MW00844562

Published by John Wiley & Sons, Inc., Hoboken, New Jersey.
Published simultaneously in Canada.

Library of Congress Cataloging-in-Publication Data:

Silvapulle, Mervyn J., 1951–
Constrained statistical inference : inequality, order, and shape restrictions / Mervyn J. Silvapulle, Pranab K. Sen.
p. cm.
Includes bibliographical references and index.
ISBN 0-471-20827-2 (cloth : acid-free paper)
1. Multivariate analysis. I. Sen, Pranab Kumar, 1937– II. Title.

QA278.S523 2004
519.5'35—dc22 2004048075

Printed in the United States of America.

10 9 8 7 6 5 4 3 2 1

To:

Param, Earlene,

and in memory of
Benedict and Anna Silvapulle

and

Anjini, Ellora, Devrani Laskar and Jacob Sen

Contents

Dedication — *v*

Preface — *xv*

1 Introduction — *1*
- *1.1 Preamble* — *1*
- *1.2 Examples* — *2*
- *1.3 Coverage and Organization of the Book* — *23*

2 Comparison of Population Means and Isotonic Regression — *25*
- *2.1 Ordered Alternative Hypotheses* — *27*
  - *2.1.1 Test of* $H_0 : \mu_1 = \mu_2 = \mu_3$ *Against an Order Restriction* — *28*
  - *2.1.2 Test of* $H_0 : \mu_1 = \ldots = \mu_k$ *Against an Order Restriction* — *31*
    - *2.1.2.1 Exact finite sample results when the error distribution is* $N(0, \sigma^2)$ — *35*
    - *2.1.2.2 Computing* $\min_{\mu \in H_1} \sum\sum(y_{ij} - \mu_i)^2$, $\bar{F}$, *and* $\bar{E}^2$. — *36*
  - *2.1.3 General Remarks* — *37*

2.2 Ordered Null Hypotheses 38
2.3 Isotonic Regression 42
2.3.1 Quasi-order and Isotonic Regression 44
2.4 Isotonic Regression: Results Related to Computational Formulas 46
2.4.1 Isotonic Regression Under Simple Order 47
2.4.2 Ordered Means of Exponential Family 50
2.5 Appendix: Proofs 53
Problems 57

3 Tests on Multivariate Normal Mean 59
3.1 Introduction 59
3.2 Statement of Two General Testing Problems 60
3.2.1 Reduction by Sufficiency 62
3.3 Theory: The Basics in Two Dimensions 63
3.4 Chi-bar-square Distribution 75
3.5 Computing the Tail Probabilities of Chi-bar-square Distributions 78
3.6 Results on Chi-bar-square Weights 81
3.7 LRT for Type A problems: $\mathbf{V}$ is Known 83
3.8 LRT for Type B problems: $\mathbf{V}$ is Known 90
3.9 Tests on the Linear Regression Parameter 95
3.9.1 Null Distributions 97
3.10 Tests When $\mathbf{V}$ is Unknown (Perlman's Test and Alternatives) 100
3.10.1 Type A Testing Problem 102
3.10.2 Type B Testing Problem 104
3.10.3 Conditional Tests of $H_0 : \boldsymbol{\theta} = \mathbf{0}$ vs $H_1 : \boldsymbol{\theta} \geq \mathbf{0}$ 105
3.11 Optimality Properties 107
3.11.1 Consistency of Tests 107
3.11.2 Monotonicity of the Power Function 109
3.12 Appendix 1: Convex Cones, Polyhedrals, and Projections 111
3.12.1 Introduction 111
3.12.2 Projections onto Convex Cones 113
3.12.3 Polyhedral Cones 122
3.13 Appendix 2: Proofs 125
Problems 133

4 *Tests in General Parametric Models* 143
4.1 *Introduction* 143
4.2 *Preliminaries* 145
4.3 *Tests of* $\boldsymbol{R\theta} = \mathbf{0}$ *Against* $\boldsymbol{R\theta} \geqslant \mathbf{0}$ 148
4.3.1 *Likelihood Ratio Test with iid Observations* 148
4.3.2 *Tests in the Presence of Nuisance Parameters* 150
4.3.3 *Wald- and Score-type Tests with iid Observations* 153
4.3.4 *Independent Observations with Covariates* 156
4.3.5 *Examples* 160
4.4 *Tests of* $\boldsymbol{h}(\boldsymbol{\theta}) = \mathbf{0}$ 164
4.4.1 *Test of* $\boldsymbol{h}(\boldsymbol{\theta}) = \mathbf{0}$ *Against* $\boldsymbol{h}(\boldsymbol{\theta}) \geqslant \mathbf{0}$ 164
4.4.2 *Test of* $\boldsymbol{h}(\boldsymbol{\theta}) = \mathbf{0}$ *Against* $\boldsymbol{h}_2(\theta) \geqslant \mathbf{0}$ 167
4.5 *An Overview of Score Tests with no Inequality Constraints* 168
4.5.1 *Test of* $H_0 : \boldsymbol{\theta} = \boldsymbol{\theta}_0$ *Against* $H_2 : \boldsymbol{\theta} \neq \boldsymbol{\theta}_0$ 169
4.5.2 *Test of* $H_0 : \boldsymbol{\psi} = \boldsymbol{\psi}_0$ *Against* $H_2 : \boldsymbol{\psi} \neq \boldsymbol{\psi}_0$ 171
4.5.3 *Tests Based on Estimating Equations* 172
4.6 *Local Score-type Tests of* $H_0 : \boldsymbol{\psi} = \mathbf{0}$ *Against* $H_1 : \boldsymbol{\psi} \in \Psi$ 175
4.6.1 *Local Score Test of* $H_0 : \boldsymbol{\theta} = \mathbf{0}$ *Against* $H_1 : \boldsymbol{\theta} \in \mathcal{C}$ 175
4.6.2 *Local Score Test of* $H_0 : \boldsymbol{\psi} = \mathbf{0}$ *Against* $H_1 : \boldsymbol{\psi} \in \mathcal{C}$ 177
4.6.3 *Local Sore-Type Test Based on Estimating Equations* 178
4.6.4 *A General Local Score-Type Test of* $H_0 : \boldsymbol{\psi} = \mathbf{0}$ 179
4.6.5 *A Simple One-Dimensional Test of* $\boldsymbol{\psi} = \mathbf{0}$ *Against* $\boldsymbol{\psi} \in \mathcal{C}$ 180
4.6.6 *A Data Example: Test Against ARCH Effect in an ARCH Model.* 181
4.7 *Approximating Cones and Tangent Cones* 183
4.7.1 *Chernoff Regularity* 186
4.8 *General Testing Problems* 194
4.8.1 *Likelihood Approach When* Θ *is Open* 195
4.8.2 *Likelihood Approach When* Θ *is not Open* 197
4.8.3 *Tests with General Objective Functions* 199
4.8.4 *Examples* 201
4.8.5 *Test of* $\boldsymbol{h}(\boldsymbol{\theta}) \geqslant \mathbf{0}$ *Against* $\boldsymbol{h}(\boldsymbol{\theta}) \ngeqslant \mathbf{0}$ 202

4.9 Properties of the mle When the True Value is on the Boundary 209
4.10 Appendix: Proofs 215

5 Likelihood and Alternatives 221
5.1 Introduction 221
5.2 The Union-Intersection Principle 222
5.2.1 UIT on Multinormal Mean 224
5.3 Intersection Union Tests (IUT) 235
5.4 Nonparametrics 243
5.4.1 Two-Sample Problem Revisited 243
5.4.2 Ordered Alternative Problem Revisited 245
5.4.3 Ordered Alternatives: Two-Way Layouts 248
5.4.4 UIP and LMPR Tests for Restricted Alternatives. 253
5.4.5 Nonparametric UIT for Restricted Alternatives 261
5.5 Restricted Alternatives and Simes-type Procedures 264
5.5.1 Probability Integral Transformation (PIT) and OSL or p-Values 267
5.5.2 Simes-Type Procedures 269
5.6 Concluding Remarks 275
Problems 276

6 Analysis of Categorical Data 283
6.1 Introduction 283
6.2 Motivating Examples 285
6.3 Independent Binomial Samples 292
6.3.1 Test of $H_0 : \boldsymbol{R}_1\boldsymbol{\pi} = \mathbf{0}$ and $\boldsymbol{R}_2\boldsymbol{\pi} = \mathbf{0}$ Against $H_1 : \boldsymbol{R}_2\boldsymbol{\pi} \geqslant \mathbf{0}$ 293
6.3.2 Test of $H_0 : \boldsymbol{A\pi} \geqslant \mathbf{0}$ Against $H_1 : \boldsymbol{A\pi} \ngeq \mathbf{0}$ 295
6.3.3 Test of $H_0 : \boldsymbol{A\pi} \ngtr \mathbf{0}$ Against $H_1 : \boldsymbol{A\pi} > \mathbf{0}$ 295
6.3.4 Pairwise Contrasts of Binomial Parameters 296
6.3.5 Discussion 297
6.4 Odds Ratios and Monotone Dependence 298
6.4.1 2×2 Tables 299
6.4.2 $2 \times c$ Tables 300
6.4.3 $r \times c$ Tables 303

6.5 Analysis of $2 \times c$ contingency tables 306
6.5.1 LRT of $\pi_1 = \pi_2$ Against Simple Stochastic Order 308
6.5.2 LRT for Simple Stochastic Order: $\boldsymbol{R\pi} \geqslant \mathbf{0}$ vs $\boldsymbol{R\pi} \ngeq \mathbf{0}$ 311
6.6 Test to Establish that Treatment is Better Than Control 313
6.7 Analysis of $\mathbf{r} \times \mathbf{c}$ Tables 315
6.7.1 LRT Against Simple Stochastic Order 317
6.7.2 LRT for and Against Stochastic Orders 320
6.8 Square Tables and Marginal Homogeneity 322
6.9 Exact Conditional Tests 324
6.9.1 Exact Conditional Tests in $2 \times c$ Tables 325
6.9.2 Comparison of Several Binomial Parameters 327
6.9.3 Exact Conditional Tests in $r \times c$ Tables 328
6.9.4 Marginal Homogeneity in Square Tables (Agresti-Coull test) 331
6.10 Discussion 335
6.11 Proofs: 335
Problems 338

7 Beyond Parametrics 345
7.1 Introduction 345
7.2 Inference on Monotone Density Function 346
7.3 Inference on Unimodal Density Function 354
7.4 Inference on Shape-Constrained Hazard Functionals 357
7.5 Inference on DMRL Functions 362
7.6 Isotonic Nonparametric Regression: Estimation 366
7.7 Shape Constraints: Hypothesis Testing 369
Problems 374

8 Bayesian Perspectives 379
8.1 Introduction 379
8.2 Statistical Decision Theory Motivations 380
8.3 Stein's Paradox and Shrinkage Estimation 384
8.4 Constrained Shrinkage Estimation 388
8.5 PCC and Shrinkage Estimation in CSI 396
8.6 Bayes Tests in CSI 400
8.7 Some Decision Theoretic Aspects: Hypothesis Testing 402
Problems 404

9 Miscellaneous Topics 407
9.1 Two-sample Problem with Multivariate Responses 408
9.1.1 One Degree of Freedom Tests of $H_0 : \boldsymbol{\mu}_1 - \boldsymbol{\mu}_2 = \mathbf{0}$ Against $H_1 : \boldsymbol{\mu}_1 - \boldsymbol{\mu}_2 \geqslant \mathbf{0}$ 411
9.1.2 O'Brien's Test of $H_0 : \boldsymbol{\mu}_1 - \boldsymbol{\mu}_2 = \mathbf{0}$ vs $H_1 : \boldsymbol{\mu}_1 - \boldsymbol{\mu}_2 \geqslant \mathbf{0}$ 412
9.1.3 A One Degree of Freedom Test for the Equality of General Parameters 414
9.1.4 Multiparameter Methods 416
9.1.4.1 $\boldsymbol{X}_1 \sim N(\boldsymbol{\mu}_1, \sigma^2 \boldsymbol{U}_1)$ and $\boldsymbol{X}_2 \sim N(\boldsymbol{\mu}_2, \sigma^2 \boldsymbol{U}_2)$ 416
9.1.4.2 $\boldsymbol{X}_1 \sim N(\boldsymbol{\mu}_1, \sigma_1^2 \boldsymbol{U}_1)$ and $\boldsymbol{X}_2 \sim N(\boldsymbol{\mu}_2, \sigma_2^2 \boldsymbol{U}_2)$ 417
9.1.4.3 $\boldsymbol{X}_1 \sim N(\boldsymbol{\mu}_1, \boldsymbol{V}_1)$ and $\boldsymbol{X}_2 \sim N(\boldsymbol{\mu}_2, \boldsymbol{V}_2)$ 418
9.1.4.4 An extension of Follmann's test 418
9.1.5 H_1 : New Treatment is Noninferior With Respect to Every Variable and Superior With Respect to Some Variables 420
9.2 Testing that an Identified Treatment is the Best: the Min Test 422
9.2.1 Monotone Critical Regions and Tests 428
9.2.2 Optimality of Min Test 429
9.2.3 Multivariate Responses 430
9.2.4 Comparison of LRT with the Min Test 431
9.3 Cross-over Interaction 434
9.3.1 A Test Against Cross-over Interaction: Two Treatments and I Groups. 437
9.3.1.1 Case 1. $\hat{\delta}_i \sim N(\delta_i, \sigma_i^2)$ where $\sigma_i^2 = \sigma^2 w_i$, σ^2 is unknown and w_i is known, $i = 1, \ldots, I$. 439
9.3.1.2 Case 2: $\hat{\delta}_i \sim N(\delta_i, \sigma_i^2)$, $\hat{\delta}_1, \ldots, \hat{\delta}_I$ are independent, and $\sigma_1, \ldots, \sigma_I$ are known 442
9.3.1.3 Case 3: $\hat{\delta}_i$ is approximately $N(\delta_i, \sigma_i^2)$ and σ_i is unknown, $i = 1, \ldots, I$. 443
9.3.2 Robustness Against Outliers 444
9.3.3 Dependent Treatment Effects 445
9.3.4 A Range Test Against Cross-over Interaction 447
9.3.5 A Test with Uniformly More Power and Less Bias 448

9.3.6 *Cross-over Interaction of Practical Significance* 449
9.3.7 *More than Two Treatments* 451
9.3.8 H_0 : *Cross-over Interaction is Present* 452
9.3.9 *Proof of Proposition 9.3.1* 453
9.4 *Directed Tests* 455
9.4.1 *Directed Test in a Simple Case* 456
9.4.2 *Directed Test Against Stochastic Order* 458
9.4.3 $r \times c$ *Tables* 460
9.4.4 *Improving Power by Enlarging the Achievable Set of* p*-Values* 462
9.4.5 *Exponential Families* 463
Problems 463

Bibliography 469

Index 525

Preface

Statistical inference has grown out of the genuine need for modeling and analysis of observational or experimental data in various interdisciplinary fields. It has its genesis in mathematical statistics and scope fortified with applications in almost all walks of life and science. In this complex domain, underlying statistical models are generally not so simple due to a complex network of constraints either in the space of the unknown parameters in a parametric way (or form of the underlying distributions in a nonparametric to semiparametric way) or even in the sample space of the observable random elements. There is therefore a growing need for statistical inference to cope with such constrained environments. This is what we mean by constrained statistical inference (CSI). In some simplest statistical models, like the two sample location or scale problem, CSI may relate to one-sided hypothesis testing or restricted estimating procedures. In a relatively more complex situation, such constraints may crop up in terms of either some inequality or ordering of associated parameters, albeit it could be more involved for functional parameters. Isotonic regression etc., have their origin in this complex. On one hand, such constraints make the statistical inference procedures more complicated. On the other hand, such constraints contain statistical information as well (viz., a reduction of the parameter or sample space), so that if properly incorporated they would be more efficient than their counterparts wherein such constraints are ignored. However, there may not be generally an optimal (decision theoretic) solution in CSI, albeit improvement is very likely. Faced with these dual aspects, CSI has been gaining attention steadily in the statistical literature, more so from the vast interdisciplinary fields of applications where such improvements matter a lot. The main objective of this treatise is to highlight the scope of CSI, though at an intermediate level of presentation, to a wider class of audience including upper undergraduate and graduate students in a broader area of statistical science, researchers, and professionals as well.

A very notable work in this direction, the first monograph on CSI, is the classical text by Barlow, Bartholomew, Bremner, and Brunk (1972), which synthesized the development of order restricted statistical inference during the 1950s and 1960s; some of their coverages have retained their glamour over 30 years. A follow-up monograph by Robertson, Wright and Dykstra (1988) captured some of the developments in the 1970s to early 1980s, albeit retaining the basic framework of its predecessor. In the past 20 years, not only a significant volume of new developments have taken place but also these have come up in much greater diversity than before. This makes our task in dealing with CSI in an updated mode much less trivial and much more compelling to meet the growing need for applied fields that cry for statistical inference; clinical trials, bioassays, biomedical sciences, genetics, and bioinformatics are all among these interdisciplinary fields. For this reason, we have organized the characterization with illustrations and motivating examples in the beginning, basic theory in the middle part, and beyond the classical likelihood based parametrics in the later part. An extensive bibliography adds more to our focus on the updating task.

The impetus for writing of this book dawned when the first author (MJS) was visiting the Department of Statistics and Actuarial Science at the University of Iowa in 1995, and continued while he was a visiting professor at the University of North Carolina at Chapel Hill (UNC-CH) in 2000. In addition, in the process of co-editing (with Tim Robertson and Richard Dykstra) a (2002) special issue of the *Journal of Statistical Inference* devoted to CSI, MJS was pleasantly surprised by the diversity of the field and the richness of the contributions made by the different researchers. Subsequently, in early 2000, while in Chapel Hill, both the current authors thought of concentrating on the theme of organizing a monograph emphasizing the broad objective as laid down above. The year at Chapel Hill (for MJS) and the support from the University and, in particular, Gary Koch, made it easier for the first author to shoulder a lion's share in this monumental task. Both the authors, wish to express their sincere appreciation to Art Cohen, Mike Perlman, and Hal Sackrowitz for their enthusiasm, encouragement, and helpful comments and discussions.

Checking for errors/ambiguities has been particularly difficult because a large part of the material presented in this book is compiled from a large number of journal publications and appears in a book for the first time. Although we have tried to detect and correct errors and eliminate ambiguities, there may well be others that escaped our scrutiny. We take responsibility for any remaining ones, and would appreciate being informed of them. Additional material that may be relevant to this book will be made available through a web site of the first author; it may be accessed from

http://www.latrobe.edu.au/mathstats/staff/silvapulle.html

The first author benefited considerably from discussions and joint research with Param Silvapulle on CSI in economics, finance, and econometrics. In addition to the intellectual input, she has also been the single most important source of support, encouragement and understanding, without which this book would have never been completed. The second author (PKS) wishes to acknowledge the understanding, patience, and open-mindedness of the first author throughout the tenure of this venture, and, in addition, to La Trobe University for hosting the two short visits in December 2001 and August 2003 that helped in bundling up the tidy ends. The support from the University of North Carolina is of course understood. Further, much of his line of thinking has culminated from his joint work with former advisees Vern Chinchilli, Azza Karmous, Mike Boyd, Cecilia Wada, and, particularly, Ming-Tien Tsai. To all of them, he owes a lot of gratitude.

We would like to thank Jeanette Varrenti and Paul Pontikis for typing some parts of the book. Ms. Margaret Tapp and Professor Bahjat F. Qaqish have been very helpful in easying the electronic transmissions of the manuscript back and forth between the two continents, (North) America and (South) Australia, throughout the past four years. We wish to thank Amy Hendrickson of TEXnology Inc. for expert and generous help with La Tex. Finally, both the authors would like to thank, from the bottom of their hearts, Steve Quigley, Susanne Steitz, Melissa Yanuzzi, and the Wiley-Interscience group for their consistent encouragement, patience, and overwhelming cooperation in bringing this project to a success.

MERVYN J. SILVAPULLE AND PRANAB K. SEN

Bundoora, Victoria, Australia and Chapel Hill, NC, USA

1

Introduction

1.1 PREAMBLE

Stochastics prevail in every walk of life and science, amidst uncertainty, unpredictability, unknowability, and improbability to shear diversity on one hand, and inequality, inexplainability, and incompatibility to impossibility on the other hand. Stochastics distorts transparency of inherent *determinacy* so that the Trojan horses of *statistical inference* are often invoked for drawing valid and interpretable conclusions from pertinent experimental or observational data. Fathoming the depth of stochastics along with the latent chaos or chance elements (i.e., uncontrollable variation) is a challenging task that statistical inference shares intricately with *statistical decision theory* (SDT), for theoretical foundation, and *statistical methodology* for effective applications. In real-life problems, often, there is an additional layer of complexity: *Constraints* (latent or not) of diverse types mar the underlying determinacy (even if that exists), and, hence, there may be a genuine need to sort out the underlying determinacy from its superimposed stochastics; we need to focus on such logical restraints, and incorporate them mathematically as well as statistically in contemplated decision making processes. *Constrained statistical inference* (CSI) has its roots in this complex.

With the ever-expanding horizons of the domain of statistical inference, from the tenuous simple parametric models (such as the binomial, Poisson, normal, Laplace, exponential laws) to more complex *functional parametric* models (such as in many *stochastic processes*), to *semiparametrics* to the omnibus *nonparametrics*, it became necessary to recharge its arsenals with the battery of compatible developments in SDT, and thereby regulate the needed methodology to meet the demand of the vast interdisciplinary fields of applications. Within this diversity, often, constraints crop up either as (partial, (non-)linear, total, or even implicit) *ordering* on the so-called parameter space(s), or as (linear or not) *inequalities* on contemplated parameters or functionals; they may also show up on the observations or experimental outcomes, termed the *responses*. In this sense, statistical constraints have their abode in the *decision* as well as the *sample* spaces. Albeit, such constrained universes could be portrayed in matching abstractions from the mathematical universe with SDT tones added for stochastic resolutions, for a more down-to-earth, comprehensive treatise of CSI, we shall adopt a pedestrian's path through the alleys of simple parametrics to functional parametrics to the vast domain, *beyond parametrics*. We specifically focus here on the constraints in the form of ordering and inequality restraints of diverse types, and keep in mind the vast field of applications. Perhaps it would be more motivating, and convincing too, to portrait first some simple and yet genuine inequality and order constrained inference problems as illustrative examples, and then step out to a more general prescription for a broader class of problems encountered in CSI. This is advocated in Section 1.2. The final section in this chapter attempts to capture the basic organization and coverage of the present treatise of CSI.

1.2 EXAMPLES

To briefly indicate an important feature of constrained statistical inference, let us first consider the simple example of testing $\mu = 0$ based on a random sample from the univariate normal distribution, $N(\mu, 1)$. Suppose that μ is known to be nonnegative. Then, we know from elementary statistics that a one-sided test with the sample mean as the test statistic is more powerful than a two-sided one. Intuitively, we would expect to do better if we know that $\mu \geqslant 0$ and if such a constraint is incorporated in the statistical inference procedure. We would expect a similar trend in multiparameter situations as well. In other words, if we know that parameters are restricted by some constraints then it is reasonable to expect that we should be able to do better by incorporating such additional information than by ignoring them. While this appears reasonable, it may or may not be an easy task.

The main theme of this book is incorporation of inequality and order restrictions on the parameters of the statistical model, and our objective is to provide a systematic development of the relevant statistical theory, motivated and illustrated by examples. In the rest of this section, we provide a broad range of motivating examples.

Example 1.2.1 *Ordered treatment means in one-way layout*

An experiment was conducted to evaluate the effect of exercise on the age at which a child starts to walk. Let Y denote the age (in months) at which a child starts to walk; the data on Y are given in Table 1.1. (The original experiment consisted of another treatment; however, here we consider only three treatments for simplicity. The data and analyzes involving all four treatments will be given in a later chapter.)

The first treatment group received a special walking exercise for 12 minutes per day beginning at age 1 week and lasting 7 weeks. The second group received daily

Table 1.1 The age at which a child first walks

Treatment (i)	Age (in months)						n_i	$\bar{y}_i$	μ_i
1	9.00	9.50	9.75	10.00	13.00	9.50	6	10.125	μ_1
2	11.00	10.00	10.00	11.75	10.50	15.00	6	11.375	μ_2
3	13.25	11.50	12.00	13.50	11.50		5	12.35	μ_3

exercises but not the special walking exercises. The third group is the control; they did not receive any exercises or other treatments. For Treatment i ($i = 1, 2, 3$), let

$$\mu_i = \text{ Mean age (in months) at which a child starts to walk.}$$

In the traditional analysis of variance (ANOVA), one would usually test

$$H_0 : \mu_1 = \mu_2 = \mu_3 \qquad \text{against} \qquad H_2 : \mu_1, \mu_2, \textit{ and } \mu_3 \textit{ are not all equal.}$$

However, let us suppose that the researcher was prepared to assume that the walking exercises would not have the negative effect of increasing the mean age at which a child starts to walk, and it was desired that this additional information be incorporated to improve on the statistical analysis. For illustrative purposes, let us suppose that the researcher wishes to incorporate the information $\mu_1 \leqslant \mu_2 \leqslant \mu_3$. In this case, the testing problem is

$$H_0 : \mu_1 = \mu_2 = \mu_3 \text{ vs } H_1 : \mu_1 \leqslant \mu_2 \leqslant \mu_3 \text{ and } \mu_1, \mu_2, \textit{ and } \mu_3 \textit{ are not all equal.}$$

Clearly, the usual ANOVA where one would test

$$H_0 : \mu_1 = \mu_2 = \mu_3 \text{ against } H_2 : \mu_1, \mu_2, \textit{ and } \mu_3 \text{ are not all equal,}$$

fails to incorporate the additional information $\mu_1 \leqslant \mu_2 \leqslant \mu_3$. Therefore, one would expect that we should be able to do better than the traditional F-test. In the next chapter, we illustrate a method that is specifically designed for this testing problem; the method to be illustrated, which we shall call the $\bar{F}$-test, is easy to understand and use because it uses essentially the same idea as that which underlies the traditional F-test and modifies it to incorporate additional information such as $\mu_1 \leqslant \mu_2 \leqslant \mu_3$.
■

Let us make a few remarks about this example. If the objective of the experiment was to establish that the special walking exercise results in a reduction in the mean age at which a child starts to walk, the testing problem needs to be formulated differently; for example, it may be formulated as test of

$$H_a : \mu_1 < \mu_2 < \mu_3 \textit{ does not hold} \qquad \text{vs} \qquad H_b : \mu_1 < \mu_2 < \mu_3.$$

This is a different statistical inference problem and it will be discussed in later examples.

To find out which treatments were effective and which were not would involve multiple tests and require Bonferroni-type adjustment to control overall type-I error; this is also likely to require large samples to achieve moderate power.

A test of H_0 against H_2 in one-way ANOVA, for example the F-test, is also a *valid* procedure for testing H_0 against H_1 as well. Here *valid* means that a 5% level test of H_0 against H_2 is also a 5% level test of H_0 against H_1. This is because the critical value and the size of a test are computed under the null hypothesis, and the null hypothesis for H_0 vs H_1 and for H_0 vs H_2 are the same. Therefore, the validity of the test is not an issue in this example.

The ordering on the μ_is in H_1 is an *assumption.* Even if the true values of $\{\mu_1, \mu_2, \mu_3\}$ do not satisfy the order restrictions in H_1, it is still possible for some tests of H_0 against H_1 to reject H_0. In this case, while rejecting H_0 would be the correct statistical decision, accepting H_1 as true would be a wrong conclusion.

In general, when we apply a test of hypothesis, it is assumed that either H_0 or H_1 is true. For this reason, $H_0 \cup H_1$ can be called the *maintained hypothesis*; in this example, the maintained hypothesis is $\mu_1 \leqslant \mu_2 \leqslant \mu_3$. It will be seen in the next chapter, that tests of H_0 against H_1 can be carried out easily.

Example 1.2.2 *Relationship between El Niño and hurricanes*

El Niño refers to unusually warm ocean currents in the Pacific that appear around Christmas time and last for several months. Monsoon rains in the central Pacific and droughts and forest fires in Indonesia and Australia have been linked to *El Niño*. A hypothesis concerning *El Niño* is the following (Kitchens (1998) page 812):

H : Warm phase of *El Niño* suppresses hurricanes and cold phase encourages.

The data[1] in Table 1.2 provide information on the numbers of hurricanes from 1950 to 1995. In this context, different types of hypothesis testing problems may arise depending on how the question of interest is formulated.

Let us first formulate the basic model as a simple one-way classification. Let *El Niño* be the factor with three levels : Cold ($i = 1$), Neutral ($i = 2$), and Warm ($i = 3$). Let Y_{ij} denote the number of hurricanes and let

$$Y_{ij} = \mu_i + e_{ij},$$

where μ_i is the expected number of hurricanes ($i = 1, 2, 3$).

Let us suppose that the hypothesis H is based on various conjectures and scientific reasoning, and it is of interest to know whether or not there is any evidence against H. In this case, the null and alternative hypotheses can be stated as

$$H_0 : \mu_1 \geqslant \mu_2 \geqslant \mu_3 \quad \text{and} \quad H_1 : \mu_1 \geqslant \mu_2 \geqslant \mu_3 \textit{ does not hold.}$$

A feature of this inference problem is that inequality constraints are present in the null hypothesis; this is different from the usual ANOVA problem where the null hypothesis would typically be of the form $\mu_1 = \mu_2 = \mu_3$.

[1]The authors are grateful to the National Hurricane Center for help with these data.

Table 1.2 Effect of *El Niño* on hurricanes

Y	50	51	52	53	54	55	56	57	58	59	60	61
H	11	8	6	6	8	9	4	3	7	7	4	8
E	c	w	n	w	c	c	n	w	n	n	n	c
Y	62	63	64	65	66	67	68	69	70	71	72	73
H	3	7	6	4	7	6	4	12	5	6	3	4
E	n	n	c	w	n	c	n	n	c	c	w	c
Y	74	75	76	77	78	79	80	81	82	83	84	85
H	4	6	6	5	5	5	9	7	2	3	5	7
E	c	c	w	w	n	n	n	n	w	w	n	c
Y	86	87	88	89	90	91	92	93	94	95		
H	4	3	5	7	8	4	3	4	3	10		
E	w	w	c	c	n	w	w	w	w	c		

Y:year (64 refers to 1964); H: number of hurricanes; E: *El Niño*; c: cold; w: warm; n: neutral.

Source: The original source of these data is the National Hurricane Center. The data in this table were extracted from the table on page 812 in Kitchens (1998).

On the other hand, if the interest is to establish that there is sufficient evidence to support the hypothesis H, then the testing problem is

$$H_0 : \mu_1 > \mu_2 > \mu_3 \textit{ does not hold} \quad \text{vs} \quad H_1 : \mu_1 > \mu_2 > \mu_3 \textit{ holds.}$$

The one-way model $Y_{ij} = \mu_i + e_{ij}$ used in the foregoing two examples is a simple one with no other covariates. If $\boldsymbol{x}_{ij}$ is a vector of covariates, then the model could be of the form[2]

$$Y_{ij} = \mu_i + \boldsymbol{x}_{ij}^T \boldsymbol{\beta} + e_{ij};$$

the hypotheses of interest are still the same. These inference problems come under the general framework of the linear model

$$\boldsymbol{Y} = \boldsymbol{X\theta} + \boldsymbol{E}.$$

Within the context of such a linear model, three testing problems of interest are:

1. $H_0 : \boldsymbol{R\theta} = \boldsymbol{0}$ against $H_1 : \boldsymbol{R\theta} \geqslant \boldsymbol{0}, \boldsymbol{R\theta} \neq \boldsymbol{0}$
2. $H_0 : \boldsymbol{R\theta} \geqslant \boldsymbol{0}$ against $H_1 : \boldsymbol{R\theta} \ngeqslant \boldsymbol{0}$
3. $H_0 : \boldsymbol{R\theta} > \boldsymbol{0}$ *does not hold* against $H_1 : \boldsymbol{R\theta} > \boldsymbol{0}$

where $\boldsymbol{R}$ is a given fixed matrix that does not depend on $\boldsymbol{\theta}$. The first two types of problems will be discussed in Chapters 3 and 4, and the third type will be discussed in later chapters under the topic *sign testing*. ■

[2] Vectors and matrices are shown in bold and the superscript T denotes transpose.

Example 1.2.3 *Comparison of two treatments when the response is multivariate (Pocock et al. (1987))*

Seventeen patients with asthma or chronic obstructive airways disease entered a trial to evaluate an inhaled active drug versus placebo. The experiment was designed as a randomized, double-blind, cross-over trial. The main purpose was to study the possibly harmful effects of drugs on lung mucociliary clearance. Analysis of those results produced no evidence of harm. In addition, standard respiratory function measures were taken at the end of both treatment periods. These were peak respiratory flow rates ($PEFR$), forced expiratory volume (FEV), and forced vital capacity (FVC), the latter two being expressed as a percentage of the predicted value for that patient's age, gender, and height in the normal population. Let us define the following variables for the changes (i.e., drug - placebo) in the three measures:

$$\begin{aligned} X_1 &= \text{ change in FEV,} \\ X_2 &= \text{ change in FVC,} \\ \text{and } X_3 &= \text{ change in PEFR.} \end{aligned}$$

Let

$$\boldsymbol{X} = (X_1, X_2, X_3)^T \quad \text{and} \quad \boldsymbol{\mu} = (\mu_1, \mu_2, \mu_3)^T$$

where $\mu_i = E(X_i)$ for $i = 1, 2, 3$. A question of interest was whether the addition of the inhaled drug could further improve respiratory function. For each measure, there was no sign of period or carry-over effects. Therefore, the univariate analyzes of drug versus placebo was performed using paired t-test (see Table 1.3).

The estimated correlation matrix of $(\bar{X}_1, \bar{X}_2, \bar{X}_3)$ reported in Pocock et al. (1987) is

$$\begin{pmatrix} 1 & 0.95 & 0.219 \\ 0.95 & 1 & 0.518 \\ 0.219 & 0.518 & 1 \end{pmatrix}.$$

All three measures showed a mean improvement on active drug but none achieved statistical significance at the 5% level. Thus, Bonferroni correction would lead to a conclusion of no evidence of improvement for the active drug. In any case, such a procedure is not the most powerful method in this setting. What is required is a

Table 1.3 Effect of drug on chronic respiratory disease

Variable	X_1	X_2	X_3
Sample Mean ($i.e. \bar{X}_i$)	7.56	4.81	2.29
$s.e.(\bar{X}_i)$	18.53	10.84	8.51
t-statistic	1.63	1.77	1.11

multivariate statistical procedure to evaluate the collective evidence of the overall benefit of the drug. Therefore, the null and alternative hypotheses take the form

$$H_0 : \boldsymbol{\mu} = \mathbf{0} \quad \text{and} \quad H_1 : \boldsymbol{\mu} \gtrsim \mathbf{0},$$

respectively, where

$$\boldsymbol{\mu} \gtrsim \mathbf{0} \text{ means that } \mu_i \geqslant 0 \text{ for every } i \text{ and } \boldsymbol{\mu} \neq \mathbf{0}.$$

It would be helpful to point out an important difference between testing $H_0 : \boldsymbol{\mu} = \mathbf{0}$ against $H_1 : \boldsymbol{\mu} \gtrsim \mathbf{0}$ and performing the three tests,

$$\begin{aligned} & H_{01} : \mu_1 = 0 \text{ vs } H_{11} : \mu_1 > 0, \\ & H_{02} : \mu_2 = 0 \text{ vs } H_{12} : \mu_2 > 0, \\ \text{and} \quad & H_{03} : \mu_3 = 0 \text{ vs } H_{13} : \mu_3 > 0 \end{aligned}$$

with Bonferroni correction. For illustrative purposes let us suppose that the standard errors, $se(\bar{\boldsymbol{X}}_i)$, in the three columns of Table 1.3 are 1.0, 10.0, and 0.2, respectively. In this case, the three Bonferroni corrected tests would conclude that the effects on FEV and PEFR are significant, but not that on FVC at 5% level of significance. A test of $H_0 : \boldsymbol{\mu} = \mathbf{0}$ against $H_1 : \boldsymbol{\mu} \gtrsim \mathbf{0}$ would also reject H_0 and accept H_1; however, this does not say which of FEV, PEFR, and FVC is significant and which is not. Therefore, the two testing procedures answer different questions.

Note that the parameter space for $\boldsymbol{\mu}$ has $2^3 = 8$ quadrants, and the Hotelling's T^2 looks for departure from the null value $\boldsymbol{\mu} = \mathbf{0}$ into all of the 2^3 quadrants, while a test against $H_1 : \boldsymbol{\mu} \gtrsim \mathbf{0}$ looks for departure into just one quadrant and hence would be more targeted to the specific objective of this study.

In a later chapter we will develop the theory for testing

$$H_0 : \boldsymbol{R\mu} = \mathbf{0} \text{ against } H_1 : \boldsymbol{R\mu} \neq \mathbf{0}, \boldsymbol{R}_1\boldsymbol{\mu} \geqslant \mathbf{0}$$

based on a sample from $N(\boldsymbol{\mu}, \boldsymbol{V})$ where $\boldsymbol{R}_1$ is a submatrix of $\boldsymbol{R}$ and $\boldsymbol{V}$ is unknown. These results can be used to deal with problems similar to those in this example. ■

Example 1.2.4 *Testing the validity of Liquidity Preference Hypothesis [an ordered hypotheses on the mean of a multivariate normal] (Richardson et al. (1992))*

Let $H(r,t)$ denote the rate of return during the time period t to $t+1$ from a *bill* with *maturity* r (i.e., at time t, the maturity date of the investment/bill is $t+r$). At time t, the *term premium* on a bill with maturity r is defined as

$$\theta_r = E\{H(r,t) - H(1,t)\}$$

where the expectation is taken with respect to the information available to the investors at time t. Thus, the term premium is the additional rate of return that the investors expect at time t for investing in a longer-term bill rather than in a short-term one that would mature at the end of next time point. The *Liquidity Preference Hypothesis*

Table 1.4 Term premia of T-bills with up to 11 months to maturity for Aug/1964-Nov/1990

Average premium		t-statistics for $\theta_r - \theta_{r-1}$		Average premium		t-statistics for $\theta_r - \theta_{r-1}$	
θ_2	0.030	$\theta_3 - \theta_2$	6.78	θ_7	0.070	$\theta_8 - \theta_7$	4.55
θ_3	0.056	$\theta_4 - \theta_3$	-0.12	θ_8	0.088	$\theta_9 - \theta_8$	1.13
θ_4	0.055	$\theta_5 - \theta_4$	3.77	θ_9	0.093	$\theta_{10} - \theta_9$	-3.97
θ_5	0.070	$\theta_6 - \theta_5$	0.80	θ_{10}	0.071	$\theta_{11} - \theta_{10}$	1.36
θ_6	0.073	$\theta_7 - \theta_6$	-0.80	θ_{11}	0.077	$\theta_{11} - \theta_9$	-2.00

(*LPH*), which plays an important role in economics, says that the term premium, θ_r, is a nonincreasing function of r. The basic argument supporting the hypothesis is that longer term bills are more risky than short term ones, and hence investors expect higher return for higher risk. There has been a series of publications in which the validity of this hypothesis was investigated (see Richardson et al. (1992) and the references therein). Table 1.4 provides summary data and some relevant statistics that have been used in the debates concerning the validity of *LPH*.

According to *LPH*, $\theta_r - \theta_s$ is nonnegative for $r > s$. Yet the estimates of $\theta_4 - \theta_3, \theta_7 - \theta_6, \theta_{10} - \theta_9$, and $\theta_{11} - \theta_9$ are negative and the last two are large compared with 5% level critical values from a standard normal distribution (see Table 1.4). This led to interesting debates about the validity of *LPH*. If we were to examine all possible pairwise differences then there is a good chance that some "large" negative estimates of $\theta_r - \theta_{r-1}$ would be observed. In any case, multiple tests of all possible pairwise differences in the θ_rs is not quite the appropriate statistical method for this problem, although the t-statistics in Table 1.4 are useful. The problem needs to be formulated as a test of

$$H_0 : \theta_{11} \geqslant \theta_{10} \geqslant \ldots \geqslant \theta_2 \quad \text{against} \quad H_1 : H_0 \text{ does not hold.} \tag{1.1}$$

As in Example 1.2.2, a feature of this testing problem is that inequalities are present in the null hypothesis.

In a later chapter, we will develop the theory for testing $\boldsymbol{\theta} \geqslant \mathbf{0}$ against $\boldsymbol{\theta} \ngeqslant \mathbf{0}$ based on observations from $\boldsymbol{X}$ where $\boldsymbol{X}$ has the multivariate normal distribution, $N(\boldsymbol{\theta}, \boldsymbol{V})$; $\boldsymbol{V}$ is either known or an estimate of it is available. These results can be used for testing (1.1). A test of (1.1) would say whether or not there is sufficient evidence against

$$H_0 : \theta_{11} \geqslant \theta_{10} \geqslant \ldots \geqslant \theta_2;$$

when there is such evidence against H_0, the test would *not* provide information as to which of the inequalities in the null hypothesis fails to hold. On the other hand, multiple tests on the pairwise differences with Bonferroni adjustment, attempt to identify which of the pairwise differences are positive; thus the later approach attempts to answer a much deeper question, one that is not really required for testing the validity of *LPH*. Therefore, in this example, it would be unnecessary to carry out multiple tests on pairwise differences; a set of multiple tests is likely to have low power compared with one that is specifically designed for testing H_0 against H_1 in (1.1). If there are a large number of observations, then multiple tests can be used for the testing problem (1.1).

In the foregoing discussions, the problem was formulated as

$$\boldsymbol{X} \sim N(\boldsymbol{\beta}, \boldsymbol{V}) \text{ where } \boldsymbol{\beta} = (\theta_2, \ldots, \theta_{11})^T,$$

and the estimate of $\boldsymbol{\beta}$ was the sample mean of 313 observations on $\boldsymbol{X}$. It is quite possible that in a more elaborate model, $(\theta_2, \ldots, \theta_{11})$ may appear as some parameters in a regression type model. The theory for such problems will also be developed in a later chapter. ■

Example 1.2.5 *Test against a set of one-sided hypotheses in a linear regression model (Wolak (1989))*

Consider the following double-log demand function:

$$\log Q_t = \alpha + \beta_1 \log PE_t + \beta_2 \log PG_t + \beta_3 \log I_t + \gamma_1 D1_t + \gamma_2 D2_t + \gamma_3 D3_t + \epsilon_t,$$

where

$$\begin{aligned} Q_t &= \text{aggregate electricity demand}, \\ PE_t &= \text{average price of electricity to the residential sector}, \\ PG_t &= \text{price of natural gas to the residential sector}, \\ I_t &= \text{income per capita}, \end{aligned}$$

and $D1_t, D2_t$, and $D3_t$ are seasonal dummy variables.

The error terms $\{\epsilon_t\}$ may or may not be independent; for example, they may satisfy an AR(1) process. Prior knowledge suggests that $(-\beta_1, \beta_2, \beta_3) \geqslant (0, 0, 0)$. A typical model selection question that arises is whether or not the foregoing model provides a better fit than the simpler model,

$$\log Q_t = \alpha + \gamma_1 D1_t + \gamma_2 D2_t + \gamma_3 D3_t + \epsilon_t.$$

This would require a test of

$$H_0 : (\beta_1, \beta_2, \beta_3) = (0, 0, 0) \text{ against } H_1 : (-\beta_1, \beta_2, \beta_3) \geqslant (0, 0, 0).$$

Strictly speaking, we should write the alternative hypothesis as $(-\beta_1, \beta_2, \beta_3) \gneqq (0, 0, 0)$. However, in what follows, we shall always interpret H_0 vs H_1 as H_0 vs $H_1 - H_0$ (i.e., $H_1 \setminus H_0$), so that the two hypothesis are disjoint. A test of

$$H_0 : (\beta_1, \beta_2, \beta_3) = (0, 0, 0) \text{ vs } H_2 : (-\beta_1, \beta_2, \beta_3) \neq (0, 0, 0)$$

is not the best for this testing/model-selection problem, although this is the one used by most standard statistical software packages. ■

Example 1.2.6 *Multivariate ANOVA (Dietz (1989))*

Vinylidene fluoride is suspected of causing liver damage. An experiment was carried out to evaluate its effects. Four groups of 10 male Fischer-344 rats received, by inhalation exposure, one of several dosages of vinylidene fluoride. Among the response variables measured on the rats were three serum enzymes: SDH, SGOT, and SGPT. Increased levels of these serum levels are often associated with liver damage. It is of interest to test whether or not these enzyme levels are affected by vinylidene fluoride. The data are given in Table 1.5 [the authors gratefully acknowledge receiving generous help from the National Toxicology Program of NIEHS regarding these data].

Let $\boldsymbol{Y}_{ij} = (Y_{ij1}, Y_{ij2}, Y_{ij3})^T$ denote the observations on the three enzymes for ith subject ($i = 1, \ldots, 10$) in treatment $j(j = 1, \ldots, 4)$. Let θ_{jk} denote the mean (or median) response for j^{th} treatment (i.e., dose) and k^{th} variable and let $\boldsymbol{\theta}_j = (\theta_{j1}, \theta_{j2}, \theta_{j3})^T$ for $j = 1, \ldots, 4$. Now, one formulation of the null and alternative hypotheses is

$$H_0 : \boldsymbol{\theta}_1 = \boldsymbol{\theta}_2 = \boldsymbol{\theta}_3 = \boldsymbol{\theta}_4 \quad \text{and} \quad H_1 : \boldsymbol{\theta}_1 \leqslant \boldsymbol{\theta}_2 \leqslant \boldsymbol{\theta}_3 \leqslant \boldsymbol{\theta}_4.$$

For this type of problem, general procedures for multivariate linear models can be used; these include methods based on likelihood, M-estimators, and ranks. The latter two methods have some robustness properties against nonnormal error distributions compared with the normal theory likelihood approach, similar to the case in the location scale model. ■

Table 1.5 Serum enzyme levels in rats

		Rat within dosage									
Dosage	Enzyme	1	2	3	4	5	6	7	8	9	10
0	SDH	18	27	16	21	26	22	17	27	26	27
	SGPOT	101	103	90	98	101	92	123	105	92	88
	SGPT	65	67	52	58	64	60	66	63	68	56
1500	SDH	25	21	24	19	21	22	20	25	24	27
	SGOT	113	99	102	144	109	135	100	95	89	98
	SGPT	65	63	70	73	67	66	58	53	58	65
5000	SDH	22	21	22	30	25	21	29	22	24	21
	SGOT	88	95	104	92	103	96	100	122	102	107
	SGPT	54	56	71	59	61	57	61	59	63	61
15000	SDH	31	26	28	24	33	23	27	24	28	29
	SGOT	104	123	105	98	167	111	130	93	99	99
	SGPT	57	61	54	56	45	49	57	51	51	48

Serum enzyme levels are in international units/liter; dosage of vinylidene in parts/million.

Reprinted from: Thirteen-week study of Vinylidene fluoride in F344 male rats, with permission from The National Toxicology Program, NIEHS; the data appeared in a report prepared by Litton Bionetics Inc. (1984). These data also appeared in Dietz (1989), where they were used to illustrate constrained inference.

Just as in the linear regression model, inequality constrained hypotheses arise in more general models such as logistic regression models, proportional hazards models, loglinear models for categorical data, Generalized Linear Models, time-series models, and threshold models. Some examples of this type are provided below.

Example 1.2.7 *Test against an ordered hypothesis on the regression parameters of a binomial response model (Piegorsch (1990), Silvapulle (1994))*

An assay was carried out with the bacterium *E. coli* strain 343/358(+) to evaluate the genotoxic effects of 9-aminoacridine (9-AA) and potassium chromate (KCr). The data are presented in Table 1.6.

An objective of the study was to evaluate the synergistic effects of KCr when it is administered simultaneously with 9-AA. The null model of interest is known as the *Simple Independent Action* (SIA) model; the reasons for choosing this particular model for the null hypothesis are discussed in Piegorsch (1990) and the references therein. They show that the complementary loglinear model,

$$-\log(1-\pi_{ij}) = \mu + \alpha_i + \beta_j + \eta_{ij}$$

with $\alpha_1 = \beta_1 = \eta_{1j} = \eta_{i1} = 0$ for every (i, j), provides an appropriate framework to test for/against *synergism*. When $\eta_{ij} = 0\ \forall(i, j)$, the SIA hypothesis holds. If η_{ij} are nonnegative and at least one of them is positive, then there is departure away from SIA towards synergism. Thus, the null and the alternative hypotheses take the forms

$$H_0 : \boldsymbol{\eta} = \mathbf{0} \quad \text{and} \quad H_1 : \boldsymbol{\eta} \geq \mathbf{0}, \tag{1.2}$$

respectively.

If we wish to test against any possible departure from the SIA model, we would test $H_0 : \boldsymbol{\eta} = \mathbf{0}$ against $H_2 : \boldsymbol{\eta} \neq \mathbf{0}$. Methods of testing H_0 against H_2 are available in standard text books on the analysis of binomial data. As was discussed in a previous example, a test of $H_0 : \boldsymbol{\eta} = \mathbf{0}$ against $H_2 : \boldsymbol{\eta} \neq \mathbf{0}$ is valid for testing $H_0 : \boldsymbol{\eta} = \mathbf{0}$ against $H_1 : \boldsymbol{\eta} \geq \mathbf{0}$, but is unlikely to be as powerful as one that is specifically developed for testing against $H_1 : \boldsymbol{\eta} \geq \mathbf{0}$.

Table 1.6 Genotoxicity of potassium chromate (KCr) and 9-aminoacridine (9-AA)

	KCr(μm)				
9-AA(μm)	0	1.4	2.9	4.3	5.7
0	0/192	9/192	3/192	1/192	3/192
40	49/102	92/192	19/192	77/192	60/192

Note: 9/192 means 9 positive responses out of 192 trials.

Reprinted from: *Mutation Research*, Vol 171, La Velle, Potasium chromate potentiates frameshift mutagenesis in *E. coli* and *S. Ttyphimurium*, Pages 1-10, Copyright (1986), with permission from Elsevier.

This type of issue arises in other areas. For example, noncarcinogenic toxic reactions to therapeutic agents have been caused by interactions of compounds that were normally safe when given alone (see Piegorsch (1990)). A formulation of this phenomenon also involves sign constraints on the interaction terms similar to (1.2).

In Chapter 4, we will consider general statistical models that include the generalized linear model in which the distribution of y_i belongs to an exponential family and that for some suitable link function, g, we have that

$$g(\mu_i) = \boldsymbol{x}_i^T \boldsymbol{\eta}$$

where μ_i is the mean of y_i. The loglikelihood functions of such generalized linear models (and some partial likelihoods) are well behaved, for example, they are concave and smooth. Consequently, we can (for example, see Silvapulle (1994)) obtain elegant results for testing

$$H_0 : \boldsymbol{R\eta} = \boldsymbol{0} \text{ against } H_1 : \boldsymbol{R\eta} \geq \boldsymbol{0}.$$

This collection of results can be used for the foregoing testing problem in the binomial response model. ■

Example 1.2.8 *Testing that a treatment is better than the control when the responses are ordinal (Grove (1980), Silvapulle (1994))*

The results of a trial involving two treatments for ulcer are given in Table 1.7. The objective is to test the hypothesis that Treatment B is better than Treatment A.

This testing problem can be formulated in several ways. One possible approach is to say that Treatment B is better than Treatment A, if the local odds ratios are all greater than or equal to 1 with at least one of them being greater than one; recall that the local odds ratios are defined as follows:

$$\phi_1 = \pi_{11}\pi_{22}/\{\pi_{21}\pi_{12}\}, \phi_2 = \pi_{12}\pi_{23}/\{\pi_{22}\pi_{13}\}, \text{ and } \phi_3 = \pi_{13}\pi_{24}/\{\pi_{23}\pi_{14}\}$$

where π_{ij} is the probability that the response of a given individual in the ith row falls in column j. Therefore, the null and alternative hypotheses are

$$H_0 : \boldsymbol{\phi} = \boldsymbol{1} \quad \text{and} \quad H_1 : \boldsymbol{\phi} \geq \boldsymbol{1}, \tag{1.3}$$

where $\boldsymbol{\phi} = (\phi_1, \phi_2, \phi_3)^T$ and $\boldsymbol{1} = (1, 1, 1)^T$. The restriction $\boldsymbol{\phi} \geqslant \boldsymbol{1}$ is known as the *likelihood ratio order.*

Table 1.7 Response of two groups to treatments to an ulcer

	Response				
Treatment group	Worse	Good	Very good	Excellent	Total
A (Control)	12	10	4	6	32
B (Treatment)	5	8	8	11	32

Table 1.8 The 2 x 2 tables formed by merging adjacent columns of Table 1.7

	1		2		3	
Treatment group	Col 1	Col 2,3,4	Col 1,2	Col 3,4	Col 1,2,3	Col 4.
A	12	20	22	10	26	6
B	5	27	13	19	21	11
Odds ratio :	γ_1		γ_2		γ_3	

Grove (1980) argues that (1.3) may be a rather stringent formulation of the hypothesis, and suggests the following formulation of the alternative hypothesis:

$$H_1 : (\sum_{j=1}^{q} \pi_{1j})(\sum_{j=q+1}^{C} \pi_{2j})\{(\sum_{j=1}^{q} \pi_{2j})(\sum_{j=q+1}^{C} \pi_{1j})\}^{-1} \geqslant 1, q = 1, \ldots, C-1 \quad (1.4)$$

where $C = 4$ is the number of columns. To interpret this, consider the collection of 2×2 tables (see Table 1.8) that may be formed by merging the adjacent columns of Table 1.7. Let $\gamma_1, \gamma_2, \gamma_3$ denote the odds ratios for the three 2×2 tables. Now the alternative hypothesis in (1.4) is the same as

$$H_1 : \gamma_1 \geqslant 1, \gamma_2 \geqslant 1, \gamma_3 \geqslant 1;$$

this is not as stringent as the H_1 in (1.3). A detailed discussion of inference when the response variable is ordinal is provided in Chapter 6 where other formulations of the alternative hypothesis such as *hazard rate order, continuation odds ratio order,* and *stochastic order* will also be studied. ■

Example 1.2.9 *A group sequential study (Lipids and cardiovascular diseases)*

The National Heart, Lung and Blood Institute (USA) planned a multicenter clinical trial in 1971 to investigate the impact of lowering the blood cholesterol level on the risk for cardiovascular diseases. The study involved 3952 males, between the ages 35 and 60, chosen from a large pool of people having no previous heart problems but having an elevated blood cholesterol level. They were randomly divided into a placebo and a treatment group, the treatment group receiving some drug to lower the cholesterol level to about 160 or less. Based on cost and other medical considerations, a 12-year (1972-1984) study was planned, and the data coordination task was given to the University of North Carolina, Chapel Hill. The primary response variable was the failure time, while there were a large number of explanatory and covariables.

Medical ethics prompted that if at any time during the study, the treatment is judged superior, the trial should be terminated, and all surviving persons should be switched to the treatment protocol. This clearly set the tone for *interim analysis,* looking at the accumulating data set regularly. It was decided to have a multiple one-sided hypothesis testing setup, in a time-sequential framework: every quarter, there would be a statistical test, one-sided alternative, and if at any time a significance is observed, the trial would be concluded along with the rejection of the null hypothesis of no treatment effect (i.e., the cholesterol level in blood has no impact on heart problems). These test statistics were not independent, nor could they be based on simple normal or exponential laws. So, the first task was to develop a time-sequential hypothesis testing procedure for one-sided alternatives. For some of these details, we refer to Sen (1981, Ch.11) and Sen (1999a). Much of these technicalities have become a part of standard statistical interim analysis, and we shall refer to that again in Chapter 5. The restraints on the parameters (survival functions) were compatible with the medical and statistical considerations, and a conclusive result was announced in the 1984 issue of the *Journal of the American Medical Association.* ■

Example 1.2.10 *Testing for/against the presence of random effects (Stram and Lee (1994))*

Consider the well-known growth curve data set of Potthoff and Roy (1964). The data consists of measurements of the distance (in *mm*) from the center pituitary to the pteryomaxillary fissure for 27 children (11 girls and 16 boys). Every subject has four measurements, taken at ages 8, 10, 12, and 14 years. The object is to model the growth in distance as a function of age and sex. To allow for individual variability in the growth function, the model being considered is

$$\boldsymbol{y}_i = (1, Age, Sex, Age.Sex)\boldsymbol{\alpha} + (1, Age)\boldsymbol{\beta}_i + \boldsymbol{\epsilon}_i$$

where $\boldsymbol{y}_i$ is 4×1 for the 4 measurements of the ith subject, $Age = (8, 10, 12, 14)^T$, $Sex = (a, a, a, a)^T$ where a is 0 for boys and 1 for girls. The parameter $\boldsymbol{\alpha}(4 \times 1)$ captures the fixed effects and $\boldsymbol{\beta}_i(2 \times 1)$ captures the random effect of subject i. Assume that

$$\boldsymbol{\epsilon}_i \sim N(\mathbf{0}, \sigma^2 \boldsymbol{I}), \quad \boldsymbol{\beta}_i \sim N(\mathbf{0}, \Psi)$$

for some 2×2 positive semi-definite matrix Ψ, and that $\boldsymbol{\epsilon}_i$ and $\boldsymbol{\beta}_i$ are independent. More generally, the regression model can be written as

$$\boldsymbol{y}_i = \boldsymbol{X}_i\boldsymbol{\alpha} + \boldsymbol{Z}_i\boldsymbol{\beta}_i + \boldsymbol{\epsilon}_i.$$

In the process of analyzing these data, one would be interested to know whether or not some random effects are present; equivalently, would a smaller model fit as well as a larger one? In this case, the parameter of interest is $\Psi = \text{cov}(\boldsymbol{\beta}_i)$, and the following testing problems are of interest:

1. $H_0 : \Psi = \begin{pmatrix} 0 & 0 \\ 0 & 0 \end{pmatrix}$ vs $H_1 : \Psi$ is positive semi-definite.

2. $H_0 : \Psi = \begin{pmatrix} \psi_{11} & 0 \\ 0 & 0 \end{pmatrix}$ and $\psi_{11} \geqslant 0 \quad vs \quad H_1 : \Psi$ is positive semi-definite.

If $\Psi = \mathbf{0}$ then there are no random effects. If $\Psi = (\psi_{11}, 0 \mid 0, 0)$ then the individual variability is captured by a random intercept, and hence the regression planes are parallel for different subjects. If Ψ is positive definite then individual variability results in not only a random intercept but also a random slope parameter for Age.

More generally, consider a generalized linear model with linear predictor of the form

$$\boldsymbol{X}_i\boldsymbol{\alpha} + \boldsymbol{U}_i\boldsymbol{\gamma}_i + \boldsymbol{Z}_i\boldsymbol{\beta}_i$$

where $\boldsymbol{\gamma}_i$ and $\boldsymbol{\beta}_i$ are random effects, for example

$$\boldsymbol{\gamma}_i \sim N(\boldsymbol{0}, \boldsymbol{D}_1) \quad \text{and} \quad \boldsymbol{\beta}_i \sim N(\boldsymbol{0}, \boldsymbol{D}_2).$$

Suppose that we wish to test

$$H_0 : \boldsymbol{D}_1 = \boldsymbol{0} \quad \text{vs} \quad H_1 : \boldsymbol{D}_1 \text{ is positive semi-definite.}$$

These hypotheses can be tested using Likelihood Ratio and Local Score tests, among others; the theory for this will be developed in Chapter 4. ■

Example 1.2.11 *Sign testing problem: testing whether an identified treatment is the best*

The problem here is to test whether or not a particular treatment, say A, is better than each of K other treatments. To illustrate an example where this arises, let A denote a drug formed by combining drugs B and C; in the medical/pharmaceutical literature, A is known as *combination drug/therapy*. Since A is a combination of drugs, it may have a higher level of risk of adverse effects. Therefore, a standard requirement of regulatory authorities to approve a combination drug, such as A, is that it must be shown that the combination drug is *better* than each of its components (see Laska and Meisner, 1989). Let X_0, X_1 and X_2 be three random variables that characterize the effects of the drugs A, B, and C, respectively. Suppose that the distribution of X_i is associated with a parameter, μ_i, and assume that larger values of μ_i are more desirable. Let

$$\theta_1 = \mu_0 - \mu_1 \text{ and } \theta_2 = \mu_0 - \mu_2.$$

Now, the null and alternative hypotheses are

$$H_0 : \theta_1 \leqslant 0 \textit{ or } \theta_2 \leqslant 0, \quad \text{and} \quad H_1 : \theta_1 > 0 \textit{ and } \theta_2 > 0$$

respectively. In each of the previous examples, the alternative parameter space included its boundary as well. By contrast, in this example, the alternative parameter space does not include the boundary, and the null parameter space is the union of the second, third, and fourth quadrants, and the boundary of the first quadrant. These types of testing problems can be solved using *intersection-union tests.*

In this example, we assumed that there is only one primary variable of interest and hence X_i was univariate. There are clinical trials in which the assessment involves several variables (called, *multiple end points*); for example, interest may be on several symptoms simultaneously. Therefore, X_i and μ_i may be vectors. ■

Example 1.2.12 *Combination drug for low-back pain and spasm (Snappin (1987))*

Dolobid(D) and Flexeril(F) are approved drugs for spasm and pain relief respectively. It was conjectured that the new drug DF, a combination of D and F, is likely to be better. As was indicated in Example 1.2.11, a requirement of the regulatory authorities for the approval of a new combination drug such as DF is that it must be better than its components D and F.

A study was conducted to evaluate the efficacy of DF relative to D only, F only, and Placebo (P). Patients entering the study were randomly assigned to one of the four treatments, DF, D, F, and P. At the end of the study period, each patient provided an evaluation of the treatment on a five-point ordinal scale ranging from Marked Improvement to Worsening of the condition (see Table 1.9). The objective is to test the claim that the combination drug is better than each of the other three treatments.

Let

$$DF \succ F \text{ denote } DF \text{is better than } F;$$

this can be formulated using different odds ratios as in Example 1.2.8. Now, the inference problem may be formulated as test of

$$H_0 : \text{Not } H_1 \quad \text{vs} \quad H_1 : DF \succ F, DF \succ D, \text{ and } DF \succ P.$$

The inference problem can also be formulated in the following simpler form. Let μ_0, μ_1, μ_2, and μ_3 denote parameters that capture the effects of the four treatments in Table 1.9; such parameters can be defined by introducing various parametric assumptions (see Snappin (1987)). Let $\theta_i = \mu_0 - \mu_i$ for $i = 1, 2, 3$. Now, one way of formulating the inference problem is test of

$$H_0 : \theta_1 \leqslant 0, \; or \; \theta_2 \leqslant 0, \; or \; \theta_3 \leqslant 0, \; \text{vs } H_1 : \theta_1 > 0, \theta_2 > 0, \; and \; \theta_3 > 0.$$

This problem will be discussed under intersection union tests. ■

Table 1.9 Comparison of Dolobid/Flexeril with Dolobid, Flexeril, and placebo.

	Degree of improvement					
Treatment	Marked	Mod	Mild	No Change	Worse	Total
Dolobid/Flexeril	50	29	19	11	5	114
Dolobid alone	41	28	23	22	2	116
Flexeril alone	38	27	18	27	3	113
Placebo	35	20	23	37	4	119

Example 1.2.13 *Cross-over interaction in a multicenter clinical trial (Ciminera et al. (1993b))*

A multicenter clinical trial was conducted to compare a drug with a placebo; the drug was a treatment to reduce blood pressure (see Table 1.10). This is an unpaired design (i.e., parallel group) within each center, and there were 12 centers. In the context of this example, we say that there is *cross-over interaction* between the treatment and center if the drug is beneficial in at least one center and harmful in at least one other. The question of interest is whether or not there is any evidence of crossover interaction.

The problem may be formulated as follows. Let δ_i denote the effect of the treatment in Center i ($i = 1, \ldots, 12$). Then the null and the alternative hypotheses may take the form

$$H_0 : \boldsymbol{\delta} \geqslant \mathbf{0} \text{ or } \boldsymbol{\delta} \leqslant \mathbf{0}, \quad \text{and} \quad H_1 : \boldsymbol{\delta} \ngeq \mathbf{0} \text{ and } \boldsymbol{\delta} \nleq \mathbf{0},$$

respectively. In this formulation, the null hypothesis says that there is no cross-over interaction. It is also possible to interchange the roles of H_0 and H_1 in the testing problem so that the null and alternative hypotheses are H_1 and H_0, respectively. In all of these, it may be of interest to replace δ_i by $\delta_i - \epsilon_i$ for some given ϵ_i so that $\delta_i - \epsilon_i > 0$ corresponds to a treatment effect that is clinically significant; such a modification through the introduction of small ϵ_is helps to accommodate the fact that a cross-over interaction may not be important if it is not clinically significant. ■

Table 1.10 Comparison of a drug with a placebo in a multicenter trial [a]

i	n_{1i}	n_{2i}	df	$\bar{z}_{1i\cdot}$	$\bar{z}_{2i\cdot}$	$\hat{\delta}_i$	ss_i	$\hat{\delta}_i/se(\hat{\delta}_i)$
1	3	3	4	-7.0	-15.3	8.3	176.67	1.53
2	4	6	8	-13.0	-6.0	-7.0	180.00	-2.29
3	7	8	13	-3.0	-15.9	12.9	1492.87	2.33
4	7	8	13	-13.6	-11.0	-2.6	693.71	-0.69
5	6	8	12	-3.0	-11.8	8.8	345.50	3.04
6	7	6	11	2.7	-4.3	7.0	236.76	2.71
7	5	8	11	-2.2	-5.0	2.8	472.80	0.75
8	7	9	14	-8.3	-17.6	9.3	1080.93	2.10
9	8	8	14	-7.4	-12.2	4.8	507.38	1.59
10	8	9	15	-7.4	-6.7	-0.7	711.87	-0.21
11	4	4	6	2.0	-8.5	10.5	519.00	1.60
12	3	3	4	0.7	-12.0	12.7	116.67	2.88
Total	69	80	125				2067.7	

[a] The within center sample means for treatments 1 and 2 are denoted by $\bar{z}_{1i\cdot}$ and $\bar{z}_{2i\cdot}$ respectively; $\hat{\delta}_i = \bar{z}_{1i\cdot} - \bar{z}_{2i\cdot}$; ss_i = within group sum of squares for group i; $se(\cdot)$ = standard error.

Example 1.2.14 *Testing against the presence of ARCH effect in ARCH or ARCH-M models (Silvapulle and Silvapulle (1995) and Beg et al. (2001))*

The AutoRegressive Conditional Heteroscedasticity (ARCH) model, introduced by Engle (1982), is one of the widely used econometric models, particularly in Financial Economics. This example involves two models in this family.

Let y_t denote the rate of return from an investment (portfolio) at time t; here we are thinking of investment in stocks or bonds. The basic form of the ARCH model is

$$y_t = \mu + \boldsymbol{\beta}^T \boldsymbol{x}_t + \epsilon_t, \tag{1.5}$$

$$\epsilon_t | \mathcal{F}_{t-1} \sim N(0, h_t^2), \qquad h_t^2 = \alpha_0 + \psi_1 \epsilon_{t-1}^2 + \ldots + \psi_p \epsilon_{t-p}^2 \tag{1.6}$$

where $\mathcal{F}_{t-1}$ is the collection of information up to time $(t-1)$ and $\boldsymbol{x}_t$ is a vector of exogenous variables. A main feature of this model is that periods of small (respectively, large) random fluctuations in y tend to be clustered. This is a frequently observed phenomenon in many high-frequency financial time series.

Another frequently observed phenomenon is that large random fluctuations in stock price would be seen as periods of high risk and this would in turn affect the stock price as investors demand higher returns for higher risk. To incorporate this, Engle et al. (1987) suggested that the specification in (1.5) be modified as

$$y_t = \mu + \boldsymbol{\beta}^T \boldsymbol{x}_t + \phi h_t + \epsilon_t. \tag{1.7}$$

Note that according to (1.7), the conditional variance h_t^2 has a direct effect on the mean of y_t; in other words, the degree of risk, measured by h_t, affects the level of stock price and rate of return. The model (1.6)-(1.7) is known as the ARCH-in-Mean (ARCH-M) model.

In many empirical applications involving the ARCH and ARCH-M models, a question of basic interest is whether or not the conditional variance h_t^2 is a constant over time; or more generally, are some of the ψ parameters equal to zero. Since h_t^2 cannot be negative, $\psi_1, \ldots, \psi_p$ are all nonnegative. Therefore, if we were to test against the alternative that h_t^2 depends on time, then the testing problem takes the form

$$H_0 : \boldsymbol{\psi} = \mathbf{0} \text{ against } H_1 : \psi_1 \geqslant 0, \ldots, \psi_p \geqslant 0. \tag{1.8}$$

For (1.5), the method that is usually applied by econometric software packages is a score test that ignores the constraints $\boldsymbol{\psi} \geqslant 0$ in the alternative hypothesis. However, it is possible to improve on this by applying a method that incorporates the constraints. The theory for this will be developed in Chapter 4; one of the methods that will be developed is a simple generalization of Rao's Score test, which we shall call a local score test (see Silvapulle and Silvapulle (1995)). For (1.7), there are some additional theoretical issues that need to be taken care of because a parameter becomes unidentified under the null hypothesis; this is a nonstandard problem (see Beg, Silvapulle and Silvapulle (2001), and Andrews (2001)). ■

Now, we provide a few more examples that fall into the categories of the previous ones.

Example 1.2.15 *Acceptance sampling (Berger (1982))*

Table 1.11 lists specifications for upholstery fabric. Similar specifications are available for other products.

It is required to develop an acceptance sampling scheme for a batch that consists of a large number of units. Suppose that a random sample of n units are to be chosen from a batch and tested. The objective is to decide whether or not to accept the batch after testing the sample of n units. Note that some of the tests would destroy or change the fabric, and therefore once a piece of fabric has been used for one test, it may not be possible to use it for another test. The inference problem can be formulated as test of

$$H_0 : \boldsymbol{\theta} \notin \Theta \quad \text{vs} \quad H_1 : \boldsymbol{\theta} \in \Theta$$

where

$$\Theta = \{\boldsymbol{\theta} : \theta_1 \geqslant 50, \theta_2 \geqslant 6, \theta_3 \leqslant 5, \theta_4 \leqslant 2, \theta_5 \in A_1, \\ \theta_6 \in B_4, \theta_7 \in C_4, \theta_8 \in D_3, \theta_9 \in E_4, \theta_{10} \in P\}.$$

First, suppose that each unit in the sample can be subjected to every test in the table of specifications. Typically, this would be a reasonable assumption if the properties of the product are uniform within each sampling unit and it can be divided into subunits for different tests. In this case, the statistical inference problem reduces to elementary inference on a single population proportion.

A different approach is required if a sampling unit cannot be subjected to all the different tests in the table of specifications. In this case, the *intersection union test*

Table 1.11 Standard specification for woven upholstery fabric–plain, tufted, or flocked

Variable	Minimum standard	
	Requirement	Formulation
Breaking strength	50 pounds	$\theta_1 \geqslant 50$
Tong tear strength	6 pounds	$\theta_2 \geqslant 6$
Dimensional change	5% shrinkage	$\theta_3 \leqslant 5$
	2% gain	$\theta_4 \leqslant 2$
Surface abrasion (heavy duty)	15000 cycles	$\theta_5 \in A_1$
Colorfastness to		
Water	class 4	$\theta_6 \in B_4$
Crocking		
Dry	class 4	$\theta_7 \in C_4$
Wet	class 3	$\theta_8 \in D_3$
Light-40 AATCCF (fading units)	class 4	$\theta_9 \in E_4$
Flammability	Pass	$\theta_{10} \in P$

Source: The original source of the information contained in this table is the Annual Book of ASTM Standards (1978), American Society for Testing and Materials, Philadelphia, part 32, page 717. This table is reproduced from *Technometrics*, Volume 24, R. L. Berger, Multiparameter hypothesis testing and acceptance sampling, pages 295–300.

provides a convenient solution to this problem. Essentially, a test at level-α accepts the batch if *each* specification is passed by a level-α test; there are no Bonferroni type adjustment because this does not come within the framework of what is usually known as multiple testing. ■

Example 1.2.16 *Analysis of* $2 \times 2 \times 2$ *tables (Cohen et al. (1983a))*

A random sample of 15 individuals who have been vaccinated against a certain disease, and another independent random sample of 15 individuals who have not been vaccinated against the same disease were chosen. The individuals were cross-classified with respect to the binary variables, Vaccination (yes, no), Immune Level (low, high), and Infection (well, sick). The data are given in Table 1.12.

Table 1.12 Effect of vaccination and immune level on infection

Infection	Not vaccinated & low immun.	Not vaccinated & high immun.	Vaccinated & low immun.	Vaccinated & high immun.
Well	4	3	3	7
Sick	7	1	3	2

Assume that vaccination affects the risk of infection only through its effect on the immune level. It is required to test the validity of the claim

H : vaccination improves immune level, which in turn reduces the risk of infection;

the terms *improves* and *reduces* lead to multiparameter one-sided hypotheses. This type of inference problem can be formulated as test of

$$H_0 : \boldsymbol{\alpha} \notin A \text{ or } \boldsymbol{\beta} \notin B \qquad \text{vs} \qquad H_1 : \boldsymbol{\alpha} \in A \text{ and } \boldsymbol{\beta} \in B$$

for some parameters $\boldsymbol{\alpha}$ and $\boldsymbol{\beta}$ and suitably chosen sets A and B. Cohen et al. (1983a) developed a methodology for this type of problem and illustrated their results using these data. This example will be studied in Chapter 6. ■

Example 1.2.17 *Meta-analysis (Cutler et al. (1991), Follmann (1996a))*

A recent meta analysis summarized the effect of sodium reduction on blood pressure from published clinical trials. An objective of the study was to provide an overall probability statement on the joint effect of sodium reduction on reducing both the systolic blood pressure (X_1) and diastolic blood pressure (X_2).

Table 1.13 provides the data for the meta analysis. In fact, the table provides considerable information that would lend to more extensive analyzes. Let $\boldsymbol{X}$ be

Table 1.13 Randomized trials of sodium reduction

Sample size	Study length (months)	Reduction in urinary sodium	Reduction in BP (mm Hg) SBP (se)	DBP (se)
		Cross-over trials; hypertensive subjects		
15	1	98	6.7 (3.76)	3.2 (3.47)
19	1	76	10 (3.06)	5.0 (1.78)
18	1	56	0.5 (1.5)	0.3 (0.8)
12	1	105	5.2 (4.1)	1.8 (3.55)
40	1.5	72	0.8 (1.8)	0.8 (1.55)
20	1	82	8.0 (2.6)	5.0 (1.62)
9	1	76	9.7 (4.33)	5.1 (2.94)
88	2	67	3.6 (0.7)	2.1 (0.4)
Parallel studies with treatment and control groups; hypertensive subjects				
31/31	24	27	1.5 (4.6)	6.9 (2.12)
6/6	2	98	*	6.0 (4.08)
6/6	2	78	*	4.0 (3.30)
10/15	12	53	8.7 (10.23)	6.3 (4.41)
15/19	1.5	117	−1.8 (4.14)	−0.5 (2.37)
15/15	2.3	89	13.3 (5.46)	6.7 (3.07)
18/12	3	171	2.0 (4.96)	−2 (2.84)
44/50	6	58	2.7 (2.2)	3.4 (1.7)
48/52	3	54	5.1 (1.42)	4.2 (0.9)
37/38	6	32	1.1 (2.15)	0.2 (1.41)
17/17	3	59	13.0 (5.99)	1.8 (3.48)
50/53	2	71	5.5 (1.48)	2.8 (0.85)
20/21	12	*	18.3 (4.35)	5.9 (2.91)
		Cross-over trials; normotensive subjects		
20	0.5	170	2.7 (2.36)	3.0 (2.03)
113	2	69	0.6 (0.84)	1.4 (1.07)
35	0.9	74	1.4 (0.74)	1.2 (0.93)
31	1	60	0.5 (0.82)	1.4 (0.90)
172	1	130	3.5 (1.25)	1.9 (0.9)
Parallel studies with treatment and control groups; normotensive subjects				
19/19	0.5	117	1.5 (3.32)	1.1 (2.03)
174/177	36	16	−0.1 (1.00)	−0.2 (0.8)

the bivariate vector that consists of X_1 and X_2. Let the mean of (X_1, X_2) be denoted by (θ_1, θ_2). Now, the statistical inference problem can be formulated as test of $H_0 : (\theta_1, \theta_2) = (0, 0)$ against $H_1 : (\theta_1, \theta_2) \geq (0, 0)$. Different rows in Table 1.13 correspond to studies with different sample sizes. Therefore, we have independent observations $\boldsymbol{X}_1, \ldots, \boldsymbol{X}_n$ with a common mean $\boldsymbol{\theta}$ but with different covariance matrices. ∎

Example 1.2.18 *Distribution-free analysis of covariance (Boyd and Sen (1986))*

The nature of the problem in this example is that there are several treatments and some covariates. We are interested to test for equality against an ordered alternative of the treatment effects. However, there was adequate evidence to suggest that the error distribution is unlikely to be normal and, further, the effects of the covariates on the response variable are unlikely to be linear. Thus, some kind of distribution-free method is required.

A trial was conducted (by A.H. Robins Company) to compare the efficacy of three treatments for the reversal of anesthesia. For confidentiality purposes, the treatments were identified as 1, 2, and 3. The response variable, Y, is *time elapsed from administration of treatment to completion of reversal.* The covariates are *Depth* = "depth of neuromuscular block at time of reversal (time treatment administered)" and *Age* = "the age (in years) of the patient"; the data are given in Table 1.14.

Based on various diagnostics such as plot of the residuals, there is adequate evidence that the distribution of Y at a fixed value of (Depth, Age) is not normal. Further, within each treatment, the dependence of Y on (Depth, Age) is unlikely to be linear. Therefore, the usual ANCOVA based on normal theory linear model is inappropriate.

Table 1.14 Patient response to treatments for reversal of anesthesia

Treatment	Y	*Depth*	*Age*	*Treatment*	Y	*Depth*	*Age*
1	4.8	17	27				
1	13.2	6	41	2	5.2	61	45
1	5.8	30	27	2	6.6	26	34
1	4.6	40	33	2	2.7	45	41
1	6	30.5	34	2	5.4	45	29
1	2.9	48	21	2	8.2	21	25
1	5.2	50	43	3	16.4	15	41
1	5.6	20	26	3	6.7	6	32
1	3.9	40	36	3	6.7	6	32
1	5.6	24.4	23	3	7.9	15	28
2	6	30	25	3	6	32	46
2	9.6	25	58	3	19.4	0	45
2	15.5	25	55	3	19	15.7	24
2	8.7	56	43	3	2.8	59	23
2	7.9	51	36	3	6.6	26	29
				3	10.4	21	35

The problem may be formulated as follows. Let $F(y - \theta_k|\boldsymbol{x})$ denote the distribution of Y for Treatment k at *(Depth, Age)* = $\boldsymbol{x}$. The functional form of F and the form of dependence of a location parameter of Y on $\boldsymbol{x}$ are unknown. The null and alternative hypotheses are $H_0 : \theta_1 = \theta_2 = \theta_3$ and $H_1 : \theta_1 \leqslant \theta_2 \leqslant \theta_3$. Some general results based on ranks can be applied for this type of problems. ■

1.3 COVERAGE AND ORGANIZATION OF THE BOOK

Statistical Decision Theory (SDT) attempts to bring both wings of statistical inference under a common shade. In that process, it often becomes more abstract than what could be more easily routed to fruitful applications. Faced with this dilemma, we plan to treat the estimation theory and hypothesis testing more or less on a complementary basis, although at the end, in Chapter 8, we would attempt to bind them together by SDT strings. Constrained statistical inference (CSI) arises in many areas of applications, and some of these have been illustrated in the preceding section. From pure theoretical perspectives, *optimal* statistical decision procedures generally exist only under very specific regularity assumptions that are not that likely to be tenable in constrained stochastic environments. As such, in the sequel, we shall first trace the rather tenuous domain of CSI problems where finite-sample optimal decisions exist. In a way to get out of this restricted scope of CSI, we would spring up in a beyond parametric scenario and illustrate the variety of competing methodologies that attempt to capture the glimpses of CSI with due emphasis on robustness, validity, and efficacy considerations in a far more general setup (where exact optimality may not percolate). In the rest of this section, we attempt to provide a motivating coverage of the book.

Ours is not the first book or monograph in this specified field. We would like to acknowledge with profound thanks the impetus we have had from the earlier two precursors: Barlow et al. (1972), to be referred to in the sequel as $B4$ (1972), and Robertson et al. (1988), those two being 16 years apart and otherwise sequentially very much related. We reverently regard $B4$ as the foundation of the CSI at a time when it was quite difficult to impart the much needed SDT to support the basic axioms. Sixteen years later, the foundation has been fortified further in Robertson et al. (1988) with annexation of some research published after 1970. In the past 16 years, there has been a phenomenal growth of research literature in CSI, not only covering the groundwork reported in the earlier two volumes, but much more so, in novel areas that have been annexed to statistical inference more intensively during the past two decades. As such, with enthusiasm, we have attempted to bring out an updated coverage of the entire integrated area of CSI, strangely after a lapse of another 16 years!

Given our motivation to integrate the CSI across the parametrics to all the way to beyond parametrics, our task has not been very easy or routine. Based on our contemplated intermediate level of presentation (for the convenience of upper-undergraduate- and lower-graduate-level students as well as researchers in applied areas), we have therefore recast the setup as follows.

In Chapters 2, 3, and 4, we consider the so called exact CSI in the parametric setting with due emphasis on the role of the classical *likelihood principle* (LP) in CSI. Chapter 2 provides an introduction to the traditional order-restricted tests in the one-way classification model. An objective of this chapter is to illustrate that, for comparing several populations in the usual one-way classification, it is easy to use a natural generalization of the traditional F-test to incorporate order restrictions on the population means. It is hoped that these methods will be incorporated into standard statistical software so that they could be used easily. Chapter 3 provides a rigorous theoretical development of the likelihood approach for testing hypotheses about the mean of a multivariate normal and the regression parameter in the normal theory linear model. These results lay the foundation on a firm footing for CSI in general statistical models that may be parametric, semiparametric, or nonparametric. In Chapter 4, we broaden the multivariate normal framework of the previous chapter and consider general parametric models which includes the generalized linear models and nonlinear time-series models. In this setting, very little is known about the exact distribution of statistics (estimators/test statistics); this is to be expected because even when there are no order restrictions on parameters, most of the available results are asymptotic in nature. Chapter 4 is devoted to developing the essentials for large sample statistical inference based mainly on the likelihood when there are constraints on parameters. Chapters 3 and 4 form the core of constrained statistical inference and open the door to a vast literature of current research interest. The remaining chapters assume familiarity with the results in these two chapters.

It would become more evident as we make progress in the subsequent chapters that in CSI, LP is often encountered with some challenges, particularly, in not so simple models, and there could be some alternative approaches that may have greater flexibility in this respect. Chapter 5 brings out this issue with the exploration of the Roy (1953) *union-intersection principle* (UIP), which has been more conveniently adapted in beyond parametrics scenarios, too. Chapters 5, 6, and 7 highlight these developments with genuine nonparametrics as well as some other semiparametric models. Functional parameters as well as more complex statistical functionals arising in survival and reliability analysis and multiple comparisons have been focused in this context. Chapter 8 deals with the essentials of SDT and attempts to tie-up some loose ends in theory that were relegated from earlier chapters to this one. In particular, shrinkage estimation and Bayes tests in CSI are unified in this coverage. Some miscellaneous CSI problems, we found hard to integrate with the presentations in earlier chapters, are relegated to the concluding one. In this way, our contention has been to present the theory and methodology at a fairly consistent and uniform (intermediate) level, albeit, for technical reasons, in latter chapters, we might have some occasional detours into higher level of abstractions, to a small extent. We therefore strive to provide more up-to-date and thorough coverage of the bibliography, particularly the post-1987 era, so that serious and advanced readers could find their desired way out through the long bibliography appended.

2

Comparison of Population Means and Isotonic Regression

The need to compare several treatments/groups/populations with respect to their means or medians or some other location parameters arises frequently in many areas of applications. Typically, we have k populations with means $\mu_1, \ldots, \mu_k$ and independent samples from the k populations. One of the statistical inference problems that often arises in this context is test of

$$H_0 : \mu_1 = \ldots = \mu_k \quad \text{vs} \quad H_2 : \mu_1, \ldots, \mu_k \text{ are not all equal.}$$

Standard methods for this include the F-test in analysis of variance (ANOVA) and nonparametric methods based on ranks. However, when several treatments are being compared, one is usually prepared to assume that certain treatments are not worse than another, for example, $\mu_1 \leq \mu_2 \leq \mu_3$. In this setup, it is of interest to incorporate the prior information $\mu_1 \leq \mu_2 \leq \mu_3$ for the purposes of improving the statistical analysis. In other words, one would like to target the analysis to detect a departure away from $\mu_1 = \mu_2 = \mu_3$ but in the direction of $\mu_1 \leq \mu_2 \leq \mu_3$. The usual unrestricted F-test in ANOVA is not particularly suitable for this because it does not incorporate prior knowledge such as $\mu_1 \leq \mu_2 \leq \mu_3$ and hence it is not specifically targeted to detect departures in any particular direction of (μ_1, μ_2, μ_3). This chapter provides an introduction to an approach for dealing with this type of statistical inference problems.

Most of the results used in this chapter are also discussed in the following chapters under more general settings. Therefore, advanced readers may start with the next chapter and refer back to this chapter when necessary, or read up to Section 2.3 and then turn to the next chapter.

In the statistics literature, a constraint/prior-knowledge of the form $\mu_i \leqslant \mu_j$ is called an *order restriction* or an *inequality constraint* on the parameters. When an order restriction appears in a hypothesis, as in the foregoing example, it is called an *ordered hypothesis* or more generally a *constrained hypothesis.*

As an example, consider Example 1.2.1 (page 2) on the comparison of two exercise programs with a control. In this example, the response variable is the age at which a child starts to walk. Let μ_i denote the population mean of the response variable for Treatment i, $i = 1, 2, 3$. An inference problem that is of interest in this context is test of

$$H_0 : \mu_1 = \mu_2 = \mu_3 \quad \text{vs} \quad H_2 : \mu_1, \mu_2, \text{ and } \mu_3 \text{ are not all equal.}$$

For illustrative purposes let us suppose that the exercise program was expected not to have detrimental effects, and that the researcher was interested to ensure that the prior knowledge, $\mu_1 \leqslant \mu_2 \leqslant \mu_3$, is incorporated in the statistical analysis when testing H_0 against H_2. This chapter introduces what we call the $\bar{F}$-test, a natural generalization of the usual F-test, to incorporate prior knowledge in the form of order restrictions. A classical test known as the $\bar{E}^2$-test is also introduced. For most practical purposes the difference between these two tests appear to be not that large. An objective of this chapter is to illustrate that it is fairly easy to apply these tests to incorporate prior information such as $\mu_1 \leqslant \mu_2 \leqslant \mu_3$. To this end, only the essential results to implement the test are presented here; a detailed study of the distribution theory in a general setting will be presented in the next chapter.

Section 2.1 provides an introduction to test of $H_0 : \mu_1 = \ldots = \mu_k$ against

$$H_1 : \mu_1, \ldots, \mu_k \text{ are not all equal and they satisfy several order restrictions.}$$

Example 1.2.2 indicated a scenario where order restrictions appear in the null hypothesis. These type of problems are discussed in Section 2.2. More specifically, this section develops a test of

$$H_1 : \{\mu_1, \ldots, \mu_k\} \textit{ satisfy several order restrictions} \quad \text{vs} \quad H_2 : \textit{not } H_1.$$

It will be shown that the tests developed in Section 2.1 for H_0 vs H_1 and those in Section 2.2 for H_1 vs H_2 can be implemented by using a simulation approach. Further, only a short computer program is required to implement this simulation approach. Thus, an important conclusion of the first two sections is that test of hypothesis involving $\mu_1, \ldots, \mu_k$ when there are order restrictions in the null or the alternative hypothesis can be implemented easily. Such a simulation approach overcomes the computational difficulties encountered in the classical approach of computing the exact critical/p-values and/or bounds for them. Consequently, the statistical results become clearer and easier to use.

Section 2.3 provides an introduction to *isotonic regression*. The foregoing example involving three treatments in which the treatment means are ordered is an example of topics studied under isotonic regression. The origin of the term isotonic regression is that the basic model used is the *regression* model $y_{ij} = \mu_i + \epsilon_{ij}$, and the parameters $\mu_1, \ldots, \mu_k$ are required to have the *same tone* (*iso tone*) as the order imposed on the treatments, for example,

$$\text{Treatment 1} \preceq \text{Treatment 2} \preceq \text{Treatment 3},$$

where Treatment 1 $\preceq$ Treatment 2 may be read as "the mean for Treatment 2 is at least as large as that for Treatment 1"; more generally, it could also be read as "Treatment 2 is at least as good as Treatment 1."

The results in Sections 2.1 and 2.2 for testing $H_0 : \mu_1 = \ldots = \mu_k$ against an order restriction are based on the linear model $y_{ij} = \mu_i + \epsilon_{ij}$. Most of these results have natural extensions to the more general linear model $\boldsymbol{Y} = \boldsymbol{X\beta} + \boldsymbol{E}$ where $\boldsymbol{E}$ has mean zero and covariance $\sigma^2\boldsymbol{U}$, σ is unknown and $\boldsymbol{U}$ is known. These will be discussed in the next chapter.

These have also been extended to incorporate robust tests based on M-estimators and those based on ranks; see Silvapulle (1992b, 1992c, 1985), Puri and Sen (1985), and Geyer (1994). The finite sample distribution theory for normal error distribution will be derived in the next chapter, and the large sample theory will be derived in the following chapter. The results have been extended to bounded influence tests in general parametric models; see Silvapulle (1997c). Section 2.3 provides an introduction to isotonic regression; most of this is related to the results in Section 2.1.

2.1 ORDERED ALTERNATIVE HYPOTHESES

In this section we adopt the same context as that is usually adopted for one-way ANOVA. Suppose that there are k treatments to be compared. Let y_{ij} denote the jth observation for Treatment i, and assume that $y_{ij} \sim N(\mu_i, \sigma^2)$ and that the observations are independent ($j = 1, \ldots, n_i, i = 1, \ldots, k$); see Table 2.1. Later we shall relax the requirement that y_{ij} be normally distributed. In the next subsection, we consider the walking exercise example (see Example 1.2.1) to motivate some of the ideas that underlie the $\bar{F}$-test and to illustrate its application when the number of order restrictions of the form $\mu_i \leqslant \mu_j$ is less than four. Then in the following subsection, we consider the more general case with an arbitrary number of treatments and test of

$$H_0 : \mu_1 = \ldots = \mu_k \quad \text{vs} \quad H_1 : \mu_i - \mu_j \geqslant 0 \text{ for } (i, j) \in B \tag{2.1}$$

where

$$B \subset \{(i, j) : i, j = 1, \ldots, k\}. \tag{2.2}$$

This particular form of the alternative hypothesis allows only pairwise contrasts; more general contrasts will be discussed in the next chapter.

Table 2.1 Normal theory one-way layout

Treatment	Independent observations	Sample mean	Population distribution
1	$y_{11}, \ldots, y_{1n_1}$	$\bar{y}_1$	$N(\mu_1, \sigma^2)$
$\vdots$	$\vdots$	$\vdots$	$\vdots$
k	$y_{k1}, \ldots, y_{kn_k}$	$\bar{y}_k$	$N(\mu_k, \sigma^2)$

2.1.1 Test of $H_0 : \mu_1 = \mu_2 = \mu_3$ Against an Order Restriction

In this subsection, we consider the special case when there are only three treatments and the null hypothesis is $H_0 : \mu_1 = \mu_2 = \mu_3$; this simple case is convenient to introduce the basic ideas. In this special case, there are three possible ordered alternatives that may be of interest in the context of one-way classification; they are

$$(1) \quad \mu_1 \leqslant \mu_2 \leqslant \mu_3, \quad (2) \quad \mu_1 \leqslant \mu_2 \textit{ and } \mu_1 \leqslant \mu_3, \quad \text{and} \quad (3) \quad \mu_1 \leqslant \mu_2.$$

For illustrative purposes, we shall consider the first; the results for the second and third are similar and are also stated in this subsection.

Consider Example 1.2.1, and assume that the setting in Table 2.1 holds. If there is no prior knowledge in the form of an order among the treatment means, then the null and alternative hypotheses would be

$$H_0 : \mu_1 = \mu_2 = \mu_3 \quad \text{and} \quad H_2 : \mu_1, \mu_2, \textit{ and } \mu_3 \textit{ are not all equal,}$$

respectively; the notation H_2, rather than H_1, is used here for the alternative hypothesis to suggest that it is "two-sided" or unrestricted. For testing H_0 against H_2, the classical F-test is usually preferred.

The purpose of the study was to test the claim that walking exercises are associated with a reduction in the mean age at which children start to walk. Let us suppose that the alternative hypothesis is

$$H_1 : \mu_1 \leqslant \mu_2 \leqslant \mu_3, \textit{ and } \{\mu_1, \mu_2, \mu_3\} \text{ are not all equal.}$$

As indicated in the previous chapter, we shall abbreviate the statement of this hypothesis to

$$H_1 : \mu_1 \leqslant \mu_2 \leqslant \mu_3.$$

More generally, we shall adopt the following convention throughout this book:

Notation: *When we refer to test of "H_0 against H_1," it should be read as "H_0 against $H_1 \setminus H_0$"; in the literature, $H_1 \setminus H_0$ is also written as $H_1 - H_0$.*

In view of this notation, test of $\mu = 0$ vs $\mu \geqslant 0$ and test of $\mu = 0$ vs $\mu > 0$ are exactly the same testing problems, where μ is a scalar.

To test H_0 against H_1, it is perfectly valid to apply the usual F-test for H_0 against H_2 on $(2, \nu)$ degrees of freedom, where ν is the error degrees of freedom. The validity is justified by the fact that both testing problems have the same null hypothesis. However, one would expect that the standard F-test for H_0 against H_2 is unlikely to have good power properties for testing H_0 against H_1 because it does not make use of the additional information $\mu_1 \leqslant \mu_2 \leqslant \mu_3$ and hence is not specifically targeted to detect departures in the direction of H_1.

Recall that the standard F-statistic is defined as

$$F = \{RSS(H_0) - RSS(H_2)\}(k-1)^{-1}/S^2 \tag{2.3}$$

where S^2 is the error mean square, k is the number of treatments, $RSS(H)$ is the abbreviation for Residual Sum of Squares under the hypothesis H,

$$RSS(H_0) = \inf_{H_0} \sum\sum (y_{ij} - \mu_i)^2 = \sum\sum (y_{ij} - \bar{y})^2$$

$$RSS(H_2) = \inf_{H_2} \sum \sum (y_{ij} - \mu_i)^2 = \sum \sum (y_{ij} - \bar{y}_i)^2$$

and $\bar{y}$ is the grand mean. The term $\{RSS(H_0) - RSS(H_2)\}$ in the numerator of F (see (2.3)) is a measure of the discrepancy between the null H_0 and the alternative H_2. The denominator, S^2, acts as a scaling factor so that the null distribution of the test statistic does not depend on the unknown scale parameter σ for the error; S^2 does not contribute to quantify the discrepancy between H_0 and H_2.

Remark: Strictly speaking, $RSS(H_2)$ should have been written as $RSS(H_0 \cup H_2)$. Since H_0 is on the boundary of H_2, this does not affect the value of $RSS(H_2)$.

These observations suggest that to test H_0 against H_1, a reasonable test statistic is obtained by modifying the foregoing F-statistic as follows:

$$\bar{F} = \{RSS(H_0) - RSS(H_1)\}/S^2 \tag{2.4}$$

where

$$RSS(H_1) = \min_{H_1} \sum \sum (y_{ij} - \mu_i)^2 = \sum \sum (y_{ij} - \tilde{\mu}_i)^2$$

and $(\tilde{\mu}_1, \tilde{\mu}_2, \tilde{\mu}_3)$ is the point at which the sum of squares $\sum \sum (y_{ij} - \mu_i)^2$ is minimized subject to the constraint in H_1. Thus, $(\tilde{\mu}_1, \tilde{\mu}_2, \tilde{\mu}_3)$ is an estimate of (μ_1, μ_2, μ_3) under H_1 and $RSS(H_1)$ is the corresponding sum of squares of the residuals; in fact, since the errors are *iid* as $N(0, \sigma^2)$, it can be shown that $(\tilde{\mu}_1, \tilde{\mu}_2, \tilde{\mu}_3)$ is the *mle* of (μ_1, μ_2, μ_3) under H_1 [see Problem 2.3]. Thus, the numerator $\{RSS(H_0) - RSS(H_1)\}$ of $\bar{F}$ is a measure of the discrepancy between H_0 and the restricted alternative H_1, just as the numerator of F is a measure of the discrepancy between H_0 and the unrestricted alternative H_2. The constant $(k-1)^{-1}$ in the numerator of the usual F statistic was not included in the definition of $\bar{F}$, but it could have been included without affecting the essential nature of any results.

To implement the foregoing procedure, we need to compute the restricted estimator $(\tilde{\mu}_1, \tilde{\mu}_2, \tilde{\mu}_3)$ under $H_1 : \mu_1 \leq \mu_2 \leq \mu_3$. Since

$$\sum_i \sum_j (y_{ij} - \mu_i)^2 = \sum_i \sum_j (y_{ij} - \bar{y}_i)^2 + \sum_i (\bar{y}_i - \mu_i)^2 n_i \tag{2.5}$$

it suffices to minimize $\sum_i (\bar{y}_i - \mu_i)^2 n_i$ subject to $H_1 : \mu_1 \leq \mu_2 \leq \mu_3$ for the purposes of computing $(\tilde{\mu}_1, \tilde{\mu}_2, \tilde{\mu}_3)$.

If the alternative hypothesis is any other order restriction, say H_1^*, then the foregoing discussions hold with trivial modifications; the main modification is that $RSS(H_1^*)$ $= \sum \sum (y_{ij} - \tilde{\mu}_i)^2$ where $(\tilde{\mu}_1, \tilde{\mu}_2, \tilde{\mu}_3)$ is the point at which $\sum \sum (y_{ij} - \mu_i)^2$ reaches its minimum subject to the constraints in H_1^*. Now we have the following result; this is a consequence of more general results in the next chapter (see (3.25)).

Proposition 2.1.1 *Let the setting be as in Table 2.1 with $k = 3$ and let the null hypothesis be $H_0 : \mu_1 = \mu_2 = \mu_3$. Then, for testing against any order restriction of the form* (2.1) *on page 27, the null distribution of $\bar{F}$ is*

$$pr(\bar{F} \leqslant c \mid H_0) = w_0 + w_1 pr(F_{1,\nu} \leqslant c) + w_2 pr(F_{2,\nu} \leqslant c/2), \quad (c > 0) \tag{2.6}$$

Table 2.2 Formulae for determining $\{w_1, w_2, w_3\}$ in (2.6)

H_1	ρ in (2.7)
$\mu_1 \leqslant \mu_2 \leqslant \mu_3$,	$\rho = -[n_1 n_3/\{(n_1+n_3)(n_2+n_3)\}]^{1/2}$
$\mu_1 \leqslant \mu_2$ *and* $\mu_1 \leqslant \mu_3$	$\rho = [n_2 n_3/\{(n_1+n_2)(n_1+n_3)\}]^{1/2}$.
$\mu_1 \leqslant \mu_2$	1.0 [i.e. $(w_0, w_1, w_2) = (0, 0.5, 0.5)$]

where

$$w_1 = 0.5,\ w_2 = (0.5 - q),\ w_0 + w_1 + w_2 = 1, q = (2\pi)^{-1}\cos^{-1}(\rho) \tag{2.7}$$

for some ρ*. The formulas for determining* ρ *are given in Table 2.2.* ■

We chose the notation $\bar{F}$ for the statistic in (2.4) because it is related to the unrestricted F-ratio and its null distribution is a weighted average of probabilities associated with F distributions. Some authors have used $\bar{F}$ for a different statistic.

It follows from (2.6) that the p-value for the $\bar{F}$-test is given by

$$p\text{-value} = w_1 \text{pr}(F_{1,\nu} \geqslant \bar{f}_{obs}) + w_2 \text{pr}(2F_{2,\nu} \geqslant \bar{f}_{obs}) \tag{2.8}$$

where $\bar{f}_{obs}$ is the sample value of $\bar{F}$.

Example 2.1.1

Now let us consider the exercise program example discussed at the beginning of this section. Suppose that we are interested to test

$$H_0 : \mu_1 = \mu_2 = \mu_3 \qquad \text{vs} \qquad H_1 : \mu_1 \leqslant \mu_2 \leqslant \mu_3.$$

The unrestricted estimate of (μ_1, μ_2, μ_3), obtained by minimizing $\sum(\bar{y}_i - \mu_i)^2 n_i$ is $(\bar{y}_1, \bar{y}_2, \bar{y}_3) = (10.125,\ 11.375,\ 12.35)$. Since this estimate satisfies the constraints in H_1, it follows that the estimate of (μ_1, μ_2, μ_3) subject to the constraint in H_1 is also equal to the unconstrained estimate. Therefore, $(\tilde{\mu}_1, \tilde{\mu}_2, \tilde{\mu}_3) = (\bar{y}_1, \bar{y}_2, \bar{y}_3) = (10.125,\ 11.375,\ 12.35)$ and, by simple substitution, we have that $\bar{F} = 5.978$. Now, let us assume that the setting in Table 2.1 holds. Then, it follows from (2.8) that for testing $H_0 : \mu_1 = \mu_2 = \mu_3$ vs $H_1 : \mu_1 \leqslant \mu_2 \leqslant \mu_3$, we have that

$$p\text{-value} = 0.5\text{pr}(F_{1,14} \geqslant 5.989) + \{(0.5 - 0.329)\}\text{pr}(F_{2,14} \geqslant 2.989) = 0.028.$$

By contrast, for the unrestricted F-statistic for testing

$$H_0 : \mu_1 = \mu_2 = \mu_3 \text{ vs } H_2 : \{\mu_1, \mu_2, \mu_3\} \text{ are not all equal}$$

p-value $= \text{pr}(F_{2,14} \geqslant 2.989) = 0.083$. Note that the p-value for the F-test is larger than that for the $\bar{F}$-test. If the alternative hypothesis is $\mu_1 \leqslant \mu_2 \leqslant \mu_3$ and the sample means satisfy the corresponding inequality $\bar{y}_1 \leq \bar{y}_2 \leq \bar{y}_3$, then it can be shown using (2.6) that the p-value for $\bar{F}$ would be smaller than that for F. Of course, it is reasonable to expect that the $\bar{F}$-test, which is specifically targeted for testing against $\mu_1 \leqslant \mu_2 \leqslant \mu_3$, would provide stronger evidence to reject H_0 than the unrestricted F-test when the sample means satisfy $\bar{y}_1 \leq \bar{y}_2 \leq \bar{y}_3$; it is reassuring to note that this, in fact, is the case. ■

In this example, the calculations were simple because there were only two order restrictions and the sample means satisfied the constraints that correspond to those in the alternative hypothesis. If the alternative hypothesis has up to three order restrictions, then closed-form expressions similar to (2.6)-(2.7) are available for computing the p-value. If the number of order restrictions is four or more then the null distribution of $\bar{F}$ is a weighted sum similar to (2.6), but in general it is quite inconvenient to use it for computing the p-value exactly. A simulation approach offers a simple and practically feasible method of computing the p-value sufficiently precisely irrespective of the number of order restrictions, for any error distribution. These are discussed in the next subsection. The derivations of the null distribution of $\bar{F}$ and that of the likelihood ratio test statistic will be provided in the next chapter.

2.1.2 Test of $H_0 : \mu_1 = \ldots = \mu_k$ Against an Order Restriction

Suppose that there are k treatments. Let the independent observations $\{y_{ij}\}$ be as in Table 2.3, where F is a cumulative distribution function. Clearly, μ_i is a location parameter for treatment i, but it does not need to be the mean ($i = 1, \ldots, k$); for example, it could be the median. Let the null and alternative hypotheses be

$$H_0 : \mu_1 = \ldots = \mu_k \quad \text{and} \quad H_1 : \mu_i - \mu_j \geqslant 0 \textit{ for } (i,j) \in B, \tag{2.9}$$

respectively, for some $B \subset \{(i,j) : i, j = 1, \ldots, k\}$. Let $\boldsymbol{\mu} = (\mu_1, \ldots, \mu_k)^T$ and H denote a hypothesis concerning $\boldsymbol{\mu}$, for example, H can be H_0 or H_1. Now, we define *the residual sum of squares under* H as

$$RSS(H) = \inf_{\boldsymbol{\mu} \in H} \sum \sum (y_{ij} - \mu_i)^2. \tag{2.10}$$

It is easily verified that this definition leads to the familiar expressions

$$RSS(H_0) = \sum \sum (y_{ij} - \bar{y})^2 \quad \text{and} \quad RSS(H_2) = \sum \sum (y_{ij} - \bar{y}_i)^2 \tag{2.11}$$

for the residual sum of squares under $H_0 : \mu_1 = \ldots = \mu_k$ and under $H_2 : \mu_1, \ldots, \mu_k$ *are not restricted*, respectively. In this sense, (2.10) is a natural extension of the two

Table 2.3 Comparison of k means

Treatment	Independent observations	Sample mean	Population distribution (cdf)
1	$y_{11}, \ldots, y_{1n_1}$	$\bar{y}_1$	$F\{(t - \mu_1)/\sigma^2\}$
$\vdots$	$\vdots$	$\vdots$	$\vdots$
k	$y_{k1}, \ldots, y_{kn_k}$	$\bar{y}_k$	$F\{(t - \mu_k)/\sigma^2\}$

familiar expressions in (2.11) for residual sums of squares. Let

$$\tilde{\mu} = (\tilde{\mu}_1, \ldots, \tilde{\mu}_k)^T = \arg\min_{\mu \in H_1} \sum\sum(y_{ij} - \mu_i)^2.$$

Thus, by (2.10), we have that

$$RSS(H_1) = \sum\sum(y_{ij} - \tilde{\mu}_i)^2 \tag{2.12}$$

for the residual sum of squares under H_1. If the error distribution is normal then $(\tilde{\mu}_1, \ldots, \tilde{\mu}_k)$ is also the *mle* of $(\mu_1, \ldots, \mu_k)$ under H_1. By motivations similar to those leading to the $\bar{F}$ in (2.4), let us define

$$\bar{F} = \{RSS(H_0) - RSS(H_1)\}/S^2, \tag{2.13}$$

where $S^2 = \nu^{-1}\sum\sum(y_{ij} - \bar{y}_i)^2$ is the mean square for error and $\nu = (n_1 + \ldots + n_k - k)$ is the error degrees of freedom. As in the previous subsection, the $\bar{F}$ in (2.13) can be considered as a modification of the usual F-statistic to incorporate the order restrictions in H_1. The exact finite sample null distribution of $\bar{F}$ for normal errors will be discussed at the end of this subsection. The p-value of $\bar{F}$ can be estimated by simulation. To this end, the following result is important.

Theorem 2.1.2 *Consider the setting in Table 2.3 and the testing problem* (2.9). *Suppose that the null hypothesis, $H_0 : \mu_1 = \ldots = \mu_k$, holds; let μ denote the common value of $\mu_1, \ldots, \mu_k$. Then the exact finite sample null distribution of $\bar{F}$ depends on the functional form of F but not on (μ, σ). Further, the asymptotic null distribution of $\bar{F}$ does not depend on (F, μ, σ), where the limit is taken as $(n_1 + \ldots + n_k) \to \infty$ such that* $0 < \lim_{n\to\infty} n_i/(n_1 + \ldots + n_k) < 1$ *for* $i = 1, \ldots, k$.

Proof: Let $y^*_{ij} = (y_{ij} - \mu)/\sigma$ for every (i, j). Let $RSS^*(H_0), RSS^*(H_1), S^*$ and $\bar{F}^*$ denote the values corresponding to $\{y^*_{ij}\}$. Then $RSS^*(H_0) = \sigma^{-2}RSS(H_0)$, $RSS^*(H_1) = \sigma^{-2}RSS(H_1)$, $(S^*)^2 = \sigma^{-2}S^2$, and hence

$$\bar{F} = \{RSS(H_0) - RSS(H_1)\}/S^2 = \{RSS^*(H_0) - RSS^*(H_1)\}/(S^*)^2 = \bar{F}^*.$$

Under H_0, the distribution of y^*_{ij} is $F(t)$, which does not depend on (μ, σ). Now, since $\bar{F}^*$ is a function of $\{y^*_{ij}\}$ only and $\bar{F} = \bar{F}^*$, it follows that the distribution of $\bar{F}$ does not depend on (μ, σ). The proof of the second part will be discussed in the next chapter. ■

This result suggests the following simulation method for computing the p-value.

Computation of the exact p-value for the $\bar{F}$-test:

Suppose that H_0 and H_1 are as in (2.9), $\bar{F}$ is as in (2.13), the setting is as in Table 2.3 and the functional form F of the error distribution is known. Then the following steps are adequate to compute the p-value for $\bar{F}$.

1. Generate independent observations $\{y_{ij} : j = 1, \ldots, n_i, i = 1, \ldots, k\}$ from $F\{(t - \mu_0)/\sigma_0\}$ where (μ_0, σ_0) can have any arbitrary values, but must be held fixed for different (i, j).

2. Compute the $\bar{F}$ statistic in (2.13) with $RSS(H_0)$ and $RSS(H_1)$ as in (2.11) and (2.12), respectively.

3. Repeat the previous two steps N times (say N = 10000), and estimate the p-value by M/N where M is the number of times the $\bar{F}$ statistic in the second step exceeded its sample value. ■

Note that in the first step of the simulation, the observations may be generated from a distribution with any value for the common location and scale parameters because, in view of Theorem 2.1.2, the null distribution of $\bar{F}$ does not depend on the common value of $\mu_1, \ldots, \mu_k$ or σ. For example, if the errors are assumed to be $N(0, \sigma^2)$ where σ is unknown, then it suffices to generate the observations from $N(0, 1)$ in the first step of the simulation for computing the p-value. Similarly, if the error distribution is a logistic with unknown variance, then it suffices to generate the observations from a logistic distribution with any values for its mean and variance.

The simulation approach can also be used to implement a bootstrap-type procedure in which the observations may be generated from the empirical distribution of the within treatment residuals. To improve robustness of validity against violation of the assumption of normal error distribution, we may estimate the p-value corresponding to several forms of error distribution and then choose their maximum as the p-value.

The second part of Theorem 2.1.2 says that if the error distribution is $F(t/\sigma)$ for some (F, σ), which may be unknown, then the asymptotic distribution of $\bar{F}$ does not depend on (F, σ). Therefore, if n_i is large for $i = 1, \ldots, k$ then the foregoing simulation for normal error would estimate the asymptotic p-value, irrespective of whether or not the true error distribution is normal. Similarly, the simulation method with any error distribution would also estimate the asymptotic p-value if n_i is large for $i = 1, \ldots, k$, irrespective of the true error distribution. It is reasonable to conjecture that closer the $F\{(. - \mu)/\sigma\}$ to the true error distribution, for some (μ, σ), the better the approximation.

Suppose that the precise form of F is unknown but it is known that F is a member of the class of distributions $\mathcal{F}$ for some $\mathcal{F}$. Let p_F denote the p-value corresponding to F in Table 2.3. Then we have

$$p\text{-value} = \sup_{F \in \mathcal{F}} p_F \tag{2.14}$$

As an example, suppose that the error distribution is known to be normal, logistic, or a t-distribution with four degrees of freedom (T_4). Then, the foregoing p-value in (2.14) for $\bar{F}$ can be computed by applying the foregoing simulation procedure to compute the p-values corresponding to normal, logistic, and T_4 and then taking their maximum as the p-value in (2.14). In most practical applications, we would not know $\mathcal{F}$, although it would not be difficult to specify a small class of distributional forms that would be well spread out in $\mathcal{F}$. In this case, a reasonable procedure is to compute the p-values corresponding to a range of choices of F (such as normal, logistic, T_r, and χ^2_r) that are well spread out in $\mathcal{F}$ and take their maximum as a reasonable approximation to the p-value.

Example 2.1.2

The experiment in Example 1.2.1 also included a fourth treatment; it was not considered in the previous worked example for simplicity. The group of children receiving the fourth treatment were checked weakly for progress but they did not receive any special exercises.

Table 2.4 Effect of exercise on the age at which a child first walks

Treatment	Age						n	$\bar{y}$
Group 1:	9.00	9.50	9.75	10.00	13.00	9.50	6	10.125
Group 2:	11.00	10.00	10.00	11.75	10.50	15.00	6	11.375
Group 3:	13.25	11.50	12.00	13.50	11.50		5	12.35
Group 4:	11.50	12.00	9.00	11.50	13.25	13.00	6	11.7

Reprinted from: *Science*, Volume 176, P. R. Zelazo, N. A. Zelazo, and S. Kolb, Walking in the newborn, Pages 314–315, Copyright (1972), with permission from AAAS.

Since Groups 1 and 2 were also checked weekly, it may be argued that Group 4 is more appropriate as a control than Group 3; we shall not pursue this issue any further here. The data for all four groups are given in Table 2.4.

Suppose that we wish to test $H_0 : \mu_1 = \mu_2 = \mu_3 = \mu_4$ against an alternative that captures our *a priori* information. There is no unique way of formulating the alternative hypothesis. For illustrative purposes, let us suppose that we wish to test

$$H_0 : \mu_1 = \mu_2 = \mu_3 = \mu_4 \text{ against } H_1 : \mu_1 \leqslant \mu_3, \mu_2 \leqslant \mu_3, \mu_1 \leqslant \mu_4, \mu_2 \leqslant \mu_4.$$

This says that Treatments 1 and 2 are at least as good as Treatments 3 and 4, but no ordering is suggested between Treatments 1 and 2, or between Treatments 3 and 4. In this case, there is no convenient formula similar to (2.8) for computing the p-value for $\bar{F}$; this is a common feature of most of the problems that involve test against inequality constraints. However, in view of Theorem 2.1.2, the simulation method on page 32 is still applicable.

Since $(\bar{y}_1, \bar{y}_2, \bar{y}_3, \bar{y}_4) = (10.1, 11.4, 12.4, 11.7)$, the sample means satisfy the restrictions in H_1. Therefore, the constrained and unconstrained estimators of $(\mu_1, \mu_2, \mu_3, \mu_4)$ are the same; in particular, $(\tilde{\mu}_1, \tilde{\mu}_2, \tilde{\mu}_3, \tilde{\mu}_4) = (\bar{y}_1, \bar{y}_2, \bar{y}_3, \bar{y}_4) = (10.1, 11.4, 12.4, 11.7)$. Now substituting directly into the formula for $\bar{F}$ in (2.13), the sample value of $\bar{F}$ is 6.43. The p-values corresponding to a range of error distributions were computed using the simulation approach. The computed values are given in the first row of Table 2.5; the figures in the second row are also p-values for a different statistic, which will be discussed later.

It is also of interest to compute the p-value by resampling; to this end, we set the error distribution equal to the empirical distribution of the residuals about the treatment means. This can be implemented by resampling with replacement from the set of within treatment residuals. The computed p-value is also given in the last column of Table 2.5. This example illustrates that simulation is a very convenient

Table 2.5 The p-values for the $\bar{F}$- and $\bar{E}^2$-tests for different error distributions

Test	Error Distribution $N(0,\sigma^2)$	T_4	T_{10}	χ_1^2	χ_2^2	χ_4^2	χ_7^2	RS
$\bar{F}$	0.052	0.051	0.058	0.050	0.048	0.049	0.051	0.048
$\bar{E}^2$	0.046	0.047	0.053	0.046	0.044	0.044	0.047	0.043

The symbol RS refers to resampling from the unrestricted within treatment residuals. To compute the p-value corresponding to $N(0, \sigma^2)$, we used $\sigma = 1$.

way of implementing tests against any order restriction in the one-way classification model when the errors are iid even if the common error distribution is not normal.

It is well known that the critical values from the F-tables are reasonably robust even if the error distribution is not normal for the unrestricted F-test of $H_0 : \mu_1 = \ldots = \mu_k$ against $H_2 : \mu_1, \ldots \mu_k$ *are unequal.* It will be argued later that a similar comment is likely to be applicable for the $\bar{F}$-test as well. Since the p-values in Table 2.5 are close for different error distributions, the results of this example are consistent with this conjecture.

The value of the standard unconstrained F-statistic for testing $H_0 : \mu_1 = \mu_2 = \mu_3 = \mu_4$ against $H_2 : \{\mu_1, \mu_2, \mu_3, \mu_4\}$ *are not all equal* is 2.14 and its p-value is 0.129 based on F-statistic $\sim F_{3,19}$. The numerical results for this example again highlight the fact that if the sample means satisfy the constraints that correspond to those in the alternative hypothesis, then the estimate of μ under H_1 and H_2 are the same and the p-value for the constrained test would be smaller than that for the unconstrained F-test. ■

2.1.2.1 Exact finite sample results when the error distribution is $N(0, \sigma^2)$

The results for normal theory likelihood ratio test of H_0 against H_1 are very similar to those for the $\bar{F}$-test. *Assume that the errors are independent and distributed as* $N(0, \sigma^2)$. Let us define

$$\bar{E}^2 = \{RSS(H_0) - RSS(H_1)\}/RSS(H_0).$$

Then the $\bar{E}^2$-test rejects H_0 for large values of $\bar{E}^2$. If the error distribution is normal then, it may be verified that

$$\bar{E}^2 = \{1 - \exp(-LRT/n)\} \tag{2.15}$$

where LRT denotes the likelihood ratio statistic (= $-2\log\Lambda$) for testing H_0 against H_1. It follows from (2.15) that LRT is an increasing function of $\bar{E}^2$, and, therefore, two tests are equivalent.

It is worth noting that $\bar{F}$ and $\bar{E}^2$ have the same numerator, namely $\{RSS(H_0) - RSS(H_1)\}$, which measures the discrepancy between H_0 and H_1. The denominators of these statistics act as scaling factors so that the null distribution of the test statistics do not depend on the unknown scale parameter of the error distribution; they do not make any contribution to quantifying the discrepancy between H_0 and H_1. Now, Theorem 2.1.2 holds with $\bar{F}$ replaced by $\bar{E}^2$ as well; simply replace $\bar{F}$ by $\bar{E}^2$ in the proof. If the error distribution is not normal then $\bar{E}^2$ is not necessarily a monotonic function of the likelihood ratio statistic, but it is still a suitable test statistic, and the three-step simulation for computing the p-value for $\bar{F}$ (see page 32) can also be applied for $\bar{E}^2$, with the latter replacing the former in the second step of the simulation.

For the Walking Exercise example discussed earlier in this section, the sample value of $\bar{E}^2$ is 0.253. The corresponding p-values for several error distributions are given in Table 2.5; they range from 0.043 for the resampling method to 0.053 for T_{10}. Note that the results for $\bar{F}$ and $\bar{E}^2$ are close; we conjecture that the difference between $\bar{F}$-test and $\bar{E}^2$-test in terms of p-value for a given set of data and in terms of power, would be small in most practical situations.

2.1.2.2 Computing $\min_{\mu \in H_1} \sum\sum(y_{ij} - \mu_i)^2$, $\bar{F}$, and $\bar{E}^2$.

The implementation of the $\bar{F}$- and $\bar{E}^2$-tests requires the computation of $RSS(H_1)$ that is equal to

$$\min_{\boldsymbol{\mu} \in H_1} \sum\sum(y_{ij} - \mu_i)^2 \quad \text{where } H_1 : \mu_i - \mu_j \geqslant 0 \text{ for } (i,j) \in B.$$

Let

$$q(\boldsymbol{\mu}) = (\bar{\boldsymbol{y}} - \boldsymbol{\mu})^T \boldsymbol{W} (\bar{\boldsymbol{y}} - \boldsymbol{\mu}) \tag{2.16}$$

where $\boldsymbol{\mu} = (\mu_1, \ldots, \mu_k)^T$, $\bar{\boldsymbol{y}} = (\bar{y}_1, \ldots, \bar{y}_k)^T$ and $\boldsymbol{W} = \text{diag}\{n_1, \ldots, n_k\}$, and let $\boldsymbol{A}$ be a matrix such that

$$\{\boldsymbol{\mu} : \mu_i - \mu_j \geqslant 0 \text{ for } (i,j) \in B\} = \{\boldsymbol{\mu} : \boldsymbol{A}\boldsymbol{\mu} \geqslant \boldsymbol{0}\}. \tag{2.17}$$

In this particular case, each row of $\boldsymbol{A}$ is a permutation of the k-vector $(1, -1, 0, \ldots, 0)$. Since

$$\sum\sum(y_{ij} - \mu_i)^2 = q(\boldsymbol{\mu}) + C(\boldsymbol{y}),$$

where $C(\boldsymbol{y}) = \sum\sum(y_{ij} - \bar{y}_i)^2$, which does not depend on $\boldsymbol{\mu}$, we have that

$$\bar{F} = \{\min_{H_0} q(\boldsymbol{\mu}) - \min_{H_1} q(\boldsymbol{\mu})\}/S^2,$$

$$\text{and} \quad \bar{E}^2 = \{\min_{H_0} q(\boldsymbol{\mu}) - \min_{H_1} q(\boldsymbol{\mu})\}/\{\min_{H_0} q(\boldsymbol{\mu}) + C(\boldsymbol{y})\}.$$

We can use the following general method for computing the minimum of $q(\boldsymbol{\mu})$ subject to $\boldsymbol{\mu} \in H_i$, $(i = 1, 2)$.

Let $\boldsymbol{x} \in \mathbb{R}^k$ be given, $\boldsymbol{B}$ be a given $k \times k$ positive definite matrix and

$$g(\boldsymbol{\mu}) = (\boldsymbol{x} - \boldsymbol{\mu})^T \boldsymbol{B} (\boldsymbol{x} - \boldsymbol{\mu}), \qquad \boldsymbol{\mu} \in \mathbb{R}^k. \tag{2.18}$$

Suppose that we wish to solve

$$\min g(\boldsymbol{\mu}) \quad \text{subject to } \boldsymbol{A}_1\boldsymbol{\mu} \geqslant \boldsymbol{0} \text{ and } \boldsymbol{A}_2\boldsymbol{\mu} = \boldsymbol{0} \tag{2.19}$$

for some matrices $\boldsymbol{A}_1$ and $\boldsymbol{A}_2$ that do not depend on $\boldsymbol{\mu}$. This constrained minimization problem in which the objective function is a quadratic in $\boldsymbol{\mu}$ and the constraints are linear equality and inequality constraints in $\boldsymbol{\mu}$ is called a *quadratic program*. There are efficient computer algorithms and software for this optimization problem. For example, we found that the subroutine QPROG in IMSL worked well; similar procedures in packages such as MATLAB, GAUSS, and SPLUS are likely to work just as well.

The foregoing quadratic programming problem is sometimes expressed in the following slightly different, but equivalent, form. Note that

$$g(\boldsymbol{\mu}) = 2f(\boldsymbol{\mu}) + \text{constant}, \quad \text{where } f(\boldsymbol{\mu}) = \boldsymbol{a}^T\boldsymbol{\mu} + (1/2)\boldsymbol{\mu}^T\boldsymbol{B}\boldsymbol{\mu}$$

and $\boldsymbol{a} = -\boldsymbol{B}^T\boldsymbol{x}$. Therefore, the minimization of $g(\boldsymbol{\mu})$ subject to some constraints on $\boldsymbol{\mu}$ is equivalent to the minimization of $f(\boldsymbol{\mu})$ subject to the same constraints on $\boldsymbol{\mu}$. Therefore, $\tilde{\boldsymbol{\mu}}$, the solution to (2.19), is also the solution to

$$\min\{\boldsymbol{a}^T\boldsymbol{\mu} + (1/2)\boldsymbol{\mu}^T\boldsymbol{B}\boldsymbol{\mu}\} \text{ subject to } \boldsymbol{A}_1\boldsymbol{\mu} \geqslant \boldsymbol{0} \text{ and } \boldsymbol{A}_2\boldsymbol{\mu} = \boldsymbol{0}. \tag{2.20}$$

2.1.3 General Remarks

It will be shown in the next chapter that, if the errors are independent and distributed as $N(0, \sigma^2)$, then the p-value for $\bar{F}$ is

$$\sum_{i=0}^{k} w_i(H_0, H_1)\text{pr}(iF_{i,\nu} \geqslant \bar{f}_{obs}), \tag{2.21}$$

where $\{w_i(H_0, H_1)\}$ are some nonnegative weights, which depend on the null hypothesis H_0 and the alternative hypothesis H_1, and $\bar{f}_{obs}$ is the sample value of $\bar{F}$. The quantities $\{w_i(H_0, H_1)\}$ are known as *chi-bar-square weights* and also as *level probabilities*; the reasons for these terms will become clear in the next chapter. They appear in the null distribution of several test statistics when there are inequality constraints on parameters.

Often, when there are order restrictions, the null distributions of various test statistics, including $\bar{F}$ and $\bar{E}^2$, take the form $w_0\text{pr}(X_0 \leqslant c) + \ldots + w_k\text{pr}(X_k \leqslant c)$ where $w_i = w_i(H_0, H_1)$ and $X_0, \ldots, X_k$ are some random variables. In this expression, computation of $\text{pr}(X_i \leqslant c)$ is usually not difficult $(i = 0, \ldots, k)$. However, simple closed-forms are not available for computing $w_0, \ldots, w_k$ exactly, in general. Consequently, there have been various attempts to obtain approximations/bounds for these quantities, so that they can be used to obtain approximations/bounds for the p-value. The details of these are not easy and/or tedious. It appears that the complicated nature of these technical details and unavailability of closed-forms for computing the p-value have led to the opinion among some authors that tests of H_0 against an order restriction are difficult to apply. Fortunately, such an opinion is ill-founded, because there is no need to compute the weights $\{w_i\}$ exactly. In view of the fact that the p-value can be computed sufficiently precisely by simulation, the aforementioned approximations/bounds on $\{w_1, \ldots, w_k\}$, and the related bounds on the exact null distributions of $\bar{F}$ and $\bar{E}^2$ are not that important for implementing tests of hypotheses involving order restrictions on $\mu_1, \ldots, \mu_k$. Even if $w_0, \ldots, w_k$ can be computed exactly, the simulation approach can be used to compute a p-value that would be more robust against violation of distributional assumptions about the error term, although for a one-way classification problem we believe that the p-values corresponding to normal and other error distributions are likely to be close. However, the bounds on $\{w_1, \ldots, w_k\}$ may be useful in nonlinear models; this will be discussed in Chapter 4.

2.2 ORDERED NULL HYPOTHESES

Now we consider testing problems in which there are inequality constraints in the null hypothesis. To introduce the general form of the testing problems, let μ_i denote the mean of population i, and let y_{ij} denote the jth observation from population i $(i = 1, \ldots, k, j = 1, \ldots, n_i)$. Assume that the observations are independent; thus the setting is the same as that in Table 2.3 for one-way classification. Suppose that we wish to test

$$H_1 : \mu_i - \mu_j \geqslant 0 \text{ for } (i,j) \in B \quad \text{vs} \quad H_2 : \textit{No restriction on } \mu_1, \ldots, \mu_k \tag{2.22}$$

where B is a given subset of $\{(i,j) : i,j = 1, \ldots, k\}$. Thus the hypothesis H_1 in (2.22) is the same as that in (2.9) although in the present context it is the null hypothesis; further, let us also note that in view of the convention introduced earlier (see page 28), the alternative parameter space is the set of all values $\{\mu_1, \ldots, \mu_k\}$ that do not satisfy the constraints H_1 in (2.22). This section provides an introduction and the essentials for carrying out a test of H_1 vs H_2. The distribution theory will be presented in the next chapter.

First, let us consider a general testing problem of the form H_a vs H_b, where H_a and H_b are given. By motivations similar to those leading to (2.13), we define

$$\bar{F} = \{RSS(H_a) - RSS(H_a \cup H_b)\}/S^2, \tag{2.23}$$

for testing H_a against H_b, where RSS is defined as in (2.10). This is a natural generalization of the usual unrestricted F-statistic for H_0 vs H_2 to a test of H_a vs H_b; clearly, the $\bar{F}$ in (2.13) is a special case of (2.23). Thus, for testing H_1 vs H_2, we have

$$\bar{F} = \{RSS(H_1) - RSS(H_1 \cup H_2)\}/S^2, \tag{2.24}$$

where, by the definition of $RSS(H)$ in (2.10), we have that

$$\begin{aligned} RSS(H_1) &= \min_{\boldsymbol{\mu}} \sum\sum (y_{ij} - \mu_i)^2, \text{subject to } (\mu_i - \mu_j) \geqslant 0 \text{ for } (i,j) \in B \\ RSS(H_1 \cup H_2) &= \min_{\boldsymbol{\mu}} \sum\sum (y_{ij} - \mu_i)^2, \text{subject to no restriction on } \boldsymbol{\mu}. \end{aligned} \tag{2.25}$$

This leads to

$$RSS(H_1) = \sum\sum(y_{ij} - \tilde{\mu}_i)^2, \quad RSS(H_1 \cup H_2) = \sum\sum(y_{ij} - \bar{y}_i)^2. \quad (2.26)$$

Now, with $q(\boldsymbol{\mu})$ as in (2.16), the $\bar{F}$ in (2.24) for the testing problem in (2.22) takes the form,

$$\bar{F} = \min_{\boldsymbol{\mu} \in H_1} q(\boldsymbol{\mu})/S^2, \quad (2.27)$$

and therefore the computational methods of subsection 2.1.2.2 are also applicable and adequate.

Let $\bar{f}_{obs}$ denote the sample value of $\bar{F}$. We need to take some care in defining the p-value and critical value for this $\bar{F}$-test of H_1 vs H_2 because pr($\bar{F} \geqslant \bar{f}_{obs}$ | *the null hypothesis,* H_1) is a function of the particular value of $(\mu_1, \ldots, \mu_k)$ in the null parameter space $\{(\mu_1, \ldots, \mu_k) : \mu_i - \mu_j \geqslant 0 \text{ for } (i, j) \in B\}$, for any $\bar{f}_{obs} > 0$. This is in sharp contrast to the testing problem studied in the previous subsection where pr($\bar{F} \geqslant \bar{f}_{obs}$ | *the null hypothesis,* H_0) does not depend on the particular value of $(\mu_1, \ldots, \mu_k)$ in the null parameter space $\{(\mu_1, \ldots, \mu_k) : \mu_1 = \ldots = \mu_k\}$.

For testing H_1 vs H_2 we define

$$p\text{-value} = \sup_{\boldsymbol{\mu} \in H_1} \text{pr}_{\boldsymbol{\mu}}(\bar{F} \geqslant \bar{f}_{obs}), \quad (2.28)$$

and a level-α test of H_1 vs H_2 rejects H_1 if this p-value $\leqslant \alpha$. An interpretation of this standard approach is that if the data are consistent with at least one point in the null parameter space then we do not reject the null hypothesis. Let $\boldsymbol{\mu}^*$ denote the value of $\boldsymbol{\mu}$ at which the supremum in (2.28) is reached. Then, $\boldsymbol{\mu}^*$ is the point in the null parameter space with which the data are most consistent, and the right-hand side of (2.28) is a measure of the degree of this consistency. If this is large, then the data are consistent with the value $\boldsymbol{\mu}^*$ in the null parameter space and hence we would not reject the null hypothesis. If, on the other hand, the right-hand side of (2.28) is small then the data are not consistent with any point in the null parameter space and hence we would reject the null hypothesis, H_1.

At first, it may seem that computing the supremum in (2.28) is a formidable task; fortunately, it is not. It can be shown that for the particular null hypothesis (i.e., H_1) in (2.22), the supremum in (2.28) is reached where $\mu_1 = \ldots = \mu_k$; further, the right-hand side of (2.28) does not depend on the common value of $\mu_1, \ldots, \mu_k$ (see the theorem below). Hence, (2.28) can be written as

$$p\text{-value} = \text{pr}(\bar{F} \geqslant \bar{f}_{obs} \mid \mu_1 = \ldots = \mu_k). \quad (2.29)$$

The main result concerning the $\bar{F}$-test of H_1 vs H_2 is given in the next result.

Theorem 2.2.1 *Suppose that the setting is as in Table 2.3, the testing problem is (2.22), and $\bar{F}$ is as in (2.24). Let μ denote the common value of $\mu_1, \ldots, \mu_k$ when they are all equal. Then we have the following:*

1. $\sup_{\boldsymbol{\mu} \in H_1} pr_{\boldsymbol{\mu}}(\bar{F} \geqslant c) = pr(\bar{F} \geqslant c \mid \mu_1 = \ldots = \mu_k)$, *for any* $c > 0$.

2. *pr*$(\bar{F} \geqslant c \mid \mu_1 = \ldots = \mu_k = \mu)$, *depends on the functional form* F *of the error distribution but not on* (μ, σ).

3. *The asymptotic distribution of* $\bar{F}$ *at* $\mu_1 = \ldots = \mu_k (= \mu$, *say*), *and hence the value of* $\lim pr(\bar{F} \geqslant c \mid \mu_1 = \ldots = \mu_k = \mu)$ *as* $n \to \infty$, *do not depend on* (F, μ, σ) *where* $c > 0$, $n = (n_1, \ldots, n_k)$ *and* $0 < \lim(n_i/n) < 1$. ■

The proof of the first part will be discussed in the next chapter; see Theorem 3.8.1 and Section 3.9. The proof of the second part follows by arguments very similar to those for the corresponding result in the previous subsection, and hence is omitted; the proof of the last part will be provided in Chapter 4. It follows from the foregoing result that the p-value and critical value for testing H_1 vs H_2 can be estimated very precisely by simulation; the main steps are given below.

Computing the exact p-value for the $\bar{F}$-test of H_1 vs H_2

Suppose that the setting is as in Table 2.3, the testing problem is (2.22) and the functional form F of the error distribution is known. Thus, $y_{ij} \sim F\{(t - \mu_i)/\sigma\}$ where F is known but σ may be unknown.

1. Generate independent observations $\{y_{ij} : j = 1, \ldots, n_i, i = 1, \ldots, k\}$ from $F\{(t - \mu_0)/\sigma_0\}$ where we may choose (μ_0, σ_0) to have any convenient value but must be the same for different values of (i, j).

2. Compute the $\bar{F}$ statistic; to this end we may either use (2.24) with RSS as in (2.26), or use (2.27).

3. Repeat the previous two steps N times (say N=10000), and estimate the p-value by M/N where M is the number of times the simulated value of $\bar{F}$ in the second step was not less than its sample value. ■

Note that in the first step of the simulation, it suffices to generate the observations from a distribution with any values for the mean and variance because, in view of Theorem 2.1.2, the null distribution of $\bar{F}$ does not depend on the common location or scale of the error distribution.

Now, suppose that the functional form F of the error distribution is unknown, but is known to be member of the class $\mathcal{F}$. Let p_F denote the p-value when the error distribution is $F(t/\sigma)$ for some σ as in Table 2.3; by Theorem 2.2.1, p_F does not depend on σ. Therefore, we have

$$p\text{-value} = \sup_{F \in \mathcal{F}} p_F. \tag{2.30}$$

As in the arguments that followed (2.14), a procedure to compute a reasonable approximation to the foregoing p-value is to compute p_F corresponding to a range of choices of F (such as normal, logistic, t_r and χ^2_r) that are well spread out in $\mathcal{F}$ and take their maximum as an approximation to the p-value in (2.30).

The results for the normal theory likelihood ratio test of H_1 against H_2 are closely related to those for the $\bar{F}$ test. For testing H_1 vs H_2, let us define

$$\bar{E}^2 = \{RSS(H_1) - RSS(H_1 \cup H_2)\}/RSS(H_1).$$

Then, it may be verified that

$$\bar{E}^2 = \{1 - \exp(-LRT/n)\}. \tag{2.31}$$

where LRT is the normal theory likelihood ratio statistic for testing H_1 vs H_2. Therefore, $\bar{E}^2$ is a monotonically increasing function of the LRT, and hence the LRT and $\bar{E}^2$-tests of H_1 vs H_2 are equivalent. Further, $\bar{E}^2$ $[= \bar{F}/(1+\bar{F})]$ is also a monotonic function of $\bar{F}$ for testing H_1 against H_2. Therefore, for this testing problem $\bar{E}^2$-test and the $\bar{F}$-test are equivalent.

The exact finite sample null distributions of $\bar{F}$ and $\bar{E}^2$ when the errors are normally distributed will be obtained in the next chapter. It is worth emphasizing that these exact results are for the specific case when the errors are normal; the exact null distributions of $\bar{F}$ and $\bar{E}^2$ are unknown for other error distributions. Hence the p-value computed using these exact results correspond to normal error only. By contrast the simulation approach to computing the p-value does not require the errors to be normally distributed.

In general, hypothesis testing problems with order restrictions in the null hypothesis arise in contexts that are more complicated than the simple one-way layout studied in this section. Nevertheless, the ideas developed in this section provide a valuable introduction and will be helpful in later sections where these are explored further. The following numerical example illustrates the main results of this subsection on test of hypothesis when there are order restrictions in the null hypothesis.

Example 2.2.1 *An example with order restrictions in the null hypothesis*

The sample mean $\bar{y}_i$ and the sample size n_i for population i are given in Table 2.6 for $i = 1, \ldots, 5$; the error mean square, $s^2 = 0.817$.

Table 2.6 Summary statistics for comparing $\mu_1, \ldots, \mu_5$

i	1	2	3	4	5
n_i	6	5	6	6	7
$\bar{y}_i$	11.825	10.811	11.751	11.837	11.231

Let the null and alternative hypothesis be

$$H_1 : \mu_1 \leqslant \mu_2, \mu_1 \leqslant \mu_3, \mu_4 \leqslant \mu_2, \mu_4 \leqslant \mu_3, \mu_1 \leqslant \mu_5, \quad \text{and} \quad H_2 : \text{ not } H_1,$$

respectively. Assume that the setting in Table 2.3 holds; in particular, the observations are assumed to be *iid*. Let us apply the $\bar{F}$-test. It follows from the definition of $\bar{F}$,

Table 2.7 Estimated p-values of $\bar{F}$ for different error distributions

Dist:	$N(0, \sigma^2)$	T_4	T_{10}	χ_1^2	χ_4^2	χ_8^2	RS
p-value:	0.091	0.091	0.092	0.074	0.086	0.085	0.091

RS : resample with replacement from the empirical distribution of the within treatment residuals, which is the same as bootstrap.

(2.27), (2.24), and (2.16) that

$$\bar{F} = \{RSS(H_1) - RSS(H_1 \cup H_2)\}/S^2 = \min_{\boldsymbol{\mu} \in H_1} q(\boldsymbol{\mu})/S^2$$
$$= [\min_{\boldsymbol{A\mu} \geqslant \mathbf{0}} (\bar{\boldsymbol{y}} - \boldsymbol{\mu})^T \boldsymbol{W} (\bar{\boldsymbol{y}} - \boldsymbol{\mu})]/S^2$$

where $\boldsymbol{A}$ is the 5×5 matrix $(-1, 1, 0, 0, 0 \mid -1, 0, 1, 0, 0 \mid 0, 1, 0, -1, 0 \mid 0, 0, 1, -1, 0 \mid -1, 0, 0, 0, 1)$; the five rows correspond to the five order restrictions in H_1. We used the subroutine QPROG in IMSL for solving this constrained minimization problem. The sample value of $\bar{F}$ is 5.04. If the error distribution is $F(t/\sigma)$, then p-value = $pr_F(\bar{F} \geqslant 5.04 \mid \mu_1 = \ldots = \mu_5)$, where the probability is computed when the error distribution is $F(t)$. In view of Theorem 2.2.1, the unknown σ can be set equal to one or any other convenient value when computing the p-value. Because the distribution function F is unknown, we approximate the p-value by the maximum of the p-values corresponding to several choices of F (see Table 2.7). In view of the results in Table 2.7, we conclude that the p-value $\geqslant$ 0.074. The results in Table 2.7 show that the p-values are not that sensitive to the form of the error distribution in this particular example; we conjecture that this is likely to be the case more generally. ■

The foregoing results and simulation approaches to computing p-values extend to tests involving order restrictions on the regression parameter $\boldsymbol{\beta}$ in the linear model, $\boldsymbol{Y} = \boldsymbol{X\beta} + \boldsymbol{E}$ where $\boldsymbol{E}$ has mean zero and covariance $\sigma^2 \boldsymbol{U}$, σ is unknown and $\boldsymbol{U}$ is known; the inequality constraints may be in the null or alternative hypothesis. This includes (1) test of $H_0 : \boldsymbol{R\beta} = \mathbf{0}$ against $H_1 : \boldsymbol{R}_1\boldsymbol{\beta} \geqslant \mathbf{0}$, where $\boldsymbol{R}_1$ is a submatrix of $\boldsymbol{R}$, and (2) test of $H_1 : \boldsymbol{R}_1\boldsymbol{\beta} \geqslant \mathbf{0}, \boldsymbol{R}_2\boldsymbol{\beta} = \mathbf{0}$ against H_2 : $\boldsymbol{\beta}$ *is unrestricted.* These will be discussed in the next chapter.

2.3 ISOTONIC REGRESSION

Consider k populations with μ_i denoting a scalar parameter of interest for group i, $i = 1, \ldots, k$. Suppose that the μ_i s are known to satisfy the *pairwise constraints,*

$$\mu_i - \mu_j \geqslant 0 \text{ for } (i, j) \in B \tag{2.32}$$

where B is a given subset of $\{(i, j) : i, j = 1, \ldots, k\}$. The objective in *isotonic regression* is to address statistical inference on $\mu_1, \ldots, \mu_k$ when they are constrained by (2.32).

Suppose that the random variable Y_i is an estimate of μ_i for $i = 1, \ldots, k$; for example, Y_i could be an observation from a population with mean μ_i or it could be the mean of several *iid* observations from a population with mean μ_i. A natural method of estimating $\{\mu_1, \ldots, \mu_k\}$ when they are known to satisfy (2.32) is to

$$\text{minimize} \sum_{i=1}^{k} (Y_i - \mu_i)^2 w_i \quad \text{subject to (2.32)} \tag{2.33}$$

where $\{w_1, \ldots, w_k\}$ are given or suitably chosen weights. For example, if $Y_i \sim N(\mu_i, \sigma^2/w_i)$ for $i = 1, \ldots, k$ and $Y_1, \ldots, Y_k$ are independent then (2.33) leads to the maximum likelihood estimate of $(\mu_1, \ldots, \mu_k)$ under (2.32). On the other hand, if $Y_1, \ldots, Y_k$ are not necessarily normal but the variance of Y_i is σ^2/w_i, where σ is unknown and $w_1, \ldots, w_k$ are known positive numbers, then (2.33) is the method of weighted least squares with weights, w_i $(i = 1, \ldots, k)$. For the time being, we shall not assume that $Y_1, \ldots, Y_k$ are normal.

Let $(\tilde{\mu}_1, \ldots, \tilde{\mu}_k)$ denote the value of $(\mu_1, \ldots, \mu_k)$, which solves the constrained minimization problem (2.33). Then we say that $(\tilde{\mu}_1, \ldots, \tilde{\mu}_k)$ is the *isotonic regression* of $(Y_1, \ldots, Y_k)$ with respect to the weights $(w_1, \ldots, w_k)$ and the order restriction (2.32). Thus, an *isotonic regression* is a weighted least squares estimate subject to a set of pairwise constraints; different weights and/or pairwise constraints correspond to different isotonic regressions.

As an example, consider the one-way ANOVA setting with k treatments: $E(X_{ij}) = \mu_i$, $\text{var}(X_{ij}) = \sigma^2$ for $j = 1, \ldots, n_i$ and $i = 1, \ldots, k$, and the X_{ij} s are independent. Let Y_i denote the sample mean $\bar{X}_i$ for $i = 1, \ldots, k$; then $E(Y_i) = \mu_i$ and $\text{var}(Y_i) = \sigma^2/n_i$. Suppose that $\mu_1, \ldots, \mu_k$ are known to satisfy

$$\textit{simple order}: \ \mu_1 \leqslant \ldots \leqslant \mu_k.$$

Clearly this order restriction is of the form (2.32). Then the isotonic regression of $(Y_1, \ldots, Y_k)$ with respect to the weights $(n_1, \ldots, n_k)$ and the simple order is the value of $(\mu_1, \ldots, \mu_k)$, which minimizes $\sum (Y_i - \mu_i)^2 n_i$ subject to $\mu_1 \leqslant \ldots \leqslant \mu_k$.

Two other important examples of pairwise constraints are:

1. *Tree order*: $\mu_1 \leqslant \mu_2, \ldots, \mu_1 \leqslant \mu_k$, also written as $\mu_1 \leqslant [\mu_2, \ldots, \mu_k]$; and
2. *Umbrella order*: $\mu_1 \leqslant \ldots \leqslant \mu_m \geqslant \ldots \geqslant \mu_k$, where m is known.

Usually, the tree order arises when a control (= group 1) is compared with several treatments. It is also worth noting that $\mu_1 \leqslant \mu_2 = \mu_3$ is also of the type in (2.32), because the constraint $\mu_2 = \mu_3$ is equivalent to $\mu_2 \leqslant \mu_3$ and $\mu_3 \leqslant \mu_2$.

As was noted in (2.17), the set of constraints in (2.32) can also be expressed as

$$\boldsymbol{A}\boldsymbol{\mu} \geqslant 0 \tag{2.34}$$

for some matrix $\boldsymbol{A}$ in which each row is a permutation of the k-vector, $(-1, 1, 0, \ldots, 0)$, and $\boldsymbol{\mu} = (\mu_1, \ldots, \mu_k)^T$. For example, for the simple and tree orders, the matrix $\boldsymbol{A}$

in (2.34) of order $(k-1) \times k$ takes the forms

$$\begin{bmatrix} -1 & 1 & 0 & \dots & & 0 \\ 0 & -1 & 1 & 0 & \dots & 0 \\ \dots & & & & & \\ 0 & \dots & & 0 & -1 & 1 \end{bmatrix} \text{and} \begin{bmatrix} -1 & 1 & 0 & \dots & & 0 \\ -1 & 0 & 1 & 0 & \dots & 0 \\ \dots & & & & & \\ -1 & 0 & \dots & 0 & 0 & 1 \end{bmatrix}, \tag{2.35}$$

respectively. Since

$$\sum (Y_i - \mu_i)^2 w_i = (\boldsymbol{Y} - \boldsymbol{\mu})^T \boldsymbol{W} (\boldsymbol{Y} - \boldsymbol{\mu}),$$

where $\boldsymbol{Y} = (Y_1, \dots, Y_k)^T$, $\boldsymbol{\mu} = (\mu_1, \dots, \mu_k)^T$, and $\boldsymbol{W} = \text{diag}\{w_1, \dots, w_k\}$, the minimization problem (2.33) can be expressed as

$$\text{minimize } (\boldsymbol{Y} - \boldsymbol{\mu})^T \boldsymbol{W} (\boldsymbol{Y} - \boldsymbol{\mu}) \quad \text{subject to } \boldsymbol{A\mu} \geqslant \boldsymbol{0}.$$

This is a standard quadratic program; see subsection 2.1.2.2 for a general method of solving it.

There are two types of inference problems that arise in isotonic regression: (i) k is fixed, and (ii) k increases with the sample size. In most parts of this text, only the first case is considered. The second case arises when a real function is to be estimated subject to shape constraints, for example, the nonparametric estimation of (i) a monotone density function, (ii) unimodal density function, (iii) monotone regression function, and (iv) convex regression function; the statistical inference problems arising in such cases are more challenging and they will be discussed later.

2.3.1 Quasi-order and Isotonic Regression

Let the labels $x_1, \dots, x_k$ denote the k groups and let $X = \{x_1, \dots, x_k\}$. Let x be a qualitative explanatory variable taking values in X; for group i, the explanatory variable x takes the value x_i. Note that $x_1, \dots, x_k$ are not necessarily real numbers. In this section, the following notation is adopted to identify a real function on X with a point in $\mathbb{R}^k$. To any given real function g on X, we associate the point $\boldsymbol{g}$ in $\mathbb{R}^k$ defined by $\boldsymbol{g} = (g_1, \dots, g_k)^T$ where $g_i = g(x_i)$ for $i = 1, \dots, k$. Conversely, to any given point $\boldsymbol{g} = (g_1, \dots, g_k)^T$ in $\mathbb{R}^k$ we associate the real function g on X defined by $g(x_i) = g_i$ for $i = 1, \dots, k$. Therefore, we shall freely interchange the roles of functions like g on X and the corresponding vector, $\boldsymbol{g} = (g_1, \dots, g_k)^T$.

Let us now consider our original problem of estimating $\mu_1, \dots, \mu_k$ subject to (2.32). Let y_i denote an estimator of μ_i, $i = 1, \dots, k$, and let

$$y(x_i) = y_i, \quad \boldsymbol{y} = (y_1, \dots, y_k)^T, \quad \mu(x_i) = \mu_i, \text{ and } \boldsymbol{\mu} = (\mu_1, \dots, \mu_k)^T.$$

Consider the regression model

$$y(x_i) = \mu(x_i) + \epsilon_i, \tag{2.36}$$

where ϵ_i is the error term, $i = 1, \dots, k$. Thus, we have a regression model in which the domain of the regression function $\mu(x)$ is the set of labels X, and the function

$\mu(x)$ satisfies the restrictions (2.32). An example of the foregoing setting is when y_i is the sample mean response of several observations from group i and μ_i is the population mean. Another example is where μ_i is the probability of success and y_i is the sample proportion of successes in group i, $i = 1, \ldots, k$.

First, let us recall some definitions on binary relations. A binary relation $\preccurlyeq$ on the set $X = \{x_1, \ldots, x_k\}$ is said to be (a) *reflexive* if $x \preccurlyeq x$ for every $x \in X$; (b) *transitive,* if $x \preccurlyeq y$ and $y \preccurlyeq z$ imply $x \preccurlyeq z$; (c) *antisymmetric* if $x \preccurlyeq y$ and $y \preccurlyeq x$ imply $x = y$; (d) a *partial order* if it is reflexive, transitive, and antisymmetric; and (e) a *quasi-order* if it is reflexive and transitive. If $\preccurlyeq$ is a quasi-order and if $x_i \preccurlyeq x_j$, then we may interpret it as "group i is not higher than group j."

Let $\preccurlyeq$ be a given quasi-order on the set of labels $X = \{x_1, \ldots, x_k\}$. A real function, μ, on X is said to be *isotonic* with respect to $\preccurlyeq$ if

$$x_i \preccurlyeq x_j \text{ implies that } \mu(x_i) \leqslant \mu(x_j), \text{ for every } x_i \text{ and } x_j \text{ in } X.$$

Thus, an isotonic function has the *same tone* as the quasi-order, and hence the term *iso tone*.

Suppose that $\{\mu_1, \ldots, \mu_k\}$ satisfy the constraints (2.32). Then, these constraints induce the order $\preccurlyeq$ on X, defined by

$$x_i \preccurlyeq x_j \quad \text{if and only if} \quad (i, j) \in B.$$

It is easily seen that the induced relation $\preccurlyeq$ is a quasi-order. Conversely, suppose that a quasi-order $\preccurlyeq$ on X is given. Then, define the corresponding set B in (2.32) as $\{(i, j) : x_i \preccurlyeq x_j\}$. Therefore, the pairwise constraints (2.32) may be referred to as a quasi-order.

It follows that estimation of $\boldsymbol{\mu}$ subject to (2.32), and estimation of the function $\mu(x)$ subject to it being isotonic with respect to the quasi-order induced by (2.32) are equivalent problems. A formal statement is given in the next proposition.

Proposition 2.3.1 *For a given a quasi-order $\preccurlyeq$ on X there exists a matrix $\boldsymbol{A}$ such that each row of $\boldsymbol{A}$ is a permutation of the k-vector (1, -1, 0, ..., 0), and*

$$\boldsymbol{A\mu} \geqslant \mathbf{0} \text{ is equivalent to } \mu \text{ is isotonic with respect to } \preccurlyeq . \tag{2.37}$$

Conversely, given a matrix $\boldsymbol{A}$ in which each row is a permutation of the k-vector (1, -1, 0, ..., 0), there exists a quasi-order $\preccurlyeq$ on X such that (2.37) *holds.* ■

Let y be a given real function on X and w be a given nonnegative function on X. Let

$$\mathcal{F} = \{\mu : \mu \text{ is an isotonic function on } X\} \text{ and } \mathcal{Q} = \{\boldsymbol{\mu} : \boldsymbol{A\mu} \geqslant 0\}$$

where $\boldsymbol{A}$ is as in Proposition 2.3.1. It follows from the foregoing proposition that $\mathcal{F}$ and $\mathcal{Q}$ are equivalent, in the sense that there is a one–one correspondence between $\mathcal{F}$ and $\mathcal{Q}$.

An isotonic function y^* on X is said to be *the least squares isotonic regression of y with respect to the weight w* if

$$\sum_{x \in X} \{y(x) - \mu(x)\}^2 w(x) \tag{2.38}$$

reaches its minimum over $\mu \in \mathcal{F}$ at $\mu = y^*$. It is also possible to consider L^p-isotonic regression with respect to w by minimizing $\sum |y(x) - \mu(x)|^p w(x)$; we shall not consider them for the time being. If there is no possibility of ambiguity, then we shall refer to y^* simply as the *isotonic regression* of y without referring to least squares explicitly.

Let

$$\boldsymbol{W} = \mathrm{diag}\{w_1, \ldots, w_k\}.$$

Then

$$\sum_{x \in X} \{y(x) - \mu(x)\}^2 w(x) = (\boldsymbol{y} - \boldsymbol{\mu})^T \boldsymbol{W} (\boldsymbol{y} - \boldsymbol{\mu}).$$

Now, since $\mathcal{F}$ and $\mathcal{Q}$ are equivalent sets, the following two minimization problems are also equivalent:

$$(A) \quad minimize \sum_{x \in X} \{y(x) - \mu(x)\}^2 w(x) \qquad \text{subject to } \mu \in \mathcal{F}, \tag{2.39}$$

$$(B) \quad minimize\, (\boldsymbol{y} - \boldsymbol{\mu})^T \boldsymbol{W} (\boldsymbol{y} - \boldsymbol{\mu}) \qquad \text{subject to } \boldsymbol{A\mu} \geqslant \boldsymbol{0}. \tag{2.40}$$

Let y^* denote the isotonic regression of y in the foregoing minimization problem (A). Then $\boldsymbol{y}^*$ is the *least square projection* of $\boldsymbol{y}$ onto $\mathcal{Q}$ with respect to the inner product $\langle \boldsymbol{x}, \boldsymbol{y} \rangle = \boldsymbol{x}^T \boldsymbol{W} \boldsymbol{y}$. Such projections will be studied in more detail in a later chapter; see Section 2.1.2.2 for some comments on computations relating to the problem (B).

2.4 ISOTONIC REGRESSION: RESULTS RELATED TO COMPUTATIONAL FORMULAS

In this section, several results relating to the computation of the isotonic regression are discussed. They are included here because the details are instructive and/or the results may be useful for studying the properties of the estimator. No attempt is made to provide a comprehensive discussion of the numerical algorithms that may be used for computing the isotonic regression; a discussion of such computational issues is outside the scope of this book. Since the computation of the isotonic regression is a problem in quadratic programming, interested readers may wish to consult the relevant literature in optimization. In this chapter, we will provide only an introduction.

2.4.1 Isotonic Regression Under Simple Order

Let y_i and μ_i denote an observation and a parameter of interest, respectively, for group i, $i = 1, \ldots, k$. Assume that $y_i = \mu_i + \epsilon_i$ and that the parameters, $\mu_1, \ldots, \mu_k$ are unknown but known to satisfy the simpler order $\mu_1 \leqslant \ldots \leqslant \mu_k$. Suppose that it is desired to estimate these parameters by the method of weighted least squares:

$$\min_{\mu_1 \leqslant \ldots \leqslant \mu_k} \sum_{i=1}^{k} (y_i - \mu_i)^2 w_i \tag{2.41}$$

where $w_1, \ldots, w_k$ are given positive numbers. This problem has been studied extensively in the early literature, and detailed discussions may be found in Barlow et al. (1972) and Robertson et al. (1988).

Let $\boldsymbol{y}^*$ denote the constrained least squares estimate of $\boldsymbol{\mu}$ resulting from (2.41); $\boldsymbol{y}^*$ is also the isotonic regression of $\boldsymbol{y}$ with respect to weights $\{w_i\}$ and the simple order. It turns out that $\boldsymbol{y}^*$ has simple and instructive explicit forms. It can be computed by a very simple algorithm known as the *Pool Adjacent Violators Algorithm*; this is described below.

Pool Adjacent Violators Algorithm (PAVA)

A set of consecutive elements of X will be called a *block*. For example, $\{x_1, x_2\}$ and $\{x_3, x_4, x_5, x_6\}$ are blocks, but $\{x_1, x_2, x_6\}$ is not a block because 1, 2, and 6 are not consecutive integers. Initially, X is partitioned into the k blocks $\{x_1\}, \ldots, \{x_k\}$ with one label in each block. Now, PAVA pools consecutive blocks as follows:
(i) If $y_1 \leqslant \ldots \leqslant y_k$ then stop the iteration and the solution is $y_j^* = y_j$ for $j = 1, \ldots, k$.
(ii) Otherwise, let i be the smallest index such that $y_i > y_{i+1}$. Now, pool the categories x_i and x_{i+1} to form a single block. Let the response variable for this block be the weighted response, $(w_i y_i + w_{i+1} y_{i+1})/(w_i + w_{i+1})$ with weight $(w_i + w_{i+1})$.
(iii) Now repeat the process for the new blocks with their weighted responses and corresponding weights until the responses are nondecreasing.
(iv) Once this process is completed, y_i^* is equal to the weighted response of the block which contains x_i.

The following example illustrates the main steps of PAVA.

Example: Let $k = 7$ and the values of $\{y_i\}$ and $\{w_i\}$ be as in the top three lines of Table 2.8. The values of y_i^* are given in the last line. The calculations in the table show that the estimator of $\boldsymbol{\mu}$ subject to the simple order $\mu_1 \leqslant \cdots \leqslant \mu_7$ is (-5/2, -5/2, 2/3, 2/3, 3, 5, 6). ■

Although the foregoing discussions are limited to the case when the objective function is a weighted sum of squares, the result has been extended to include the case when it is of the form $(\boldsymbol{y} - \boldsymbol{\mu})^T \boldsymbol{G} (\boldsymbol{y} - \boldsymbol{\mu})$ where $\boldsymbol{G}$ is a positive definite matrix (Diaz and Salvador (1988)) and the simple order set is replaced by an *acute* cone. The PAVA has also been extended to include more general quadratic programming problems and concave regression problems (Tang and Lin (1991), Qian (1994b))

The isotonic regression under simple order has another form that is instructive. Although the variable x is treated as qualitative in isotonic regression, let us consider

Table 2.8 Illustration of PAVA

i	1		2	3		4	5	6	7
w_i	1		1	2		1	2	2	1
y_i	-1		-4	1		0	3	5	6
First pool:									
Blocks		{ 1, 2 }		{ 3 }		{ 4 }	{ 5 }	{ 6 }	{ 7 }
Pooled w		2		2		1	2	2	1
Pooled y		-5/2		1		0	3	5	6
Second pool:									
Blocks		{ 1, 2 }			{ 3, 4 }		{ 5 }	{ 6 }	{ 7 }
Pooled w		2			3		2	2	1
Pooled y		-5/2			2/3		3	5	6
y_i^*	-5/2		-5/2	2/3		2/3	3	5	6

the special case when it also has a quantitative interpretation. Let $\mu(x)$ be a nondecreasing function of x and let $F(x) = \int^x \mu(t)dt$. Then $F(x)$ is a convex function and $\mu(x)$ is the left slope of $F(x)$. This suggests that an alternative approach to estimating $\mu(x)$ subject to the constraint that μ is nondecreasing, is to estimate F subject to the constraint that it is convex and then estimate $\mu(x_i)$ as the left slope of the estimate of F at x_i. In fact a version of this approach and PAVA are equivalent. Such an equivalence holds, with suitable modifications, even when $x_1, \ldots, x_k$ are not quantitative but k categories. These are discussed below.

Let P_0 denote the origin in two-dimensions, and let the coordinates of P_i be defined by

$$P_i \equiv (w_1 + \ldots + w_i, w_1 y_1 + \ldots + w_i y_i), \qquad i = 1, \ldots, k.$$

Let the *Cumulative Sum Diagram* (*CSD*) be the curve which joins the consecutive points, $P_0, P_1, \ldots, P_k$ by straight line segments. The *Greatest Convex Minorant* (*GCM*) of *CSD* is defined as

$$GCM(t) = \max\{g(t) : g \text{ is convex and } g(t) \leqslant CSD(t) \text{ for every } t\}. \qquad (2.42)$$

Therefore, *GCM* of *CSD* is the maximum of the convex functions that do not lie above $CSD(t)$ at any t. Alternatively, $GCM(t)$ can be seen as the path of a taut string that joins P_0 and P_k such that $P_0, \ldots, P_k$ are on or above the string (see Figures 2.1 and 2.2). Now we have the following; see pages 9-10 in Barlow et al. (1972) or Theorem 1.2.1 in Robertson et al. (1988) for a proof.

Proposition 2.4.1 *For the simple increasing order, $\mu_1 \leqslant \ldots \leqslant \mu_k$, we have that*

$$y_i^* = \textit{left slope of } GCM(t) \textit{ at } t = (w_1 + \ldots + w_i),\ i = 1, \ldots, k. \qquad \blacksquare$$

The solution based on *GCM* can be given an explicit form using min-max and max-min functions. To this end, let

$$Av(s,t) = (w_s y_s + \ldots + w_t y_t)/(w_s + \ldots + w_t) \qquad (2.43)$$

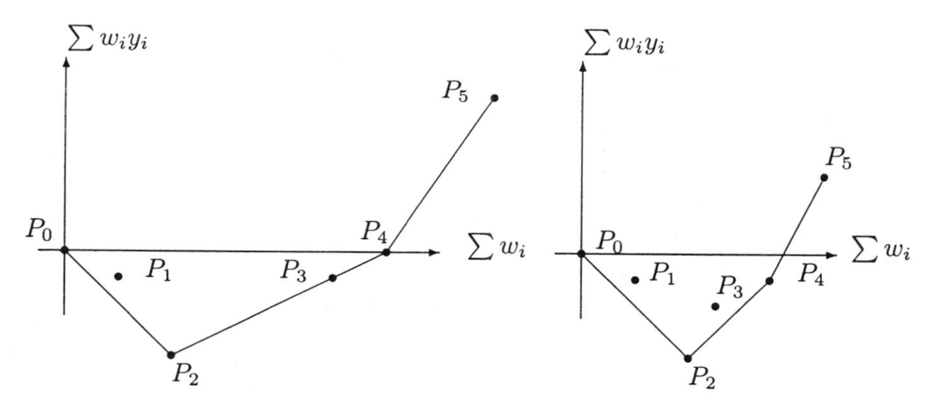

Fig. 2.1 Greatest convex minorant of $P_0P_1P_2P_3P_4P_5$

Fig. 2.2 Greatest convex minorant of $P_0P_1P_2P_3P_4P_5$

for $t \geqslant s$. Thus, $Av(s,t)$ is the weighted mean of the observations $y_s, \ldots, y_t$ and it is also the left slope of the line segment $P_{s-1}P_t$. Then equivalent closed-forms of y_i^* are given in the next result, and it is followed by an illustrative example; for more details see Barlow et al. (1972), Robertson et al. (1988), and Qian (1992).

Proposition 2.4.2 *For the simple increasing order, $\mu_1 \leqslant \ldots \leqslant \mu_k$, we have that*

$$\begin{aligned} y_i^* &= \min_{t\geqslant i}\max_{s\leqslant i} Av(s,t) &= \min_{t\geqslant i}\max_{s\leqslant t} Av(s,t) \\ &= \max_{s\leqslant i}\min_{t\geqslant i} Av(s,t) &= \max_{s\leqslant i}\min_{t\geqslant s} Av(s,t) \end{aligned}$$ ■

Example: Let $k = 6$. The values of (w_i, y_i) and the coordinates of the points P_i are given in Table 2.9, and Fig. 2.1 shows the points. Let us denote the slope of P_iP_j by $[P_iP_j]$. Now, by the foregoing Proposition, we have that

$$\begin{aligned} y_2^* &= \max_{s\leqslant 2}\min_{t\geqslant 2} Av(s,t) = \max_{s\leqslant 2}\min_{t\geqslant 2}[P_{s-1}P_t] \\ &= \max\{\min_{t\geqslant 2}[P_0P_t], \min_{t\geqslant 2}[P_1P_t]\} \\ &= \max\{ \quad \min\{[P_0P_2],[P_0P_3],[P_0P_4],[P_0P_5]\}, \\ &\qquad\qquad \min\{[P_1P_2],[P_1P_3],[P_1P_4],[P_1P_5]\} \quad \} \\ &= \max\{[P_0P_2],[P_1P_2]\} = [P_0P_2]. \end{aligned}$$

This simple example illustrates how the max-min formula is related to the left slope of the *GCM(t)* at $t = w_1 + w_2$. Similarly, it would be instructive to verify that y_3^* is equal to the left slope of the *GCM(t)* at $t = (w_1 + \ldots + w_3)$. Fig. 2.2 shows another example of a greatest convex minorant of $P_0P_1P_2P_3P_4P_5$. ■

There are more general versions of the *min max* and *max min* formulas; see Section 1.4 in Robertson et al. (1988).

Table 2.9 Greatest convex minorant

i	1		2	3	4	5	6
w_i	1		1	3	1	2	2
y_i	-1		-3	1	1	3	4
Pooled blocks		{ 1, 2 }		{ 3 }	{ 4 }	{ 5 }	{ 6 }
Pooled w		2		3	1	2	2
Pooled y		-2		1	1	3	4
Final Estimate							
y_i^*	-2		-2	1	1	3	4
Greatest convex minorant :							
$\sum_{j=1}^{i} w_j$	1		2	5	6	8	10
$\sum_{j=1}^{i} w_j y_j$	-1		-4	- 1	0	6	14
P_i	(1,-1)		(2,-4)	(5,-1)	(6,0)	(8,6)	(10,14)

2.4.2 Ordered Means of Exponential Family

This section contains more advanced topics and it can be ignored at first reading. Further, most of the results in this section will not be required until Chapter 7.

There are different forms of what is known as the exponential family. First, we shall consider the form that is used in Generalized Linear Models where there may be an unknown scale parameter (for example, see McCullagh and Nelder (1989)); another form in which there are no unknown scale parameters will be considered at the end of this subsection. Suppose that the variable y has probability function,

$$f(y;\theta,\phi) = \exp\{\frac{y\theta - b(\theta)}{a(\phi)} + c(y;\phi)\}, \tag{2.44}$$

where θ and ϕ are scalars and $a(\cdot) > 0, b(\cdot)$ and $c(\cdot)$ are some functions. The function f may be the probability density of a continuous random variable such as normal or gamma, or it may the probability function of a discrete random variable such as poisson or binomial. Let $\ell(\theta,\phi)$ denote $\log f(y;\theta,\phi)$. Then, it is well known that (for example, see McCullagh and Nelder (1989), p 22)

$$(\partial/\partial\theta)\ell(\theta,\phi) = \{y - \dot{b}(\theta)\}/a(\phi), \quad (\partial^2/\partial\theta^2)\ell(\theta,\phi) = -\ddot{b}(\theta)/a(\phi),$$
$$E(y) = \mu = \dot{b}(\theta) \text{ and } \mathrm{var}(y) = \ddot{b}(\theta)a(\phi).$$

Let $\hat{\theta}$ and $\hat{\mu}$ denote the unconstrained *mle*s of θ and μ based on a single observation of y. Then, we have $\hat{\mu} = \dot{b}(\hat{\theta}) = y$.

Now, suppose that we have k groups and assume that the variable y_i in group i has probability function,

$$f(y_i;\theta_i,\phi_i) = \exp\{\frac{y_i\theta_i - b(\theta_i)}{a(\phi_i)} + c(y_i;\phi_i)\}, \tag{2.45}$$

where θ_i and ϕ_i are scalars and $a(\cdot) > 0, b(\cdot)$ and $c(\cdot)$ are some functions. Let μ_i denote the population mean of y_i. The function $a(\phi_i)$ usually takes the form ϕ/m_i;

ϕ and m_i are known as the dispersion parameter and prior weight, respectively. As an example, if y_i is the mean of m_i independent observations from $N(\mu_i, \sigma^2)$ then $a(\phi_i) = \sigma^2/m_i$. Let $y_{i1}, \ldots, y_{in_i}$ denote n_i independent observations from group $i, i = 1, \ldots, k$. Let $\boldsymbol{\theta} = (\theta_1, \ldots, \theta_k)^T$, $\boldsymbol{\phi} = (\phi_1, \ldots, \phi_k)^T$, $\bar{y}_i = n_i^{-1}(y_{i1} + \ldots + y_{in_i})$, $\bar{\boldsymbol{y}} = (\bar{y}_1, \ldots, \bar{y}_k)^T$. We are interested to obtain the *mle* of $\boldsymbol{\mu}$ subject to

$$\mu_i - \mu_j \geqslant 0 \text{ for } (i, j) \in B \tag{2.46}$$

where B is a given subset of $\{(i, j) : i, j = 1, \ldots, k\}$. Let $\boldsymbol{A}$ be defined by

$$\{\boldsymbol{\mu} : \boldsymbol{A}\boldsymbol{\mu} \geqslant 0\} = \{\boldsymbol{\mu} : \mu_i - \mu_j \geqslant 0 \text{ for } (i, j) \in B\}.$$

It follows from (2.45) that the loglikelihood $L(\boldsymbol{\theta}, \boldsymbol{\phi})$ can be expressed as

$$L(\boldsymbol{\theta}, \boldsymbol{\phi}) = \sum_{i=1}^{k} \frac{n_i\{\bar{y}_i\theta_i - b(\theta_i)\}}{a(\phi_i)} + H(\boldsymbol{y}; \boldsymbol{\phi}) \tag{2.47}$$

where H is some function that does not depend on $\boldsymbol{\theta}$. Let $(\hat{\boldsymbol{\theta}}, \hat{\boldsymbol{\phi}})$ denote the unconstrained *mle* of $(\boldsymbol{\theta}, \boldsymbol{\phi})$; thus $(\hat{\boldsymbol{\theta}}, \hat{\boldsymbol{\phi}})$ is the unconstrained maximizer of $L(\boldsymbol{\theta}, \boldsymbol{\phi})$. Let the corresponding estimator of $\boldsymbol{\mu}$ be denoted by $\hat{\boldsymbol{\mu}}$. If $\boldsymbol{\mu}(\boldsymbol{\theta}, \boldsymbol{\phi})$ expresses $\boldsymbol{\mu}$ as a function of $(\boldsymbol{\theta}, \boldsymbol{\phi})$ then by the invariance of *mle*, we have that $\hat{\boldsymbol{\mu}} = \boldsymbol{\mu}(\hat{\boldsymbol{\theta}}, \hat{\boldsymbol{\phi}})$. Since the exponential family was expressed in the canonical form, $\hat{\boldsymbol{\mu}} = \bar{\boldsymbol{y}}$. Similarly, let $(\tilde{\boldsymbol{\phi}}, \tilde{\boldsymbol{\mu}})$ denote the *mle* of $(\boldsymbol{\phi}, \boldsymbol{\mu})$ subject to $\boldsymbol{\mu} \in \mathcal{Q}$ where $\mathcal{Q} = \{\boldsymbol{\mu} : \boldsymbol{A}\boldsymbol{\mu} \geqslant \mathbf{0}\}$. Since the *mle* is invariant under transformations, the constrained *mle* of $\boldsymbol{\theta}$ is $\tilde{\boldsymbol{\theta}} = \boldsymbol{\theta}(\tilde{\boldsymbol{\mu}})$. It turns out that, under some conditions concerning the dispersion parameter $\boldsymbol{\phi}$, $\tilde{\boldsymbol{\mu}}$ is equal to the isotonic regression of $\hat{\boldsymbol{\mu}}$ with respect to some weights, and, therefore, $\tilde{\boldsymbol{\mu}}$ can be obtained by minimizing a quadratic function of $\bar{\boldsymbol{y}}$. This is stated in the next result; the proof is given in the Appendix.

Proposition 2.4.3 *Suppose that the parameter $\boldsymbol{\theta}$ is to be estimated by maximizing the function $L(\boldsymbol{\theta}, \boldsymbol{\phi})$ in (2.47) subject to* (2.46), *which we write as $\boldsymbol{A}\boldsymbol{\mu} \geqslant \mathbf{0}$ where $\mu_i = E(\bar{y}_i)$. Let $(\tilde{\boldsymbol{\theta}}, \tilde{\boldsymbol{\mu}})$ denote the resulting estimator of $(\boldsymbol{\theta}, \boldsymbol{\mu})$.*
(a) Assume that $\phi_1, \ldots, \phi_k$ are known constants. Then $\tilde{\boldsymbol{\mu}}$ is also the value of $\boldsymbol{\mu}$ at which $\sum(\bar{y}_i - \mu_i)^2\{n_i/a(\phi_i)\}$ reaches its minimum subject to $\boldsymbol{A}\boldsymbol{\mu} \geqslant \mathbf{0}$.
(b) Assume that $\phi_1 = \ldots = \phi_k (= \phi$, say), and ϕ may be an unknown nuisance parameter. Then $\tilde{\boldsymbol{\mu}}$ is also the value of $\boldsymbol{\mu}$ at which $\sum(\bar{y}_i - \mu_i)^2 n_i$ reaches its minimum subject to $\boldsymbol{A}\boldsymbol{\mu} \geqslant \mathbf{0}$. ■

To consider important special cases of this result, let $y_{i1}, \ldots, y_{in_i}$ denote n_i independent observations from any one of the following (here we use the same notation as in McCullagh and Nelder (1989) page 30):

1. Normal with mean μ_i and unknown variance σ^2.

2. Poisson with mean μ_i.

3. Binary with $\text{pr}(y_{ij} = 1) = \mu_i$ and $\text{pr}(y_{ij} = 0) = 1 - \mu_i$.

4. Gamma with parameters (μ_i, ν), where $\text{var}(y_i) = \mu_i^2/\nu$ and ν is unknown.

5. Inverse gaussian with parameter (μ_i, σ^2) where σ is unknown.

All of these cases belong to the canonical exponential family in (2.44). Therefore, for each of the five cases just listed, it follows from Proposition 2.4.3 that

$$\tilde{\boldsymbol{\mu}} = \arg \min_{\boldsymbol{A\mu} \geqslant \mathbf{0}} \sum (\bar{y}_i - \mu_i)^2 n_i.$$

If the constraint $\boldsymbol{A\mu} \geqslant \mathbf{0}$ is the same as $\mu_1 \leqslant \ldots \leqslant \mu_k$ then it follows from Proposition 2.4.2 that

$$\tilde{\mu}_i = \min_{t \geqslant i} \max_{s \leqslant i} (n_s \bar{y}_s + \ldots + n_t \bar{y}_t)/(n_s + \ldots + n_t); \tag{2.48}$$

further, $\tilde{\mu}_i$ can also be expressed as the left slope of a greatest convex minorant of a particular piecewise linear function as shown in Proposition 2.4.1.

Now, let us consider the case when $\phi_1, \ldots, \phi_k$ are unknown nuisance parameters; this was not considered in Proposition 2.4.3. Let $(\tilde{\boldsymbol{\theta}}, \tilde{\boldsymbol{\phi}})$ denote the solution of

$$\min L(\boldsymbol{\theta}, \boldsymbol{\phi}) \text{ subject to } \boldsymbol{A\mu} \geqslant \mathbf{0}.$$

It can be shown using Lemmas 2.5.3 and 2.5.2 in the appendix to this chapter that $\tilde{\boldsymbol{\mu}}$ is still the isotonic regression of $\bar{\boldsymbol{y}}$ with respect to the weights $\tilde{w}_i = n_i/a(\tilde{\phi}_i)$ as in part (a) of Proposition 2.4.3. However, since the weights depend on $\tilde{\boldsymbol{\phi}}$, which in turn may depend on $\tilde{\boldsymbol{\mu}}$, it follows that $\tilde{\boldsymbol{\mu}}$ is not just a function of $\bar{\boldsymbol{y}}$ alone, as is the case in parts (a) and (b) of Proposition 2.4.3. Therefore, the foregoing results do not provide a simple closed-form for computing $\tilde{\boldsymbol{\mu}}$ when $\phi_1, \ldots, \phi_k$ are unknown.

If the conditions of Proposition 2.4.3 are satisfied, then it provides a convenient way of computing $\tilde{\boldsymbol{\mu}}$ starting from the unconstrained estimator $\hat{\boldsymbol{\mu}}$. Further, it provides analytically simpler forms for the solution. This may be helpful in the study of the properties of the solutions. For example, to study the asymptotic properties of $\tilde{\boldsymbol{\mu}}$, it may be convenient to use the fact that $\tilde{\boldsymbol{\mu}}$ is obtained by minimizing a quadratic function of $\hat{\boldsymbol{\mu}}$; since properties of $\hat{\boldsymbol{\mu}}$ are easier to obtain, and minimizers of sum of squares are more tractable, this approach may turn out to be easier.

If there are no unknown scale parameters, then a result similar to the foregoing proposition holds even when the exponential family is not in the canonical form (2.44). Suppose that the density function of y_{ij} is

$$f(y; \theta_i, \phi_i) = \exp\{p_1(\theta_i)p_2(\phi_i)K(y; \phi_i) + S(y; \phi_i) + q(\theta_i, \phi_i)\},$$

where $(\partial/\partial\theta_i)q(\theta_i, \phi_i) = -\theta_i(\partial/\partial\theta_i)p_1(\theta_i)p_2(\phi_i)$ and ϕ_i is known $(i = 1, \ldots, k)$. Then, we have $E\{K(y_i; \phi_i)\} = \theta_i$ and $var\{K(Y_i; \phi_i)\} = \{(\partial/\partial\theta_i)p_1(\theta_i)p_2(\phi_i)\}^{-1}$. Further, the unconstrained *mle* $\hat{\theta}_i$ of θ_i is $(1/n_i)\{K(y_{i1}; \phi_i) + \ldots + K(y_{in_i}; \phi_i)\}$; let $\hat{\boldsymbol{\theta}} = (\hat{\theta}_1, \ldots, \hat{\theta}_k)^T$. Now, the *mle* of $\boldsymbol{\theta}$ subject to $\boldsymbol{A\theta} \geqslant \mathbf{0}$ is the isotonic regression of $\hat{\boldsymbol{\theta}}$ with weights $w_i = n_i p_2(\phi_i)$ (see Theorem 1.5.2 in Robertson et al. (1988); they attribute this result to the student, Zehua Chen).

The next result (proof is given in the appendix to this chapter) appears on page 45 in Barlow et al. (1972); also on page 38 in Robertson et al. (1988). This result will be used for obtaining simpler formulas for the constrained *mle* of a multinomial parameter, and also the constrained nonparametric *mle* of a monotonic density function in a later chapter.

Corollary 2.4.4 *(Maximum of a product). Let c_i and y_i be given positive numbers, s be a given real number,*

$$L(\boldsymbol{\theta}) = \theta_1^{y_1} \ldots \theta_k^{y_k},$$

$\boldsymbol{A}$ be a given matrix in which every row is a permutation of the k-vector $(-1, 1, 0, \ldots, 0)$, and $z_i = (y_i / \sum y_i)(s/c_i)$ $(i = 1, \ldots, k)$. Then

$$\textit{the maximum of } L(\boldsymbol{\theta}) \textit{ subject to } \boldsymbol{A\theta} \geqslant \mathbf{0} \textit{ and } \sum c_i\theta_i = s$$

and

$$\textit{the minimum of } \sum (z_i - \theta_i)^2 c_i \textit{ subject to } \boldsymbol{A\theta} \geqslant \mathbf{0}$$

are reached at the same value of $\boldsymbol{\theta}$. ■

Example 2.4.1 *Multinomial Parameters*

Consider a multinomial distribution with k categories, and probability function

$$n!(y_1! \ldots y_k!)^{-1}\theta_1^{y_1} \ldots \theta_k^{y_k},$$

where $n = (y_1 + \ldots + y_k)$ is the total number of observations. Let $\tilde{\boldsymbol{\theta}}$ denote the *mle* of $\boldsymbol{\theta}$ subject to $\boldsymbol{A\theta} \geqslant \mathbf{0}$ and $\sum \theta_i = 1$. Then, we have that

$$\tilde{\boldsymbol{\theta}} = \arg \min_{\boldsymbol{A\theta} \geqslant \mathbf{0}} \sum_{i=1}^{k} (\hat{\theta}_i - \theta_i)^2 \text{ where } \hat{\theta}_i = y_i/n \qquad \text{for } i = 1, \ldots, k.$$

This can be deduced from the previous corollary as follows.

Since $n!(y_1! \ldots y_k!)^{-1}$ does not depend on $\boldsymbol{\theta}$ and $\sum \theta_i = 1$ is equivalent to $\sum c_i\theta_i = s$ where $c_i = 1 = s$, the results of the previous corollary on maximization of a product are applicable. Note that $z_i = \{y_i/(\sum y_i)\}(s/c_i) = (y_i/n) = \hat{\theta}_i$, which is the unrestricted *mle* of θ_i $(i = 1, \ldots, k)$. Further, $w_i = (y_i/z_i) = n$. Now, the desired result follows from the previous corollary. ■

2.5 APPENDIX: PROOFS

The proof of Proposition 2.4.3 involves several intermediate results that are of independent interest. These are stated below as lemmas. The proofs of these lemmas, given below, use various mathematical results on projections onto polyhedrals. A detailed account of the relevant results and definitions are given in the next chapter. Therefore, these proofs are meant to be read only if the reader is familiar with the basic results on projections on to polyhedrals discussed in the Appendix to the next chapter. Proofs of the following lemmas are also given from first principles in Barlow et al. (1972); see also Section 1.7 in Robertson et al. (1988).

Lemma 2.5.1 *Let $\boldsymbol{A}$ be a matrix in which every row is a permutation of $(-1, 1, 0, 0, \ldots, 0)$, $\mathcal{Q} = \{\boldsymbol{\mu} : \boldsymbol{A\mu} \geqslant \boldsymbol{0}\}$, and let $\boldsymbol{y}^*$ denote the value of $\boldsymbol{\mu}$ at which $\sum\{y_i - \mu_i\}^2 w_i$ reaches its minimum over $\boldsymbol{A\mu} \geqslant \boldsymbol{0}$. For any real function ϕ on $\mathbb{R}$, define $\boldsymbol{\phi}^* = (\phi(y_1^*), \ldots, \phi(y_k^*))^T$. Then we have the following:*

1. *Let F be the unique face of $\mathcal{Q}$ such that $\boldsymbol{y}^*$ is in the relative interior of F. Then, $\boldsymbol{\phi}^*$ is in the linear space spanned by F.*
2. *$(\boldsymbol{y} - \boldsymbol{y}^*)^T \boldsymbol{W} \boldsymbol{\phi}^* = 0$ where $\boldsymbol{W} = \operatorname{diag}\{w_1, \ldots, w_k\}$.*

Proof: Let $\boldsymbol{a}_i$ denote a typical row of $\boldsymbol{A}$. Let I and J be such that

$$F = \{\boldsymbol{\mu} : \boldsymbol{a}_i^T \boldsymbol{\mu} = 0 \text{ for } i \in I, \text{ and } \boldsymbol{a}_j^T \boldsymbol{\mu} \geq 0 \text{ for } j \in J\}.$$

Then

$$ri(F) = \{\boldsymbol{\mu} : \boldsymbol{a}_i^T \boldsymbol{\mu} = 0 \text{ for } i \in I, \text{ and } \boldsymbol{a}_j^T \boldsymbol{\mu} > 0 \text{ for } j \in J\},$$

where $ri()$ denotes the relative interior. Let $\langle F \rangle = \{\boldsymbol{\mu} : \boldsymbol{a}_i^T \boldsymbol{\mu} = 0 \text{ for } i \in I\}$, the linear space spanned by F. Then $(\boldsymbol{y} - \boldsymbol{y}^*) \perp_{\boldsymbol{W}^{-1}} \langle F \rangle$. Suppose that $\boldsymbol{a}_i^T \boldsymbol{\mu} = 0$ for some $i \in I$ and that the nonzero elements of $\boldsymbol{a}_i$ occur at positions p and q. Then $\mu_p = \mu_q$. Consequently, $\phi(\mu_p) = \phi(\mu_q)$ and $\boldsymbol{a}_i^T \boldsymbol{\phi}_0 = 0$ where $\boldsymbol{\phi}_0 = (\phi(\mu_1), \ldots, \phi(\mu_k))^T$. Since $\boldsymbol{y}^* \in F \subset \langle F \rangle$, it follows that $\boldsymbol{a}_i^T \boldsymbol{y}^* = 0$ and hence $\boldsymbol{a}_i^T \boldsymbol{\phi}^* = 0$ for every $i \in I$; hence, $\boldsymbol{\phi}^*$ also belongs to $\langle F \rangle$. This completes the proof of the first part. The proof of the second part follows from $(\boldsymbol{y} - \boldsymbol{y}^*) \perp_{\boldsymbol{W}^{-1}} \langle F \rangle$. ■

The first part of the lemma provides a geometric picture and the second part follows from the first. The next lemma says that the minimum of a weighted sum of squares and that of a more general function, both subject to the same constraints, are reached at the same point.

Lemma 2.5.2 *Let $y_1, \ldots, y_k$ be given real numbers, and let I be an interval containing $\{y_1, \ldots, y_k\}$. Let Φ be a convex function that is finite on I and ∞ elsewhere. Let ϕ be the derivative of Φ (at a corner, choose a value between the left and right derivatives); assume that ϕ is finite on I. Let*

$$\Delta(u, v) = \begin{cases} \Phi(u) - \Phi(v) - (u - v)\phi(v), & \text{if } u, v \in I \\ \infty, & \text{if } u \notin I \text{ or } v \notin I. \end{cases}$$

For any two points, $\boldsymbol{y}$ and $\boldsymbol{\mu}$ in $\mathbb{R}^k$, let

$$\Delta[\boldsymbol{y}, \boldsymbol{\mu}] = \sum_{i=1}^{k} \Delta(y_i, \mu_i) w_i.$$

Let $\boldsymbol{A}$ be a matrix in which every row is a permutation of $(-1, 1, 0, 0 \ldots, 0)$, $\mathcal{Q} = \{\boldsymbol{\mu} : \boldsymbol{A\mu} \geqslant \boldsymbol{0}\}$, and let $\boldsymbol{y}^*$ denote the value of $\boldsymbol{\mu}$ at which $\sum\{y_i - \mu_i\}^2 w_i$ reaches its minimum over $\boldsymbol{A\mu} \geqslant \boldsymbol{0}$. Then, *(i)* $\Delta[\boldsymbol{y}, \boldsymbol{\mu}] \geqslant \Delta[\boldsymbol{y}, \boldsymbol{y}^*] + \Delta[\boldsymbol{y}^*, \boldsymbol{\mu}]$, *for any* $\boldsymbol{\mu} \in \mathcal{Q}$.

(ii) $\boldsymbol{y}^*$ *solves* $\min_{\boldsymbol{\mu} \in \mathcal{Q}} \Delta[\boldsymbol{y}, \boldsymbol{\mu}]$.

(iii) $\boldsymbol{y}^*$ *also solves* $\max_{\boldsymbol{\mu} \in \mathcal{Q}} \sum[\Phi(\mu_i) + (y_i - \mu_i)\phi(\mu_i)]w_i$.

(iv) If Φ *is strictly convex then* $\Delta[\boldsymbol{y}, \boldsymbol{\mu}]$ *has a unique minimum over* $\boldsymbol{\mu} \in \mathcal{Q}$.

Proof:
[This proof is essentially the same as that in Barlow et al. (1972), page 42]. It follows from the definition of Δ that, for real numbers r, s and t,

$$\Delta(r, t) = \Delta(r, s) + \Delta(s, t) - (r - s)\{\phi(s) - \phi(t)\}.$$

Let $\boldsymbol{\mu}$ be a point in $\mathcal{Q}$, $\boldsymbol{\phi}_0 = (\phi(\mu_1), \ldots, \phi(\mu_k))^T$, and $\boldsymbol{\phi}^* = (\phi(y_1^*), \ldots, \phi(y_k^*))^T$. It follows from (2.49) that

$$\Delta[\boldsymbol{y}, \boldsymbol{\mu}] = \Delta[\boldsymbol{y}, \boldsymbol{t}^*] + \Delta[\boldsymbol{y}^*, \boldsymbol{\mu}] - (\boldsymbol{y} - \boldsymbol{y}^*)^T \boldsymbol{W}(\boldsymbol{\phi}^* - \boldsymbol{\phi}_0). \tag{2.49}$$

Since $\boldsymbol{A\mu} \geqslant \mathbf{0}$ and ϕ is nondecreasing, it follows that $\boldsymbol{A\phi}_0 \geqslant \mathbf{0}$ and hence $\boldsymbol{\phi}_0 \in \mathcal{Q}$. Therefore, it follows from Lemma 3.12.3 (see page 114) that

$$(\boldsymbol{y} - \boldsymbol{y}^*)^T \boldsymbol{W} \boldsymbol{\phi}_0 \leqslant 0. \tag{2.50}$$

Further, by Lemma 2.5.1, we have that

$$(\boldsymbol{y} - \boldsymbol{y}^*)^T \boldsymbol{W} \boldsymbol{\phi}^* = 0. \tag{2.51}$$

Now, part (i) of the theorem follows from (2.49), (2.50), and (2.51). Part (ii) follows from part (i) because $\Delta[\cdot, \cdot] \geqslant 0$. Part (iii) follows from

$$\Delta[\boldsymbol{y}, \boldsymbol{\mu}] = \sum_{i=1}^{k} \Phi(y_i)w_i - \sum_{i=1}^{k} \{\Phi(\mu_i) + (y_i - \mu_i)\phi(\mu_i)\}w_i]$$

and the fact that the first term, $\sum \Phi(y_i)w_i$, does not depend on $\boldsymbol{\mu}$. Part (iv) follows from $\Delta[\boldsymbol{y}^*, \boldsymbol{\mu}] > 0$ if $\boldsymbol{y}^* \neq \boldsymbol{\mu}$. ■

Lemma 2.5.3 *Let* $L(\boldsymbol{\theta}, \boldsymbol{\phi})$ *denote the loss function in (2.47). Then,*

$$\begin{aligned} L(\boldsymbol{\theta}, \boldsymbol{\phi}) &= \sum_{i=1}^{k} \{\bar{y}_i\theta_i - b(\theta_i)\}\{n_i/a(\phi_i)\} + \sum_{i,j} c(y_{ij}; \phi_i) \\ &= \sum_{i=1}^{k} \{\Phi(\mu_i) + \theta(\mu_i)(\bar{y}_i - \mu_i)\}\{n_i/a(\phi_i)\} + H(\boldsymbol{y}; \boldsymbol{\phi}, \mu_0), \end{aligned}$$

for some function H, $\boldsymbol{\phi} = (\phi_1, \ldots, \phi_k)^T$, *and* $\Phi(t) = \int_{\mu_0}^{t} \theta(\mu)d\mu$.

Proof:

Let θ, μ and ϕ denote scalar parameters. Let $\dot{b}(\boldsymbol{\theta})$ and $\ddot{b}(\boldsymbol{\theta})$ denote the first and second derivatives of $b(\boldsymbol{\theta})$. For the exponential family in (2.47), we have that

$$\mu = \dot{b}(\theta) \text{ and } var(Y) = \ddot{b}(\theta)a(\phi);$$

for example, see McCullagh and Nelder (1989), page 29). Since $\ddot{b}(\theta) > 0$, it follows that $\dot{b}(\theta)$ is a strictly increasing function. Therefore, $\mu = \dot{b}(\theta)$ can be inverted, and hence we have that

$$\theta(\mu) = \dot{b}^{-1}(\mu), \mu = \dot{b}\{\theta(\mu)\}, \text{ and } (d/d\mu)b\{\theta(\mu)\} = \dot{b}(\theta)\dot{\theta}(\mu) = \mu\dot{\theta}(\mu).$$

Now, to obtain the unconstrained *mle*, $\hat{\boldsymbol{\theta}}$, we solve $(\partial/\partial\boldsymbol{\theta})L(\boldsymbol{\theta}, \boldsymbol{\phi}) = 0$. This leads to

$$\hat{\mu}_i = \bar{y}_i = \dot{b}(\hat{\theta}_i) = \dot{b}\{\theta(\hat{\mu}_i)\}. \tag{2.52}$$

Let μ_0 be a value in the allowable range of μ. Then

$$\begin{aligned} b\{\theta(\mu)\} &= \int_{\mu_0}^{\mu} (d/dt)b\{\theta(t)\}dt + C(\mu_0) \\ &= \int_{\mu_0}^{\mu} t\dot{\theta}(t)dt + C(\mu_0) \\ &= \mu\theta(\mu) - \int_{\mu_0}^{\mu} \theta(t)dt + C(\mu_0). \end{aligned} \tag{2.53}$$

Substituting (2.52) and (2.53) in (2.47), we have

$$Loglikelihood = \sum \{n_i/a(\phi_i)\}\{\Phi(\mu_i) + \theta(\mu_i)(\hat{\mu}_i - \mu_i)\} + H(\boldsymbol{y}, \boldsymbol{\phi})$$

where $\Phi(x) = \int_{\mu_0}^{x} \theta(t)dt$. ■

Proof of Proposition 2.4.3:

Follows from the previous two lemmas.

Proof of Corollary 2.4.4:

Let $w_i = (y_i/z_i)$ for $i = 1, \ldots, k$, and $\ell(\boldsymbol{\theta}) = \log L(\boldsymbol{\theta})$. Then

$$\ell(\boldsymbol{\theta}) = \sum y_i \log \theta_i = \sum (z_i \log \theta_i)w_i. \tag{2.54}$$

Since $\sum(z_i - \theta_i)w_i = s^{-1}(\sum y_i)(s - \sum c_i\theta_i)$, it follows that

$$\sum c_i\theta_i = s \text{ if and only if } \sum (z_i - \theta_i)w_i = 0. \tag{2.55}$$

Let $\Phi(t) = t\log(t)$ and let Δ be defined as in Lemma 2.5.2. Then

$$\begin{aligned} \Delta[a, b] &= \Phi(a) - \Phi(b) - (a - b)\phi(b) \\ &= a\log a - a\log b - (a - b) \end{aligned}$$

and hence

$$\Delta[z_i, \theta_i]w_i = \sum(z_i \log z_i)w_i - \{\ell(\boldsymbol{\theta}) + r(\boldsymbol{\theta})\} \tag{2.56}$$

where $r(\boldsymbol{\theta}) = \sum(z_i - \theta_i)w_i$. Since $\sum(z_i \log z_i)w_i$ does not depend on $\boldsymbol{\theta}$, we can restrict our attention to $\{\ell(\boldsymbol{\theta}) + r(\boldsymbol{\theta})\}$ for the purposes of finding the minimum of $\sum \Delta[z_i, \theta_i]w_i$. Let $\mathcal{A} = \{\boldsymbol{\theta} : \boldsymbol{A\theta} \geqslant \mathbf{0}\}$ and $\mathcal{R} = \{\boldsymbol{\theta} : r(\boldsymbol{\theta}) = 0\}$. Let

$$\tilde{\boldsymbol{\theta}} = \arg\min_{\boldsymbol{\theta}\in\mathcal{A}} \sum(z_i - \theta_i)^2 w_i.$$

Let $\phi(t) = 1$ for every $t \in \mathbb{R}$. Then, by Lemma 2.5.1, we have that $\boldsymbol{\phi}^* = (\phi(\tilde{\theta}_1), \ldots, \phi(\tilde{\theta}_k))^T = \mathbf{1}^T$, and $(\boldsymbol{z} - \tilde{\boldsymbol{\theta}})^T \boldsymbol{W}\mathbf{1} = 0$; hence $r(\tilde{\boldsymbol{\theta}}) = \sum(z_i - \tilde{\theta}_i)w_i = 0$. Further,

$$\begin{aligned} \tilde{\boldsymbol{\theta}} &= \arg\min\nolimits_{\boldsymbol{\theta}\in\mathcal{A}} \sum \Delta[z_i, \theta_i]w_i \quad \text{by Lemma 2.5.2, part(ii)} && (2.57)\\ &= \arg\max\nolimits_{\boldsymbol{\theta}\in\mathcal{A}}\{\ell(\boldsymbol{\theta}) + r(\boldsymbol{\theta})\} \quad \text{by (2.56)} && (2.58) \end{aligned}$$

Now,

$$\begin{aligned} \max_{\boldsymbol{\theta}\in\mathcal{A}\cap\mathcal{R}} \ell(\boldsymbol{\theta}) &= \max_{\boldsymbol{\theta}\in\mathcal{A}\cap\mathcal{R}} \{\ell(\boldsymbol{\theta}) + r(\boldsymbol{\theta})\}, \quad \text{because } r(\boldsymbol{\theta}) = 0 \text{ for } \boldsymbol{\theta} \in \mathcal{R}.\\ &\leqslant \max_{\boldsymbol{\theta}\in\mathcal{A}}\{\ell(\boldsymbol{\theta}) + r(\boldsymbol{\theta})\}, \quad \text{because } \mathcal{A}\cap\mathcal{R} \subset \mathcal{A}\\ &= \ell(\tilde{\boldsymbol{\theta}}) + r(\tilde{\boldsymbol{\theta}}) \quad \text{by (2.57) and (2.58)}\\ &= \ell(\tilde{\boldsymbol{\theta}}) \quad \text{because } r(\tilde{\boldsymbol{\theta}}) = 0. \end{aligned}$$

Since $r(\tilde{\boldsymbol{\theta}}) = 0$, we have that $\tilde{\boldsymbol{\theta}} \in \mathcal{R}$, and hence $\max_{\boldsymbol{\theta}\in\mathcal{A}\cap\mathcal{R}} \ell(\boldsymbol{\theta}) = \ell(\tilde{\boldsymbol{\theta}})$. Since Φ is strictly convex, $\tilde{\boldsymbol{\theta}}$ is uniquely determined. ■

Problems

2.1

In Example 1.2.2, assume that the error distribution is normal. Now, test the following hypotheses.

1. $H_0 : \mu_1 = \mu_2 = \mu_3$ vs $H_1 : \mu_1 \geqslant \mu_2 \geqslant \mu_3$.

2. $H_0 : \mu_1 \geqslant \mu_2 \geqslant \mu_3$ vs $H_1 : \mu_1 \geqslant \mu_2 \geqslant \mu_3$ *does not hold.*

[Hint: Apply the $\bar{F}$-test and follow the steps as in the worked example on walking exercise].

2.2

For the Example 2.1.2 on page 33, compute the p-values when the error distribution is (a) Normal, (b) Logistic, and (c) χ^2_5 for each of the following:

1. $H_0 : \mu_1 = \mu_2 = \mu_3 = \mu_4$ vs $H_1 : \mu_1 \leqslant \mu_2 \leqslant \mu_3, \mu_2 \leqslant \mu_4$.

2. $H_1 : \mu_1 \leqslant \mu_2 \leqslant \mu_3, \mu_2 \leqslant \mu_4$ vs $H_2 :$ *No restriction on* $(\mu_1, \ldots, \mu_4)$.

2.3 Consider the normal theory one-way ANOVA context in Table 2.1. Let H_1 be as in (2.9), and $(\tilde{\mu}_1, \ldots, \tilde{\mu}_k)$ be the solution of

$$\min \sum\sum (y_{ij} - \mu_i)^2 \text{ subject to } H_1.$$

Show that $\tilde{\mu}_1, \ldots, \tilde{\mu}_k)$ is the constrained *mle* of $(\mu_1, \ldots, \mu_k)$ under H_1.
[Hint: Let $L(\mu_1, \ldots, \mu_k, \sigma)$ denote the log likelihood corresponding to $y_{ij} \sim N(\mu_i, \sigma^2)$. Then $L(\mu_1, \ldots, \mu_k, \sigma) = -(2\sigma^2)^{-1} \sum\sum (y_{ij} - \mu_i)^2 - N \log \sigma^2$. For any given, $(\mu_1, \ldots, \mu_k)$, L reaches its minimum at $\tilde{\sigma}^2 = N^{-1} \sum\sum (y_{ij} - \mu_i)^2$. Therefore, the *concentrated* loglikelihood obtained by eliminating σ is

$$L^*(\mu_1, \ldots, \mu_k) = L(\mu_1, \ldots, \mu_k, \tilde{\sigma}) = -\sum\sum (y_{ij} - \mu_i)^2 - 2N^{-1}.$$

The *mle* of $(\mu_1, \ldots, \mu_k)$ under H_1 is its value at which $\sum\sum (y_{ij} - \mu_i)^2$ reaches its minimum.]

2.4 Let y be a binary random variable taking the values 1 and 0 with probability μ and $(1 - \mu)$, respectively, $(0 < \mu < 1)$. Let $f(y)$ denote the probability function of y. Verify that $f(y) = \mu^y (1 - \mu)^{1-y} = \exp[y \log\{\mu/(1 - \mu)\} + \log(1 - \mu)]$, and deduce that the probability function of y is of the form (2.45) with $\theta(\mu) = \log\{\mu/(1 - \mu)\}, a(\phi) = 1$ and $b(\theta) = -\log(1 - \mu)$.

Suppose that there are k groups and for group i, let $y_{i1}, \ldots, y_{in_i}$ denote n_i independent binary random variables with $\text{pr}(y_{ij} = 1) = \mu_i (i = 1, \ldots, k)$. Verify that

$$\text{likelihood} = \Pi_{i=1}^{k} \Pi_{j=1}^{n_i} \mu_i^{y_{ij}} (1 - \mu_i)^{1-y_{ij}}.$$

Let $\boldsymbol{A}$ be a matrix with k columns in which every row is a permutation of $(1, -1, 0, \ldots, 0)$. Use Proposition 2.4.3 and show that the *mle* $\tilde{\boldsymbol{\mu}}$ of $\boldsymbol{\mu}$ subject to $\boldsymbol{A\mu} \geqslant \boldsymbol{0}$ is also the value of $\boldsymbol{\mu}$ which

$$\text{maximizes} \sum (\bar{y}_i - \mu_i)^2 n_i \text{ subject to } \boldsymbol{A\mu} \geqslant \boldsymbol{0}. \tag{2.59}$$

Now, suppose that the constraint $\boldsymbol{A\mu} \geqslant \boldsymbol{0}$ is the same as $\mu_1 \leqslant \ldots \leqslant \mu_k$. Show that

$$\tilde{\mu}_i = \min_{t \geqslant i} \max_{s \leqslant i} (n_s \bar{y}_s + \ldots + n_t \bar{y}_t)/(n_s + \ldots + n_t).$$

Suppose that $k = 5, (n_1, \ldots, n_5) = (10, 15, 25, 12, 16)$ and $\bar{\boldsymbol{y}} = (3/10, 5/15, 8/25, 8/12, 14/16)$. Use a diagram to illustrate that $\tilde{\mu}_i$ can also be expressed as the left slope of a greatest convex minorant of a particular piece-wise linear function. [Hint: see Proposition 2.4.1].

3

Tests on Multivariate Normal Mean

3.1 INTRODUCTION

In this chapter we consider two broad classes of hypothesis testing problems when the population distribution is normal. For example, the observations may be independently and identically distributed from the multivariate normal distribution, $N(\boldsymbol{\theta}, \boldsymbol{V})$, or they may satisfy a linear model of the form $\boldsymbol{Y} = \boldsymbol{X\theta} + \boldsymbol{E}$ with normal errors. For the standard problem of testing $H_0 : \boldsymbol{R\theta} = \boldsymbol{0}$ against $H_2 : \boldsymbol{R\theta} \neq \boldsymbol{0}$, where $\boldsymbol{R}$ is a given fixed matrix, it is easy to apply the likelihood ratio test because the (likelihood ratio) test statistic can be computed easily, and the statistical tables for its null distribution, χ^2_q where $q = rank(\boldsymbol{R})$, are easily available. In what follows, we shall use the acronym LRT for the likelihood ratio statistic and the test as well.

If the hypotheses involve inequalities in $\boldsymbol{\theta}$, then the theory becomes more complicated, and several difficulties in applying the results are also encountered. One such difficulty is caused by the fact that the null distribution of LRT depends on the particular inequalities. For example, when $\boldsymbol{V}$ is known, the null distribution of LRT for testing

$$H_0 : \boldsymbol{R\theta} = \boldsymbol{0} \text{ against } H_1 : \boldsymbol{R\theta} \geqslant \boldsymbol{0}, \boldsymbol{R\theta} \neq \boldsymbol{0} \tag{3.1}$$

depends on the matrix $\boldsymbol{R}$ through $\boldsymbol{RVR}^T$, not just on $rank(\boldsymbol{R})$, as is the case for testing $H_0 : \boldsymbol{R\theta} = \boldsymbol{0}$ against $H_2 : \boldsymbol{R\theta} \neq \boldsymbol{0}$. Consequently, it is not practicable to have statistical tables of critical values because we would need one table for each $\boldsymbol{RVR}^T$ that may arise in practice, and there are infinite number of them. Another difficulty is that, except in some very special cases, it is difficult to compute the critical values exactly.

Fortunately, the p-values and critical values of the tests discussed in this chapter can be computed easily by simulation. A simple example illustrated this in the previous chapter. Its extension to the setting of this chapter is easy. The required computer programs can be developed so that they are applicable to most practical situations in which the constraints on $\boldsymbol{\theta}$ are linear. The main requirement for this is a computer program for solving a quadratic program; such programs are available in IMSL, NAG and several other software packages. See Bazaraa et al. (1993) for relevant theory and references.

Sometimes, constrained inference problems may lead to unexpected results; the simple example in Silvapulle (1997a) illustrated this. For example, the sample means of a bivariate response may appear to suggest that a treatment performed substantially worse than the control, but a likelihood ratio test may conclude to accept that the treatment is better than the control. The crux of these issues are also discussed.

We adopt the general approach of computing the p-values approximately by simulation. For most practical purposes, the simulation approach is simple, adequate, and easy to implement.

The next section provides a discussion of the two types of testing problems that are discussed in this chapter, which we call Type A and Type B problems. Subsection 3.2.1 shows that for a large class of problems considered in this chapter, considerations may be restricted to a single observation from a normal distribution; the reasoning is based on sufficiency of the sample mean for a random sample from a multivariate normal. Some examples in two dimensions are discussed in detail in Section 3.3 to introduce the essential basics. A distribution, called chi-bar-square, arises naturally in Type A and Type B testing problems. This class of distributions is defined in Section 3.4. Computing the critical values of a chi-bar-square distribution is not as easy as reading a number from a table of critical values. In fact, except in some very special cases, it is virtually impossible to compute the tail probability of a chi-bar-square distribution using a hand calculator. Various results relating to the tail probability of a chi-bar-square distribution are discussed in Section 3.5. Then, in Sections 3.7 and 3.8 we consider the two types of testing problems in multivariate analysis with known covariance matrices. These results are extended in the following section to deal with the same type of testing problems in the linear model when the covariance matrix is $\sigma^2\boldsymbol{U}$ where σ^2 is unknown but $\boldsymbol{U}$ is known. Then, we consider the same two types of testing problems when the observations are from a multivariate normal and the covariance matrix is completely unknown. The proofs of several results are demanding, and therefore they are given in an Appendix to ensure that the main ideas and results are presented without the interruption of lengthy technical details.

3.2 STATEMENT OF TWO GENERAL TESTING PROBLEMS

Let us first define some terms. A set $\mathcal{A} \subset \mathbb{R}^p$ is said to be *convex* if $\{\lambda\boldsymbol{x}+(1-\lambda)\boldsymbol{y}\} \in \mathcal{A}$ whenever $\boldsymbol{x}, \boldsymbol{y} \in \mathcal{A}$ and $0 < \lambda < 1$. Therefore $\mathcal{A}$ is a convex set if the line segment joining $\boldsymbol{x}$ and $\boldsymbol{y}$ is in $\mathcal{A}$ whenever the points $\boldsymbol{x}$ and $\boldsymbol{y}$ are in $\mathcal{A}$. A set $\mathcal{A}$ is said to be a *cone with vertex* $\boldsymbol{x}_0$ if $\boldsymbol{x}_0 + k(\boldsymbol{x} - \boldsymbol{x}_0) \in \mathcal{A}$ for every $\boldsymbol{x} \in \mathcal{A}$ and $k \geqslant 0$; if the vertex is the origin, then we shall simply refer to it as a *cone*. Therefore, a cone is a set that consists of infinite straight lines starting from the origin. A *polyhedral* $\mathcal{P}$ in $\mathbb{R}^p$ is a set of the form $\{\boldsymbol{\theta} \in \mathbb{R}^p : \boldsymbol{a}_1^T\boldsymbol{\theta} \geqslant 0, \ldots, \boldsymbol{a}_k^T\boldsymbol{\theta} \geqslant 0\}$ where $\boldsymbol{a}_1, \ldots, \boldsymbol{a}_k$ are given elements of $\mathbb{R}^p$. Thus, $\mathcal{P}$ is the intersection of the *half-spaces*, $\{\boldsymbol{a}_1^T\boldsymbol{\theta} \geqslant 0\}, \ldots$ $\{\boldsymbol{a}_k^T\boldsymbol{\theta} \geqslant 0\}$. It is easily seen that a polyhedral is a closed convex cone. A more detailed discussion of polyhedrals is given in the first Appendix to this chapter.

Notation: Throughout this chapter, we shall adopt the following notation without any further comment:
(a) $\mathcal{C}_a, \mathcal{C}_b, \mathcal{C}$, and $\mathcal{M}$ are subsets of an Euclidean space, for example $\mathbb{R}^p$.
(b) $\mathcal{C}$ is a closed convex cone, $\mathcal{M}$ is a linear space, $\mathcal{M} \subset \mathcal{C}$, and $\mathcal{C}_a \subset \mathcal{C}_b$.

We shall consider two types of inequality constrained testing problems. These are stated below in their most general forms, where $\boldsymbol{\theta} \in \mathbb{R}^p$.

Type A : Test $H_0 : \boldsymbol{\theta} \in \mathcal{M}$ against $H_1 : \boldsymbol{\theta} \in \mathcal{C}$,
Type B : Test $H_1 : \boldsymbol{\theta} \in \mathcal{C}$ against $H_2 : \boldsymbol{\theta} \notin \mathcal{C}$.

Clearly, the null hypothesis in Type A is of the form $\boldsymbol{R\theta} = \mathbf{0}$, for some matrix $\boldsymbol{R}$. Strictly speaking, the alternative hypothesis in the foregoing Type A problem should be stated as $H_1 : \boldsymbol{\theta} \in \mathcal{C}, \boldsymbol{\theta} \notin \mathcal{M}$. However, we shall continue to adopt the convention introduced in the previous chapter (see page 28): *"Test of H_0 against H_1" is to be interpreted as "test of H_0 against $H_1 - H_0$."* Therefore, the foregoing Type A and Type B testing problems may also be stated in the following equivalent forms:

Type A : Test $H_0 : \boldsymbol{\theta} \in \mathcal{M}$ against $H_1 : \boldsymbol{\theta} \in \mathcal{C}$ and $\boldsymbol{\theta} \notin \mathcal{M}$.
Type B : Test $H_1 : \boldsymbol{\theta} \in \mathcal{C}$ against $H_2 : \boldsymbol{\theta} \in \mathbb{R}^p$.

Some examples of Type A problems:

1. $H_0 : \boldsymbol{\theta} = \mathbf{0}$ against $H_1 : \boldsymbol{\theta} \geqslant \mathbf{0}$.
2. $H_0 : (\boldsymbol{\theta}_a^T, \boldsymbol{\theta}_b^T)^T = \mathbf{0}$ against $H_1 : \boldsymbol{\theta}_a \geqslant \mathbf{0}$.
3. $H_0 : \boldsymbol{R\theta} = \mathbf{0}$ against $H_1 : \boldsymbol{R}_1\boldsymbol{\theta} \geqslant \mathbf{0}$, where $\boldsymbol{R}_1$ is a submatrix of $\boldsymbol{R}$.
4. $H_0 : \boldsymbol{A}(\boldsymbol{\theta}) = \mathbf{0}$ against $H_1 : \boldsymbol{A}(\boldsymbol{\theta})$ *is negative semi-definite*, where $\boldsymbol{\theta} = (\theta_1, \theta_2, \theta_3)^T$ and $\boldsymbol{A}(\boldsymbol{\theta})$ is the 2×2 matrix, $(\theta_1, \theta_3 \mid \theta_3, \theta_2)$; this is equivalent to $H_0 : (\theta_1, \theta_2, \theta_3) = \mathbf{0}$ against $H_1 : \theta_1 \leq 0, \theta_2 \leq 0, \theta_1\theta_2 - \theta_3^2 \geqslant 0$.

In the last example, although the parameter space $\{\boldsymbol{\theta} : \theta_1 \leq 0, \theta_2 \leq 0, \theta_1\theta_2 - \theta_3^2 \geqslant 0\}$ involves nonlinear inequalities, it turns out to be a closed convex cone. In view of the convention stated above, $\boldsymbol{\theta} \neq \mathbf{0}, (\boldsymbol{\theta}_a^T, \boldsymbol{\theta}_b^T)^T \neq \mathbf{0}$, $\boldsymbol{R\theta} \neq \mathbf{0}$, and $\boldsymbol{A}(\boldsymbol{\theta}) \neq \mathbf{0}$ are assumed to be implicit in the alternative hypothesis of the foregoing examples (1), (2), (3), and (4), respectively.

Some examples of Type B problems:

1. $H_1 : \boldsymbol{\theta} \geqslant \mathbf{0}$ against $H_2 : \boldsymbol{\theta} \ngeqslant \mathbf{0}$.
2. $H_1 : \boldsymbol{\theta}_a \geqslant \mathbf{0}$ and $\boldsymbol{\theta}_b = \mathbf{0}$ against $H_2 : \boldsymbol{\theta} \in \mathbb{R}^p$, where $\boldsymbol{\theta} = (\boldsymbol{\theta}_a^T, \boldsymbol{\theta}_b^T)^T$.

3. $H_1 : \boldsymbol{R}_1\boldsymbol{\theta} \geqslant \mathbf{0}$ and $\boldsymbol{R}_2\boldsymbol{\theta} = \mathbf{0}$ against H_2 : H_1 *is not true*, where $\boldsymbol{R}_1$ and $\boldsymbol{R}_2$ are some known fixed matrices.

4. H_1 : $\boldsymbol{A}(\boldsymbol{\theta})$ *is negative semi-definite* against H_2 : H_1 *is not true*, where $\boldsymbol{\theta} = (\theta_1, \theta_2, \theta_3)$ and $\boldsymbol{A}(\boldsymbol{\theta})$ is the 2×2 matrix $(\theta_1, \theta_3 \mid \theta_3, \theta_2)$; this is equivalent to $H_1 : \theta_1 \leq 0, \theta_2 \leq 0, \theta_1\theta_2 - \theta_3^2 \geqslant 0$, and H_2 : *no restriction on* $\boldsymbol{\theta}$.

These examples provide some idea of the type of testing problems studied in this chapter. The main difference between Type A and Type B problems is that inequalities may be present in the alternative hypothesis of Type A and the null hypothesis of Type B problems.

Here are some examples that are not Type A or Type B testing problems:

1. $H_0 : \theta_1 \leq 0$ or $\theta_2 \leq 0$ against $H_1 : \theta_1 > 0$ and $\theta_2 > 0$.

2. $H_0 : \boldsymbol{\theta} \geqslant \mathbf{0}$ or $\boldsymbol{\theta} \leq \mathbf{0}$ against $H_1 : \boldsymbol{\theta} \ngeqslant \mathbf{0}$ and $\boldsymbol{\theta} \nleq \mathbf{0}$.

3. $H_0 : \theta_1 = \theta_2 = 0$ against $H_1 : \theta_2^2 \geqslant |\theta_1|$.

The first of these arises in clinical trials involving *combination drugs*; this testing problem is known as the *sign testing* problem. The second one arises in tests against *cross-over interactions* between treatments and groups. For the first two examples, some exact finite sample results are available. These will be studied in Chapter 9. The third is an artificial example; this example resembles a Type A problem but the alternative parameter space is not a closed convex cone. For such problems, large sample results will be studied in Chapter 4.

When the population distribution is $N(\boldsymbol{\theta}, \boldsymbol{V})$, the nature of the solutions to inequality constrained testing problems also depend on what is known about the covariance matrix $\boldsymbol{V}$ in addition to the structure of the null and alternative parameter spaces. We shall consider three cases: (i) $\boldsymbol{V}$ is a known positive definite matrix, (ii) $\boldsymbol{V} = \sigma^2\boldsymbol{U}$, where $\boldsymbol{U}$ is a known positive definite matrix and σ is unknown, and (iii) $\boldsymbol{V}$ is unknown. In the next chapter, we shall study asymptotic results when $\boldsymbol{V}$ is a function of some nuisance parameters.

Almost all of the results in this chapter are for Type A or Type B problems. Occasionally we shall consider test of $H_a : \boldsymbol{\theta} \in \mathcal{C}_a$ against $H_b : \boldsymbol{\theta} \in \mathcal{C}_b$ because it incorporates Type A and Type B problems and it is instructive to introduce some results within this general context.

3.2.1 Reduction by Sufficiency

Often, we shall consider test of hypotheses based on a single observation from a multivariate normal distribution. Although this case is considered for simplicity, it can also be justified using the sufficiency of sample mean. To illustrate this, let $\boldsymbol{X}_1, \ldots, \boldsymbol{X}_n$ be n independent and identically distributed observations from $N(\boldsymbol{\theta}, \boldsymbol{V})$, where $\boldsymbol{V}$ is a known positive definite matrix and $\boldsymbol{\theta} \in \mathbb{R}^p$. Suppose that we wish to test

$$H_a : \boldsymbol{\theta} \in \mathcal{C}_a \text{ against } H_b : \boldsymbol{\theta} \in \mathcal{C}_b,$$

where, as was indicated earlier, $\mathcal{C}_a \subset \mathcal{C}_b$. In this case, the likelihood ratio test can be formulated in terms of a single observation of $\bar{\boldsymbol{X}}$, where $\bar{\boldsymbol{X}}$ is the sample mean. As is shown below, this is a consequence of the fact that $\bar{\boldsymbol{X}}$ is sufficient for $\boldsymbol{\theta}$.

Let $L(\boldsymbol{\theta})$ denote the loglikelihood for the n observations $\boldsymbol{X}_1, \ldots, \boldsymbol{X}_n$. Since

$$L(\boldsymbol{\theta}) = -(n/2)^{-1} \log |\boldsymbol{V}| - (np/2) \log(2\pi) - 2^{-1} \sum_{i=1}^{n} (\boldsymbol{X}_i - \boldsymbol{\theta})^T \boldsymbol{V}^{-1} (\boldsymbol{X}_i - \boldsymbol{\theta}),$$

we have that

$$L(\boldsymbol{\theta}) = \ell(\boldsymbol{\theta}) + g(\boldsymbol{X}_1, \ldots, \boldsymbol{X}_n, \boldsymbol{V}),$$

where

$$\ell(\boldsymbol{\theta}) = -2^{-1} (\bar{\boldsymbol{X}} - \boldsymbol{\theta})^T (n^{-1}\boldsymbol{V})^{-1} (\bar{\boldsymbol{X}} - \boldsymbol{\theta})$$

and $g(\boldsymbol{X}_1, \ldots, \boldsymbol{X}_n, \boldsymbol{V})$ does not depend on $\boldsymbol{\theta}$. Therefore, $\ell(\boldsymbol{\theta})$ is the *kernel* of the loglikelihood for $\{\boldsymbol{X}_1, \ldots, \boldsymbol{X}_n\}$. Now, the *LRT* for testing $H_a : \boldsymbol{\theta} \in \mathcal{C}_a$ against $H_b : \boldsymbol{\theta} \in \mathcal{C}_b$, based on $\{\boldsymbol{X}_1, \ldots, \boldsymbol{X}_n\}$ is

$$2[\max\{L(\boldsymbol{\theta}) : \boldsymbol{\theta} \in \mathcal{C}_b\} - \max\{L(\boldsymbol{\theta}) : \boldsymbol{\theta} \in \mathcal{C}_a\}].$$

Therefore, it follows that

$$LRT = 2[\max\{\ell(\boldsymbol{\theta}) : \boldsymbol{\theta} \in \mathcal{C}_b\} - \max\{\ell(\boldsymbol{\theta}) : \boldsymbol{\theta} \in \mathcal{C}_a\}]. \tag{3.2}$$

On the other hand, since $\bar{\boldsymbol{X}} \sim N(\boldsymbol{\theta}, n^{-1}\boldsymbol{V})$, the kernel of the loglikelihood for the single observation $\bar{\boldsymbol{X}}$ is $\ell(\boldsymbol{\theta})$. Therefore, the maximum likelihood estimator (*mle*) and the LRT based on $\bar{\boldsymbol{X}}$ and those based on $\boldsymbol{X}_1, \ldots, \boldsymbol{X}_n$ are identical. Hence, in much of the developments in this chapter where the covariance matrix $\boldsymbol{V}$ is known, we will consider statistical inference based on a single observation that may be thought of as $\bar{\boldsymbol{X}}$.

3.3 THEORY: THE BASICS IN TWO DIMENSIONS

Before we study general testing problems, it would be instructive to consider some simple special cases. We shall illustrate the derivation of the null distribution of the LRT for some special cases and then use the ideas developed to introduce the general results.

Example 3.3.1 *Maximum Likelihood Estimation*

Let $\boldsymbol{X} = (X_1, X_2)^T \sim N(\boldsymbol{\theta}, \boldsymbol{I})$, where $\boldsymbol{I}$ is the 2×2 identity matrix and $\boldsymbol{\theta} = (\theta_1, \theta_2)^T$. Consider the maximum likelihood estimation of $\boldsymbol{\theta}$ based on one observation on $\boldsymbol{X}$ and subject to the constraint $\boldsymbol{\theta} \in \mathcal{C}$ where $\mathcal{C}$ is the nonnegative orthant $\{(\theta_1, \theta_2)^T : \theta_1 \geqslant 0, \theta_2 \geqslant 0\}$. For the single observation $\boldsymbol{X}$, the kernel $\ell(\boldsymbol{\theta})$ of the loglikelihood is given by

$$-2\ell(\boldsymbol{\theta}) = \{(X_1 - \theta_1)^2 + (X_2 - \theta_2)^2\} = \|\boldsymbol{X} - \boldsymbol{\theta}\|^2$$

where $\|\cdot\|$ is the Euclidean norm defined by $\|\boldsymbol{x}\|^2 = (x_1^2 + x_2^2)$. Let $\tilde{\boldsymbol{\theta}}$ be the *mle* of $\boldsymbol{\theta}$ subject to $\boldsymbol{\theta} \geqslant \mathbf{0}$. Since $-2\ell(\boldsymbol{\theta})$ is equal to the squared distance between $\boldsymbol{X}$ and $\boldsymbol{\theta}$, $\tilde{\boldsymbol{\theta}}$ is the point in $\mathcal{C}$ that is closest to $\boldsymbol{X}$; in other words, $\tilde{\boldsymbol{\theta}}$ is the *projection* of $\boldsymbol{X}$ onto $\mathcal{C}$ (see Fig. 3.1). With Π denoting the projection function, we can write

$$\tilde{\boldsymbol{\theta}} = \Pi(\boldsymbol{X} \mid \mathbb{R}^{+2}).$$

Let Q_1, Q_2, Q_3, Q_4 denote the four quadrants in the two-dimensional plane as in Fig. 3.1. Then the *mle* $\tilde{\boldsymbol{\theta}}$ is given by

$$(\tilde{\theta}_1, \tilde{\theta}_2) = \begin{cases} (X_1, X_2) & \text{if } \boldsymbol{X} \in Q_1 \\ (0, X_2) & \text{if } \boldsymbol{X} \in Q_2 \\ (0, 0) & \text{if } \boldsymbol{X} \in Q_3 \\ (X_1, 0) & \text{if } \boldsymbol{X} \in Q_4. \end{cases} \tag{3.3}$$

■

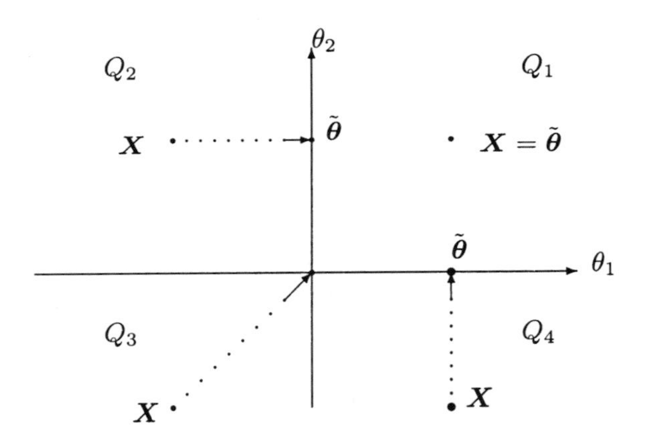

Fig. 3.1 The maximum likelihood estimator $\tilde{\boldsymbol{\theta}}$ of $\boldsymbol{\theta}$ subject to $\boldsymbol{\theta} \geqslant \mathbf{0}$, based on a single observation of $\boldsymbol{X}$, where $\boldsymbol{X} \sim N(\boldsymbol{\theta}, \boldsymbol{I})$.

Since the distribution function of $\boldsymbol{X}$ is known and $\tilde{\boldsymbol{\theta}}$ is a function of $\boldsymbol{X}$ only, we can write down explicit expressions for the distribution of $\tilde{\boldsymbol{\theta}}$. In contrast to the familiar situation where the parameter space for $\boldsymbol{\theta}$ is $\mathbb{R}^2$, here we observe that the distribution of the constrained *mle*, $\tilde{\boldsymbol{\theta}}$, is not normal. Further, the distribution of $(\tilde{\boldsymbol{\theta}} - \boldsymbol{\theta})$ and that of any scaled form of it depend on $\boldsymbol{\theta}$. Therefore, we cannot construct a confidence region for $\boldsymbol{\theta}$ in the usual way based on the distribution of $(\tilde{\boldsymbol{\theta}} - \boldsymbol{\theta})$. Consequently, the explicit form of the distribution of $(\tilde{\boldsymbol{\theta}} - \boldsymbol{\theta})$ has not found much use in statistical inference.

Example 3.3.2 *Type A testing problem (Example 3.3.1 continued).*

Let $\boldsymbol{X} = (X_1, X_2)^T \sim N(\boldsymbol{\theta}, \boldsymbol{I})$, where $\boldsymbol{I}$ is the 2×2 identity matrix and $\boldsymbol{\theta} = (\theta_1, \theta_2)^T$. Consider the likelihood ratio test of

$$H_0 : \theta_1 = \theta_2 = 0 \text{ against } H_1 : \theta_1 \geqslant 0, \theta_2 \geqslant 0,$$

based on a single observation of $\boldsymbol{X}$. Since $-2\ell(\boldsymbol{\theta}) = \|\boldsymbol{X} - \boldsymbol{\theta}\|^2$ and $LRT = 2[\max\{\ell(\boldsymbol{\theta}) : \boldsymbol{\theta} \in H_1\} - \max\{\ell(\boldsymbol{\theta}) : \boldsymbol{\theta} \in H_0\}]$ we have that

$$LRT = \|\boldsymbol{X}\|^2 - \|\boldsymbol{X} - \tilde{\boldsymbol{\theta}}\|^2 = \|\tilde{\boldsymbol{\theta}}\|^2.$$

The last step follows since $(\boldsymbol{X} - \tilde{\boldsymbol{\theta}})^T \tilde{\boldsymbol{\theta}} = 0$ as is clear from Figure 3.1; it may also be verified by considering each of the four cases in (3.3) (see Fig. 3.1). Since we are interested in the distribution of LRT under the null hypothesis, *let us suppose that the null hypothesis is true* for the rest of these derivations. To obtain an expression for $\text{pr}(LRT \leq c)$, note that

$$\begin{aligned} \text{pr}(LRT \leq c) &= \sum_{i=1}^{4} \text{pr}(LRT \leq c \text{ and } \boldsymbol{X} \in Q_i) \\ &= \sum_{i=1}^{4} \text{pr}(LRT \leq c \mid \boldsymbol{X} \in Q_i)\text{pr}(\boldsymbol{X} \in Q_i). \end{aligned} \tag{3.4}$$

Let us evaluate each of the four summands in the last expression.

Using the circular symmetry of $N(\boldsymbol{0}, \boldsymbol{I})$, it can be shown that the direction $(= \boldsymbol{X}/\|\boldsymbol{X}\|)$ and the length $(= \|\boldsymbol{X}\|)$ of $\boldsymbol{X}$ are statistically independent (see Exercise 3.1). Consequently, the conditional distribution of $(X_1^2 + X_2^2)$, given that the direction of $\boldsymbol{X}$ is in Q_1, is the same as that of its unconditional distribution. Therefore,

$$\begin{aligned} \text{pr}(LRT \leqslant c \mid \boldsymbol{X} \in \mathcal{Q}_1) &= \text{pr}(X_1^2 + X_2^2 \leqslant c \mid \boldsymbol{X} \in \mathcal{Q}_1) \\ &= \text{pr}(X_1^2 + X_2^2 \leqslant c) = \text{pr}(\chi_2^2 \leqslant c). \end{aligned} \tag{3.5}$$

Now, consider the second summand in (3.4) :

$$\begin{aligned} \text{pr}(LRT \leq c \mid \boldsymbol{X} \in \mathcal{Q}_2) &= \text{pr}(X_2^2 \leq c \mid X_2 \geqslant 0, X_1 \leq 0) \\ &= \text{pr}(X_2^2 \leq c \mid X_2 \geqslant 0) \quad \text{since } X_1 \text{ and } X_2^2 \text{ are independent} \\ &= \text{pr}(X_2^2 \leq c) \quad \text{since } X_2 \text{ is symmetric} \\ &= \text{pr}(\chi_1^2 \leq c) \quad \text{since } X_2 \sim N(0,1). \end{aligned} \tag{3.6}$$

Similarly, $\text{pr}(LRT \leq c \mid \boldsymbol{X} \in Q_4) = \text{pr}(\chi_1^2 \leq c)$. Therefore, we have that

$$LRT = \begin{cases} X_1^2 + X_2^2 & \text{given } \boldsymbol{X} \in Q_1 \\ X_2^2 & \text{given } \boldsymbol{X} \in Q_2 \\ 0 & \text{given } \boldsymbol{X} \in Q_3 \\ X_1^2 & \text{given } \boldsymbol{X} \in Q_4 \end{cases} \sim \begin{cases} \chi_2^2 \\ \chi_1^2 \\ 0 \\ \chi_1^2. \end{cases} \tag{3.7}$$

Now, it follows from (3.4)-(3.7) that the null distribution of the *LRT* is the following weighted sum of chi-square distributions:

$$\text{pr}(LRT \leq c) = (1/4) + (1/2)\text{pr}(\chi_1^2 \leq c) + (1/4)\text{pr}(\chi_2^2 \leq c) \tag{3.8}$$

for $c > 0$. For notational convenience, we define χ_0^2, the *chi-square distribution with zero degrees of freedom,* to be the distribution that takes the value zero with probability one. With this notation, the foregoing result is usually written as

$$\text{pr}(LRT \leq c \mid H_0) = \sum_{i=0}^{2} w_i \text{pr}(\chi_i^2 \leq c),$$

where $(w_0, w_1, w_2) = (0.25, 0.5, 0.25)$. For later use, let us note that w_0, w_1, and w_3 are the probabilities that $\boldsymbol{X}$ falls in the cones $Q_3, Q_2 \cup Q_4$, and Q_1 respectively. It follows from (3.7) that the critical region, $\{LRT \geqslant c\}$, is the region to the upper-right of the curve $PQRS$ in Fig. 3.2; PQ is parallel to the x_1-axis, RS is parallel to x_2-axis and QR is a circular arc of radius $\sqrt{c}$. ■

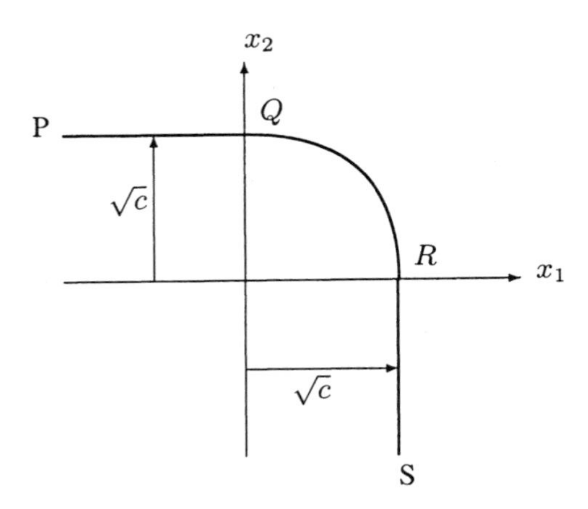

Fig. 3.2 The likelihood ratio test of $\boldsymbol{\theta} = \mathbf{0}$ against $\boldsymbol{\theta} \geqslant \mathbf{0}$ based on a single observation of $\boldsymbol{X} \sim N(\boldsymbol{\theta}, \boldsymbol{I})$; the critical region, $\{\boldsymbol{x} : LRT \geqslant c\}$ is the upper-right region bounded by the curve $PQRS$.

The foregoing two examples show, in a very simple situation, how the constrained *MLE* over $\mathcal{C}$ behaves like a projection of a normal random variate onto $\mathcal{C}$ (see Fig. 3.1), and how a mixture of chi-square distributions arise when the alternative hypothesis involves inequality constraints. In fact, more general mixtures of chi-square distributions arise when the hypotheses involve other forms of inequalities. The ideas underlying such general results are introduced below using simple bivariate examples.

Example 3.3.3 *Likelihood ratio test of* $H_0 : \boldsymbol{\theta} = \mathbf{0}$ *against* $H_1 : \boldsymbol{R\theta} \geqslant \mathbf{0}$.

Let $\boldsymbol{X} = (X_1, X_2)^T \sim N(\boldsymbol{\theta}, \boldsymbol{I})$. Consider the LRT of

$$H_0 : \boldsymbol{\theta} = \mathbf{0} \quad \text{against} \quad H_1 : \boldsymbol{R\theta} \geqslant \mathbf{0}$$

where $\boldsymbol{R}$ is the 2×2 nonsingular matrix $(1, 4 \mid 1, -2)$ (i.e., the two rows of $\boldsymbol{R}$ are $(1, 4)$ and $(1, -2)$ respectively). Let $\mathcal{C} = \{\boldsymbol{\theta} : \boldsymbol{R\theta} \geqslant \mathbf{0}\}$. Then, $\mathcal{C}$ is a closed convex cone (see Fig. 3.3).

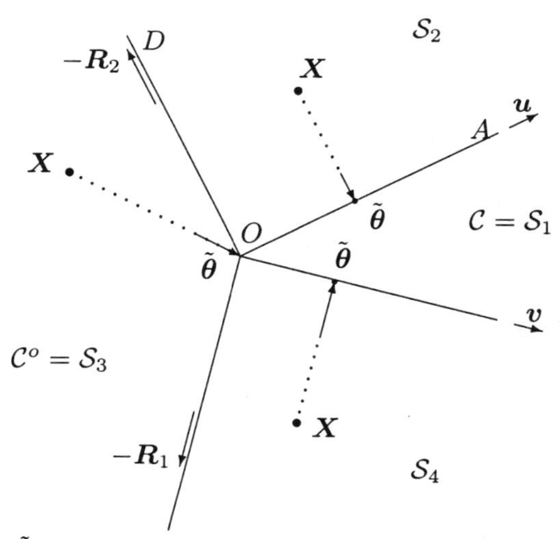

Fig. 3.3 The *mle* $\tilde{\boldsymbol{\theta}}$ of $\boldsymbol{\theta}$ based on a single observation of $\boldsymbol{X}$ when $\boldsymbol{X} \sim N(\boldsymbol{\theta}, \boldsymbol{I})$ and $\boldsymbol{\theta}$ is constrained to lie in the cone $\mathcal{C}$.

Let $\mathcal{C}^o = \{\boldsymbol{\alpha} : \boldsymbol{\alpha}^T\boldsymbol{\theta} \leq \boldsymbol{0}$ for every $\boldsymbol{\theta} \in \mathcal{C}\}$; $\mathcal{C}^o$ is called the *negative dual* or *polar cone* of $\mathcal{C}$ with respect to the inner product, $\langle \boldsymbol{\alpha}, \boldsymbol{\theta} \rangle = \boldsymbol{\alpha}^T\boldsymbol{\theta} = \alpha_1\theta_1 + \alpha_2\theta_2$. Thus, $\mathcal{C}^o$ is the collection of vectors which do not form an acute angle with any vector in $\mathcal{C}$. It may be verified that the boundaries of $\mathcal{C}^o$ are the perpendiculars to the boundaries of $\mathcal{C}$, and that $\mathcal{C}^o$ is the closed convex cone formed by these perpendiculars (see Fig.3.3). Let $\boldsymbol{R}_1^T$ and $\boldsymbol{R}_2^T$ denote the two rows of $\boldsymbol{R}$. It may also be verified that $-\boldsymbol{R}_1$ and $-\boldsymbol{R}_2$ are parallel to the boundaries of $\mathcal{C}^o$ as shown in Fig. 3.3. Clearly, $\mathcal{C}$ and $\mathcal{C}^o$ partition the plane into 4 cones; let us denote these by $\mathcal{S}_1, \mathcal{S}_2, \mathcal{S}_3, \mathcal{S}_4$ with $\mathcal{S}_1 = \mathcal{C}$ and $\mathcal{S}_3 = \mathcal{C}^o$ (see Fig. 3.3). The partition $\{\mathcal{S}_1, \mathcal{S}_2, \mathcal{S}_3, \mathcal{S}_4\}$ in this example corresponds to $\{\mathcal{Q}_1, \mathcal{Q}_2, \mathcal{Q}_3, \mathcal{Q}_4\}$ in Example 3.3.2.

Let $\boldsymbol{u}$ and $\boldsymbol{v}$ denote unit vectors parallel to the upper and lower boundaries of $\mathcal{C}$. Since the kernel of the loglikelihood is equal to $-0.5\|\boldsymbol{X} - \boldsymbol{\theta}\|^2$, the *mle* of $\boldsymbol{\theta}$ subject to $\boldsymbol{\theta} \in \mathcal{C}$ is the point in $\mathcal{C}$ closest to $\boldsymbol{X}$. In other words, the *mle* is the projection of $\boldsymbol{X}$ onto $\mathcal{C}$, which we write as $\Pi(\boldsymbol{X} \mid \mathcal{C})$. As in Examples 3.3.1 and 3.3.2, we obtain the following for the *mle*:

$$\tilde{\boldsymbol{\theta}} = \begin{cases} \boldsymbol{X} & \text{given } \boldsymbol{X} \in \mathcal{S}_1 \\ (\boldsymbol{u}^T\boldsymbol{X})\boldsymbol{u} & \text{given } \boldsymbol{X} \in \mathcal{S}_2 \\ \boldsymbol{0} & \text{given } \boldsymbol{X} \in \mathcal{S}_3 \\ (\boldsymbol{v}^T\boldsymbol{X})\boldsymbol{v} & \text{given } \boldsymbol{X} \in \mathcal{S}_4. \end{cases} \tag{3.9}$$

By considering each of the cases, $\boldsymbol{X} \in \mathcal{S}_i, i = 1, 2, 3, 4$ separately, it may be verified that $(\boldsymbol{X} - \tilde{\boldsymbol{\theta}})$ and $\tilde{\boldsymbol{\theta}}$ are orthogonal when $\tilde{\boldsymbol{\theta}}$ is not zero; Fig. 3.3 shows this clearly. Therefore, it follows from (3.2) that

$$LRT = \|\boldsymbol{X}\|^2 - \|\boldsymbol{X} - \tilde{\boldsymbol{\theta}}\|^2 = \|\tilde{\boldsymbol{\theta}}\|^2.$$

To derive the null distribution of *LRT*, first note that

$$\text{pr}(LRT \le c) = \sum_{i=1}^{4} \text{pr}(LRT \le c \mid \boldsymbol{X} \in \mathcal{S}_i)\text{pr}(\boldsymbol{X} \in \mathcal{S}_i). \tag{3.10}$$

Suppose that H_0 holds. Then $\boldsymbol{X} \sim N(\boldsymbol{0}, \boldsymbol{I})$. Therefore, the length and direction of $\boldsymbol{X}$ are independent (see Exercise 3.1). Hence $\|\boldsymbol{X}\|^2 \leqslant c$ and $\boldsymbol{X} \in \mathcal{S}_1$ are independent, and

$$\begin{aligned} \text{pr}(LRT \le c \mid \boldsymbol{X} \in \mathcal{S}_1) &= \text{pr}(X_1^2 + X_2^2 \le c \mid \boldsymbol{X} \in \mathcal{S}_1) \\ &= \text{pr}(X_1^2 + X_2^2 \le c) = \text{pr}(\chi_2^2 \le c). \end{aligned}$$

To obtain an expression for $\text{pr}(LRT \le c \mid \boldsymbol{X} \in \mathcal{S}_2)$, note that we can think of OA and OD as the first and second axes. In this new orthogonal coordinate system, the *LRT* given $\boldsymbol{X} \in \mathcal{S}_2$ is the squared length of the first coordinate and hence is distributed as χ_1^2; essentially the same argument was used in the previous example leading to (3.6). Similarly, the *LRT* given $\boldsymbol{X} \in \mathcal{S}_4$ is distributed as χ_1^2. Thus, as in Examples 3.3.1 and 3.3.2, we obtain the following for the conditional null distribution of *LRT* :

$$LRT = \begin{cases} X_1^2 + X_2^2 & \text{given } \boldsymbol{X} \in \mathcal{S}_1 \\ (\boldsymbol{u}^T\boldsymbol{X})^2 & \text{given } \boldsymbol{X} \in \mathcal{S}_2 \\ 0 & \text{given } \boldsymbol{X} \in \mathcal{S}_3 \\ (\boldsymbol{v}^T\boldsymbol{X})^2 & \text{given } \boldsymbol{X} \in \mathcal{S}_4. \end{cases} \sim \begin{cases} \chi_2^2 \\ \chi_1^2 \\ \chi_0^2 \\ \chi_1^2. \end{cases} \tag{3.11}$$

Collecting these results and substituting in (3.10), we have that

$$\text{pr}(LRT \le c \mid H_0) = q\text{pr}(\chi_0^2 \le c) + 0.5\text{pr}(\chi_1^2 \le c) + (0.5 - q)\text{pr}(\chi_2^2 \le c), \tag{3.12}$$

where $q = \text{pr}(\boldsymbol{X} \in \mathcal{S}_3 \mid H_0) = \text{pr}(\boldsymbol{X} \in \mathcal{C}^o \mid H_0) = (2\pi)^{-1}\gamma$ and γ is the angle (in radians) of the cone $\mathcal{C}^o$ at its vertex; therefore,

$$q = (2\pi)^{-1} \cos^{-1}[\boldsymbol{R}_1^T \boldsymbol{R}_2 / \{\boldsymbol{R}_1^T \boldsymbol{R}_1)(\boldsymbol{R}_2^T \boldsymbol{R}_2)\}^{1/2}], \tag{3.13}$$

where $\boldsymbol{R}_1^T$ and $\boldsymbol{R}_2^T$ are the two rows of $\boldsymbol{R}$. The critical region $\{LRT \geqslant c\}$ is the region to the right of the curve $PQRS$ in Figure 3.4; PQ is orthogonal to the upper boundary of $\mathcal{C}$, RS is orthogonal to the lower boundary of $\mathcal{C}$ and QR is a circular arc of radius $\sqrt{c}$. ■

Example 3.3.4 *Likelihood ratio test of* $H_0 : \boldsymbol{\theta} = \boldsymbol{0}$ *against* $H_1 : \boldsymbol{R\theta} \geqslant \boldsymbol{0}$ *when* $\boldsymbol{X} \sim N(\boldsymbol{\theta}, \boldsymbol{V})$ *and* $\boldsymbol{V}$ *is arbitrary.*

Let $\boldsymbol{X} = (X_1, X_2)^T \sim N(\boldsymbol{\theta}, \boldsymbol{V})$ where $\boldsymbol{V}$ is an arbitrary positive definite matrix. Consider the LRT of $H_0 : \boldsymbol{\theta} = \boldsymbol{0}$ against $H_1 : \boldsymbol{\theta} \in \mathcal{C}$ where $\mathcal{C} = \{\boldsymbol{\theta} : \boldsymbol{R\theta} \geqslant \boldsymbol{0}\}$ is a closed convex cone and $\boldsymbol{R}$ is a 2×2 nonsingular matrix. We can transform this problem to the set up in Example 3.3.3. Let $\boldsymbol{V}^{-1}$ be factorized as $\boldsymbol{V}^{-1} = \boldsymbol{A}^T\boldsymbol{A}$, for some $\boldsymbol{A}$ (for example, use Cholesky decomposition; see Anderson (1984) page 586);

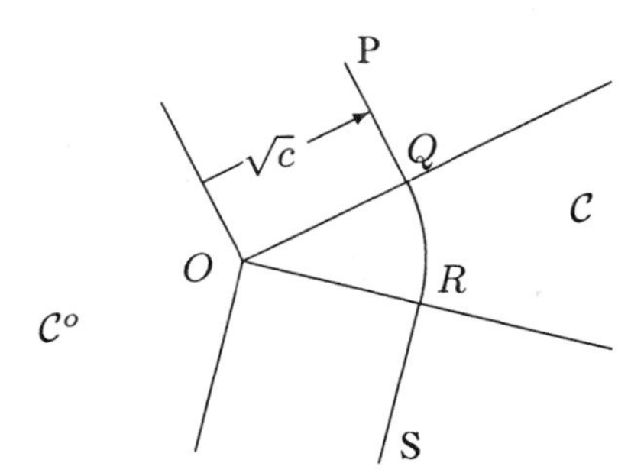

Fig. 3.4 The region to the right of $PQRS$ is the critical region $\{\boldsymbol{x} : \|\tilde{\boldsymbol{\theta}}(\boldsymbol{x})\|^2 \geqslant c\}$ for the *LRT* of $H_0 : \boldsymbol{\theta} = \mathbf{0}$ against $H_1 : \boldsymbol{\theta} \in \mathcal{C}$ based on a single observation of $\boldsymbol{X}$ where $\boldsymbol{X} \sim N(\boldsymbol{\theta}, \boldsymbol{I})$; $\tilde{\boldsymbol{\theta}}(\boldsymbol{x})$ is the *mle* of $\boldsymbol{\theta}$ subject to $\boldsymbol{\theta} \in \mathcal{C}$ when $\boldsymbol{X} = \boldsymbol{x}$, as shown in Figure 3.3

if $\boldsymbol{V} = [v_{ij}]$ then $\boldsymbol{A} = (v_{11}v_{22} - v_{12}^2)^{-1/2}[\sqrt{v_{22}}, -v_{12}/\sqrt{v_{22}} \mid 0, v_{11} - (v_{12}^2/v_{22})]$. Now, transform $\boldsymbol{X}$ and the associated quantities, $\boldsymbol{\theta}$ and $\mathcal{C}$, by $\boldsymbol{A}$:

$$\boldsymbol{Y} = \boldsymbol{A}\boldsymbol{X}, \boldsymbol{\alpha} = \boldsymbol{A}\boldsymbol{\theta}, \mathcal{P} = \boldsymbol{A}\mathcal{C}$$

where $\boldsymbol{A}\mathcal{C} = \{\boldsymbol{A}\boldsymbol{\theta} : \boldsymbol{\theta} \in \mathcal{C}\}$. Then $\mathcal{P} = \{\boldsymbol{\alpha} : \boldsymbol{R}\boldsymbol{A}^{-1}\boldsymbol{\alpha} \geqslant \mathbf{0}\}$, $\boldsymbol{Y} \sim N(\boldsymbol{\alpha}, \boldsymbol{I})$, and the testing problem is equivalent to the test of $H_0 : \boldsymbol{\alpha} = \mathbf{0}$ against $H_1 : \boldsymbol{\alpha} \in \mathcal{P}$ based on $\boldsymbol{Y}$. This is precisely the set-up in the last example with $\{\boldsymbol{Y}, \mathcal{P}\}$ in place of $\{\boldsymbol{X}, \mathcal{C}\}$. Since the *mle* and LRT are invariant under nonsingular transformations, we have that

$$\tilde{\boldsymbol{\theta}} = \boldsymbol{A}^{-1}\tilde{\boldsymbol{\alpha}} \quad \text{and} \quad LRT = \|\tilde{\boldsymbol{\alpha}}\|^2 = \tilde{\boldsymbol{\theta}}^T \boldsymbol{V}^{-1}\tilde{\boldsymbol{\theta}}.$$

Now, by (3.13) and (3.12), we have that

$$\text{pr}(LRT \leq c \mid H_0) = q\text{pr}(\chi_0^2 \leq c) + 0.5\text{pr}(\chi_1^2 \leq c) + (0.5 - q)\text{pr}(\chi_2^2 \leq c), \quad (3.14)$$

where

$$q = (2\pi)^{-1}\cos^{-1}[(\boldsymbol{R}_1^T\boldsymbol{V}\boldsymbol{R}_2)/\{(\boldsymbol{R}_1^T\boldsymbol{V}\boldsymbol{R}_1)(\boldsymbol{R}_2^T\boldsymbol{V}\boldsymbol{R}_2)\}^{1/2}], \quad (3.15)$$

and $\boldsymbol{R}_1^T$ and $\boldsymbol{R}_2^T$ are the two rows of $\boldsymbol{R}$. ■

This example shows that we do not have to consider the case for a general $\boldsymbol{V}$ separately because the inference problem can be restated in terms of the identity covariance matrix and therefore can be handled by the method for the previous example. However, it is instructive to review the details in terms of the original variables and parameters without using such a transformation because some important features tend to be masked by the transformation.

Geometry of *mle* and LRT when $\boldsymbol{X} \sim N(\boldsymbol{\theta}, \boldsymbol{V}), \boldsymbol{V} \neq \boldsymbol{I}$.

Since $\boldsymbol{V}^{-1}$ is positive definite, it induces the inner product defined by $\langle \boldsymbol{x}, \boldsymbol{y} \rangle_V = \boldsymbol{x}^T\boldsymbol{V}^{-1}\boldsymbol{y}$. This further induces the distance function defined by $\|\boldsymbol{x} - \boldsymbol{y}\|_V = \{(\boldsymbol{x} -$

$\boldsymbol{y})^T\boldsymbol{V}^{-1}(\boldsymbol{x}-\boldsymbol{y})\}^{1/2}$. Two vectors $\boldsymbol{x}$ and $\boldsymbol{y}$ are said to be *orthogonal* with respect to $\langle . \rangle_{\boldsymbol{V}}$ if $\langle \boldsymbol{x}, \boldsymbol{y} \rangle_{\boldsymbol{V}} = 0$ and denote it by $\boldsymbol{x} \perp_{\boldsymbol{V}} \boldsymbol{y}$. If necessary, we shall use the abbreviation $\boldsymbol{V}$-orthogonal, $\boldsymbol{V}$-distance, etc. to indicate that the relevant inner product is $\langle \cdot \rangle_{\boldsymbol{V}}$.

Let $\mathcal{B}$ be a given subset of $\mathbb{R}^2$ and let $\tilde{\boldsymbol{\theta}}$ denote the *mle* of $\boldsymbol{\theta}$ over $\boldsymbol{\theta} \in \mathcal{B}$ based on a single observation of $\boldsymbol{X}$ where $\boldsymbol{X} \sim N(\boldsymbol{\theta}, \boldsymbol{V})$. Then, since $\ell(\boldsymbol{\theta}) = (-1/2)(\boldsymbol{X} - \boldsymbol{\theta})^T\boldsymbol{V}^{-1}(\boldsymbol{X}-\boldsymbol{\theta}) = (-1/2)\|\boldsymbol{X}-\boldsymbol{\theta}\|^2_{\boldsymbol{V}}$, it follows that $\tilde{\boldsymbol{\theta}} = \arg\min_{\boldsymbol{\theta}\in\mathcal{B}} \|\boldsymbol{X}-\boldsymbol{\theta}\|^2_{\boldsymbol{V}}$ and hence $\tilde{\boldsymbol{\theta}}$ is the point in $\mathcal{B}$ that is $\boldsymbol{V}$-closest to $\boldsymbol{X}$. In this case, we say that $\tilde{\boldsymbol{\theta}}$ is the $\boldsymbol{V}$-projection of $\boldsymbol{X}$ onto $\mathcal{B}$ and denote it by $\Pi_{\boldsymbol{V}}(\boldsymbol{X} \mid \mathcal{B})$ or equivalently by $\Pi_{\boldsymbol{V}}(\boldsymbol{X}, \mathcal{B})$. More details about such projections are given in the appendix to this chapter. The following result relating to the contours of $N(\mathbf{0}, \boldsymbol{V})$ is useful.

Proposition 3.3.1 *The ellipse $\boldsymbol{x}^T\boldsymbol{V}^{-1}\boldsymbol{x} = Const$ and any line through the center of the ellipse are $\boldsymbol{V}$-orthogonal at their points of intersection.*

Proof: Let us write

$$\boldsymbol{V} = \begin{pmatrix} \sigma_1^2 & \rho\sigma_1\sigma_2 \\ \rho\sigma_1\sigma_2 & \sigma_2^2 \end{pmatrix}, \text{ and hence } \boldsymbol{V}^{-1} = \Delta^{-1}\begin{pmatrix} \sigma_2^2 & -\rho\sigma_1\sigma_2 \\ -\rho\sigma_1\sigma_2 & \sigma_1^2 \end{pmatrix}$$

where $\Delta = \{\sigma_1^2\sigma_2^2(1-\rho^2)\}$. Now, the ellipse $\boldsymbol{x}^T\boldsymbol{V}^{-1}\boldsymbol{x} = Constant$ can be expressed as

$$\sigma_2^2x_1^2 - 2\rho\sigma_1\sigma_2x_1x_2 + \sigma_1^2x_2^2 = Constant. \tag{3.16}$$

By implicit differentiation, we have that

$$dx_2/dx_1 = (\sigma_2/\sigma_1)(\rho\sigma_1x_2 - \sigma_2x_1)/(\sigma_1x_2 - \rho\sigma_2x_1); \tag{3.17}$$

this is the slope of the tangent to the contour (3.16) at (x_1, x_2). Now consider the line ℓ defined by $x_2 = mx_1$ (see Figure 3.5). For the time being, assume that $\sigma_1 m \neq \sigma_2\rho$ to avoid division by zero; this can be relaxed by taking the limit $\sigma_1 m \to \sigma_2\rho$.

Note that the slope, a, of the tangent to the contour (3.16) at the point of intersection with the line ℓ is given by

$$a = (\sigma_2/\sigma_1)(\rho\sigma_1m - \sigma_2)/(\sigma_1m - \rho\sigma_2). \tag{3.18}$$

Further, $(1, m)$, a vector parallel to the line ℓ, and $(1, a)$, a vector parallel to the tangent of the ellipse at the point of intersection with ℓ, are $\boldsymbol{V}$-orthogonal. Therefore, the line ℓ and any contour (3.16) are $\boldsymbol{V}$-orthogonal at their points of intersections. ■

To illustrate this further, let C_1 and C_2 be two contours defined by $\boldsymbol{x}^T\boldsymbol{V}^{-1}\boldsymbol{x} = c_1^2$ and $\boldsymbol{x}^T\boldsymbol{V}^{-1}\boldsymbol{x} = c_2^2$ respectively, for some c_1 and c_2 (see Fig. 3.5). Let O denote the origin and, as shown in Fig. 3.5, let F and Q be the points at which ℓ intersects C_1 and C_2, respectively. Let T_Q be the tangent to C_2 at Q; similarly, let T_F be the tangent to C_1 at F. Let P and E be points on T_Q and T_F respectively. Now, we have the following:

1. For a given i, $(i = 1, 2)$, every point on C_i is at $\boldsymbol{V}$-equidistant from O.

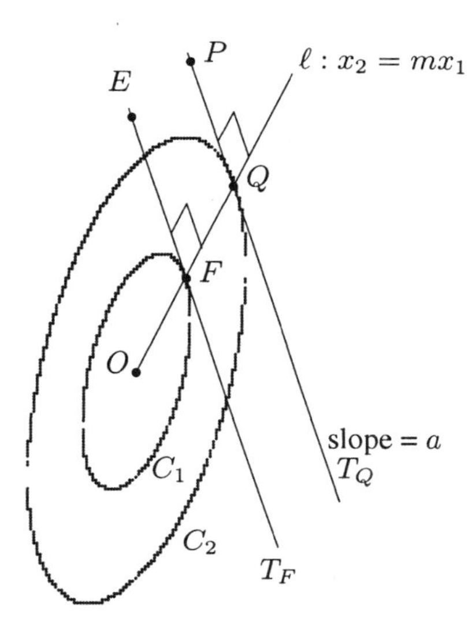

Fig. 3.5 $\boldsymbol{V}$-projection onto a radial line ℓ through the center O of the ellipses C_1 and C_2; the point on ℓ that is $\boldsymbol{V}$-closest to P is Q; the point on T_F that is $\boldsymbol{V}$-closest to O is F.

2. T_Q and T_F are parallel and their common slope a is given by (3.18).

3. T_Q and T_F are $\boldsymbol{V}$-orthogonal to ℓ.

By Pythagoras theorem, the point on a line that is $\boldsymbol{V}$-closest to a point, say P, is obtained by dropping a $\boldsymbol{V}$-perpendicular to the line from P. Therefore, in Fig. 3.5 the point on ℓ that is $\boldsymbol{V}$-closest to P is Q; the point on T_Q that is $\boldsymbol{V}$-closest to O is also Q. If $\boldsymbol{X} \sim N(\boldsymbol{\theta}, \boldsymbol{V})$, the observed value of $\boldsymbol{X}$ is at O, and $\boldsymbol{\theta}$ is restricted to T_Q then the *mle* of $\boldsymbol{\theta}$ is at Q; if the observation is at P and $\boldsymbol{\theta}$ is restricted to ℓ then the *mle* of $\boldsymbol{\theta}$ is at Q.

The foregoing are also helpful to illustrate the form of the *mle* of $\boldsymbol{\theta}$ when it is constrained to lie in a cone. Let $\boldsymbol{X} \sim N(\boldsymbol{\theta}, \boldsymbol{V})$ and let us consider the *mle* of $\boldsymbol{\theta}$ based on a single observation of $\boldsymbol{X}$ when $\boldsymbol{\theta}$ is constrained to lie on a convex cone. Let $\mathcal{C}$ be the convex cone AOB in Fig. 3.6 and suppose that $\boldsymbol{\theta}$ is constrained to lie in it. Fig. 3.6 shows the *mle* when $\boldsymbol{\theta}$ is restricted to $\mathcal{C}$ and Fig. 3.7 shows the critical region for testing $H_0 : \boldsymbol{\theta} = \boldsymbol{0}$ against $H_1 : \boldsymbol{\theta} \in \mathcal{C}$. Let $OC \perp_V OA$ and $OD \perp_V OB$. Then, COD is the polar cone of $\mathcal{C}$ with respect to $\langle , \rangle_V$. Let Q and R be the points of intersection of an arbitrary contour ($\boldsymbol{x}^T \boldsymbol{V}^{-1} \boldsymbol{x} = constant$) with OA and OB, respectively. Let PQ and SR be the tangents to the contour at Q and R, respectively. Thus, $PQRS$ is a smooth curve with continuous slope everywhere. If $\boldsymbol{X} \in AOB$ then $\tilde{\boldsymbol{\theta}} = \boldsymbol{X}$; if $\boldsymbol{X} \in COA$, say $\boldsymbol{X} = \overline{OP}$, then $\tilde{\boldsymbol{\theta}} = \overline{OQ}$; if $\boldsymbol{X} \in DOC$ then $\tilde{\boldsymbol{\theta}} = \boldsymbol{0}$; and if $\boldsymbol{X} \in DOB$, say $\boldsymbol{X} = \overline{OS}$, then $\tilde{\boldsymbol{\theta}} = \overline{OR}$. Now, the boundary of a typical critical region is $PQRS$, where the QR is the segment of the contour C (ie. $\boldsymbol{x}^T \boldsymbol{V}^{-1} \boldsymbol{x} = constant$) that lies in AOB.

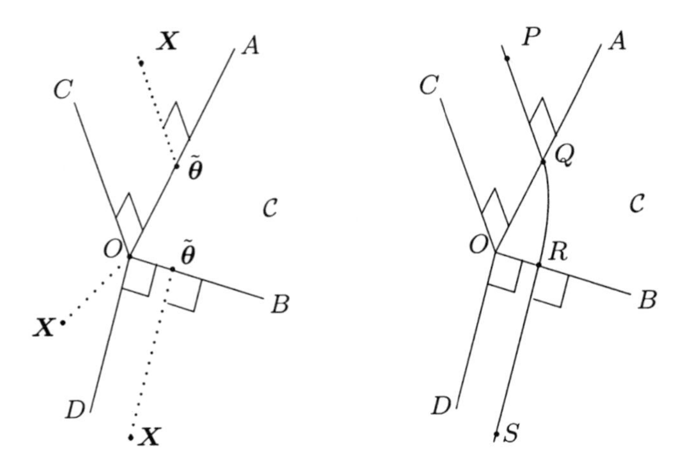

Fig. 3.6 The constrained *mle* of $\boldsymbol{\theta}$ subject to $\boldsymbol{\theta} \in \mathcal{C}$ when $\boldsymbol{X} \sim N(\boldsymbol{\theta}, \boldsymbol{V})$.

Fig. 3.7 The *LRT* of $H_0 : \boldsymbol{\theta} = \mathbf{0}$ vs $H_1 : \boldsymbol{\theta} \in \mathcal{C}$ when $\boldsymbol{X} \sim N(\boldsymbol{\theta}, \boldsymbol{V})$ and $\boldsymbol{V} \neq \boldsymbol{I}$; a typical boundary of the critical region is $PQRS$

Example 3.3.5 *A numerical example: likelihood ratio test against inequality constraints*

Let $\boldsymbol{X}$ be a bivariate random variable distributed as $N(\boldsymbol{\theta}, \boldsymbol{V})$ where $\boldsymbol{V} = (1, \rho; \rho, 1)$ with $\rho = 0.9$. Now, consider the estimation of $\boldsymbol{\theta}$ and test of $H_0 : \boldsymbol{\theta} = \mathbf{0}$ against $H_1 : \boldsymbol{\theta} \geqslant \mathbf{0}$ based on the following five observations on $\boldsymbol{X}$: (-4,-3), (-4, -3), (-3, -2), (-2, -1), and (-2, -1). For these data, the mean $\bar{\boldsymbol{x}}$ is (-3, -2); as was pointed out in Section 3.2.1, the estimation and tests depend on the data only through this mean because $\boldsymbol{V}$ is known. The kernel, $\ell(\boldsymbol{\theta})$, of the loglikelihood is $\{-0.5n(\bar{\boldsymbol{X}} - \boldsymbol{\theta})^T \boldsymbol{V}^{-1}(\bar{\boldsymbol{X}} - \boldsymbol{\theta})\}$. To estimate $\boldsymbol{\theta}$, we minimize $(\bar{\boldsymbol{x}} - \boldsymbol{\theta})^T \boldsymbol{V}^{-1}(\bar{\boldsymbol{x}} - \boldsymbol{\theta})$ over $\{\theta_1 \geqslant 0, \theta_2 \geqslant 0\}$; this achieves its minimum at $(0, 0.7)$. This may be verified by using the geometric method in the previous example. Alternatively, since the dimension of $\boldsymbol{\theta}$ is only two, the minimum can be computed as explained in the next paragraph; while we would not use this method in high dimensions, it is instructive in two dimensions.

Note that $(\bar{\boldsymbol{x}} - \boldsymbol{\theta})^T \boldsymbol{V}^{-1}(\bar{\boldsymbol{x}} - \boldsymbol{\theta})$ is proportional to $f(\theta_1, \theta_2)$ defined by

$$f(\theta_1, \theta_2) = \theta_1^2 + \theta_2^2 - 1.8\theta_1\theta_2 + 2.4\theta_1 - 1.4\theta_2 + 2.2.$$

This function is convex because its second derivative is proportional to $\boldsymbol{V}^{-1}$, which is positive definite. The minimum of $f(\boldsymbol{\theta})$, without any restrictions on (θ_1, θ_2), is achieved at $\boldsymbol{\theta} = \bar{\boldsymbol{x}} = (-3, -2)$, which is outside the region $\{\boldsymbol{\theta} \geq \mathbf{0}\}$. Because $f(\boldsymbol{\theta})$ is convex, the minimum of $f(\boldsymbol{\theta})$ subject to $\{\boldsymbol{\theta} \geq \mathbf{0}\}$ is achieved on the boundary of $\{\boldsymbol{\theta} \geq \mathbf{0}\}$, which consists of the nonnegative θ_1- and θ_2-axes. Now, to find the absolute minimum of $f(\boldsymbol{\theta})$ over $\boldsymbol{\theta} \geq \mathbf{0}$, it suffices to find the minima of $f(\boldsymbol{\theta})$ on the nonnegative segments of the θ_1- and θ_2-axes separately; the lowest of the two minima is the absolute minimum of $f(\boldsymbol{\theta})$ subject to $\boldsymbol{\theta} \geqslant \mathbf{0}$.

The null distribution of LRT is given by (see (3.14) and (3.15)):

$$pr(LRT \geq c \mid H_0) = 0.5pr(\chi_1^2 \geq c) + 0.5(1 - \pi^{-1}\cos^{-1}\rho)pr(\chi_2^2 \geq c).$$

Substituting $\rho = 0.9$, the sample value of the likelihood ratio statistic, $n\tilde{\boldsymbol{\theta}}^T \boldsymbol{V}^{-1}\tilde{\boldsymbol{\theta}}$, is 12.9 for which the p-value is

$$0.5\text{pr}(\chi_1^2 \geqslant 12.9) + 0.5(1 - \pi^{-1}\cos^{-1} 0.9)\text{pr}(\chi_2^2 \geqslant 12.9),$$

which is less than 0.05. ■

Now, let us consider a curious phenomenon of the LRT based on the foregoing example.

Example 3.3.6 *A curious example involving the LRT; the data are inconsistent with the the null and the alternative hypotheses (Silvapulle (1997a))*

Consider an experiment to establish that a new treatment for pain relief is better than a placebo. For each patient, let X_1 and X_2 denote the reductions in pain at two locations (say neck and back) due to the treatment. For illustrative purposes, suppose that the distribution of $\boldsymbol{X}$ and the data are the same as in the previous example. Suppose also that the new treatment is expected to be at least as good as the placebo in controlling pain. Thus, it is assumed *a priori* that $\boldsymbol{\theta} \geqslant \boldsymbol{0}$. Suppose that the inference problem for establishing that the new treatment is better than the placebo be formulated as test of $H_0 : \boldsymbol{\theta} = \boldsymbol{0}$ against $H_1 : \boldsymbol{\theta} \geqslant \boldsymbol{0}$. Since every observed value is negative, the pain has increased under the new treatment for each patient in the sample; clearly, the new treatment has performed worse than the placebo. For these data, the *mle* of (θ_1, θ_2) under $H_1 : \boldsymbol{\theta} \geqslant \boldsymbol{0}$ is $(0, 0.7)$ and the LRT of H_0 vs H_1 would reject H_0 in favor of H_1 at 5% level (LRT=12.9, p-value < 0.05).

At first, it would appear that there is something wrong because there is no way that we could claim that the new treatment is better than the placebo when, in fact, the pain under the new treatment turned out to be worse for each patient. In fact, there is nothing wrong as far as the statistical calculations are concerned. However, several issues arise in the interpretation of the results.

To provide an insight into why the *mle* of θ_2 is positive and the LRT rejects H_0, first recall that the maximum likelihood estimator and the likelihood ratio statistic are based on the distribution of $\bar{\boldsymbol{X}}$ being $N(\boldsymbol{\theta}, n^{-1}\boldsymbol{V})$. The contours of $N(\boldsymbol{\theta}, n^{-1}\boldsymbol{V})$ passing through the observed value $\bar{\boldsymbol{x}} = (-3, -2)$ for $\boldsymbol{\theta} = (0, 0)$ and $\boldsymbol{\theta} = (0, 0.7)$ are shown in Fig. 3.8; these are indicated by C_2 and C_1, respectively. Note that C_1 is smaller than C_2. Therefore, if $g(\boldsymbol{x}|\boldsymbol{\theta})$ denotes the density function of $\bar{\boldsymbol{X}}$ then $g((-3, -2)|(0, 0.7)) > g((-3, -2)|(0, 0))$. Thus, the likelihood at $\boldsymbol{\theta} = (0, 0.7)$ is larger than that at $\boldsymbol{\theta} = (0, 0)$. Therefore, the data are more likely to have arisen from $N(\boldsymbol{\theta}, \boldsymbol{V})$ with $\boldsymbol{\theta} = (0, 0.7)$ than $\boldsymbol{\theta} = \boldsymbol{0}$.

A small p-value for $H_0 : \boldsymbol{\theta} = \boldsymbol{0}$ vs $H_1 : \boldsymbol{\theta} \geqslant \boldsymbol{0}$ says that the data are considerably more inconsistent with $\boldsymbol{\theta} = \boldsymbol{0}$ than with $\boldsymbol{\theta} \gtrsim \boldsymbol{0}$. If we accept the formulation of the problem as given above, then there are no contradictions; the LRT does exactly what we expect it to do.

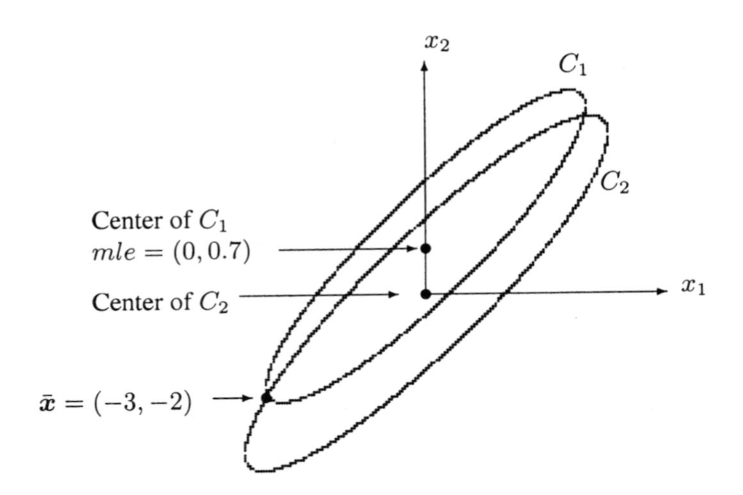

Fig. 3.8 The *mle* of $\boldsymbol{\theta}$ subject to $\boldsymbol{\theta} \geqslant \mathbf{0}$ when $\boldsymbol{X} \sim N(\boldsymbol{\theta}, \boldsymbol{V})$, $\boldsymbol{V} = (1, 0.9 \mid 0.9, 1)$ and the sample mean is $(-3, -2)$.

Up to this point there are no difficulties in the statistical calculations. However, recently, there have been debates about the interpretation of these results. If the assumption $\boldsymbol{X} \sim N(\boldsymbol{\theta}, \boldsymbol{V})$ where $\boldsymbol{V} = (1, 0.9|0.9, 1)$ is correct and $\bar{\boldsymbol{x}} = (-3, -2)^T$, should we accept that the new treatment is better? Some authors argue that *LRT* is deficient, while others argue there is nothing wrong with *LRT*. It is not possible to discuss all the relevant issues concerning this. Our objective here has been to draw attention to the phenomenon. We refer the readers to Silvapulle (1997a), Cohen et al. (2000), Cohen and Sackrowitz (2004), Perlman and Wu (2002b), and Chaudhuri and Perlman (2004). Although these authors discuss different examples, the statistical issues therein are the same as those in the foregoing example.

If we wish to establish that the new treatment is better than the placebo, then the testing problem could have been formulated as $H_0 : \boldsymbol{\theta} \notin A$ vs $H_1 : \boldsymbol{\theta} \in A$ where A is the set of all values of $\boldsymbol{\theta}$ that correspond to the treatment being better than the placebo. For example, one possible choice is $A = \{\theta_1 > 0, \theta_2 > 0\}$; for this particular choice of A, if the test rejects H_0, then clearly there is statistical evidence not only to reject H_0 but also to support H_1. This particular testing problems will be discussed in later chapters under the topic, *sign testing*. More generally, in a test of H_a vs H_b for some H_a and H_b, if we wish to interpret the rejection of H_a as statistical evidence in support of H_b, it may be necessary to ensure that $H_a \cup H_b$ is the full parameter space. ■

The phenomenon in the simple example just discussed manifests in other settings as well; see Problem 3.9 for an example. In the analysis of ordinal response data, this property of LRT translates to it being not *concordant monotone*; for a discussion of this see Cohen et al. (2000). For further discussions and different points, see Cohen and Sackrowitz (2004), Chaudhuri and Perlman (2004), and Perlman and Wu (1999, 2002b).

It is possible to construct tests of $\boldsymbol{\theta} = \mathbf{0}$ vs $\boldsymbol{\theta} \geqslant \mathbf{0}$ in which points such as (-3,-2) do not lie in the critical region. Cohen and Sackrowitz (1998) introduced a class of tests that they called the *Directed Tests* for testing this type of hypotheses. They developed their ideas on precise mathematical formulations of the problem; this will be introduced in Chapter 9. However, at the time of writing this book, there is a continuing debate concerning the advantages and disadvantages of various tests for constrained inference problems, in particular, as to whether or not *LRT* is deficient, and as to whether or not the directed tests are themselves deficient (see the references in the previous paragraph).

3.4 CHI-BAR-SQUARE DISTRIBUTION

The family of chi-bar-square distributions, to be introduced in this section, plays an important role when the null and/or alternative hypothesis involve parameter inequalities. The null distributions of likelihood ratio statistics for Type A and Type B problems with multivariate normal data turn out to be chi-bar-square. The role of chi-bar-square distributions in Type A and Type B testing problems is similar to that of chi-square distributions in the application of likelihood ratio for testing $\boldsymbol{\theta} = \mathbf{0}$ against $\boldsymbol{\theta} \neq \mathbf{0}$ in small and in large samples. In this section, we introduce the general form of chi-bar-square distribution and show how it relates to an important fundamental result on likelihood ratio test.

Let $\mathcal{C} \subset \mathbb{R}^p$ ($\mathcal{C}$ is a closed convex cone) and let $\boldsymbol{Z}_{p\times 1} \sim N(\mathbf{0}, \boldsymbol{V})$, where $\boldsymbol{V}$ is a positive definite matrix. We define $\bar{\chi}^2(\boldsymbol{V}, \mathcal{C})$ to be the random variable, which has the same distribution as $[\boldsymbol{Z}^T\boldsymbol{V}^{-1}\boldsymbol{Z} - \min_{\boldsymbol{\theta}\in\mathcal{C}}(\boldsymbol{Z}-\boldsymbol{\theta})^T\boldsymbol{V}^{-1}(\boldsymbol{Z}-\boldsymbol{\theta})]$. If there is no possibility of any ambiguity, we shall also write

$$\bar{\chi}^2(\boldsymbol{V}, \mathcal{C}) = \boldsymbol{Z}^T\boldsymbol{V}^{-1}\boldsymbol{Z} - \min_{\boldsymbol{\theta}\in\mathcal{C}}(\boldsymbol{Z}-\boldsymbol{\theta})^T\boldsymbol{V}^{-1}(\boldsymbol{Z}-\boldsymbol{\theta}). \tag{3.19}$$

It would be instructive to provide a geometric interpretation of $\bar{\chi}^2$ and its related quantities. Fig. 3.9 provides a schematic diagram of $\mathcal{C}$ and a typical value of $\boldsymbol{Z}$ represented by OA. Let the inner product be defined as $\langle \boldsymbol{x}, \boldsymbol{y}\rangle_V = \boldsymbol{x}^T\boldsymbol{V}^{-1}\boldsymbol{y}$, for $\boldsymbol{x}, \boldsymbol{y} \in \mathbb{R}^p$; this induces the norm $\|\boldsymbol{x}\|_V = \{\boldsymbol{x}^T\boldsymbol{V}^{-1}\boldsymbol{x}\}^{1/2}$ and the distance $\|\boldsymbol{x} - \boldsymbol{y}\|_V = \{(\boldsymbol{x}-\boldsymbol{y})^T\boldsymbol{V}^{-1}(\boldsymbol{x}-\boldsymbol{y})\}^{1/2}$ between $\boldsymbol{x}$ and $\boldsymbol{y}$. Let B be the point in $\mathcal{C}$ that is $\boldsymbol{V}$-closest to A, and let $\tilde{\boldsymbol{Z}}$ denote the vector OB. Therefore, $\tilde{\boldsymbol{Z}}$ is the value of $\boldsymbol{x}$ at which

$$(\boldsymbol{Z}-\boldsymbol{x})^T\boldsymbol{V}^{-1}(\boldsymbol{Z}-\boldsymbol{x}), \quad \boldsymbol{x} \in \mathcal{C}$$

is the minimum. In other words, $\tilde{\boldsymbol{Z}}$ is the $\boldsymbol{V}$-projection of $\boldsymbol{Z}$ onto $\mathcal{C}$; let us denote it by $\Pi_{\boldsymbol{V}}(\boldsymbol{Z}|\mathcal{C})$ or equivalently $\Pi_{\boldsymbol{V}}(\boldsymbol{Z}, \mathcal{C})$. It is shown in the Appendix that $\boldsymbol{Z} - \tilde{\boldsymbol{Z}}$ is $\boldsymbol{V}$-orthogonal to $\tilde{\boldsymbol{Z}}$; in other words, OB is $\boldsymbol{V}$-orthogonal to AB. Therefore, by

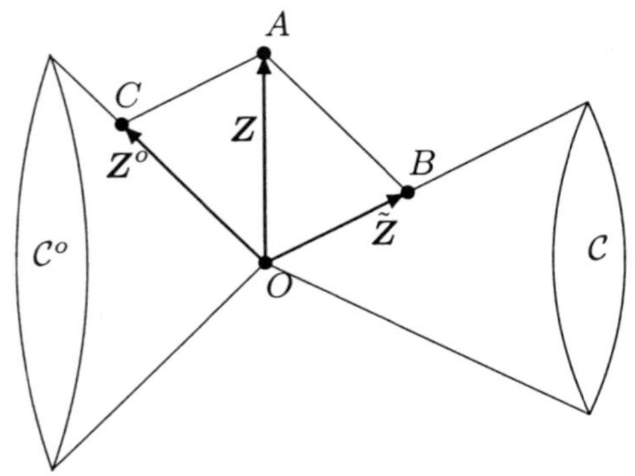

Fig. 3.9 $\boldsymbol{V}$-projection: OB and OC are the $\boldsymbol{V}$-projections of OA onto $\mathcal{C}$ and $\mathcal{C}^o$ respectively.

applying the Pythagoras theorem to OAB, we have that

$$\|OA\|_{\boldsymbol{V}}^2 = \|OB\|_{\boldsymbol{V}}^2 + \|BA\|_{\boldsymbol{V}}^2,$$

and hence

$$\boldsymbol{Z}^T\boldsymbol{V}^{-1}\boldsymbol{Z} = \tilde{\boldsymbol{Z}}^T\boldsymbol{V}^{-1}\tilde{\boldsymbol{Z}} + \min_{\boldsymbol{x}\in\mathcal{C}}(\boldsymbol{Z}-\boldsymbol{x})^T\boldsymbol{V}^{-1}(\boldsymbol{Z}-\boldsymbol{x}).$$

Therefore,

$$\bar{\chi}^2(\boldsymbol{V},\mathcal{C}) = \|OB\|_{\boldsymbol{V}}^2.$$

We denote the $\boldsymbol{V}$-distance between the point $\boldsymbol{Z}$ and the set $\mathcal{C}$ by $\|\boldsymbol{Z}-\mathcal{C}\|_{\boldsymbol{V}}$ and it is defined by

$$\|\boldsymbol{Z}-\mathcal{C}\|_{\boldsymbol{V}}^2 = \inf\{(\boldsymbol{Z}-\boldsymbol{x})^T\boldsymbol{V}^{-1}(\boldsymbol{Z}-\boldsymbol{x}) : \quad \boldsymbol{x}\in\mathcal{C}\};$$

in Fig. 3.9, $\|\boldsymbol{Z}-\mathcal{C}\|_{\boldsymbol{V}}$ is equal to $\|AB\|_{\boldsymbol{V}}$.

The *polar cone* $\mathcal{C}^o$ of $\mathcal{C}$ with respect to the inner product $\langle\boldsymbol{x},\boldsymbol{y}\rangle_{\boldsymbol{V}} = \boldsymbol{x}^T\boldsymbol{V}^{-1}\boldsymbol{y}$ is defined by

$$\mathcal{C}^o = \{\boldsymbol{x} : \boldsymbol{x}^T\boldsymbol{V}^{-1}\boldsymbol{y} \le 0 \text{ for every } \boldsymbol{y}\in\mathcal{C}\}.$$

Because $\mathcal{C}$ is a closed convex cone, it may be verified easily that $\mathcal{C}^o$ is also a closed convex cone. It consists of the vectors that form obtuse angles with every vector in $\mathcal{C}$; here obtuse means that $\boldsymbol{x}^T\boldsymbol{V}^{-1}\boldsymbol{y}\le 0$.

Let $\boldsymbol{Z}^o$ denote the point in $\mathcal{C}^o$ that is closest to $\boldsymbol{Z}$. In other words, $\boldsymbol{Z}^o$ is the projection of $\boldsymbol{Z}$ onto $\mathcal{C}^o$ and is defined as the value of $\boldsymbol{x}$ at which

$$(\boldsymbol{Z}-\boldsymbol{x})^T\boldsymbol{V}^{-1}(\boldsymbol{Z}-\boldsymbol{x}), \quad \boldsymbol{x}\in\mathcal{C}^o$$

is a minimum. Let C in $\mathcal{C}^o$ be the point such that $\boldsymbol{Z}^o = OC$. Again, the Pythagoras theorem is applicable to OAC. In fact, $OBAC$ is a rectangle, and hence

$$OC = BA, CA = OB, \|OA\|_{\boldsymbol{V}}^2 = \|OC\|_{\boldsymbol{V}}^2 + \|CA\|_{\boldsymbol{V}}^2 = \|OB\|_{\boldsymbol{V}}^2 + \|BA\|_{\boldsymbol{V}}^2.$$

It would also be useful to note that

$$\bar{\chi}^2(\boldsymbol{V}, \mathcal{C}^o) = \|OC\|_{\boldsymbol{V}}^2 = \|\boldsymbol{Z} - \mathcal{C}\|^2.$$

These results are summarized below.

Proposition 3.4.1 *Let $\boldsymbol{V}$ be a $p \times p$ positive definite matrix. Then*

1. $\|\boldsymbol{Z}\|_{\boldsymbol{V}}^2 = \|\Pi_{\boldsymbol{V}}(\boldsymbol{Z} \mid \mathcal{C})\|_{\boldsymbol{V}}^2 + \|\boldsymbol{Z} - \Pi_{\boldsymbol{V}}(\boldsymbol{Z} \mid \mathcal{C})\|_{\boldsymbol{V}}^2$.
2. $\Pi_{\boldsymbol{V}}(\boldsymbol{Z} \mid \mathcal{C}^o) = \boldsymbol{Z} - \Pi_{\boldsymbol{V}}(\boldsymbol{Z} \mid \mathcal{C})$.
3. *If $\boldsymbol{Z} \sim N(\boldsymbol{0}, \boldsymbol{V})$ then*

$$\|\Pi_{\boldsymbol{V}}(\boldsymbol{Z} \mid \mathcal{C})\|_{\boldsymbol{V}}^2 \sim \bar{\chi}^2(\boldsymbol{V}, \mathcal{C}) \quad \textit{and} \quad \|\boldsymbol{Z} - \Pi_{\boldsymbol{V}}(\boldsymbol{Z} \mid \mathcal{C})\|_{\boldsymbol{V}}^2 \sim \bar{\chi}^2(\boldsymbol{V}, \mathcal{C}^o). \blacksquare$$

Now we have the following fundamental result; proof is given in the Appendix.

Theorem 3.4.2 *Let $\mathcal{C}$ be a close convex cone in $\mathbb{R}^p$ and $\boldsymbol{V}$ be a $p \times p$ positive definite matrix. Then the distribution of $\bar{\chi}^2(\boldsymbol{V}, \mathcal{C})$ is given by*

$$pr\{\bar{\chi}^2(\boldsymbol{V}, \mathcal{C}) \leq c\} = \sum_{i=0}^{p} w_i(p, \boldsymbol{V}, \mathcal{C}) pr(\chi_i^2 \leq c) \tag{3.20}$$

where $w_i(p, \boldsymbol{V}, \mathcal{C}), i = 0, \ldots, p$ are some nonnegative numbers and $\sum_{i=0}^{p} w_i(p, \boldsymbol{V}, \mathcal{C}) = 1$. $\blacksquare$

More details about the quantities $w_i(p, \boldsymbol{V}, \mathcal{C})\}$ and their computation are discussed in the next section. Several proofs of Theorem 3.4.2 have appeared in the literature for progressively more general cases. The proofs in Gourieroux et al. (1982) and Shapiro (1985) are noteworthy. The proof in Gourieroux et al. (1982) is for the special case when $\mathcal{C} = \{\boldsymbol{\theta} : \boldsymbol{R\theta} \geqslant \boldsymbol{0}\}$ and $\boldsymbol{R}$ is a full row-rank matrix. This proof is instructive because it is a direct extension of the ideas presented earlier. The proof in Shapiro (1985) is for general closed convex cones and is the most general known; further, Shapiro's proof is based on results relating to projections on to polyhedrals. The proof given in the Appendix to this chapter is based on the proof due to Shapiro (1985). The special case of the theorem for $p = 2$ was discussed earlier in Section 3.3. The main idea underlying the proof for the general case is similar although the technical details are more involved.

In the special case of two dimensions, we saw that the sum of the three weights is 1. Theorem 3.4.2 shows that this generalizes to the case in (3.20). Thus the right-hand side of (3.20) is a weighted mean of several tail probabilities of χ^2-distributions and hence is known as a *chi-bar-square distribution.* We shall refer to $\{w_i(p, \boldsymbol{V}, \mathcal{C})\}$ as *chi-bar-square weights* or simply as *weights*. Another term used for these weights is *level probabilities.*

Now, suppose that we wish to test $H_0 : \boldsymbol{\theta} = \boldsymbol{0}$ against $H_1 : \boldsymbol{\theta} \in \mathcal{C}$ based on a single observation of the p-dimensional vector $\boldsymbol{X}$ where $\boldsymbol{X} \sim N(\boldsymbol{\theta}, \boldsymbol{V})$ and $\boldsymbol{V}$ is a known positive definite matrix. It follows from the definition, that the LRT takes the form

$$\boldsymbol{X}^T\boldsymbol{V}^{-1}\boldsymbol{X} - \min\{(\boldsymbol{X}-\boldsymbol{\theta})^T\boldsymbol{V}^{-1}(\boldsymbol{X}-\boldsymbol{\theta}) : \boldsymbol{\theta} \in \mathcal{C}\}, \tag{3.21}$$

which is the expression on the right-hand side of (3.19) with $\boldsymbol{X}$ in place of $\boldsymbol{Z}$. Therefore, the distribution of LRT under $H_0 : \boldsymbol{\theta} = \boldsymbol{0}$ is $\bar{\chi}^2(\boldsymbol{V}, \mathcal{C})$. In view of this, (3.21) is sometimes called a $\bar{\chi}^2$-statistic. It is worth noting that the $\bar{\chi}^2$-statistic is based on principles of generalized least squares, and therefore it is a reasonable test statistic even if the distribution of $\boldsymbol{X}$ is not normal. The null distribution of the $\bar{\chi}^2$-statistic in (3.21) is a $\bar{\chi}^2$-distribution if $\boldsymbol{X} \sim N(\boldsymbol{0}, \boldsymbol{V})$.

3.5 COMPUTING THE TAIL PROBABILITIES OF CHI-BAR-SQUARE DISTRIBUTIONS

As was indicated earlier, the null distribution of several test statistics for or against inequality constraints turn out to be $\bar{\chi}^2$. Therefore, we need to be able to compute its tail probability to obtain the p-value and/or critical value. This would be easy if the chi-bar-square weights $\{w_i\}$ are known. Unfortunately, the *exact* computation of $\{w_i\}$ is quite difficult in general. The difficulties in computing $\{w_i\}$ may give the false impression that likelihood ratio tests of hypotheses for Type A and Type B testing problems are too complicated to implement. However, we can compute the tail probability of a chi-bar-square distribution by simulation; this is easy to use and fast. Therefore, for most practical needs, the fact that w_i has a complicated form does not cause any serious difficulties. Later, we will see situations where the chi-bar-square weights depend on unknown nuisance parameters; in those cases our inability to compute the weights accurately and fast may cause difficulties.

Simulation 1: To compute pr$\{\bar{\chi}^2(\boldsymbol{V}, \mathcal{C}) \geqslant c\}$

(1) Generate $\boldsymbol{Z}$ from $N(\boldsymbol{0}, \boldsymbol{V})$. (2) Compute $\bar{\chi}^2(\boldsymbol{V}, \mathcal{C})$ in (3.19). (3) Repeat the first two steps N times (say, N = 10000). (4) Estimate pr$\{\bar{\chi}^2(\boldsymbol{V}, \mathcal{C}) \geqslant c\}$ by (n/N) where n is the number of times $\bar{\chi}^2(\boldsymbol{V}, \mathcal{C})$ in the second step turned to be greater than or equal to c. ■

If $\mathcal{C}$ involves only linear constraints, then a quadratic program can be used for computing $\bar{\chi}^2$ (see Section 2.1.2.2). We can also try to compute the tail probability of a $\bar{\chi}^2$ distribution by computing the weights first. However, the weights of a chi-bar-square distribution are likely to be required only for computing the tail probability; if this is the case, then it is likely to be just as easy to compute the tail probability directly by the foregoing simulation method.

Now, let us consider the case when

$$\mathcal{C} = \{\boldsymbol{\theta} \in \mathbb{R}^p : \boldsymbol{a}_i^T\boldsymbol{\theta} \geqslant 0, i = 1, \ldots, k\}. \tag{3.22}$$

If the weights $w_i(p, \boldsymbol{V}, \mathcal{C}\}$ are of interest, then formula (3.29) given in the next section can be used for a simulation-based method for computing them. The simulation steps are easy to implement. It requires only a quadratic program. Such programs are available in several software packages; a simple and easy to implement algorithm for solving quadratic programs is also given in Wollan and Dykstra (1987).

Simulation 2: To compute $w_i(p, \boldsymbol{V}, \mathcal{C})$ **when** $\mathcal{C}$ **is the polyhedral in** (3.22)

(1) Generate $\boldsymbol{Z}$ from $N(\boldsymbol{0}, \boldsymbol{V})$. (2) Compute $\tilde{\boldsymbol{Z}}$, the point at which $(\boldsymbol{Z}-\boldsymbol{\theta})^T\boldsymbol{V}^{-1}(\boldsymbol{Z}-\boldsymbol{\theta})$ is a minimum over $\boldsymbol{\theta} \in \mathcal{C}$. (3) Let $J = \{j : \boldsymbol{a}_j^T\tilde{\boldsymbol{Z}} = 0\}$ and $\phi = \{\boldsymbol{\theta} : \boldsymbol{a}_j^T\boldsymbol{\theta} = 0$ for every $j \in J\}$ and s be the dimension of ϕ; compute s. (4) Repeat the previous steps N times (say $N = 10000$). (5) Estimate $w_i(p, \boldsymbol{V}, \mathcal{C})$ by the proportion of times s turned out to be equal to i, $(i = 1, \ldots, p)$. ■

For more details and proofs relating to the foregoing simulation method see Silvapulle (1996) and Shapiro (1985).

The case when $\mathcal{C}$ is the nonnegative orthant $\mathbb{R}^{+p}$ arises frequently in inequality constrained statistical inference. When the $\boldsymbol{a}_j$'s defining the polyhedral $\mathcal{C} = \{\boldsymbol{x} : \boldsymbol{a}_j^T\boldsymbol{x} \geqslant 0\}$ are linearly independent, a result in the next section shows that $\{w_i(p, \boldsymbol{V}, \mathcal{C})\}$ can be obtained from $\{w_i(k, \boldsymbol{W}, \mathbb{R}^{+k})\}$ for some $k \times k$ positive definite matrix $\boldsymbol{W}$. Thus, the case when the constrained set is the nonnegative orthant is of particular interest. If the weights take the form $w_i(k, \boldsymbol{W}, \mathbb{R}^{+k})$, then the foregoing Simulation 2 reduces to the following simpler form:

Simulation 3: To compute $w_i(p, \boldsymbol{W}, \mathbb{R}^{+p}), i = 1, \ldots, p$

(1) Generate $\boldsymbol{Z}$ from $N(\boldsymbol{0}, \boldsymbol{W})$. (2) Compute $\tilde{\boldsymbol{Z}}$, the point at which $(\boldsymbol{Z}-\boldsymbol{\theta})^T\boldsymbol{W}^{-1}(\boldsymbol{Z}-\boldsymbol{\theta})$ is the minimum over $\boldsymbol{\theta} \geqslant \boldsymbol{0}$. (3) Count the number of positive components of $\tilde{\boldsymbol{Z}}$ (this is equal to s in Step 3 of Simulation 2). (4) Repeat the previous three steps N times (say $N = 10000$). (5) Estimate $w_i(p, \boldsymbol{W}, \mathbb{R}^{+p})$ by the proportion of times $\tilde{\boldsymbol{Z}}$ turned out to have exactly i positive components, $i = 1, \ldots, p$. ■

This approach for the case when the polyhedral is $\mathbb{R}^{+k}$ appears in Wolak (1987). The general idea of using simulation to compute the weights has been around for sometime, for example, see Perlman (1969) and Gourieroux et al. (1982). The authors have a FORTRAN program that implements the foregoing procedure using subroutines from IMSL; this works quite fast.

If $pr\{\bar{\chi}^2(\boldsymbol{V}, \mathcal{C}) \geqslant c\}$ is very small, then Simulation 1 would require a large number of simulations. In this case, we can overcome the problem by computing the weights of the chi-bar-square distribution separately, for example, by simulation. Once this is done, the tail probabilities of the chi-square distributions in $\sum w_i(p, \boldsymbol{V}, \mathcal{C})\text{pr}(\chi_i^2 \geqslant c)$ may be computed using methods that are already available. The advantage of this method is that the value of c has no bearing on $w_i(p, \boldsymbol{V}, \mathcal{C})$ and standard statistical software packages have programs for computing $\text{pr}(\chi_i^2 \geqslant c)$ very accurately for large values of c.

In most practical applications, $\mathcal{C}$ is a polyhedral. However, for completeness, let us also consider the case when $\mathcal{C}$ is not a polyhedral. In this general case, closed-form expressions for $w_i(p, \boldsymbol{V}, \mathcal{C})$ are quite complicated. They can be expressed as some integrals (see Kuriki (1993), and Lin and Lindsay (1997)). At this stage, it is unclear whether or not such integral representations can be used for computing them. However, a simulation approach can be used for computing the weights even in this general case.

Simulation 4: To compute $w_i(p, \boldsymbol{V}, \mathcal{C})$ when $\mathcal{C}$ is not necessarily a polyhedral

(1) Let $0 < c_1 < \ldots < c_{p+1} < \infty$ and $a_{ij} = \text{pr}(\chi_i^2 \leq c_{j+1})$ for $i, j = 0, \ldots, p$. (2) Generate $\boldsymbol{Z}$ from $N(\boldsymbol{0}, \boldsymbol{V})$, and compute $\bar{\chi}^2(\boldsymbol{V}, \mathcal{C})$ in (3.19). (3) Repeat the previous step a large number of times and let p_j be the proportion of times $\bar{\chi}^2(\boldsymbol{V}, \mathcal{C})$ computed in the previous step fell in the interval $(0, c_{j+1}), j = 0, \ldots, p$. (4) Estimate $w_j(p, \boldsymbol{V}, \mathcal{C})$ by solving the system of $(p+1)$ equations $p_j = a_{0j}w_0 + \ldots + a_{pj}w_p$, $(j = 0, \ldots, p)$. ■

For most of the Type A and Type B testing problems, the tail probability P of the likelihood ratio statistic takes the form

$$P = \sum_{i=0}^{p} w_i(p, \boldsymbol{V}, \mathcal{C}) \text{pr}(\chi_{i+k}^2 \geqslant c),$$

for some k. Since $\text{pr}(\chi_i^2 \geqslant c) \leq \text{pr}(\chi_j^2 \geqslant c)$ for $i < j$, $\sum_0^p w_i(p, \boldsymbol{V}, \mathcal{C}) = 1$ and $0 \leq w_i(p, \boldsymbol{V}, \mathcal{C}) \leq (1/2)$ by part 4 in Proposition 3.6.1, we have the following lower and upper bounds for P :

$$(1/2)\{\text{pr}(\chi_k^2 \geqslant c) + \text{pr}(\chi_{k+1}^2 \geqslant c)\} \leq P \leq (1/2)\{\text{pr}(\chi_{p+k-1}^2 \geqslant c) + \text{pr}(\chi_{p+k}^2 \geqslant c)\}. \quad (3.23)$$

These bounds are easy to compute and may turn out to be adequate for practical purposes in some cases; for example, if the upper bound is small then there may not be any need to compute the exact p-value to reject the null hypothesis. It is worth pointing out that these may be crude bounds if p is large.

The proof of Theorem 3.4.2 shows that $\text{pr}\{\bar{\chi}^2(\boldsymbol{V}, \mathcal{C}) \geqslant c\}$ and $w_i(p, \boldsymbol{V}, \mathcal{C})$ are continuous in $(\boldsymbol{V}, \mathcal{C})$. Therefore, if $\mathcal{C}$ is close to $\mathbb{R}^{+p}$ and $\boldsymbol{V}$ is close to $\boldsymbol{I}$, then $w_i(p, \boldsymbol{V}, \mathcal{C})$ is close to $w_i(p, \boldsymbol{I}, \mathbb{R}^{+p})$, and an approximation to $\text{pr}(\bar{\chi}^2(\boldsymbol{V}, \mathcal{C}) \geqslant c)$ is $\sum_{i=0}^{p} 2^{-p}p!/\{i!(p-i)!\}\text{pr}(\chi_i^2 \geqslant c)$ because $w_i(p, \boldsymbol{I}, \mathbb{R}^{+p}) = 2^{-p}p!/\{i!(p-i)!\}$ by part 2 of Proposition 3.6.1. It may be verified that this approximation lies between the lower and the upper bounds given in (3.23). This provides a quick and rough estimate of the tail probability $\text{pr}(\bar{\chi}^2(\boldsymbol{V}, \mathcal{C}) \geqslant c)$. The closer the $(\boldsymbol{V}, \mathcal{C})$ to $(\boldsymbol{I}, \mathbb{R}^{+p})$, the better the approximation; for some comparisons on this, see Piegorsch (1990).

The following closed-form expressions for $w_i(p, \boldsymbol{V}, \mathbb{R}^{+p})$ are useful.

1. Let $p = 1$. Then

$$w_0(1, \boldsymbol{V}) = w_1(1, \boldsymbol{V}) = 0.5. \quad (3.24)$$

Therefore, $\text{pr}(\bar{\chi}^2 \geqslant t) = 0.5\text{pr}(\chi_1^2 \geqslant t)$ for $t > 0$. This is related to the familiar result where the p-value for a two-sided alternative concerning a scalar parameter is equal to twice the p-value for the corresponding one-sided alternative.

2. Let $p = 2$. Then (see Example 3.3.4)

$$\begin{aligned} &w_0(2, \boldsymbol{V}) = (1/2)\pi^{-1}\cos^{-1}(\rho_{12}), \qquad w_1(2, \boldsymbol{V}) = (1/2), \text{ and} \\ &w_2(2, \boldsymbol{V}) = (1/2) - (1/2)\pi^{-1}\cos^{-1}(\rho_{12}) \end{aligned} \quad (3.25)$$

where ρ_{12} is the correlation coefficient $v_{12}\{v_{11}v_{22}\}^{-1/2}$.

3. Let $p = 3$. Then, (for example, see Kudo, 1963, pp 414 - 415)

$$\begin{aligned} w_3(3, \boldsymbol{V}) &= (4\pi)^{-1}(2\pi - \cos^{-1}\rho_{12} - \cos^{-1}\rho_{13} - \cos^{-1}\rho_{23}), \\ w_2(3, \boldsymbol{V}) &= (4\pi)^{-1}(3\pi - \cos^{-1}\rho_{12.3} - \cos^{-1}\rho_{13.2} - \cos^{-1}\rho_{23.1}), \\ w_1(3, \boldsymbol{V}) &= (1/2) - w_3(3, \boldsymbol{V}), \\ w_0(3, \boldsymbol{V}) &= (1/2) - w_2(3, \boldsymbol{V}), \end{aligned} \tag{3.26}$$

   where ρ_{ij} is the correlation coefficient $v_{ij}\{v_{ii}v_{jj}\}^{-1/2}$, and $\rho_{ij.k}$ is the partial correlation coefficient $(\rho_{ij} - \rho_{ik}\rho_{jk})/\{(1-\rho_{ik}^2)(1-\rho_{jk}^2)\}^{1/2}$.

4. Let $\boldsymbol{V} = \boldsymbol{I}$. Then, by part 2 of Proposition 3.6.1,

$$w_i(p, \boldsymbol{I}) = 2^{-p}p!/\{i!(p-i)!\} \text{ for every } p \text{ and } i. \tag{3.27}$$

When $p = 4$, all the weights can be expressed as functions of $w_4(4, \boldsymbol{V})$ and $\boldsymbol{V}$ (for example, see Wolak (1987)). However, numerical integration or Monte Carlo simulation needs to be used to compute $w_4(4, \boldsymbol{V})$. Therefore, these formulas may not be that helpful, because it may be just as easy to compute all the weights or the tail probability of the $\bar{\chi}^2$ distribution by simulation.

The simulation approaches discussed in this section require a computer program for computing $\min\{(\boldsymbol{Z} - \boldsymbol{\theta})^T\boldsymbol{V}^{-1}(\boldsymbol{Z} - \boldsymbol{\theta}) : \boldsymbol{\theta} \in \mathcal{C}\}$. If $\mathcal{C}$ involves only linear inequalities, we can use a standard quadratic program (see Section 2.1.2.2). If $\mathcal{C}$ involves nonlinear inequalities in $\boldsymbol{\theta}$, then a general-purpose nonlinear optimization program may be used; such programs are available in IMSL, MATLAB, and NAG. If $\mathcal{C}$ involves nonlinear inequality constraints and minimization of $(\boldsymbol{Z}-\boldsymbol{\theta})^T\boldsymbol{V}^{-1}(\boldsymbol{Z}-\boldsymbol{\theta})$ subject to the constraint $\boldsymbol{\theta} \in \mathcal{C}$ is difficult to be incorporated in a simulation program, then another approach worth considering is to compute bounds for the critical values and/or p-values, for example, by the following method. Let $\mathcal{P}_1$ and $\mathcal{P}_2$ be two polyhedrals, such that $\mathcal{P}_1 \subset \mathcal{C} \subset \mathcal{P}_2$. Since $\bar{\chi}^2(\boldsymbol{V}, \mathcal{P}_1) \leqslant \bar{\chi}^2(\boldsymbol{V}, \mathcal{C}) \leqslant \bar{\chi}^2(\boldsymbol{V}, \mathcal{P}_2)$ we have that

$$\text{pr}\{\bar{\chi}^2(\boldsymbol{V}, \mathcal{P}_1) \geqslant c\} \leqslant \text{pr}\{\bar{\chi}^2(\boldsymbol{V}, \mathcal{C}) \geqslant c\} \leqslant \text{pr}\{\bar{\chi}^2(\boldsymbol{V}, \mathcal{P}_2) \geqslant c\}.$$

This approach is useful for obtaining lower and upper bounds for the tail probability of a chi-bar-square distribution; this method was used in Silvapulle (1992b).

3.6 RESULTS ON CHI-BAR-SQUARE WEIGHTS

Often, the constraints defining $\mathcal{C}$ are linear and independent. In this case, $w_i(p, \boldsymbol{V}, \mathcal{C})$ can be expressed as $w_i(k, \boldsymbol{W}, \mathbb{R}^{+k})$ for some $\boldsymbol{W}$ and k. Therefore, to simplify notation, we denote $w_i(p, \boldsymbol{V}, \mathcal{C})$ by $w_i(p, \boldsymbol{V})$ when $\mathcal{C}$ is the positive orthant:

$$w_i(p, \boldsymbol{V}) = w_i(p, \boldsymbol{V}, \mathbb{R}^{+p}). \tag{3.28}$$

The main results concerning $w_i(p, \boldsymbol{V}, \mathcal{C})$ are stated in this section; the proofs are not given here but may be found in Kudo (1963), Perlman (1969), Wolak (1987b), Shapiro (1985, 1988), and Silvapulle (1996a).

Proposition 3.6.1 *Let $\mathcal{C}$ be a closed convex cone in $\mathbb{R}^p$ and $\boldsymbol{V}$ be a $p \times p$ nonsingular covariance matrix. Then we have the following:*

1. *Let $\mathcal{C} = \{\boldsymbol{x} : \boldsymbol{a}_i^T \boldsymbol{x} \geqslant 0, i = 1, \ldots, k\}$ for some $\boldsymbol{a}_1, \ldots, \boldsymbol{a}_k$, $J = \{j : \boldsymbol{a}_j^T \Pi_{\boldsymbol{V}}(\boldsymbol{Z}|\mathcal{C}) = \boldsymbol{0}\}$, and $\phi = \{\boldsymbol{\theta} : \boldsymbol{a}_j^T \boldsymbol{\theta} = 0, \forall j \in J\}$. Then*

$$w_i(p, \boldsymbol{V}, \mathcal{C}) = pr(\textit{The linear space } \phi \textit{ is of dimension } i). \tag{3.29}$$

   This can be restated as follows: Suppose that $\mathcal{C}$ is a polyhedral, $\boldsymbol{Z} \sim N(\boldsymbol{0}, \boldsymbol{V})$ and let $F(\boldsymbol{Z}, \boldsymbol{V}, \mathcal{C})$ denote the face of $\mathcal{C}$ such that $\Pi_{\boldsymbol{V}}(\boldsymbol{Z}|\mathcal{C})$ lies in the relative interior of $F(\boldsymbol{Z}, \boldsymbol{V}, \mathcal{C})$; for a definition of a face of a cone and relative interior of a face see the Appendix to this chapter (page 123). Then

$$w_i(p, \boldsymbol{V}, \mathcal{C}) = pr\{ \textit{ dimension of the linear space spanned by} F(\boldsymbol{Z}, \boldsymbol{V}, \mathcal{C}) \textit{ is } i\}.$$

2. *Let $\boldsymbol{Z} \sim N(\boldsymbol{0}, \boldsymbol{V})$ and $\mathcal{C}$ be the nonnegative orthant. Then, $w_i(p, \boldsymbol{V}) = pr\{\Pi_{\boldsymbol{V}}(\boldsymbol{Z}|\mathcal{C})$ has exactly i positive components$\}$.*

3. *$\sum_0^p (-1)^i w_i(p, \boldsymbol{V}, \mathcal{C}) = 0$.*

4. *$0 \leq w_i(p, \boldsymbol{V}, \mathcal{C}) \leq 0.5$.*

5. *Let $\mathcal{C}^o$ denote the polar cone, $\{\boldsymbol{x} \in \mathbb{R}^p : \boldsymbol{x}^T \boldsymbol{V}^{-1} \boldsymbol{y} \leq 0$ for every $\boldsymbol{y} \in \mathcal{C}\}$, of $\mathcal{C}$ with respect to the inner product $\langle \boldsymbol{x}, \boldsymbol{y} \rangle = \boldsymbol{x}^T \boldsymbol{V}^{-1} \boldsymbol{y}$. Then $w_i(p, \boldsymbol{V}, \mathcal{C}^o) = w_{p-i}(p, \boldsymbol{V}, \mathcal{C})$.*

6. *Let $\mathcal{C} = \{\boldsymbol{\theta} \in \mathbb{R}^p : \boldsymbol{R\theta} \geqslant \boldsymbol{0}\}$ where $\boldsymbol{R}$ is a $p \times p$ nonsingular matrix. Then $\bar{\chi}^2(\boldsymbol{V}, \mathcal{C}) = \bar{\chi}^2(\boldsymbol{RVR}^T, \mathbb{R}^{+p})$ and $w_i(p, \boldsymbol{V}, \mathcal{C}) = w_i(p, \boldsymbol{RVR}^T)$.*

7. *$w_i(p, \boldsymbol{V}) = w_{p-i}(p, \boldsymbol{V}^{-1})$.*

8. *Let $\mathcal{C} = \{\boldsymbol{\theta} \in \mathbb{R}^p : \boldsymbol{R\theta} \geqslant \boldsymbol{0}\}$ where $\boldsymbol{R}$ is a $k \times p$ matrix of rank $k (\leq p)$. Then*

$$w_{p-k+i}(p, \boldsymbol{V}, \mathcal{C}) = \begin{cases} w_i(k, \boldsymbol{RVR}^T) & \textit{for } i = 0, \ldots, k \\ 0 & \textit{otherwise} \end{cases}$$

9. *Let $\mathcal{C} = \{\boldsymbol{\theta} \in \mathbb{R}^p : \boldsymbol{R}_1 \boldsymbol{\theta} \geqslant \boldsymbol{0}, \boldsymbol{R}_2 \boldsymbol{\theta} = \boldsymbol{0}\}$ where $\boldsymbol{R}_1$ is $s \times p$, $\boldsymbol{R}_2$ is $t \times p$, $s + t \leq p$, $[\boldsymbol{R}_1^T, \boldsymbol{R}_2^T]$ is of full rank, and*

$$\boldsymbol{A} = \boldsymbol{R}_1 \boldsymbol{V} \boldsymbol{R}_1^T - (\boldsymbol{R}_1 \boldsymbol{V} \boldsymbol{R}_2^T)(\boldsymbol{R}_2 \boldsymbol{V} \boldsymbol{R}_2^T)^{-1}(\boldsymbol{R}_2 \boldsymbol{V} \boldsymbol{R}_1^T).$$

   Then

$$w_{p-s-t+j}(p, \boldsymbol{V}, \mathcal{C}) = \begin{cases} w_j(s, \boldsymbol{A}) & \textit{for } j = 0, \ldots, s \\ 0 & \textit{otherwise.} \end{cases}$$

10. *Let $\boldsymbol{R}$ be a $r \times p$ matrix of rank r, $\boldsymbol{R}_1$ be a $q \times p$ submatrix of $\boldsymbol{R}$, $\mathcal{M} = \{\boldsymbol{\beta} : \boldsymbol{R\beta} = \boldsymbol{0}\}$, $\mathcal{C} = \{\boldsymbol{\beta} : \boldsymbol{R}_1 \boldsymbol{\beta} \geqslant \boldsymbol{0}\}$, and $\mathcal{M}^\perp$ be the orthogonal complement of $\mathcal{M}$ with respect to $\langle \boldsymbol{x}, \boldsymbol{y} \rangle_{\boldsymbol{V}} = \boldsymbol{x}^T \boldsymbol{V}^{-1} \boldsymbol{y}$. Then*

$$w_{r-q+j}(p, \boldsymbol{V}, \mathcal{C} \cap \mathcal{M}^\perp) = \begin{cases} w_j(q, \boldsymbol{R}_1 \boldsymbol{V} \boldsymbol{R}_1^T)) & \textit{for } j = 0, \ldots, q \\ 0 & \textit{otherwise} \end{cases}$$

11. *Let $\boldsymbol{C}$ denote the correlation matrix corresponding to $\boldsymbol{V}$. Then $\bar{\chi}^2(\boldsymbol{V},\mathcal{C}) = \bar{\chi}^2(\boldsymbol{C},\mathcal{C})$ and $w_i(p,\boldsymbol{V},\mathcal{C}) = w_i(p,\boldsymbol{C},\mathcal{C})$ for every i.*

Proof: For (1) see Shapiro (1985); (2) follows from (1) but it has also been known for quite some time (for example, see Perlman (1969)). See (5.3), (5.4), (5.5), (5.8) and (5.10) in Shapiro (1988) for (6), (7), (8), (9), and (10) respectively. ■

Kudo (1963) showed that $w_i(p,\boldsymbol{V}) = \sum p\{(\Lambda_N)^{-1}\}p(\Lambda_{M:N})$ where $M \subset \{1,\ldots,p\}$, the sum extends over all M with exactly i elements, $N = \{1,\ldots,p\}\setminus M$, Λ_M is the covariance matrix of the vector $\{Z_i : i \in M\}$, $\Lambda_{M:N}$ is the same under $Z_j = 0$ for $j \notin M$, and $p(\boldsymbol{A})$ is $\text{pr}(\boldsymbol{X} \geqslant \boldsymbol{0})$ when $\boldsymbol{X} \sim N(\boldsymbol{0},\boldsymbol{A})$. Thus, $w_i(p,\boldsymbol{V})$ can be expressed in terms of orthant probabilities. Schervish (1983) provided a program for computing probabilities of rectangles corresponding to multivariate normal; this can also be used for computing the orthant probabilities. Bohrer and Chow (1978) provided a computer program based on this result for computing $w_i(p,\boldsymbol{V})$, for $p \leq 10$. Another FORTRAN program is given in Sun (1988a). Kudo (1963) also discussed its use for computations. Formulas for the chi-bar square weights when $\boldsymbol{V}$ is a diagonal matrix are given in Section 2.4 of Robertson et al. (1988) for some special cases.

The following useful corollary to part 5 of the foregoing proposition follows from Theorem 3.4.2; it says that when $\mathcal{C}$ is replaced by its polar cone the weights appear in the reverse order.

Corollary 3.6.2 *Let $\mathcal{C}$ and $\boldsymbol{V}$ be as in Proposition 3.6.1. Then*

$$pr\{\bar{\chi}^2(\boldsymbol{V},\mathcal{C}^o) \leq c\} = \sum_{i=0}^{p} w_{p-i}(p,\boldsymbol{V},\mathcal{C})pr(\chi_i^2 \leq c). \quad ■$$

3.7 LRT FOR TYPE A PROBLEMS: V IS KNOWN

Section 3.3 provided detailed discussions on the distribution of *LRT* in two dimensions, and the discussions therein showed how chi-bar-square distributions arise. The main result in Section 3.4 showed that the *LRT* for testing $\boldsymbol{\theta} = \boldsymbol{0}$ against $\boldsymbol{\theta} \in \mathcal{C}$ based on a single observation of $\boldsymbol{X}$, where $\boldsymbol{X} \sim N(\boldsymbol{\theta},\boldsymbol{V})$, is the $\bar{\chi}^2$ statistic

$$\boldsymbol{X}^T\boldsymbol{V}^{-1}\boldsymbol{X} - \min_{\boldsymbol{\theta}\in\mathcal{C}}(\boldsymbol{X}-\boldsymbol{\theta})^T\boldsymbol{V}^{-1}(\boldsymbol{X}-\boldsymbol{\theta})$$

and that its null distribution is the chi-bar-square distribution

$$\sum_{i=0}^{p} w_i(p,\boldsymbol{V},\mathcal{C})\text{pr}(\chi_i^2 \leq c).$$

A similar result holds even if the null parameter space, $\{\boldsymbol{0}\}$, is replaced by a linear space; this section provides a discussion of such results.

Let $\mathcal{M}$ be a linear space contained in $\mathcal{C}$. Suppose that we are interested to test

$$H_0 : \boldsymbol{\theta} \in \mathcal{M} \text{ against } H_1 : \boldsymbol{\theta} \in \mathcal{C}$$

based on a single observation of $\boldsymbol{X}$, where $\boldsymbol{\theta}$ is a location parameter of $\boldsymbol{X}$, for example, it could be the mean of $\boldsymbol{X}$. A least squares statistic for testing H_0 against H_1 is

$$L_A = \min\{q(\boldsymbol{a}) : \boldsymbol{a} \in \mathcal{M}\} - \min\{q(\boldsymbol{a}) : \boldsymbol{a} \in \mathcal{C}\}, \tag{3.30}$$

where

$$q(\boldsymbol{a}) = (\boldsymbol{X} - \boldsymbol{a})^T \boldsymbol{V}^{-1}(\boldsymbol{X} - \boldsymbol{a})$$

and $\boldsymbol{V}$ is a known positive definite matrix. If $\boldsymbol{X} \sim N(\boldsymbol{\theta}, \boldsymbol{V})$ then L_A is the LRT for testing H_0 against H_1. A general result on the distribution of LRT for a Type A testing problem is given in the next result.

Theorem 3.7.1 *Let $\boldsymbol{X} \sim N(\boldsymbol{\theta}, \boldsymbol{V})$ where $\boldsymbol{V}$ is a positive definite matrix. Then, the LRT for the Type A testing problem, $H_0 : \boldsymbol{\theta} \in \mathcal{M}$ against $H_1 : \boldsymbol{\theta} \in \mathcal{C}$, is similar*[1] *and its null distribution is given by*

$$pr(LRT \leq c \mid H_0) = \sum_{i=0}^{p} w_i(p, \boldsymbol{V}, \mathcal{C} \cap \mathcal{M}^{\perp}) pr(\chi_i^2 \leq c) \tag{3.31}$$

where $\mathcal{M}^{\perp} = \{\boldsymbol{x} : \boldsymbol{x}^T \boldsymbol{V}^{-1} \boldsymbol{y} = 0$ for every $\boldsymbol{y} \in \mathcal{M}\}$ is the orthogonal complement of $\mathcal{M}$ with respect to the inner product $\langle \boldsymbol{x}, \boldsymbol{y} \rangle_{\boldsymbol{V}} = \boldsymbol{x}^T \boldsymbol{V}^{-1} \boldsymbol{y}$.

Proof: The proof of this theorem is based on deeper results about projections onto to convex cones. These are discussed in the Appendix to this chapter. Here, we shall provide an intuitive explanation of the main ideas after a short proof.

It follows from parts (h) and (i) of Proposition 3.12.6 in the Appendix that

$$\begin{aligned} LRT &= \min\{\|\boldsymbol{X} - \boldsymbol{\theta}\|_{\boldsymbol{V}}^2 : \boldsymbol{\theta} \in \mathcal{M}\} - \min\{\|\boldsymbol{X} - \boldsymbol{\theta}\|_{\boldsymbol{V}}^2 : \boldsymbol{\theta} \in \mathcal{C}\} \\ &= \|\boldsymbol{X}\|_{\boldsymbol{V}}^2 - \min\{\|\boldsymbol{X} - \boldsymbol{\theta}\|_{\boldsymbol{V}}^2 : \boldsymbol{\theta} \in \mathcal{C} \cap \mathcal{M}^{\perp}\}. \end{aligned}$$

The last expression on the right-hand side is also the LRT for testing $\boldsymbol{\theta} = \boldsymbol{0}$ against $\boldsymbol{\theta} \in \mathcal{C} \cap \mathcal{M}^{\perp}$. Therefore, the proof follows from Theorem 3.4.2. ■

Figure 3.10 provides an intuitive picture of the ideas underlying the foregoing proof in three-dimensions. Let O be the origin and OR be the θ_1 axis; θ_2 and θ_3 axes are not shown. Let $\mathcal{M}$ be the linear space spanned by the line OR; thus, $\mathcal{M}$ is simply the θ_1 axis. The rectangle $OPQR$ spans a plane; similarly, the rectangle $OSTR$ spans another plane. Let $\mathcal{C}$ be the convex cone generated by these two planes as shown in the figure; thus, the cone $\mathcal{C}$ has a wedge shape with its *spine* being OR. Clearly, $\mathcal{M}^{\perp}$ is the plane perpendicular to the θ_1 axis and passes through O. Therefore, $\mathcal{M}^{\perp}$ is spanned by the θ_2-axis and the θ_3-axis, and $\mathcal{C} \cap \mathcal{M}^{\perp}$ is the cone SOP. Let OA represent the vector $\boldsymbol{X}$. Let B be the projection of A onto $\mathcal{C}$; more precisely, let OB be the projection of OA onto $\mathcal{C}$. Suppose that B is an interior point of OPQR. Then B is also the projection of A onto the plane spanned by $OPQR$. Therefore,

[1] Here, similar means that the null distribution of the test statistic is the same at every point in the null parameter space.

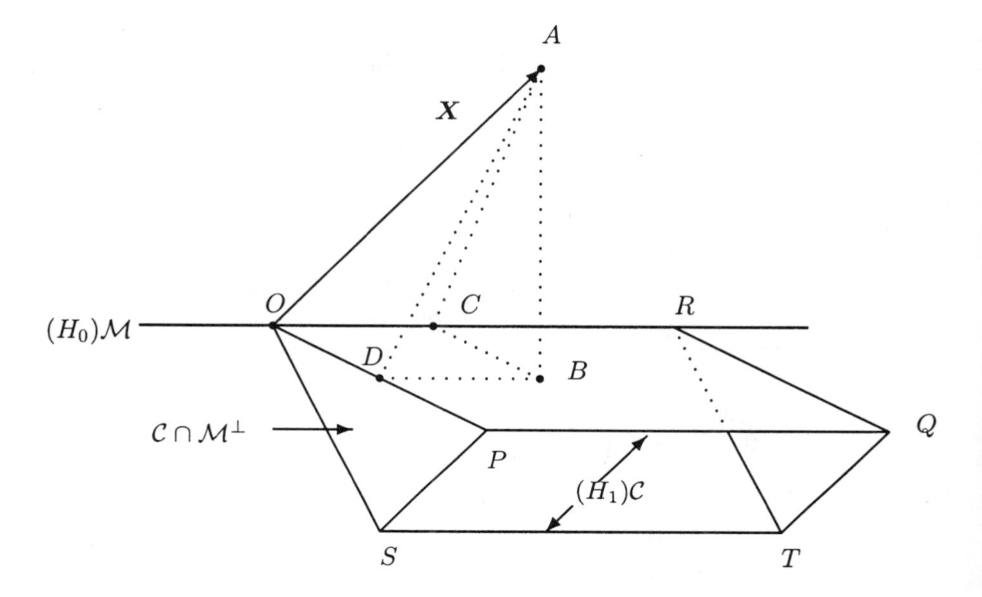

Fig. 3.10 Projections onto $\mathcal{C}$, $\mathcal{M}$, and $\mathcal{C} \cap \mathcal{M}^{\perp}$ where $\mathcal{M} \subset \mathcal{C}$.

AB is orthogonal to $OPQR$. Let D and C be the projections of A onto $\mathcal{C} \cap \mathcal{M}^{\perp}$ and $\mathcal{M}$ respectively. Then D and C are also the projections of B onto $\mathcal{C} \cap \mathcal{M}^{\perp}$ and $\mathcal{M}$ respectively. Now, we have the following :

$$AB \perp BD, AB \perp BC, AC \perp OC, AD \perp OD.$$

It follows that

$$\begin{aligned}
&LRT \text{ of } \boldsymbol{\theta} \in \mathcal{M} \text{ vs } \boldsymbol{\theta} \in \mathcal{C} \\
&\quad = \min\{\|\boldsymbol{X} - \boldsymbol{\theta}\|^2_{\boldsymbol{V}} : \boldsymbol{\theta} \in \mathcal{M}\} - \min\{\|\boldsymbol{X} - \boldsymbol{\theta}\|^2_{\boldsymbol{V}} : \boldsymbol{\theta} \in \mathcal{C}\} \\
&\quad = \|AC\|^2 - \|AB\|^2 = \|BC\|^2 = \|DO\|^2 = \|OA\|^2 - \|AD\|^2 \\
&\quad = \|\boldsymbol{X}\|^2_{\boldsymbol{V}} - \min\{\|\boldsymbol{X} - \boldsymbol{\theta}\|^2_{\boldsymbol{V}} : \boldsymbol{\theta} \in \mathcal{C} \cap \mathcal{M}^{\perp}\}.
\end{aligned}$$

This is also the *LRT* for testing $\boldsymbol{\theta} = \mathbf{0}$ against $\boldsymbol{\theta} \in \mathcal{C} \cap \mathcal{M}^{\perp}$. Since $\mathcal{C} \cap \mathcal{M}^{\perp}$ is a closed convex cone, its null distribution is a chi-bar-square and is given by (3.31).

In general, the weights of the chi-bar-square distribution in (3.31) depend on the parameter spaces $\mathcal{M}$ and $\mathcal{C}$. There is no easy way to compute these weights for an arbitrary $\{\mathcal{C}, \mathcal{M}, \boldsymbol{V}\}$. The simulation procedure indicated in Section 3.5 is available as a general purpose procedure for computing $\text{pr}(LRT \leq c \mid H_0)$. Fortunately, often in practice, the constraints defining $\mathcal{C}$ turn out to be linear and independent. In this case, we have the following simpler result, which follows from Proposition 3.6.1.

Corollary 3.7.2 *Let $\boldsymbol{X} \sim N(\boldsymbol{\theta}, \boldsymbol{V})$ where $\boldsymbol{V}$ is a positive definite matrix, $\boldsymbol{R}$ be a matrix of order $r \times p$, $rank(\boldsymbol{R}) = r \leq p$, and let $\boldsymbol{R}_1$ be a submatrix of $\boldsymbol{R}$ of order*

$q \times p$. Let the null and alternative hypothesis be $H_0 : \boldsymbol{R\theta} = \mathbf{0}$ and $H_1 : \boldsymbol{R}_1\boldsymbol{\theta} \geqslant \mathbf{0}$ respectively. Then, the LRT is similar and its null distribution is given by

$$pr(LRT \leq c|H_0) = \sum_{i=0}^{q} w_i(q, \boldsymbol{R}_1\boldsymbol{V}\boldsymbol{R}_1^T)pr(\chi^2_{r-q+i} \leq c) \qquad \blacksquare$$

Note that the chi-bar-square weights appearing in this corollary take the simpler form $w_i(q, \boldsymbol{R}_1\boldsymbol{V}\boldsymbol{R}_1^T)$, which by definition is equal to $w_i(q, \boldsymbol{R}_1\boldsymbol{V}\boldsymbol{R}_1^T, \mathbb{R}^{+q})$. Further, the number of terms in the foregoing $\bar{\chi}^2$ distribution depends on the number of inequalities in H_1 only, not on the dimension p of $\boldsymbol{\theta}$. If the alternative hypothesis has only independent linear inequalities, and the number of them does not exceed 3, then we can use the explicit formulas for weights given in (3.24), (3.25), and (3.26). If the alternative hypothesis does not have any inequality constraints then $q = 0$ and hence we recover the classical result, $\text{pr}(LRT \leq c|H_0) = \text{pr}(\chi^2_r \leq c)$.

If $\boldsymbol{X}$ is not normal but the distribution of $\boldsymbol{X} - \boldsymbol{\theta}$ does not depend on any unknown parameters, then L_A in (3.30) is still a suitable statistic for testing $H_0 : \boldsymbol{\theta} \in \mathcal{M}$ against $H_1 : \boldsymbol{\theta} \in \mathcal{C}$; it is a generalized least squares based statistic. For this we have the following:

Proposition 3.7.3 *Suppose that the distribution of $\boldsymbol{X} - \boldsymbol{\theta}$ does not depend on any unknown parameters. Then the null distribution of L_A in* (3.30) *does not depend on the particular value of $\boldsymbol{\theta}$ in the null parameter space; ie. the test is similar.*

Proof: It may be verified easily that $\mathcal{C} + \boldsymbol{\alpha} = \mathcal{C}$ for any $\boldsymbol{\alpha} \in \mathcal{M}$, where $\mathcal{C} + \boldsymbol{\alpha}$ is defined as $\{\boldsymbol{x}+\boldsymbol{\alpha} : \boldsymbol{x} \in \mathcal{C}\}$; and $\mathcal{M}+\boldsymbol{\alpha} = \mathcal{M}$ for any $\boldsymbol{\alpha} \in \mathcal{M}$. Let $\boldsymbol{\theta}$ be an element of $\mathcal{M}$ and let $\boldsymbol{Y} = (\boldsymbol{X} - \boldsymbol{\theta})$. Then $L_A = \min\{q^*(\boldsymbol{a}) : \boldsymbol{a} \in \mathcal{M}\} - \min\{q^*(\boldsymbol{a}) : \boldsymbol{a} \in \mathcal{C}\}$, where $q^*(\boldsymbol{a}) = (\boldsymbol{Y} - \boldsymbol{a})^T\boldsymbol{V}^{-1}(\boldsymbol{Y} - \boldsymbol{a})$, and the distribution of $\boldsymbol{Y}$ does not depend on any unknown parameters. Therefore, if H_0 is true then the distribution of L_A is the same at every value in the null parameter space, $\mathcal{M}$. ■

In view of the foregoing Proposition, even if $\boldsymbol{X}$ is not normal, we can talk about the null distribution of L_A without specifying the null value of $\boldsymbol{\theta}$, and for the purposes of computing the p-value/critical value of L_A, we may assume that $\boldsymbol{\theta} = \mathbf{0}$. With t denoting the observed value of L_A, we have that

$$p\text{-value} = \text{pr}(L_A \geqslant t \mid H_0) = \text{pr}(L_A \geqslant t \mid \boldsymbol{\theta} = \mathbf{0}). \tag{3.32}$$

A consequence of this is that, even if $\boldsymbol{X}$ is not normal but the distribution of $\boldsymbol{X} - \boldsymbol{\theta}$ is known and does not depend on any unknown parameters, then exact finite sample p-value corresponding to L_A can be computed by simulation with pseudo-random observations generated at $\boldsymbol{\theta} = \mathbf{0}$.

The foregoing results are useful in large samples even if the population distribution is not normal. Let us suppose that $\boldsymbol{Z}_1, \ldots, \boldsymbol{Z}_n$ are independently and identically distributed as $\boldsymbol{Z}$, with $E(\boldsymbol{Z}) = \boldsymbol{\theta}$ and $\text{cov}(\boldsymbol{Z}) = \boldsymbol{W}$; it is assumed that the distribution of $\boldsymbol{Z}$ may not be normal, $\boldsymbol{W}$ is known, and that the distribution of $(\boldsymbol{Z} - \boldsymbol{\theta})$ does not depend on any unknown parameters. Suppose that we are interested to test $H_0 : \boldsymbol{\theta} \in \mathcal{M}$ against $H_1 : \boldsymbol{\theta} \in \mathcal{C}$. Then, with $\boldsymbol{X}$ denoting $n^{-1}\sum \boldsymbol{Z}_i$ and assuming that n is large, we have that $\boldsymbol{X}$ is approximately $N(\boldsymbol{\theta}, \boldsymbol{V})$, where $\boldsymbol{V} = n^{-1}\boldsymbol{W}$. Therefore, L_A in (3.30) is still a suitable statistic for testing H_0 against H_1, although it may not be the likelihood ratio statistic. Further, for large samples, its null distribution is approximately $\bar{\chi}^2$. In particular, the results in Theorem 3.7.1 and Corollary 3.7.2 are applicable as large sample approximations.

Another large sample application of the foregoing may take the following form. Let $\boldsymbol{\theta}$ denote a parameter associated with a general statistical model, and let $\boldsymbol{\theta}_0$ denote its true value. For large samples, we typically have $\sqrt{n}(\hat{\boldsymbol{\theta}} - \boldsymbol{\theta}_0) \approx N(\boldsymbol{0}, \boldsymbol{V})$. Assume that $\hat{\boldsymbol{V}}$ is a consistent estimator of $\boldsymbol{V}$. Let

$$\hat{T} = \min_{\boldsymbol{\theta} \in \mathcal{M}} n(\hat{\boldsymbol{\theta}} - \boldsymbol{\theta})^T \hat{\boldsymbol{V}}^{-1} (\hat{\boldsymbol{\theta}} - \boldsymbol{\theta}) - \min_{\boldsymbol{\theta} \in \mathcal{C}} n(\hat{\boldsymbol{\theta}} - \boldsymbol{\theta})^T \hat{\boldsymbol{V}}^{-1} (\hat{\boldsymbol{\theta}} - \boldsymbol{\theta}).$$

If the $\hat{\boldsymbol{V}}$ in the foregoing definition of $\hat{T}$ is replaced by its probability limit, $\boldsymbol{V}$, then the resulting change on $\hat{T}$ is of order $o_p(1)$. Let $\boldsymbol{Z}_n = \sqrt{n}(\hat{\boldsymbol{\theta}} - \boldsymbol{\theta}_0)$ and $\boldsymbol{Z} \sim N(\boldsymbol{0}, \boldsymbol{V})$. Suppose that $\boldsymbol{\theta}_0 \in \mathcal{M}$ and $\boldsymbol{Z}_n \xrightarrow{d} \boldsymbol{Z}$. Then, we have that

$$\begin{aligned} \hat{T} &= \min_{\boldsymbol{\theta} \in \mathcal{M}} (\boldsymbol{Z}_n - \boldsymbol{\theta})^T \boldsymbol{V}^{-1} (\boldsymbol{Z}_n - \boldsymbol{\theta}) - \min_{\boldsymbol{\theta} \in \mathcal{C}} (\boldsymbol{Z}_n - \boldsymbol{\theta})^T \boldsymbol{V}^{-1} (\boldsymbol{Z}_n - \boldsymbol{\theta}) + o_p(1) \\ &\xrightarrow{d} \min_{\boldsymbol{\theta} \in \mathcal{M}} (\boldsymbol{Z} - \boldsymbol{\theta})^T \boldsymbol{V}^{-1} (\boldsymbol{Z}_n - \boldsymbol{\theta}) - \min_{\boldsymbol{\theta} \in \mathcal{C}} (\boldsymbol{Z} - \boldsymbol{\theta})^T \boldsymbol{V}^{-1} (\boldsymbol{Z} - \boldsymbol{\theta}). \end{aligned} \tag{3.33}$$

The distribution of this limit is given by (3.31); if the constraints in H_1 are linear and independent then the limiting distribution is given by Corollary 3.7.2. A difficulty in applying this general result will arise if the weights, $w_i(p, \boldsymbol{V}, \mathcal{M}^\perp \cap \mathcal{C})$ in (3.31) depend on the unknown parameters defining $\boldsymbol{V}$. Methods of dealing with such nuisance parameters will be discussed in more detail in the next chapter. For now, we note that if $\boldsymbol{V}$ depends on unknown nuisance parameters then a standard procedure in statistical inference is to define the p-value as the supremum of the tail probability over the nuisance parameter:

$$p\text{-value} = \sup_{\boldsymbol{V}} \text{pr}(\hat{T} \geqslant t_{obs} \mid H_0)$$

where t_{obs} is the observed value of $\hat{T}$; this is discussed at the beginning of the next section. Now, if adopt the large sample arguments leading to (3.33), we have

$$p\text{-value} \simeq \sup_{\boldsymbol{V}} \lim \text{pr}(\hat{T} \geqslant t_{obs} \mid H_0) = \sup_{\boldsymbol{V}} \text{pr}\{\bar{\chi}^2(\boldsymbol{V}, \mathcal{C} \cap \mathcal{M}^\perp) \geqslant t_{obs}\}.$$

Example 3.7.1 *Effect of in-breeding on growth (Kudo (1963))*

In a large scale survey of child health carried out in Hiroshima and Nagasaki (see Neel and Schull (1954)), approximately 10000 children born of related as well as unrelated parents were studied. One facet of the study involved obtaining four anthropometric measurements as indices of growth and development. These were weight (y_1 gm), height (y_2 mm), head girth (y_3 mm), and chest girth (y_4 mm). The multivariate linear regression model, $E(\boldsymbol{Y}) = \boldsymbol{\mu} + t\boldsymbol{\alpha} - c\boldsymbol{\beta}$ was estimated where $\boldsymbol{Y} = (Y_1, Y_2, Y_3, Y_4)^T$, t is the age in months at examination, c is the in-breeding coefficient (see Neel and Schull (1954), p71) of the individual, and $\{\boldsymbol{\mu}, \boldsymbol{\alpha}, \boldsymbol{\beta}\}$ are unknown parameters. This analysis does not allow for difference between the two cities, Hiroshima and Nagasaki; preliminary analysis failed to detect any significant difference between the two cities. For now, we assume that there is no difference between the two cities; the method of analysis given below would not be affected even if there is a difference. The null and alternative hypotheses of interest may be stated as

$$\begin{aligned} H_0 &: \text{In-breeding has no effect on growth,} \\ and \quad H_1 &: \text{In-breeding has a negative effect on growth.} \end{aligned}$$

respectively. The foregoing statement of the testing problem is broad, and it needs to be formulated in terms of some parameters so that it can be tested statistically. We formulate the problem as

$$H_0 : (\beta_1, \beta_2, \beta_3, \beta_4) = (0, 0, 0, 0) \text{ and } H_1 : (\beta_1, \beta_2, \beta_3, \beta_4) \geqslant (0, 0, 0, 0). \quad (3.34)$$

The null hypothesis will be rejected if there is evidence that at least one β_i is positive. The least squares estimate of $\boldsymbol{\beta}$ is $\hat{\boldsymbol{\beta}} = (2.710, 0.775.0.068, 0.329)^T$; the upper triangle of an estimate of the covariance matrix of $\hat{\boldsymbol{\beta}}$ is [1.199, 0.2164, 0.0839, 0.1841 | 0.0850, 0.0179, 0.0338 | 0.0232, 0.0137 | 0.0532], where "|" is used to separate different rows. The estimated covariance matrix is based on 10483 degrees of freedom. Now, since $\hat{\boldsymbol{\beta}}$ is approximately $N(\boldsymbol{\beta}, \boldsymbol{V})$, we wish to apply Corollary 3.7.2 with $\boldsymbol{X}$ replaced by $\hat{\boldsymbol{\beta}}$. Note that we have only an estimate of the covariance matrix of $\hat{\boldsymbol{\beta}}$. However, because the estimate is based on very large number of observations (more than 10000 degrees of freedom), it is reasonable to apply the results of the earlier sections with the estimated covariance matrix of $\hat{\boldsymbol{\beta}}$ being treated as the exact covariance matrix of $\hat{\boldsymbol{\beta}}$; essentially we are applying (3.33). Since each component of $\hat{\boldsymbol{\beta}}$ is positive, the estimate of $\boldsymbol{\beta}$ based on the single observation of $\hat{\boldsymbol{\beta}}$ subject to $\boldsymbol{\beta} \geqslant \mathbf{0}$ is also $\hat{\boldsymbol{\beta}}$. The sample value of the test statistic is

$$L_A = \|\hat{\boldsymbol{\beta}}\|_{\boldsymbol{V}}^2 - \min\{(\hat{\boldsymbol{\beta}} - \boldsymbol{\beta})^T \boldsymbol{V}^{-1} (\hat{\boldsymbol{\beta}} - \boldsymbol{\beta}) : \boldsymbol{\beta} \geqslant \mathbf{0}\} = \|\hat{\boldsymbol{\beta}}\|_{\boldsymbol{V}}^2 = 9.3088.$$

Therefore, the p-value, $\text{pr}(L_A \geqslant 9.3088 \mid \boldsymbol{\beta} = \mathbf{0})$, is equal to

$$w_0(4, \boldsymbol{V})\text{pr}(\chi_0^2 \geqslant 9.3088) + \ldots + w_4(4, \boldsymbol{V})\text{pr}(\chi_4^2 \geqslant 9.3088).$$

Now, we compute the weights $w_0(4, \boldsymbol{V}), \ldots, w_4(4, \boldsymbol{V})$ by simulation (see page 79). The computed values are $w_0 = 0.0062$, $w_1 = 0.0682$, $w_2 = 0.2700$, $w_3 = 0.4355$, and $w_4 = 0.2201$; a FORTRAN computer program took a few seconds to compute these weights by simulation using 50000 samples. Now,

$$\begin{aligned} p\text{-value } &= 0.0682 \times \text{pr}(\chi_1^2 \geqslant 9.3088) + 0.27 \times \text{pr}(\chi_2^2 \geqslant 9.3088) \\ &+ 0.4355 \times \text{pr}(\chi_3^2 \geqslant 9.3088) + 0.2201 \times \text{pr}(\chi_4^2 \geqslant 9.3088) = 0.026. \end{aligned}$$

Therefore, it appears that there is sufficient evidence to accept that in-breeding has a negative effect on growth. Let us note that the p-value could have been computed using Simulation 1 without computing the weights first.

Another approach to formulating and testing the foregoing hypothesis is to carry out the multiple hypotheses tests of $\beta_1 = 0$ vs $\beta_1 > 0$, $\beta_2 = 0$ vs $\beta_2 > 0$, $\beta_3 = 0$ vs $\beta_3 > 0$, and $\beta_4 = 0$ vs $\beta_4 > 0$. There are advantages and disadvantages of this approach. The main advantage is that the multiple tests attempt to identify which of $\beta_1, \ldots, \beta_4$ are positive, and hence which aspects of growth are affected and which are not by in-breeding. However, a disadvantage of the multiple testing formulation is that they require an adjustment to the individual levels of significance, such as Bonferroni-type adjustment, which is known to result in low power for the combined tests. ■

Example 3.7.2 *Meta analysis: Effect of sodium on blood pressure (Follmann (1996b)).*

A large number of studies have been carried out to evaluate the effect of sodium reduction in reducing blood pressure. Let $\boldsymbol{X}^T = (X_1, X_2)$ = (systolic blood pressure, diastolic blood pressure). The extent of evidence in favor of the positive effect of sodium reduction in reducing the blood pressure vary considerably across various studies. What is desired is an overall probability statement that summarizes the strength of evidence in favor of the hypothesis that sodium reduction reduces blood pressure. To this end, Cutler et al. (1991) collected summary information available in a large number of published studies. It is worth pointing out that meta analysis requires considerable care in selecting the sample and ensuring that the observations are independent and that they are measuring the same parameter; we shall not consider such issues here. The relevant data are given in Table 1.13 (Cutler et al. (1991)).

Let us consider the data for the normotensive subjects. There are $n = 7$ observations. Let $\boldsymbol{\theta} = (\theta_1, \theta_2)^T$ denote the mean reduction in blood pressure. Let us formulate the null and alternative hypotheses as $H_0 : \boldsymbol{\theta} = \boldsymbol{0}$ and $H_1 : \boldsymbol{\theta} \geqslant \boldsymbol{0}$, respectively. This is a Type A testing problem with two inequality constraints in H_1. Because the sample size in each of the seven studies with normotensive subjects is reasonably large, let us assume that the estimated standard error of $\hat{\theta}_{ij}$ is its true standard deviation. From ith published study we obtain an estimate $\hat{\boldsymbol{\theta}}_i$ and an estimated standard error for each component of $\hat{\boldsymbol{\theta}}_i$. Because these standard errors are not the same across different studies, we apply weighted least squares to estimate $\boldsymbol{\theta}$; each $\hat{\theta}_{ij}$ is weighted by the inverse of its variance. This leads to the following estimates:

$$\hat{\boldsymbol{\theta}} = (0.17, 1.04)^T; \hat{\boldsymbol{V}} = (0.1524, 0.0974 \mid 0.0974, 0.1510).$$

Now, the projection of $\hat{\boldsymbol{\theta}}$, with respect to $\|.\|_{\hat{\boldsymbol{V}}}$, onto H_0 is $\boldsymbol{0}$ and that onto the positive orthant is $\hat{\boldsymbol{\theta}}$ itself because both components of $\hat{\boldsymbol{\theta}}$ are positive. Direct substitution provides $\hat{\boldsymbol{\theta}}^T \hat{\boldsymbol{V}}^{-1} \hat{\boldsymbol{\theta}} = 9.97$. Let ρ denote the correlation coefficient corresponding to $\boldsymbol{V}$; thus it is the correlation coefficient between systolic and diastolic blood pressures. If ρ were known, a large sample p-value would be

$$[0.5\text{pr}\{\chi_1^2 \geqslant 9.97\} + (2\pi)^{-1}\cos^{-1}(\rho)\text{pr}\{\chi_2^2 \geqslant 9.97\}].$$

Since ρ is unknown,

$$p\text{-value} \simeq \sup_{\rho}[0.5\text{pr}\{\chi_1^2 \geqslant 9.97\} + (2\pi)^{-1}\cos^{-1}(\rho)\text{pr}\{\chi_2^2 \geqslant 9.97\}].$$

Let us make the reasonable assumption that the reductions in systolic and diastolic blood pressures are nonnegatively correlated; therefore, $\cos^{-1}\rho \leq \pi/2$. Hence

$$\sup_{\rho>0}[(1/2)\text{pr}\{\chi_1^2 \geqslant 9.97\} + \{(2\pi)^{-1}\cos^{-1}\rho\}\text{pr}\{\chi_1^2 \geqslant 9.97\}] \leq 0.001.$$

Therefore, a large sample estimate of the p-value does not exceed 0.001; there is sufficient evidence to reject H_0.

The results presented so far for testing $H_0 : \boldsymbol{\theta} = \mathbf{0}$ against $H_1 : \boldsymbol{\theta} \geqslant \mathbf{0}$ say that there is sufficient statistical evidence to reject $\boldsymbol{\theta} = \mathbf{0}$ but this alone does not mean that there is sufficient statistical evidence to accept $\boldsymbol{\theta} \gtrapprox \mathbf{0}$. Suppose that we are prepared to assume *a priori* that a reduction in sodium intake cannot increase mean systolic or diastolic blood pressure, which translates to $\boldsymbol{\theta} \geqslant \mathbf{0}$. Now, since we have already noted that there is sufficient statistical evidence to reject $\boldsymbol{\theta} = \mathbf{0}$ we can conclude that there is sufficient statistical evidence to accept $\boldsymbol{\theta} \gtrapprox \mathbf{0}$.

Now, let us suppose that we do not wish to assume *a priori* that $\boldsymbol{\theta} \geqslant \mathbf{0}$ but would like to establish that $\boldsymbol{\theta} \gtrapprox \mathbf{0}$. This type of problems will be studied in a later chapter under the topic *sign testing*; see also *combination therapy*. It will be seen that we can formulate the inference problem as test of

$$H_0 : \theta_1 \leq 0 \text{ or } \theta_2 \leq 0 \quad \text{vs} \quad H_1 : \theta_1 > 0 \text{ and } \theta_2 > 0.$$

Once the problem is formulated in this form, the so-called *min test* can be applied. In this formulation if the p-value turns out to be small then we can conclude that there is sufficient statistical evidence not only to reject $H_0 : \theta_1 \leq 0$ or $\theta_2 \leq 0$ but also to accept $H_1 : \theta_1 > 0$ and $\theta_2 > 0$. ■

3.8 LRT FOR TYPE B PROBLEMS: V IS KNOWN

Let $\boldsymbol{X} \sim N(\boldsymbol{\theta}, \boldsymbol{V})$, where $\boldsymbol{V}$ is a given positive definite matrix. We consider the Type B problem of testing

$$H_1 : \boldsymbol{\theta} \in \mathcal{C} \text{ against } H_2 : \boldsymbol{\theta} \notin \mathcal{C}. \tag{3.35}$$

In contrast to Type A problems, a main feature of Type B problems is that the null hypothesis involves inequalities. In this case, some important issues arise. To illustrate some of these, we first consider a simple bivariate case.

Let $\boldsymbol{X} = (X_1, X_2)^T$ be a bivariate random variable with density function $f(\boldsymbol{x} - \boldsymbol{\theta})$ for some f and $\boldsymbol{\theta}$, and let the null and alternative hypotheses be

$$H_1 : (\theta_1, \theta_2) \geqslant (0,0) \quad \text{and} \quad H_2 : (\theta_1, \theta_2) \not\geqslant (0,0),$$

respectively. This is a special case of (3.35) with $\mathcal{C} = \{\boldsymbol{\theta} : \boldsymbol{\theta} \geqslant \mathbf{0}\}$. Let

$$L_B = \min\{\|\boldsymbol{X} - \boldsymbol{\theta}\|^2 : \boldsymbol{\theta} \geqslant \mathbf{0}\} - \min\{\|\boldsymbol{X} - \boldsymbol{\theta}\|^2 : \boldsymbol{\theta} \in \mathbb{R}^2\}, \tag{3.36}$$

where $\|\boldsymbol{x}\|^2 = \boldsymbol{x}^T\boldsymbol{x}$. Let $\tilde{\boldsymbol{\theta}}$ be the projection of $\boldsymbol{X}$ onto the positive orthant. Then

$$L_B = \|\boldsymbol{X} - \tilde{\boldsymbol{\theta}}\|^2.$$

If $X \sim N(\theta, I)$ then $\tilde{\theta}$ is the *mle* of θ and L_B is the *LRT*; however, for now we shall not assume that X is $N(\theta, I)$. With ℓ_B denoting the sample value of L_B, one is tempted to define the p-value as $\mathrm{pr}(L_B \geqslant \ell_B \mid H_0)$. A difficulty that arises with this is that this probability depends on the particular null value, θ, which may be anywhere in the null parameter space $\{\theta : \theta \geqslant 0\}$. Thus, $\mathrm{pr}_{\theta}(L_B \geqslant \ell_B \mid \theta \in H_0)$ is not just a fixed number on the null parameter space, but a function of θ, and hence does not define a p-value. In this case, a reasonable approach to overcome the difficulty appears to be not to reject null hypothesis if there is at least one value in the null parameter space with which the data are consistent; or equivalently reject H_0 if the data are inconsistent with every value of θ in the null parameter space. In fact, the usual procedure is to define the p-value as $sup_{\theta \in H_0} \mathrm{pr}_{\theta}(L_B \geqslant \ell_B)$. At first, it may seem to be a formidable task to evaluate this supremum. Fortunately, it is reached at $\theta = \mathbf{0}$, and therefore

$$\text{p-value} = \sup_{\theta \in H_0} \mathrm{pr}_{\theta}(L_B \geqslant \ell_B) = \mathrm{pr}_{\mathbf{0}}(L_B \geqslant \ell_B). \qquad (3.37)$$

The proof of this is easy, and is given below in Theorem 3.8.1.

It is implicit in the statement (3.37) that the strength of evidence against H_0 and in favor of H_1 based on $L_B = \ell_B$ depends of the assumed true value of θ in the null parameter space and that it is *least* when $\theta = \mathbf{0}$. For this reason, $\theta = \mathbf{0}$ is called the *least favorable null value* of L_B (also, the *least favorable null configuration of* L_B), and the distribution of L_B at this value is called the *least favorable null distribution* of L_B. The foregoing result holds in higher dimensions as well for Type B testing problems even if the distribution of X is not normal; this is stated below.

Theorem 3.8.1 *Assume that the distribution of $X - \theta$ does not depend on any unknown parameters. Let the null and alternative hypotheses be $H_1 : \theta \in \mathcal{C}$ and $H_2 : \theta \notin \mathcal{C}$, respectively, where $\mathcal{C}$ is a closed convex cone. Let V be a given positive definite matrix, and let $L_B = \min\{(X - a)^T V^{-1}(X - a) : a \in \mathcal{C}\}$. Then, $pr_{\theta}(L_B \geqslant c \mid \theta \in \mathcal{C}) \leq pr_{\mathbf{0}}(L_B \geqslant c)$, and therefore the least favorable null value of L_B is $\mathbf{0}$ and*

$$p\text{-}value = sup_{\theta \in \mathcal{C}} pr_{\theta}(L_B \geqslant \ell_B) = pr_{\mathbf{0}}(L_B \geqslant \ell_B) \qquad (3.38)$$

where ℓ_B is the sample value of L_B. If $\mathcal{C}$ contains a linear space $\mathcal{M}$ then every point in $\mathcal{M}$ is also a least favorable null value for L_B.

Proof : The proof follows a set containment argument. The following proof is illustrated in Fig. 3.6 for the special case, $p = 2$, $V = I$ and $\mathcal{C} = \mathbb{R}^{+2}$. Suppose that $\theta \in \mathcal{C}$ and let $Z = X - \theta$. Then, for $c > 0$,

$$\begin{aligned}\mathrm{pr}_{\theta}(L_B \geqslant c) &= \mathrm{pr}_{\theta}[\|X - \mathcal{C}\|_V^2 \geqslant c] \\ &= \mathrm{pr}[\|(Z + \theta) - \mathcal{C}\|_V^2 \geqslant c] = \mathrm{pr}[\|Z - (\mathcal{C} - \theta)\|_V^2 \geqslant c] \\ &\leq \mathrm{pr}[\|Z - \mathcal{C}\|_V^2 \geqslant c];\end{aligned}$$

the last step follows because $\mathcal{C} \subset \mathcal{C} - \theta$ and hence the distance from Z to $\mathcal{C}$ is at least as large as that to $\mathcal{C} - \theta$. Therefore, we have that $\mathrm{pr}_{\theta}(L_B \geqslant c | \theta \in \mathcal{C}) \leq \mathrm{pr}_{\mathbf{0}}(L_B \geqslant c)$. If $\theta \in \mathcal{M}$ then $\mathcal{C} = \mathcal{C} - \theta$ and hence the last part of the theorem follows. ■.

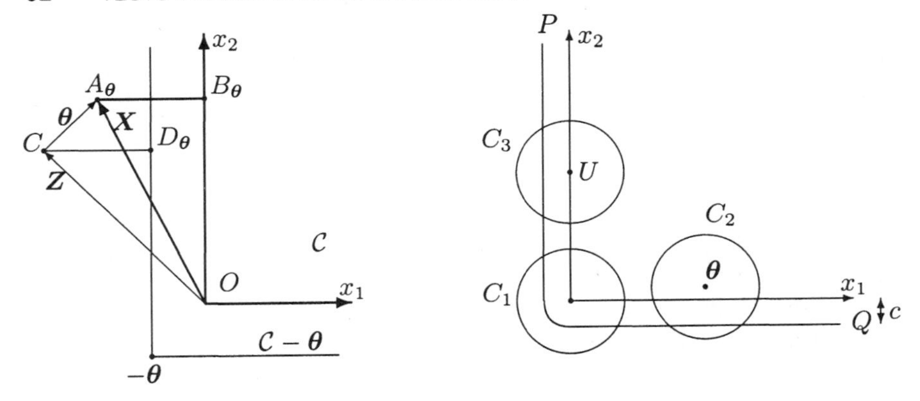

Fig. 3.11 Least favorable null value for $H_1 : \boldsymbol{\theta} \geqslant \mathbf{0}$ vs $H_2 : \boldsymbol{\theta} \ngeqslant \mathbf{0}$ is the origin

Fig. 3.12 Least favorable null value for $H_1 : \boldsymbol{\theta} \geqslant \mathbf{0}$ vs $H_2 : \boldsymbol{\theta} \ngeqslant \mathbf{0}$ is the origin

Least favorable null value when $p = 2$

The foregoing proof is illustrated in Fig. 3.11 for the special case $p = 2, \boldsymbol{V} = \boldsymbol{I}$ and $\mathcal{C} = \mathbb{R}^{+2}$. Let $\boldsymbol{Z}$ be a random variable and let it be represented by OC. Let us think of $\boldsymbol{X}$ as the sum of the deterministic part $\boldsymbol{\theta}$ in $\mathbb{R}^{+2}$ and the random part $\boldsymbol{Z}$ that is unrelated to $\boldsymbol{\theta}$. Thus, the position of the point C is not related to $\boldsymbol{\theta}$. The vector $\boldsymbol{X}$ is represented by $OA_{\boldsymbol{\theta}}$ where the position of $A_{\boldsymbol{\theta}}$ depends on $\boldsymbol{\theta}$. Let $B_{\boldsymbol{\theta}}$ and $D_{\boldsymbol{\theta}}$ be the points in $\mathbb{R}^{+2}$ and $\mathbb{R}^{+2} - \boldsymbol{\theta}$ that are closest to $A_{\boldsymbol{\theta}}$ and C, respectively. Then $L_B = \|A_{\boldsymbol{\theta}} B_{\boldsymbol{\theta}}\|^2 = \|CD_{\boldsymbol{\theta}}\|^2$; because the position of C is unrelated $\boldsymbol{\theta}$, it follows that $\|CD_{\boldsymbol{\theta}}\|^2$ is a maximum when $\boldsymbol{\theta} = \mathbf{0}$. Therefore, $\boldsymbol{\theta} = \mathbf{0}$ is the least favorable null value.

It would be instructive to provide a different geometric interpretation of the foregoing proof. To this end, let us rewrite the proof of the previous theorem slightly differently. Let c and d be positive numbers. Suppose that $\boldsymbol{\theta} \in \mathcal{C}$ and let $\boldsymbol{Z} = \boldsymbol{X} - \boldsymbol{\theta}$. Now, the critical region is of the form $\mathcal{A} = \{\boldsymbol{x} \in \mathbb{R}^p : \|\boldsymbol{x} - \mathcal{C}\| \geqslant c\}$. Again, $\mathcal{C} \subset \mathcal{C} - \boldsymbol{\theta}$ for any $\boldsymbol{\theta} \in \mathcal{C}$, and we have that

$$\text{pr}_{\boldsymbol{\theta}}(\|\boldsymbol{X} - \mathcal{C}\|_{\boldsymbol{V}} \geqslant c \text{ and } \|\boldsymbol{Z}\|_{\boldsymbol{V}} \leqslant d) = \text{pr}(\|\boldsymbol{Z} - (\mathcal{C} - \boldsymbol{\theta})\|_{\boldsymbol{V}} \geqslant c \text{ and } \|\boldsymbol{Z}\|_{\boldsymbol{V}} \leqslant d)$$

$$\leqslant \text{pr}(\|\boldsymbol{Z} - \mathcal{C}\|_{\boldsymbol{V}} \geqslant c \text{ and } \|\boldsymbol{Z}\|_{\boldsymbol{V}} \leqslant d) = \text{pr}_{\mathbf{0}}(\|\boldsymbol{X} - \mathcal{C}\|_{\boldsymbol{V}} \geqslant c \text{ and } \|\boldsymbol{Z}\|_{\boldsymbol{V}} \leqslant d).$$

Now, the proof follows by taking the limit $d \to \infty$.

This provides a different geometric interpretation of the main idea that underlies Theorem 3.8.1. Let us illustrate this using the same simple bivariate example. Since the test statistic is $L_B = \|\boldsymbol{X} - \mathbb{R}^{+2}\|^2$, the critical region is of the form $\mathcal{A} = \{\boldsymbol{x} \in \mathbb{R}^2 : \|\boldsymbol{x} - \mathbb{R}^{+2}\| \geqslant c\}$ for some $c > 0$; this is the region to the left/bottom of the curve PQ in Fig. 3.12. Let C_1 be a circle of radius d with center at the origin. Let $\boldsymbol{\theta}$ be the point in $\mathcal{C}$ as shown in Figure 3.12. Let the circle C_2 be obtained by shifting the center of C_1 to $\boldsymbol{\theta}$. Note that the part of the interior of C_1 in the critical region is larger than that corresponding to C_2, for any $\boldsymbol{\theta}$ in $\mathcal{C}$. Now, it is easily seen

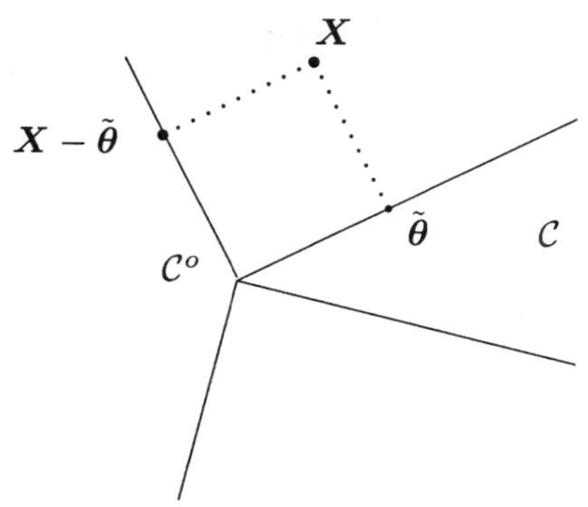

Fig. 3.13 Likelihood ratio test of $\boldsymbol{\theta} \in \mathcal{C}$ against $\boldsymbol{\theta} \notin \mathcal{C}$.

that the probability of $\boldsymbol{X}$ falling in the critical region and $\|\boldsymbol{Z}\| \leq d$ decreases as $\boldsymbol{\theta}$ moves along any straight line from the origin into the nonnegative orthant. This is true for any $d > 0$ and therefore the probability of $\boldsymbol{X}$ falling in the critical region is a maximum when $\boldsymbol{\theta} = \mathbf{0}$, and hence the least favorable null value of L_B is $\mathbf{0}$.

Least favorable null distribution of LRT when $p = 2$

Now, let us derive the least favorable null distribution of the LRT in two dimensions. Some important results do not become clear when $\mathcal{C}$ is the nonnegative orthant, $\mathbb{R}^{+2}$. Therefore, we consider an arbitrary $\mathcal{C}$. Let $\boldsymbol{X} \sim N(\boldsymbol{\theta}, \boldsymbol{I})$, $H_1 : \boldsymbol{\theta} \in \mathcal{C}$ and $H_2 : \boldsymbol{\theta} \notin \mathcal{C}$; this set-up is the same as that in Example 3.3.3 in Section 3.3 and the details therein would be helpful here as well. For testing H_1 vs H_2, we have that

$$LRT = \min\{\|\boldsymbol{X} - \boldsymbol{\theta}\|^2 : \boldsymbol{\theta} \in \mathcal{C}\} - \min\{\|\boldsymbol{X} - \boldsymbol{\theta}\|^2 : \boldsymbol{\theta} \in \mathbb{R}^2\} = \|\boldsymbol{X} - \tilde{\boldsymbol{\theta}}\|^2,$$

where $\tilde{\boldsymbol{\theta}}$ is the *mle* of $\boldsymbol{\theta}$ subject to $\boldsymbol{\theta} \in \mathcal{C}$. Note that $(\boldsymbol{X} - \tilde{\boldsymbol{\theta}})$ is also the point in $\mathcal{C}^o$ that is closest to $\boldsymbol{X}$. Fig. 3.13 shows this for the present two dimensional case; this also holds in $\mathbb{R}^p$ (see Proposition 3.12.4). Since $\mathcal{C}^o$ is a closed convex cone in $\mathbb{R}^2$, $\|\boldsymbol{X} - \tilde{\boldsymbol{\theta}}\|^2$ is the LRT for testing

$$H_0^* : \boldsymbol{\theta} = \mathbf{0} \text{ against } H_1^* : \boldsymbol{\theta} \in \mathcal{C}^o.$$

We just noted that $\|\boldsymbol{X} - \tilde{\boldsymbol{\theta}}\|^2$ is also the LRT for testing $H_1 : \boldsymbol{\theta} \in \mathcal{C}$ against $H_2 : \boldsymbol{\theta} \notin \mathcal{C}$ based on a single observation of $\boldsymbol{X}$. Therefore, the least favorable null distribution of the LRT for testing $H_1 : \boldsymbol{\theta} \in \mathcal{C}$ against $H_2 : \boldsymbol{\theta} \notin \mathcal{C}$, and the null distribution of the LRT for testing $H_0^* : \boldsymbol{\theta} = \mathbf{0}$ against $H_1^* : \boldsymbol{\theta} \in \mathcal{C}^o$ are the same.

Since the angles (in radians) of $\mathcal{C}$ and $\mathcal{C}^o$ at their vertices sum to π, we conclude from Example 3.3.3 in Section 3.3 that

$$\begin{aligned} \text{pr}(LRT \geqslant c \mid \boldsymbol{\theta} \in \mathcal{C}) &\leq \text{pr}(LRT \geqslant c \mid \boldsymbol{\theta} = \mathbf{0}) \\ &= (0.5 - q)\text{pr}(\chi_0^2 \geqslant c) + 0.5\text{pr}(\chi_1^2 \geqslant c) + q\text{pr}(\chi_2^2 \geqslant c) \end{aligned} \tag{3.39}$$

where $q = \gamma/(2\pi)$, and γ is the angle (in radians) of the cone $\mathcal{C}^o$ at the vertex. Note that for this Type B testing problem, the weights in the relevant chi-bar-square distribution appear in the reverse order to that of the corresponding Type A problem of testing $\boldsymbol{\theta} = \mathbf{0}$ against $\boldsymbol{\theta} \in \mathcal{C}$ (see Example 3.3.3).

In the foregoing example, if we were to assume that the covariance matrix of $\boldsymbol{X}$ is $\boldsymbol{V}$ rather than $\boldsymbol{I}$, then the details of Example 3.3.4 are relevant and (3.39) would still hold with q as in (3.15).

This example in two-dimensions illustrates some important results that hold in higher dimensions as well; this is stated in the next theorem.

Theorem 3.8.2 *Let $\boldsymbol{X} \sim N(\boldsymbol{\theta}, \boldsymbol{V})$ where $\boldsymbol{V}$ is a $p \times p$ positive definite matrix, and let the null and alternative hypotheses be*

$$H_1 : \boldsymbol{\theta} \in \mathcal{C} \quad \text{and} \quad H_2 : \boldsymbol{\theta} \notin \mathcal{C},$$

respectively. Then the least favorable null value of LRT is $\boldsymbol{\theta} = \mathbf{0}$. The least favorable null distribution of LRT is $\bar{\chi}^2(\boldsymbol{V}, \mathcal{C}^o)$ and

$$pr(LRT \leq c \mid \boldsymbol{\theta} = \mathbf{0}) = \sum_{i=0}^{p} w_{p-i}(p, \boldsymbol{V}, \mathcal{C}) pr(\chi_i^2 \leq c).$$

Proof: It follows from Propositions 3.4.1 (page 77) and 3.12.4 (page 116) that

$$\begin{aligned} LRT \quad &= \min\{(\boldsymbol{X} - \boldsymbol{a})^T \boldsymbol{V}^{-1} (\boldsymbol{X} - \boldsymbol{a}) : \boldsymbol{a} \in \mathcal{C}\} = \|\Pi_{\boldsymbol{V}}(\boldsymbol{X} \mid \mathcal{C}^o)\|_{\boldsymbol{V}}^2 \\ &= \boldsymbol{X}^T \boldsymbol{V}^{-1} \boldsymbol{X} - \min\{(\boldsymbol{X} - \boldsymbol{\theta})^T \boldsymbol{V}^{-1} (\boldsymbol{X} - \boldsymbol{\theta}) : \boldsymbol{\theta} \in \mathcal{C}^o\} \\ &= \bar{\chi}^2(\boldsymbol{V}, \mathcal{C}^o). \end{aligned}$$

Now, the proof follows from Corollary 3.6.2 on page 83. ■

As in Type A testing problems, if the inequalities defining $\mathcal{C}$ are all linear and independent, then the weights of the $\bar{\chi}^2$ distribution in the previous theorem take the simpler forms corresponding to the nonnegative orthant. The following corollary can be deduced from Proposition 3.6.1 (page 82).

Corollary 3.8.3 *Let $\boldsymbol{X} \sim N(\boldsymbol{\theta}, \boldsymbol{V})$. Let the null and alternative hypotheses be*

$$H_1 : \boldsymbol{R}_1 \boldsymbol{\theta} \geqslant \mathbf{0}, \boldsymbol{R}_2 \boldsymbol{\theta} = \mathbf{0} \quad \text{and} \quad H_2 : \boldsymbol{\theta} \text{ is not restricted},$$

where $\boldsymbol{R}_1$ is $s \times m$, $\boldsymbol{R}_2$ is $t \times m$, and the rank of $[\boldsymbol{R}_1^T, \boldsymbol{R}_2^T]$ is $(s+t)$. Then

$$LRT = \min\{(\boldsymbol{X} - \boldsymbol{a})^T \boldsymbol{V}^{-1} (\boldsymbol{X} - \boldsymbol{a}) : \boldsymbol{R}_1 \boldsymbol{a} \geqslant \mathbf{0}, \boldsymbol{R}_2 \boldsymbol{a} = \mathbf{0}\},$$

and the least favorable null distribution of LRT is

$$pr(LRT \leq c \mid \boldsymbol{\theta} = \mathbf{0}) = \sum_{0}^{s} w_{s-i}(s, \boldsymbol{A}) pr(\chi_{t+i}^2 \leq c),$$

where $\boldsymbol{A} = \boldsymbol{R}_1 \boldsymbol{V} \boldsymbol{R}_1^T - (\boldsymbol{R}_1 \boldsymbol{V} \boldsymbol{R}_2^T)(\boldsymbol{R}_2 \boldsymbol{V} \boldsymbol{R}_2^T)^{-1}(\boldsymbol{R}_2 \boldsymbol{V} \boldsymbol{R}_1^T)$. ■

In some situations, it is possible that linear equality constraints may be present in both the null and the alternative hypotheses as well. For example, suppose that the testing problem is

$$H_1 : \boldsymbol{\theta} \in \mathcal{C} \quad \text{vs} \quad H_2 : \boldsymbol{\theta} \in \mathcal{L}$$

where $\mathcal{L}$ is a linear subspace of $\mathbb{R}^p$ and $\mathcal{C} \subset \mathcal{L}$. In this case, the testing problem can be reformulated in the form discussed in this section. To this end, let $k = dim(\mathcal{L})$ and $\boldsymbol{A}$ be a matrix such that $rank(\boldsymbol{A}) = k(\leq p)$ and $\mathcal{L} = \{\boldsymbol{A\beta} : \boldsymbol{\beta} \in \mathbb{R}^k\}$. Then the hypotheses take the form $H_1 : \boldsymbol{\beta} \in \mathcal{P}$ and $H_2 : \boldsymbol{\beta} \notin \mathcal{P}$, where $\mathcal{P}$ is the closed convex cone, $\{\boldsymbol{\beta} : \boldsymbol{A\beta} \in \mathcal{C}\}$. Now, with L denoting $-2\log(likelihood)$, we have the following for $\boldsymbol{\theta} \in \mathcal{L}$:

$$\begin{aligned} L &= (\boldsymbol{X} - \boldsymbol{\theta})^T \boldsymbol{V}^{-1} (\boldsymbol{X} - \boldsymbol{\theta}) = (\boldsymbol{X} - \boldsymbol{A\beta})^T \boldsymbol{V}^{-1} (\boldsymbol{X} - \boldsymbol{A\beta}) \\ &= (\boldsymbol{X} - \boldsymbol{A}\hat{\boldsymbol{\beta}})^T \boldsymbol{V}^{-1} (\boldsymbol{X} - \boldsymbol{A}\hat{\boldsymbol{\beta}}) + (\hat{\boldsymbol{\beta}} - \boldsymbol{\beta})^T \boldsymbol{W}^{-1} (\hat{\boldsymbol{\beta}} - \boldsymbol{\beta}) \end{aligned}$$

where $\hat{\boldsymbol{\beta}} = (\boldsymbol{A}^T \boldsymbol{V}^{-1} \boldsymbol{A})^{-1} \boldsymbol{A}^T \boldsymbol{V}^{-1} \boldsymbol{X}$ and $\boldsymbol{W} = (\boldsymbol{A}^T \boldsymbol{V}^{-1} \boldsymbol{A})^{-1}$. Therefore, the problem is equivalent to testing $H_1 : \boldsymbol{\beta} \in \mathcal{P}$ against $H_2 : \boldsymbol{\beta} \notin \mathcal{P}$ based on a single observation of $\hat{\boldsymbol{\beta}}$ where $\hat{\boldsymbol{\beta}} \sim N(\boldsymbol{\beta}, \boldsymbol{W})$. This is of the same form as that considered in the foregoing theorem and the corollary.

Finally, let L_A denote the LRT for testing $\boldsymbol{\theta} = \boldsymbol{0}$ vs $\boldsymbol{\theta} \in \mathcal{C}$ and L_B denote the LRT for testing $\boldsymbol{\theta} \in \mathcal{C}$ vs $\boldsymbol{\theta} \notin \mathcal{C}$ based on a single observation of $\boldsymbol{X}$ from $N(\boldsymbol{\theta}, \boldsymbol{V})$. Then, by using arguments similar to the proof of Lemma 3.13.6 (page 129), it can be deduced that

$$\text{pr}(L_A \in A \text{ and } L_B \in B \mid \boldsymbol{\theta} = \boldsymbol{0}) = \sum_{i=0}^{p} w_i(p, \boldsymbol{V}, \mathcal{C}) \text{pr}(\chi_i^2 \in A) \text{pr}(\chi_{p-i}^2 \in B).$$

This was obtained by Raubertas et al. (1986). There may be some scenarios in which this may be useful. For example, if one wishes to reject $\boldsymbol{\theta} = \boldsymbol{0}$ and accept $\boldsymbol{\theta} \in \mathcal{C} \setminus \{\boldsymbol{0}\}$ if L_A is large and L_B is small, this result may have some use. However, a sketch of the critical region would show that it has an unappealing shape when $\boldsymbol{V} \neq \boldsymbol{I}$.

3.9 TESTS ON THE LINEAR REGRESSION PARAMETER

In this section it will be shown that most of the normal theory results in the previous sections extend to the normal theory linear model as well. It will be seen that most of the results concerning the LRT on the comparison of normal means studied in the previous chapter are special cases of the results presented in this section.

Let us consider the linear model,

$$\boldsymbol{Y} = \boldsymbol{X\theta} + \boldsymbol{E} \tag{3.40}$$

where $\boldsymbol{Y}$ is an $N \times 1$ vector of observations, $\boldsymbol{X}$ is a known $N \times p$ matrix of rank p with $p < N$, $\boldsymbol{\theta}$ is a $p \times 1$ vector of unknown parameters and $\boldsymbol{E}$ has mean $\boldsymbol{0}$ and covariance matrix $\sigma^2 \boldsymbol{U}$, $\boldsymbol{U}$ is known and σ is unknown. Assume that the distribution of $\boldsymbol{E}$ does not depend any unknown parameters. Let

$$\begin{array}{ll} \hat{\boldsymbol{\theta}} = (\boldsymbol{X}^T\boldsymbol{U}^{-1}\boldsymbol{X})^{-1}\boldsymbol{X}^T\boldsymbol{U}^{-1}\boldsymbol{Y}, & \boldsymbol{W} = (\boldsymbol{X}^T\boldsymbol{U}^{-1}\boldsymbol{X})^{-1}, \\ Q(\boldsymbol{\theta}) = (\boldsymbol{Y} - \boldsymbol{X}\boldsymbol{\theta})^T\boldsymbol{U}^{-1}(\boldsymbol{Y} - \boldsymbol{X}\boldsymbol{\theta}), & q(\boldsymbol{\theta}) = (\hat{\boldsymbol{\theta}} - \boldsymbol{\theta})^T\boldsymbol{W}^{-1}(\hat{\boldsymbol{\theta}} - \boldsymbol{\theta}). \end{array} \tag{3.41}$$

Using the normal equation, it is easily seen that

$$Q(\boldsymbol{\theta}) = (\boldsymbol{Y} - \boldsymbol{X}\hat{\boldsymbol{\theta}})^T\boldsymbol{U}^{-1}(\boldsymbol{Y} - \boldsymbol{X}\hat{\boldsymbol{\theta}}) + q(\boldsymbol{\theta}). \tag{3.42}$$

Let the null and alternative hypotheses be

$$H_a : \boldsymbol{\theta} \in \mathcal{C}_a \text{ and } H_b : \boldsymbol{\theta} \in \mathcal{C}_b,$$

respectively, where $\mathcal{C}_a \subset \mathcal{C}_b$. Let $\boldsymbol{\theta}^a$ and $\boldsymbol{\theta}^b$ be the points at which $Q(\boldsymbol{\theta})$ is minimized over $\mathcal{C}_a$ and $\mathcal{C}_b$, respectively. To define a test statistic based on generalized least squares without assuming normality, first note that $Q(\boldsymbol{\theta})$ is a generalized sum of squares. Therefore, $\{Q(\boldsymbol{\theta}^a) - Q(\boldsymbol{\theta}^b)\}$ is the reduction in this generalized sum of squares; it is a measure of the discrepancy between H_a and H_b. However, as it stands, this reduction cannot be used as a statistic for testing H_a against H_b because its null distribution depends on the unknown scale parameter σ^2; but, the null distribution of $\{Q(\boldsymbol{\theta}^a) - Q(\boldsymbol{\theta}^b)\}/\sigma^2$ does not depend on σ^2. Therefore, it provides a good starting point for constructing a test statistic. As in the classical least squares theory, the idea is to replace the σ^2 in the denominator of $\{Q(\boldsymbol{\theta}^a) - Q(\boldsymbol{\theta}^b)\}/\sigma^2$ by a suitable estimate of it so that the distribution of the resulting statistic is independent of σ^2. One such estimator turns out to be $Q(\boldsymbol{\theta}^a)$, and we define the following statistic:

$$\bar{E}^2 = \{Q(\boldsymbol{\theta}^a) - Q(\boldsymbol{\theta}^b)\}/Q(\boldsymbol{\theta}^a). \tag{3.43}$$

In the analysis of the classical normal theory linear models with no inequality constraints, tests on regression parameters are usually carried out using F-tests. A similar test statistic can also be developed when the hypotheses involve inequality constraints. Let

$$S^2 = \nu^{-1}(\boldsymbol{Y} - \boldsymbol{X}\hat{\boldsymbol{\theta}})^{-1}\boldsymbol{U}^{-1}(\boldsymbol{Y} - \boldsymbol{X}\hat{\boldsymbol{\theta}})$$

with $\nu = (N - p)$, the usual error mean square. Motivated by $\{Q(\boldsymbol{\theta}^a) - Q(\boldsymbol{\theta}^b)\}/\sigma^2$ as a measure of discrepancy between H_a and H_b, we define

$$\bar{F} = \{Q(\boldsymbol{\theta}^a) - Q(\boldsymbol{\theta}^b)\}/S^2. \tag{3.44}$$

In the definition of the traditional unrestricted/two-sided F-statistic for testing equality constraints, the numerator of the F-statistic is $\{Q(\boldsymbol{\theta}^a) - Q(\boldsymbol{\theta}^b)\}/k$ where k is the number of equality constraints imposed by the null hypothesis. However, we did not introduce the divisor k in the definition of $\bar{F}$ because it does not really simplify anything.

The difference between the definitions of $\bar{E}^2$ and $\bar{F}$ is that $\bar{E}^2$ uses the dispersion of the *restricted* residuals, $\boldsymbol{Y} - \boldsymbol{X}\boldsymbol{\theta}^a$, to estimate σ^2 while $\bar{F}$ uses the dispersion of the *unrestricted* residuals, $\boldsymbol{Y} - \boldsymbol{X}\hat{\boldsymbol{\theta}}$. Because of the close similarity between these statistics, we would not expect the differences between them to be substantial, at least in large samples. Wright (1988) provided some calculations to compare the powers of $\bar{E}^2$ with $\bar{F}$ in one-way lay-out models.

The basic idea of the $\bar{F}$ statistic appears to be due to Kudo (1963) who suggested it when the observations are *iid* from $N(\boldsymbol{\theta}, \sigma^2 \boldsymbol{U})$ and stated its null distribution. Wolak (1987a) applied it for inference with linearly independent constraints in the linear model, Wright (1988) considered it for ANOVA models, and Silvapulle (1996a) obtained results for the case of testing for or against inequality constraints in a more general context. Dufour (1989) obtained bounds when the parameter spaces may not be cones. Wolak (1987a) also provided a detailed discussion of $\bar{F}$ and how it relates to Kuhn-Tucker multipliers. Large sample results relating to $\bar{F}$ will be obtained in the next chapter.

So far, the discussions have been motivated by generalized least squares, without assuming that $\boldsymbol{E}$ has any known distributional form. Now, suppose that $\boldsymbol{E} \sim N(\boldsymbol{0}, \sigma^2 \boldsymbol{U})$. Then the loglikelihood $L(\boldsymbol{\theta}, \sigma^2)$ takes the form

$$L(\boldsymbol{\theta}, \sigma^2) = -(1/2)\sigma^{-2} Q(\boldsymbol{\theta}) - (1/2) np \log(\sigma^2) + Const. \qquad (3.45)$$

For any given $\boldsymbol{\theta}$, $L(\boldsymbol{\theta}, \sigma^2)$ is maximized at $\sigma^2 = (np)^{-1} Q(\boldsymbol{\theta})$. Therefore, the profile (or concentrated) loglikelihood $\tilde{L}(\boldsymbol{\theta})$ is obtained by substituting $\sigma^2 = (np)^{-1} Q(\boldsymbol{\theta})$ in $L(\boldsymbol{\theta}, \sigma^2)$. This leads to

$$\tilde{L}(\boldsymbol{\theta}) = \max_{\sigma} L(\boldsymbol{\theta}, \sigma^2) = -(1/2) np \log Q(\boldsymbol{\theta}) + Const.$$

Therefore, $\boldsymbol{\theta}^a$ and $\boldsymbol{\theta}^b$, which were introduced as generalized least squares estimates, are also the mle's of $\boldsymbol{\theta}$ under H_a and H_b, respectively. Now, the likelihood ratio statistic for testing H_a against H_b is given by

$$LRT = np \log\{Q(\boldsymbol{\theta}^a)/Q(\boldsymbol{\theta}^b)\}. \qquad (3.46)$$

Since

$$\bar{E}^2 = [1 - \exp\{-LRT/(np)\}],$$

LRT is an increasing function of $\bar{E}^2$. Therefore, the $\bar{E}^2$-test, which rejects H_a for large values of $\bar{E}^2$, is equivalent to the likelihood ratio test when $\boldsymbol{E} \sim N(\boldsymbol{0}, \sigma^2 \boldsymbol{U})$.

3.9.1 Null Distributions

Many of the results that we established for $\bar{\chi}^2$ statistics also have corresponding results for $\bar{E}^2$ and $\bar{F}$. In what follows, the suffices A and B indicate that the test statistics correspond to Type A and Type B problems, respectively. The proof of the next result is given in the Appendix (page 130).

Theorem 3.9.1 *Let the linear model be as in (3.40), and assume that the distribution of $\boldsymbol{E}/\sigma$ does not depend on any unknown parameters. Then we have the following:*
(a) For the Type A testing problem,

$$H_0 : \boldsymbol{\theta} \in \mathcal{M} \quad against \quad H_1 : \boldsymbol{\theta} \in \mathcal{C},$$

the null distributions of $\bar{E}^2$ and $\bar{F}$ do not depend on σ or on the value of $\boldsymbol{\theta}$ in the null parameter space; in other words, the tests are similar.
(b) For the Type B testing problem,

$$H_0 : \boldsymbol{\theta} \in \mathcal{C} \quad \textit{against} \quad H_1 : \boldsymbol{\theta} \notin \mathcal{C},$$

a least favorable null value of $\bar{E}^2$ and of $\bar{F}$ is $\boldsymbol{\theta} = \mathbf{0}$; if the null hypothesis also includes the linear equality constraint, $\boldsymbol{R\theta} = \mathbf{0}$, then any $\boldsymbol{\theta}$ satisfying $\boldsymbol{R\theta} = \mathbf{0}$ is also a least favorable null value. ∎

Suppose that the testing problem is as in part (a) or part (b) of the foregoing Theorem. An important consequence of this result is that the critical values and p-values can be computed by simulation for a large class of error distributions that includes the normal distribution; the simulations can be carried out at any fixed value of $(\boldsymbol{\theta}, \sigma)$. The simulation steps for computing the p-values of $\bar{F}$ and $\bar{E}^2$ corresponding to any error distribution are given below.

Computation of the p-values for $\bar{F}$ and $\bar{E}^2$ by simulation
Suppose that the conditions of Theorem 3.9.1 are satisfied. Let $F(\boldsymbol{\epsilon}/\sigma)$ denote the *cdf* of the error $\boldsymbol{E}$ where σ may be unknown, but F is assumed known; for example, the distribution of $\boldsymbol{E}$ may be $N(\mathbf{0}, \sigma^2\boldsymbol{U})$. Now, the following steps would compute the p-value for $\bar{F}$ (respectively, for $\bar{E}^2$).

1. Generate one observation of $\boldsymbol{Y}$ from $F(\boldsymbol{\epsilon})$; this is same as generating one observation $\boldsymbol{E}$ from $F(\boldsymbol{\epsilon})$ and then computing $\boldsymbol{Y}$ as in (3.40) with $\boldsymbol{\theta} = \mathbf{0}$.

2. Compute the $\bar{F}$ (respectively, $\bar{E}^2$) statistic .

3. Repeat the previous two steps N times (say $N = 10000$), and estimate the p-value by M/N where M is the number of times the $\bar{F}$ (respectively, $\bar{E}^2$) statistic in the second step exceeded its sample value. ∎

Note that in the first step of the simulation, the observations may be generated from a distribution with any value for the common scale parameter because, in view of Theorem 3.9.1, the null distributions of $\bar{F}$ and $\bar{E}^2$ do not depend on σ. Thus, if $\boldsymbol{E} \sim N(\mathbf{0}, \sigma^2\boldsymbol{U})$ where σ is unknown, then it suffices to generate $\boldsymbol{Y}$ from $N(\mathbf{0}, \boldsymbol{U})$ in the first step of the simulation for computing the p-value. The level-α critical value can also be computed by ordering the N pseudo-random values of the statistic computed in step 2, and then finding upper level-α quantile of the empirical distribution of the N values.

The forgoing results would be particularly useful in model selection. For example, suppose that we wish to test H_0 vs H_1 where

$$\begin{aligned} & H_0 : y = \theta_0 + \theta_1 x_1 + \ldots + \theta_q x_q + e, \\ \textit{and} \quad & H_1 : y = \theta_0 + \theta_1 x_1 + \ldots + \theta_q x_q + \theta_{q+1} x_{q+1} + \ldots + \theta_p x_p + e. \end{aligned}$$

Suppose also that some of $\{\theta_{q+1}, \ldots, \theta_p\}$ are known to be nonnegative. Then the task of comparing the smaller model (H_0) to the larger one (H_1) reduces to a testing problem of the type just discussed.

If we assume that the error distribution is normal, then we can derive the null distributions of $\bar{E}_A^2$ and $\bar{F}_A$. The main result is given in the next theorem, and the proof is given in the Appendix (see page 131). The derivations use the same basic approach and constructions as for deriving the $\bar{\chi}^2$ distribution.

Theorem 3.9.2 *Let the linear model be as in (3.40), and assume that* $\boldsymbol{E} \sim N(\boldsymbol{0}, \sigma^2\boldsymbol{U})$.
(a) Let the null and alternative hypotheses be

$$H_0 : \boldsymbol{\theta} \in \mathcal{M} \quad \textit{and} \quad H_1 : \boldsymbol{\theta} \in \mathcal{C}$$

respectively, where $dim(\mathcal{M}) = q$. *Then the null distributions of* $\bar{E}^2$ *and* $\bar{F}$ *are given by*

$$pr(\bar{E}_A^2 \leq c \mid H_0) = \textstyle\sum_{i=0}^{p} w_i(p, \boldsymbol{W}, \mathcal{C} \cap \mathcal{M}^{\perp}) pr[\beta\{i/2, (N-q-i)/2\} \leq c], \tag{3.47}$$

$$pr(\bar{F}_A \leq c \mid H_0) = \textstyle\sum_{i=0}^{p} w_i(p, \boldsymbol{W}, \mathcal{C} \cap \mathcal{M}^{\perp}) pr(iF_{i,\nu} \leq c) \tag{3.48}$$

where $\beta(a, b)$ *is the beta distribution with parameters* a *and* b.
(b) Let the null and alternative hypotheses be

$$H_0 : \boldsymbol{\theta} \in \mathcal{C} \quad \textit{and} \quad H_1 : \boldsymbol{\theta} \notin \mathcal{C}$$

respectively. Then $\boldsymbol{\theta} = \boldsymbol{0}$ *is a least favorable null value for* $\bar{E}_B^2$ *and* $\bar{F}_B$. *Further, the least favorable null distributions of* $\bar{E}_B^2$ *and* $\bar{F}_B$ *are:*

$$pr(\bar{E}_B^2 \leq c \mid \boldsymbol{\theta} = \boldsymbol{0}) = \textstyle\sum_{i=0}^{p} w_{p-i}(p, \boldsymbol{W}, \mathcal{C}) pr[\beta\{i/2, (N-p)/2\} \leq c],$$

$$pr(\bar{F}_B \leq c \mid \boldsymbol{\theta} = \boldsymbol{0}) = \textstyle\sum_{i=0}^{p} w_{p-i}(p, \boldsymbol{W}, \mathcal{C}) pr(iF_{i,\nu} \leq c). \qquad \blacksquare$$

As in Sections 3.7 and 3.8 on results concerning $\bar{\chi}^2$ statistics, if the constraints are linear and independent then the foregoing results take simpler forms because the weights can be expressed in terms of those corresponding to the nonnegative orthant using the results in Proposition 3.6.1; this is stated in the next corollary.

Corollary 3.9.3 *Let the linear model be as in (3.40), and assume that* $\boldsymbol{E} \sim N(\boldsymbol{0}, \sigma^2\boldsymbol{U})$.
(a) Let $\boldsymbol{R}$ *be a matrix of order* $r \times p$, $r \leq p$, $rank(\boldsymbol{R}) = r$, $\boldsymbol{R}_1$ *be a submatrix of* $\boldsymbol{R}$ *of order* $q \times p$. *Let the null and alternative hypotheses be*

$$H_0 : \boldsymbol{R\theta} = \boldsymbol{0} \textit{ and } H_1 : \boldsymbol{R}_1\boldsymbol{\theta} \geqslant \boldsymbol{0}.$$

Then we have the following for the null distributions of $\bar{E}_A^2$ *and* $\bar{F}_A$*:*

$$pr(\bar{E}_A^2 \leq c | \boldsymbol{R\theta} = \boldsymbol{0}) = \textstyle\sum_{i=0}^{q} w_i(q, \boldsymbol{R}_1\boldsymbol{V}\boldsymbol{R}_1^T) pr[\beta\{(r-q+i)/2, (N-p+q-i)/2\} \leq c]$$

$$pr(\bar{F}_A \leq c | \boldsymbol{R\theta} = \boldsymbol{0}) = \textstyle\sum_{i=0}^{q} w_i(q, \boldsymbol{R}_1\boldsymbol{V}\boldsymbol{R}_1^T) pr[(r-q+i)F_{r-q+i,\nu} \leq c].$$

(b) Now, let the null and alternative hypotheses be

$$H_1 : \boldsymbol{R}_1\boldsymbol{\theta} \geqslant \boldsymbol{0}, \boldsymbol{R}_2\boldsymbol{\theta} = \boldsymbol{0} \textit{ and } H_2 : \boldsymbol{\theta} \textit{ is not restricted},$$

where $\boldsymbol{R}_1$ is $s \times p$, $\boldsymbol{R}_2$ is $t \times p$, and $[\boldsymbol{R}_1^T, \boldsymbol{R}_2^T]$ has full rank. Then, any $\boldsymbol{\theta}$ satisfying $\boldsymbol{R}_1\boldsymbol{\theta} = \mathbf{0}$ and $\boldsymbol{R}_2\boldsymbol{\theta} = \mathbf{0}$ is a least favorable null value for $\bar{E}_B^2$ and $\bar{F}_B$. Further,

$$pr(\bar{E}_B^2 \leq c|\boldsymbol{\theta} = \mathbf{0}) = \textstyle\sum_{i=0}^{s} w_{s-i}(s, \boldsymbol{A})pr[\beta\{(t+i)/2, (N-k)/2\} \leq c],$$

$$pr(\bar{F}_B \leq c|\boldsymbol{\theta} = \mathbf{0}) = \textstyle\sum_{i=0}^{s} w_{s-i}(s, \boldsymbol{A})pr[(t+i)F_{t+i,\nu} \leq c]. \qquad \blacksquare$$

The foregoing results for Type B testing problems are useful even if some equality constraints on the regression parameters are present under both the null and alternative hypotheses. For example, consider the Type B testing problem and suppose that some linear equality constraints are present in the null and alternative hypotheses. Then the hypotheses take the form $H_1 : \boldsymbol{\theta} \in \mathcal{C}$ and $H_2 : \boldsymbol{\theta} \in \mathcal{L}$, where $\mathcal{L} = \{\boldsymbol{A\beta} : \boldsymbol{\beta} \in \mathbb{R}^k\}$ for some matrix $\boldsymbol{A}$ of order $p \times k$ and $rank(\boldsymbol{A}) = k$. Let $\mathcal{P} = \{\boldsymbol{\beta} : \boldsymbol{A\beta} \in \mathcal{C}\}$. Then the linear model takes the form $\boldsymbol{Y} = \boldsymbol{B\beta} + \boldsymbol{E}$, where $\boldsymbol{B} = \boldsymbol{XA}$, and the null and alternative hypotheses take the standard form, $H_0 : \boldsymbol{\beta} \in \mathcal{P}$ and $H_1 : \boldsymbol{\beta} \notin \mathcal{P}$. Now we can apply the foregoing results.

3.10 TESTS WHEN V IS UNKNOWN (PERLMAN'S TEST AND ALTERNATIVES)

In the foregoing sections we considered the cases when the covariance matrix $\boldsymbol{V}$ is either known or of the form $\sigma^2\boldsymbol{U}$ where $\boldsymbol{U}$ is known and σ is unknown. Now we consider the case when $\boldsymbol{V}$ is completely unknown.

Let $\boldsymbol{X}_1, \ldots, \boldsymbol{X}_n$ be n independently and identically distributed observations from $N(\boldsymbol{\theta}, \boldsymbol{V})$ where $\boldsymbol{V}$ is unknown. Exact results for the likelihood ratio test were obtained by Perlman in his seminal paper, Perlman (1969). Corresponding results for a closely related test were obtained by Silvapulle (1995); conditional version of this test was introduced by Perlman and Wu (2002a). Other procedures include Tang (1994), Wang and McDermott (1998a, 1998b), Tang et al. (1989a), O'Brien (1984), and Follmann (1996b).

Let us consider the general problem of testing

$$H_a : \boldsymbol{\theta} \in \mathcal{C}_a \text{ against } H_b : \boldsymbol{\theta} \in \mathcal{C}_b \tag{3.49}$$

where $\mathcal{C}_a$ and $\mathcal{C}_b$ are *one-sided* closed cones; a set A is said to be *one-sided* if there exists $\boldsymbol{a}$ such that $A \subset \{\boldsymbol{x} : \boldsymbol{a}^T\boldsymbol{x} \geqslant 0\}$. In this section, the cones considered are all *one-sided*. Since $\boldsymbol{V}$ is unknown, the loglikelihood up to a constant is

$$\ell(\boldsymbol{\theta}, \boldsymbol{V}) = \sum_{i=1}^{n}(-1/2)\{\log|\boldsymbol{V}| + (\boldsymbol{X}_i - \boldsymbol{\theta})^T\boldsymbol{V}^{-1}(\boldsymbol{X}_i - \boldsymbol{\theta})\}. \tag{3.50}$$

Further, we have that

$$LRT = 2[\max\{\ell(\boldsymbol{\theta}, \boldsymbol{V}) : \boldsymbol{\theta} \in \mathcal{C}_b, \boldsymbol{V} > 0\} - \max\{\ell(\boldsymbol{\theta}, \boldsymbol{V}) : \boldsymbol{\theta} \in \mathcal{C}_a, \boldsymbol{V} > 0\}],$$

where $\boldsymbol{V} > 0$ means that $\boldsymbol{V}$ is positive definite. Note that the maximization needs to be carried out over $\boldsymbol{\theta}$ and over all the positive definite matrices, $\boldsymbol{V} > 0$, because now

V is an unknown parameter; therefore, this *LRT* is not the same as that for the case when V is known. Let

$$S = n^{-1} \sum (X_i - \bar{X})(X_i - \bar{X})^T$$

denote the sample covariance matrix, and let

$$\mathcal{U}(\mathcal{C}_a, \mathcal{C}_b) = \{\|\Pi_S(\bar{X}|\mathcal{C}_b)\|_S^2 - \|\Pi_S(\bar{X}|\mathcal{C}_a)\|_S^2\}\{1 + \|\bar{X} - \|\Pi_S(\bar{X}|\mathcal{C}_b)\|_S^2\}^{-1}. \tag{3.51}$$

Now, we have the following that simplifies the *LRT*.

Proposition 3.10.1 *(Perlman (1969), Theorem 5.1). $\mathcal{U}(\mathcal{C}_a, \mathcal{C}_b)$ is a strictly increasing function of LRT for testing $H_a : \theta \in \mathcal{C}_a$ against $H_b : \theta \in \mathcal{C}_b$.* ■

It follows that the likelihood ratio test is equivalent to rejecting H_a for large values of $\mathcal{U}(\mathcal{C}_a, \mathcal{C}_b)$. Therefore, to apply the *LRT*, we do not have to compute the maximum of the function, $\ell(\theta, V)$, or any other function of V over the complicated parameter space, $\{V > 0\}$. If the constraints in $\mathcal{C}_a$ and $\mathcal{C}_b$ are linear, then $\mathcal{U}(\mathcal{C}_a, \mathcal{C}_b)$ can be computed using only a quadratic program. The foregoing result does not require observations to be *iid* but it does require an observation of Y from $N(\sqrt{n}\theta, V)$, and an observation S from the Wishart distribution $W(n-1, V)$ where Y and S are independent.

Now, let us introduce *Perlman's test* as follows:

Reject $H_0 : \theta \in \mathcal{C}_a$ in favor of $H_1 : \theta \in \mathcal{C}_b$ for large values of $\mathcal{U}(\mathcal{C}_a, \mathcal{C}_b)$.

Before we study properties of this test, let us introduce another class of tests. In classical multivariate analysis, tests of $\theta = \mathbf{0}$ against $\theta \neq \mathbf{0}$ are carried out using the well-known Hotelling's T^2 statistic, $\bar{X}^T S^{-1} \bar{X}$. For this unrestricted testing problem, the LRT reduces to $\bar{X}^T S^{-1} \bar{X}$, the Hotelling's T^2. To motivate another test, let us note that, since S is positive definite, $\|.\|_S^2 = \bar{X}^T S^{-1} \bar{X}$ is the squared distance between $\bar{X}$ and the null parameter space. A similar statistic is available for the two-sample problem. Based on an idea of Shorack (1967), generalizations of the statistic $\bar{X}^T S^{-1} \bar{X}$ have been proposed when there are inequalities in θ (see Silvapulle (1995)). The basic form of the statistic is quite simple:

$$T^{*2} = \|\bar{X} - \mathcal{C}_a\|_S^2 - \|\bar{X} - \mathcal{C}_b\|_S^2. \tag{3.52}$$

In other words, T^{*2} is the difference between "the squared distance from $\bar{X}$ to the null parameter space" and "the squared distance from $\bar{X}$ to the alternative parameter space." Let us remark that the foregoing T^{*2} and the Hotelling T^2 have similarities with respect to their algebraic forms; but, Hotelling's T^2 is based on likelihood ratio and the T^{*2} in (3.52) is not.

For the special case of testing $H_0 : \theta = \mathbf{0}$ against $H_1 : \theta \geqslant \mathbf{0}$, Sen and Tsai (1999), obtained a union-intersection test. Their test statistic, which they denote by T_n^*, is the same as T^{*2}; hence the latter can also be motivated from the union-intersection principle.

Perlman (1969) obtained the null distribution of $\mathcal{U}(\mathcal{C}_a, \mathcal{C}_b)$ for some important testing problems. Several corresponding results for the T^{*2} were obtained by Silvapulle (1995). The important ones are discussed below. Let U denote either the $\mathcal{U}(\mathcal{C}_a, \mathcal{C}_b)$ in (3.51) or the T^{*2} in (3.52). Let u denote the sample value of U. In general, the null distribution of U depends on $(\boldsymbol{\theta}, \boldsymbol{V})$. Therefore, $\text{pr}(U \geqslant u \mid H_0, \boldsymbol{V})$ is not an operational p-value. A suitable test procedure is,

$$\text{reject } H_0 \text{ if } \sup\{\text{pr}(U \geqslant u \mid \boldsymbol{\theta}, \boldsymbol{V}) : \boldsymbol{\theta} \in H_0, \boldsymbol{V} > 0\} \text{ is small,} \tag{3.53}$$

where the supremum is taken over $\boldsymbol{\theta}$ in H_0 and over all positive definite matrices $\boldsymbol{V}$; for example, see Bickel and Doksum (1977) page 170. Similarly, if $\inf\{\text{pr}(U \geqslant u|\boldsymbol{\theta}, \boldsymbol{V}) : \boldsymbol{\theta} \in H_0, \boldsymbol{V} > 0\}$ is large then do not reject H_0. Otherwise, the test is not conclusive. In some cases, for example, in Type A testing problems, $\text{pr}(U \geqslant u|H_0, \boldsymbol{V})$ does not depend on the value of $\boldsymbol{\theta}$ in the null parameter space but it depends on $\boldsymbol{V}$. In such cases, we shall also consider the possibility of estimating $\text{pr}(U \geqslant u|H_0, \boldsymbol{V})$ by substituting $\hat{\boldsymbol{V}}$ for $\boldsymbol{V}$, but it does require caution (see Section 4.3.2).

The null distributions of *LRT* and T^{*2} for inequality constrained problems, involve the following two random variables:

$$G(i, j) = \chi_i^2/\chi_j^2 \quad \text{and} \quad H(k, r, n) = (\chi_k^2/\chi_{n-r}^2)(1 + \chi_{r-k}^2/\chi_{n-r+k}^2),$$

where the different χ^2 variates are independent. For any given c, the tail probabilities $\text{pr}\{G(i, j) \geqslant c\}$ and $\text{pr}\{H(k, r, n) \geqslant c\}$ can be computed easily by simulation because generation of random numbers from independent χ^2 variates is straight forward. Consequently, it will be seen that, in terms of computational demands, the difference between T^{*2} and LRT is small; the choice between the two would need to be based on their statistical properties.

It is also worth noting that $(j/i)G(i, j)$ is the usual F distribution with (i, j) degrees of freedom which we denote by $F_{i,j}$. Therefore,

$$\text{pr}\{G(i, j) \geqslant c\} = \text{pr}((i/j)F_{i,j} \geqslant c),$$

and hence the F-distribution can be used to compute the tail probabilities of $G(i, j)$.

3.10.1 Type A Testing Problem

To consider a special Type A problem, let us partition $\boldsymbol{\theta}$ as $(\boldsymbol{\theta}_1^T, \boldsymbol{\theta}_2^T)^T$ where $\boldsymbol{\theta}_1$ is $q \times 1$ and $\boldsymbol{\theta}_2$ is $r \times 1$; let $\boldsymbol{V}, \boldsymbol{S}$ and $\boldsymbol{X}$ be partitioned conformably. Now, consider the test of

$$H_0 : \boldsymbol{\theta}_2 = \mathbf{0} \text{ against } H_1 : \boldsymbol{\theta}_2 \in \mathcal{P} \tag{3.54}$$

where $\mathcal{P}$ is a one-sided closed convex cone. Then, it is easily shown that the null distributions of *LRT* and T^{*2} do not depend on $\boldsymbol{\theta}$ in the null parameter space; the argument is virtually the same as for Proposition 3.7.3. A bounds test can be carried out using the bounds given in the next result.

Proposition 3.10.2 *(Perlman (1969) Theorem 6.3, and Silvapulle (1995) Theorem 3.) Let the testing problem be as in (3.54), and let $\mathcal{U}$ denote $\mathcal{U}(\mathcal{C}_a, \mathcal{C}_b)$ for this testing* *problem. Then under* H_0,

$$\begin{array}{ll}\inf_{\theta,V} pr(\mathcal{U} \geqslant u \mid H_0, V) & = (1/2)pr\{G(1, n-r) \geqslant u\}, \\ \inf_{\theta,V} pr(T^{*2} \geqslant u \mid H_0, V) & = (1/2)pr\{H(1, r, n) \geqslant u\}.\end{array} \tag{3.55}$$

If $\mathcal{P}$ *contains an open set of dimension* r *then*

$$\begin{array}{l}\sup_{\theta,V} pr(\mathcal{U} \geqslant u \mid H_0, V) = \\ \quad (1/2)[pr\{G(r-1, n-r) \geqslant u\} + pr\{G(r, n-r) \geqslant u\}], \\ \sup_{\theta,V} pr(T^{*2} \geqslant u \mid H_0, V) = \\ \quad (1/2)[pr\{H(r-1, r, n) \geqslant u\} + pr\{G(r, n-r) \geqslant u\}].\end{array} \tag{3.56}$$

■

The results in (3.55) and (3.56) are adequate to carry out the test as in (3.53). These two results are still quite general because the parameter space $\mathcal{P}$ in (3.54) may include nonlinear inequalities in θ. In most practical situations, the inequalities in θ are linear and independent. In this case we can obtain closed forms for the null distributions of $\mathcal{U}$ and T^{*2}; these are given in the next result.

Proposition 3.10.3 *(Perlman (1969) Corollary 7.6, and Silvapulle (1995)). Let the null and alternative hypotheses be*

$$H_0 : \theta_2 = \mathbf{0} \text{ and } H_1 : A\theta_2 \geqslant \mathbf{0} \tag{3.57}$$

respectively, where A *is a given nonsingular matrix. Let* $\mathcal{U}$ *denote the* $\mathcal{U}(\mathcal{C}_a, \mathcal{C}_b)$ *for this testing problem. Then*

$$pr(\mathcal{U} \geqslant u \mid H_0, V) = \sum_{i=1}^{r} pr\{G(i, n-r) \geqslant u\} w_i(r, W), \tag{3.58}$$

$$pr(T^{*2} \geqslant u \mid H_0, V) = \sum_{i=1}^{r} pr\{H(k, r, n) \geqslant u\} w_i(r, W). \tag{3.59}$$

where $W = A^{-1}V_{22}(A^{-1})^T$, *and* V_{22} *is the bottom-right submatrix of* V *obtained by partitioning* V *to conform with the partitioning* $\theta = (\theta_1^T, \theta_2^T)$. ■

Note that the expressions on the right-hand side of (3.58) depend on W but not on θ in the null parameter space. Therefore, we have to use its infimum and supremum over $V > 0$, given in (3.55) and (3.56), to obtain bounds on the p-value and perform the test as in (3.53). These bounds can be quite far from those corresponding to the true W if the sample size is small. This is the price that we pay for not knowing anything about the covariance matrix, V. One approach to improve the situation is to narrow down the class of matrices allowed for V based on nonsample information, and then take the supremum of (3.58) over that smaller class. Finding and evaluating the *exact* supremum over such a restricted set of matrices would typically be difficult, but it should be possible to generate a well-spread finite number of matrices within that smaller class, evaluate (3.58) for each V and take the maximum as an estimate of the p-value. If the sample size is large, an estimate of (3.58) is obtained by substituting an estimate $\hat{W}$ of W for W in $w_i(r, W)$. One needs to be cautious in using this as a p-value because if $\hat{W}$ is far from W, the resulting estimate of the p-value could be a poor estimate. A more desirable approach would be to compute the expressions on the right-hand side of (3.58) for a range of possible values of $\boldsymbol{W}$ and take the maximum of these values. Even this has a certain degree of subjective element, but it is still better than simply substituting an estimate for $\boldsymbol{W}$. A formal way of doing this is to take the supremum over a confidence region for $\boldsymbol{W}$ and make some adjustments for using only a confidence region rather than all possible $\boldsymbol{W}$. A rigorous development of this procedure is provided in the next chapter (see Silvapulle (1996b) and Berger and Boos (1994)).

The *LRT* is biased when $\mathcal{P}$ contains a non-empty open set; see Perlman (1969), p 558. There are points in the alternative parameter space that are close to the null space at which the restricted test, (3.53), has less power than the unrestricted one. However, for the restricted testing problem, we would still prefer the restricted tests to the unrestricted one because the former would have better overall performance (see Silvapulle (1995) and Perlman and Wu (2002a)). The results of a simulation study shows that *LRT* is neither dominated by nor dominates T^{*2}. For *Type A* problems, some calculations show that the T^{*2} is likely to be more powerful than the *LRT* for values for $\boldsymbol{\theta}$ near the boundary; for values of $\boldsymbol{\theta}$ in the central direction, *LRT* is likely to be more powerful than the T^{*2} (see Silvapulle (1995)). At this stage we do not have sufficient results to recommend one over the other.

3.10.2 Type B Testing Problem

Now, consider the Type B problem of testing

$$H_1 : \boldsymbol{\theta}_2 \in \mathcal{P} \text{ against } H_2 : \boldsymbol{\theta}_2 \notin \mathcal{P}, \tag{3.60}$$

where as in the previous subsection we assume that $\boldsymbol{\theta}$ is partitioned as $(\boldsymbol{\theta}_1^T, \boldsymbol{\theta}_2^T)^T$ and $\mathcal{P}$ is assumed to be a one-sided closed cone. In the notation introduced at the beginning of this section, $\mathcal{C}_a = \mathcal{P} \subset \mathcal{C}_b = \mathbb{R}^r$. Let $\mathcal{U}$ denote $\mathcal{U}(\mathcal{C}_a, \mathcal{C}_b)$ for this testing problem and let u denote the sample value of $\mathcal{U}$. It follows from (3.51) and (3.52) that T^{*2} and $\mathcal{U}$ are equal. The null distribution of $\mathcal{U}$ depends on $(\boldsymbol{\theta}_2, \boldsymbol{V})$ and therefore, the test is

$$\text{reject } H_0 \text{ if } \sup_{\boldsymbol{\theta}_2 \in \mathcal{P}, \boldsymbol{V} > 0} \text{pr}(\mathcal{U} \geqslant u \mid \boldsymbol{\theta}, \boldsymbol{V}) \text{ is small.} \tag{3.61}$$

The main result on the null distribution of $\mathcal{U}$ is the following:

Proposition 3.10.4 *(Perlman (1969), section 8). Let $\mathcal{U}$ denote $\mathcal{U}(\mathcal{C}_a, \mathcal{C}_b)$ for the testing problem (3.60). Then we have that,*

1. $\mathcal{U} = min\{(\bar{\boldsymbol{X}}_2 - \boldsymbol{a})^T \boldsymbol{S}_{22}^{-1} (\bar{\boldsymbol{X}}_2 - \boldsymbol{a}) : \boldsymbol{a} \in \mathcal{P}\}$.

2. $pr(U \geqslant u \mid H_0, \boldsymbol{V}) \leq pr(U \geqslant u \mid \boldsymbol{\theta} = \boldsymbol{0}, \boldsymbol{V})$, *for any* $\boldsymbol{V}$.

3. $\sup_{\boldsymbol{\theta}_2 \in \mathcal{C}, \boldsymbol{V} > 0} pr(U \geqslant u \mid \boldsymbol{\theta}, \boldsymbol{V}) \leq$

$$(1/2)[pr\{G(r-1, n-r-1) \geqslant u\} + pr\{G(r, n-r) \geqslant u\}].$$

4. *If $\mathcal{P}$ contains an open set of dimension r, then*

$$\inf_{\theta_2 \in \mathcal{P}, V>0} pr(LRT \geqslant t \mid H_0, V) = 0.$$ ■

There are other procedures for testing when V is unknown. None of the tests is uniformly best. A conditional version of the T^{*2} was suggested by Perlman and Wu (2002a); they also compared the performance of several statistics for this testing problem. Likelihood based conditional tests have been proposed by Wang and McDermott(1998a,b) ; the implementation of this procedure is difficult and there appears to be some issues that need to be resolved (see Perlman and Wu (1999)). Other procedures have also been suggested, such as the simple test in Tang (1994). Although the test in Tang (1994) is uniformly more powerful than the likelihood ratio test, it does not follow that this test is better than the likelihood ratio test or the T^{*2} test. These will be discussed later.

3.10.3 Conditional Tests of $H_0 : \theta = 0$ vs $H_1 : \theta \geq 0$

An approach to dealing with the unknown nuisance parameter V is to eliminate it from the null distribution by conditioning on a sufficient statistic for V or modify the test so that the probability of Type I error, as a function of V, is as flat as possible. Tests of the former type were developed by Wang and McDermott(1998a,b), and those of the latter type were developed by Perlman and Wu (2002a). These two types of tests are discussed below in turn.

Wang-McDermott test:

Let $X_1, \ldots, X_n$ be *iid* as X where $X \sim N(\theta, V)$. Let the null and alternative hypotheses be

$$H_0 : \theta = 0 \qquad \text{and} \qquad H_1 : \theta \geq 0$$

respectively. Let $W = \sum X_i X_i^T$, $S = W - n\bar{X}\bar{X}^T$, $\tilde{\theta}$ be the *mle* of θ subject to $\theta \geqslant 0$,

$$\mathcal{U} = n\tilde{\theta}^T S^{-1}\tilde{\theta}\{1 + (\bar{X} - \tilde{\theta})^T S^{-1}(\bar{X} - \tilde{\theta})\}^{-1},$$

and u_{obs} be the observed value of $\mathcal{U}$. It was shown earlier that the *unconditional* LRT of H_0 against H_1 is, *reject H_0 for large values of $\mathcal{U}$*. For this test, the p-value is $\sup_{V>0} \text{pr}_V(\mathcal{U} \geqslant u_{obs})$.

Under H_0, W is a complete sufficient statistic for V, and hence the distribution of $\mathcal{U}$ conditional on $W = w$ does not depend on V or any other unknown parameters. Now, the level-α *conditional* LRT of Wang and McDermott is the following:

$$\text{reject } H_0 \text{ if } \mathcal{U} \geqslant c_\alpha(w) \text{ where } \text{pr}_0\{\mathcal{U} \geqslant c_\alpha(w) \mid W = w\} = \alpha.$$

Note that the probability on the LHS does not depend on V. Therefore, the p-value for this conditional test is

$$\text{pr}_0(\mathcal{U} \geqslant u_{obs} \mid W = w). \tag{3.62}$$

It turns out that the density of $\bar{\boldsymbol{X}}$ under H_0, conditional on $\boldsymbol{W} = \boldsymbol{w}$, has a simple closed form. Further, $\mathcal{U}$ has a simple closed-form for the purposes of computing it. Therefore, the p-value in (3.62) can be computed by using a simple brute force numerical integration or Monte Carlo; an easy-to-implement algorithm for computing the foregoing p-value is given in Wang and McDermott (1998a).

The power function of this conditional test approaches 1 uniformly in $(\boldsymbol{\theta}, \boldsymbol{V})$ as $\boldsymbol{\theta}^T \boldsymbol{V}^{-1}\boldsymbol{\theta} \to \infty$; further, the test is consistent. However, it is biased; see Sen and Tsai (1999).

It also turns out that $c^*_\alpha \geqslant c_\alpha(\boldsymbol{w})$ where c^*_α and $c_\alpha(\boldsymbol{w})$ are the level-α critical values of the unconditional and conditional LRTs, respectively. Since the test statistics for the two tests are the same, it follow that the conditional LRT is uniformly more powerful than the unconditional test. It does not automatically follow that the unconditional test is better because the probability of Type I error is higher for the unconditional test although both tests have the same size.

Conditional versions of the T^{*2} discussed in the previous subsections were studied by Wang and McDermott (1998b). Essentially the same ideas are applicable. Briefly the test is
reject H_0 *if* $T^{*2} > c'_\alpha$ *where* c'_α *satisfies* $pr_{\mathbf{0},\boldsymbol{V}}\{T^{*2} > c'_\alpha \mid \boldsymbol{W} = \boldsymbol{w}\} = \alpha$.

Perlman-Wu test:

Now, let us consider the class of conditional tests proposed by Perlman and Wu (2002a). These authors proposed a class of tests in which they condition on K, where K is the number of positive components of $\tilde{\boldsymbol{\theta}}$. In contrast to the statistic $\boldsymbol{W}$ in the foregoing discussions, the statistic K is not sufficient for $\boldsymbol{V}$, and the resulting conditional test turns out to be not similar. However, Perlman and Wu (2002a) identify a particular member of this class of conditional tests that is nearly similar, in a sense to be made precise in the theorems stated below.

Let C_α denote the class of tests of H_0 against H_1 for which the test statistic is $\mathcal{U}$ and they are conditional on K. Let $\mathbf{c} = (c_1, \ldots, c_k)$ where $c_i > 0$ for $i = 1, \ldots, p$. Now, define the conditional test of Perlman and Wu as follows:

Do not reject H_0 if $K = 0$, and reject H_0 if $K = k$ and $\mathcal{U} > c_k$ for $k = 1, \ldots, p$.

Let $T(\boldsymbol{c})$ denote this test. The unconditional LRT is a special case of this corresponding to $c_1 = \ldots = c_k$. Each choice of $\mathbf{c}$ defines a conditional test and it is not similar. One way of choosing an optimal $\mathbf{c}$ is to find that value for which the minimum probability of Type I error over the null parameter space is a maximum over all possible values of $\{\boldsymbol{c}\}$. Perlman and Wu (2002a) obtained results in this direction.

Theorem 3.10.5 *(Perlman and Wu (2002a)). For any* α, $0 < \alpha < 1$, *we have that*

$$\max_{T(\boldsymbol{c}) \in C_\alpha} \inf_{\boldsymbol{V}>0} pr_{\mathbf{0},\boldsymbol{V}}\{T(\boldsymbol{c}) \text{ rejects } H_0\} = \alpha/2.$$

Further, this maximum is attained only by the test $T(\mathbf{c}_\alpha)$ *where* $\mathbf{c}_\alpha = (c_{1,\alpha}, \ldots, c_{p,\alpha})^T$ *and* $pr(\chi^2_k/\chi^2_{n-p} \geqslant c_{k,\alpha}) = \alpha$ *where* χ^2_k *and* χ^2_{n-p} *are independent (* $k = 1, \ldots, p$*).*
■

The same conditioning approach is applicable to the statistic T^{*2}. The Perlman and Wu (2002a) type conditional version of T^{*2} is the following:

Do not reject H_0 if $K = 0$, and reject H_0 if $K = k$ and $T^{*2}(\boldsymbol{c}) > c_k$

for $k = 1, \ldots, p$. The optimal choice is given in the next theorem.

Theorem 3.10.6 *(Perlman and Wu (2002a)). For any α, $0 < \alpha < 1$, we have that*

$$\max_{T^{*2}(\boldsymbol{c}) \in C_\alpha} \inf_{\boldsymbol{V}>0} pr_{\boldsymbol{0},\boldsymbol{V}}\{T^{*2}(\boldsymbol{c}) \text{ rejects } H_0\} = \alpha/2.$$

Further, this maximum is attained only by the test $T^{*2}(\boldsymbol{c}_\alpha^*)$ where $\boldsymbol{c}_\alpha^* = (c_{1,\alpha}^*, \ldots, c_{p,\alpha}^*)^T$ and $pr[(\chi_k^2/\chi_{n-p}^2)\{1+(\chi_{p-k}^2/\chi_{n-p-k}^2)\} \geqslant c_{k,\alpha}^*] = \alpha$ where the various χ^2 variates are independent ($k = 1, \ldots, p$). ∎

The forgoing conditional tests are unlikely to be admissible. The acceptance regions of $T(\boldsymbol{c})$ and $T^{*2}(\boldsymbol{c})$ are neither monotonic nor convex. In fact, Perlman and Wu (2002a) conjecture that the conditional tests of Wang and McDermott (1998a,b) and the unconditional tests, T^{*2} and LRT, are also inadmissible. A simulation study in Perlman and Wu (2002a) compared the performance of the foregoing conditional and unconditional tests; they observed that overall, $T(\boldsymbol{c})$ and $T^{*2}(\boldsymbol{c})$ performed better.

3.11 OPTIMALITY PROPERTIES

In the classical unrestricted setting, the parameter spaces tend to be open and several tests tend to have various optimality properties such as UMP, UMPI, unbiasedness, consistency, monotonic power function, etc. However, these optimality properties do not automatically carry over when the parameter space is not open. Further, often different optimality/desirable properties cannot be achieved without sacrificing some other desirable properties. In this section we shall consider consistency and monotonicity of power function. A thorough investigation of all these relevant issues is outside the scope of this book.

3.11.1 Consistency of Tests

Let n denote the sample size and let a level-α test of $H_0 : \boldsymbol{\theta} \in \Theta_0$ against $H_1 : \boldsymbol{\theta} \notin \Theta_0$ be " reject H_0 if $T_n \geqslant c_n$" where T_n is a test statistic and c_n is the critical value. As the sample size increases, one would expect that the information in the sample about $\boldsymbol{\theta}$ would also increase, and, hence, if T_n is a good test statistic then the probability of rejecting H_0 would tend 1 as $n \to \infty$ for $\boldsymbol{\theta} \notin \Theta_0$. This leads to the following definition. A test of $H_0 : \boldsymbol{\theta} \in \Theta_0$ is said to be *consistent* at $\boldsymbol{\theta}$ ($\notin \Theta_0$) if the probability of rejecting the null hypothesis tends to 1 as $n \to \infty$ when the true value of the parameter is $\boldsymbol{\theta}$. Tests under inequality constraints are usually consistent at points in the alternative parameter space, but not necessarily at every point in the parameter space. Let us consider a simple example in two-dimensions to illustrate the main idea.

Example 3.11.1 *Consistency of the LRT in 2-dimensions.*

Let $\boldsymbol{X}_1, \ldots, \boldsymbol{X}_n$ be *iid* as $N(\boldsymbol{\theta}, \boldsymbol{I})$, and let the null and alternative hypotheses be defined as

$$H_0 : \boldsymbol{\theta} = \boldsymbol{0} \text{ and } H_1 : \boldsymbol{\theta} \geqslant \boldsymbol{0},$$

respectively, where $\boldsymbol{\theta} \in \mathbb{R}^2$. Then, $LRT = n\{\|\bar{\boldsymbol{X}}\|^2 - \|\bar{\boldsymbol{X}} - \mathbb{R}^{+2}\|^2\}$. First, let us consider the case when the true value of $\boldsymbol{\theta}$ satisfies $\theta_1 > 0$ and $\theta_2 > 0$. Since $\bar{\boldsymbol{X}} \xrightarrow{p} \boldsymbol{\theta}$, it follows that, with probability tending to 1, $\bar{\boldsymbol{X}}$ lies in an arbitrarily small neighborhood of $\boldsymbol{\theta}$ contained in $\mathbb{R}^{+2}$ and $LRT = n\|\bar{\boldsymbol{X}}\|^2$; therefore, $LRT \xrightarrow{p} \infty$ as $n \to \infty$. Hence, LRT is consistent at $\boldsymbol{\theta}$ when $\theta_1 > 0$ and $\theta_2 > 0$. Similarly, by considering the interior and the boundary of each of the other three quadrants separately, it is easily seen that

$$\text{LRT is consistent at } \boldsymbol{\theta} \text{ if and only if } \boldsymbol{\theta} \notin \{\boldsymbol{\theta} \in \mathbb{R}^2 : \theta_1 \leq 0, \theta_2 \leq 0\}.$$

In fact, it may be verified that $lim_{n\to\infty}pr\{LRT$ rejects $H_0|\boldsymbol{\theta}\}$ is 1, less than 1, or zero according as $\boldsymbol{\theta}$ is not in Q_3, lies on the boundary of Q_3, or lies in the interior of Q_3, where $Q_3 = \{\boldsymbol{\theta} \in \mathbb{R}^2 : \theta_1 \leq 0, \theta_2 \leq 0\}$.

Now consider the LRT of $H_1 : \boldsymbol{\theta} \geqslant \boldsymbol{0}$ against $H_2 : \boldsymbol{\theta} \not\geqslant \boldsymbol{0}$. Again, by considering the interior and boundary of each quadrant separately, we have that

$$\text{LRT is consistent at any } \boldsymbol{\theta} \notin \{\boldsymbol{\theta} \in \mathbb{R}^2 : \theta_1 \geqslant 0, \theta_2 \geqslant 0\};$$

further, $lim_{n\to\infty}pr\{LRT$ rejects $H_1|\boldsymbol{\theta}\}$ is 1, less than 1, or zero according as $\boldsymbol{\theta}$ is not in $\mathbb{R}^{+2}$, lies on the boundary of $\mathbb{R}^{+2}$, or lies in the interior of $\mathbb{R}^{+2}$. ■

Essentially the same arguments can be used to study the consistency properties of more general tests of inequality constrained hypotheses (see Problem 3.29). For the rest of this section all inner products will be assumed to be with respect to a given positive definite matrix, $\boldsymbol{V}$. Let $\boldsymbol{X}_1, \ldots \boldsymbol{X}_n$ be *iid* as $\boldsymbol{X}$ where $\boldsymbol{X} \sim N(\boldsymbol{\theta}, \boldsymbol{V})$ and $\boldsymbol{V}$ is known. Then we have the following:

(1) Let the null and alternative hypotheses be $H_0 : \boldsymbol{\theta} \in \mathcal{M}$ and $H_1 : \boldsymbol{\theta} \in \mathcal{C}$, respectively. Then $LRT = n\|\bar{\boldsymbol{X}} - \mathcal{P}\|^2$ where $\mathcal{P} = (\mathcal{C}^o \oplus \mathcal{M}) = (\mathcal{C} \cap \mathcal{M}^{\perp})^o$; see Theorem 3.7.1 on page 84 and Proposition 3.12.6 on page 118. Since $(\bar{\boldsymbol{X}} - \boldsymbol{\theta}) = o_p(1)$ and $\mathcal{P}$ is a closed set, it follows that LRT is consistent if and only if $\boldsymbol{\theta} \notin \mathcal{P}$. As a special case, let $\mathcal{M} = \{\boldsymbol{0}\}$. Then, since $LRT = n\|\bar{\boldsymbol{X}} - \mathcal{C}^o\|^2$, $(\bar{\boldsymbol{X}} - \boldsymbol{\theta}) = o_p(1)$ and $\mathcal{C}^o$ is a closed set, it follows that LRT is consistent if and only if $\boldsymbol{\theta} \notin \mathcal{C}^o$.

(2) Let the null and alternative hypotheses be $H_1 : \boldsymbol{\theta} \in \mathcal{C}$ and $H_2 : \boldsymbol{\theta} \notin \mathcal{C}$, respectively. Then $LRT = n\|\bar{\boldsymbol{X}} - \mathcal{C}\|^2$, $(\bar{\boldsymbol{X}} - \boldsymbol{\theta}) = o_p(1)$ and $\mathcal{C}$ is a closed set. Therefore, LRT is consistent at any $\boldsymbol{\theta} \notin \mathcal{C}$.

Now, let us consider the case when $\boldsymbol{V}$ is unknown, and the null and alternative hypotheses are $H_0 : \boldsymbol{\theta} = \boldsymbol{0}$ and $H_1 : \boldsymbol{\theta} \in \mathcal{C}$, respectively. In this case, the *LRT* statistic is consistent (see Wang and McDermott (1998a) p 383). Sen and Tsai (1999) note that essentially the same arguments are applicable to T^{*2} because $T^{*2} \geqslant LRT$; hence T^{*2} is also consistent. Wang and McDermott (1998) also showed that their conditional *LRT* is consistent.

It may also be of interest to study the behavior of the power when the effect of the sample size is absorbed into the value of $\boldsymbol{\theta}$. Problem 3.30 is a result along this line.

3.11.2 Monotonicity of the Power Function

For the simple univariate case of testing $H_0 : \mu = 0$ against $H_1 : \mu > 0$ based on a single observation of X where $X \sim N(\mu, 1)$, the power of a test with critical region $\{x \in \mathbb{R} : x \geqslant c\}$, for any fixed $c > 0$, increases to 1 as μ increases to ∞. Similarly, it is also known that for testing $\boldsymbol{\theta} = \mathbf{0}$ against $\boldsymbol{\theta} \neq \mathbf{0}$ based on a single observation of $\boldsymbol{X}$ from $N(\boldsymbol{\theta}, \boldsymbol{V})$, where $\boldsymbol{V}$ is a given positive definite matrix, the power of the LRT ($= \|\boldsymbol{X}\|_{\boldsymbol{V}}^2$) at $k\boldsymbol{\Delta}$ increases to 1 as k increases from zero to ∞, for any fixed $\boldsymbol{\Delta} \neq \mathbf{0}$ (see Anderson (1955)); this result holds even when the distribution of $\boldsymbol{X}$ is not normal but symmetric and unimodal (see Theorem 3.11.4).

The result of Anderson (1955) (stated as Theorem 3.11.4 on page 111) is applicable when the acceptance region is convex and symmetric. Therefore, it is usually not applicable when there are order restrictions on the parameters because the acceptance region is usually not symmetric and further it may not be even convex. However, it turns out that the power function of several tests when there are order restrictions on $\boldsymbol{\theta}$ are monotonic along certain straight lines. The main results concerning the monotonicity of power are stated below, and the proofs are given for some of them; for the others, references are given where detailed proofs may be found.

There are essentially two approaches to study the monotonicity of the power function. The simplest one is called the *set containment argument*, similar to that in Lemma 8.2 of Perlman (1969). This uses the fact $\mathcal{C} + \boldsymbol{\theta} \subset \mathcal{C}$ when $\boldsymbol{\theta} \in \mathcal{C}$ and hence the distance from any given point to $\mathcal{C} + \boldsymbol{\theta}$ is not more than the distance from the same point to $\mathcal{C}$. The second method is *geometric*. This approach starts with the fact that the power of a test is equal to the volume of a region under the probability density function and above the critical region and uses geometric properties of this volume. So far, general results using such a geometric approach have assumed that the density function is elliptically symmetric and unimodal; the set containment argument does not require the density function to be elliptically symmetric or unimodal. Therefore, the results based on set containment arguments are applicable to a broader class of distributions. However, there are cases in which the geometric approach works but not the set containment argument; in some cases the converse is true. Therefore, both approaches are useful and essential.

For the rest of this section, the inner product (and hence $\|.\|, \Pi()$, etc) will be assumed to be with respect to a given positive definite matrix $\boldsymbol{V}$, for tests based on a single observation.

Proposition 3.11.1 *Let $\boldsymbol{X} = \boldsymbol{\theta} + \boldsymbol{E}$ where the distribution of $\boldsymbol{E}$ does not depend on any unknown parameters. Let $\boldsymbol{V}$ be a given positive definite matrix, $\pi_{01}(\boldsymbol{\theta})$ denote the power function of $\|\boldsymbol{X} - (\mathcal{C} \cap \mathcal{M}^{\perp})^o\|_{\boldsymbol{V}}^2$ for testing $H_0 : \boldsymbol{\theta} \in \mathcal{M}$ against $H_1 : \boldsymbol{\theta} \in \mathcal{C}$; similarly, let $\pi_{12}(\boldsymbol{\theta})$ denote the power function of $\|\boldsymbol{X} - \mathcal{C}\|_{\boldsymbol{V}}^2$ for testing $H_1 : \boldsymbol{\theta} \in \mathcal{C}$ against $H_2 : \boldsymbol{\theta} \notin \mathcal{C}$. Then, $\pi_{01}(\boldsymbol{\theta}) \leq \pi_{01}(\boldsymbol{\theta}')$ for $\boldsymbol{\theta} - \boldsymbol{\theta}' \in \mathcal{C}^o$; and $\pi_{12}(\boldsymbol{\theta}) \geqslant \pi_{12}(\boldsymbol{\theta}')$ for $\boldsymbol{\theta}' - \boldsymbol{\theta} \in \mathcal{C}$.*

Proof: The proof follows a set containment argument. To prove the first, let $\boldsymbol{\theta} - \boldsymbol{\theta}' \in \mathcal{C}^o$. Then $\mathcal{C}^o + (\boldsymbol{\theta} - \boldsymbol{\theta}') \subset \mathcal{C}^o$, and hence $\mathcal{C}^o - \boldsymbol{\theta}' \subset \mathcal{C}^o - \boldsymbol{\theta}$. Now, $\pi_{01}(\boldsymbol{\theta}) =$

$\text{pr}\{\|\boldsymbol{X}-\mathcal{C}^o\|_{\boldsymbol{V}}^2 \geqslant c|\boldsymbol{\theta}\} = \text{pr}\{\|\boldsymbol{E}+\boldsymbol{\theta}-\mathcal{C}^o\|_{\boldsymbol{V}}^2 \geqslant c\} = \text{pr}\{\|\boldsymbol{E}-(\mathcal{C}^o-\boldsymbol{\theta})\|_{\boldsymbol{V}}^2 \geqslant c\} \leq \text{pr}\{\|\boldsymbol{E}-(\mathcal{C}^o-\boldsymbol{\theta}')\|_{\boldsymbol{V}}^2 \geqslant c\} = \pi_{01}(\boldsymbol{\theta}')$.

To prove the second, let $\boldsymbol{\theta}'-\boldsymbol{\theta} \in \mathcal{C}$. Then $\mathcal{C}+(\boldsymbol{\theta}'-\boldsymbol{\theta}) \subset \mathcal{C}$, and hence $\mathcal{C}-\boldsymbol{\theta} \subset \mathcal{C}-\boldsymbol{\theta}'$. Now, $\pi_{12}(\boldsymbol{\theta}) = \text{pr}\{\|\boldsymbol{X}-\mathcal{C}\|_{\boldsymbol{V}}^2 \geqslant c|\boldsymbol{\theta}\} = \text{pr}\{\|\boldsymbol{E}-(\mathcal{C}-\boldsymbol{\theta})\|_{\boldsymbol{V}}^2 \geqslant c\} \geqslant \text{pr}\{\|\boldsymbol{E}-(\mathcal{C}-\boldsymbol{\theta}')\|_{\boldsymbol{V}}^2 \geqslant c\} = \pi_{12}(\boldsymbol{\theta}')$. ■

A simple diagram can be used to illustrate this result. The first part of the result says that for testing $H_0 : \boldsymbol{\theta} = \mathbf{0}$ against $H_1 : \boldsymbol{\theta} \in \mathcal{C}$, the power of the test at any point, say $\boldsymbol{\theta}_0$, increases in any direction of $-\mathcal{C}^o$ and decreases in any direction of $\mathcal{C}^o$. Similarly, the second part of the result says that for testing $H_1 : \boldsymbol{\theta} \in \mathcal{C}$ against $H_2 : \boldsymbol{\theta} \notin \mathcal{C}$, the power of the test at any point, say $\boldsymbol{\theta}_0$, increases in any direction of $-\mathcal{C}$ and decreases in any direction of $\mathcal{C}$.

Now, we state the main results that have been obtained using the geometric approach. Let h be a nonincreasing function on $[0,\infty)$, $\boldsymbol{V}$ be a positive definite matrix of order $p \times p$, and $\boldsymbol{X}$ be a random variable with a density function $|\boldsymbol{V}|^{-1/2}h(\|\boldsymbol{x}-\boldsymbol{\theta}\|^2)$; a density function of this form is said to be *elliptically symmetric* and *unimodal* about $\boldsymbol{\theta}$. Let

$$p(\boldsymbol{\theta}) = 1 - \int_A |\boldsymbol{V}|^{-1/2}h(\|\boldsymbol{x}-\boldsymbol{\theta}\|_{\boldsymbol{V}}^2)d\boldsymbol{x}, \qquad A \subset \mathbb{R}^p, \boldsymbol{\theta} \in \mathbb{R}^p.$$

For several tests, including the LRT, of $H_0 : \boldsymbol{\theta} \in \mathcal{M}$ against $H_1 : \boldsymbol{\theta} \in \mathcal{C}$ and of $H_1 : \boldsymbol{\theta} \in \mathcal{C}$ against $H_2 : \boldsymbol{\theta} \notin \mathcal{C}$, the power of the test at $\boldsymbol{\theta}$ takes the form $p(\boldsymbol{\theta})$, where A is the acceptance region. Therefore, monotonicity properties of functions of the form $p(\boldsymbol{\theta})$ are of interest. Anderson's theorem is applicable to establish the monotonicity of $p(\boldsymbol{\theta})$ when A is symmetric and convex. When there are inequality constraints, the acceptance region A is usually not symmetric and it may not be even convex. Mukerjee et al. (1986) studied the monotonicity properties of $p(\boldsymbol{\theta})$ when A satisfies certain conditions that include that A be convex but not necessarily symmetric. They showed that $p(k\boldsymbol{\theta}_0)$ is monotonic in $|k|$ for $\boldsymbol{\theta}_0 \in \mathbb{R}^p$; their main result (stated as Theorem 3.11.5 on page 111) leads to the following useful corollary.

Corollary 3.11.2 *(Mukerjee et al. (1986)). Let the distribution of $\boldsymbol{X}$ be elliptically symmetric and unimodal about $\boldsymbol{\theta}$. Let $\boldsymbol{\theta}_1 \in \mathbb{R}^p$ and $\boldsymbol{\theta}_2 \in \mathcal{C}$. Then as a function of k, $\pi_{12}(\boldsymbol{\theta}_1 + k\boldsymbol{\theta}_2)$ is nondecreasing for $k \in (-\infty,\infty)$ and $\pi_{12}(\boldsymbol{\theta}_1 + k\tilde{\boldsymbol{\theta}}_1)$ is nondecreasing for $k \geqslant -1$ where $\tilde{\boldsymbol{\theta}}_1 = \Pi(\boldsymbol{\theta}_1|\mathcal{C}^o)$.* ■

It is also shown in Robertson et al. (1988), page 106, that (a) $\pi_{01}(k\boldsymbol{\Delta})$ is nondecreasing in k, where $(k > 0)$ and $\boldsymbol{\Delta} \in \mathcal{C}$, and (b) $\pi_{12}(k\boldsymbol{\Delta})$ and $\pi_{12}^*(k\boldsymbol{\Delta})$ are nondecreasing in k, where $k > 0$, $\boldsymbol{\Delta} \in \mathcal{C}^o \oplus \mathcal{M}$ and π_{12}^* is the power function of the $\bar{E}^2$-test.

For some $\bar{E}^2$-tests, the acceptance region is not convex and hence the results of Mukerjee et al. (1986) are not applicable. Hu and Wright (1994b) studied the monotonicity of $p(\boldsymbol{\theta}_0 + t\boldsymbol{\theta}_1)$ in t for some given $\boldsymbol{\theta}_0$ and $\boldsymbol{\theta}_1$ and when A takes a special form; their regularity conditions do not require A to be convex or symmetric. Their result extends some of the results in Mukerjee et al. (1986), and is applicable for the $\bar{E}^2$-test of H_0 against H_1. A useful corollary of their main result (see Theorem 3.11.6) is the following:

Corollary 3.11.3 *(Hu and Wright (1994b)). Suppose that $\boldsymbol{X}_1, \ldots, \boldsymbol{X}_n$ are iid as $N(\boldsymbol{\theta}, \sigma^2 \boldsymbol{U})$ where $\boldsymbol{U}$ is known and σ is unknown. Then, the power function of the LRT (namely, $\bar{E}^2$-test) of $H_0 : \boldsymbol{\theta} \in \mathcal{M}$ against $H_1 : \boldsymbol{\theta} \in \mathcal{C}$ is nondecreasing on each line segment starting at a point in $\mathcal{M}$ and continuing in the direction of a vector in $\mathcal{C}$. Consequently, this test is unbiased as well.* ■

Since $\bar{F} = \bar{\chi}^2 / S^2$ and S^2 is independent of $\boldsymbol{\theta}$, monotonicity of $\bar{\chi}^2$ carry over to $\bar{F}$ as well; hence, monotonicity of $\bar{F}$ is not studied separately.

The main results concerning the monotonicity of $p(\boldsymbol{\theta})$ are stated below for completeness.

Theorem 3.11.4 *(Anderson (1955)). Let A be a convex and symmetric set in $\mathbb{R}^p$. Let f be a function such that (i) $f(\boldsymbol{x}) = f(-\boldsymbol{x})$, (ii) $\{\boldsymbol{x} : f(\boldsymbol{x}) \geqslant u\}$ is convex for every $u > 0$, and (iii) $\int_A f(\boldsymbol{x}) d\boldsymbol{x} < \infty$ (in the Lebesgue sense). Then $\int_A f(\boldsymbol{x} + k\boldsymbol{y}) \geqslant \int_A f(\boldsymbol{x} + \boldsymbol{y}) d\boldsymbol{y}$ for $0 \leq k \leq 1$.* ■

Theorem 3.11.5 *(Mukerjee et al. (1986)). Let $\boldsymbol{\theta} \in \mathbb{R}^p$ and $S_{\boldsymbol{\theta}} = \{b\boldsymbol{\theta} : -\infty < b < \infty\}$ be the linear space spanned by $\boldsymbol{\theta}$; for a given convex set $A \subset \mathbb{R}^p$ let the positive part of A in the direction of $\boldsymbol{\theta}$ be defined by*

$$A^+ = \{\boldsymbol{x} \in A : \Pi_{\boldsymbol{V}}(\boldsymbol{x} | S_{\boldsymbol{\theta}}) = b\boldsymbol{\theta} \text{ where } b > 0\}.$$

Let f be an elliptically symmetric and unimodal density function. Suppose that $\boldsymbol{x} - 2\Pi_{\boldsymbol{V}}(\boldsymbol{x} | S_{\boldsymbol{\theta}} \in A$ for each $\boldsymbol{x} \in A^+$. Then, $\int_{A - k\boldsymbol{\theta}} f(\boldsymbol{x}) d\boldsymbol{x}$ is a nonincreasing function of k for $k > 0$. ■

Theorem 3.11.6 *(Hu and Wright (1994b)). Suppose that $\boldsymbol{X}$ has the unimodal and elliptically symmetric density $|\boldsymbol{V}|^{-1/2} h(\|\boldsymbol{x}\|_{\boldsymbol{V}}^2)$ where h is nonincreasing on $[0, \infty)$. Let $\mathcal{C} \subset \mathbb{R}^p$ be a closed convex cone, $\mathcal{M}$ be a linear subspace such that $\mathcal{M} \subset \mathcal{C}$, and*

$$A = \{\boldsymbol{x} \in \mathbb{R}^p : \|\Pi_{\boldsymbol{V}}(\boldsymbol{x}|\mathcal{M}) - \Pi_{\boldsymbol{V}}(\boldsymbol{x}|\mathcal{C})\|_{\boldsymbol{V}}^2 \leq a + b\|\boldsymbol{x} - \Pi_{\boldsymbol{V}}(\boldsymbol{x}|\mathcal{C})\|_{\boldsymbol{V}}^2\},$$

where $a > 0$ and $b > 0$. If $\boldsymbol{\mu}_0 \in \mathcal{M}$ and $\boldsymbol{\nu}_0 \in \mathcal{C}$ then $pr(\boldsymbol{X} + (\boldsymbol{\mu}_0 + t\boldsymbol{\nu}_0) \in A)$, is nonincreasing in t for $t > 0$. ■

Now let us consider the case when the covariance matrix $\boldsymbol{V}$ is completely unknown, and the null and alternative hypotheses are $H_0 : \boldsymbol{\theta} = \boldsymbol{0}$ and $H_1 : \boldsymbol{\theta} \in \mathcal{C}$, respectively. The power function of *LRT* and T^{*2} are monotonic. In particular, the set containment argument can be used to establish that the power functions of *LRT* and T^{*2} at $\boldsymbol{\theta}_0$ increase in any direction of $\mathcal{C}$ (see Sen and Tsai (1999)).

3.12 APPENDIX 1: CONVEX CONES, POLYHEDRALS, AND PROJECTIONS

3.12.1 Introduction

This Appendix provides a brief account of the main results on projections onto convex cones in finite dimensional spaces as they relate to the statistical inference problems discussed in this book. An excellent account of projections onto convex cones is given in Stoer and Witzgall (1970)(SW); other references include Bazaraa et al. (1993) and Hiriart-Urruty and Lemaréchal (1993). Zarantonello (1998) provides an extensive account of projections onto convex sets in Hilbert spaces; while some of the results and ideas therein are relevant, such general results are not required at this stage. In this Appendix, we provide a reasonably self-contained discussion that would be adequate for this book.

Let $\mathbb{R}^p$ denote the p-dimensional Euclidean space, $\boldsymbol{V}$ be a $p \times p$ symmetric positive definite matrix, $\boldsymbol{x} \in \mathbb{R}^p$, and $\boldsymbol{y} \in \mathbb{R}^p$. Then, $\langle \boldsymbol{x}, \boldsymbol{y} \rangle_{\boldsymbol{V}} = \boldsymbol{x}^T \boldsymbol{V}^{-1} \boldsymbol{y}$ defines an *inner product* on $\mathbb{R}^p$. This induces the corresponding *norm* $\|\boldsymbol{x}\|_{\boldsymbol{V}} = \langle \boldsymbol{x}, \boldsymbol{x} \rangle_{\boldsymbol{V}}^{1/2}$; the corresponding *distance* between $\boldsymbol{x}$ and $\boldsymbol{y}$ is $\|\boldsymbol{x} - \boldsymbol{y}\|_{\boldsymbol{V}}$. If $\boldsymbol{x}^T \boldsymbol{V}^{-1} \boldsymbol{y} = 0$ then we say that $\boldsymbol{x}$ and $\boldsymbol{y}$ are *orthogonal* with respect to $\boldsymbol{V}$, which we denote by $\boldsymbol{x} \perp_{\boldsymbol{V}} \boldsymbol{y}$. We may abbreviate "orthogonal with respect to $\boldsymbol{V}$", "distance with respect to $\boldsymbol{V}$" to $\boldsymbol{V}$-orthogonal, $\boldsymbol{V}$-distance, etc. However, if $\boldsymbol{V}$ is obvious we will drop the reference to $\boldsymbol{V}$.

If $\boldsymbol{x} \perp_{\boldsymbol{V}} \boldsymbol{y}$ then a version of the Pythagoras theorem holds: namely, $\|\boldsymbol{x} + \boldsymbol{y}\|_{\boldsymbol{V}}^2 = \|\boldsymbol{x}\|_{\boldsymbol{V}}^2 + \|\boldsymbol{y}\|_{\boldsymbol{V}}^2$. A consequence of this is that the shortest distance between a point and a plane is the distance from the point to the plane along a line that is orthogonal to the plane; here, distance and orthogonality are with respect to $\langle , \rangle_{\boldsymbol{V}}$.

Let C be a closed convex set in $\mathbb{R}^p$ and $\boldsymbol{x} \in \mathbb{R}^p$. Let $\tilde{\boldsymbol{x}}$ in C be the point in C that is closest to $\boldsymbol{x}$ with respect to the distance $\|\cdot\|_{\boldsymbol{V}}$; thus,

$$\|\boldsymbol{x} - \tilde{\boldsymbol{x}}\| = \min_{\boldsymbol{\theta} \in C} (\boldsymbol{x} - \boldsymbol{\theta})^T \boldsymbol{V}^{-1} (\boldsymbol{x} - \boldsymbol{\theta}).$$

The vector $\tilde{\boldsymbol{x}}$ is called the *projection* of $\boldsymbol{x}$ onto C, and is denoted by $\Pi_{\boldsymbol{V}}(\boldsymbol{x} \mid C)$ (SW p 47); thus

$$\tilde{\boldsymbol{x}} = \Pi_{\boldsymbol{V}}(\boldsymbol{x} \mid C) = \arg \min_{\boldsymbol{\theta} \in C} (\boldsymbol{x} - \boldsymbol{\theta})^T \boldsymbol{V}^{-1} (\boldsymbol{x} - \boldsymbol{\theta}).$$

Almost all the results on projections onto convex cones discussed here are generalizations of the corresponding results for projections onto linear spaces. Therefore, the basic results on the latter are essential for the rest of this appendix. For detailed discussions of relevant results on projections onto linear spaces, see textbooks on linear models, such as Arnold (1981), Chapter 2, or Rao (1972).

We may assume, without loss of generality, that $\boldsymbol{V} = \boldsymbol{I}$ when considerations are restricted to $\langle \boldsymbol{x}, \boldsymbol{y} \rangle_{\boldsymbol{V}}$ where $\boldsymbol{x}$ and $\boldsymbol{y}$ are elements of $\mathbb{R}^p$. To show this, let $\boldsymbol{V}^{-1} = \boldsymbol{U}^T \boldsymbol{U}$ be a factorization of $\boldsymbol{V}^{-1}$, for example, the Cholesky factorization. Now apply the invertible linear transformation $\boldsymbol{U}$ on $\mathbb{R}^p$. Then, any two points $\boldsymbol{x}$ and $\boldsymbol{y}$ in $\mathbb{R}^p$ are mapped to $\boldsymbol{p}$ and $\boldsymbol{q}$ in $\mathbb{R}^p$ where $\boldsymbol{p} = \boldsymbol{U}\boldsymbol{x}$ and $\boldsymbol{q} = \boldsymbol{U}\boldsymbol{y}$. Since $\langle \boldsymbol{x}, \boldsymbol{y} \rangle_{\boldsymbol{V}} = \boldsymbol{x}^T \boldsymbol{V}^{-1} \boldsymbol{y} = (\boldsymbol{U}\boldsymbol{x})^T (\boldsymbol{U}\boldsymbol{y}) = \boldsymbol{p}^T \boldsymbol{q} = \langle \boldsymbol{p}, \boldsymbol{q} \rangle_{\boldsymbol{I}}$, the distance between $\boldsymbol{x}$ and $\boldsymbol{y}$ in $\{\mathbb{R}^p, \langle \cdot \rangle_{\boldsymbol{V}}\}$ is equal to the distance between their images $\boldsymbol{p}$ and $\boldsymbol{q}$ in $\{\mathbb{R}^p, \langle \cdot \rangle_{\boldsymbol{I}}\}$. Therefore, for the purposes of studying properties that depend only on distances between points, we may identify $\{\mathbb{R}^p, \langle \cdot \rangle_{\boldsymbol{V}}\}$ with $\{\mathbb{R}^p, \langle \cdot \rangle_{\boldsymbol{I}}\}$ and, hence, without loss of generality, assume that $\boldsymbol{V}$ is the identity matrix. Instead of writing $\langle \boldsymbol{p}, \boldsymbol{q} \rangle_{\boldsymbol{I}}$ and $\|\boldsymbol{p} - \boldsymbol{q}\|_{\boldsymbol{I}}$ we shall use the simpler notation, $\boldsymbol{p}^T \boldsymbol{q}$ and $\|\boldsymbol{p} - \boldsymbol{q}\|$, respectively.

Remark: Since $\boldsymbol{U}$ and $\boldsymbol{U}^{-1}$ are one-to-one, onto and, continuous, $\boldsymbol{U} : \{\mathbb{R}^p, \langle\cdot\rangle_{\boldsymbol{V}}\} \to \{\mathbb{R}^p, \langle\cdot\rangle_{\boldsymbol{I}}\}$ is a *homeomorphism* and hence $\{\mathbb{R}^p, \langle\cdot\rangle_{\boldsymbol{V}}\}$ and $\{\mathbb{R}^p, \langle\cdot\rangle_{\boldsymbol{I}}\}$ are *homeomorphic*. Thus, these two spaces have the same topological structure. Since $\langle \boldsymbol{x}, \boldsymbol{y}\rangle_{\boldsymbol{V}} = \langle \boldsymbol{Ux}, \boldsymbol{Uy}\rangle_{\boldsymbol{I}}$, $\boldsymbol{U}$ is an *isomorphism* and $\{\mathbb{R}^p, \langle\cdot\rangle_{\boldsymbol{V}}\}$ and $\{\mathbb{R}^p, \langle\cdot\rangle_{\boldsymbol{I}}\}$ are *isomorphic*.

A set S is said to be the *orthogonal sum* of S_1 and S_2 if $S = S_1 + S_2 (= \{s_1 + s_2 : s_1 \in S_1, s_2 \in S_2\})$ and every element $\boldsymbol{x} \in S$ admits a unique *orthogonal decomposition* of the form $\boldsymbol{x} = \boldsymbol{y} + \boldsymbol{z}$ with $\boldsymbol{y} \in S_1, \boldsymbol{z} \in S_2$ and $\boldsymbol{y}^T\boldsymbol{z} = 0$. Such an orthogonal sum will be denoted by $S_1 \oplus S_2$. As an example, the *direct sum* of two orthogonal linear subspaces is an orthogonal sum.

For any set $S \subset \mathbb{R}^p$, its *orthogonal complement* $S^\perp$ is defined as $\{\boldsymbol{y} \in \mathbb{R}^p : \boldsymbol{y}^T\boldsymbol{x} = 0, \text{ for all } \boldsymbol{x} \in S\}$. Clearly, $S^\perp$ is a linear space. For a set $S \subset \mathbb{R}^p$, we define the *polar cone* S^o as

$$S^o = \{\boldsymbol{x} : \boldsymbol{x}^T\boldsymbol{y} \le 0 \text{ for all } \boldsymbol{y} \in S\}.$$

The next result states a few important but elementary properties of cones and polar cones; the proofs are easy and hence omitted.

Proposition 3.12.1 *Let $\mathcal{C}$ be a closed convex cone and P and Q be subsets of $\mathbb{R}^p$. Then we have the following:*

(1) *$\mathcal{C}$ is closed under addition: $\boldsymbol{x} \in \mathcal{C}$ and $\boldsymbol{y} \in \mathcal{C} \Rightarrow \boldsymbol{x} + \boldsymbol{y} \in \mathcal{C}$.*
(2) $\mathcal{C} + \boldsymbol{x} \subset \mathcal{C} \subset \mathcal{C} - \boldsymbol{x}, \quad \forall \boldsymbol{x} \in \mathcal{C}$.
(3) *P^o is a closed convex cone.*
(4) *If $P \subset Q$ then $Q^o \subset P^o$.*
(5) *If P is a linear subspace then $P^\perp = P^o$.* ■

Clearly, we can also define orthogonal sum, orthogonal complement, polar cone, etc. with respect to any symmetric and positive definite matrix, $\boldsymbol{V}$; all that it requires is to replace $\boldsymbol{x}^T\boldsymbol{y}$ by $\boldsymbol{x}^T\boldsymbol{V}^{-1}\boldsymbol{y}$ in the foregoing definitions. For example, the *polar cone of S with respect to $\boldsymbol{V}$* is $\{\boldsymbol{x} : \boldsymbol{x}^T\boldsymbol{V}^{-1}\boldsymbol{y} \le 0 \text{ for all } \boldsymbol{y} \in S\}$. This can be denoted by $S^o_{\boldsymbol{V}}$; however, we will write S^o instead if the suffix $\boldsymbol{V}$ is obvious. Similarly, the *orthogonal complement of S with respect to $\boldsymbol{V}$* is defined as $\{\boldsymbol{y} \in \mathbb{R}^p : \boldsymbol{y}^T\boldsymbol{V}^{-1}\boldsymbol{x} = 0, \text{ for all } \boldsymbol{x} \in S\}$. This can be denoted by $S^\perp_{\boldsymbol{V}}$, although we would typically write $S^\perp$ if the suffix $\boldsymbol{V}$ is obvious. In what follows, the main results are presented for the case $\boldsymbol{V} = \boldsymbol{I}$; however, the corresponding results for a general $\boldsymbol{V}$ would be obvious.

For statistical inference based on the linear model, projections onto linear spaces play an important role. The corresponding results on projections onto convex cones play a similar important role in constrained statistical inference in the exact normal and large sample theory.

3.12.2 Projections onto Convex Cones

Before we consider projections onto convex cones, let us recall some results on projections onto linear spaces. Let $\boldsymbol{P}$ be a $p \times p$ matrix. We say that $\boldsymbol{P}$ is a *projection matrix* if $\boldsymbol{P} = \boldsymbol{P}^T$ and $\boldsymbol{P}^2 = \boldsymbol{P}$.

Proposition 3.12.2 *Let $\boldsymbol{P}$ be $p \times p$ projection matrix, $\boldsymbol{Q} = \boldsymbol{I} - \boldsymbol{P}$, $\mathcal{L}_P = \{\boldsymbol{P}\boldsymbol{x} : \boldsymbol{x} \in \mathbb{R}^p\}$ and $\mathcal{L}_Q = \{\boldsymbol{Q}\boldsymbol{x} : \boldsymbol{x} \in \mathbb{R}^p\}$. Then we have the following: (1) $\boldsymbol{P}$ is positive semi-definite. (2) $\boldsymbol{Q}$ is a projection matrix. (3) $\mathcal{L}_P$ and $\mathcal{L}_Q$ are orthogonal linear subspaces. (4) Any $\boldsymbol{x} \in \mathbb{R}$ can be represented uniquely as $\boldsymbol{x} = \boldsymbol{p} + \boldsymbol{q}$ where $\boldsymbol{p} \in \mathcal{L}_P$ and $\boldsymbol{q} \in \mathcal{L}_Q$.*

Proof: See Bazaraa et al. (1993), page 448, for example. ■

Let $\mathcal{L}$ be a linear subspace of $\mathbb{R}^p$. Some results on projections onto linear spaces that are important for inference in linear models, include the following:

1. The projection $\Pi(\boldsymbol{x}|\mathcal{L})$ of $\boldsymbol{x}$ onto $\mathcal{L}$ exists and is unique.

2. Let $\mathcal{L} = \{\boldsymbol{X}\boldsymbol{\beta} : \boldsymbol{\beta} \in \mathbb{R}^k\}$, and $\boldsymbol{P} = \boldsymbol{X}(\boldsymbol{X}^T\boldsymbol{X})^{-1}\boldsymbol{X}^T$, where $k \leqslant p$, $\boldsymbol{X}$ is $p \times k$ and $\text{rank}(\boldsymbol{X}) = k$. Then, $\Pi(\boldsymbol{x}|\mathcal{L}) = \boldsymbol{P}\boldsymbol{x}$ for every $\boldsymbol{x}$ and $\boldsymbol{P}$ is called the *projection matrix* onto $\mathcal{L}$.

3. $\boldsymbol{x} - \Pi(\boldsymbol{x}|\mathcal{L})$ is the projection of $\boldsymbol{x}$ onto $\mathcal{L}^{\perp}$, and the projection matrix onto $\mathcal{L}^{\perp}$ is $(\boldsymbol{I} - \boldsymbol{P})$.

4. If $\boldsymbol{x}$ is projected onto $\mathcal{L}$ and then the projection itself is projected onto a subspace $\mathcal{M}$ of $\mathcal{L}$, the resulting projection is the projection of $\boldsymbol{x}$ onto $\mathcal{M}$ [i.e., $\Pi\{\Pi(\boldsymbol{x}|\mathcal{L})|\mathcal{M}\} = \Pi(\boldsymbol{x}|\mathcal{M})$].

The essence of many of these results carries over to projections onto convex cones and plays similar important roles in constrained statistical inference. The main difference that causes technical difficulties is that the projection of $\boldsymbol{x}$ onto a convex cone is not a linear function of $\boldsymbol{x}$ and hence it cannot be expressed as $\boldsymbol{Q}\boldsymbol{x}$ where $\boldsymbol{x}$ is arbitrary and $\boldsymbol{Q}$ does not depend on $\boldsymbol{x}$. The geometric nature of projections onto linear spaces and their relevance in the context of statistical inference in linear models appear in standard textbooks on linear models. Therefore, we do not discuss such results here, but provide a discussion of the corresponding results for closed convex cones and schematic diagrams to help interpret the results.

Just as in the case for linear spaces, projection of a point onto a closed convex set exists uniquely, and it can be characterized by the angle between the line of projection and the set. These results are given below.

Proposition 3.12.3 *Let C be a nonempty closed convex set in $\mathbb{R}^p$ and $\boldsymbol{a} \in \mathbb{R}^p \setminus C$. Then we have the following:*

(a) *There exists a unique $\boldsymbol{a}^* \in C$ that is closest to $\boldsymbol{a}$ in the sense that, for $\boldsymbol{x} \in C$, $\|\boldsymbol{a} - \boldsymbol{x}\|$ is a minimum at $\boldsymbol{x} = \boldsymbol{a}^*$.*

(b) *The point $\boldsymbol{a}^* \in C$ is the unique point in C closest to $\boldsymbol{a}$ if and only if $(\boldsymbol{a} - \boldsymbol{a}^*)^T(\boldsymbol{x} - \boldsymbol{a}^*) \leq 0$ for all $\boldsymbol{x} \in C$.*

(c) *There exists a vector $\mathbf{p}$ and a scalar α such that $\mathbf{p}^T\boldsymbol{a} > \alpha$ and $\mathbf{p}^T\boldsymbol{x} \leq \alpha$ for every $\boldsymbol{x} \in C$.*

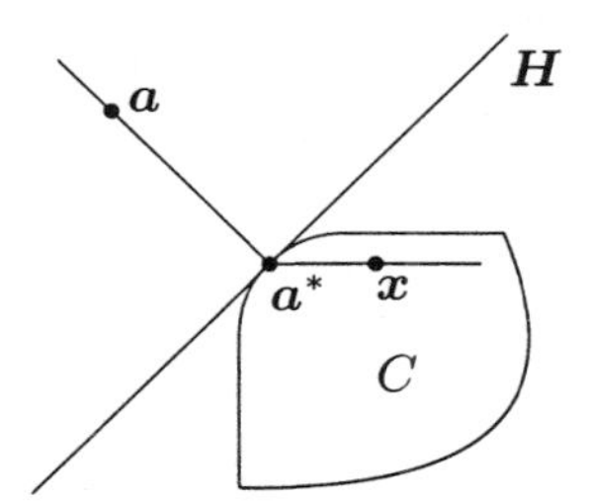

Fig. 3.14 The projection $\boldsymbol{a}^*$ of $\boldsymbol{a}$ onto the convex set C; the plane $\boldsymbol{H}$, which is orthogonal to $\boldsymbol{a} - \boldsymbol{a}^*$, separates C and the point $\boldsymbol{a}$

[For (a), see SW 3.3.1 or Bazaraa et al. (1993), page 43; for (b) see Bazaraa et al. (1993), page 43 or Hiriart-Urutty and Lemaréchal (1993), page 117]

Proof of (a): Let $f : C \to \mathbb{R}$ be defined by $f(\boldsymbol{x}) = \|\boldsymbol{a} - \boldsymbol{x}\|$. Let $\boldsymbol{x}_0 \in C$ and $S = \{\boldsymbol{y} \in \mathbb{R}^p : f(\boldsymbol{y}) \leq f(\boldsymbol{x}_0)\}$. Then $S \cap C$ is nonempty and compact. Since f is continuous, it has a minimum over $S \cap C$, say at $\boldsymbol{a}^*$. Clearly, $f(\boldsymbol{a}^*) < f(\boldsymbol{x})$ for $\boldsymbol{x} \notin S$. Therefore, $\boldsymbol{a}^*$ is also the global minimum of $f(\boldsymbol{x})$ over C. Now, to show that $\boldsymbol{a}^*$ is unique, suppose that $\tilde{\boldsymbol{a}} \in C, \tilde{\boldsymbol{a}} \neq \boldsymbol{a}^*$, and $\|\boldsymbol{a} - \tilde{\boldsymbol{a}}\| = \|\boldsymbol{a} - \boldsymbol{a}^*\|$. Let $\bar{\boldsymbol{a}} = 0.5(\boldsymbol{a}^* + \tilde{\boldsymbol{a}})$. Clearly, $\bar{\boldsymbol{a}} \in C$ because C is convex. Further, we have that $\|\boldsymbol{a} - \bar{\boldsymbol{a}}\|^2 = 0.5\|\boldsymbol{a} - \boldsymbol{a}^*\|^2 + 0.5\|\boldsymbol{a} - \tilde{\boldsymbol{a}}\|^2 - 0.25\|\tilde{\boldsymbol{a}} - \boldsymbol{a}^*\|^2 < \|\boldsymbol{a} - \boldsymbol{a}^*\|^2$. Thus, $\bar{\boldsymbol{a}}$ is closer to $\boldsymbol{a}$ than $\boldsymbol{a}^*$; this is a contradiction. Therefore, $\boldsymbol{a}^*$ is unique.

Proof of (b): Assume that $\boldsymbol{a}^* \in C$ and $(\boldsymbol{a} - \boldsymbol{a}^*)^T(\boldsymbol{x} - \boldsymbol{a}^*) \leq 0$ for every $\boldsymbol{x} \in C$. Let $\boldsymbol{x} \in C$. Then $\|\boldsymbol{a} - \boldsymbol{x}\|^2 = \|\boldsymbol{a} - \boldsymbol{a}^*\|^2 + \|\boldsymbol{a}^* - \boldsymbol{x}\|^2 + 2(\boldsymbol{a} - \boldsymbol{a}^*)^T(\boldsymbol{a}^* - \boldsymbol{x})$. Since $(\boldsymbol{a} - \boldsymbol{a}^*)^T(\boldsymbol{a}^* - \boldsymbol{x}) \geqslant 0$ by assumption, we have that $\|\boldsymbol{a} - \boldsymbol{x}\|^2 \geqslant \|\boldsymbol{a} - \boldsymbol{a}^*\|^2$. Therefore, $\boldsymbol{a}^*$ is the point in C closest to $\boldsymbol{a}$. To prove the converse, assume that $\boldsymbol{a}^* \in C$ is the point in C closest to $\boldsymbol{a}$. Let $\boldsymbol{x} \in C$. Then, $\boldsymbol{a}^* + \lambda(\boldsymbol{x} - \boldsymbol{a}^*) \in C$ for $0 < \lambda \leq 1$ because C is convex. Now, since $\boldsymbol{a}^*$ is closer to $\boldsymbol{a}$ than $\boldsymbol{a}^* + \lambda(\boldsymbol{x} - \boldsymbol{a}^*)$, we have $\|\boldsymbol{a} - \boldsymbol{a}^*\|^2 \leq \|\boldsymbol{a} - \{\boldsymbol{a}^* + \lambda(\boldsymbol{x} - \boldsymbol{a}^*)\}\|^2 = \|\boldsymbol{a} - \boldsymbol{a}^*\|^2 + \lambda^2\|\boldsymbol{x} - \boldsymbol{a}^*\|^2 - 2\lambda(\boldsymbol{a} - \boldsymbol{a}^*)^T(\boldsymbol{x} - \boldsymbol{a}^*)$. Therefore, $2\lambda(\boldsymbol{a} - \boldsymbol{a}^*)^T(\boldsymbol{x} - \boldsymbol{a}^*) \leq \lambda^2\|\boldsymbol{a} - \boldsymbol{a}^*\|^2$. Now, divide by $\lambda(> 0)$ and take the limit as $\lambda \to 0^+$. This leads to $(\boldsymbol{a} - \boldsymbol{a}^*)^T(\boldsymbol{x} - \boldsymbol{a}^*) \leq 0$ for every $\boldsymbol{x} \in C$.

Proof of (c): Let $\boldsymbol{a}^* = \Pi(\boldsymbol{a} \mid \mathcal{C})$, $\boldsymbol{p}$ be a unit vector parallel to $(\boldsymbol{a} - \boldsymbol{a}^*)$ and $\alpha = \boldsymbol{p}^T\boldsymbol{a}^*$. Now, the proof follows from part (b). ■

Fig. 3.14 illustrates the nature of the foregoing results. It shows that if $\Pi(\boldsymbol{a}|\mathcal{C})$ is treated as the origin then the angle between $\boldsymbol{a}$ and $\boldsymbol{x}$ must be obtuse for any $\boldsymbol{x} \in \mathcal{C}$ (the *angle* θ between the vectors $\boldsymbol{x}$ and $\boldsymbol{y}$ in $\mathbb{R}^p$ is defined by $\cos\theta = \boldsymbol{x}^T\boldsymbol{y}\{\|\boldsymbol{x}\|\|\boldsymbol{y}\|\}^{-1/2}$ and this angle is said to be *obtuse* if $\cos\theta < 0$). Further, the hyperplane $H = \{\boldsymbol{x} \in \mathbb{R}^p : \boldsymbol{x}^T\boldsymbol{p} = \alpha\}$ is orthogonal to $(\boldsymbol{a} - \boldsymbol{a}^*)$, passes through $\boldsymbol{a}^*$, and *separates* $\boldsymbol{a}$ and C; here separates means that the point $\boldsymbol{a}$ and the set C lie on the opposite sides of the hyperplane.

An important result on projections onto linear spaces is that, for any $\boldsymbol{x} \in \mathbb{R}^p$ and linear space $\mathcal{L}$, we have a unique orthogonal decomposition of the form $\boldsymbol{x} =$

$\Pi(\boldsymbol{x}|\mathcal{L}) + \Pi(\boldsymbol{x}|\mathcal{L}^{\perp})$ that corresponds to the orthogonal sum $\mathbb{R}^p = \mathcal{L} \oplus \mathcal{L}^{\perp}$. For a closed convex cone $\mathcal{C}$, essentially the same results hold with the polar cone $\mathcal{C}^o$ playing the role of $\mathcal{L}^{\perp}$. These are given in the next proposition

Proposition 3.12.4 *Let $\mathcal{C}$ be a closed convex cone and $\boldsymbol{x} \in \mathbb{R}^p$.*

(a) Assume that $\boldsymbol{x} = \boldsymbol{y} + \boldsymbol{z}$ with $\boldsymbol{y} \in \mathcal{C}$, $\boldsymbol{z} \in \mathcal{C}^o$ and $\boldsymbol{y}^T\boldsymbol{z} = 0$. Then $\boldsymbol{y} = \Pi(\boldsymbol{x}|\mathcal{C})$ and $\boldsymbol{z} = \Pi(\boldsymbol{x}|\mathcal{C}^o)$.

(b) Conversely, $\boldsymbol{x} = \Pi(\boldsymbol{x}|\mathcal{C}) + \Pi(\boldsymbol{x}|\mathcal{C}^o)$ and $\Pi(\boldsymbol{x}|\mathcal{C})^T\Pi(\boldsymbol{x}|\mathcal{C}^o) = 0$.

(c) $\mathbb{R}^p = \mathcal{C} \oplus \mathcal{C}^0$.

(d) $\mathcal{C}^{oo} = \mathcal{C}$.

(For (a) and (b) see Hiriart-Urruty (1993), pp120-121 or SW 2.7.5.)

Proof of (a): Let $\boldsymbol{a}$ be an arbitrary point in $\mathcal{C}$. Then, $\boldsymbol{a}^T\boldsymbol{z} \leq 0$, and $\|\boldsymbol{x} - \boldsymbol{a}\|^2 = \|\boldsymbol{x} - \boldsymbol{y}\|^2 + \|\boldsymbol{y} - \boldsymbol{a}\|^2 - 2\boldsymbol{a}^T\boldsymbol{z} \geqslant \|\boldsymbol{x} - \boldsymbol{y}\|^2$. Therefore, $\boldsymbol{y}$ is the point in $\mathcal{C}$ closest to $\boldsymbol{x}$ and hence $\boldsymbol{y} = \Pi(\boldsymbol{x}|\mathcal{C})$. Similarly, let $\boldsymbol{b} \in \mathcal{C}^o$. Then $\boldsymbol{b}^T\boldsymbol{y} \leq 0$, and $\|\boldsymbol{x} - \boldsymbol{b}\|^2 = \|\boldsymbol{x} - \boldsymbol{z}\|^2 + \|\boldsymbol{z} - \boldsymbol{b}\|^2 - 2\boldsymbol{b}^T\boldsymbol{y} \geqslant \|\boldsymbol{x} - \boldsymbol{z}\|^2$. Therefore, $\boldsymbol{z}$ is the point in $\mathcal{C}^o$ closest to $\boldsymbol{x}$ and hence $\boldsymbol{z} = \Pi(\boldsymbol{x}|\mathcal{C}^o)$.

Proof of (b): Let $\boldsymbol{x}^*$ denote $\Pi(\boldsymbol{x}|\mathcal{C})$ and $\boldsymbol{y} = (1+\delta)\boldsymbol{x}^*$ for $|\delta| < 1$. Then, by Proposition 3.12.3 (on page 114), $(\boldsymbol{x} - \boldsymbol{x}^*)^T(\boldsymbol{y} - \boldsymbol{x}^*) \leq 0$ for $|\delta| < 1$. Therefore, $\delta(\boldsymbol{x} - \boldsymbol{x}^*)^T\boldsymbol{x}^* \leq 0$ for $|\delta| < 1$. Since δ may be positive or negative, we have that $(\boldsymbol{x} - \boldsymbol{x}^*)^T\boldsymbol{x}^* = 0$. Let $\boldsymbol{u}$ be an arbitrary point in $\mathcal{C}$ and $\boldsymbol{y} = \boldsymbol{x}^* + \boldsymbol{u}$. Then, $\boldsymbol{y} \in \mathcal{C}$ because $\mathcal{C}$ is a convex cone. Now, $0 \geqslant (\boldsymbol{x} - \boldsymbol{x}^*)^T(\boldsymbol{y} - \boldsymbol{x}^*) = (\boldsymbol{x} - \boldsymbol{x}^*)^T\boldsymbol{u}$. Since $\boldsymbol{u}$ is arbitrary, we have that $\boldsymbol{x} - \boldsymbol{x}^* \in \mathcal{C}^o$. Thus, we have shown that $\boldsymbol{x} = \boldsymbol{x}^* + (\boldsymbol{x} - \boldsymbol{x}^*)$, $\boldsymbol{x}^* \in \mathcal{C}$, $(\boldsymbol{x} - \boldsymbol{x}^*) \in \mathcal{C}^o$ and $(\boldsymbol{x} - \boldsymbol{x}^*)^T\boldsymbol{x}^* = 0$. Therefore, by part (a), $\boldsymbol{x} - \boldsymbol{x}^* = \Pi(\boldsymbol{x}|\mathcal{C}^o)$, and hence $\boldsymbol{x} = \Pi(\boldsymbol{x}|\mathcal{C}) + \Pi(\boldsymbol{x}|\mathcal{C}^o)$ and $\Pi(\boldsymbol{x}|\mathcal{C})^T\Pi(\boldsymbol{x}|\mathcal{C}^o) = 0$.

Proof of (c); Follows directly from parts (a) and (b).

Proof of (d): It follows from the definition of polar cones that $\mathcal{C} \subset \mathcal{C}^{oo}$. To prove the inclusion in the other direction, let $\boldsymbol{x} \in \mathcal{C}^{oo}$ and let $\boldsymbol{x} = \boldsymbol{y} + \boldsymbol{z}$ be the orthogonal decomposition corresponding to $\mathbb{R}^p = \mathcal{C} \oplus \mathcal{C}^o$. Now, $\boldsymbol{z}^T\boldsymbol{z} = \boldsymbol{z}^T\boldsymbol{z} + \boldsymbol{y}^T\boldsymbol{z} = (\boldsymbol{y} + \boldsymbol{z})^T\boldsymbol{z} = \boldsymbol{x}^T\boldsymbol{z} \leq 0$, since $\boldsymbol{x} \in \mathcal{C}^{oo}$ and $\boldsymbol{z} \in \mathcal{C}^o$. It follows that $\boldsymbol{z}^T\boldsymbol{z} = 0$ and hence $\boldsymbol{x} = \boldsymbol{y} \in \mathcal{C}$. Therefore, $\mathcal{C}^{oo} \subset \mathcal{C}$. ■

This result is illustrated in Figures 3.15 and 3.16; the first one shows the projections with respect to an arbitrary matrix $\boldsymbol{V}$ and the second shows those for the identity matrix. Let A be an arbitrary point, and $\boldsymbol{x}$ denote OA. Let $\boldsymbol{y} = OB$ and $\boldsymbol{z} = OC$ be the projections of OA onto $\mathcal{C}$ and $\mathcal{C}^o$, respectively. Now, the foregoing proposition says that OC and OB are orthogonal, $OBAC$ is a rectangle (with respect to $\langle . \rangle_{\boldsymbol{V}}$), and $OA = OB + OC$. Further, the orthogonal decomposition of OA into OB and OC where B and C lie $\mathcal{C}$ and $\mathcal{C}^o$, respectively, is unique.

The following result appears in Rockafellar (1970) on page 146:

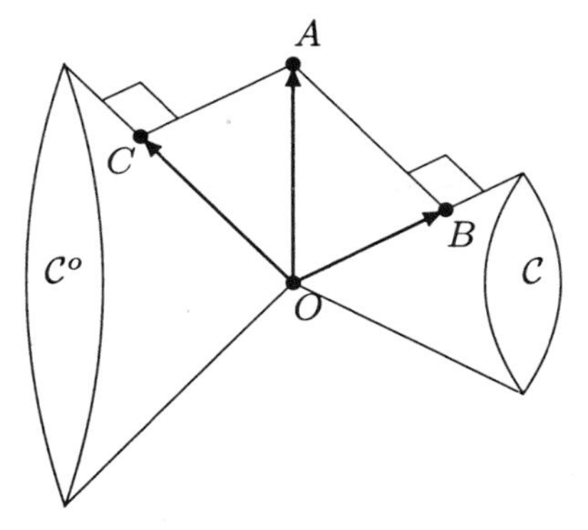

Fig. 3.15 The orthogonal projections of OA onto $\mathcal{C}$ and $\mathcal{C}^o$ when $\boldsymbol{V} \neq \boldsymbol{I}$

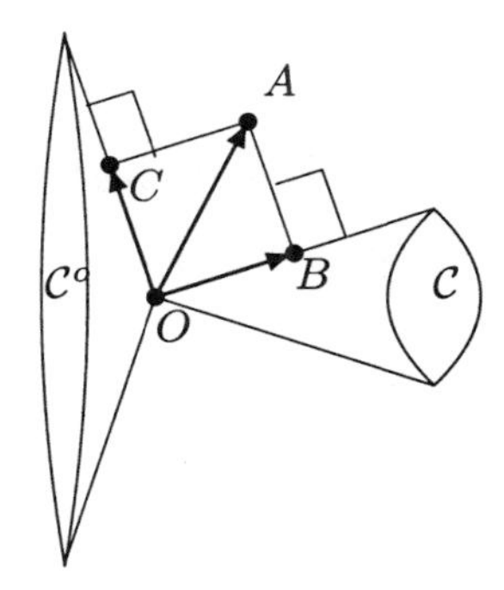

Fig. 3.16 The orthogonal projections of OA onto $\mathcal{C}$ and $\mathcal{C}^o$ when $\boldsymbol{V} = \boldsymbol{I}$.

Proposition 3.12.5 . *Let $\mathcal{C}_1, \ldots, \mathcal{C}_k$ be k convex cones in $\mathbb{R}^p$. Then*

$$(\mathcal{C}_1 + \ldots + \mathcal{C}_k)^o = \mathcal{C}_1^o + \ldots + \mathcal{C}_k^o,$$

$$\text{and} \quad (\mathcal{C}_1 \cap \ldots \cap \mathcal{C}_k)^o = \text{closure of } (\mathcal{C}_1^o + \ldots + \mathcal{C}_k^o). \qquad \blacksquare$$

Example: Let $X, P, Q \subset \mathbb{R}^p$, $P \subset Q^o$, and $X = P \oplus Q$. Show that for any $\boldsymbol{x} \in X$, there exist unique $\boldsymbol{p} \in P$ and $\boldsymbol{q} \in Q$ such that $\boldsymbol{x} = \boldsymbol{p} + \boldsymbol{q}$ and $\boldsymbol{p}^T\boldsymbol{q} = 0$; further, $\boldsymbol{p} = \Pi(\boldsymbol{x}|P)$ and $\boldsymbol{q} = \Pi(\boldsymbol{x}|Q)$.
Solution: Let $\boldsymbol{a} \in P$. Then

$$\begin{aligned}
\|\boldsymbol{x} - \boldsymbol{a}\|^2 = \|\boldsymbol{p} + \boldsymbol{q} - \boldsymbol{a}\|^2 &= \|\boldsymbol{p} - \boldsymbol{a}\|^2 + \|\boldsymbol{q}\|^2 + 2(\boldsymbol{p} - \boldsymbol{a})^T\boldsymbol{q} \\
&= \|\boldsymbol{p} - \boldsymbol{a}\|^2 + \|\boldsymbol{q}\|^2 - 2\boldsymbol{a}^T\boldsymbol{q}, \qquad \text{because } \boldsymbol{p}^T\boldsymbol{q} = 0 \qquad (3.63) \\
&\geqslant \|\boldsymbol{p} - \boldsymbol{a}\|^2 + \|\boldsymbol{q}\|^2;
\end{aligned}$$

the last step follows because $\boldsymbol{a} \in P$ and $\boldsymbol{q} \in Q \subset P^o$ and hence $\boldsymbol{a}^T\boldsymbol{q} \leqslant 0$. Therefore, $\|\boldsymbol{x} - \boldsymbol{a}\|^2$ reaches its minimum over $\boldsymbol{a} \in P$ when $\boldsymbol{a} = \boldsymbol{p}$. Hence, $\boldsymbol{p} = \Pi(\boldsymbol{x}|P)$. Similarly, $\boldsymbol{q} = \Pi(\boldsymbol{x}|Q)$. $\blacksquare$

Another important feature of projections onto linear spaces is that if $\mathcal{M}$ and $\mathcal{L}$ are linear subspaces and $\mathcal{M} \subset \mathcal{L}$ then projecting $\boldsymbol{x}$ onto $\mathcal{L}$ first and then onto $\mathcal{M}$ leads to the same point as projecting $\boldsymbol{x}$ directly onto $\mathcal{M}$; further, this two-stage projection leads to the orthogonal decomposition, $\Pi(\boldsymbol{x}|\mathcal{L}) = \Pi(\boldsymbol{x}|\mathcal{M}) + \Pi(\boldsymbol{x}|\mathcal{L} \cap \mathcal{M}^{\perp})$, and hence $\|\Pi(\boldsymbol{x}|\mathcal{L})\|^2 = \|\Pi(\boldsymbol{x}|\mathcal{M})\|^2 + \|\Pi(\boldsymbol{x}|\mathcal{L} \cap \mathcal{M}^{\perp})\|^2$. This plays an important role for statistical inference in normal theory linear models. For example, consider the likelihood ratio statistic for testing $\boldsymbol{\theta} \in \mathcal{M}$ against $\boldsymbol{\theta} \in \mathcal{L}$ based on $\boldsymbol{X} \sim N(\boldsymbol{\theta}, \boldsymbol{V})$. This LRT statistic is equal to the LRT statistic for testing $\boldsymbol{\theta} = \boldsymbol{0}$ against $\boldsymbol{\theta} \in \mathcal{L} \cap \mathcal{M}^{\perp}$, and hence its null distribution is χ^2_q where $q = dim(\mathcal{L} \cap \mathcal{M}^{\perp}) = dim(\mathcal{L}) - dim(\mathcal{M})$. Similar results hold for projections onto closed convex cones, $\mathcal{C}_1$ and $\mathcal{C}_2$, as well provided that $\mathcal{C}_1 \subset \mathcal{C}_2$ and at least one of them is a linear space. These results are stated below. It is worth noting that if neither $\mathcal{C}_1$ nor $\mathcal{C}_2$ is a linear space then we cannot obtain orthogonal decompositions in general; it will be seen that this is the root of the difficulties in obtaining least favorable null distributions when the null and alternative parameter spaces are $\mathcal{C}_1$ and $\mathcal{C}_2$ with neither of them being a linear space.

Proposition 3.12.6 *Let $\mathcal{L}$ and $\mathcal{M}$ be linear spaces, $\mathcal{C}$ be a closed convex cone, $\mathcal{M} \subset \mathcal{C} \subset \mathcal{L} \subset \mathbb{R}^p$, and $\boldsymbol{y} \in \mathbb{R}^p$. Then*

(a) $\Pi\{\Pi(\boldsymbol{y}|\mathcal{L})|\mathcal{C}\} = \Pi(\boldsymbol{y}|\mathcal{C})$.

(b) $\Pi\{\Pi(\boldsymbol{y}|\mathcal{C})|\mathcal{M}\} = \Pi(\boldsymbol{y}|\mathcal{M})$.

(c) $\mathcal{C} = \mathcal{M} \oplus (\mathcal{C} \cap \mathcal{M}^{\perp})$.

(d-1) *Given $\boldsymbol{x} \in \mathbb{R}^p$, there exist unique $\boldsymbol{m} \in \mathcal{M}, \boldsymbol{m}_o \in \mathcal{M} \cap \mathcal{C}^{\perp}$, and $\boldsymbol{c}_o \in \mathcal{C}^o$ such that $\boldsymbol{x} = \boldsymbol{m} + \boldsymbol{m}_o + \boldsymbol{c}_o$ and the components are pairwise orthogonal, namely $\boldsymbol{m}^T\boldsymbol{m}_o = \boldsymbol{m}^T\boldsymbol{c}_o = \boldsymbol{m}_o^T\boldsymbol{c}_o = 0$. Further, the components $\boldsymbol{m}, \boldsymbol{m}_o$ and $\boldsymbol{c}_0$ are the projections of $\boldsymbol{x}$ onto $\mathcal{M}, \mathcal{C} \cap M^{\perp}$ and $\mathcal{C}^o$, respectively.*

(d-2) $\Pi\{\Pi(\boldsymbol{x} \mid \mathcal{C}) \mid \mathcal{C} \cap \mathcal{M}^{\perp}\} = \Pi(\boldsymbol{x} \mid \mathcal{C} \cap \mathcal{M}^{\perp})$.

(d-3) *For any $\boldsymbol{x} \in \mathcal{C}$, we have*

$$\boldsymbol{x} = \Pi(\boldsymbol{x} \mid \mathcal{M}) + \Pi(\boldsymbol{x} \mid \mathcal{C} \cap \mathcal{M}^{\perp}) + \Pi(\boldsymbol{x} \mid \mathcal{C}^o)$$

and the three projections in this decomposition are pairwise orthogonal.

(d-4) $\mathbb{R}^p = \mathcal{M} \oplus (\mathcal{C} \cap \mathcal{M}^{\perp}) \oplus \mathcal{C}^o$

(d-5) $(\mathcal{C} \cap \mathcal{M}^{\perp})^o = \mathcal{C}^o \oplus \mathcal{M}$.

(e) $\mathcal{L} = \mathcal{C} \oplus (\mathcal{L} \cap \mathcal{C}^o)$.

(f) $\mathbb{R}^p = \mathcal{C} \oplus (\mathcal{C}^o \cap \mathcal{L}) \oplus \mathcal{L}^{\perp}$.

(g) $\boldsymbol{y} - \Pi(\boldsymbol{y}|\mathcal{C}) \perp \Pi(\boldsymbol{y}|\mathcal{M}) - \Pi(\boldsymbol{y}|\mathcal{C})$.

(h) $\|\boldsymbol{y} - \Pi(\boldsymbol{y}|\mathcal{M})\|^2 - \|\boldsymbol{y} - \Pi(\boldsymbol{y}|\mathcal{C})\|^2 = \|\Pi(\boldsymbol{y}|\mathcal{M}) - \Pi(\boldsymbol{y}|\mathcal{C})\|^2$.

(i) $\Pi(\boldsymbol{y}|\mathcal{C}) - \Pi(\boldsymbol{y}|\mathcal{M}) = \Pi(\boldsymbol{y}|\mathcal{C} \cap \mathcal{M}^{\perp})$.

Proof of (a): Let $\mathcal{L} = \{\boldsymbol{X\beta} : \boldsymbol{\beta} \in \mathbb{R}^k\}$ where $\boldsymbol{X}$ is a $p \times k$ matrix of rank k, $\mathcal{K} \subset \mathbb{R}^k$ be a closed convex cone such that $\mathcal{C} = \boldsymbol{X}\mathcal{K}$, and $\hat{\boldsymbol{\beta}} = (\boldsymbol{X}^T\boldsymbol{X})^{-1}\boldsymbol{X}^T\boldsymbol{y}$. Then, by using the normal equation, we have that

$$\|\boldsymbol{y} - \boldsymbol{X\beta}\|^2 = \|\boldsymbol{y} - \boldsymbol{X}\hat{\boldsymbol{\beta}}\|^2 + \|\boldsymbol{X}\hat{\boldsymbol{\beta}} - \boldsymbol{X\beta}\|^2$$

and $\Pi(\boldsymbol{X}|\mathcal{L}) = \boldsymbol{X}\hat{\boldsymbol{\beta}}$. Therefore, the value of $\boldsymbol{\beta} \in \mathcal{K}$, say $\tilde{\boldsymbol{\beta}}$, at which $\|\boldsymbol{y} - \boldsymbol{X\beta}\|^2$ is a minimum is also the same as that at which $\|\boldsymbol{X}\hat{\boldsymbol{\beta}} - \boldsymbol{X\beta}\|^2$ is a minimum. Since $\Pi(\boldsymbol{y}|\mathcal{L}) = \boldsymbol{X}\hat{\boldsymbol{\beta}}$, $\Pi(\boldsymbol{X}\hat{\boldsymbol{\beta}}|\mathcal{C}) = \boldsymbol{X}\tilde{\boldsymbol{\beta}}$ and $\Pi(\boldsymbol{y}|\mathcal{C}) = \boldsymbol{X}\tilde{\boldsymbol{\beta}}$, the proof follows. One could also show this directly using $\|\boldsymbol{y} - \boldsymbol{x}\|^2 = \|\boldsymbol{y} - \boldsymbol{Py}\|^2 + \|\boldsymbol{Py} - \boldsymbol{x}\|^2$ where $\boldsymbol{P}$ is the projection matrix $\boldsymbol{X}(\boldsymbol{X}^T\boldsymbol{X})^{-1}\boldsymbol{X}^T$ onto $\mathcal{L}$, $\boldsymbol{x} \in \mathcal{L}$, and $\boldsymbol{y} \in \mathbb{R}^p$.

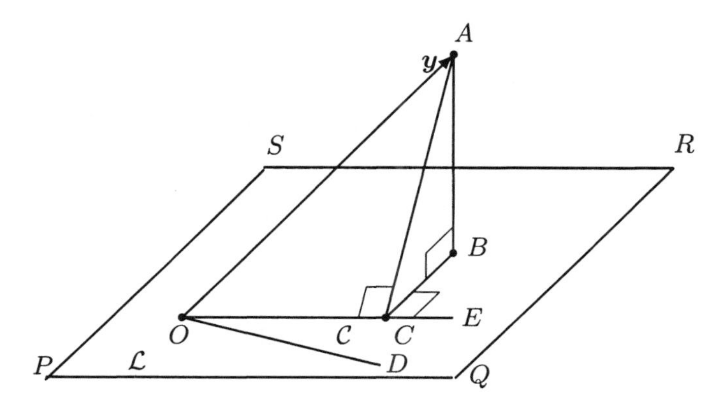

Fig. 3.17 Projections of OA onto the cone, $\mathcal{C}$, and the linear space, $\mathcal{L}$, where $\mathcal{C} \subset \mathcal{L}$.

Fig. 3.17 illustrates this result; $\mathcal{L}$ is the two-dimensional plane containing $PQRS$; O is the origin; $\mathcal{C}$ is the cone EOD contained in $\mathcal{L}$; B is the projection of A onto $\mathcal{L}$; and C is the projection of B onto $\mathcal{C}$. Then C is also the projection of A onto $\mathcal{C}$.
Proof of (b) (Raubertas et al. (1986), p 2815) : Since $\Pi(\boldsymbol{y}|\mathcal{C}) = \boldsymbol{y} - \Pi(\boldsymbol{y}|\mathcal{C}^o)$, we have that $\Pi\{\Pi(\boldsymbol{y}|\mathcal{C})|\mathcal{M}\} = \Pi\{\boldsymbol{y} - \Pi(\boldsymbol{y}|\mathcal{C}^o)|\mathcal{M}\} = \Pi(\boldsymbol{y}|\mathcal{M}) - \Pi\{\Pi(\boldsymbol{y}|\mathcal{C}^o)|\mathcal{M}\}$; the last equality follows since any projection onto any linear space is a linear function. Since $\mathcal{M} \subset \mathcal{C}$, we have that $\mathcal{C}^o \subset \mathcal{M}^\perp$. Therefore, $\Pi\{\Pi(\boldsymbol{y}|\mathcal{C}^o)|\mathcal{M}\} = \mathbf{0}$, and we conclude that $\Pi\{\Pi(\boldsymbol{y}|\mathcal{C})|\mathcal{M}\} = \Pi(\boldsymbol{y}|\mathcal{M})$ (see Fig. 3.10, page 85).

Proof of (c): Let $\mathcal{C}^* = \mathcal{M}^\perp \cap \mathcal{C}$. First let us show that $\mathcal{C} = \{\boldsymbol{y}+\boldsymbol{z} : \boldsymbol{y} \in \mathcal{M}, \boldsymbol{z} \in \mathcal{C}^*\}$. Recall that $\mathbb{R}^p = \mathcal{M} \oplus \mathcal{M}^\perp$. Given $\boldsymbol{x} \in \mathcal{C}$, we can write $\boldsymbol{x} = \boldsymbol{y} + \boldsymbol{z}$, where $\boldsymbol{y} \in \mathcal{M}, \boldsymbol{z} \in \mathcal{M}^\perp$ and $\boldsymbol{y}^T\boldsymbol{z} = 0$. Since $-\boldsymbol{y} \in \mathcal{M} \subset \mathcal{C}$ and $\mathcal{C}$ is a closed convex cone, it follows that $\boldsymbol{z} = \boldsymbol{x} + (-\boldsymbol{y}) \in \mathcal{C}$. Therefore, $\boldsymbol{z} \in \mathcal{C}^*$ and hence $\mathcal{C} \subset \mathcal{M} + \mathcal{C}^*$. Since $\mathcal{M} \subset \mathcal{C}$, $\mathcal{C}^* = \mathcal{M}^\perp \cap \mathcal{C} \subset \mathcal{C}$ and $\mathcal{C}$ is a closed convex cone it follows that $\mathcal{M} + \mathcal{C}^* \subset \mathcal{C}$. Therefore, $\mathcal{C} = \mathcal{M} + \mathcal{C}^*$.

The claim that $\mathcal{M}+\mathcal{C}^*$ is an orthogonal sum follows directly from $\mathbb{R}^p = \mathcal{M} \oplus \mathcal{M}^\perp$. Let $\boldsymbol{x} \in \mathcal{C}$ and let us start with the decomposition, $\boldsymbol{x} = \boldsymbol{y}+\boldsymbol{z}$ in the previous paragraph where $\boldsymbol{y} \in \mathcal{M}, \boldsymbol{z} \in \mathcal{M}^\perp$ and $\boldsymbol{y}^T\boldsymbol{z} = 0$; the pair $\{\boldsymbol{y}, \boldsymbol{z}\}$ is uniquely defined. Since we just proved that $\boldsymbol{z}$ also belongs to $\mathcal{C} \cap \mathcal{M}^\perp$, we have shown the following: $\mathcal{C} = \mathcal{M}+\mathcal{C}^*$ and given $\boldsymbol{x} \in \mathcal{C}$ there exist unique $\boldsymbol{y} \in \mathcal{M}$ and $\boldsymbol{z} \in \mathcal{C}^*$ such that $\boldsymbol{x} = \boldsymbol{y} + \boldsymbol{z}$ and $\boldsymbol{y}^T\boldsymbol{z} = 0$. Therefore, $\mathcal{C} = \mathcal{M} \oplus \mathcal{C}^*$.
Remark: Note that $\boldsymbol{z}$ is the point in $\mathcal{M}^\perp$ closest to $\boldsymbol{x}$, and further since $\boldsymbol{z} \in \mathcal{C}^* \subset \mathcal{M}^\perp$ it is also the point in $\mathcal{C}^*$ closest to $\boldsymbol{x}$ as well. Therefore, $\boldsymbol{z} = \Pi(\boldsymbol{x} \mid \mathcal{C}^*)$ and $\boldsymbol{y} = \Pi(\boldsymbol{x}|\mathcal{M})$.

Proof of (d-1): Let $\boldsymbol{x} \in \mathbb{R}^p$. Since $\mathbb{R}^p = \mathcal{C} \oplus \mathcal{C}^o$, there exist unique $\boldsymbol{c} \in \mathcal{C}$ and $\boldsymbol{c}_0 \in \mathcal{C}^o$ such that $\boldsymbol{x} = \boldsymbol{c} + \boldsymbol{c}_o$ and $\boldsymbol{c}^T\boldsymbol{c}_o = 0$. We also have $\boldsymbol{c} = \Pi(\boldsymbol{x}|\mathcal{C})$ and $\boldsymbol{c}_0 = \Pi(\boldsymbol{x}|\mathcal{C}^o)$. Since $\mathcal{C} = \mathcal{M} \oplus (\mathcal{C} \cap \mathcal{M}^\perp)$, there exist unique $\boldsymbol{m} \in \mathcal{M}$ and $\boldsymbol{m}_0 \in \mathcal{C} \cap \mathcal{M}^\perp$ such that $\boldsymbol{c} = \boldsymbol{m} + \boldsymbol{m}_0$ and $\boldsymbol{m}^T\boldsymbol{m}_0 = 0$. We also have

$$\boldsymbol{m} = \Pi(\boldsymbol{c} \mid \mathcal{M}) = \Pi\{\Pi(\boldsymbol{x} \mid \mathcal{C}) \mid \mathcal{M}\} = \Pi(\boldsymbol{x} \mid \mathcal{M}) \qquad \text{by part (b)}, \qquad (3.64)$$

$$\boldsymbol{m}_0 = \Pi(\boldsymbol{c} \mid \mathcal{C} \cap \mathcal{M}^\perp) = \Pi\{\Pi(\boldsymbol{x} \mid \mathcal{C}) \mid \mathcal{C} \cap \mathcal{M}^\perp\}. \tag{3.65}$$

Since $\mathcal{M} \subset \mathcal{C}$, we have that $\mathcal{C}^o \subset \mathcal{M}^\perp$ and hence $\boldsymbol{c}_o^T \boldsymbol{m} = 0$. Now

$$0 = \boldsymbol{c}_o^T \boldsymbol{c} = \boldsymbol{c}_0^T(\boldsymbol{m} + \boldsymbol{m}_o) = \boldsymbol{c}_o^T \boldsymbol{m} + \boldsymbol{c}_o^T \boldsymbol{m}_o = \boldsymbol{c}_o^T \boldsymbol{m}_o.$$

Therefore, given $\boldsymbol{x} \in \mathbb{R}^p$, we have shown that there exist $\boldsymbol{m} \in \mathcal{M}, \boldsymbol{m}_o \in \mathcal{M} \cap \mathcal{C}^\perp$, and $\boldsymbol{c}_o \in \mathcal{C}^o$ such that $\boldsymbol{x} = \boldsymbol{m} + \boldsymbol{m}_o + \boldsymbol{c}_o$ and the three components are pairwise orthogonal, (i.e., $\boldsymbol{m}^T \boldsymbol{m}_o = \boldsymbol{m}^T \boldsymbol{c}_o = \boldsymbol{m}_o^T \boldsymbol{c}_o = 0$).

To prove uniqueness of the decomposition let $\boldsymbol{x} \in \mathbb{R}^p$. Suppose that $\boldsymbol{x} = \boldsymbol{m}' + \boldsymbol{m}_o' + \boldsymbol{c}_o'$ where $\boldsymbol{m}' \in \mathcal{M}, \boldsymbol{m}_o' \in \mathcal{C} \cap \mathcal{M}^\perp, \boldsymbol{c}_o' \in \mathcal{C}^o$, and $\{\boldsymbol{m}', \boldsymbol{m}_o', \boldsymbol{c}_o'\}$ are pairwise orthogonal. Then, since $\boldsymbol{m}' \in \mathcal{C}$ and $\boldsymbol{m}_o' \in \mathcal{C}$ it follows that $\boldsymbol{m}' + \boldsymbol{m}_o' \in \mathcal{C}$. It also follows from the pairwise orthogonality of the components that $(\boldsymbol{m}' + \boldsymbol{m}_o')^T \boldsymbol{c}_o' = 0$. Since $\mathbb{R}^p = \mathcal{C} \oplus \mathcal{C}^o$ it follows from the orthogonal decomposition $\boldsymbol{x} = (\boldsymbol{m}' + \boldsymbol{m}_0') + \boldsymbol{c}_0'$ that $\boldsymbol{c}_o' = \Pi(\boldsymbol{x} \mid \mathcal{C}^o) = \boldsymbol{c}_0$ and $(\boldsymbol{m}' + \boldsymbol{m}_o') = \Pi(\boldsymbol{x} \mid \mathcal{C})$. Since $\mathcal{C} = \mathcal{M} \oplus (\mathcal{C} \cap \mathcal{M}^\perp)$, $\boldsymbol{m}' + \boldsymbol{m}_o' \in \mathcal{C}, \boldsymbol{m}' \in \mathcal{M}, \boldsymbol{m}_o' \in \mathcal{C} \cap \mathcal{M}^\perp$, and $\boldsymbol{m}'^T \boldsymbol{m}_o' = 0$ it follows that $\boldsymbol{m}' = \Pi\{\boldsymbol{m}' + \boldsymbol{m}_o' \mid \mathcal{M}) = \Pi\{\Pi(\boldsymbol{x} \mid \mathcal{C}) \mid \mathcal{M}\} = \Pi(\boldsymbol{x} \mid \mathcal{M})$ and $\boldsymbol{m}_o' = \Pi\{\boldsymbol{m}' + \boldsymbol{m}_o' \mid \mathcal{C} \cap \mathcal{M}^\perp) = \Pi\{\Pi(\boldsymbol{x} \mid \mathcal{C}) \mid \mathcal{C} \cap \mathcal{M}^\perp\} = \boldsymbol{m}_0$. Therefore, the decomposition $\boldsymbol{x} = \boldsymbol{m} + \boldsymbol{m}_o + \boldsymbol{c}_o$ is unique.

We already noted that $\boldsymbol{c}_o = \Pi(\boldsymbol{x} \mid \mathcal{C}^o)$ and $\boldsymbol{m} = \Pi\{\Pi(\boldsymbol{x} \mid \mathcal{C}) \mid \mathcal{M}\} = \Pi(\boldsymbol{x} \mid \mathcal{M})$. Therefore, to complete the proof of (d-1) it suffices to prove that $\boldsymbol{m}_o = \Pi(\boldsymbol{x} \mid \mathcal{C} \cap \mathcal{M}^\perp)$. To this end, let $\boldsymbol{b} \in \mathcal{C} \cap \mathcal{M}^\perp$. Now,

$$\begin{aligned}(\boldsymbol{m} + \boldsymbol{c}_o)^T(\boldsymbol{m}_o - \boldsymbol{b}) &= \boldsymbol{m}^T(\boldsymbol{m}_0 - \boldsymbol{b}) + \boldsymbol{c}_o^T(\boldsymbol{m}_0 - \boldsymbol{b}) \\ &= 0 + \boldsymbol{c}_o^T(\boldsymbol{m}_0 - \boldsymbol{b}), \qquad \text{because } \boldsymbol{m}_o - \boldsymbol{b} \in \mathcal{M}^\perp \text{ and } m \in \mathcal{M} \\ &= \boldsymbol{c}_o^T \boldsymbol{m} - \boldsymbol{c}_o^T \boldsymbol{b} \\ &= 0 - \boldsymbol{c}_o^T \boldsymbol{b} \qquad \text{because } \boldsymbol{c}_o \in \mathcal{C}^o \subset \mathcal{M}^\perp \text{ and } \boldsymbol{m} \in \mathcal{M}\end{aligned}$$

The last term is nonnegative because $\boldsymbol{c}_o \in \mathcal{C}^o$ and $\boldsymbol{b} \in \mathcal{C}$. Therefore, $(\boldsymbol{m} + \boldsymbol{c}_o)^T(\boldsymbol{m}_o - \boldsymbol{b})$ is nonnegative and reaches its minimum of zero when $\boldsymbol{b} = \boldsymbol{m}_0$. Now,

$$\begin{aligned}\|\boldsymbol{x} - \boldsymbol{b}\|^2 &= \|\boldsymbol{m} + \boldsymbol{m}_o + \boldsymbol{c}_o - \boldsymbol{b}\|^2 \\ &= \|\boldsymbol{m} + \boldsymbol{c}_o\|^2 + \|\boldsymbol{m}_o - \boldsymbol{b}\|^2 + 2(\boldsymbol{m} + \boldsymbol{c}_o)^T(\boldsymbol{m}_o - \boldsymbol{b}). \\ &\geqslant \|\boldsymbol{m} + \boldsymbol{c}_o\|^2 + \|\boldsymbol{m} - \boldsymbol{b}\|^2 \qquad \forall \boldsymbol{b} \in \mathcal{C} \cap \mathcal{M}^\perp,\end{aligned}$$

equality holding when $\boldsymbol{b} = \boldsymbol{m}_0$. Hence, $\|\boldsymbol{x} - \boldsymbol{b}\|^2$ reaches its minimum over $\boldsymbol{b} \in \mathcal{C} \cap \mathcal{M}^\perp$ when $\boldsymbol{b} = \boldsymbol{m}_o$. Therefore, $\Pi(\boldsymbol{x} \mid \mathcal{C} \cap \mathcal{M}^\perp) = \boldsymbol{m}_o = \Pi\{\Pi(\boldsymbol{x} \mid \mathcal{C}) \mid \mathcal{C} \cap \mathcal{M}^\perp\}$.

Proofs of part (d-2), (d-3), and (d-4): Contained in the proof of (d-1).

Proof of part (d-5): Since $\mathcal{C} \cap \mathcal{M}^\perp \subset \mathcal{C}$ we have $\mathcal{C}^o \subset (\mathcal{C} \cap \mathcal{M}^\perp)^o$. Since $\mathcal{C} \cap \mathcal{M}^\perp \subset \mathcal{M}^\perp$ it follows that $\mathcal{M} = (\mathcal{M}^\perp)^\perp = (\mathcal{M}^\perp)^o \subset (\mathcal{C} \cap \mathcal{M}^\perp)^o$. Since $(\mathcal{C} \cap \mathcal{M}^\perp)^o$ is a closed convex cone, it is closed under addition, and hence $\mathcal{C}^o + \mathcal{M} \subset (\mathcal{C} \cap \mathcal{M}^\perp)^o$.

To prove the set containment in the reverse order, first note that $\mathcal{C}^o \subset \mathcal{C}^o + \mathcal{M}$, and hence $(\mathcal{C}^o + \mathcal{M})^o \subset \mathcal{C}$. Since $\mathcal{M} \subset \mathcal{C}^o + \mathcal{M}$, it follows that $(\mathcal{C}^o + \mathcal{M})^o \subset \mathcal{M}^\perp$. Therefore, $(\mathcal{C}^o + \mathcal{M})^o \subset \mathcal{C} \cap \mathcal{M}^\perp$ and $(\mathcal{C} \cap \mathcal{M}^\perp)^o \subset \mathcal{C}^o + \mathcal{M}$.

Therefore, we have proved $(\mathcal{C} \cap \mathcal{M}^{\perp})^{o} = \mathcal{C}^{o} + \mathcal{M}$. The proof that this is an orthogonal sum follows from part (d-4).

Proof of (e): The proof is similar to that for part (c) and is based on $\mathbb{R}^{p} = \mathcal{C} \oplus \mathcal{C}^{o}$.

Proof of (f): Given $\boldsymbol{y} \in \mathbb{R}^{p}$ we have the unique orthogonal decomposition, $\boldsymbol{y} = \Pi(\boldsymbol{y}|\mathcal{L}) + \Pi(\boldsymbol{y}|\mathcal{L}^{\perp})$. Since $\mathcal{L} = \mathcal{C} \oplus (\mathcal{C}^{o} \cap \mathcal{L})$, we also have the unique orthogonal decomposition

$$\begin{aligned} \Pi(\boldsymbol{y}|\mathcal{L}) &= \Pi\{\Pi(\boldsymbol{y}|\mathcal{L})|\mathcal{C})\} + \Pi\{\Pi(\boldsymbol{y}|\mathcal{L})|\mathcal{C}^{o} \cap \mathcal{L})\} \\ &= \Pi(\boldsymbol{y}|\mathcal{C}) + \Pi(\boldsymbol{y}|\mathcal{C}^{o} \cap \mathcal{L}) \qquad \text{by part (a)}. \end{aligned} \tag{3.66}$$

Therefore, we have the orthogonal decomposition,

$$\boldsymbol{y} = \Pi(\boldsymbol{y}|\mathcal{C}) + \Pi(\boldsymbol{y}|\mathcal{C}^{o} \cap \mathcal{L}) + \Pi(\boldsymbol{y}|\mathcal{L}^{\perp}).$$

Uniqueness of the decomposition can also be established as for part (d-1). Hence, $\mathbb{R}^{p} = \mathcal{C} \oplus (\mathcal{C}^{o} \cap \mathcal{L}) \oplus \mathcal{L}^{\perp}$.

Proofs of (g) and (h): It is easily verified that $\mathcal{M}^{\perp} = \mathcal{M}^{o}$, and if $\mathcal{C}_1 \subset \mathcal{C}_2$ then $\mathcal{C}_2^{o} \subset \mathcal{C}_1^{o}$. Since $\mathcal{M} \subset \mathcal{C}$ we have that $\mathcal{C}^{o} \subset \mathcal{M}^{\perp}$ and $\Pi(\boldsymbol{y}|\mathcal{C}^{o})^{T}\Pi(\boldsymbol{y}|\mathcal{M}) = 0$. Note that $\Pi(\boldsymbol{y}|\mathcal{C}^{o}) \in \mathcal{C}^{o} \subset \mathcal{M}^{\perp}$. Therefore, $\Pi(\boldsymbol{y}|\mathcal{C}^{o}) \perp \mathcal{M}$. Now,

$$\begin{aligned} &\{\boldsymbol{y} - \Pi(\boldsymbol{y}|\mathcal{C})\}^{T}\{\Pi(\boldsymbol{y}|\mathcal{C}) - \Pi(\boldsymbol{y}|\mathcal{M})\} \\ &\quad = \Pi(\boldsymbol{y}|\mathcal{C}^{o})^{T}\Pi(\boldsymbol{y}|\mathcal{C}) - \Pi(\boldsymbol{y}|\mathcal{C}^{o})^{T}\Pi(\boldsymbol{y}|\mathcal{M}) = \mathbf{0}. \end{aligned} \tag{3.67}$$

Therefore, $\|\boldsymbol{y} - \Pi(\boldsymbol{y}|\mathcal{M})\|^{2} = \|\boldsymbol{y} - \Pi(\boldsymbol{y}|\mathcal{C})\|^{2} + \|\Pi(\boldsymbol{y}|\mathcal{C}) - \Pi(\boldsymbol{y}|\mathcal{M})\|^{2}$.

Proof of (i):

$$\begin{aligned} \Pi(\boldsymbol{x} \mid \mathcal{C}) &= \Pi\{\Pi(\boldsymbol{x} \mid \mathcal{C}) \mid \mathcal{M}\} + \Pi\{\Pi(\boldsymbol{x} \mid \mathcal{C}) \mid \mathcal{C} \cap \mathcal{M}^{\perp}\} \\ &= \Pi(\boldsymbol{x} \mid \mathcal{M}) + \Pi(\boldsymbol{x} \mid \mathcal{C} \cap \mathcal{M}^{\perp}) \qquad \text{by parts (b) and (c)} \end{aligned} \tag{3.68}$$

The Fig. 3.17 illustrates the main ideas of this proof in three-dimensions. ■

A consequence of the orthogonal sum $\mathcal{C} = \mathcal{M} \oplus (\mathcal{C} \cap \mathcal{M}^{\perp})$ in part(c) of the foregoing proposition is that any $\boldsymbol{y} \in \mathcal{C}$ has a unique decomposition of the form $\boldsymbol{y} = \boldsymbol{a} + \boldsymbol{b}$, where $\boldsymbol{a} \in \mathcal{M}, \boldsymbol{b} \in \mathcal{C} \cap \mathcal{M}^{\perp}, \boldsymbol{a}^{T}\boldsymbol{b} = 0$; further for such a decomposition, $\boldsymbol{a} = \Pi(\boldsymbol{y} \mid \mathcal{M})$, and $\boldsymbol{b} = \Pi(\boldsymbol{y} \mid \mathcal{C} \cap \mathcal{M}^{\perp})$. Similarly, a consequence of the orthogonal sum $\mathbb{R}^{p} = \mathcal{M} \oplus (\mathcal{C} \cap \mathcal{M}^{\perp}) \oplus \mathcal{C}^{o}$ is that any $\boldsymbol{y} \in \mathbb{R}^{p}$ has a unique decomposition of the form $\boldsymbol{y} = \boldsymbol{a} + \boldsymbol{b} + \boldsymbol{c}$, where $\boldsymbol{a} \in \mathcal{M}, \boldsymbol{b} \in \mathcal{C} \cap \mathcal{M}^{\perp}, \boldsymbol{c} \in \mathcal{C}^{o}$, $\boldsymbol{a}^{T}\boldsymbol{b} = 0, \boldsymbol{a}^{T}\boldsymbol{c} = 0, \boldsymbol{b}^{T}\boldsymbol{c} = 0$; and, further, for such a decomposition we have that $\boldsymbol{a} = \Pi(\boldsymbol{y} \mid \mathcal{M})$, $\boldsymbol{b} = \Pi(\boldsymbol{y} \mid \mathcal{C} \cap \mathcal{M}^{\perp})$ and $\boldsymbol{c} = \Pi(\boldsymbol{y} \mid \mathcal{C}^{o})$.

In contrast to the decompositions in linear subspaces, the decomposition $\boldsymbol{x} = \boldsymbol{x}_1 + \boldsymbol{x}_2$, with $\boldsymbol{x}_1 \in \mathcal{C}$ and $\boldsymbol{x}_2 \in \mathcal{C}^{o}$ is not unique when $\mathcal{C}$ is a closed convex cone unless we impose the requirement $\boldsymbol{x}_1^{T}\boldsymbol{x}_2 = 0$; however, the decomposition is optimal in the sense stated below (Hiriart-Urruty and Lemaréchal (1993), pp 118, 121).

Proposition 3.12.7 *Let $\mathcal{C}$ be a closed convex cone $\boldsymbol{x} \in \mathbb{R}^p$. Then we have the following:*
(a) $\boldsymbol{x} = \boldsymbol{x}_1 + \boldsymbol{x}_2$, $\boldsymbol{x}_1 \in \mathcal{C}$, $\boldsymbol{x}_2 \in \mathcal{C}^o \Rightarrow \|\boldsymbol{x}_1\| \geqslant \|\Pi(\boldsymbol{x}|\mathcal{C})\|$ and $\|\boldsymbol{x}_2\| \geqslant \|\Pi(\boldsymbol{x}|\mathcal{C}^o)\|$.
(b) $\|\Pi(\boldsymbol{x}_1|\mathcal{C}) - \Pi(\boldsymbol{x}_2|\mathcal{C})\|^2 \leq \{\Pi(\boldsymbol{x}_1|\mathcal{C}) - \Pi(\boldsymbol{x}_2|\mathcal{C})\}^T(\boldsymbol{x}_1 - \boldsymbol{x}_2)$, for $\boldsymbol{x}_1, \boldsymbol{x}_2 \in \mathbb{R}^p$.
(c) $\|\Pi(\boldsymbol{x}_1|\mathcal{C}) - \Pi(\boldsymbol{x}_2|\mathcal{C})\| \leq \|(\boldsymbol{x}_1 - \boldsymbol{x}_2)\|$, for $\boldsymbol{x}_1, \boldsymbol{x}_2 \in \mathbb{R}^p$. ■

3.12.3 Polyhedral Cones

Let $\boldsymbol{a}_1, \ldots, \boldsymbol{a}_q$ be q points in $\mathbb{R}^p$ and $\mathcal{P} = \{\boldsymbol{x} \in \mathbb{R}^p : \boldsymbol{a}_i^T\boldsymbol{x} \geqslant 0 \text{ for } i = 1, \ldots, q\}$. Then $\mathcal{P}$ is a closed convex cone and it is called a *polyhedral cone.* Note that $\mathcal{P}$ is the intersection of the half-spaces, $\{\boldsymbol{x} : \boldsymbol{a}_1^T\boldsymbol{x} \geqslant 0\}, \ldots,$ and $\{\boldsymbol{x} : \boldsymbol{a}_q^T\boldsymbol{x} \geqslant 0\}$. With the $p \times q$ matrix $\boldsymbol{A}$ defined as $[\boldsymbol{a}_1, \ldots, \boldsymbol{a}_q]$, we may express $\mathcal{P}$ as $\{\boldsymbol{x} \in \mathbb{R}^p : \boldsymbol{A}^T\boldsymbol{x} \geqslant \boldsymbol{0}\}$. Because the equality constraint $\boldsymbol{a}^T\boldsymbol{x} = 0$ is equivalent to the two inequality constraints $\boldsymbol{a}^T\boldsymbol{x} \geqslant 0$ and $(-\boldsymbol{a})^T\boldsymbol{x} \geqslant 0$, the set of constraints $\boldsymbol{A}^T\boldsymbol{x} \geqslant \boldsymbol{0}$ may contain equality constraints as well. We will say that the set of constraints $\boldsymbol{A}^T\boldsymbol{x} \geqslant \boldsymbol{0}$ defining $\mathcal{P}$ is *tight* if $\mathcal{P}$ cannot be defined using a submatrix of $\boldsymbol{A}$ with fewer columns than $\boldsymbol{A}$; in other words, $\boldsymbol{A}$ has no redundant columns, and hence no redundant constraints. Without loss of generality, we shall always assume that the set of constraints defining $\mathcal{P}$ is tight.

Let S be a subset of $\mathbb{R}^p$. Then the *cone generated* by S consists of elements of the form $w_1\boldsymbol{b}_1 + \ldots + w_m\boldsymbol{b}_m$ where $\{\boldsymbol{b}_1, \ldots, \boldsymbol{b}_m\} \subset S$, $w_i \geqslant 0$, for $i = 1, \ldots, m$; it is also called the *conical hull* or *positive hull* of S. The elements of S are called *generators* of the cone generated by S. The conical hull is the smallest cone that contains S. The set consisting of elements of the form $w_1\boldsymbol{b}_1 + \ldots + w_m\boldsymbol{b}_m$ where $\{\boldsymbol{b}_1, \ldots, \boldsymbol{b}_m\} \subset S$, $\sum w_i = 1$, and $w_i \geqslant 0$, for $i = 1, \ldots, m$, is called the *convex hull* of S. The convex hull is the smallest convex set containing S.

An explicit formula for the relationship between a polyhedral cone and its polar cone is given in the next result.

Proposition 3.12.8 *Let $\boldsymbol{a}_1, \ldots, \boldsymbol{a}_q$ be q points in $\mathbb{R}^p$ and let $\mathcal{C} = \{w_1\boldsymbol{a}_1 + \ldots + w_q\boldsymbol{a}_q : w_i \geqslant 0, i = 1, \ldots, q\}$, the cone generated by $\{\boldsymbol{a}_1, \ldots, \boldsymbol{a}_q\}$. Then $\mathcal{C}^o = \{\boldsymbol{x} \in \mathbb{R}^p : \boldsymbol{a}_i^T\boldsymbol{x} \leq 0 \text{ for } i = 1, \ldots, q\}$. Consequently, if $\mathcal{P} = \{\boldsymbol{x} : \boldsymbol{a}_i^T\boldsymbol{x} \geqslant 0, \text{ for } i = 1, \ldots, q\}$ then $\mathcal{P}^o = \{w_1\boldsymbol{a}_1 + \ldots + w_q\boldsymbol{a}_q : w_i \leq 0, i = 1, \ldots, q\}$.*
[For example see, Hiriart-Urutty and Lemaréchal (1993), p 119].

Proof: Let $\boldsymbol{x} \in \mathcal{C}^o$. Then $\boldsymbol{x}^T(w_1\boldsymbol{a}_1 + \ldots + w_q\boldsymbol{a}_q) \leq 0$ for $(w_1, \ldots, w_q) \geqslant 0$. Now, by choosing $w_i > 0$ and the $w_j = 0$ for $j \neq i$, we have that $\boldsymbol{a}_i^T\boldsymbol{x} \leq 0$ for $i = 1, \ldots, q$. Therefore, $\boldsymbol{x} \in \{\boldsymbol{x} : \boldsymbol{a}_i^T\boldsymbol{x} \leq 0, \text{ for } i = 1, \ldots, q\}$. To prove the set containment in the opposite direction, let $\boldsymbol{x} \in \{\boldsymbol{x} : \boldsymbol{a}_i^T\boldsymbol{x} \leq 0, \text{ for } i = 1, \ldots, q\}$ and $\boldsymbol{z} \in \mathcal{C}$. Then $\boldsymbol{z} = (w_1\boldsymbol{a}_1 + \ldots + w_q\boldsymbol{a}_q)$ for some $w_i \geqslant 0$ for $i = 1, \ldots, q$. Now, $\boldsymbol{x}^T\boldsymbol{z} = w_1\boldsymbol{a}_1^T\boldsymbol{x} + \ldots + w_q\boldsymbol{a}_q^T\boldsymbol{x} \leq 0$. Therefore, $\boldsymbol{x} \in \mathcal{C}^o$. The second part of the proposition follows from the first by substituting $\mathcal{P} = -\mathcal{C}^o$ and noting that $\mathcal{P}^o = -\mathcal{C}^{oo} = -\mathcal{C}$. ■

The following restatement of the foregoing proposition in matrix notation is useful.

Corollary 3.12.9 *Let $\boldsymbol{a}_1, \ldots, \boldsymbol{a}_q$ be q points in $\mathbb{R}^p$ and let $\boldsymbol{A} = [\boldsymbol{a}_1, \ldots, \boldsymbol{a}_q]$.*

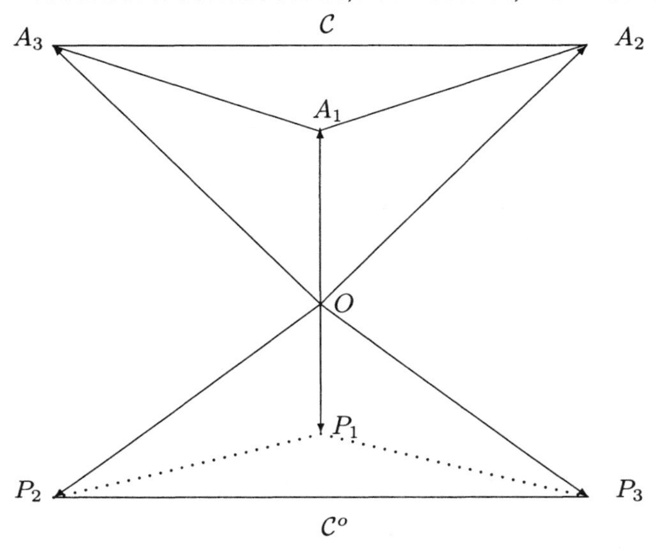

Fig. 3.18 A polyhedral, $\mathcal{C}$, and its polar cone $\mathcal{C}^o$

1. *If* $\mathcal{P} = \{\boldsymbol{x} : \boldsymbol{A}^T\boldsymbol{x} \geqslant \boldsymbol{0}\}$ *then* $\mathcal{P}^o = \{\boldsymbol{Ay} : \boldsymbol{y} \leq \boldsymbol{0}\}$.
2. *If* $\mathcal{C} = \{\boldsymbol{Ax} : \boldsymbol{x} \geqslant \boldsymbol{0}\}$ *then* $\mathcal{C}^o = \{\boldsymbol{y} : \boldsymbol{A}^T\boldsymbol{y} \leq \boldsymbol{0}\}$. ■

Fig. 3.18 provides a representation of this relationship between a polyhedral and its polar in three-dimensions. Let OA_1, OA_2, and OA_3 correspond to $\boldsymbol{a}_1, \boldsymbol{a}_2$, and $\boldsymbol{a}_3$, respectively, and $\mathcal{C} = \{\boldsymbol{Ax} : \boldsymbol{x} \geqslant \boldsymbol{0}\}$. Let $OA_1 \perp OP_2P_3$, $OA_2 \perp OP_1P_3$, and $OA_3 \perp OP_1P_2$. Then $OP_1 \perp OA_2A_3$, $OP_2 \perp OA_1A_3$, and $OP_3 \perp OA_1A_2$. The polyhedrals $OA_1A_2A_3$ and $OP_1P_2P_3$ are $\mathcal{C}$ and $\mathcal{C}^o$, respectively. Note that every boundary plane of $\mathcal{C}$ corresponds to an edge of $\mathcal{C}^o$, the two being orthogonal to each other; similarly, every edge of $\mathcal{C}$ corresponds to a boundary plane of $\mathcal{C}^o$, again the two being orthogonal to each other. This correspondence defines the function Φ in Proposition 3.12.11.

Now we state two useful results on polyhedral cones; for example see SW 2.8.6 and 2.8.8.

Proposition 3.12.10 *(a) Minkowski's theorem: For every polyhedral cone* $\mathcal{P}$ *there exist* $\boldsymbol{a}_1, \ldots, \boldsymbol{a}_q$ *in* $\mathbb{R}^p$ *such that* $\mathcal{P} = \{w_1\boldsymbol{a}_1 + \ldots + w_q\boldsymbol{a}_q : w_i \geqslant 0\, i = 1, \ldots, q\}$. *(b) Weyl's theorem: Every closed convex cone with finitely many generators is a polyhedral cone.* ■

Let $\boldsymbol{A} = (\boldsymbol{a}_1, \ldots, \boldsymbol{a}_q)$ be a $p \times q$ matrix, and $\mathcal{P} = \{\boldsymbol{x} : \boldsymbol{A}^T\boldsymbol{x} \geqslant \boldsymbol{0}\}$. Let $J = \{1, \ldots, m\}$ be a subset of $\{1, \ldots, q\}$; J may be empty. Let $I = \{1, \ldots, q\} \setminus J$. Let $\boldsymbol{A}_J$ and $\boldsymbol{A}_I$ denote the matrices with their columns being $\{\boldsymbol{a}_j : j \in J\}$ and $\{\boldsymbol{a}_i : i \in I\}$ respectively. The set $F_J = \{\boldsymbol{x} : \boldsymbol{A}_J^T\boldsymbol{x} = \boldsymbol{0}, \boldsymbol{A}_I^T\boldsymbol{x} \geqslant \boldsymbol{0}\}$ is called a *face* of $\mathcal{P}$; without loss of generality, it is assumed that the set of constraints defining F_J

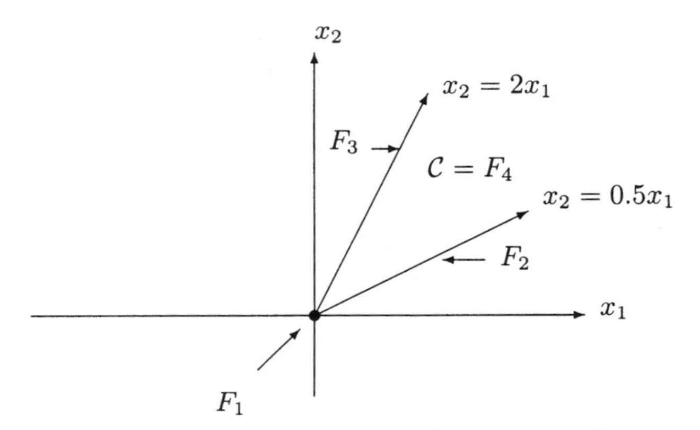

Fig. 3.19 The faces of the polyhedral $\{\boldsymbol{x} : 2x_1 - \frac{1}{2}x_2 \geqslant 0, x_2 - \frac{1}{2}x_1 \geqslant 0\}$.

is tight. Now, the *relative interior* of F_J, denoted, $ri(F_J)$, is defined as the interior of F_J with respect to the relative topology induced in the linear space spanned by F_J; it may be verified that the relative interior of a face is obtained by replacing the inequality constraints defining F_J by their corresponding strict inequalities; thus, $ri(F_J) = \{\boldsymbol{x} : \boldsymbol{A}_J^T\boldsymbol{x} = \boldsymbol{0}, \boldsymbol{A}_I^T\boldsymbol{x} > \boldsymbol{0}\}$. The next example illustrates these definitions.

Example: Let $C = \{\boldsymbol{x} : 2x_1 - x_2 \geqslant 0, x_2 - \frac{1}{2}x_1 \geqslant 0\}$.
Then C is a polyhedral cone and its faces are (see Fig. 3.19)

$$\begin{aligned}
F_1 &= \{\boldsymbol{x} : 2x_1 - x_2 = 0, x_2 - 2^{-1}x_1 = 0\}, \\
F_2 &= \{\boldsymbol{x} : 2x_1 - x_2 \geqslant 0, x_2 - 2^{-1}x_1 = 0\}, \\
F_3 &= \{\boldsymbol{x} : 2x_1 - x_2 = 0, x_2 - 2^{-1}x_1 \geqslant 0\}, \\
F_4 &= \{\boldsymbol{x} : 2x_1 - x_2 \geqslant 0, x_2 - 2^{-1}x_1 \geqslant 0\},
\end{aligned}$$

Descriptions of the faces of C and their relative interiors are:

$F_1 =$ the vertex,	$ri(F_1) =$ the vertex,
$F_2 =$ the lower boundary of C	$ri(F_2) = F_2$ excluding the vertex
$F_3 =$ the upper boundary of C	$ri(F_3) = F_3$ excluding the vertex
$F_4 = C$	$ri(F_4) = C$ excluding its boundary, $F_2 \cup F_3$.■

Note that the polyhedral $\mathcal{P}$ is the union of sets of the form $ri(F_J)$, each of which is an open set in a linear subspace, and hence does not include its boundary. A main feature of statistical inference problems under inequality constraints is that the parameter space includes the boundary. An advantage of $ri(F_J)$ is that the parameter space $\mathcal{P}$ that includes its boundary is partitioned into sets of the form $ri(F_J)$, each of which is open with respect to the linear space spanned by $ri(F_J)$ and has a linear structure. Consequently, we can apply well-known results for open parameter spaces to a polyhedral $\mathcal{P}$ as well by considering each member of the partition of $\mathcal{P}$ separately.

There is an important relationship between the faces of a polyhedral cone and those of its polar cone. This was useful for deriving the chi-bar-square distribution for a general closed convex cone; however, these relationships are not required to understand the proofs in the form they are presented in this monograph. For completeness, we state these results below; for example, see SW 2.13.2, 2.13.3.

Proposition 3.12.11 *Let $\mathcal{F}(\mathcal{P}) = \{F : F \text{ is a face of } \mathcal{P} \text{ and } F \text{ is nonempty}\}$; it is the collection of faces of $\mathcal{P}$. Then*
1. $F^{\perp} \cap \mathcal{P}^o$ is a face of $\mathcal{P}^o$.
2. For any given face F of $\mathcal{P}$, let Φ denote the operation of choosing $F^{\perp} \cap \mathcal{P}^o$. Then $\mathcal{F}(\mathcal{P}) \xrightarrow{\Phi} \mathcal{F}(\mathcal{P}^o) \xrightarrow{\Phi} \mathcal{F}(\mathcal{P}^{oo}) = \mathcal{F}(\mathcal{P})$ and $\Phi\Phi(F) = F$.
3. $< F^{\perp} \cap \mathcal{P}^o > = < F >^{\perp}$ where $< A >$ denotes the linear space spanned by A. Further, every face of $\mathcal{P}$ is uniquely associated with the face $F^{\perp} \cap \mathcal{P}^o$ of $\mathcal{P}^o$, and any face of $\mathcal{P}^o$ is of the form $F^{\perp} \cap \mathcal{P}^o$ for some face F of $\mathcal{P}$. ■

3.13 APPENDIX 2: PROOFS

Proof of Theorem 3.4.2: To prove this theorem it is convenient to establish several lemmas that are of independent interest.

Lemma 3.13.1 .

1. *Suppose that $\boldsymbol{Z} \sim N(\boldsymbol{0}, \boldsymbol{I})$ and let $\mathcal{P} \subset \mathbb{R}^p$ be a cone. Then the direction and length of $\boldsymbol{Z}$ are independent. Consequently, $\|\boldsymbol{Z}\|$ and $\boldsymbol{Z}/\|\boldsymbol{Z}\|$ are independent random variables and $\{\|\boldsymbol{Z}\| \geqslant c\}$ and $\{\boldsymbol{Z} \in \mathcal{P}\}$ are independent events.*

2. *More generally, suppose that the distribution of $\boldsymbol{X}$ is orthogonally invariant (i.e., $\boldsymbol{X}$ and $\boldsymbol{PX}$ have the same distribution for any orthogonal matrix $\boldsymbol{P}$). Then $\{\|\boldsymbol{Z}\| \geqslant c\}$ and $\{\boldsymbol{Z} \in \mathcal{P}\}$ are independent events and $\boldsymbol{Z}/\|\boldsymbol{Z}\|$ is distributed uniformly on the unit sphere.*

Proof: First, let us transform $\boldsymbol{Z}$ into polar coordinates: $Z_1 = R\sin\theta_1$, $Z_2 = R\cos\theta_1\sin\theta_2$, $Z_3 = R\cos\theta_1\cos\theta_2\sin\theta_3, \ldots, Z_{p-1} = R\cos\theta_1\cos\theta_2\ldots\cos\theta_{p-2}\sin\theta_{p-1}$, $Z_p = R\cos\theta_1\cos\theta_2\ldots\cos\theta_{p-2}\cos\theta_{p-1}$, where $-(\pi/2) < \theta_i \leqslant (\pi/2)$ for $i = 1, \ldots, p-2$ and $-\pi < \theta_{p-1} \leqslant \pi$. Then $R^2 = \|\boldsymbol{Z}\|^2$ and the Jacobian of the transformation is $R^{p-1}\cos^{p-2}\theta_1\cos^{p-3}\theta_2\ldots\cos\theta_{p-2}$ (see Anderson (1984), page 280). Now, since the density of $\boldsymbol{Z}$ at $\boldsymbol{z}$ is a function of $\|\boldsymbol{z}\|$ only, it follows that the density of $(R, \theta_1, \ldots, \theta_{p-1})$ factors into two components, one is a function of R only and the other is a function of $(\theta_1, \ldots, \theta_{p-1})$. Now the proof follows because $\{\boldsymbol{Z}/\|\boldsymbol{Z}\|$ depends on $(\theta_1, \ldots, \theta_{p-1})$ only. A similar result also appeared in Kudo (1963).

The second result is stated in Perlman (1969) and is essentially the same as the first. ■

Lemma 3.13.2 *Let $\mathcal{P} \subset \mathbb{R}^p$ be a polyhedral cone, $\boldsymbol{y} \in \mathbb{R}^p$, F be a face of $\mathcal{P}$, $\mathcal{L}$ denote the linear space spanned by F, and $\boldsymbol{P}$ denote the projection matrix onto $\mathcal{L}$. Let the statements (A), (B) and (C) be defined as follows:*

(A) $\Pi(\boldsymbol{y}|\mathcal{P}) \in ri(F)$.

(B) $\boldsymbol{Py} \in ri(F)$ *and* $(\boldsymbol{I} - \boldsymbol{P})\boldsymbol{y} \in F^{\perp} \cap \mathcal{P}^o$.

(C) $\boldsymbol{Py} = \Pi(\boldsymbol{y}|\mathcal{P})$ *and* $(\boldsymbol{I} - \boldsymbol{P})\boldsymbol{y} = \Pi(\boldsymbol{y}|\mathcal{P}^o)$.

Then, (A) is true if and only if (B) is true. If (B) is true then (C) is true.

Proof: Let us recall that $\mathbb{R}^p = \mathcal{L} \oplus \mathcal{L}^{\perp}$ where $\oplus$ is an orthogonal sum. Consequently, $\boldsymbol{y} \in \mathbb{R}^p$ has a unique orthogonal decomposition of the form $\boldsymbol{y} = \boldsymbol{a} + \boldsymbol{b}$, where $\boldsymbol{a} \in \mathcal{L}, \boldsymbol{b} \in \mathcal{L}^{\perp}$ and $\boldsymbol{a}^T\boldsymbol{b} = 0$; further, $\boldsymbol{a} = \Pi(\boldsymbol{y}|\mathcal{L}) = \boldsymbol{Py}$ and $\boldsymbol{b} = \Pi(\boldsymbol{y}|\mathcal{L}^{\perp}) = (\boldsymbol{I} - \boldsymbol{P})\boldsymbol{y}$. A similar result holds for closed convex cones as well: $\mathbb{R}^p = \mathcal{P} \oplus \mathcal{P}^o$ is an orthogonal sum. Consequently, $\boldsymbol{y} \in \mathbb{R}^p$ has a unique orthogonal decomposition of the form $\boldsymbol{y} = \boldsymbol{p} + \boldsymbol{q}$, where $\boldsymbol{p} \in \mathcal{P}, \boldsymbol{q} \in \mathcal{P}^o$, and $\boldsymbol{p}^T\boldsymbol{q} = 0$; further, $\boldsymbol{p} = \Pi(\boldsymbol{y}|\mathcal{P})$ and $\boldsymbol{q} = \Pi(\boldsymbol{y}|\mathcal{P}^o)$. Consequently, if $\boldsymbol{y} = \boldsymbol{p} + \boldsymbol{q}$ where $\boldsymbol{p} \in \mathcal{P}, \boldsymbol{q} \in \mathcal{P}^o$, and $\boldsymbol{p}^T\boldsymbol{q} = 0$, then $\boldsymbol{p} = \Pi(\boldsymbol{y}|\mathcal{P})$ and $\boldsymbol{q} = \Pi(\boldsymbol{y}|\mathcal{P}^o)$. Let us write down the two unique orthogonal decompositions of $\boldsymbol{y}$:

$$\boldsymbol{y} = \Pi(\boldsymbol{y}|\mathcal{L}) + \Pi(\boldsymbol{y}|\mathcal{L}^{\perp}) = \Pi(\boldsymbol{y}|\mathcal{P}) + \Pi(\boldsymbol{y}|\mathcal{P}^o) \tag{3.69}$$

Proof of (B) $\Rightarrow$ (C) : Assume that (B) is true. Since $\Pi(\boldsymbol{y}|\mathcal{L}) = \boldsymbol{Py} \in ri(F) \subset \mathcal{P}$, $(\boldsymbol{I} - \boldsymbol{P})\boldsymbol{y} = \Pi(\boldsymbol{y}|\mathcal{L}^{\perp}) = F^{\perp} \cap \mathcal{P}^o \subset \mathcal{P}^o$ and $\Pi(\boldsymbol{y}|\mathcal{L})^T\Pi(\boldsymbol{y}|\mathcal{L}^{\perp}) = 0$ it follows that the two orthogonal decompositions in (3.69) are identical, and hence $\boldsymbol{Py} = \Pi(\boldsymbol{y}|\mathcal{L}) = \Pi(\boldsymbol{y}|\mathcal{P})$ and $(\boldsymbol{I} - \boldsymbol{P})\boldsymbol{y} = \Pi(\boldsymbol{y}|\mathcal{L}^{\perp}) = \Pi(\boldsymbol{y}|\mathcal{P}^o)$; this completes the proof of (C).
Proof of (B) $\Rightarrow$ (A): Follows since $\Pi(\boldsymbol{y}|\mathcal{P}) = \boldsymbol{Py} \in ri(F)$. (see Fig. 3.20).
Proof of (A) $\Rightarrow$ (B): Assume that (A) is true. Let $\boldsymbol{y}^*$ denote $\Pi(\boldsymbol{y}|\mathcal{P})$. Then $(\boldsymbol{y} - \boldsymbol{y}^*)^T(\boldsymbol{x} - \boldsymbol{y}^*) \leq 0$ for every $\boldsymbol{x} \in \mathcal{P}$ (by Proposition 3.12.3 on page 114). Suppose that $(\boldsymbol{y} - \boldsymbol{y}^*)^T(\boldsymbol{x} - \boldsymbol{y}^*)$ is negative for some $\boldsymbol{x} \in ri(F)$. Then $g(\boldsymbol{t}) = (\boldsymbol{y} - \boldsymbol{y}^*)^T(\boldsymbol{t} - \boldsymbol{y}^*)$ is negative for values of $\boldsymbol{t}$ in a small neighborhood of $\boldsymbol{x}$ contained in $ri(F)$. Therefore, there is an $\boldsymbol{x}_1 \in ri(F)$ such that $(\boldsymbol{y} - \boldsymbol{y}^*)^T(\boldsymbol{x}_1 - \boldsymbol{y}^*) < 0$, and hence $\boldsymbol{y}$ is closer to $\boldsymbol{x}_1$ in $\mathcal{P}$ than to $\boldsymbol{y}^*$; this is a contradiction because $\boldsymbol{y}^*$ is the closest point in $\mathcal{P}$. Therefore, $(\boldsymbol{y} - \boldsymbol{y}^*)^T(\boldsymbol{x} - \boldsymbol{y}^*) = 0$ for every $\boldsymbol{x}$ in $ri(F)$ and $\boldsymbol{y} - \boldsymbol{y}^*$ is orthogonal to $\mathcal{L}$. Since $\boldsymbol{y}^* \in ri(F)$ it follows that $\boldsymbol{y}^* \in \mathcal{L}$. Therefore, $\boldsymbol{Py} = \Pi(\boldsymbol{y}|\mathcal{L}) = \boldsymbol{y}^* \in ri(F)$. Further, $\boldsymbol{y} - \boldsymbol{y}^* = \Pi(\boldsymbol{y}|\mathcal{P}^o) \in \mathcal{P}^o$ by (3.69) and $\boldsymbol{y} - \boldsymbol{y}^* = (\boldsymbol{I} - \boldsymbol{P})\boldsymbol{y} = \Pi(\boldsymbol{y}|\mathcal{L}^{\perp}) \in \mathcal{L}^{\perp} \subset F^{\perp}$. Therefore, $\boldsymbol{y} - \boldsymbol{y}^* \in F^{\perp} \cap \mathcal{P}^o$. This completes the proof of (A) $\Rightarrow$ (B).
■

Lemma 3.13.3 *Let* $\boldsymbol{Y} \sim N(\boldsymbol{0}, \boldsymbol{I})$, $\boldsymbol{P}$ *be a projection matrix,* $r = rank(\boldsymbol{P})$ *and* $\mathcal{C} \subset \mathbb{R}^p$ *be a nonempty cone. Then the distribution of* $\|\boldsymbol{PY}\|^2$, *conditional on* $\boldsymbol{PY} \in \mathcal{C}$, *is* χ^2_r.

Proof : Let $\boldsymbol{P} = \boldsymbol{UDU}^T$ where $\boldsymbol{U}$ is an orthogonal matrix and $\boldsymbol{D}$ is a diagonal matrix that we assume, without loss of generality, to be $diag(1, \ldots, 1, 0, \ldots, 0)$ with each of the first r diagonal elements equal to 1 and the rest being zeros. Let $\boldsymbol{Z} = \boldsymbol{U}^T\boldsymbol{Y}$, and $\boldsymbol{Z}_a = (Z_1, \ldots, Z_r)^T$. Then $\boldsymbol{Z} \sim N_p(\boldsymbol{0}, \boldsymbol{I})$ and $\boldsymbol{Z}_a \sim N_r(\boldsymbol{0}, \boldsymbol{I})$. Further,

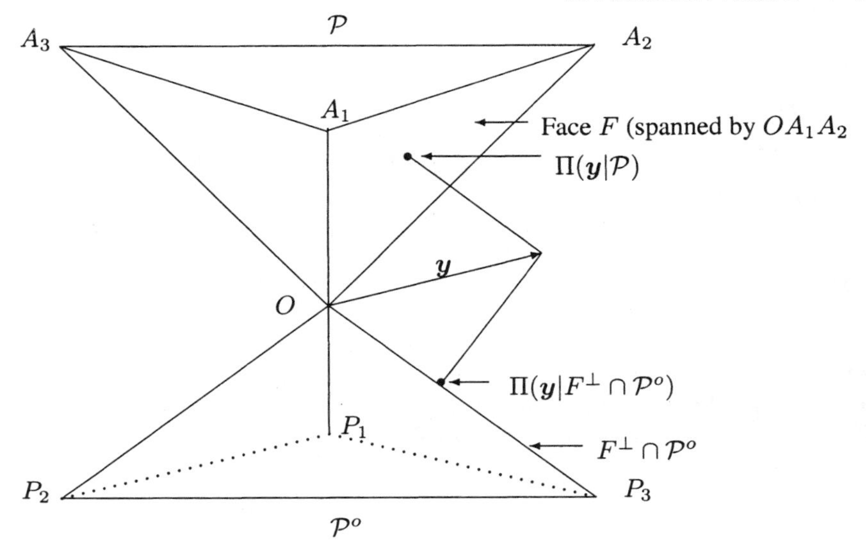

Fig. 3.20 Projections onto F and $F^{\perp} \cap \mathcal{P}^o$.

since $\boldsymbol{U}$ is a rotation matrix, we have that $\|\boldsymbol{PY}\|^2 = \|\boldsymbol{UDU}^T\boldsymbol{Y}\|^2 = \|\boldsymbol{UDZ}\|^2 = \|\boldsymbol{DZ}\|^2 = Z_1^2 + \ldots + Z_r^2$. Now, with $\{\boldsymbol{PY} \in \mathcal{C}\}$ denoting the event that $\boldsymbol{PY} \in \mathcal{C}$, we have that $\{\boldsymbol{PY} \in \mathcal{C}\} = \{\boldsymbol{UDU}^T\boldsymbol{Y} \in \mathcal{C}\} = \{\boldsymbol{U}(\boldsymbol{Z}_a : \boldsymbol{0})^T \in \mathcal{C}\} = \{\boldsymbol{Z}_a \in \mathcal{K}\}$, where $\mathcal{K}$ is a cone in $\mathbb{R}^r$. Since the length and direction of $N(\boldsymbol{0}, \boldsymbol{I})$ are independent, we have that $\text{pr}(\|\boldsymbol{PY}\|^2 \geqslant c \mid \boldsymbol{PY} \in \mathcal{C}) = \text{pr}(Z_1^2 + \ldots + Z_r^2 \geqslant c \mid \boldsymbol{Z}_a \in \mathcal{K}) = \text{pr}(\chi_r^2 \geqslant c)$. ■

Lemma 3.13.4 *Let $\boldsymbol{Y} \sim N(\boldsymbol{0}, \boldsymbol{I})$, $\mathcal{P} \subset \mathbb{R}^p$ be a polyhedral cone, F be a face of $\mathcal{P}$, $\boldsymbol{P}$ be the projection matrix onto the linear space spanned by F, and $r = rank(\boldsymbol{P})$. Then we have the following.*
(i) $pr\{\|\Pi(\boldsymbol{Y}|\mathcal{P})\|^2 \geqslant c \mid \Pi(\boldsymbol{Y}|\mathcal{P}) \in ri(F)\} = pr(\chi_r^2 \geqslant c)$.
(ii) Conditional on $\Pi(\boldsymbol{Y}|\mathcal{P}) \in ri(F)$, $\|\Pi(\boldsymbol{y}|\mathcal{P})\|^2$ and $\|\Pi(\boldsymbol{y}|\mathcal{P}^o)\|^2$, are independent and are distributed as χ_r^2 and χ_{p-r}^2, respectively. Hence

$$
\begin{aligned}
&pr\{\|\Pi(\boldsymbol{Y}|\mathcal{P})\|^2 \geqslant c_1, \|\Pi(\boldsymbol{Y}|\mathcal{P}^o)\|^2 \geqslant c_2 \mid \Pi(\boldsymbol{Y}|\mathcal{P}) \in ri(F)\} \\
&\quad = pr(\chi_r^2 \geqslant c_1) pr(\chi_{p-r}^2 \geqslant c_2).
\end{aligned}
\tag{3.70}
$$

Proof:
(i) Since $\boldsymbol{PY}$ and $(\boldsymbol{I} - \boldsymbol{P})\boldsymbol{Y}$ are independent and $ri(F)$ is a cone, it follows from the previous two lemmas that $\text{pr}\{\|\Pi(\boldsymbol{y}|\mathcal{P})\|^2 \geqslant c \mid \Pi(\boldsymbol{Y}|\mathcal{P}) \in ri(F)\} = \text{pr}\{\|\boldsymbol{PY}\|^2 \geqslant c \mid \Pi(\boldsymbol{Y}|\mathcal{P}) \in ri(F)\} = \text{pr}\{\|\boldsymbol{PY}\|^2 \geqslant c \mid \boldsymbol{PY} \in ri(F), (\boldsymbol{I} - \boldsymbol{P})\boldsymbol{Y} \in F^{\perp} \cap \mathcal{P}^o\} = \text{pr}\{\|\boldsymbol{PY}\|^2 \geqslant c \mid \boldsymbol{PY} \in ri(F)\} = \text{pr}(\chi_r^2 \geqslant c)$.
(ii) By arguments similar to those for the previous part and the fact that $F^{\perp} \cap \mathcal{P}^o$ is

a cone, we have the following:

$$\begin{aligned}
&\text{pr}\{\|\Pi(\boldsymbol{Y}|\mathcal{P})\|^2 \geqslant c_1, \|\Pi(\boldsymbol{Y}|\mathcal{P}^o)\|^2 \geqslant c_2 \mid \Pi(\boldsymbol{Y}|\mathcal{P}) \in ri(F)\} \\
&= \text{pr}\{\|\boldsymbol{PY}\|^2 \geqslant c_1, \|(\boldsymbol{I}-\boldsymbol{P})\boldsymbol{Y}\|^2 \geqslant c_2 \mid \boldsymbol{PY} \in ri(F), (\boldsymbol{I}-\boldsymbol{P})\boldsymbol{Y} \in F^{\perp} \cap \mathcal{P}^o\} \\
&= \text{pr}\{\|\boldsymbol{PY}\|^2 \geqslant c_1 \mid \boldsymbol{PY} \in ri(F)\} \times \\
&\qquad \text{pr}\{\|(\boldsymbol{I}-\boldsymbol{P})\boldsymbol{Y}\|^2 \geqslant c_2 \mid (\boldsymbol{I}-\boldsymbol{P})\boldsymbol{Y} \in F^{\perp} \cap \mathcal{P}^o\} \\
&= \text{pr}(\chi_r^2 \geqslant c_1)\text{pr}(\chi_{p-r}^2 \geqslant c_2). \qquad \blacksquare
\end{aligned}$$

Lemma 3.13.5 *Let $\mathcal{P}$ be a polyhedral cone in $\mathbb{R}^p$. Then there exist a collection of faces of $\mathcal{P}$, say $\{F_1, \ldots, F_K\}$, such that the collection of their relative interiors, $\{ri(F_1), \ldots, ri(F_K)\}$, forms a partition of $\mathcal{P}$. Further,*

$$\|\Pi(\boldsymbol{y}|\mathcal{P})\|^2 = \sum_{i=0}^{K} I\{\Pi(\boldsymbol{y}|\mathcal{P}) \in ri(F_i)\}\|P_i\boldsymbol{y}\|^2$$

where $I(.)$ is the indicator function and $\boldsymbol{P}_i$ is the projection matrix onto the linear space spanned by F_i.

Proof: Since $\mathcal{P}$ is a closed convex set, $\Pi(\boldsymbol{Y}|\mathcal{P})$ exists and is unique (see Proposition 3.12.3 on page 114), or Stoer and Witzgall (1970), p 51). If F and G are two different faces of $\mathcal{P}$, then $ri(F) \cap ri(G)$ is the empty set (see Stoer and Witzgall (1970), p 43, 2.4.14). The polyhedral cone $\mathcal{P}$, a typical face F, and its relative interior $ri(F)$ can be expressed as follows :

$$\begin{aligned}
\mathcal{P} &= \{\boldsymbol{x} : \boldsymbol{a}_i^T\boldsymbol{x} \leq 0 \text{ for } i = 1, \ldots, m\}, \\
F &= \{\boldsymbol{x} : \boldsymbol{a}_i^T\boldsymbol{x} = 0 \text{ for } i \in I, \boldsymbol{a}_j^T\boldsymbol{x} \leq 0 \text{ for } j \in J\}, \\
ri(F) &= \{\boldsymbol{x} : \boldsymbol{a}_i^T\boldsymbol{x} = 0 \text{ for } i \in I, \boldsymbol{a}_j^T\boldsymbol{x} < 0 \text{ for } j \in J\}
\end{aligned} \tag{3.71}$$

where $\{I, J\}$ is a partition of $\{1, \ldots, m\}$; it is assumed that the equality constraints defining F are contained in $\{\boldsymbol{a}_i^T\boldsymbol{x} = 0, i \in I\}$ and hence the set of constraints $\{\boldsymbol{a}_j^T\boldsymbol{x} \leq 0, j \in J\}$ does not include a constraint of the form $\boldsymbol{a}^T\boldsymbol{x} = 0$ for some $\boldsymbol{a} \neq \boldsymbol{0}$. Thus, it is clear that the collection of all relative interiors of faces of $\mathcal{P}$ covers $\mathcal{P}$. Let $\{F_1, \ldots, F_K\}$ denote a collection of faces of $\mathcal{P}$ such that $\{ri(F_1), \ldots, ri(F_K)\}$ forms a partition of $\mathcal{P}$. Let $\boldsymbol{P}_i$ denote the projection matrix onto the linear space spanned by F_i, $i = 1, \ldots, K$. Now, if $\Pi(\boldsymbol{y}|\mathcal{P}) \in ri(F_i)$ then $\Pi(\boldsymbol{y}|\mathcal{P}) = \boldsymbol{P}_i\boldsymbol{y}$, by Lemma 3.13.2. This completes the proof. $\blacksquare$

The next lemma is really a restatement of the results in the previous lemmas, but it is stated here because this alternative statement is helpful.

Lemma 3.13.6 *Let $\mathcal{P}$ be a convex polyhedral in $\mathbb{R}^p$, and $\boldsymbol{X}_{p\times 1} \sim N(\boldsymbol{0}, \boldsymbol{I})$. Then there exists a collection of faces of $\mathcal{P}$, say $\{F_1, \ldots, F_k\}$, such that their relative interiors, $\{ri(F_1), \ldots, ri(F_k)\}$, are a partition of $\mathcal{P}$. Let $\mathcal{S}_i = \{\boldsymbol{x} : \Pi(\boldsymbol{x}|\mathcal{P}) \in ri(F_i)\}$, $\boldsymbol{P}_i$ denote the projection matrix onto the linear space spanned by F_i, and $\bar{\chi}^2(\boldsymbol{I}, \mathcal{P}) = \|\boldsymbol{X}\|^2 - \min\{\|\boldsymbol{X} - \boldsymbol{a}\|^2 : \boldsymbol{a} \in \mathcal{P}\}$. Then, we have the following :*

(i) $\bar{\chi}^2(\boldsymbol{I}, \mathcal{P}) = \sum_{i=1}^{k} I(\boldsymbol{X} \in \mathcal{S}_i)\|\boldsymbol{P}_i\boldsymbol{X}\|^2$.
(ii) If $\boldsymbol{X} \in \mathcal{S}_i$, *then* $\boldsymbol{P}_i\boldsymbol{X}$ *and* $(\boldsymbol{I} - \boldsymbol{P}_i)\boldsymbol{X}$ *are the projections of* $\boldsymbol{X}$ *onto* $\mathcal{P}$ *and* $\mathcal{P}^o$, *respectively; hence* $\bar{\chi}^2(\boldsymbol{I}, \mathcal{P}) = \|\boldsymbol{P}_i\boldsymbol{X}\|^2$ *and* $\bar{\chi}^2(\boldsymbol{I}, \mathcal{P}^o) = \|(\boldsymbol{I} - \boldsymbol{P}_i)\boldsymbol{X}\|^2$.
(iii) If $\boldsymbol{X} \in \mathcal{S}_i$, *then* $\|\boldsymbol{P}_i\boldsymbol{X}\|^2$ *and* $\|(\boldsymbol{I}-\boldsymbol{P}_i)\boldsymbol{X}\|^2$ *are independent and are distributed as* χ^2_ν *and* $\chi^2_{p-\nu}$, *respectively, where* $\nu = rank(\boldsymbol{P}_i)$. ■

Lemma 3.13.7 *Let* $\boldsymbol{Y} \sim N_p(\boldsymbol{0}, \boldsymbol{I}), \mathcal{P}$ *be a polyhedral cone, and let* $\bar{\chi}^2$ *denote* $\|\Pi(\boldsymbol{Y}|\mathcal{P})\|^2$. *Then*

$$pr(\bar{\chi}^2 \geqslant c) = \sum_{i=0}^{p} w_i pr(\chi_i^2 \geqslant c), \text{ where } w_i = \sum_{rank(\boldsymbol{P}_j)=i} pr\{\Pi(\boldsymbol{Y}|\mathcal{P}) \in ri(F_j)\}.$$

Thus, $w_i = pr\{\Pi(\boldsymbol{Y}|\mathcal{P})$ *lies in the relative interior of a face of* $\mathcal{P}$ *of dimension* i $\}$, *for* $i = 0, \ldots, p$. *Further,* $w_0 + \ldots + w_p = 1$.

Proof: In view of the previous lemma, we have the following:

$$\begin{aligned} &\text{pr}(\bar{\chi}^2 \geqslant c) \\ &= \sum_{j=0}^{k} \text{pr}\{\|\Pi(\boldsymbol{Y}|\mathcal{P})\|^2 \geqslant c \mid \Pi(\boldsymbol{Y}|\mathcal{P}) \in ri(F_j)\} \times \text{pr}\{\Pi(\boldsymbol{Y}|\mathcal{P}) \in ri(F_j)\} \\ &= \sum_{i=0}^{p} w_i \text{pr}(\chi_i^2 \geqslant c), \text{ where } w_i = \sum_{rank(\boldsymbol{P}_j)=i} \text{pr}\{\Pi(\boldsymbol{Y}|\mathcal{P}) \in ri(F_j)\}. \end{aligned} \tag{3.72}$$

Thus, $w_i = \text{pr}\{\Pi(\boldsymbol{Y}|\mathcal{P})$ *lies on the relative interior of a face of dimension* i $\}$, the dimension of a face being defined as the dimension of the linear space spanned by the face. ■

Proof of Theorem 3.4.2 continued: We will prove the theorem for a polyhedral cone first, and then deduce the result for a general cone. Assume that $\mathcal{C}$ is a polyhedral cone. Let $\boldsymbol{V}^{-1} = \boldsymbol{A}^T\boldsymbol{A}$ be a factorization of $\boldsymbol{V}^{-1}$, $\boldsymbol{X} = \boldsymbol{A}\boldsymbol{Y}$, and $\mathcal{P}$ be the polyhedral cone $\boldsymbol{A}\mathcal{C} = \{\boldsymbol{A}\boldsymbol{x} : \boldsymbol{x} \in \mathcal{C}\}$. Then the faces of $\mathcal{C}$ take the form $\boldsymbol{A}F$ for some face F of $\mathcal{P}$ and $dim(\boldsymbol{A}F) = dim(F) =$ dimension of the linear space spanned by F. Now,

$$\begin{aligned} LRT &= \boldsymbol{Y}^T\boldsymbol{V}^{-1}\boldsymbol{Y} - \min\{(\boldsymbol{Y} - \boldsymbol{\alpha})^T\boldsymbol{V}^{-1}(\boldsymbol{Y} - \boldsymbol{\alpha}) : \boldsymbol{\alpha} \in \mathcal{C}\} \\ &= \boldsymbol{X}^T\boldsymbol{X} - \min\{(\boldsymbol{X} - \boldsymbol{\beta})^T(\boldsymbol{X} - \boldsymbol{\beta}) : \boldsymbol{\beta} \in \mathcal{P}\}. \end{aligned} \tag{3.73}$$

It follows from Lemma 3.13.7 that $\text{pr}(LRT \leq c \mid H_0) = \sum_{i=0}^{p} w_i(p, \boldsymbol{V}, \mathcal{P})\text{pr}(\chi_i^2 \leq c)$, where $w_i(p, \boldsymbol{V}, \mathcal{P}) = \text{pr}\{\Pi(\boldsymbol{X}||\mathcal{P})$ lies on the relative interior of a face G of $\mathcal{P}$ with $dim(G) = i\} = \text{pr}\{\Pi_{\boldsymbol{V}}(\boldsymbol{Y}|\mathcal{C})$ lies on the relative interior of a face F of $\mathcal{C}$ with $dim(F) = i\}$.

Now let us relax the assumption that $\mathcal{C}$ is a polyhedral and assume only that it is a closed convex cone. Let $\mathcal{C}_1 \subset \mathcal{C}_2 \subset \ldots$ be a sequence of polyhedral cones such that it converges to $\mathcal{C}$. Then, for any $c > 0$, $\text{pr}(\|\Pi_{\boldsymbol{V}}(\boldsymbol{Y}, \mathcal{C}_n)\|^2_{\boldsymbol{V}} \leq c)$ converges to $\text{pr}(\|\Pi_{\boldsymbol{V}}(\boldsymbol{Y}, \mathcal{C})\|^2_{\boldsymbol{V}} \leq c)$ as $n \to \infty$. For any fixed $(i, p, \boldsymbol{V})$, the sequence

$\{w_i(p, \boldsymbol{V}, \mathcal{C}_n)\}$ is bounded, and hence has limit points. Therefore, $\text{pr}(\|\Pi_{\boldsymbol{V}}(\boldsymbol{Y}, \mathcal{C}_n)\|^2_{\boldsymbol{V}} \le c)$ has limit points of the form $\sum_{i=0}^{p} w_i \text{pr}(\chi_i^2 \le c)$ for some $\{w_i\}$. Since all such limit points are equal to $\text{pr}(\|\Pi_{\boldsymbol{V}}(\boldsymbol{Y}, \mathcal{C})\|^2_{\boldsymbol{V}} \le c)$, for any c, it follows that limit points of $\{w_i(p, \boldsymbol{V}, \mathcal{C}_n)\}$ are unique. Therefore, we define $w_i(p, \boldsymbol{V}, \mathcal{C}) = \lim_{n\to\infty} w_i(p, \boldsymbol{V}, \mathcal{C}_n)$. ■

Proof of Proposition 3.6.1: The proofs of virtually all of these are based on the details of the proof of the previous theorem. The proof of part 1 is contained in the proof of Theorem 3.4.2 just given. Part 2 is a corollary to part 1. It was also proved by Nuesch (1964); see also Perlman (1969) and Wolak (1987). To our knowledge, part 3 has been a known result but it is not clear whether or not a proof has been published. This was posed as a conjecture (see Shapiro (1987)); Professor Alexander Shapiro informed us that he has received more than one proof of the conjecture, one by Professor J. Kinoses and another by Professor J. Lawrence; Professor Tim Robertson also informed us that he is aware of other proofs. Part 4 follows from part 3. Part 5 is shown in Shapiro (1985); see also Shapiro (1988, equation 3.4). Part 6 can be established easily by applying the transformation $\boldsymbol{Y} = \boldsymbol{RZ}$ where $\boldsymbol{Z} \sim N(\boldsymbol{0}, \boldsymbol{V})$; see also Shapiro (1988, equation 5.3). See equations (5.4), (5.5), (5.8), and (5.10) in Shapiro (1988) for parts 7, 8, 9, and 10, respectively. ■

Proof of Theorem 3.9.1: (a) Let $\boldsymbol{Z} = \boldsymbol{E}/\sigma$ and assume that the null hypothesis holds. Then, we have that

$$\bar{E}_A^2 = \{\min_{\boldsymbol{a}\in\mathcal{M}} Q^*(\boldsymbol{a}) - \min_{\boldsymbol{a}\in\mathcal{C}} Q^*(\boldsymbol{a})\}/\min_{\boldsymbol{a}\in\mathcal{M}} Q^*(\boldsymbol{a}) \tag{3.74}$$

where $Q^*(\boldsymbol{a}) = (\boldsymbol{AZ} - \boldsymbol{a})^T \boldsymbol{W}^{-1}(\boldsymbol{AZ} - \boldsymbol{a})$ and $\boldsymbol{A} = \boldsymbol{W}^{-1}\boldsymbol{X}^T\boldsymbol{U}^{-1}$. The distribution of $\boldsymbol{Z}$ and hence that of (3.74) does not depend on any unknown parameters. Consequently, the null distribution of $\bar{E}_A^2$ is the same for any value of $\boldsymbol{\theta}$ in $\mathcal{M}$ and any value of $\sigma > 0$; by similar arguments, the corresponding result holds for $\bar{F}_A$ as well.

(b) Let $\boldsymbol{\theta} \in \mathcal{C}$. Let

$$\begin{aligned} g_1(\boldsymbol{\theta}, \boldsymbol{E}) &= \min\{\|\boldsymbol{X\theta} + \boldsymbol{E} - \boldsymbol{Xa}\|^2_{\boldsymbol{U}} : \boldsymbol{a} \in \mathcal{C}\} \\ \text{and} \quad g_2(\boldsymbol{\theta}, \boldsymbol{E}) &= \min\{\|\boldsymbol{X\theta} + \boldsymbol{E} - \boldsymbol{Xa}\|^2_{\boldsymbol{U}} : \boldsymbol{a} \in \mathbb{R}^p\}. \end{aligned} \tag{3.75}$$

To show the dependence of $\bar{E}_B^2$ on $\boldsymbol{\theta}$, let us write

$$\bar{E}_B^2(\boldsymbol{\theta}, \boldsymbol{E}) = \{g_1(\boldsymbol{\theta}, \boldsymbol{E}) - g_2(\boldsymbol{\theta}, \boldsymbol{E})\}/g_1(\boldsymbol{\theta}, \boldsymbol{E}).$$

Now,

$$\begin{aligned} &\min\{\|\boldsymbol{Y} - \boldsymbol{Xa}\|^2_{\boldsymbol{U}} : \boldsymbol{a} \in \mathcal{C}\} = \min\{\|\boldsymbol{X\theta} + \boldsymbol{E} - \boldsymbol{Xa}\|^2_{\boldsymbol{U}} : \boldsymbol{a} \in \mathcal{C}\} \\ &= \min\{\|\boldsymbol{E} - \boldsymbol{X}(\boldsymbol{a} - \boldsymbol{\theta})\|^2_{\boldsymbol{U}} : \boldsymbol{a} \in \mathcal{C}\} = \min\{\|\boldsymbol{E} - \boldsymbol{Xb}\|^2_{\boldsymbol{U}} : \boldsymbol{b} \in \mathcal{C} - \boldsymbol{\theta}\} \\ &\le \min\{\|\boldsymbol{E} - \boldsymbol{Xb}\|^2_{\boldsymbol{U}} : \boldsymbol{b} \in \mathcal{C}\} \end{aligned}$$

the last step follows since $\mathcal{C} \subset \mathcal{C} - \boldsymbol{\theta}$. Similarly,

$$\min\{\|\boldsymbol{Y} - \boldsymbol{Xa}\|^2_{\boldsymbol{U}} : \boldsymbol{a} \in \mathbb{R}^p\} = \min\{\|\boldsymbol{E} - \boldsymbol{Xb}\|^2_{\boldsymbol{U}} : \boldsymbol{b} \in \mathbb{R}^p\}.$$

Thus, $g_1(\boldsymbol{\theta}, \boldsymbol{E}) \leq g_1(\mathbf{0}, \boldsymbol{E})$ and $g_2(\boldsymbol{\theta}, \boldsymbol{E}) = g_2(\mathbf{0}, \boldsymbol{E})$. Now,

$$\begin{aligned}\bar{E}_B^2(\boldsymbol{\theta}, \boldsymbol{E}) &= 1 - \{g_2(\boldsymbol{\theta}, \boldsymbol{E})/g_1(\boldsymbol{\theta}, \boldsymbol{E})\} = 1 - \{g_2(\mathbf{0}, \boldsymbol{E})/g_1(\boldsymbol{\theta}, \boldsymbol{E})\} \\ &\leqslant 1 - \{g_2(\mathbf{0}, \boldsymbol{E})/g_1(\mathbf{0}, \boldsymbol{E})\} = \bar{E}_B^2(\mathbf{0}, \boldsymbol{E}).\end{aligned}$$

Therefore, $\bar{E}_B^2(\boldsymbol{\theta}, \boldsymbol{E}) \leq \bar{E}_B^2(\mathbf{0}, \boldsymbol{E})$ and

$$\text{pr}\{\bar{E}_B^2(\boldsymbol{\theta}, \boldsymbol{E}) \geqslant c\} \leq \text{pr}\{\bar{E}_B^2(\mathbf{0}, \boldsymbol{E}) \geqslant c\}, \quad \boldsymbol{\theta} \in \mathcal{C}.$$

Hence the least favorable null value for $\bar{E}_B^2$ is $\mathbf{0}$.

To prove the corresponding result for $\bar{F}$, note that $S^2(\boldsymbol{\theta}, \boldsymbol{E}) = S^2(\mathbf{0}, \boldsymbol{E})$. Now,

$$\begin{aligned}\bar{F}(\boldsymbol{\theta}, \boldsymbol{E}) &= \{g_1(\boldsymbol{\theta}, \boldsymbol{E}) - g_2(\boldsymbol{\theta}, \boldsymbol{E})\}/S^2(\boldsymbol{\theta}, \boldsymbol{E}) \\ &= \{g_1(\boldsymbol{\theta}, \boldsymbol{E}) - g_2(\mathbf{0}, \boldsymbol{E})\}/S^2(\mathbf{0}, \boldsymbol{E}) \\ &\leqslant \{g_1(\mathbf{0}, \boldsymbol{E}) - g_2(\mathbf{0}, \boldsymbol{E})\}/S^2(\mathbf{0}, \boldsymbol{E}) = \bar{F}(\mathbf{0}, \boldsymbol{E}).\end{aligned}$$

Therefore, $\text{pr}\{\bar{F}_B(\boldsymbol{\theta}, \boldsymbol{E}) \geqslant c\} \leq \text{pr}\{\bar{F}_B(\mathbf{0}, \boldsymbol{E}) \geqslant c\}$ for $\boldsymbol{\theta} \in \mathcal{C}$. ■

Proof of Theorem 3.9.2 : (a) Without loss of generality assume that $\sigma = 1$, $\boldsymbol{U}$ is the identity matrix, and the null hypothesis is $\boldsymbol{R\theta} = \mathbf{0}$ where $\boldsymbol{R}$ is a $r \times p$ matrix of rank r. Now

$$\bar{E}_B^2 = \{q(\boldsymbol{\theta}^0) - q(\boldsymbol{\theta}^1)\}/Q(\boldsymbol{\theta}^0).$$

Note that

$$Q(\boldsymbol{\theta}^0) = [\{q(\boldsymbol{\theta}^0) - q(\boldsymbol{\theta}^1)\} + \|\boldsymbol{Y} - \boldsymbol{X}\hat{\boldsymbol{\theta}}\|^2 + q(\boldsymbol{\theta}^1)].$$

Since $\hat{\boldsymbol{\theta}}$ is independent of $(\boldsymbol{Y} - \boldsymbol{X}\hat{\boldsymbol{\theta}})$, we have that $\boldsymbol{Y} - \boldsymbol{X}\hat{\boldsymbol{\theta}}$ is also independent of $\{q(\boldsymbol{\theta}^0) - q(\boldsymbol{\theta}^1), q(\boldsymbol{\theta}^1)\}$. Let $\{F_1, \ldots, F_k\}$ be a collection of faces of $\mathcal{C} \cap \mathcal{M}^{\perp}$ as in Lemma 3.13.4.

Now, suppose that $\Pi_{\boldsymbol{W}}(\hat{\boldsymbol{\theta}}|\mathcal{C} \cap \mathcal{M}^{\perp}) \in ri(F)$ for some $F \in \{F_1, \ldots, F_k\}$. Let $dim\{ri(F)\} = j$. Then $q(\boldsymbol{\theta}^0) - q(\boldsymbol{\theta}^1) = \|\boldsymbol{P}\hat{\boldsymbol{\theta}}\|^2$ where $\boldsymbol{P}$ is the projection matrix onto the linear space spanned by F. Now, since $q(\boldsymbol{\theta}^0) \sim \chi_r^2, q(\boldsymbol{\theta}^0) - q(\boldsymbol{\theta}^1) = \|\boldsymbol{P}\hat{\boldsymbol{\theta}}\|^2 \sim \chi_j^2$, and $q(\boldsymbol{\theta}^0) - q(\boldsymbol{\theta}^1)$ and $q(\boldsymbol{\theta}^1)$ are independent by Proposition 3.12.6 part (g), we have that $q(\boldsymbol{\theta}^1) \sim \chi_{r-j}^2$. Therefore,

$$\begin{aligned}\bar{E}_A^2 &= \{q(\boldsymbol{\theta}^0) - q(\boldsymbol{\theta}^1)\}/[\{q(\boldsymbol{\theta}^0) - q(\boldsymbol{\theta}^1)\} + \|\boldsymbol{Y} - \boldsymbol{X}\hat{\boldsymbol{\theta}}\|^2 + q(\boldsymbol{\theta}^1)] \\ &\sim \chi_j^2/(\chi_j^2 + \chi_{N-p+r-j}^2),\end{aligned} \tag{3.76}$$

where the two terms in the denominator are independent. Therefore,

$$\bar{E}_A^2 \sim \beta\{2^{-1}j, 2^{-1}(N - p + r - j)\}.$$

Let B_i denote the event $\Pi(\hat{\boldsymbol{\theta}}|\mathcal{C}\cap\mathcal{M}^{\perp})\in ri(F_i)\}$. Now, by adopting an argument similar to the derivation of $\bar{\chi}^2$ we have the following:

$$\text{pr}(\bar{E}_A^2 \leq c|H_0) = \sum_i \text{pr}\{\bar{E}_A^2 \leq c \mid B_i\}\text{pr}\{B_i\}.$$

$$= \sum_{j=0}^{p} \sum_{dim(F_i)=j} \text{pr}\{\bar{E}_A^2 \leq c \mid B_i\}\text{pr}(B_i)$$

$$= \sum_{j=0}^{p} w_j(p, \boldsymbol{W}, \mathcal{C}\cap\mathcal{M}^{\perp})\text{pr}[\beta\{2^{-1}j, 2^{-1}(N-p+r-j)\} \leq c].$$

The distribution of $\bar{F}$ is obtained similarly.

$$\text{pr}(\bar{F}_A \leq c|H_0) = \sum_{i=1}^{K} \text{pr}\{\|\Pi(\hat{\boldsymbol{\theta}}|\mathcal{C}\cap\mathcal{M}^{\perp})\|^2 S^{-2} \leq c \mid B_i\}\text{pr}(B_i)$$

$$= \sum_{j=0}^{p} \sum_{dim(F_i)=j} \text{pr}\{\|\boldsymbol{P}_i\hat{\boldsymbol{\theta}}\|^2 S^{-2} \leq c \mid B_i)\text{pr}(B_i)$$

$$= \sum_{j=0}^{p} w_j(p, \boldsymbol{W}, \mathcal{C}\cap\mathcal{M}^{\perp})\text{pr}(jF_{j,\nu} \leq c). \quad \blacksquare$$

Proof of part (b):

$$\bar{E}_B^2 = \{Q(\boldsymbol{\theta}^1) - Q(\boldsymbol{\theta}^2)\}/Q(\boldsymbol{\theta}^1) = q(\boldsymbol{\theta}^1)/[q(\boldsymbol{\theta}^1) + \|\boldsymbol{Y} - \boldsymbol{X}\hat{\boldsymbol{\theta}}\|_{\boldsymbol{U}}^2].$$

As in Lemma 3.13.4, let $\{F_1, \ldots, F_k\}$ be a collection of faces of $\mathcal{C}$ so that their relative interiors are a partition of $\mathcal{C}$. Let A_i be the event $\Pi_{\boldsymbol{W}}(\hat{\boldsymbol{\theta}}|\mathcal{C}) \in ri(F_i)$. Now, we have

$$\begin{aligned}\text{pr}(\bar{E}_B^2 \leq c \mid \boldsymbol{\theta} = \boldsymbol{0}) &= \sum_{i=0}^{K} \text{pr}[q(\boldsymbol{\theta}^1)/\{q(\boldsymbol{\theta}^1) + \|\boldsymbol{Y} - \boldsymbol{X}\hat{\boldsymbol{\theta}}\|_{\boldsymbol{U}}^2\} \leq c \mid A_i]\text{pr}(A_i)\\ &= \sum_{j=0}^{p} \sum_{dim(F_i)=j} \text{pr}[\chi_{p-j}^2/\{\chi_{p-j}^2 + \chi_{N-p}^2\} \leq c]\text{pr}(A_i)\\ &= \sum_{j=0}^{p} w_{p-j}(p, \boldsymbol{W}, \mathcal{C})\text{pr}[\beta\{2^{-1}j, 2^{-1}(N-p)\} \leq c].\end{aligned}$$

The proof for the $\bar{F}$ is similar.

$$\text{pr}(\bar{F}_B \leq c|\boldsymbol{\theta} = \boldsymbol{0}) = \sum_{i=0}^{K} \text{pr}[q(\boldsymbol{\theta}^1)/S^2 \leq c \mid A_i]\text{pr}(A_i)$$

$$= \sum_{i=0}^{k}\sum_{j=0}^{p} \text{pr}[\chi_{p-j}^2/S^2 \leq c]\text{pr}(A_i) = \sum_{j=0}^{p} w_{p-j}(p, \boldsymbol{W}, \mathcal{C})\text{pr}[jF_{j,\nu} \leq c]. \quad \blacksquare$$

Problems

Section 3.3

3.1 *The length and direction of $N(\mathbf{0}, \boldsymbol{I})$ are independent; two-dimensional case:* Let $\boldsymbol{X} = (X_1, X_2)^T \sim N(\mathbf{0}, \boldsymbol{I})$, where $\boldsymbol{I}$ is the 2×2 identity matrix. Let (r, α) be the polar coordinates of $\boldsymbol{X}$ given by $(X_1, X_2) = r(\sin\alpha, \cos\alpha)$. Obtain the joint density of (r, α), and deduce that r and α are independent and that $\|\boldsymbol{X}\|$ and $\boldsymbol{X}/\|\boldsymbol{X}\|$ are independent [see Anderson (1984), p 279].

3.2 *Derivation of a $\bar{\chi}^2$ distribution in three dimensions:* Let $\boldsymbol{X} = (X_1, X_2, X_3)^T \sim N(\boldsymbol{\theta}, \boldsymbol{I})$ and let the null and alternative hypotheses be $H_0 : \boldsymbol{\theta} = \mathbf{0}$ and $H_1 : \boldsymbol{\theta} \geqslant \mathbf{0}$. Let $\tilde{\boldsymbol{\theta}}$ be the *mle* of $\boldsymbol{\theta}$ under H_1 based on a single observation of $\boldsymbol{X}$; the test of H_0 against H_1 is also to be based on a single observation of $\boldsymbol{X}$. Show that $\tilde{\boldsymbol{\theta}} = (X_1 I\{X_1 > 0\}, X_2 I\{X_2 > 0\}, X_3 I\{X_3 > 0\}$, where I is the indicator function and $LRT = X_1^2 I(X_1 > 0) + X_2^2 I(X_2 > 0) + X_3^2 I(X_3 > 0)$. Show that $\boldsymbol{X}$ has exactly 3, 2, 1, or 0 positive components with probabilities (1/8), (3/8), (3/8), or (1/8), respectively. Deduce that, for $c > 0$, the null distribution of LRT is given by, $\text{pr}(LRT \leq c) = \sum_{i=0}^{3} w_i \text{pr}(\chi_i^2 \leq c)$, where $(w_0, w_1, w_2, w_3) = (1/8, 3/8, 3/8, 1/8)$. Verify that $w_i = \text{pr}(\text{exactly } i \text{ components of } \boldsymbol{Z} \text{ are positive })$, where $\boldsymbol{Z}_{3\times 1} \sim N(\mathbf{0}, \boldsymbol{I})$, and $i = 0, \ldots, 3$.

3.3 *Derivation of a $\bar{\chi}^2$ distribution in three dimensions:* Let $\boldsymbol{X} \sim N(\boldsymbol{\theta}, \boldsymbol{I})$, where $\boldsymbol{X}$ is 3×1, and let $\mathcal{C} = \{\boldsymbol{\theta} \in \mathbb{R}^3 : \theta_2 \geqslant 0, \theta_3 \geqslant 0\}$. Show that the *mle* of $\boldsymbol{\theta}$ subject to $\boldsymbol{\theta} \in \mathcal{C}$ and the LRT of $H_0 : \boldsymbol{\theta} = \mathbf{0}$ against $H_1 : \boldsymbol{\theta} \in \mathcal{C}$ are given by

$$(\tilde{\theta}_1, \tilde{\theta}_2, \tilde{\theta}_3) = \begin{cases} (X_1, X_2, X_3) & \text{if } X_2 \geqslant 0 \text{ and } X_3 \geqslant 0 \\ (X_1, X_2, 0) & \text{if } X_3 < 0 \text{ and } X_2 > 0 \\ (X_1, 0, X_3) & \text{if } X_2 < 0 \text{ and } X_3 > 0 \\ (X_1, 0, 0) & \text{if } X_2 < 0 \text{ and } X_3 < 0 \end{cases}$$

and

$$LRT = \begin{cases} X_1^2 + X_2^2 + X_3^2 & \sim \chi_3^2 \quad \text{if } X_2 \geqslant 0 \text{ and } X_3 \geqslant 0 \\ X_1^2 + X_2^2 & \sim \chi_2^2 \quad \text{if } X_3 < 0 \text{ and } X_2 > 0 \\ X_1^2 + X_3^2 & \sim \chi_2^2 \quad \text{if } X_2 < 0 \text{ and } X_3 > 0 \\ X_1^2 & \sim \chi_1^2 \quad \text{if } X_2 < 0 \text{ and } X_3 < 0, \end{cases}$$

respectively. Deduce that the null distribution of LRT is given by

$$\text{pr}(LRT \leq c | H_0) = (1/4)\text{pr}(\chi_1^2 \leq c) + (1/2)\text{pr}(\chi_2^2 \leq c) + (1/4)\text{pr}(\chi_3^2 \leq c).$$

3.4 *Derivation of a $\bar{\chi}^2$ distribution in four dimensions:* Let $\boldsymbol{X} \sim N(\boldsymbol{\theta}, \boldsymbol{I})$ where $\boldsymbol{X}$ is 4×1. Let the null and alternative hypothesis be $H_0 : \boldsymbol{\theta} = \mathbf{0}$ and $H_1 : \theta_3 \geqslant 0, \theta_4 \geqslant 0$. Show that

$$\text{pr}(LRT \leq c | H_0) = (1/4)\text{pr}(\chi_2^2 \leq c) + (1/2)\text{pr}(\chi_3^2 \leq c) + (1/4)\text{pr}(\chi_4^2 \leq c).$$

3.5 *Derivation of a $\bar{\chi}^2$ distribution in p dimensions:* Let $\boldsymbol{X} \sim N(\boldsymbol{\theta}, \boldsymbol{I})$ where $\boldsymbol{X}$ is $p \times 1$. Let the null and alternative hypothesis be $H_0 : \boldsymbol{\theta} = \mathbf{0}$ and $H_1 : \boldsymbol{\theta}_2 \geqslant 0$

where θ_2 is a subvector of θ consisting of k elements. Show that $\text{pr}(LRT \leq c|H_0) = w_{p-k}\text{pr}(\chi^2_{p-k} \leq c) + \ldots + w_p\text{pr}(\chi^2_p \leq c)$ where $w_{p-k+i} = 2^{-k}k!/\{(k-i)!i!\}$.

3.6 *Application of LRT in two dimensions:* Let $\boldsymbol{X} = (X_1, X_2)^T \sim N(\boldsymbol{\theta}, \boldsymbol{I})$. The mean of a sample of five observations on $\boldsymbol{X}$ is $(-1, 2)$. Compute the *mle* under the constraint $\theta_1 \geqslant 0, \theta_2 \geqslant 0$ and the p-value for testing $H_0 : \theta_1 = \theta_2 = 0$ against $H_1 : \theta_1 \geqslant 0, \theta_2 \geqslant 0$.

3.7 *Derivation of a $\bar{\chi}^2$ distribution in two dimensions:* Let $\boldsymbol{X} \sim N(\boldsymbol{\theta}, \boldsymbol{I}), H_0 : \theta_1 = \theta_2 = 0, H_1 : \theta_1 \geqslant 0, \theta_2 - m\theta_1 \geqslant 0$, where m is known. Show from first principles that the distribution of the LRT under H_0 is given by $\text{pr}(LRT \leq c|H_0) = \sum_{i=0}^{i=2} w_i \text{pr}(\chi^2_i \leq c)$ where $w_0 = q, w_1 = 0.5, w_2 = 0.5 - q$, and $q = (2\pi)^{-1}cos^{-1}\{-m/(1+m^2)^{1/2}\}$. Indicate the critical region corresponding to the critical value 4. [The details of the derivations are also given in (Gourieroux et al., 1982, section 4.2).]

3.8 *Derivation of the critical region of LRT in two dimensions:* For Example 3.3.5, prove the following: (i) $(1, \rho) \perp_{\boldsymbol{V}} (0, 1)$ and $(1, \rho^{-1}) \perp_{\boldsymbol{V}} (1, 0)$. (ii) The boundaries of the polar cone of the nonnegative orthant are $\theta_2 = \rho^{-1}\theta_1$ and $\theta_2 = \rho\theta_1$. (iii) For $\rho = 0.9$ and $\bar{\boldsymbol{x}} = (-3, 2)$, the *mle* of $\boldsymbol{\theta}$ subject to $\boldsymbol{\theta} \geqslant \boldsymbol{0}$ is the point of intersection of the θ_2 axis and the line of slope ρ through $\bar{\boldsymbol{x}}$. (iv) For $\rho = 0.9$ and $\bar{\boldsymbol{x}} = (3, -2)$, the *mle* of $\boldsymbol{\theta}$ subject to $\boldsymbol{\theta} \geqslant \boldsymbol{0}$ is the point of intersection of the θ_1 axis and the line of slope ρ^{-1} through $\bar{\boldsymbol{x}}$. Indicate the critical region $\{LRT \geqslant 6\}$.

3.9 Let X_i be an observation from the ith treatment and assume that $X_i \sim N(\mu_i, 1)$ for $i = 1, 2, 3$. Let $H_0 : \mu_1 = \mu_2 = \mu_3$ and $H_1 : \mu_1 \geq \mu_2$ and $\mu_1 \geq \mu_3$. (i) Verify that the LRT of H_0 against H_1 rejects H_0 at 5% level when $(X_1, X_2, X_3) = (50, 150, 70)$. (ii) Verify that the phenomenon in part (i) is essentially the same the same as that in Example 3.3.6.
[Hint: Let $Y_1 = X_1 - X_2, Y_2 = X_1 - X_3, \boldsymbol{Y} = (Y_1, Y_2)^T$ and $\boldsymbol{X} = (X_1, X_2)^T$. Then $\boldsymbol{Y} = \boldsymbol{AX}, \boldsymbol{Y} \sim N(\boldsymbol{\theta}, \boldsymbol{V})$, and $\boldsymbol{\theta} = \boldsymbol{A\mu}$ where $\boldsymbol{V} = \boldsymbol{AA}^T, \boldsymbol{A} = (1, -1, 0; 1, 0, -1)$, and $\boldsymbol{V} = 2(1, 0.5; 0.5, 1)$. Verify that the LRT of H_0 against H_1 is equivalent to the LRT of $H_0^* : \boldsymbol{\theta} = \boldsymbol{0}$ against $H_1^* : \boldsymbol{\theta} \geq \boldsymbol{0}$ based on a single observation of $\boldsymbol{Y}$. Verify (by using the result in Problem 8) that $\boldsymbol{Y} = (-100, -20)^T$, which corresponds to $\boldsymbol{X} = (50, 150, 70)^T$, is in the critical region.]

Section 3.4

3.10 *Pythagoras theorem in $\mathbb{R}^p$:* Let $\boldsymbol{V}$ be a $p \times p$ positive definite matrix. Define the inner product $\langle \boldsymbol{x}, \boldsymbol{y} \rangle_{\boldsymbol{V}} = \boldsymbol{x}^T\boldsymbol{V}^{-1}\boldsymbol{y}$, and the distance function $\|\boldsymbol{x} - \boldsymbol{y}\|_{\boldsymbol{V}} = \{\boldsymbol{x}^T\boldsymbol{V}^{-1}\boldsymbol{y}\}^{1/2}$. We say that $\boldsymbol{x}$ is *orthogonal* to $\boldsymbol{y}$ with respect to $\boldsymbol{V}$, denoted $\boldsymbol{x} \perp_{\boldsymbol{V}} \boldsymbol{y}$, if $\boldsymbol{x} \neq \boldsymbol{0}, \boldsymbol{y} \neq \boldsymbol{0}$, and $\langle \boldsymbol{x}, \boldsymbol{y} \rangle_{\boldsymbol{V}} = 0$. If A, B, C are three points in $\mathbb{R}^p$ such that $AB \perp_{\boldsymbol{V}} AC$, show that $\|AB\|^2_{\boldsymbol{V}} + \|AC\|^2_{\boldsymbol{V}} = \|BC\|^2_{\boldsymbol{V}}$.

3.11 Let $\mathcal{L}$ be a subspace of $\mathbb{R}^p$, $r = dim(\mathcal{L})$, and $\boldsymbol{x} \in \mathbb{R}^p$. Let O be the origin and A be the point such $\boldsymbol{x} = OA$. Let $\boldsymbol{V}$ be a $p \times p$ positive definite matrix. Show that the point in $\mathcal{L}$ that is $\boldsymbol{V}$-closest to A is obtained by dropping a $\boldsymbol{V}$-perpendicular to $\mathcal{L}$ from A. [Hint: Use the Pythagoras theorem.]

3.12 Let $X \sim N(\theta, V), H_0 : \theta = 0$, and $H_1 : \theta \in \mathcal{C}$. Suppose that $V^{-1} = A^T A = B^T B$ for some A and B. Show that BA^{-1} is an orthogonal matrix. Now, transform the setting from $\{X, \theta, V, \mathcal{C}\}$ by A as in Example 3.3.4 on page 68; also, consider the transformation by B. Show that the geometry of the image under transformation A is the same as that under transformation B except for a rotation defined by the orthogonal matrix BA^{-1}.

3.13 Let $Y_1, \ldots, Y_n$ be independent standard normal rv's, and let $Y = (Y_1, \ldots, Y_n)^T$. Let B be a $k \times n$ matrix of given constants such that $\{y : By \geqslant 0\}$ is nonempty. Show that the conditional distribution of $\|Y\|^2$ given $BY \geqslant 0$ is χ^2_n.

3.14 A function $f : \mathbb{R}^p \to \mathbb{R}$ is said to be Lipchitz continuous if $|f(x) - f(y)| \leqslant k\|x - y\|$ for some $k > 0$. Let $\mathcal{A} \subset \mathbb{R}^p$ and $d_{\mathcal{A}}$ be the distance function $d_{\mathcal{A}} : \mathbb{R}^p \to \mathbb{R}$ defined by $d_{\mathcal{A}}(x) = \|x - \mathcal{A}\|$. Show that $d_{\mathcal{A}}$ is Lipchitz continuous with $k = 1$.

3.15 Let A be an $m \times n$ matrix and $rank(A) = m$. Let $P = I - A^T(AA^T)^{-1}A$. Then, show that P is the projection matrix onto the null space of A.

3.16 Let V be a positive definite matrix, and let $V^{-1} = P^T D^{-1} P$ for some D and P, where P is orthogonal. Consider the problem of testing $H_0 : \theta = 0$ against $H_1 : \theta \in \mathcal{C}$ based on a single observation of X where $X \sim N(\theta, V)$. Show that the LRT of H_0 against H_1 based on a single observation of X is equivalent to testing $H_0^* : \alpha = 0$ against $H_1^*; \alpha \in \mathcal{P}$ based on a single observation of Y, where $Y = PX$ and $\mathcal{P} = P\mathcal{C} = \{Px : x \in \mathcal{C}\}$ which is also a closed convex cone. Deduce that $\bar{\chi}^2(V, \mathcal{C}) \sim \bar{\chi}^2(D, P\mathcal{C})$.

Similarly, let $V^{-1} = A^T A$; thus A can be the Cholesky factor or $D^{-1/2}P$. Let $Z = AX$ and $\mathcal{Q} = A\mathcal{C} = \{Ax : x \in \mathcal{C}\}$. Show that $\mathcal{Q}$ is also a closed convex cone, the testing problem is equivalent to test of $H_0' : \gamma = 0$ vs $H_1' : \gamma \in \mathcal{Q}$ based on a single observation of $Z \sim N(\gamma, I)$ and $\bar{\chi}^2(V, \mathcal{C}) \sim \bar{\chi}^2(I, \mathcal{Q})$.

3.17 *The $\bar{\chi}^2(V, \mathcal{C})$ distribution and the weights depend on V only through the correlation matrix corresponding to V:* Let $V = (v_{ij})$ be a positive definite matrix of order $p \times p$. Consider the problem of testing $H_0 : \theta = 0$ against $H_1 : \theta \in \mathcal{C}$ based on a single observation of X where $X \sim N(\theta, V)$. Let Ω denote the correlation matrix corresponding to V. Let A be the diagonal matrix $diag(a_1, \ldots, a_p)$ where $a_i = v_{ii}^{1/2}$. Verify that $V = A\Omega A$. Show that $A\mathcal{C} = \mathcal{C}$. Deduce that $\bar{\chi}^2(V, \mathcal{C}) \sim \bar{\chi}^2(\Omega, \mathcal{C})$ and hence the $\bar{\chi}^2$-weights depend on V only through its correlation matrix.

3.18 Let R be a $k \times p$ matrix of rank k, $k \leq p$, and $y = Ra$. Verify that

$$\min\{(x - a)^T V^{-1}(x - a) : Ra \geqslant 0\} = \min\{(y - b)^T(RVR^T)^{-1}(y - b) : b \geqslant 0\}.$$

Deduce that

$$\bar{\chi}^2(V, \mathcal{C}) \sim \bar{\chi}^2(RVR^T, \mathbb{R}^{+k})$$

where $\mathcal{C} = \{\theta : R\theta \geqslant 0\}$.

[Hint: Find a matrix Q such that $RVQ^T = 0$ and $C^T = [R^T, Q^T]$ is nonsingular; then show that $\min\{(x - a)^T V^{-1}(x - a) : Ra \geqslant 0\} = \min\{(z - c)^T(CVC^T)^{-1}(z - c) : c \geqslant 0\}$. Now use the fact that CVC^T is block diagonal.]

Section 3.8

3.19 *Derivation of the null distribution of LRT for a Type B problem in two dimensions:* Let $\boldsymbol{X} \sim N(\boldsymbol{\theta}, \boldsymbol{I}), H_1 : \theta_1 \geqslant 0, \theta_2 - m\theta_1 \geqslant 0$ and H_2 : *not* H_1, where m is known. Verify the following: (a) $LRT = \min\{(\boldsymbol{X} - \boldsymbol{a})^T(\boldsymbol{X} - \boldsymbol{a}) : a_1 \geqslant 0, a_2 - ma_1 \geqslant 0\}$. (b) $\text{pr}(LRT \leq c|\boldsymbol{\theta} = \boldsymbol{0}) = (0.5 - q)\text{pr}(\chi_0^2 \leq c) + 0.5\text{pr}(\chi_1^2 \leq c) + q\text{pr}(\chi_2^2 \leq c)$, where $q = (2\pi)^{-1}\cos^{-1}(-m(1 + m^2)^{-1/2})$.

3.20 Let $\boldsymbol{X} \sim N(\boldsymbol{\theta}, \boldsymbol{I})$, where $\boldsymbol{X}$ is 2×1, $H_0 : \boldsymbol{\theta} \in \mathcal{C}$ and $H_2 : \boldsymbol{\theta} \notin \mathcal{C}$. Let $\mathcal{C}^o$ denote the polar cone of $\mathcal{C}$ with respect to $\boldsymbol{I}$. Let $\mathcal{S}_1, \ldots, \mathcal{S}_4$ denote the four cones as in Example 3 in Section 3.3. (a) Show that, conditional on $\boldsymbol{X} \in \mathcal{S}_i$, $\tilde{\boldsymbol{\theta}}$ and $\boldsymbol{X} - \tilde{\boldsymbol{\theta}}$ are independent, $i = 1, \ldots, 4$. (b) Let L_A and L_B denote the LRT for the Type A testing problem, $H_0 : \boldsymbol{\theta} = \boldsymbol{0}$ against $H_1 : \boldsymbol{\theta} \geqslant \boldsymbol{0}$, and for the Type B testing problem, $H_1 : \boldsymbol{\theta} \in \mathcal{C}$ against $H_2 : \boldsymbol{\theta} \notin \mathcal{C}$, respectively. Show that the joint distribution of L_A and L_B at $\boldsymbol{\theta} = \boldsymbol{0}$ is given by

$$\begin{aligned}\text{pr}(L_A \geqslant c_1, L_B \geqslant c_2|\boldsymbol{\theta} = \boldsymbol{0}) = (0.5 - q)\text{pr}(\chi_0^2 \geqslant c_1)\text{pr}(\chi_2^2 \geqslant c_2) + \\ 0.5\text{pr}(\chi_1^2 \geqslant c_1)\text{pr}(\chi_1^2 \geqslant c_2) + q\text{pr}(\chi_2^2 \geqslant c_1)\text{pr}(\chi_0^2 \geqslant c_2).\end{aligned}$$

3.21

3.21.1. Let $\boldsymbol{X} \sim N(\boldsymbol{\theta}, \boldsymbol{V})$. Let $\tilde{\boldsymbol{\theta}}$ denote the *mle* subject to $\boldsymbol{\theta} \in \mathcal{C}$, L_A and L_B denote the LRT s for $H_0 : \boldsymbol{\theta} = \boldsymbol{0}$ against $H_1 : \boldsymbol{\theta} \in \mathcal{C}$ and $H_1 : \boldsymbol{\theta} \in \mathcal{C}$ against $H_2 : \boldsymbol{\theta} \notin \mathcal{C}$, respectively. Then the joint distribution of L_A and L_B at $\boldsymbol{\theta} = \boldsymbol{0}$ is given by the following:

$$\text{pr}(L_A \leq c_1, L_B \leq c_2|\boldsymbol{\theta} = \boldsymbol{0}) = \sum_{i=0}^{p} w_i(p, \boldsymbol{V}, \mathcal{C})\text{pr}(\chi_i^2 \leq c_1)\text{pr}(\chi_{p-i}^2 \leq c_2).$$

3.21.2. Let the hypotheses H_0, H_1, and H_2 be defined as $H_0 : \boldsymbol{\theta} \in \mathcal{M}$, $H_1 : \boldsymbol{\theta} \in \mathcal{C}$, and $H_2 : \boldsymbol{\theta} \in \mathcal{L}$ where $\mathcal{M}$ and $\mathcal{L}$ are linear spaces, $\mathcal{C}$ is a closed convex cone, $\mathcal{M} \subset \mathcal{C} \subset \mathcal{L}$, and $dim(\mathcal{L}) = k \leq p$. Obtain an expression for $\text{pr}(L_A \leq c_1, L_B \leq c_2)$.

3.21.3. Suppose that $\boldsymbol{X} \sim N(\boldsymbol{\theta}, \sigma^2\boldsymbol{U})$ where σ is unknown and $\boldsymbol{U}$ is known. Using similar arguments obtain expressions for $\text{pr}_{\boldsymbol{0}}(\bar{E}_A^2 \geqslant c, \bar{E}_B^2 \geqslant d)$ and $\text{pr}_{\boldsymbol{0}}(\bar{F}_A \geqslant c, \bar{F}_B \geqslant d)$.

3.21.4. Consider the linear model $\boldsymbol{Y} = \boldsymbol{X\beta} + \boldsymbol{E}$ where $\boldsymbol{E} \sim N(\boldsymbol{0}, \sigma^2\boldsymbol{U})$, σ is unknown and $\boldsymbol{U}$ is known. Using similar arguments obtain expressions for $\text{pr}_{\boldsymbol{0}}(\bar{E}_A^2 \geqslant c, \bar{E}_B^2 \geqslant d)$ and $\text{pr}_{\boldsymbol{0}}(\bar{F}_A \geqslant c, \bar{F}_B \geqslant d)$.
[Hint:

$$\begin{aligned}&\text{pr}_{\boldsymbol{0}}(L_A \leq c_1, L_B \leq c_2) \\ &= \sum \text{pr}_{\boldsymbol{0}}(L_A \leq c_1, L_B \leq c_2 \mid \boldsymbol{X} \in \mathcal{S}_i)\text{pr}(\boldsymbol{X} \in \mathcal{S}_i) \\ &= \sum \text{pr}_{\boldsymbol{0}}(\|\boldsymbol{P}_i\boldsymbol{X}\|^2 \leq c_1, \|(\boldsymbol{I} - \boldsymbol{P})_i\boldsymbol{X}\|^2 \leq c_2 \mid \boldsymbol{X} \in \mathcal{S}_i)\text{pr}(\boldsymbol{X} \in \mathcal{S}_i).\end{aligned}$$

Now use Lemma 3.13.6.]

Section 3.9

3.22 For Example 1.2.2 in Chapter 1 on *El Niño*, recall that the hypothesis of interest is that

the warm phase of El Niño tends to suppress hurricanes whereas cold phase encourages hurricanes.

(a) Test the foregoing hypothesis as a null hypothesis.
(b) Test the same hypothesis as the alternative.

3.23 *Connections among $\bar{E}^2, \bar{F}$ and LRT:* Let $\boldsymbol{X} = (X_1, X_2)^T$ and assume that $\boldsymbol{X} \sim N(\boldsymbol{\theta}, \sigma^2 \boldsymbol{U})$. Let $\boldsymbol{X}_1, \ldots, \boldsymbol{X}_n$ be n independently and identically distributed observations on $\boldsymbol{X}$. Show that this set up is a special case of the linear model in 3.40. Let the null and alternative hypotheses be $H_0 : \boldsymbol{\theta} = \boldsymbol{0}$ and $H_1 : \boldsymbol{\theta} \geqslant \boldsymbol{0}$. Let $Q_1(\boldsymbol{\theta}) = \sum(\boldsymbol{X}_i - \boldsymbol{\theta})^T \boldsymbol{U}^{-1}(\boldsymbol{X}_i - \boldsymbol{\theta})$ and $q_1(\boldsymbol{\theta}) = n(\bar{\boldsymbol{X}} - \boldsymbol{\theta})^T \boldsymbol{U}^{-1}(\bar{\boldsymbol{X}} - \boldsymbol{\theta})$. Show that $Q_1(\boldsymbol{\theta}) = \sum(\boldsymbol{X}_i - \bar{\boldsymbol{X}})^T \boldsymbol{U}^{-1}(\boldsymbol{X}_i - \bar{\boldsymbol{X}}) + q_1(\boldsymbol{\theta})$ and that the loglikelihood $L(\boldsymbol{\theta}, \sigma^2)$ takes the form $L(\boldsymbol{\theta}, \sigma^2) = -(1/2)\sigma^{-2} Q_1(\boldsymbol{\theta}) - \frac{1}{2} np \log(\sigma^2) + Const.$ Let the null and alternative hypotheses be $H_a : \boldsymbol{\theta} \in \mathcal{C}_a$ and $H_b : \boldsymbol{\theta} \in \mathcal{C}_b$, respectively, where $\mathcal{C}_a \subset \mathcal{C}_b$. Let $\boldsymbol{\theta}^a$ and $\boldsymbol{\theta}^b$ be the points at which $Q_1(\boldsymbol{\theta})$ is minimized over $\mathcal{C}_a$ and $\mathcal{C}_b$, respectively. Show that, the likelihood ratio statistic for testing H_a against H_b is given by $LRT = np \log\{Q_1(\boldsymbol{\theta}^a)/Q_1(\boldsymbol{\theta}^b)\}$. Let $\bar{E}^2 = \{Q_1(\boldsymbol{\theta}^a) - Q_1(\boldsymbol{\theta}^b)\}/Q_1(\boldsymbol{\theta}^a)$ and $\bar{F} = \{Q_1(\boldsymbol{\theta}^a) - Q_1(\boldsymbol{\theta}^b)\}/S^2$, where $S^2 = \nu^{-1} \sum(\boldsymbol{X}_i - \boldsymbol{X})^T \boldsymbol{U}^{-1}(\boldsymbol{X}_i - \boldsymbol{X})$ and $\nu = 2(n-1)$. Show that $\bar{E}^2 = [1 - \exp\{-LRT/(np)\}]$. Deduce that the $\bar{E}^2$-test that rejects H_a for large values of $\bar{E}^2$, is equivalent to the likelihood ratio test.

(a) Now let the null and alternative hypotheses be $H_0 : \boldsymbol{\theta} = \boldsymbol{0}$ and $H_1 : \boldsymbol{\theta} \geqslant \boldsymbol{0}$. Let the four quadrants $Q_1, \ldots, Q_4$ of $\mathbb{R}^2$ be defined as in Examples 1 and 2 of Section 3.3. Show that $\bar{X}_1, \bar{X}_2, \sum \|\boldsymbol{X}_i - \bar{\boldsymbol{X}}\|^2$ are mutually independent. By arguments similar to those for (3.7) show that

$$\bar{E}_A^2 = \begin{cases} T/\{T + \sum \|\boldsymbol{X}_i - \bar{\boldsymbol{X}}\|^2\} & \sim \beta(\frac{2}{2}, \frac{n-2}{2}) \quad \text{given } \bar{\boldsymbol{X}} \in Q_1 \\ (n\bar{X}_1^2)/\{T + \sum \|\boldsymbol{X}_i - \bar{\boldsymbol{X}}\|^2\} & \sim \beta(\frac{1}{2}, \frac{n-1}{2}) \quad \text{given } \bar{\boldsymbol{X}} \in Q_4 \\ (n\bar{X}_2^2)/\{T + \sum \|\boldsymbol{X}_i - \bar{\boldsymbol{X}}\|^2\} & \sim \beta(\frac{1}{2}, \frac{n-1}{2}) \quad \text{given } \bar{\boldsymbol{X}} \in Q_2 \\ 0/\{0 + T + \sum \|\boldsymbol{X}_i - \bar{\boldsymbol{X}}\|^2\} & \sim \beta(\frac{0}{2}, \frac{n-0}{2}) \quad \text{given } \bar{\boldsymbol{X}} \in Q_3 \end{cases}$$

where $T = n\bar{X}_1^2 + n\bar{X}_2^2$.

By applying $\text{pr}(\bar{E}_A^2 \leq c | H_0) = \sum_1^4 \text{pr}(\bar{E}_A^2 \leq c | \bar{\boldsymbol{X}} \in Q_i)\text{pr}(\bar{\boldsymbol{X}} \in Q_i)$ or otherwise, deduce that $\text{pr}(\bar{E}_A^2 \leq c | H_0)$ is equal to

$$\frac{1}{4}\text{pr}\{\beta(\frac{0}{2}, \frac{n-0}{2}) \leq c\} + \frac{1}{2}\text{pr}\{\beta(\frac{1}{2}, \frac{n-1}{2}) \leq c) + \frac{1}{4}\text{pr}\{\beta(\frac{2}{2}, \frac{n-2}{2}) \leq c\}.$$

(b) Show that

$$\bar{F}_A = \begin{cases} (n\bar{X}_1^2 + n\bar{X}_2^2)/S^2 & \sim 2F_{2,\nu} \quad \text{conditional on } \bar{\boldsymbol{X}} \in Q_1 \\ n\bar{X}_1^2/S^2 & \sim 1F_{1,\nu} \quad \text{conditional on } \bar{\boldsymbol{X}} \in Q_4 \\ n\bar{X}_2^2/S^2 & \sim 1F_{1,\nu} \quad \text{conditional on } \bar{\boldsymbol{X}} \in Q_2 \\ 0/S^2 & \sim 0F_{0,\nu} \quad \text{conditional on } \bar{\boldsymbol{X}} \in Q_3. \end{cases}$$

Show that, under the null hypothesis,

$$\text{pr}(\bar{F}_A \leq c) = (1/4)\text{pr}(2F_{2,\nu} \leq c) + (1/2)\text{pr}(1F_{1,\nu} \leq c) + (1/4)\text{pr}(0F_{0,\nu} \leq c). \blacksquare$$

3.24 Let $\boldsymbol{X} \sim N(\boldsymbol{\theta}, \sigma^2\boldsymbol{U})$, where σ is unknown. Let $\boldsymbol{X}_1, \ldots, \boldsymbol{X}_n$ be independently and identically distributed as $\boldsymbol{X}$. Let the null and alternative hypotheses be $H_1 : \boldsymbol{R}_1\boldsymbol{\theta} \geqslant \boldsymbol{0}, \boldsymbol{R}_2\boldsymbol{\theta} = \boldsymbol{0}$ and $H_2 : \boldsymbol{\theta}$ *is not restricted*, where $\boldsymbol{R}_1^T$ is $s \times m$ and $\boldsymbol{R}_2^T$ is $t \times m$, and $[\boldsymbol{R}_1^T, \boldsymbol{R}_2^T]$ has full rank. Show that the least favorable null value of $\bar{E}^2$ and of $\bar{F}$ is $\boldsymbol{\theta} = \boldsymbol{0}$, and their least favorable null distributions are

$$\text{pr}_0(\bar{E}^2 \leq c|\boldsymbol{\theta} = \boldsymbol{0}) = \textstyle\sum_{i=0}^{s} w_{s-i}(s, \boldsymbol{A})\text{pr}[\beta\{\frac{1}{2}(t+i), \frac{1}{2}(np-p)\}\} \leq c].$$

$$\text{pr}_0(\bar{F} \leq c|\boldsymbol{\theta} = \boldsymbol{0}) = \textstyle\sum_{i=0}^{s} w_{s-i}(s, \boldsymbol{A})\text{pr}\{(t+i)F_{t+i,\nu} \leq c\}$$

where $\boldsymbol{A} = \boldsymbol{R}_1\boldsymbol{V}\boldsymbol{R}_1^T - (\boldsymbol{R}_1\boldsymbol{V}\boldsymbol{R}_2^T)(\boldsymbol{R}_2\boldsymbol{V}\boldsymbol{R}_2^T)^{-1}(\boldsymbol{R}_2\boldsymbol{V}\boldsymbol{R}_1^T)$.

3.25 Let $\boldsymbol{X}_1, \ldots, \boldsymbol{X}_n$ be independently and identically distributed as $\boldsymbol{X}$ where $\boldsymbol{X} \sim N(\boldsymbol{\theta}, \sigma^2\boldsymbol{U})$. Let the null and alternative hypotheses be $H_0 : \boldsymbol{R\theta} = \boldsymbol{0}$ and $H_1 : \boldsymbol{R}_1\boldsymbol{\theta} \geqslant \boldsymbol{0}$, where $\boldsymbol{R}$ is a full row-rank matrix of order $r \times p$ and $\boldsymbol{R}_1$ is a submatrix of $\boldsymbol{R}$ of order $q \times p$. Show that $\bar{E}^2$ and $\bar{F}$ tests are similar and that their null distributions are given by

$$\text{pr}(\bar{E}^2 \leq c|H_0) = \textstyle\sum_{j=0}^{q} w_j(q, \boldsymbol{R}_1\boldsymbol{V}\boldsymbol{R}_1^T)\text{pr}\{\beta(2^{-1}(r-q+j), 2^{-1}(np-p+q-j) \leq c\},$$

$$\text{pr}(\bar{F} \leq c) = \textstyle\sum_{j=0}^{q} w_j(q, \boldsymbol{R}_1\boldsymbol{V}\boldsymbol{R}_1^T)\text{pr}\{(r-q+j)F_{r-q+j,\nu} \leq c\}.$$

Section 3.10

3.26 For the hypertensive patients in the sodium reduction example (see Example 1.2.17 in Chapter 1), obtain an overall p-value for testing the hypothesis that a reduction in sodium intake leads to a reduction in blood pressure. State the assumptions that you made and discuss their limitations.

3.27 Let $\boldsymbol{X} \sim N(\boldsymbol{\theta}, \boldsymbol{V})$ where $\boldsymbol{V}$ is unknown. Let $\boldsymbol{\theta}_2$ and $\boldsymbol{S}$ be as in Section 3.10. Consider the test of $H_0 : \boldsymbol{\theta}_2 = \boldsymbol{0}$ against $H_1 : \boldsymbol{\theta}_2 \in \mathcal{P}$, where $\mathcal{P}$ is a one-sided closed cone. Show that the *LRT* takes the simpler form (see Perlman (1969), equation (6.13))

$$LRT = \{\|\boldsymbol{Y}\|_{\boldsymbol{T}}^2 - \|\boldsymbol{Y} - \Pi_{\boldsymbol{T}}(\boldsymbol{Y}|\mathcal{P})\|_{\boldsymbol{T}}^2\{1 + \|\boldsymbol{Y} - \Pi_{\boldsymbol{T}}(\boldsymbol{Y}|\mathcal{P})\|_{\boldsymbol{T}}^2\}^{-1}$$

where $\boldsymbol{Y} = \bar{\boldsymbol{X}}_{22}$ and $\boldsymbol{T} = \boldsymbol{S}_{22}$.

3.28 Let $\boldsymbol{A}$ be the $k \times (k-1)$ matrix such that $\boldsymbol{A\mu} \geqslant \boldsymbol{0}$ is equivalent to $\mu_1 \leqslant \ldots \leqslant \mu_k$. Let $\boldsymbol{y}^*$ be the point at which $(\boldsymbol{y} - \boldsymbol{\mu})^T\boldsymbol{W}(\boldsymbol{y} - \boldsymbol{\mu})$ reaches its minimum over $\boldsymbol{A\mu} \geqslant \boldsymbol{0}$, where $\boldsymbol{W}$ is a diagonal matrix with positive diagonal elements. Let $\mathcal{Q} = \{\boldsymbol{\mu} : \boldsymbol{A\mu} \geqslant \boldsymbol{0}\}$. Let $\mathcal{M}$ be the linear space spanned by $(1, \ldots, 1)$. Deduce the following [Hint: Use the results relating to Fig. 3.10, in particular those in Section 3.12.2. Proofs of these from first principles are provided in Barlow et al. (1972); see also Robertson et al. (1988)].

1. $\mathcal{M}$ is the largest linear space contained in $\mathcal{Q}$.
2. $(\boldsymbol{y} - \boldsymbol{y}^*) \perp_{\boldsymbol{W}^{-1}} \mathcal{M}$.

3. $(\boldsymbol{y}-\boldsymbol{y}^*)\boldsymbol{W}\boldsymbol{1}=0$. (Hint: $\boldsymbol{1}\in\mathcal{M}$).
4. $(\boldsymbol{y}-\boldsymbol{\mu})^T\boldsymbol{W}(\boldsymbol{y}-\boldsymbol{\mu})\geqslant(\boldsymbol{y}-\boldsymbol{y}^*)^T\boldsymbol{W}(\boldsymbol{y}-\boldsymbol{y}^*)+(\boldsymbol{y}^*-\boldsymbol{\mu})^T\boldsymbol{W}(\boldsymbol{y}^*-\boldsymbol{\mu})$ for every $\boldsymbol{\mu}\in\mathcal{Q}$.
5. $(\boldsymbol{y}-\boldsymbol{y}^*)^T\boldsymbol{W}\boldsymbol{y}^*=0$.
6. $(\boldsymbol{y}-\boldsymbol{y}^*)^T\boldsymbol{W}\boldsymbol{\mu}\leqslant 0$ for every $\boldsymbol{\mu}\in\mathcal{Q}$.
7. If $\tilde{\boldsymbol{y}}\in\mathcal{Q}$ and $(\boldsymbol{y}-\tilde{\boldsymbol{y}})^T\boldsymbol{W}(\tilde{\boldsymbol{y}}-\boldsymbol{\mu})\geqslant 0$ for every $\boldsymbol{\mu}\in\mathcal{Q}$ then $\tilde{\boldsymbol{y}}=\boldsymbol{y}^*$.
8. Let $\tilde{\boldsymbol{y}}\in\mathcal{Q}$. Then $\tilde{\boldsymbol{y}}=\boldsymbol{y}^*$ if and only if $(\boldsymbol{y}-\tilde{\boldsymbol{y}})^T\boldsymbol{W}\tilde{\boldsymbol{y}}=0$ and $(\boldsymbol{y}-\tilde{\boldsymbol{y}})^T\boldsymbol{W}\boldsymbol{\mu}\leqslant 0$ for every $\boldsymbol{\mu}\in\mathcal{Q}$.

3.29 Let $\hat{\boldsymbol{\theta}}$ be an estimator of $\boldsymbol{\theta}$ where $\boldsymbol{\theta}\in\mathbb{R}^p$. Suppose that $\sqrt{n}(\hat{\boldsymbol{\theta}}-\boldsymbol{\theta})$ converges in distribution to $N(\boldsymbol{0},\boldsymbol{V})$ where $\boldsymbol{V}$ is positive definite. Let $\hat{\boldsymbol{V}}$ be a consistent estimator of $\boldsymbol{V}$ (i.e., $\hat{\boldsymbol{V}}\xrightarrow{p}\boldsymbol{V}$), $H_0:\boldsymbol{\theta}\in\mathcal{M}$, $H_1:\boldsymbol{\theta}\in\mathcal{C}$, and $H_2:\boldsymbol{\theta}\notin\mathcal{C}$. $L_A=n\{\|\hat{\boldsymbol{\theta}}-\mathcal{M}\|^2-\|\hat{\boldsymbol{\theta}}-\mathcal{C}\|^2\}$ and $L_B=n\|\hat{\boldsymbol{\theta}}-\mathcal{C}\|^2$, where $\|\cdot\|$ is defined with respect to $\langle\cdot\rangle_{\hat{\boldsymbol{V}}}$. Following the notation introduced in this chapter, it is implicitly assumed that $\mathcal{C}$ is a closed convex cone, $\mathcal{M}$ is a linear space, and $\mathcal{M}\subset\mathcal{C}$. Show that L_A is consistent at any $\boldsymbol{\theta}\notin(\mathcal{C}\cap\mathcal{M}^\perp)^o$ and that L_B is consistent at any $\boldsymbol{\theta}\notin\mathcal{C}$.

Now assume that $\mathcal{C}$ is a polyhedral. Explain why the limiting probability of rejecting the null hypothesis may depend on the particular face of the polyhedral on which the true value lies.

3.30

3.30.1. Let $H_0:\boldsymbol{\theta}=\boldsymbol{0}$ and $H_1:\boldsymbol{\theta}\in\mathcal{C}$ be the null and alternative hypotheses where $\boldsymbol{\theta}\in\mathbb{R}^p$. Assume that $\boldsymbol{X}=\boldsymbol{\theta}+\boldsymbol{E}$, where the distribution of $\boldsymbol{E}$ does not depend on any unknown parameters. Let $\boldsymbol{V}$ be a positive definite matrix, the inner product be defined with respect to $\langle\cdot\rangle_{\boldsymbol{V}}$, and let

$$L_A=\|\boldsymbol{X}\|^2-\|\boldsymbol{X}-\mathcal{C}\|^2.$$

Then prove the following: (a) $L_A=\|\boldsymbol{X}-\mathcal{C}^o\|^2=\|\boldsymbol{E}-\mathcal{P}_{\boldsymbol{\theta}}\|^2$, where $\mathcal{P}_{\boldsymbol{\theta}}=\mathcal{C}^o-\boldsymbol{\theta}$. (b) $\|\boldsymbol{\theta}-\mathcal{C}^o\|=\|\boldsymbol{0}-\mathcal{P}_{\boldsymbol{\theta}}\|$; in other words, the distance between $\boldsymbol{\theta}$ and $\mathcal{C}^o$ is equal to the distance between $\boldsymbol{0}$ and $\mathcal{P}_{\boldsymbol{\theta}}$. (c) For any given $K>0$, if the distance between $\boldsymbol{\theta}$ and $\mathcal{C}^o$ is more than K then $B_K(\boldsymbol{0})\cap\mathcal{P}_{\boldsymbol{\theta}}$ is the empty set, where $B_K(\boldsymbol{0})$ is the ball of radius K centered at $\boldsymbol{0}$. (d) $L_A\xrightarrow{p}\infty$ as $\|\boldsymbol{\theta}-\mathcal{C}^o\|\to\infty$. (e) For the test that rejects the H_0 for large values of L_A, show that the power tends to 1 as $\|\boldsymbol{\theta}-\mathcal{C}^o\|\to\infty$.

3.30.2. Let $\boldsymbol{X},\boldsymbol{\theta},\boldsymbol{E},\boldsymbol{V}$ and the inner product be as in the previous part. Let $H_0:\boldsymbol{\theta}\in\mathcal{M}$ and $H_1:\boldsymbol{\theta}\in\mathcal{C}$, and let

$$L_A=\|\boldsymbol{X}-\mathcal{M}\|^2-\|\boldsymbol{X}-\mathcal{C}\|^2;$$

$\mathcal{M}$ is a linear space, $\mathcal{C}$ is a closed convex cone, and $\mathcal{M}\subset\mathcal{C}$. Consider the test that rejects the H_0 for large values of L_A. Prove that

$$\text{the power of } L_A \text{ tends to 1 as } \|\boldsymbol{\theta}-(\mathcal{C}\cap\mathcal{M}^\perp)^o\|\to\infty.$$

[Hint: Recall that $L_A = \|\boldsymbol{X} - (\mathcal{C} \cap \mathcal{M}^{\perp})^o\|^2$ and hence L_A has the same numerical value as the L_A in the previous part with $\mathcal{C}$ replaced by $\mathcal{C} \cap \mathcal{M}^{\perp}$. Now apply the previous part.]

3.30.3. Let $\boldsymbol{X}, \boldsymbol{\theta}, \boldsymbol{E}, \boldsymbol{V}$ and the inner product be as in the previous part. Let $H_1 : \boldsymbol{\theta} \in \mathcal{C}$ and $H_2 : \boldsymbol{\theta} \notin \mathcal{C}$, and let $L_B = \|\boldsymbol{X} - \mathcal{C}\|^2$. Show that $L_B \xrightarrow{p} \infty$ and that the power of the test that rejects H_1 for large values of L_B tends to 1 as $\|\boldsymbol{\theta} - \mathcal{C}\| \to \infty$. [Hint: Apply a method similar to that for the first part].

3.31 Let $\boldsymbol{A}$ be a nonsingular matrix of order $p \times p$, $\boldsymbol{x} \in \mathbb{R}^p$ and $\mathcal{P}$ be a cone. Then prove the following:

1. $\boldsymbol{A}\Pi(\boldsymbol{x}|\mathcal{P}) = \Pi_{\boldsymbol{AVA}^T}(\boldsymbol{Ax}|\boldsymbol{A}\mathcal{P})$.

2. $\|\Pi_{\boldsymbol{V}}(\boldsymbol{x}|\mathcal{P})\|_{\boldsymbol{V}} = \|\Pi_{\boldsymbol{AVA}^T}(\boldsymbol{Ax}|\boldsymbol{A}\mathcal{P})\|_{\boldsymbol{AVA}^T}$.

3. $\|\boldsymbol{x} - \Pi_{\boldsymbol{V}}(\boldsymbol{x}|\mathcal{P})\|_{\boldsymbol{V}} = \|\boldsymbol{Ax} - \Pi_{\boldsymbol{AVA}^T}(\boldsymbol{Ax}|\boldsymbol{A}\mathcal{P})\|_{\boldsymbol{AVA}^T}$.

4. Let $\mathcal{P}_1 \supset \mathcal{P}_2 \ldots \supset$ be s sequence of decreasing cones in $\mathbb{R}^p$. Then $\|\boldsymbol{x} - \Pi_{\boldsymbol{V}}(\boldsymbol{x}|\mathcal{P}_n)\|_{\boldsymbol{V}}$ is a nondecreasing sequence and converges to $\|\boldsymbol{x} - \Pi_{\boldsymbol{V}}(\boldsymbol{x}|\mathcal{P})\|_{\boldsymbol{V}}$ where $\mathcal{P} = \lim \mathcal{P}_n$; a similar result holds for increasing sequences.

[Perlman (1969)]

3.32 (a) Let $\boldsymbol{x}, \boldsymbol{y} \in \mathbb{R}^p$, $\mathcal{C}$ be a closed convex cone, $\boldsymbol{V}$ be a positive definite matrix of order $p \times p$, Π denote $\Pi_{\boldsymbol{V}}$, $\|\cdot\|$ denote $\|\cdot\|_{\boldsymbol{V}}$, and $\mathcal{C}^o$ denote the polar cone of $\mathcal{C}$ with respect to $\langle\cdot\rangle_{\boldsymbol{V}}$. Show that [see Lemma 2.2 in Mukerjee et al. (1986)]

$$\Pi(\boldsymbol{x} + \boldsymbol{y} \mid \mathcal{C}) \leq \|\Pi(\boldsymbol{x} \mid \mathcal{C}) + \Pi(\boldsymbol{y} \mid \mathcal{C})\| \leq \|\Pi(\boldsymbol{x} \mid \mathcal{C})\| + \|\Pi(\boldsymbol{y} \mid \mathcal{C})\|.$$

(b) $\|\Pi(\boldsymbol{y} + \boldsymbol{x} \mid \mathcal{C})\| \leq \|\Pi(\boldsymbol{y} \mid \mathcal{C})\|, \quad \boldsymbol{x} \in \mathcal{C}^o$.
(c) $\|\boldsymbol{y} - \boldsymbol{x}\|^2 \geq \|\Pi(\boldsymbol{y} \mid \mathcal{C}) - \Pi(\boldsymbol{x} \mid \mathcal{C})\|^2, \quad \boldsymbol{x}, \boldsymbol{y} \in \mathbb{R}^p$.

3.33 Let $\boldsymbol{Y} = (Y_1, \ldots, Y_n)^T$ where $Y_1, \ldots, Y_n$ are independent standard normal random variables. Let $\mathcal{L}$ be a linear subspace of $\mathbb{R}^n, dim(\mathcal{L}) = r$, $\boldsymbol{A}$ be an $m \times n$ matrix such that each row of $\boldsymbol{A}$ is orthogonal to $\mathcal{L}$, and $\tilde{\boldsymbol{Y}}$ denote the projection of $\boldsymbol{Y}$ onto $\mathcal{L}$. Assume that $\{\boldsymbol{y} : \boldsymbol{Ay} \geqslant \boldsymbol{0}\}$ is nonempty. Then, show that the conditional distribution of $\|\tilde{\boldsymbol{Y}}\|^2$, given $\boldsymbol{AY} \geqslant \boldsymbol{0}$, is χ^2_r. [Meyer (2003b)]
[Hint: Use Lemma 3.13.3. Let $\boldsymbol{P}$ be the symmetric projection matrix onto $\mathcal{L}$. Then $\mathcal{L} = \{\boldsymbol{Py} : \boldsymbol{y} \in \mathbb{R}^n\}$. Verify that if $\boldsymbol{a}_i^T$ is the ith row of $\boldsymbol{A}$ then $\boldsymbol{a}_i = \boldsymbol{Py}_i$ for some $\boldsymbol{y}_i$ $(i = 1, \ldots, m)$. Deduce that $\boldsymbol{A} = \boldsymbol{BP}$ where $\boldsymbol{B}^T = [\boldsymbol{y}_1, \ldots, \boldsymbol{y}_m]$, and hence $\boldsymbol{AY} \geq \boldsymbol{0}$ is equivalent to $\boldsymbol{PY} \in \mathcal{C} = \{\boldsymbol{x} : \boldsymbol{Bx} \geqslant \boldsymbol{0}\}$. For a different proof, see Meyer (2003b)].

3.34 Let $\boldsymbol{Y} = (Y_1, \ldots, Y_n)^T$ where $Y_1, \ldots, Y_n$ are independent standard normal random variables. Let $\mathcal{L}$ be a linear subspace of $\mathbb{R}^n$, $dim(\mathcal{L}) = r$. Let $\boldsymbol{A}$ be an $m \times n$ matrix such that each row of $\boldsymbol{A}$ is a vector in $\mathcal{L}$. Let $\tilde{\boldsymbol{Y}}$ denote the projection of $\boldsymbol{Y}$ onto $\mathcal{L}$. Assume that $\{\boldsymbol{Ay} \geqslant \boldsymbol{0}\}$ is non-empty. Then, show that the conditional distribution of $\|\tilde{\boldsymbol{Y}}\|^2$, given $\boldsymbol{AY} \geqslant \boldsymbol{0}$, is χ^2_r. [Meyer (2003b)]

[Hint: Use Lemma 3.13.3. Let $\boldsymbol{P}$ be a projection matrix onto $\mathcal{L}$. Let the ith row of $\boldsymbol{A}$ be $\boldsymbol{a}_i^T$. Then $\boldsymbol{a}_i = (\boldsymbol{I} - \boldsymbol{P})\boldsymbol{y}_i$ for some $\boldsymbol{y}_i$ $(i = 1, \ldots, m)$. Deduce that $\boldsymbol{A} = \boldsymbol{Y}_0^T(\boldsymbol{I} - \boldsymbol{P})$, where $\boldsymbol{Y}_0^T = [\boldsymbol{y}_1, \ldots, \boldsymbol{y}_m]$, and hence $\boldsymbol{AZ} \geqslant \mathbf{0}$ is equivalent to $(\boldsymbol{I} - \boldsymbol{P})\boldsymbol{Z} \in \mathcal{P}$ where $\mathcal{P} = \{\boldsymbol{x} : \boldsymbol{Y}_0^T\boldsymbol{x} \geqslant 0\}$. Now, use the fact that $\boldsymbol{PZ}$ and $(|\boldsymbol{I} - \boldsymbol{P})\boldsymbol{Z}$ are independent to verify that $\text{pr}(\|\boldsymbol{PZ}\|^2 \leqslant c \mid \boldsymbol{BZ} \geqslant 0) = \text{pr}(\|\boldsymbol{PZ}\|^2 \leqslant c \mid (\boldsymbol{I} - \boldsymbol{P})\boldsymbol{Z} \in \mathcal{P}) = \text{pr}(\|\boldsymbol{PZ}\|^2 \leqslant c)$.]

4

Tests in General Parametric Models

4.1 INTRODUCTION

In Chapter 3, tests of hypotheses concerning the mean of a multivariate normal were considered. Often, the parameter of interest in a hypothesis is more than just the mean of a multivariate normal. Some examples of this type were discussed in Chapter 1. In this chapter, the relevant theory is developed in a general context. To illustrate the nature of the problem, consider Example 1.2.8. It describes the results of a study to establish that a new treatment is better than an old one when the response is an ordinal variable. In this example, the inference problem can be formulated as test of $\boldsymbol{g}(\boldsymbol{\pi}) = \mathbf{0}$ against $\boldsymbol{g}(\boldsymbol{\pi}) \geqslant \mathbf{0}$ where $\boldsymbol{\pi}$ is the vector consisting of all the $\{\pi_{ij}\}$ and π_{ij} is the probability that the response for Treatment i is the jth ordinal category of the response variable, and $\boldsymbol{g}$ is a nonlinear function; in fact, every component of $\boldsymbol{g}$ is an odds ratio. This example does not fit into the setting in Chapter 3. The theory developed in this chapter is applicable to this and much more general settings. A sample of models to which the theory of this chapter is applicable include the following:

(1) nonlinear regression
(2) generalized linear models (logistic regression, log-linear models, etc.)
(3) proportional hazards model, Cox's regression model
(4) time series model (for example, ARCH models used in finance)
(5) mixed effects models (positive semi-definite covariance matrices)
(6) quasi-likelihood
(7) generalized estimating equations (GEE)

(8) generalized method of moments (GMM).

It is thus clear that the results of this chapter are applicable to a broad range of contexts.

Let $\boldsymbol{\theta}$ denote the parameter of a statistical model and Θ denote the parameter space. A common feature of most of the inference problems with no inequality constraints on $\boldsymbol{\theta}$ is that Θ is open and the likelihood function is smooth over Θ. This feature plays a crucial role in ensuring that (i) $\sqrt{n}(\hat{\boldsymbol{\theta}} - \boldsymbol{\theta}) \approx N\{\mathbf{0}, \mathcal{I}(\boldsymbol{\theta})^{-1}\}$, and (ii) the asymptotic null distribution of the likelihood ratio statistic for testing $\boldsymbol{\theta}_a = \mathbf{0}$ against $\boldsymbol{\theta}_a \neq \mathbf{0}$ is a chi-square where $\boldsymbol{\theta}_a$ is a subvector of $\boldsymbol{\theta}$. These two results form the basis of statistical inference in these models when there are no inequality constraints on $\boldsymbol{\theta}$.

In this chapter, we will build on these and obtain corresponding results when there are inequality constraints on $\boldsymbol{\theta}$ or the parameter space for $\boldsymbol{\theta}$ is constrained so that the constrained space is not open (topologically). These results will form the basis of inference in this setting. Most of the details in this chapter are given for the case when the observations are independently and identically distributed. However, corresponding results hold for non-*iid* cases, which include regression type models, time series models, and stochastic processes. As a general guide, with $\hat{\boldsymbol{\theta}}$ denoting an unconstrained estimator of $\boldsymbol{\theta}$, if $\sqrt{n}(\hat{\boldsymbol{\theta}} - \boldsymbol{\theta})$ is asymptotically normal and the asymptotic null distribution of the likelihood ratio statistic for testing $\boldsymbol{\theta}_a = \mathbf{0}$ against $\boldsymbol{\theta}_a \neq \mathbf{0}$ is a chi-square, where $\boldsymbol{\theta}_a$ is a subvector of $\boldsymbol{\theta}$, then it is very likely that most of the results in this chapter for likelihood ratio tests would be applicable for constrained statistical inference on $\boldsymbol{\theta}$ as well.

The plan of this chapter is as follows. Section 4.2 provides a brief discussion of regularity conditions relating to the likelihood function that would be relevant to the rest of the chapter. Section 4.3 considers statistical models, which are usually referred to as *regular models,* and develops the likelihood ratio, Wald and score-type tests of $\boldsymbol{R\theta} = \mathbf{0}$ against $\boldsymbol{R\theta} \geqslant \mathbf{0}$ where $\boldsymbol{R}$ is a given matrix of constants. These results are of particular importance because a large number of real-life examples with inequality constraints on $\boldsymbol{\theta}$ are of this type. In any case, a detailed study of the case when the constraints are linear is essential before we consider nonlinear constraints. Section 4.4 extends the results of the previous section to the case when the constraints are nonlinear in $\boldsymbol{\theta}$. In particular, tests of $\boldsymbol{h}(\boldsymbol{\theta}) = \mathbf{0}$ against $\boldsymbol{h}(\boldsymbol{\theta}) \geqslant \mathbf{0}$ where $\boldsymbol{h}$ is a vector function will be discussed; it will be seen that there are several tests that are asymptotically equivalent to the LRT. Subsection 4.4.2 provides a statistic for testing $\boldsymbol{h}(\boldsymbol{\theta}) = \mathbf{0}$ against $\boldsymbol{h}_2(\boldsymbol{\theta}) \geqslant \mathbf{0}$, where $\boldsymbol{h}(\boldsymbol{\theta})$ is a smooth vector function of $\boldsymbol{\theta}$ and $\boldsymbol{h}_2(\boldsymbol{\theta})$ is a subvector of $\boldsymbol{h}(\boldsymbol{\theta})$. The main requirement for this is that $\sqrt{n}(\hat{\boldsymbol{\theta}} - \boldsymbol{\theta})$ be asymptotically normal; thus $\hat{\boldsymbol{\theta}}$ may not be the *mle.* In Section 4.5, a brief overview of likelihood ratio, Wald and score tests is provided with particular reference to the results that would be required for developing a class of score tests against inequality constraints. In Section 4.6, the Rao's score and Neyman's C(α) tests are generalized to tests against inequality constraints. These are further extended to develop tests against inequality constraints when the model is estimated by solving a set of estimating equations; for example, the class of *Generalized Estimating Equation*(GEE) methods and the *Generalized Method of Moments* (GMM) that is widely used in economics, fall into this category.

The results in Sections 4.3–4.6 provide a good coverage of the standard results required for testing $\boldsymbol{R\theta} = \mathbf{0}$ against $\boldsymbol{R\theta} \geqslant \mathbf{0}$. These results are extended further in the next two sections where the null and alternative hypotheses are stated in the general form, $H_0 : \boldsymbol{\theta} \in \Theta_0$ and $H_1 : \boldsymbol{\theta} \in \Theta_1$, respectively, where $\Theta_0 \subset \Theta_1 \subset \Theta$. In this case, the asymptotic null distributions of test statistics such as the likelihood ratio/Wald/ score, depend on the local shapes of Θ_0 and Θ_1 at the assumed true value in the null parameter space. The relevant local shapes of Θ_0 and Θ_1 are characterized by cones that approximate them at $\boldsymbol{\theta}_0$; these are called *approximating cones*. These will be discussed in detail later in this chapter.

General results for testing $H_0 : \boldsymbol{\theta} \in \Theta_0$ against $H_1 : \boldsymbol{\theta} \in \Theta_1$ are discussed in Section 4.8. A topic of particular interest in economics involves test of $\boldsymbol{h}(\boldsymbol{\theta}) \geqslant \mathbf{0}$ against $\boldsymbol{h}(\boldsymbol{\theta}) \ngeqslant \mathbf{0}$ where $\boldsymbol{h}$ is a given nonlinear function. An example of this type is test of the hypothesis that a certain function (for example, a production function or a cost function) is concave against that it is not concave. The general theory for this is discussed in Section 4.8. The asymptotic distribution of the constrained local and global *mle* are studied in the last section. Some of the proofs are relegated to the Appendix at the end of this chapter.

4.2 PRELIMINARIES

Let $Y_1, \ldots, Y_n$ be independently and identically distributed with common density function $f(y; \boldsymbol{\theta})$, $\boldsymbol{\theta} \in \Theta \subset \mathbb{R}^p$ and y can be univariate or multivariate. Let

$$\begin{aligned} \ell(\boldsymbol{\theta}) &= \sum \log f(Y_i; \boldsymbol{\theta}), \qquad \boldsymbol{S}(\boldsymbol{\theta}) = (\partial/\partial\boldsymbol{\theta})\ell(\boldsymbol{\theta}) \\ \text{and} \quad \mathcal{I}_{\boldsymbol{\theta}} &= \mathcal{I}(\boldsymbol{\theta}) = E_{\boldsymbol{\theta}}\{(\partial/\partial\boldsymbol{\theta}) \log f(Y; \boldsymbol{\theta})(\partial/\partial\boldsymbol{\theta}^T) \log f(Y; \boldsymbol{\theta})\} \end{aligned} \tag{4.1}$$

denote the loglikelihood, score function, and the information matrix for one observation, respectively. Let $\boldsymbol{\theta}_0$ denote the *true value* and $\hat{\boldsymbol{\theta}}$ denote the global *mle* over Θ.

Let A be a subset of Θ containing the true value of $\boldsymbol{\theta}$. Let $\hat{\boldsymbol{\theta}}_A$ denote the *mle* over A defined by

$$\hat{\boldsymbol{\theta}}_A = \arg \max_{\boldsymbol{\theta} \in A} \ell(\boldsymbol{\theta});$$

equivalently, $\ell(\hat{\boldsymbol{\theta}}_A) = \max_{\boldsymbol{\theta} \in A} \ell(\boldsymbol{\theta})$. Strictly speaking, it is better to define the *mle* over A by $\ell(\hat{\boldsymbol{\theta}}_A) = \sup_{\boldsymbol{\theta} \in A} \ell(\boldsymbol{\theta}) + o_p(1)$; if we do so, $\hat{\boldsymbol{\theta}}_A$ would be well defined even if $\sup_{\boldsymbol{\theta} \in A} \ell(\boldsymbol{\theta})$ is reached at a point on the boundary of A that is not in A. However, we shall not be concerned with this technical detail. Every result in this chapter requires $\hat{\boldsymbol{\theta}}_A$ to be consistent, where A may the natural parameter space Θ or the ones defined by null/alternative hypotheses. For the case when the true value of $\boldsymbol{\theta}$ is an interior point of A, consistency of the *mle* has been discussed in great detail in the literature; for example, see Lehmann (1983, section 6.3). For a discussion of the case when the true parameter may be on the boundary of the parameter space, see Self and Liang (1987, Theorem 1) and Andrews (1999, Section 3.1); van der Vaart and Wellner (1996, Section 3.2) consider the more general case, which includes M-estimation.

Consistency of $\hat{\boldsymbol{\theta}}_A$ would follow from the following two, albeit strong, conditions (see Andrews (1999)): There exists a function $\ell_0(\boldsymbol{\theta})$ such that
(i) $\sup_{\boldsymbol{\theta}\in A} |n^{-1}\ell(\boldsymbol{\theta}) - \ell_0(\boldsymbol{\theta})| \xrightarrow{p} 0$, and
(ii) $\sup_{\boldsymbol{\theta}\in A, \|\boldsymbol{\theta}-\boldsymbol{\theta}_0\|>\epsilon} \ell_0(\boldsymbol{\theta}_0) < \ell_0(\boldsymbol{\theta}_0), \quad \epsilon > 0.$
The first condition says that $n^{-1}\ell(\boldsymbol{\theta})$ converges uniformly over the parameter space, and the second says that $\boldsymbol{\theta}_0$ is the global maximum of $\ell_0(\boldsymbol{\theta})$ and it is separated from all the other maxima. The uniform law of large numbers of Andrews (1992) may be helpful in establishing the uniform convergence of $n^{-1}\ell(\boldsymbol{\theta})$. In any case, let us note that if the estimator is the global maximizer of a function, such as the likelihood, then constraints on $\boldsymbol{\theta}$ do not generally introduce additional difficulties for establishing consistency. Throughout this chapter, we shall assume the following:

Assumption A1: *The mle,* $\hat{\boldsymbol{\theta}}_A$, *is consistent where* $\boldsymbol{\theta}_0 \in A \subset \Theta$. ■

Further, we shall also typically assume that a set of regularity conditions are satisfied to ensure that the asymptotic null distribution of the *LRT* of $\boldsymbol{\theta} = \boldsymbol{\theta}_0$ against $\boldsymbol{\theta} \neq \boldsymbol{\theta}_0$ is a chi-square, and $\sqrt{n}(\hat{\boldsymbol{\theta}} - \boldsymbol{\theta}_0)$ is asymptotically normal. If this were not true, it is quite unlikely that the results stated in this section about the asymptotic null distribution of LRT for test against inequality restrictions would hold. It is also of interest to note that such a set of regularity conditions is sufficient to derive the asymptotic distributions of LRT under inequality constraints. With this in mind, let us define the following set of regularity conditions:

Condition Q:

1. Distinct values of $\boldsymbol{\theta}$ correspond to distinct distributions.
2. The first three partial derivatives of $\log f(y; \boldsymbol{\theta})$ with respect to $\boldsymbol{\theta}$ exist almost everywhere.
3. There exists a $G(y)$ such that $\int G(y)dy < \infty$ and the absolute values of the first three partial derivatives of $\log f(y; \boldsymbol{\theta})$ with respect to $\boldsymbol{\theta}$ are bounded by $G(y)$ in a neighborhood of $\boldsymbol{\theta}_0$.
4. The Fisher information matrix, $\mathcal{I}(\boldsymbol{\theta})$, is finite and positive definite. ■

Remark: In most cases, it would be possible to replace the condition involving the third derivative of $\log f(y; \boldsymbol{\theta})$ by one that involves only the second derivative. However, we shall not concern ourselves with this technical detail.

In what follows we shall assume that Condition Q is satisfied. Under this condition, we have the following standard results; for example, see Cox and Hinkley (1974), Sen and Singer (1993), Lehmann (1991), and Ferguson (1996).

Proposition 4.2.1 *Suppose that Condition Q is satisfied. Then we have the following:*

1. $n^{-1/2}\boldsymbol{S}(\boldsymbol{\theta}_0) \xrightarrow{d} N\{\boldsymbol{0}, \mathcal{I}(\boldsymbol{\theta}_0)\}$.
2. $n^{-1/2}\mathcal{I}(\boldsymbol{\theta}_0)^{-1}\boldsymbol{S}(\boldsymbol{\theta}_0) = n^{1/2}(\hat{\boldsymbol{\theta}} - \boldsymbol{\theta}_0) + o_p(1)$.
3. $\sqrt{n}(\hat{\boldsymbol{\theta}} - \boldsymbol{\theta}_0) \xrightarrow{d} N\{\boldsymbol{0}, \mathcal{I}(\boldsymbol{\theta}_0)^{-1}\}$.

4. *The LRT for testing* $\boldsymbol{R\theta} = 0$ *against* $\boldsymbol{R\theta} \neq \mathbf{0}$ *is asymptotically* χ^2_r *under* H_0 *where* $r = rank(\boldsymbol{R})$. ■

This proposition states important first order results that are useful for inference when there are no inequality constraints on $\boldsymbol{\theta}$. Typically, these results are proved using a quadratic approximation of $\ell(\boldsymbol{\theta})$ and a linear approximation of $\nabla\ell(\boldsymbol{\theta})$ near $\boldsymbol{\theta}_0$. Since we will be making use of such quadratic approximations, they are stated in the next proposition.

Proposition 4.2.2 *Assume that Condition Q is satisfied. Let* $\boldsymbol{u} = \sqrt{n}(\boldsymbol{\theta} - \boldsymbol{\theta}_0)$ *and* $K > 0$ *be given. Then we have the following:*

$$(A)\quad \ell(\boldsymbol{\theta}) = \ell(\boldsymbol{\theta}_0) + n^{-1/2}\boldsymbol{u}^T\boldsymbol{S}(\boldsymbol{\theta}_0) - (1/2)\boldsymbol{u}^T\mathcal{I}_{\boldsymbol{\theta}_0}\boldsymbol{u} + r_n(\boldsymbol{u}), \tag{4.2}$$

where $\sup_{\|\boldsymbol{u}\|<K} |r_n(\boldsymbol{u})| = o_p(1)$.

$$(B)\quad \ell(\boldsymbol{\theta}) = \ell(\boldsymbol{\theta}_0) + (1/2)n^{-1}\boldsymbol{S}(\boldsymbol{\theta}_0)^T\mathcal{I}_{\boldsymbol{\theta}_0}^{-1}\boldsymbol{S}(\boldsymbol{\theta}_0) - (1/2)(\boldsymbol{Z}_n - \boldsymbol{u})^T\mathcal{I}_{\boldsymbol{\theta}_0}(\boldsymbol{Z}_n - \boldsymbol{u}) + \delta_n(\boldsymbol{u}), \tag{4.3}$$

where $\boldsymbol{Z}_n = n^{-1/2}\mathcal{I}_{\boldsymbol{\theta}_0}^{-1}\boldsymbol{S}(\boldsymbol{\theta}_0)$ *and* $\sup_{\|\boldsymbol{u}\|<K} |\delta_n(\boldsymbol{u})| = o_p(1)$.

$$(C)\quad \ell(\boldsymbol{\theta}) = \ell(\hat{\boldsymbol{\theta}}) - (1/2)(\boldsymbol{Z}_n - \boldsymbol{u})^T\mathcal{I}_{\boldsymbol{\theta}_0}(\boldsymbol{Z}_n - \boldsymbol{u}) + \epsilon_n(\boldsymbol{u}), \tag{4.4}$$

where $\boldsymbol{Z}_n = \sqrt{n}(\hat{\boldsymbol{\theta}} - \boldsymbol{\theta}_0)$, $\hat{\boldsymbol{\theta}}$ *is* $\sqrt{n}$*-consistent, and* $\sup_{\|\boldsymbol{u}\|<K} |\epsilon_n(\boldsymbol{u})| = o_p(1)$.

Further, $\delta_n(\boldsymbol{u}) = n^{-1/2}\|\boldsymbol{u}\|^3 O_p(1)$. ■

These three quadratic approximations are very important in constrained inference, just as in unconstrained inference. A detailed proof of the first approximation is given in Sen and Singer (1993, Section 5.3); the second approximation is used in Self and Liang (1987), and the third one is used in Silvapulle (1994). For a detailed account see Andrews (1999).

Practically all the asymptotic results of this chapter are based on the assumption that the *mle* is $\sqrt{n}$-consistent–i.e., $\sqrt{n}(\hat{\boldsymbol{\theta}}_A - \boldsymbol{\theta}_0) = O_p(1)$, where $\boldsymbol{\theta}_0 \in A \subset \Theta$. Fortunately, $\sqrt{n}$-consistency of *mle* follows from the consistency of the *mle* and the quadratic approximations in the previous proposition; this is stated in the next result.

Lemma 4.2.3 *Suppose that Condition Q is satisfied and* $\boldsymbol{\theta}_0 \in A \subset \Theta$*. Then, the consistency of* $\hat{\boldsymbol{\theta}}_A$ *ensures that* $\sqrt{n}(\hat{\boldsymbol{\theta}}_A - \boldsymbol{\theta}_0) = O_p(1)$.

Proof: For the case when the observations are *iid*, see Lemma 1 in Chernoff (1954). For the non-*iid* case, see Andrews (1999). For a proof of a similar result, see page 141 in Pollard (1984); see also Lemma 4.3 in Geyer (1994). ■

The $\sqrt{n}$-consistency of $\hat{\boldsymbol{\theta}}_A$ enables us to restrict our attention to neighborhoods of the form $\{\boldsymbol{\theta} : \sqrt{n}\|\boldsymbol{\theta} - \boldsymbol{\theta}_0\| < K\}$, where K is arbitrary but fixed, for the purposes of evaluating the global maximum of $\ell(\boldsymbol{\theta})$. Once we restrict attention to such a neighborhood, the remainder terms in Proposition 4.2.2 become negligible because they would converge to zero uniformly over the neighborhood. This is the approach that will be used throughout.

A result similar to Lemma 4.2.3 is also available for local *mle*. Except for the very last section of this chapter, the results in this chapter relate to test of hypotheses and hence local *mle* would not play a role. However, for completeness let us state the result; the following is Lemma 1 in Self and Liang (1987)

Proposition 4.2.4 *Suppose that Condition Q is satisfied and that, for some* $\delta_0 > 0$, $B = A \cap \{\boldsymbol{\theta} : \|\boldsymbol{\theta} - \boldsymbol{\theta}_0\| \leq \delta\}$ *is a closed set for* $0 < \delta < \delta_0$. *Then, there exists a sequence of points* $\boldsymbol{\theta}^\dagger$ *in* B *at which* $\ell(\boldsymbol{\theta})$ *has a local maximum and* $\boldsymbol{\theta}^\dagger \xrightarrow{p} \boldsymbol{\theta}_0$. *Further,* $\sqrt{n}(\boldsymbol{\theta}^\dagger - \boldsymbol{\theta}_0) = O_p(1)$. ■

4.3 TESTS OF $R\theta = 0$ AGAINST $R\theta \geqslant 0$

Let the setting be as in the previous section; in particular, $\boldsymbol{\theta} \in \mathbb{R}^p$ denotes the parameter associated with a statistical model, Θ denotes the parameter space for $\boldsymbol{\theta}$, $\boldsymbol{\theta}_0$ denotes the true parameter, and $\{\ell(\boldsymbol{\theta}), \boldsymbol{S}(\boldsymbol{\theta}), \mathcal{I}(\boldsymbol{\theta})\}$ be as in (4.1). Let $\boldsymbol{R}$ be a known full-rank matrix of constants of order $r \times p$ $(r \leq p)$. In this section, we consider test of

$$H_0 : \boldsymbol{R\theta} = \boldsymbol{0} \quad \text{vs} \quad H_1 : \boldsymbol{R\theta} \geqslant \boldsymbol{0}. \tag{4.5}$$

In this section, we will assume that Θ *is open.* The results for the case when Θ is open also hold for many cases when it is not open, but the technical details are more involved. The main advantage of considering the case when Θ is open is that the classical arguments based on Taylor series, for example, as in Cox and Hinkley (1974) and Lehmann (1991), can also be used. For most of this section, *the observations will be assumed to be independently and identically distributed.*

For the time being, let us note that practically every result in this section for testing (4.5) also holds, with minor modifications, for testing

$$H_0^* : \boldsymbol{\theta} \in \mathcal{M} \quad \text{vs} \quad H_1^* : \boldsymbol{\theta} \in \mathcal{C} \tag{4.6}$$

as well where $\mathcal{M}$ is a linear space, $\mathcal{C}$ is a closed convex cone, and $\mathcal{M} \subset \mathcal{C}$. These results also have generalizations when the $\mathcal{C}$ in H_1^* is replaced by a set that may be neither concave nor conical; such generalizations will be discussed in a later section.

Notation: *Unless the contrary is clear, the following notation will be adopted:* $\hat{\boldsymbol{\theta}}$ *is the unconstrained estimator over* Θ, *and* $\tilde{\boldsymbol{\theta}}$ *and* $\bar{\boldsymbol{\theta}}$ *denote the estimators under* H_1 *and* H_0, *respectively.*

4.3.1 Likelihood Ratio Test with *iid* Observations

For testing $H_0 : \boldsymbol{R\theta} = \boldsymbol{0}$ against $H_1 : \boldsymbol{R\theta} \geqslant \boldsymbol{0}$, we have

$$LRT = 2[\sup_{\boldsymbol{R\theta} \geqslant \boldsymbol{0}} \ell(\boldsymbol{\theta}) - \sup_{\boldsymbol{R\theta} = \boldsymbol{0}} \ell(\boldsymbol{\theta})]; \tag{4.7}$$

its asymptotic null distribution is given in the next result.

Proposition 4.3.1 *Let $Y_1, \ldots, Y_n$ be independently and identically distributed with common density function $f(y; \boldsymbol{\theta})$. Assume that Condition Q is satisfied. Let H_0 and H_1 be as in* (4.5). *Suppose that H_0 is satisfied and let $\boldsymbol{\theta}_0$ denote the true value in H_0. Let the function $D(\boldsymbol{z})$, $\boldsymbol{z} \in \mathbb{R}^p$, be defined by*

$$D(\boldsymbol{z}) = \min_{\boldsymbol{R\theta}=\boldsymbol{0}} (\boldsymbol{z} - \boldsymbol{\theta})^T \mathcal{I}_{\boldsymbol{\theta}_0} (\boldsymbol{z} - \boldsymbol{\theta}) - \min_{\boldsymbol{R\theta} \geqslant \boldsymbol{0}} (\boldsymbol{z} - \boldsymbol{\theta})^T \mathcal{I}_{\boldsymbol{\theta}_0} (\boldsymbol{z} - \boldsymbol{\theta}). \tag{4.8}$$

Let $\boldsymbol{Z}_n = n^{-1/2} \mathcal{I}_{\boldsymbol{\theta}_0}^{-1} \boldsymbol{S}(\boldsymbol{\theta}_0)$ and $\boldsymbol{Z} \sim N(\boldsymbol{0}, \mathcal{I}_{\boldsymbol{\theta}_0}^{-1})$. Then, we have the following:

1. *$LRT = D(\boldsymbol{Z}_n) + o_p(1)$, and hence $LRT \xrightarrow{d} D(\boldsymbol{Z})$.*
2. *The asymptotic null distribution of the LRT is the chi-bar square given by*

$$\lim_{n \to \infty} pr_{\boldsymbol{\theta}_0}(LRT \geqslant c | H_0) = \sum_{i=0}^{r} w_i(r, \boldsymbol{R} \mathcal{I}_{\boldsymbol{\theta}_0}^{-1} \boldsymbol{R}^T) pr(\chi_i^2 \geqslant c). \tag{4.9}$$

Proof: The large sample arguments used here are similar to those used for the case when there are no inequality constraints. Let the notation be as in (4.3). By the same result, we have that

$$\ell(\boldsymbol{\theta}) = A_n - (1/2) q(\boldsymbol{Z}_n - \boldsymbol{u}) + \delta_n(\boldsymbol{u}), \tag{4.10}$$

where $\boldsymbol{Z}_n = n^{-1/2} \mathcal{I}_{\boldsymbol{\theta}_0}^{-1} \boldsymbol{S}(\boldsymbol{\theta}_0)$, $q(\boldsymbol{Z}_n - \boldsymbol{u}) = (\boldsymbol{Z}_n - \boldsymbol{u})^T \mathcal{I}_{\boldsymbol{\theta}_0} (\boldsymbol{Z}_n - \boldsymbol{u})$, A_n does not depend on $\boldsymbol{u}$, and $\sup\{|\delta_n(\boldsymbol{u})| : \|\boldsymbol{u}\| < K\} = o_p(1)$. Since $\boldsymbol{R\theta}_0 = \boldsymbol{0}$, it follows that $\boldsymbol{R\theta} \geq \boldsymbol{0}$ if and only if $\boldsymbol{Ru} \geq \boldsymbol{0}$, and $\boldsymbol{R\theta} = \boldsymbol{0}$ if and only if $\boldsymbol{Ru} = \boldsymbol{0}$. Now, it follows from the foregoing quadratic approximation that

$$\begin{aligned} LRT &= 2[\sup\{\ell(\boldsymbol{\theta}) : \boldsymbol{R\theta} \geqslant \boldsymbol{0}\} - \sup\{\ell(\boldsymbol{\theta}) : \boldsymbol{R\theta} = \boldsymbol{0}\}] \\ &= \inf\{q(\boldsymbol{Z}_n - \boldsymbol{u}) : \boldsymbol{Ru} = \boldsymbol{0}\} - \inf\{q(\boldsymbol{Z}_n - \boldsymbol{u}) : \boldsymbol{Ru} \geq \boldsymbol{0}\} + o_p(1) \\ &= D(\boldsymbol{Z}_n) + o_p(1). \end{aligned}$$

Note that $\boldsymbol{Z}_n \xrightarrow{d} \boldsymbol{Z}$, where $\boldsymbol{Z} \sim N(\boldsymbol{0}, \mathcal{I}_{\boldsymbol{\theta}_0}^{-1})$, and $D(\boldsymbol{z})$ is continuous in $\boldsymbol{z}$ (in fact, it can be shown, using the triangular inequality, that $|D(\boldsymbol{z}) - D(\boldsymbol{x})| \leq \|\boldsymbol{z} - \boldsymbol{x}\|$ and hence $D(\boldsymbol{z})$ is Lipchitz-continuous). It follows that $D(\boldsymbol{Z}_n)$ and the LRT have a common limiting distribution and it is equal to the distribution of $D(\boldsymbol{Z})$. By Corollary 3.7.2 (on page 85), the distribution of $D(\boldsymbol{Z})$ is the chi-bar-square in (4.9). ■

This result is a direct generalization of the classical result that the asymptotic distribution of the likelihood ratio statistic for testing $\boldsymbol{R\theta} = \boldsymbol{0}$ against $\boldsymbol{R\theta} \neq \boldsymbol{0}$ is χ_r^2. In fact, for testing (4.5) and (4.6) the family of chi-bar-square distributions plays a role similar to that of the chi-squared distributions in inference under no inequality constraints.

It would be helpful to note that the main steps in the proof of Proposition 4.3.1 are very similar to those for proving that the *LRT* of $\boldsymbol{R\theta} = \mathbf{0}$ against $\boldsymbol{R\theta} \neq \mathbf{0}$ is asymptotically χ_r^2 under H_0. In fact, the main idea of the proof carries over to much more general contexts, although some of the technical details would be different.

Let t_{obs} denote the observed sample value of *LRT*. If

$$\sum_{i=0}^{r} w_i(r, \boldsymbol{R}\mathcal{I}_{\boldsymbol{\theta}_0}^{-1}\boldsymbol{R}^T)\text{pr}(\chi_i^2 \geqslant t_{obs}) \tag{4.11}$$

does not depend on the assumed true value $\boldsymbol{\theta}_0$ in the null parameter space, then it would be the large sample p-value. However, in general, it depends on $\boldsymbol{\theta}_0$ through $\mathcal{I}_{\boldsymbol{\theta}_0}^{-1}$ in $w_i(q, \boldsymbol{R}\mathcal{I}_{\boldsymbol{\theta}_0}^{-1}\boldsymbol{R}^T)$. Thus, $\boldsymbol{\theta}_0$ is a nuisance parameter and hence (4.11) is not a usable definition of a p-value. Some methods of dealing with this are discussed in the next subsection.

4.3.2 Tests in the Presence of Nuisance Parameters

Tests of statistical hypotheses in the presence of nuisance parameters are encountered frequently in statistical analysis. The presence of such parameters usually causes difficulties for statistical inference. Let $\boldsymbol{\theta}$ be partitioned as $(\boldsymbol{\lambda} : \boldsymbol{\psi})$ where $\boldsymbol{\psi} \in \mathbb{R}^r$; here, and in what follows, we shall use the following notation for partitioning a vector such as $\boldsymbol{\theta}$ to avoid writing the superscript "T" several times:

$$(\boldsymbol{\lambda} : \boldsymbol{\psi}) \quad \text{denotes} \quad (\boldsymbol{\lambda}^T, \boldsymbol{\psi}^T)^T.$$

For simplicity, we shall consider the special case,

$$H_0 : \boldsymbol{\psi} = \mathbf{0} \text{ and } H_1 : \boldsymbol{\psi} \geqslant \mathbf{0}; \tag{4.12}$$

it will be clear that the main procedures discussed below are applicable for testing $\boldsymbol{\psi} = \mathbf{0}$ against $\boldsymbol{\psi} \in C$ where $\mathcal{C} \subset \mathbb{R}^r$ and for more general testing problems as well.

Let T be a test statistic such that large values of T provide evidence against H_0, t denote the sample value of T, and $\boldsymbol{\lambda}_0$ denote the true value of $\boldsymbol{\lambda}$. Since the discussion of this subsection is centered around p-value and probability of Type I error, let us suppose that $H_0 : \boldsymbol{\psi} = \mathbf{0}$ holds; therefore, we let $\boldsymbol{\theta}_0 = (\boldsymbol{\lambda}_0 : \mathbf{0})$ denote the true value. Suppose that $\text{pr}(T \geqslant t \mid \boldsymbol{\psi} = \mathbf{0}, \boldsymbol{\lambda})$ depends on $\boldsymbol{\lambda}$. If $\boldsymbol{\lambda}_0$ were known, then the p-value would be $\text{pr}(T \geqslant t \mid \boldsymbol{\psi} = \mathbf{0}, \boldsymbol{\lambda} = \boldsymbol{\lambda}_0)$. Because $\boldsymbol{\lambda}_0$ is unknown, it is not usable as a p-value.

Under the null hypothesis, $H_0 : \boldsymbol{\psi} = \mathbf{0}$, the only unknown parameter is $\boldsymbol{\lambda}$, and therefore the notation $\text{pr}_{\boldsymbol{\lambda}}(\cdot|H_0)$ will be used for the probability evaluated under H_0 at $\boldsymbol{\theta} = (\boldsymbol{\lambda} : \mathbf{0})$. Let

$$p_n(c; \boldsymbol{\lambda}) = \text{pr}_{\boldsymbol{\lambda}}(LRT \geqslant c|H_0) \qquad \text{and} \qquad p_\infty(c; \boldsymbol{\lambda}) = \lim_{n \to \infty} p_n(c; \boldsymbol{\lambda}).$$

In many cases, closed-form expressions are available for $p_\infty(t; \boldsymbol{\lambda})$, but not for $p_n(t; \boldsymbol{\lambda})$. In this case, our main interest would be on large sample tests based on $p_\infty(t, \boldsymbol{\lambda})$.

By Proposition 4.3.1,

$$p_\infty(t;\boldsymbol{\lambda}) = \sum_{i=0}^{r} w_i\{r, \mathcal{I}^{\psi\psi}(\boldsymbol{\lambda})\}\text{pr}(\chi_i^2 \geqslant t) \tag{4.13}$$

where $\mathcal{I}^{\psi\psi}(\boldsymbol{\lambda})$ is the $(\boldsymbol{\psi}, \boldsymbol{\psi})$ block of $\mathcal{I}^{-1}(\boldsymbol{\theta})$ evaluated at $\boldsymbol{\theta} = (\boldsymbol{\lambda} : \mathbf{0})$. If $p_\infty(t;\boldsymbol{\lambda})$ does not depend on $\boldsymbol{\lambda}$ then the large sample p-value corresponding to $LRT = t$ is $p_\infty(t;\boldsymbol{\lambda})$. Therefore, for the following discussions, we shall consider only the case when $p_\infty(c;\boldsymbol{\lambda})$ does depend on $\boldsymbol{\lambda}$.

Bounds test:

Let $p_{n,L}(t)$ and $p_{n,U}(t)$ be positive numbers such that

$$p_{n,L}(t) \leq p_n(t;\boldsymbol{\lambda}) \leq p_{n,U}(t) \qquad \text{for any } \boldsymbol{\lambda}.$$

Then, the *bounds test* at level α is the following: reject H_0 if $p_{n,U}(t) \leq \alpha$, do not reject H_0 if $p_{n,L}(t) > \alpha$ and the test is inconclusive if $p_{n,L}(t) \leq \alpha < p_{n,U}(t)$. Usually, it is difficult to obtain sharp bounds for $p_n(t;\boldsymbol{\lambda})$. If the sample size is large then $p_n(t;\boldsymbol{\lambda})$ can be approximated by the tail probability of the chi-bar-square distribution in (4.13), and hence the bounds in (3.23) can be used for a large sample bounds test.

Supremum of pr(Type I Error) over the null parameter space (p_{sup}-test):

A test, which we call the p_{sup}-test, is based on

$$p\text{-value} = \sup_{\boldsymbol{\lambda}\in\Lambda} p_n(t;\boldsymbol{\lambda}) \tag{4.14}$$

where t is the sample value of LRT and Λ is the parameter space for $\boldsymbol{\lambda}$, for example, see Lehmann (1983). This is the generally accepted procedure for exact inference. The argument underlying this procedure is that if the value of t for LRT is consistent with at least one point in the null parameter space then we should not reject H_0. A large sample approximation of (4.14) is

$$\text{reject } H_0 \text{ if } \sup_{\boldsymbol{\lambda}\in\Lambda} p_\infty(t;\boldsymbol{\lambda}) \leq \alpha. \tag{4.15}$$

Ideally, it would be better to approximate $lim_n sup_{\boldsymbol{\lambda}} p_n(t;\boldsymbol{\lambda})$ than $sup_{\boldsymbol{\lambda}} lim_n p_n(t;\boldsymbol{\lambda})$. It does not appear that this has been investigated for tests against inequality constraints.

In some cases, $p_\infty(t;\boldsymbol{\lambda})$ may vary substantially as a function of $\boldsymbol{\lambda}$ in Λ and hence this test may be too conservative. Let us illustrate this using an example. Consider the Perlman's test of $\boldsymbol{\psi} = \mathbf{0}$ against $\boldsymbol{\psi} \geqslant \mathbf{0}$ based on a random sample from $N(\boldsymbol{\psi}, \boldsymbol{\lambda})$, where $\boldsymbol{\lambda}$ is an unknown $q \times q$ covariance matrix. For Perlman's statistic, $\mathcal{U}$, which is equivalent to the *LRT*, the finite sample null distribution is the distribution given by (see Proposition 3.10.3 on page 103)

$$p_n(t;\boldsymbol{\lambda}) = \text{pr}_{\boldsymbol{\lambda}}(\mathcal{U} \geqslant t|H_0) = \sum_{i=1}^{r} w_i(r,\boldsymbol{\lambda})\text{pr}(\chi_i^2/\chi_{n-r}^2 \geqslant t); \tag{4.16}$$

it was shown in (3.56) that

$$p_{sup} = \sup_{\boldsymbol{\lambda}\in\Lambda} p_n(t;\boldsymbol{\lambda}) = 0.5\{\text{pr}(\chi_{r-1}^2/\chi_{n-r}^2 \geqslant t) + \text{pr}(\chi_r^2/\chi_{n-r}^2 \geqslant t)\}. \tag{4.17}$$

It can be shown that the supremum of $p_n(t; \boldsymbol{\lambda})$ over $\boldsymbol{\lambda}$ is achieved when the correlation matrix corresponding to $\boldsymbol{\lambda}$ is equal to $\boldsymbol{J}$, where $\boldsymbol{J}$ is the matrix in which every element is equal to 1 [Silvapulle (1996b), Perlman (1969)]. Clearly, $\boldsymbol{J}$ is an extreme form for a correlation matrix. The true correlation matrix in most practical situations is unlikely to be close to $\boldsymbol{J}$. If we restrict the parameter space for $\boldsymbol{\lambda}$ away from a neighborhood of $\boldsymbol{J}$, then the supremum of $p_n(t; \boldsymbol{\lambda})$ would be smaller than the expression on the right-hand side of (4.17). Consequently, the test in (4.14) with p_{sup} as in (4.17) appears to be conservative for most practical applications. For a discussion of this example, see Silvapulle (1996b).

There are some values of $\boldsymbol{\lambda}$ that one would be able to eliminate without looking at the data. After looking at the data, one may be tempted to eliminate values of $\boldsymbol{\lambda}$ that are far from an estimate of $\boldsymbol{\lambda}$; however, this needs to be done carefully to ensure that the probability of Type I error does not exceed the nominal level. One approach that may overcome the conservative nature of p_{sup}-test is to approximate the p-value by the supremum of $\text{pr}(LRT \geqslant t|\boldsymbol{\lambda})$ over a confidence region for $\boldsymbol{\lambda}$. The p^*-test discussed below provides such an approach.

Supremum of pr(Type I Error) over a confidence region(p^*-test):

In this method, the test is carried out in two stages. For illustrative purposes, suppose that we wish to test H_0 vs H_1 at level 0.05. In the first stage, we construct a confidence region for $\boldsymbol{\lambda}$, say a 99 % confidence region; let this be denoted by $\mathcal{A}$. Let $p^* = \{(1 - 0.99) + \sup_{\boldsymbol{\lambda}\in\mathcal{A}} p_n(t; \boldsymbol{\lambda})\}$. Now, in the second stage, a test at level 0.05 is

$$\text{reject } H_0 \text{ if } p^* \leq 0.05.$$

An interpretation of this is the following: In the first stage, when we replace the full parameter space for $\boldsymbol{\lambda}$ by a 99% confidence region for it, the "probability" of an error is $(1 - 0.99)$. In the second stage, we reject H_0 if $\sup_{\boldsymbol{\lambda}\in\mathcal{A}} p_n(t; \boldsymbol{\lambda}) \leq 0.04$, so that the combined probability of Type I error is not more than $(1 - 0.99) + 0.04 = 0.05$. We call this, the *p^*-test,* and the foregoing p^* is an upper bound for the p-value of this test. Obviously, it is also possible to use a 98% confidence region for Stage 1 and carry out the test in Stage 2 with $p^* = \{(1 - 0.98) + \sup_{\boldsymbol{\lambda}\in\mathcal{A}} p_n(t; \boldsymbol{\lambda})\}$. A rigorous statement of this is given below in Proposition 4.3.2.

Suppose that we wish to test

$$H_0 : \boldsymbol{\psi} = \mathbf{0} \text{ against } H_1 : \boldsymbol{\psi} \in C$$

for some C. Let $T_{\boldsymbol{\lambda}}$ be a test statistic for a given $\boldsymbol{\lambda}$, and let $t_{\boldsymbol{\lambda}}$ denote the sample value of $T_{\boldsymbol{\lambda}}$; note that the test statistic may depend on $\boldsymbol{\lambda}$. For simplicity, assume that large values of $T_{\boldsymbol{\lambda}}$ favor H_1. Let α_1 be a given small number in the range $0 \leq \alpha_1 < 1$. Let $\mathcal{A}$ be a random subset of Λ such that

$$\text{pr}(\boldsymbol{\lambda}_0 \in \mathcal{A} \mid H_0) \geqslant 1 - \alpha_1.$$

Thus, once the sample values have been substituted, $\mathcal{A}$ becomes a $100(1 - \alpha_1)\%$ confidence region for $\boldsymbol{\lambda}_0$ under H_0. Let

$$p^* = \alpha_1 + \sup_{\boldsymbol{\lambda}\in\mathcal{A}} \text{pr}\{T_{\boldsymbol{\lambda}} \geqslant t_{\boldsymbol{\lambda}} \mid \boldsymbol{\theta} = (\boldsymbol{\lambda} : \mathbf{0})\}.$$

Now, a level-α *p^*-test* is

$$\text{reject } H_0 \text{ if } p^* \leq \alpha. \tag{4.18}$$

This is stated in the next result, and the proof is given in the appendix to this chapter (see page 215).

Proposition 4.3.2 *The size of the p^*-test in* (4.18) *does not exceed α. Consequently, an upper bound for the p-value of this test is p^*.* ■

Although the p^*-test was motivated by the need for a test with better power than the p_{sup}-test, there is no guarantee that the former would have such a power advantage. However, it is worth noting that the power of p^*-test at level α is no less than that of the p_{sup}-test at level $(\alpha - \alpha_1)$. This suggests that the power of the p^*-test cannot be much worse than that of the p_{sup} test provided α_1 is small. Simulations in Silvapulle (1996b) suggest that the p^*-test has better power than the p_{sup}-test in the specific cases considered; for illustrative examples of this method and further discussions, see Silvapulle (1996b) and Berger and Boos (1994).

Point estimate of the p-value ($\hat{p}$-test):

Because $\boldsymbol{\lambda}_0$ is unknown, one procedure is to use an estimated value of $p_n(t; \boldsymbol{\lambda}_0)$ under H_0. Let $\hat{p}$ be a consistent estimate of $p_n(t; \boldsymbol{\lambda}_0)$ under H_0; for example, it could be $p_n(t; \hat{\boldsymbol{\lambda}})$ or $p_\infty(t; \hat{\boldsymbol{\lambda}})$ where $\hat{\boldsymbol{\lambda}}$ is a consistent estimate of $\boldsymbol{\lambda}$ under H_0. Now the large sample test, which we call the $\hat{p}$-test, is

$$\text{reject } H_0 \text{ if } \hat{p} \leq \alpha. \tag{4.19}$$

In small samples, this procedure must be used with caution because there is a possibility that the size of the test may exceed the nominal level α substantially. For example, for testing $\boldsymbol{\psi} = \mathbf{0}$ against $\boldsymbol{\psi} \geqslant \mathbf{0}$ based on a sample from $N(\boldsymbol{\psi}, \boldsymbol{\lambda})$, the tail probability

$$\sum w_i(r, \hat{\boldsymbol{\lambda}}) \text{pr}(\chi_i^2/\chi_{n-r}^2 \geqslant t)$$

with $\hat{\boldsymbol{\lambda}}$ being a consistent estimator of $\boldsymbol{\lambda}$ may turn out to be a poor estimate of

$$\sum w_i(r, \boldsymbol{\lambda}_0) \text{pr}(\chi_i^2/\chi_{n-r}^2 \geqslant t)$$

if the chi-bar-square weights, $w_i(r, \boldsymbol{\lambda})$, are sensitive to the form of the covariance matrix $\boldsymbol{\lambda}$ and $\hat{\boldsymbol{\lambda}}$ is not close to $\boldsymbol{\lambda}_0$; this was illustrated in Silvapulle (1996b).

4.3.3 Wald- and Score-type Tests with *iid* Observations

In the standard inference literature, Wald and score tests of $\boldsymbol{R\theta} = \mathbf{0}$ against $\boldsymbol{R\theta} \neq \mathbf{0}$ are known to be asymptotically equivalent to the likelihood ratio test. These tests may be generalized in different directions when there are inequality constraints. In this section, we will introduce Wald- and score-type tests that are asymptotically equivalent to the likelihood ratio test for testing against inequality constraints.

Wald-type tests

The usual Wald statistic for testing $\boldsymbol{R\theta} = \mathbf{0}$ against $\boldsymbol{R\theta} \neq \mathbf{0}$ is

$$n(\boldsymbol{R}\hat{\boldsymbol{\theta}})^T(\boldsymbol{R}\hat{\mathcal{I}}^{-1}\boldsymbol{R}^T)^{-1}(\boldsymbol{R}\hat{\boldsymbol{\theta}}) \tag{4.20}$$

where $\hat{\mathcal{I}}$ is a consistent estimator of $\mathcal{I}(\boldsymbol{\theta})$. The following interpretations of this statistic are helpful; they are based on $\sqrt{n}(\hat{\boldsymbol{\theta}} - \boldsymbol{\theta}) \approx N\{\mathbf{0}, \mathcal{I}(\boldsymbol{\theta})^{-1}\}$ and $\sqrt{n}(\boldsymbol{R}\hat{\boldsymbol{\theta}} - \boldsymbol{R\theta}) \approx N(\mathbf{0}, \boldsymbol{A})$ where $\boldsymbol{A} = \{\boldsymbol{R}\mathcal{I}(\boldsymbol{\theta})^{-1}\boldsymbol{R}^T\}$:

1. It [i.e., (4.20)] is a measure of $\|\boldsymbol{R}\hat{\boldsymbol{\theta}}\|^2$. Since $\boldsymbol{R}\hat{\boldsymbol{\theta}}$ is an estimate of $\boldsymbol{R\theta}$ obtained without imposing the constraints in the null hypothesis, it is reasonable to expect that $\|\boldsymbol{R}\hat{\boldsymbol{\theta}}\|^2$ would be small if $\boldsymbol{R\theta} = \mathbf{0}$; and it would be large if $\boldsymbol{R\theta} \neq \mathbf{0}$.

2. First, construct the LRT (=$n(\boldsymbol{R}\hat{\boldsymbol{\theta}})^T(\boldsymbol{RVR}^T)^{-1}(\boldsymbol{R}\hat{\boldsymbol{\theta}})$) for testing $\boldsymbol{R\theta} = \mathbf{0}$ against $\boldsymbol{R\theta} \neq \mathbf{0}$ based on a single observation of $\hat{\boldsymbol{\theta}}$ under the assumption that $\hat{\boldsymbol{\theta}}$ is exactly $N(\boldsymbol{\theta}, \boldsymbol{V})$, where $\boldsymbol{V}$ is assumed known. Now, (4.20) is a large sample approximation to this statistic obtained by substituting $\boldsymbol{V} = n^{-1}\hat{\mathcal{I}}^{-1}$.

3. It is a measure of $\{d(\hat{\boldsymbol{\theta}}, H_0) - d(\hat{\boldsymbol{\theta}}, H_2)\}$, where $d(\hat{\boldsymbol{\theta}}, H_i)$ is a measure of the distance between $\hat{\boldsymbol{\theta}}$ and the parameter space defined by the hypothesis H_i $(i = 0, 1)$; (4.20) is obtained by choosing

$$d(\hat{\boldsymbol{\theta}}, H) = \min\{n(\hat{\boldsymbol{\theta}} - \boldsymbol{\theta})^T\hat{\mathcal{I}}(\hat{\boldsymbol{\theta}} - \boldsymbol{\theta}) : \boldsymbol{\theta} \in H\}.$$

For testing $\boldsymbol{R\theta} = \mathbf{0}$ against $\boldsymbol{R\theta} \geqslant \mathbf{0}$, a statistic that resembles (4.20) and is along the lines of the first interpretation (of the foregoing three) is

$$W = n(\boldsymbol{R}\tilde{\boldsymbol{\theta}})^T(\boldsymbol{R}\tilde{\mathcal{I}}^{-1}\boldsymbol{R}^T)^{-1}(\boldsymbol{R}\tilde{\boldsymbol{\theta}}) \tag{4.21}$$

where $\tilde{\mathcal{I}}$ is an estimator of $\mathcal{I}$ under $H_1 : \boldsymbol{R\theta} \geqslant \mathbf{0}$; thus, $\tilde{\mathcal{I}}$ can be $\mathcal{I}(\tilde{\boldsymbol{\theta}})$ or $\mathcal{I}(\hat{\boldsymbol{\theta}})$. We shall call this a Wald-type statistic. Note that the definition in (4.21) does not follow from the asymptotic distribution of $\boldsymbol{R}\tilde{\boldsymbol{\theta}}$ in the same way as (4.20) does from that of $\boldsymbol{R}\hat{\boldsymbol{\theta}}$ because $\boldsymbol{R}\tilde{\boldsymbol{\theta}}$ is not asymptotically normal. It will be seen that the W in (4.21) is asymptotically equivalent to *LRT.*

A statistic along the lines of the third interpretation is

$$D = \inf_{\boldsymbol{R\theta}=\mathbf{0}}\{n(\hat{\boldsymbol{\theta}} - \boldsymbol{\theta})^T\hat{\mathcal{I}}(\hat{\boldsymbol{\theta}} - \boldsymbol{\theta})\} - \inf_{\boldsymbol{R\theta}\geqslant\mathbf{0}}\{n(\hat{\boldsymbol{\theta}} - \boldsymbol{\theta})^T\hat{\mathcal{I}}(\hat{\boldsymbol{\theta}} - \boldsymbol{\theta})\}. \tag{4.22}$$

We shall call this a *Distance Statistic.*

Because the parameter $\boldsymbol{\theta}$ enters the hypotheses only through $\boldsymbol{R\theta}$, we can think of $\boldsymbol{R\theta}$ as the parameter of interest and define a statistic along the lines of the second interpretation with $\boldsymbol{R\theta}$ playing the role of the parameter of interest. This leads to a statistic which is equal to D in (4.22) although its functional form is different:

$$\begin{aligned} D = &\inf_{\boldsymbol{R\theta}=\mathbf{0}}\{n(\boldsymbol{R}\hat{\boldsymbol{\theta}} - \boldsymbol{R\theta})^T(\boldsymbol{R}\hat{\mathcal{I}}^{-1}\boldsymbol{R}^T)^{-1}(\boldsymbol{R}\hat{\boldsymbol{\theta}} - \boldsymbol{R\theta}) - \\ &\qquad \inf_{\boldsymbol{R\theta}\geqslant\mathbf{0}}\{n(\boldsymbol{R}\hat{\boldsymbol{\theta}} - \boldsymbol{R\theta})^T(\boldsymbol{R}\hat{\mathcal{I}}^{-1}\boldsymbol{R}^T)^{-1}(\boldsymbol{R}\hat{\boldsymbol{\theta}} - \boldsymbol{R\theta}). \end{aligned} \tag{4.23}$$

It will be seen later that D is also asymptotically equivalent to the *LRT.*

A Global score statistic

Recall that the score statistic for testing $\theta = \theta_0$ against $\theta \neq \theta_0$ is

$$n^{-1}S(\theta_0)^T \mathcal{I}_{\theta_0}^{-1} S(\theta_0) \tag{4.24}$$

where $S(\theta) = (\partial/\partial\theta)\ell(\theta)$ is the score function. The motivation for this is that if θ_0 is the true value then $E\{\ell(\theta)\}$ has a global maximum and a stationary value at θ_0, $n^{-1/2}S(\theta_0) \xrightarrow{d} N\{0, \mathcal{I}_{\theta_0}\}$, and $S(\theta_0)$ tends to be close to (respectively, far from) $\mathbf{0}$ when θ_0 is the true (respectively, not true) value. One possible way of extending this to the inequality constrained testing problem is to assess the extent to which $\{S(\bar{\theta}) - S(\tilde{\theta})\}$ is different from zero, where $\bar{\theta}$ and $\tilde{\theta}$ are estimates of θ under the null and alternative hypotheses, respectively. If H_0 is true, then $\bar{\theta}$ and $\tilde{\theta}$ are expected to be close and hence $\{S(\bar{\theta}) - S(\tilde{\theta})\}$ is expected to be close to zero; on the other hand, if H_0 is not true then $S(\tilde{\theta})$ is expected to be close to zero but not $S(\bar{\theta})$ and therefore, $\{S(\bar{\theta}) - S(\tilde{\theta})\}$ is expected to be away from zero. This suggests that a test can be based on the difference $\{S(\bar{\theta}) - S(\tilde{\theta})\}$. Instead of the score function $S(\theta)$ we may use the effective score, $R\mathcal{I}(\theta_0)^{-1}S(\theta_0)$. This suggests the following:

$$S_G = [R\hat{\mathcal{I}}^{-1}\{S(\bar{\theta}) - S(\tilde{\theta})\}]^T (R\hat{\mathcal{I}}^{-1}R^T)^{-1}[R\hat{\mathcal{I}}^{-1}\{S(\bar{\theta}) - S(\tilde{\theta})\}]. \tag{4.25}$$

The following form has also been suggested (see Robertson et al. (1988)):

$$\{S(\bar{\theta}) - S(\tilde{\theta})\}^T \mathcal{I}(\bar{\theta})^{-1}\{S(\bar{\theta}) - S(\tilde{\theta})\}.$$

The reason for referring to these as "Global" score statistics is that they use information about the shape of the score function over the whole parameter space. This is in contrast to the Rao score statistic and Neyman's $C(\alpha)$ statistic, which are based only on the properties of the score function in a *local neighborhood* of the assumed true null value.

It is well known that the likelihood ratio, Wald, and score tests are asymptotically equivalent for testing $R\theta = \mathbf{0}$ against $R\theta \neq \mathbf{0}$. The next proposition says that a corresponding result also holds for the foregoing Wald-type statistic in (4.21), the distance statistic (4.22), and the score statistic in (4.25). In fact, this result holds even in some regression-type models such as the generalized linear models (see Silvapulle (1994)); this will be discussed in the next subsection.

Proposition 4.3.3 *Suppose that Condition Q is satisfied. Then for testing* $H_0 : R\theta = \mathbf{0}$ *vs* $H_1 : R\theta \geq \mathbf{0}$, *we have the following:*

$$LRT = W + o_p(1) = D + o_p(1) = S_G + o_p(1) \text{ under } H_0. \tag{4.26}$$

Consequently, the common asymptotic null distribution of LRT, W, D, *and* S_G *is the chi-bar square in* (4.9).

Proof: Follows from the local quadratic approximations of loglikelihood in Proposition 4.2.2 (see Gourieroux and Monfort (1995) and Silvapulle (1994)). ■

It follows that, under some reasonable regularity conditions, the test statistics LRT, W, D, and S_G are asymptotically equivalent in the sense that they have the same asymptotic local power, more precisely Pitman asymptotic efficiency against local alternatives. Although the four statistics in (4.26) are asymptotically equivalent, very little is known about their relative advantages. From a computational point of view, D is easier to use than the others because it requires only the unconstrained estimator, $\hat{\boldsymbol{\theta}}$, which is usually easier to compute than the constrained estimators. The statistics S_G and W do not have any significant computational advantage over LRT because all of them require the constrained estimator. In contrast to S_G, W, and D, *LRT* has the advantage that it is invariant under a reparameterization. If a test that is computationally simpler than the LRT is required, then D is worth considering. In view of these, at this stage, S_G and W do not appear to be serious competitors to *LRT* or D.

For testing $\boldsymbol{R\theta} = \mathbf{0}$ against $\boldsymbol{R\theta} \neq \mathbf{0}$, the classical score statistic, also known as Rao's score statistic and the Lagrange multiplier statistic, provides a simple way of testing when the full model is complicated; the implementation requires estimation of the model only under H_0. This is important when the full model is significantly more complicated than the null model; this particular feature of the score statistic has been crucial for its popularity. Unfortunately, the global score statistic, S_G, does not have this simplicity. A natural generalization of the Rao's score statistic, which preserves the attractive feature (i.e., requires only the null model to be estimated) of the classical Rao's score statistic will be introduced in a later section.

4.3.4 Independent Observations with Covariates

Most of the foregoing results for independently and identically distributed observations extend in a natural way to other more general cases with non-*iid* observation, for example, linear and nonlinear regression models, generalized linear models, and some stochastic processes.

Let $\boldsymbol{Y}_1, \ldots, \boldsymbol{Y}_n$ be independent, $f_i(\boldsymbol{y}; \boldsymbol{\theta})$ denote the density function of $\boldsymbol{Y}_i$ and $\ell(\boldsymbol{\theta})$ denote the loglikelihood, $\sum \log f_i(\boldsymbol{Y}_i; \boldsymbol{\theta})$. This includes the setting where the observations are $(\boldsymbol{y}_i, \boldsymbol{x}_i), i = 1, \ldots, n$, with $\boldsymbol{x}$ denoting a covariate; thus, regression-type models are included. Recall that in the previous subsections, we defined $\mathcal{I}(\boldsymbol{\theta})$ as the information per observation; it was the same for every observation because they were identically distributed. Now, since the observations are not identically distributed, we need to make some modifications. Let

$$\nabla = (\partial/\partial\boldsymbol{\theta}) \text{ and } \nabla^2 = (\partial^2/\partial\boldsymbol{\theta}\partial\boldsymbol{\theta}^T).$$

Suppose that

$$n^{-1/2}\nabla\ell(\boldsymbol{\theta}) \xrightarrow{d} N(\mathbf{0}, \mathcal{V}_{\boldsymbol{\theta}}), \qquad \text{and} \qquad n^{-1}\nabla^2\ell(\boldsymbol{\theta}) \xrightarrow{a.s.} -\mathcal{V}_{\boldsymbol{\theta}} \tag{4.27}$$

for some positive definite matrix $\mathcal{V}_{\boldsymbol{\theta}}$. As was introduced at the beginning of this section, let $\hat{\boldsymbol{\theta}}, \tilde{\boldsymbol{\theta}}$ and $\bar{\boldsymbol{\theta}}$ denote the maximizers of $\ell(\boldsymbol{\theta})$ over Θ, $\{\boldsymbol{R\theta} \geqslant \mathbf{0}\}$ and $\{\boldsymbol{R\theta} =$

$\mathbf{0}\}$, respectively. For testing

$$H_0 : \boldsymbol{R\theta} = \mathbf{0} \text{ against } H_1 : \boldsymbol{R\theta} \geqslant \mathbf{0}, \tag{4.28}$$

the likelihood ratio statistic is

$$LRT = 2\{\ell(\tilde{\boldsymbol{\theta}}) - \ell(\bar{\boldsymbol{\theta}})\}. \tag{4.29}$$

We shall continue to assume that Assumption A1 on page 146 holds with appropriate modifications. Assume that (a) the quadratic approximation (4.3) in Proposition 4.2.2 (page 147) holds with $\mathcal{I}(\boldsymbol{\theta})$ replaced by $\mathcal{V}(\boldsymbol{\theta})$, (b) $\mathcal{V}(\boldsymbol{\theta})$ is continuous in $\boldsymbol{\theta}$ and (c) $\tilde{\boldsymbol{\theta}}$ and $\bar{\boldsymbol{\theta}}$ are $n^{1/2}$-consistent. The proof of the quadratic approximation may involve lengthy technical details; for example, for the case of generalized linear models, see Fahrmeir and Kaufmann (1985). The $\sqrt{n}$-consistency of $\tilde{\boldsymbol{\theta}}$ and $\bar{\boldsymbol{\theta}}$ would follow from the consistency of the global *mle* and the quadratic approximation, in much the same way as in the *iid* case. Thus, conceptually, the conditions required for the regression type setting is not that different from that for the *iid* setting but the technical details could be complicated.

Now, the proofs of Propositions 4.3.1 and 4.3.3 still hold with appropriate modifications; see Silvapulle (1994) for details.

Proposition 4.3.4 *The asymptotic null distribution of LRT for testing*

$$H_0 : \boldsymbol{R\theta} = \mathbf{0} \textit{ against } H_1 : \boldsymbol{R\theta} \geqslant \mathbf{0}$$

is equal to the distribution of

$$\inf_{\boldsymbol{R\theta}=\mathbf{0}}\{(\boldsymbol{Z} - \boldsymbol{\theta})^T\mathcal{V}_{\boldsymbol{\theta}_0}(\boldsymbol{Z} - \boldsymbol{\theta})\} - \inf_{\boldsymbol{R\theta}\geqslant\mathbf{0}}\{(\boldsymbol{Z} - \boldsymbol{\theta})^T\mathcal{V}_{\boldsymbol{\theta}_0}(\boldsymbol{Z} - \boldsymbol{\theta})\}$$

where $\boldsymbol{Z} \sim N(\mathbf{0}, \mathcal{V}_{\boldsymbol{\theta}_0}^{-1})$. *Consequently, as* $n \to \infty$,

$$pr_{\boldsymbol{\theta}_0}(LRT \geqslant c \mid H_0) \to \sum_{i=0}^{r} w_i(r, \boldsymbol{R}\mathcal{V}_{\boldsymbol{\theta}_0}^{-1}\boldsymbol{R}^T)pr(\chi_i^2 \geqslant c). \qquad \blacksquare$$

It appears that, if

1. $\sqrt{n}(\hat{\boldsymbol{\theta}} - \boldsymbol{\theta}_0)$ converges to $N\{\mathbf{0}, \mathcal{V}_{\boldsymbol{\theta}_0}^{-1}\}$, and

2. the asymptotic null distribution of the likelihood ratio statistic, $2\{\ell(\hat{\boldsymbol{\theta}}) - \ell(\bar{\boldsymbol{\theta}})\}$, for testing $\boldsymbol{R\theta} = \mathbf{0}$ against $\boldsymbol{R\theta} \neq \mathbf{0}$ is a chi-square,

then it is almost certain that the conclusions of the foregoing proposition also hold. If the parameter space for $\boldsymbol{\theta}$ is open and the asymptotic null distribution of $2\{\ell(\hat{\boldsymbol{\theta}}) - \ell(\bar{\boldsymbol{\theta}})\}$ for testing $H_0 : \boldsymbol{R\theta} = \mathbf{0}$ against $H_2 : \boldsymbol{R\theta} \neq \mathbf{0}$ is not a chi-square, then the results of Proposition 4.3.4 are unlikely to hold.

It is also possible to construct Wald-type, score-type, and distance-based tests as follows:

$$\begin{aligned} W &= n(\boldsymbol{R}\tilde{\boldsymbol{\theta}})^T(\boldsymbol{R}\hat{\mathcal{V}}^{-1}\boldsymbol{R}^T)^{-1}(\boldsymbol{R}\tilde{\boldsymbol{\theta}}) \\ S_G &= n^{-1}\boldsymbol{U}^T(\boldsymbol{R}\hat{\mathcal{V}}^{-1}\boldsymbol{R}^T)^{-1}\boldsymbol{U}, \quad \text{where } \boldsymbol{U} = \boldsymbol{R}\hat{\mathcal{V}}^{-1}\{\boldsymbol{S}(\tilde{\boldsymbol{\theta}}) - \boldsymbol{S}(\bar{\boldsymbol{\theta}})\} \\ D &= \{d(\hat{\boldsymbol{\theta}}, H_0) - d(\hat{\boldsymbol{\theta}}, H_1)\}, \quad \text{where } d(\hat{\boldsymbol{\theta}}, H) = \inf_{\boldsymbol{\theta}\in H} n(\hat{\boldsymbol{\theta}} - \boldsymbol{\theta})^T\hat{\mathcal{V}}(\hat{\boldsymbol{\theta}} - \boldsymbol{\theta}). \end{aligned} \tag{4.30}$$

Now, it can be shown that (see Silvapulle (1994), Gourieroux and Monfort (1995))

$$LRT = W + o_p(1) = S_G + o_p(1) = D + o_p(1). \tag{4.31}$$

Therefore, these tests have the same asymptotic null distribution. Further, under some weak conditions, (4.31) also hold under Pitman-type local alternatives and hence the tests in (4.31) also have the same asymptotic local power.

Least squares in linear regression

The foregoing results extend in a natural way to some quasi-likelihood models as well. To illustrate the main ideas, let us first consider the linear regression model,

$$y_i = \boldsymbol{x}_i^T\boldsymbol{\theta} + \epsilon_i,$$

where the error terms are *iid* with common variance σ^2. Let

$$R_n(\boldsymbol{\theta}) = (-1/2)\sum(y_i - \boldsymbol{x}_i^T\boldsymbol{\theta})^2 \qquad \text{and} \qquad \boldsymbol{W} = \lim(n^{-1}\boldsymbol{X}^T\boldsymbol{X}),$$

where $\boldsymbol{X}$ is the design matrix $(\boldsymbol{x}_1, \ldots, \boldsymbol{x}_n)^T$. Let $\hat{\boldsymbol{\theta}}, \tilde{\boldsymbol{\theta}}$, and $\bar{\boldsymbol{\theta}}$ be the estimators corresponding to the objective function $R_n(\boldsymbol{\theta})$; thus $\hat{\boldsymbol{\theta}}$ is simply the unconstrained ordinary least squares estimator. Let $\boldsymbol{S}(\boldsymbol{\theta}) = \nabla R_n(\boldsymbol{\theta})$. Then we have that $\sqrt{n}(\hat{\boldsymbol{\theta}} - \boldsymbol{\theta}_0) \xrightarrow{d} N(\boldsymbol{0}, \sigma^2\boldsymbol{W}^{-1})$,

$$n^{-1/2}\boldsymbol{S}(\boldsymbol{\theta}_0) \xrightarrow{d} N(\boldsymbol{0}, \sigma^2\boldsymbol{W}) \quad \text{and} \quad -n^{-1}\nabla^2 R_n(\boldsymbol{\theta}) \xrightarrow{p} \boldsymbol{W}.$$

Note that the asymptotic covariance of $\boldsymbol{S}(\boldsymbol{\theta}_0)$ and the limit of $-n^{-1}\nabla^2 R_n(\boldsymbol{\theta}_0)$ are proportional, but not equal; if R_n were the loglikelihood then they would have been equal. Now, for testing $H_0 : \boldsymbol{R\theta} = \boldsymbol{0}$ against $H_2 : \boldsymbol{R\theta} \neq \boldsymbol{0}$, an LR-type statistic and its asymptotic null distribution are given by

$$T_n = 2\{R_n(\hat{\boldsymbol{\theta}}) - R_n(\bar{\boldsymbol{\theta}})\}/\hat{\sigma}^2 \xrightarrow{d} \chi_r^2 \quad \text{under } H_0,$$

where $\hat{\sigma}$ is a consistent estimator of σ.

Note that

$$T_n = \{\|(\boldsymbol{Y} - \boldsymbol{X}\bar{\boldsymbol{\theta}})/\sigma\|^2 - \|(\boldsymbol{Y} - \boldsymbol{X}\hat{\boldsymbol{\theta}})/\sigma\|^2\}(\sigma^2/\hat{\sigma}^2),$$

$(\sigma^2/\hat{\sigma}^2) \xrightarrow{p} 1$, and the distribution of the term in $\{\ldots\}$ does not depend on σ. Thus, essentially σ^2 factors out and cancels with $\hat{\sigma}^{-2}$ in the limit. This illustrates that in the linear model, we do not need to know the exact population distribution to construct tests of $H_0 : \boldsymbol{R\theta} = \mathbf{0}$ against $H_1 : \boldsymbol{R\theta} \neq \mathbf{0}$.

In what follows we extend Proposition 4.3.4 to incorporate this type of setting.

A quasi-likelihood method

Consider a general parametric model with unknown parameter $\boldsymbol{\theta}$. Let $R_n(\boldsymbol{\theta})$ denote an objective function which will be maximized for estimating $\boldsymbol{\theta}$. Let $\hat{\boldsymbol{\theta}}, \tilde{\boldsymbol{\theta}}$, and $\bar{\boldsymbol{\theta}}$ be the estimators corresponding to this objective function. Let $\boldsymbol{S}_n(\boldsymbol{\theta}) = \nabla R_n(\boldsymbol{\theta})$. Suppose that

$$n^{-1/2}\boldsymbol{S}(\boldsymbol{\theta}_0) \xrightarrow{d} N\{\mathbf{0}, \sigma^2 \boldsymbol{W}(\boldsymbol{\theta}_0)\} \quad \text{and} \quad n^{-1}\nabla^2 R_n(\boldsymbol{\theta}) \xrightarrow{p} -\boldsymbol{W}(\boldsymbol{\theta}).$$

Let $\boldsymbol{Z}_n(\boldsymbol{\theta}_0) = n^{-1/2}\boldsymbol{W}(\boldsymbol{\theta}_0)^{-1}\boldsymbol{S}(\boldsymbol{\theta}_0)$. Assume that R_n satisfies regularity conditions similar to those required for the *iid* setting. Therefore, we have $\sqrt{n}(\hat{\boldsymbol{\theta}} - \boldsymbol{\theta}_0) \xrightarrow{d} N\{\mathbf{0}, \sigma^2\boldsymbol{W}(\boldsymbol{\theta}_0)^{-1}\}$, $\boldsymbol{Z}_n(\boldsymbol{\theta}_0) \xrightarrow{d} N\{\mathbf{0}, \sigma^2\boldsymbol{W}(\boldsymbol{\theta}_0)^{-1}\}$, and a quadratic approximation similar to (4.3) holds for $R_n(\boldsymbol{\theta})$ with the information matrix $\mathcal{I}$ replaced by $\boldsymbol{W}$. Now consider test of $H_0 : \boldsymbol{R\theta} = \mathbf{0}$ vs $H_1 : \boldsymbol{R\theta} \geqslant \mathbf{0}$. A suitable test statistic for this is

$$T_n = 2\{R_n(\tilde{\boldsymbol{\theta}}) - R_n(\bar{\boldsymbol{\theta}})\}/\hat{\sigma}^2,$$

where $\hat{\sigma}^2$ is a consistent estimator of σ. Then we have

$$\begin{aligned} T_n &= 2\{R_n(\tilde{\boldsymbol{\theta}}) - R_n(\bar{\boldsymbol{\theta}})\}/\sigma^2 + o_p(1) \\ &= D(\boldsymbol{Z}_n)/\sigma^2 + o_p(1) = D(\boldsymbol{Z}_n/\sigma) + o_p(1) \\ &\xrightarrow{d} D(\boldsymbol{U}) \quad \text{under } H_0 \end{aligned} \tag{4.32}$$

where $D(\boldsymbol{z})$ is as in (4.8) with $\mathcal{I}_{\boldsymbol{\theta}_0}$ replaced by $\boldsymbol{W}(\boldsymbol{\theta}_0)$, and $\boldsymbol{U} \sim N\{\mathbf{0}, \boldsymbol{W}(\boldsymbol{\theta}_0)^{-1}\}$. Therefore,

$$T_n \xrightarrow{d} \bar{\chi}^2\{\boldsymbol{R}\boldsymbol{W}(\boldsymbol{\theta}_0)^{-1}\boldsymbol{R}^T, \mathbb{R}^{+r}\} \quad \text{under } H_0.$$

We can also define Wald, score, and distance type test statistics as in (4.30). All that we need to do is to introduce $\hat{\sigma}^{-2}$ on the RHS of the definitions of W, S_G, and D in (4.30) and replace $\hat{\mathcal{V}}$ by $\hat{\boldsymbol{W}}$, a consistent estimator of $\boldsymbol{W}(\boldsymbol{\theta}_0)$:

$$\begin{aligned} W &= \hat{\sigma}^{-2} n(\boldsymbol{R}\tilde{\boldsymbol{\theta}})^T(\boldsymbol{R}\hat{\boldsymbol{W}}^{-1}\boldsymbol{R}^T)^{-1}(\boldsymbol{R}\tilde{\boldsymbol{\theta}}) \\ S_G &= \hat{\sigma}^{-2} n^{-1}\boldsymbol{U}^T(\boldsymbol{R}\hat{\boldsymbol{W}}^{-1}\boldsymbol{R}^T)^{-1}\boldsymbol{U}, \quad \text{where } \boldsymbol{U} = \boldsymbol{R}\hat{\boldsymbol{W}}^{-1}\{\boldsymbol{S}(\tilde{\boldsymbol{\theta}}) - \boldsymbol{S}(\bar{\boldsymbol{\theta}})\} \\ D &= \hat{\sigma}^{-2}\{d(\hat{\boldsymbol{\theta}}, H_0) - d(\hat{\boldsymbol{\theta}}, H_1)\}, \quad \text{where } d(\hat{\boldsymbol{\theta}}, H) = \inf_{\boldsymbol{\theta} \in H} n(\hat{\boldsymbol{\theta}} - \boldsymbol{\theta})^T\hat{\boldsymbol{W}}(\hat{\boldsymbol{\theta}} - \boldsymbol{\theta}). \end{aligned} \tag{4.33}$$

The rest of the details for the asymptotic equivalence of T_n, W, S_G, and D remain essentially the same; see Silvapulle (1994) and Gourieroux and Monfort (1995) for details. Finally, let us note that, for a generalized linear model with canonical link, the foregoing results are adequate to apply quasi-likelihood ratio test in the presence of an overdispersion parameter, σ.

Robust test in the linear model

The foregoing results can be used for developing *robust tests* of $\boldsymbol{R\theta} = \mathbf{0}$ vs $\boldsymbol{R\theta} \geqslant \mathbf{0}$ in the linear regression model $\boldsymbol{Y} = \boldsymbol{X\theta} + \boldsymbol{E}$. Suppose that the objective function is

$$R_n(\boldsymbol{\theta}) = -\sum \rho\{(y_i - \boldsymbol{x}_i^T\boldsymbol{\theta})/s\}s,$$

where ρ is a suitably chosen convex function and s is an estimate of scale, for example, a multiple of the median of the absolute values of the least square residuals. Now, it may be verified that the foregoing tests of $\boldsymbol{R\theta} = \mathbf{0}$ vs $\boldsymbol{R\theta} \geqslant \mathbf{0}$ are applicable. See Silvapulle (1992b) and Silvapulle (1992c) for details where proofs of the asymptotic equivalence of the tests are also given; these provide details for the statistics of the form T_n and D, respectively. It is shown that the robustness properties of the M-estimator carry over to the corresponding test statistics T_n and D. These results also hold in *group sequential analysis*; see Silvapulle and Sen (1993).

4.3.5 Examples

Example 4.3.1

An assay was carried out with the bacterium *E. coli* to evaluate the genotoxic effects of 9-AA and potassium chromate. Further background and the data are given in Example 1.2.7 of Chapter 1 (page 11). The objective is to test whether potassium chromate and 9-AA have a synergistic effect. The *simple independent action* (SIA) model has attracted interest for studies of the type in this example. To test SIA against synergism, let $\pi_{ij} = \text{pr}\{$ positive response for a test unit in cell $(i, j)\}$, and

$$\log(1 - \pi_{ij}) = \mu + \alpha_i + \tau_j + \eta_{ij} \tag{4.34}$$

where $i = 1, 2$ and $j = 1, \ldots, 5$. To ensure that the parameters in (4.34) are identified, let us impose the constraints, $\alpha_1 = \tau_1 = \eta_{i1} = \eta_{1j} = 0$ for all (i, j). Let $\boldsymbol{\theta} = (\mu, \alpha_2, \tau_2, \tau_3, \tau_4, \tau_5, \eta_{22}, \eta_{23}, \eta_{24}, \eta_{25})^T$, $\boldsymbol{\lambda} = (\mu, \alpha_2, \tau_2, \tau_3, \tau_4, \tau_5)^T$ and $\boldsymbol{\psi} = (\eta_{22}, \eta_{23}, \eta_{24}, \eta_{25})^T$. It is easily seen that the number of distinct π_{ij} parameters, namely ten, is also the number of distinct parameters on the right-hand side of (4.34), and that (4.34) is a one-to-one transformation from $\boldsymbol{\pi}$ to $\boldsymbol{\theta}$, where $\boldsymbol{\pi} = (\pi_{11}, \pi_{12}, \pi_{13}, \pi_{14}, \pi_{15}, \pi_{21}, \pi_{22}, \pi_{23}, \pi_{24}, \pi_{25})^T$. The SIA model is equivalent to $\boldsymbol{\psi} = \mathbf{0}$. If $\boldsymbol{\psi} \geqslant \mathbf{0}$ and $\boldsymbol{\psi} \neq \mathbf{0}$, then there is synergism; if $\boldsymbol{\psi} \leq \mathbf{0}$ and $\boldsymbol{\psi} \neq \mathbf{0}$, then there is antagonism.

In this example, the question of interest is whether or not there is any synergism. Therefore, the statistical inference problem may be formulated as a test of

$$H_0 : \boldsymbol{\psi} = \mathbf{0} \text{ against } H_1 : \boldsymbol{\psi} \geqslant \mathbf{0}. \tag{4.35}$$

This is within the general framework considered in this section; it is a special case of (4.28) with $\boldsymbol{R} = [\mathbf{0} : \boldsymbol{I}_4]$, where $\boldsymbol{I}_4$ is the 4×4 identity matrix and $\mathbf{0}$ is the 4×6 matrix of zeroes. Now, the null and alternative hypotheses are $H_0 : \boldsymbol{R\theta} = \mathbf{0}$ and $H_1 : \boldsymbol{R\theta} \geqslant \mathbf{0}$, respectively.

For cell (i, j), let n_{ij} and y_{ij} denote the number of units tested and the number of positive responses, respectively. Then the likelihood is the product of all the binomial probabilities, and therefore

$$\ell(\boldsymbol{\theta}) = \sum_{i=1}^{2}\sum_{j=1}^{5}[y_{ij}\log\pi_{ij} + (n_{ij} - y_{ij})\log(1 - \pi_{ij}) + c_{ij}]$$

where π_{ij} as a function of $\boldsymbol{\theta}$ is given by (4.34) and the constant c_{ij} is the logarithm of the binomial coefficient, $n_{ij}!\{y_{ij}!(n_{ij} - y_{ij})!\}^{-1}$. By definition,

$$\begin{aligned} LRT &= 2\{\max_{\boldsymbol{\theta}\in H_1}\ell(\boldsymbol{\theta}) - \max_{\boldsymbol{\theta}\in H_0}\ell(\boldsymbol{\theta})\} \\ &= 2\{\max_{\boldsymbol{\psi}\geqslant\mathbf{0}}\ell(\boldsymbol{\theta}) - \max_{\boldsymbol{\psi}=\mathbf{0}}\ell(\boldsymbol{\theta})\}. \end{aligned} \tag{4.36}$$

In this model, the loglikelihood is concave, and by modifying the arguments in Silvapulle (1981), it can be verified that the *mle* exists at a finite point with probability one. The value of $\max\{\ell(\boldsymbol{\theta}) : \boldsymbol{\psi} \geqslant \mathbf{0}\}$ under H_1 can be computed using a nonlinear inequality constrained optimization program in software libraries, such as *IMSL*, *MATLAB*, or *NAG*. In this example, we did not need to use such a constrained optimization program, for reasons explained below.

Note that the equation (4.34) is a reparameterization of $\boldsymbol{\pi}$ in terms of $\boldsymbol{\theta}$. Therefore, to compute the global *mle*, $\hat{\boldsymbol{\theta}}$, it suffices to solve (4.34) with $\boldsymbol{\pi} = \hat{\boldsymbol{\pi}}$ where $\hat{\boldsymbol{\pi}}$, the vector of sample proportions of successes, is the global *mle* of $\boldsymbol{\pi}$. The computed value of $\hat{\boldsymbol{\psi}}$ turned to be nonnegative for each component. Therefore, the *mle* of $\boldsymbol{\theta}$ under the constraint $\boldsymbol{\psi} \geqslant \mathbf{0}$ is also the unconstrained estimator, $\hat{\boldsymbol{\theta}}$. Therefore, the sample value of the *LRT* for testing $\boldsymbol{\psi} = \mathbf{0}$ against $\boldsymbol{\psi} \geqslant \mathbf{0}$ is equal to that for testing $\boldsymbol{\psi} = \mathbf{0}$ against $\boldsymbol{\psi} \neq \mathbf{0}$. Although the *LRT* for the two tests have the same numerical value for these data, they have different p-values, because their null distributions are different.

The computed value of *LRT* is 60.5. A large sample approximation of the p-value for testing $\boldsymbol{\psi} = \mathbf{0}$ against $\boldsymbol{\psi} \geqslant \mathbf{0}$ is

$$\sup_{\boldsymbol{\lambda}}\sum_{i=1}^{4} w_i\{4, \mathcal{I}^{\psi\psi}(\boldsymbol{\lambda})\}\mathrm{pr}(\chi_i^2 \geqslant 60.5).$$

Because this supremum is not easy to compute, it is worth applying a bounds test first. An upper bound for the foregoing large sample p-value is (see, (3.23))

$$0.5[\mathrm{pr}(\chi_3^2 \geqslant 60.5) + \mathrm{pr}(\chi_4^2 \geqslant 60.5)]; \tag{4.37}$$

this is less than 0.001. Therefore, the large sample p-value corresponding to a sample value of 60.5 for the *LRT* is less than 0.001. Therefore, there is sufficient evidence to reject H_0 and accept H_1; in other words, reject the SIA model and accept that there is synergism between potassium chromate and 9-AA.

For illustrative purposes, let us apply the *LRT* for testing $\boldsymbol{\psi} = \mathbf{0}$ against $\boldsymbol{\psi} \neq \mathbf{0}$ as well. It was noted that the *LRT* for this is also 60.5. The large sample p-value for this unrestricted test is $\mathrm{pr}(\chi_4^2 \geqslant 60.5)$, which is also less than 0.001. Therefore, assuming that the nominal level is 0.05, we would reject $\psi = \mathbf{0}$ and accept $\psi \neq \mathbf{0}$. It is valid to apply this unrestricted test for testing $\psi = \mathbf{0}$ against $\psi \geqslant \mathbf{0}$ as well. Recall that the reason that we prefer to apply a one-sided/restricted test as opposed to a two-sided/unrestricted test for testing $\psi = \mathbf{0}$ against $\psi \geqslant \mathbf{0}$ is for power advantage; validity is not the issue because a 5% level test of $\psi = \mathbf{0}$ against $\psi \neq \mathbf{0}$ is also a 5% level test of $\psi = \mathbf{0}$ against $\psi \geqslant \mathbf{0}$. Since the p-value for $\psi = \mathbf{0}$ against $\psi \neq \mathbf{0}$ is small, we would reject $\psi = \mathbf{0}$ and accept $\psi \geqslant \mathbf{0}$. Therefore, in this example, the evidence against $\psi = \mathbf{0}$ is sufficiently strong and we need not have carried out a one-sided test to decide whether or not to reject H_0. However, if it is known that $\psi \geqslant \mathbf{0}$ then a test of $H_0 : \psi = \mathbf{0}$ against $H_1 : \psi \geqslant \mathbf{0}$ would in general summarize the statistical evidence better than a test of $H_0 : \psi = \mathbf{0}$ against $H_2 : \psi \neq \mathbf{0}$.

The foregoing computations can be modified to incorporate an overdispersion parameter as well. ■

Example 4.3.2 *Comparison of treatments with ordinal response data.*

Example 1.2.8 of Chapter 1 (page 12) provides details of a trial to compare two treatments for ulcer; the data are given in Table 1.7. The objective of the study is to show that Treatment B is better than Treatment A. The hypotheses can be formulated in different ways. One possibility is to say that Treatment B is better than Treatment A if the *local odds ratios* are all greater than or equal to 1 with at least one of them being greater than 1; for interpretations and discussions about odds ratios, see Chapter 6. However, this may be an unnecessarily stringent requirement. A less stringent, and hence perhaps more desirable, formulation of the problem is

$$H_0 : \gamma_1 = \gamma_2 = \gamma_3 = 1 \quad \text{and} \quad H_1 : \gamma_1 \geqslant 1, \gamma_2 \geqslant 1, \gamma_3 \geqslant 1, \tag{4.38}$$

$$\text{where } \gamma_q = (\sum_{j=1}^{q} \pi_{1j})(\sum_{j=q+1}^{C} \pi_{2j})\{(\sum_{j=1}^{q} \pi_{2j})(\sum_{j=q+1}^{C} \pi_{1j})\}^{-1}$$
$$\text{and } \pi_{ij} = \text{pr(the response of a subject in row } i \text{ falls in column } j). \tag{4.39}$$

The γ_i s are also certain odds ratios of interest as discussed in Example 1.2.8 of Chapter 1. The parameter $\boldsymbol{\pi} = (\pi_{11}, \pi_{12}, \pi_{13}, \pi_{21}, \pi_{22}, \pi_{23})^T$ defines the unrestricted model completely. The likelihood is simply the product of the two multinomial probabilities for Treatment A and Treatment B:

$$\text{Likelihood at } \boldsymbol{\pi} \propto \Pi_{i=1,j=1}^{i=2,j=4} \pi_{ij}^{y_{ij}}. \tag{4.40}$$

To apply the results of this section, it is important that the hypotheses involve only linear functions of the parameter that appears in the likelihood. However, in the foregoing formulation, this is not the case because the likelihood is expressed as a function of $\boldsymbol{\pi}$ (see 4.40) and the hypothesis involve nonlinear functions of $\boldsymbol{\pi}$. General methods of testing hypotheses involving nonlinear functions of parameters will be discussed later. Such methods can be applied for testing (4.38). However, in this particular example, there is a simple reparameterization such that the hypotheses involve only linear functions and hence the results of this section are applicable.

Let us parameterize the π_{ij} s as

$$\log(\pi_{i1} + \ldots + \pi_{iq}) = \lambda_q - \psi_{iq} \tag{4.41}$$

where $i = 1, 2$ and $q = 1, 2, 3$. Let $\boldsymbol{\theta} = (\lambda_1, \lambda_2, \lambda_3, \psi_{21}, \psi_{22}, \psi_{23})^T, \boldsymbol{\lambda} = (\lambda_1, \lambda_2, \lambda_3)^T$ and $\boldsymbol{\psi} = (\psi_{21}, \psi_{22}, \psi_{23})^T$. Then $\gamma_1 \geqslant 1, \gamma_2 \geqslant 1, \gamma_3 \geqslant 1$ is equivalent to $\boldsymbol{\psi} \geqslant \mathbf{0}$. Now,

$$\text{Likelihood at } \boldsymbol{\theta} = L(\boldsymbol{\theta}) = \text{ Const } \Pi_{i=1,j=1}^{i=2,j=4}\{\pi_{ij}(\boldsymbol{\theta})\}^{y_{ij}}$$

where the value of $\pi_{ij}(\boldsymbol{\theta})$ in terms of $\boldsymbol{\theta}$ is given by (4.41). For the data in Table 1.7, $\hat{\gamma}_1 = (12 \times 27)/(5 \times 20) = 3.24$, $\hat{\gamma}_2 = (22 \times 19)/(13 \times 10) = 3.22, 4$ and $\hat{\gamma}_3 = (26 \times 11)/(21 \times 6) = 2.27$. Thus, each $\hat{\gamma}_i$ is greater than 1 and hence $\hat{\boldsymbol{\psi}} \geqslant \mathbf{0}$. Again, as in the previous example, the unrestricted estimator of $\boldsymbol{\theta}$ satisfies the inequalities in H_1. Consequently, the estimator of $\boldsymbol{\theta}$ under H_1 is also the unrestricted estimator of $\boldsymbol{\theta}$. Now, the *LRT* for testing $\boldsymbol{\psi} = \mathbf{0}$ against $\boldsymbol{\psi} \geqslant \mathbf{0}$ is 6.0. Under H_0, we have that $\pi_{1j} = \pi_{2j}$ for $j = 1, \ldots, 4$. Therefore, the *mle* of $(\pi_{i1}, \pi_{i2}, \pi_{i3}, \pi_{i4})$ under H_0 is the vector of sample proportions for column totals $(i = 1, 2)$. Hence, the *mle* of $\boldsymbol{\pi}$ under H_0 is $\bar{\boldsymbol{\pi}} = (\bar{\boldsymbol{\pi}}_1^T, \bar{\boldsymbol{\pi}}_2^T)^T = (17/64, 18/64, 12/64, 17/64, 17/64, 18/64, 12/64, 17/64)^T$. Because the *mle* is invariant under transformations, the *mle* of $\boldsymbol{\lambda}$ under H_0 is obtained by solving (4.41) with $\boldsymbol{\pi} = \bar{\boldsymbol{\pi}}$ and $\boldsymbol{\psi} = \mathbf{0}$; this leads to $\bar{\boldsymbol{\lambda}} = \{\log(17/64), \log(35/64), \log(47/64)\}$. Now, an estimate of the large sample p-value for *LRT* is (see $\hat{p}$-test)

$$\hat{p} = \sum_{i=1}^{3} w_i\{3, \mathcal{I}^{\psi\psi}(\bar{\boldsymbol{\lambda}})\}\text{pr}(\chi_i^2 \geqslant 6.0) = 0.054;$$

the formulas in (3.26) were used for computing $w_i\{3, \mathcal{I}^{\psi\psi}(\bar{\boldsymbol{\lambda}})\}$. Using the same closed-form for the weights, the supremum of the tail probability over $\boldsymbol{\lambda}$ was computed as

$$\sup_{\boldsymbol{\lambda}} \sum_{i=1}^{3} w_i\{3, \mathcal{I}^{\psi\psi}(\boldsymbol{\lambda})\}\text{pr}(\chi_i^2 \geqslant 6.0) = 0.083.$$

Therefore, a large sample approximation of the p-value is 0.083; as expected, this is larger than $\hat{p}$ (= 0.054) and the difference between the two is not negligible. In this example, some of the observed cell frequencies (in Table 1.7) are not that large, and, therefore, it may be prudent to be conservative and use $0.083(= p_{sup})$ rather than $0.054(= \hat{p})$ as a large sample p-value.

It would be of interest to compare these results with a test against the unrestricted alternative that Treatments A and B are different for which the null and alternative hypotheses take the form

$$H_0 : \boldsymbol{\psi} = \mathbf{0} \text{ and } H_2 : \boldsymbol{\psi} \neq \mathbf{0},$$

respectively. Because the unrestricted *mle* of $\boldsymbol{\psi}$ satisfies the constraint $\boldsymbol{\psi} \geqslant \mathbf{0}$, the *LRT* of $\boldsymbol{\psi} = \mathbf{0}$ against $\boldsymbol{\psi} \neq \mathbf{0}$ and that of $\boldsymbol{\psi} = \mathbf{0}$ against $\boldsymbol{\psi} \geqslant \mathbf{0}$ have the same numerical value, namely 6.0. However, the p-values are different because the null distributions are different. For $H_0 : \boldsymbol{\psi} = \mathbf{0}$ against $H_2 : \boldsymbol{\psi} \neq \mathbf{0}$, the large sample p-value is $\text{pr}(\chi_3^2 \geqslant 6.0) = 0.112$. As expected (because the components of $\boldsymbol{U}$ are all positive), this p-value is larger than those obtained earlier for testing against $\boldsymbol{R\theta} \geq \mathbf{0}$.

As has been noted earlier, if the unrestricted estimate satisfies the inequalities in H_1, then the *LRT* against $H_1 : \boldsymbol{\psi} \geqslant \mathbf{0}$ and that against $H_2 : \boldsymbol{\psi} \neq \mathbf{0}$ would have the same numerical value but the latter would have a larger p-value. This follows easily because the chi-bar-square distribution corresponding to the test against $\boldsymbol{\psi} \geqslant \mathbf{0}$ has shorter tails than the χ^2 distribution corresponding to the test against $\boldsymbol{\psi} \neq \mathbf{0}$. This is consistent with our intuition: if the unrestricted estimate, $\hat{\boldsymbol{\psi}}$ is nonnegative, then the data are pointing us in the direction of $\boldsymbol{\psi} \geqslant \mathbf{0}$ and the evidence against $\boldsymbol{\psi} = \mathbf{0}$ from a test against $\boldsymbol{\psi} \geqslant \mathbf{0}$ is likely to be stronger than that from a global test against $\boldsymbol{\psi} \neq \mathbf{0}$. The corresponding phenomenon is well-known in the classical one-sided t-test on the mean of a normal distribution. ■

4.4 TESTS OF $h(\theta) = 0$

The results in Section 4.1 for linear constraints of $\boldsymbol{\theta}$ can be extended to nonlinear constraints as well. For simplicity we shall first consider the case when the observations are *iid*. Throughout this section we shall assume that the natural parameter space Θ is open and that Condition Q is satisfied. Most of the results of this section can be extended to the setting when the observations are not *iid*. Some of these will be discussed later in this chapter.

Let $\boldsymbol{h}(\boldsymbol{\theta}) = (h_1(\boldsymbol{\theta}), \ldots, h_r(\boldsymbol{\theta}))^T$ be a continuously differentiable vector function. In the next subsection, we will define several tests of $\boldsymbol{h}(\boldsymbol{\theta}) = \mathbf{0}$ vs $\boldsymbol{h}(\boldsymbol{\theta}) \geqslant \mathbf{0}$. Then in subsection 4.4.2 we will consider a test against $\boldsymbol{h}_2(\boldsymbol{\theta}) \geqslant \mathbf{0}$ where $\boldsymbol{h}_2$ is a subvector of $\boldsymbol{h}$.

4.4.1 Test of $h(\theta) = 0$ Against $h(\theta) \geqslant 0$

Let us consider the hypotheses

$$H_0 : \boldsymbol{h}(\boldsymbol{\theta}) = \mathbf{0} \qquad \text{vs} \qquad H_1 : \boldsymbol{h}(\boldsymbol{\theta}) \geqslant \mathbf{0}. \tag{4.42}$$

Let $\boldsymbol{H}(\boldsymbol{\theta}) = (\partial/\partial\boldsymbol{\theta}^T)\boldsymbol{h}(\boldsymbol{\theta})$. Assume that

$$\boldsymbol{H}(\boldsymbol{\theta}_0) \text{ has full row rank.} \tag{4.43}$$

As usual, let $\hat{\boldsymbol{\theta}}$ denote the unconstrained *mle*, and $\tilde{\boldsymbol{\theta}}$ and $\bar{\boldsymbol{\theta}}$ denote the *mle*'s under H_1 and H_0, respectively. The definition of *LRT* does not require any new ideas: $LRT = 2\{\ell(\tilde{\boldsymbol{\theta}}) - \ell(\bar{\boldsymbol{\theta}})\}$. The asymptotic null distribution of this turns out to be a chi-bar-square. Another convenient statistic for testing (4.42) is the distance statistic based on $\hat{\boldsymbol{\theta}}$. Since $\sqrt{n}(\hat{\boldsymbol{\theta}} - \boldsymbol{\theta}_0) \xrightarrow{d} N\{\mathbf{0}, \mathcal{I}(\boldsymbol{\theta}_0)^{-1}\}$, we define

$$D = d(\hat{\boldsymbol{\theta}}, H_0) - d(\hat{\boldsymbol{\theta}}, H_1), \text{ where } \quad d(\hat{\boldsymbol{\theta}}, H) = \inf_{\boldsymbol{\theta} \in H} n(\hat{\boldsymbol{\theta}} - \boldsymbol{\theta})^T \hat{\mathcal{I}}(\hat{\boldsymbol{\theta}} - \boldsymbol{\theta})$$

where $\hat{\mathcal{I}}$ is a consistent estimator of $\mathcal{I}(\boldsymbol{\theta}_0)$. It will be seen that the *LRT* and D are asymptotically equivalent and have the common asymptotic null distribution, $\bar{\chi}^2\{\boldsymbol{H}(\boldsymbol{\theta}_0)\mathcal{I}(\boldsymbol{\theta}_0)^{-1}\boldsymbol{H}(\boldsymbol{\theta}_0)^T, \mathbb{R}^{+r}\}$. Even if $\boldsymbol{H}(\boldsymbol{\theta}_0)$ has rank less than the dimension of $\boldsymbol{h}$, the asymptotic null distribution of *LRT* is still a chi-bar-square, but it has a different form.

Much of the existing literature concerning tests of $\boldsymbol{h}(\boldsymbol{\theta}) = \mathbf{0}$ usually assumes that $\boldsymbol{H}(\boldsymbol{\theta}_0)$ is of full row-rank. It is important to note that this may fail to be satisfied even in what appears to be well-behaved models.

An example where $\boldsymbol{H}(\boldsymbol{\theta}_0)$ is not of full rank:
Let $\boldsymbol{A}(\boldsymbol{\theta})$ be the 2×2 matrix $(\theta_1, \theta_3 \mid \theta_3, \theta_2)$,

$$H_0 : \boldsymbol{A}(\boldsymbol{\theta}) = \mathbf{0} \quad \text{and} \quad H_1 : \boldsymbol{A}(\boldsymbol{\theta}) \text{ is positive semi-definite.}$$

This may be restated in the equivalent form $H_0 : \boldsymbol{h}(\boldsymbol{\theta}) = \mathbf{0}$ and $H_1 : \boldsymbol{h}(\boldsymbol{\theta}) \geqslant \mathbf{0}$, where $\boldsymbol{h}(\boldsymbol{\theta}) = (-\theta_1, \theta_1\theta_2 - \theta_3^2)^T$. Now, $(\partial/\partial\boldsymbol{\theta}^T)\boldsymbol{h}(\boldsymbol{\theta}) = [1, 0, 0|\theta_2, \theta_1, -2\theta_3]$; the last row becomes a zero vector at the null value, $\boldsymbol{\theta} = \mathbf{0}$, and hence $(\partial/\partial\boldsymbol{\theta}^T)\boldsymbol{h}(\boldsymbol{\theta})$ is not of full rank. Therefore, condition (4.43) is not satisfied.

To provide a more concrete example, we consider the regression model

$$y = g(\boldsymbol{x}; \boldsymbol{\theta}) + \epsilon, \quad \text{where} \quad g(\boldsymbol{x}; \boldsymbol{\theta}) = \theta_1 x_1^2 + 2\theta_3 x_1 x_2 + \theta_2 x_2^2 + \theta_4 x_1 + \theta_5 x_2 + \theta_6.$$

Let the hypotheses be

$$H_0 : g(\boldsymbol{x}; \boldsymbol{\theta}) = \text{linear in } \boldsymbol{x} \quad \text{and} \quad H_1 : g(\boldsymbol{x}; \boldsymbol{\theta}) \text{ is concave in } \boldsymbol{x}.$$

First note that the Hessian matrix of $g(\boldsymbol{x}; \boldsymbol{\theta})$ with respect to $\boldsymbol{x}$ is $2[\theta_1, \theta_3|\theta_3, \theta_2]$. Now, concavity of $g(\boldsymbol{x}; \boldsymbol{\theta})$ is equivalent to the Hessian matrix, $(1/2)\boldsymbol{A}(\boldsymbol{\theta})$, of $g(\boldsymbol{x}; \boldsymbol{\theta})$ being negative semi-definite. Therefore, the hypotheses take the form $H_0 : \boldsymbol{A}(\boldsymbol{\theta}) = \mathbf{0}$ and $H_1 : \boldsymbol{A}(\boldsymbol{\theta})$ *is positive semi-definite.*

Another example where this arises is when $\boldsymbol{A}$ is the covariance matrix of a collection of random effects; see Example 1.2.10 on page 14. ■

We can also define tests that resemble the W and S_G introduced earlier for linear constraints of $\boldsymbol{\theta}$. All of these test statistics differ only by a $o_p(1)$ term under the null hypothesis. Thus, they have the same asymptotic null distribution. Further, they also have the same asymptotic local power against Pitman-type local alternatives; this would require some conditions, for example, contiguity. Other types of asymptotic efficiencies of these tests have not been studied in any detail. As noted earlier, not much is known about their relative advantages and disadvantages. For completeness, they are introduced here.

For simplicity, we shall adopt the following notation. A function, say $h(\boldsymbol{\theta})$, evaluated at $\tilde{\boldsymbol{\theta}}$ and $\bar{\boldsymbol{\theta}}$ will be denoted by $\tilde{h}$ and $\bar{h}$, respectively; thus $\tilde{h} = h(\tilde{\boldsymbol{\theta}})$ and $\bar{h} = h(\bar{\boldsymbol{\theta}})$. Let $\check{\boldsymbol{H}}$ and $\check{\mathcal{I}}$ denote consistent estimators of $\boldsymbol{H}(\boldsymbol{\theta})$ and $\mathcal{I}(\boldsymbol{\theta})$ respectively, under the null hypothesis; therefore, they could be obtained by substituting $\hat{\boldsymbol{\theta}}, \tilde{\boldsymbol{\theta}}$ or $\bar{\boldsymbol{\theta}}$ for $\boldsymbol{\theta}$.

(1) Wald-type test

This is very similar to the Wald-type test introduced earlier for linear constraints. Recall that the Wald-type statistic for testing $\boldsymbol{h}(\boldsymbol{\theta}) = \mathbf{0}$ against $\boldsymbol{h}(\boldsymbol{\theta}) \neq \mathbf{0}$ is

$n\hat{\boldsymbol{h}}^T(\check{\boldsymbol{H}}\check{\mathcal{I}}^{-1}\check{\boldsymbol{H}}^T)^{-1}\hat{\boldsymbol{h}}$. This suggests the statistic

$$W = n\tilde{\boldsymbol{h}}^T(\check{\boldsymbol{H}}\check{\mathcal{I}}^{-1}\check{\boldsymbol{H}}^T)^{-1}\tilde{\boldsymbol{h}}.$$

(2) Hausman-Wald type test

This simply evaluates the distance between the *mle*'s under H_0 and H_1.

$$W_H = n(\tilde{\boldsymbol{\theta}} - \bar{\boldsymbol{\theta}})^T\check{\mathcal{I}}(\tilde{\boldsymbol{\theta}} - \bar{\boldsymbol{\theta}}).$$

(3) Global score test:

Motivated by arguments given for the linear constraints setting, we define

$$S_G = n^{-1}\boldsymbol{U}^T(\check{\boldsymbol{H}}\check{\mathcal{I}}^{-1}\check{\boldsymbol{H}}^T)^{-1}\boldsymbol{U}, \quad \text{where } \boldsymbol{U} = \check{\boldsymbol{H}}\check{\mathcal{I}}^{-1}(\tilde{\boldsymbol{S}} - \bar{\boldsymbol{S}}).$$

Another, possible global score statistic is (see Robertson et al. (1988)),

$$n^{-1}(\tilde{\boldsymbol{S}} - \bar{\boldsymbol{S}})^T\check{\mathcal{I}}^{-1}(\tilde{\boldsymbol{S}} - \bar{\boldsymbol{S}}).$$

(4) Kühn-Tücker multiplier test

First, it would be instructive to provide an insight into the so called *Lagrange multiplier test*. Let us consider the optimization problem

$$\text{Problem P1:} \qquad \max \ell(\boldsymbol{\theta}) \qquad \text{subject to } \boldsymbol{h}(\boldsymbol{\theta}) = \boldsymbol{b}$$

where $\boldsymbol{b}$ is a given set of constraints. Let the Lagrangian function be defined as $\{\ell(\boldsymbol{\theta}) + \boldsymbol{\lambda}^T(\boldsymbol{h}(\boldsymbol{\theta}) - \boldsymbol{b})\}$, where $\boldsymbol{\lambda} = (\lambda_1, \ldots \lambda_r)^T$. Let the optimal solution be $(\bar{\boldsymbol{\theta}}_b : \bar{\boldsymbol{\lambda}}_b)$; $\bar{\boldsymbol{\lambda}}_b$ is the Lagrange multiplier. Now, we have (for example, see Intriligator (1971) page 36)

$$\bar{\boldsymbol{\lambda}}_b = -(\partial/\partial \boldsymbol{b})\ell(\bar{\boldsymbol{\theta}}_b).$$

Thus, $\bar{\boldsymbol{\lambda}}_b$ is a measure of the sensitivity of the constrained maximum of $\ell(\boldsymbol{\theta})$ over $\{\boldsymbol{\theta} : \boldsymbol{h}(\boldsymbol{\theta}) = \boldsymbol{b}\}$ to the value $\boldsymbol{b}$. If the unconstrained maximum of $\ell(\boldsymbol{\theta})$ is achieved over $\{\boldsymbol{\theta} : \boldsymbol{h}(\boldsymbol{\theta}) = \boldsymbol{b}\}$ then $\bar{\boldsymbol{\lambda}}_b = \boldsymbol{0}$.

For brevity of notation, let us assume that $\boldsymbol{b} = \boldsymbol{0}$ and suppress the suffix $\boldsymbol{b}$ in $\boldsymbol{\lambda}_b$. Aichison and Silvey (1958) showed that, under $\boldsymbol{h}(\boldsymbol{\theta}) = \boldsymbol{0}$, $\bar{\boldsymbol{\lambda}}/\sqrt{n}$ is asymptotically $N\{\boldsymbol{0}, \boldsymbol{V}(\boldsymbol{\theta}_0)\}$ where $\boldsymbol{V}(\boldsymbol{\theta}) = [\boldsymbol{H}\mathcal{I}^{-1}\boldsymbol{H}^T]_{\boldsymbol{\theta}}^{-1}$; it is asymptotically normal with nonzero mean under local alternatives. Therefore, it is reasonable to expect that the Lagrange multipliers have the potential to be building blocks for constructing a test of $\boldsymbol{h}(\boldsymbol{\theta}) = \boldsymbol{0}$ against $\boldsymbol{h}(\boldsymbol{\theta}) \neq \boldsymbol{0}$. The so called *Lagrange multiplier* statistic for testing $\boldsymbol{h}(\boldsymbol{\theta}) = \boldsymbol{0}$ vs $\boldsymbol{h}(\boldsymbol{\theta}) \neq \boldsymbol{0}$ is $n^{-1}\bar{\boldsymbol{\lambda}}^T\boldsymbol{V}(\bar{\boldsymbol{\theta}})^{-1}\bar{\boldsymbol{\lambda}}$. Since $\bar{\boldsymbol{\lambda}}$ satisfies $[(\partial/\partial\boldsymbol{\theta})\ell(\boldsymbol{\theta}) + \boldsymbol{H}(\boldsymbol{\theta})^T\bar{\boldsymbol{\lambda}}]_{\boldsymbol{\theta}=\bar{\boldsymbol{\theta}}} = \boldsymbol{0}$ we have

$$n^{-1}\bar{\boldsymbol{\lambda}}^T\boldsymbol{V}(\bar{\boldsymbol{\theta}})^{-1}\bar{\boldsymbol{\lambda}} = n^{-1}[\boldsymbol{S}(\boldsymbol{\theta})^T\mathcal{I}(\boldsymbol{\theta})^{-1}\boldsymbol{S}(\boldsymbol{\theta})]_{\boldsymbol{\theta}=\bar{\boldsymbol{\theta}}};$$

the expression on the right-hand side is Rao's score statistic.

The same idea can be extended to construct tests when there are inequality constraints. For example, consider the optimization problem

$$\text{Problem P2:} \qquad \max \ell(\boldsymbol{\theta}) \qquad \text{subject to } \boldsymbol{h}(\boldsymbol{\theta}) \geqslant \boldsymbol{b}$$

Let the Lagrangian function be defined as above. The first-order Kühn-Tücker conditions (also known as Karush-Kühn-Tücker conditions) are

$$\nabla \ell(\tilde{\boldsymbol{\theta}}) + \nabla \boldsymbol{h}(\tilde{\boldsymbol{\theta}})^T \tilde{\boldsymbol{\lambda}} = \mathbf{0}, \boldsymbol{h}(\tilde{\boldsymbol{\theta}}) \geqslant \mathbf{0}, \tilde{\boldsymbol{\lambda}} \geqslant \mathbf{0}, \lambda_j h_j(\tilde{\boldsymbol{\theta}}) = 0 \text{ for } j = 1, \ldots, r.$$

Let the solution be $(\tilde{\boldsymbol{\theta}}_b, \tilde{\boldsymbol{\lambda}}_b)$; $\tilde{\boldsymbol{\lambda}}_b$ are known as Kühn-Tücker multipliers. Again $\tilde{\boldsymbol{\lambda}}_b$ can also be interpreted as a measure of sensitivity of the constrained maximum of $\ell(\boldsymbol{\theta})$ to $\boldsymbol{b}$, in much the same way as for the foregoing equality constrained optimization problem (for details about this interpretation and its relevance, see Intriligator (1971) page 60).

Let $\bar{\boldsymbol{\lambda}}$ and $\tilde{\boldsymbol{\lambda}}$ denote the Kühn-Tücker multipliers associated with the maximization of $\ell(\boldsymbol{\theta})$ subject to $\boldsymbol{h}(\boldsymbol{\theta}) = \mathbf{0}$ and $\boldsymbol{h}(\boldsymbol{\theta}) \geqslant \mathbf{0}$, respectively. Now, the KT-statistic is defined as

$$KT = n(\tilde{\boldsymbol{\lambda}} - \bar{\boldsymbol{\lambda}})^T (\check{\boldsymbol{H}} \check{\mathcal{I}}^{-1} \check{\boldsymbol{H}})(\tilde{\boldsymbol{\lambda}} - \bar{\boldsymbol{\lambda}}).$$

Now, the next result says that the foregoing tests are asymptotically equivalent.

Proposition 4.4.1 *Under the null hypothesis,* $H_0 : \boldsymbol{h}(\boldsymbol{\theta}) = \mathbf{0}$, *we have*

$$LRT = W + o_p(1) = W_H + o_p(1) = S_G + o_p(1) = D + o_p(1) = KT + o_p(1).$$

Proof: The proofs of these, except that for D, are given in Gourieroux and Monfort (1995, Chapter, 21); the equivalence of D to the *LRT* follows from the quadratic approximation (C) in Proposition 4.2.2. ■

An unattractive feature of W, W_H, S_G, and KT is that they depend on $\boldsymbol{H}(\boldsymbol{\theta})$ explicitly; further, some of the technical arguments also assume that $\boldsymbol{H}$ is of full rank. By contrast, *LRT* and distance-based tests do not depend on the choice of the function $\boldsymbol{h}$ to define a given set of constraints.

In some cases, it maybe difficult to compute the *mle* and/or the information matrix. In such cases, it may be of interest to construct a test based on an estimator that may be easier to obtain than the *mle*. Such a method is given in the next subsection.

4.4.2 Test of $h(\theta) = 0$ Against $h_2(\theta) \geqslant 0$

Let us consider the setting in Section 4.2. In particular, the parameter space Θ is open and Condition Q is satisfied. As in the previous subsection, let $\boldsymbol{h}$ be an $r \times 1$ vector function of $\boldsymbol{\theta}$; it is assumed that $\boldsymbol{h}$ is continuously differentiable. Let $\boldsymbol{H}(\boldsymbol{\theta})_{r \times p} = (\partial/\partial \boldsymbol{\theta}^T)\boldsymbol{h}(\boldsymbol{\theta})$ and assume that its rank is r for values of $\boldsymbol{\theta}$ in the null parameter space.

Let $\boldsymbol{h}$ be partitioned as $(\boldsymbol{h}_1 : \boldsymbol{h}_2)$ where $\boldsymbol{h}_2$ is $q \times 1$; recall that $(\boldsymbol{h}_1 : \boldsymbol{h}_2)$ denotes the partitioned vector $(\boldsymbol{h}_1^T, \boldsymbol{h}_2^T)^T$. In this section, we shall introduce a test of

$$H_0 : \boldsymbol{h}(\boldsymbol{\theta}) = \mathbf{0} \text{ against } H_1 : \boldsymbol{h}_2(\boldsymbol{\theta}) \geqslant \mathbf{0} \tag{4.44}$$

based on the basic idea that underlies the Wald test.

Let $\hat{\boldsymbol{\theta}}$ denote an estimator of $\boldsymbol{\theta}$ such that

$$n^{1/2}(\hat{\boldsymbol{\theta}} - \boldsymbol{\theta}_0) \xrightarrow{d} N\{\mathbf{0}, \boldsymbol{W}(\boldsymbol{\theta}_0)\}$$

for some $\boldsymbol{W}(\boldsymbol{\theta})$. For example, $\hat{\boldsymbol{\theta}}$ can be the unrestricted maximum likelihood estimator. It follows that

$$n^{1/2}\{\boldsymbol{h}(\hat{\boldsymbol{\theta}}) - \boldsymbol{h}(\boldsymbol{\theta}_0)\} \xrightarrow{d} N\{\mathbf{0}, \boldsymbol{V}(\boldsymbol{\theta}_0)\}, \tag{4.45}$$

where $\boldsymbol{V}(\boldsymbol{\theta}) = \boldsymbol{H}(\boldsymbol{\theta})\boldsymbol{W}(\boldsymbol{\theta})\boldsymbol{H}(\boldsymbol{\theta})^T$. Now, we can use some of the arguments used in subsection 4.3.3 for defining Wald-type and distance statistics. Let $\hat{\boldsymbol{V}}$ be a consistent estimator of $\boldsymbol{V}(\boldsymbol{\theta})$ under the null hypothesis, for example, $\hat{\boldsymbol{V}} = \boldsymbol{H}(\hat{\boldsymbol{\theta}})\boldsymbol{W}(\hat{\boldsymbol{\theta}})\boldsymbol{H}(\hat{\boldsymbol{\theta}})^T$.

A suitable statistic for testing H_0 against H_1 is

$$D = n[\boldsymbol{h}(\hat{\boldsymbol{\theta}})^T\hat{\boldsymbol{V}}^{-1}\boldsymbol{h}(\hat{\boldsymbol{\theta}}) - \min_{\boldsymbol{a}}\{(\boldsymbol{h}(\hat{\boldsymbol{\theta}}) - \boldsymbol{a})^T\hat{\boldsymbol{V}}^{-1}(\boldsymbol{h}(\hat{\boldsymbol{\theta}}) - \boldsymbol{a}) : \boldsymbol{a}_2 \geqslant \mathbf{0}\}] \tag{4.46}$$

where $\boldsymbol{a}$ in $\mathbb{R}^r$ is partitioned as $(\boldsymbol{a}_1 : \boldsymbol{a}_2)$ to conform with $\boldsymbol{h} = (\boldsymbol{h}_1 : \boldsymbol{h}_2)$; see Kodde and Palm (1986). Note that $D = d\{\boldsymbol{h}(\hat{\boldsymbol{\theta}}), H_0\} - d\{\boldsymbol{h}(\hat{\boldsymbol{\theta}}), H_1\}$, where $d\{\boldsymbol{h}(\hat{\boldsymbol{\theta}}), H\}$ is a measure of the distance between $\boldsymbol{h}(\hat{\boldsymbol{\theta}})$ and H. The test rejects H_0 if D is large.

Suppose that H_0 is satisfied. It follows from $\{\boldsymbol{V}(\boldsymbol{\theta}_0) - \hat{\boldsymbol{V}})\} = o_p(1)$ that, under H_0, the value of D would change by $o_p(1)$ if $\hat{\boldsymbol{V}}$ is replaced by $\boldsymbol{V}(\boldsymbol{\theta}_0)$ (see Lemma 4.10.2 in the Appendix to this chapter). Therefore, the limiting distribution of D in (4.46) is equal to the distribution of

$$T = \boldsymbol{X}^T\boldsymbol{V}(\boldsymbol{\theta}_0)^{-1}\boldsymbol{X} - \min_{\boldsymbol{a}}\{(\boldsymbol{X} - \boldsymbol{a})^T\boldsymbol{V}(\boldsymbol{\theta}_0)^{-1}(\boldsymbol{X} - \boldsymbol{a}) : \boldsymbol{a}_2 \geqslant \mathbf{0}\}$$

where $\boldsymbol{X} \sim N\{\mathbf{0}, \boldsymbol{V}(\boldsymbol{\theta}_0)\}$. Consequently, the asymptotic null distribution of D is given by

$$\mathrm{pr}_{\boldsymbol{\theta}_0}\{D \geqslant c \mid H_0\} \to \sum_{i=0}^{q} w_i\{q, \boldsymbol{V}_{22}(\boldsymbol{\theta}_0)\}\mathrm{pr}(\chi^2_{r-q+i} \geqslant c), \tag{4.47}$$

where $\boldsymbol{V}_{(22)}$ is the diagonal block of $\boldsymbol{V}$ corresponding to $\boldsymbol{h}_2$. Note that the large sample null distribution of D depends on the assumed true value $\boldsymbol{\theta}_0$ of the parameter $\boldsymbol{\theta}$ in the null parameter space. Approaches to dealing with such nuisance parameter problems were discussed in an earlier section.

If $\hat{\boldsymbol{\theta}}$ is the *mle*, then it follows from the quadratic approximations of the loglikelihood, that *LRT* and the other tests introduced in this subsection are asymptotically equivalent and hence their asymptotic null distribution is given by (4.47) with $\mathcal{I}(\boldsymbol{\theta})^{-1}$ substituted for $\boldsymbol{W}(\boldsymbol{\theta})$ in the formula for $\boldsymbol{V}(\boldsymbol{\theta})$.

4.5 AN OVERVIEW OF SCORE TESTS WITH NO INEQUALITY CONSTRAINTS

In this section, a brief summary of the relevant asymptotic results on likelihood ratio, Wald- and score-type tests is provided when there are no inequality constraints.

Although formal proofs are not given, the essential arguments are indicated. Details of the arguments may be found in Basawa (1985), Cox and Hinkley (1974), Sen and Singer (1993), and van der Vaart (2000). Hall and Mathiason (1990) provide a thorough account of many of these results at an intermediate level. These will be used in Section 4.6 to introduce local score and score-type tests against inequality constraints.

Let $f(y;\boldsymbol{\theta})$ denote the density of Y where $\boldsymbol{\theta} \in \Theta \subset \mathbb{R}^p$, and $Y_1, \ldots, Y_n$ denote n *iid* observations on the random variable/vector Y. It is assumed that Θ is open in $\mathbb{R}^p$. First, we shall consider the case of testing $\boldsymbol{\theta} = \boldsymbol{\theta}_0$ against $\boldsymbol{\theta} \neq \boldsymbol{\theta}_0$ and then consider the more general case when there are nuisance parameters. In this section we will be studying the distribution of statistics when H_0 is not true. Therefore, the true value may not be $\boldsymbol{\theta}_0$.

4.5.1 Test of $H_0 : \boldsymbol{\theta} = \boldsymbol{\theta}_0$ Against $H_2 : \boldsymbol{\theta} \neq \boldsymbol{\theta}_0$

Let

$$\ell(\boldsymbol{\theta}) = \sum_{i=1}^{n} \log f(Y_i;\boldsymbol{\theta}), \qquad \boldsymbol{S}(\boldsymbol{\theta}) = (\partial/\partial\boldsymbol{\theta})\ell(\boldsymbol{\theta})$$
$$\text{and} \quad \mathcal{I}(\boldsymbol{\theta}) = E_{\boldsymbol{\theta}}\{(\partial/\partial\boldsymbol{\theta}) \log f(Y;\boldsymbol{\theta})(\partial/\partial\boldsymbol{\theta}^T) \log f(Y;\boldsymbol{\theta})\} \tag{4.48}$$

denote the loglikelihood, score function, and the information matrix, respectively. Assume that Condition Q in subsection 4.2 is satisfied. Let the null and alternative hypotheses be

$$H_0 : \boldsymbol{\theta} = \boldsymbol{\theta}_0 \quad \text{and} \quad H_2 : \boldsymbol{\theta} \neq \boldsymbol{\theta}_0, \tag{4.49}$$

respectively. Then, because $\log x \leq (x - 1)$ (or by Jensen's inequality), it follows that

$$E_{\boldsymbol{\theta}} \log\{f(Y;\boldsymbol{\theta}^*)/f(Y;\boldsymbol{\theta})\} \leq E_{\boldsymbol{\theta}}\{f(Y;\boldsymbol{\theta}^*)/f(Y;\boldsymbol{\theta})\} - 1 = 0,$$

and hence

$$\text{plim}_{\boldsymbol{\theta}} n^{-1}\ell(\boldsymbol{\theta}^*) = E_{\boldsymbol{\theta}} \log f(Y;\boldsymbol{\theta}^*) \leq E_{\boldsymbol{\theta}} \log f(Y;\boldsymbol{\theta}) = \text{plim}_{\boldsymbol{\theta}} n^{-1}\ell(\boldsymbol{\theta})$$

where $\text{plim}_{\boldsymbol{\theta}}$ is the probability limit when $\boldsymbol{\theta}$ is the true value. Therefore, $\text{plim}_{\boldsymbol{\theta}} n^{-1}\ell(\boldsymbol{\theta}^*)$ achieves its global maximum at $\boldsymbol{\theta}^* = \boldsymbol{\theta}$. Further, $\boldsymbol{\theta}^* = \boldsymbol{\theta}$ is a stationary point of $\text{plim}_{\boldsymbol{\theta}} n^{-1}\ell(\boldsymbol{\theta}^*)$. This suggests that it may be possible to use the slope of loglikelihood at $\boldsymbol{\theta}_0$ to test $H_0 : \boldsymbol{\theta} = \boldsymbol{\theta}_0$. To this end, note that

$$n^{-1}\boldsymbol{S}(\boldsymbol{\theta}_0) \xrightarrow{p} \mathbf{0} \text{ under } H_0, \text{ and } n^{-1}\boldsymbol{S}(\boldsymbol{\theta}_0) \xrightarrow{p} \boldsymbol{a}(\boldsymbol{\theta}) \text{ under } H_2 \tag{4.50}$$

where $\boldsymbol{a}(\boldsymbol{\theta}) = E_{\boldsymbol{\theta}}\{(\partial/\partial\boldsymbol{\theta}) \log f(Y;\boldsymbol{\theta}_0)\}$; assume that $\boldsymbol{a}(\boldsymbol{\theta}) \neq \mathbf{0}$ for $\boldsymbol{\theta} \neq \boldsymbol{\theta}_0$. We also have $n^{-1/2}\boldsymbol{S}(\boldsymbol{\theta}_0) \xrightarrow{d} N\{\mathbf{0}, \mathcal{I}(\boldsymbol{\theta}_0)\}$ under H_0. This suggests the *score statistic*

$$S_L = n^{-1}\boldsymbol{S}(\boldsymbol{\theta}_0)^T \mathcal{I}(\boldsymbol{\theta}_0)^{-1}\boldsymbol{S}(\boldsymbol{\theta}_0) \tag{4.51}$$

for testing $H_0 : \boldsymbol{\theta} = \boldsymbol{\theta}_0$ vs $H_2 : \boldsymbol{\theta} \neq \boldsymbol{\theta}_0$. Because $S_L \xrightarrow{d} \chi^2_p$ under H_0 and $S_L \xrightarrow{p} \infty$ under H_2, the score test rejects H_0 if S_L is large.

The *Wald statistic* for $H_0 : \boldsymbol{\theta} = \boldsymbol{\theta}_0$ against $H_2 : \boldsymbol{\theta} \neq \mathbf{0}$ is defined as

$$T_W = n(\hat{\boldsymbol{\theta}} - \boldsymbol{\theta}_0)^T \hat{\mathcal{I}}(\hat{\boldsymbol{\theta}} - \boldsymbol{\theta}_0), \tag{4.52}$$

where $\hat{\mathcal{I}}$ is a consistent estimator of $\mathcal{I}(\boldsymbol{\theta}_0)$. In the usual definition of a Wald statistic, the normalizing matrix $\hat{\mathcal{I}}$ is chosen so that it would be consistent irrespective of whether the null hypothesis is true or not; however, for the validity of the test, it suffices $\hat{\mathcal{I}}$ to be consistent under the null hypothesis only. Clearly, $T_W \xrightarrow{d} \chi^2_p$ under $H_0 : \boldsymbol{\theta} = \boldsymbol{\theta}_0$ and $T_W \xrightarrow{p} \infty$ under $H_2 : \boldsymbol{\theta} \neq \boldsymbol{\theta}_0$, and therefore the Wald test rejects H_0 if T_W is large.

Under $H_2 : \boldsymbol{\theta} \neq \boldsymbol{\theta}_0$, the probability of rejecting H_0 tends to 1 as $n \to \infty$ for the foregoing score and Wald tests at a fixed level. Similarly, at any fixed level, the power of many other tests at $\boldsymbol{\theta} \neq \boldsymbol{\theta}_0$ also tend to 1. Therefore, to compare the performance of different tests, the limiting power of a test at a fixed level and at a fixed point in the alternative parameter space is not particularly helpful. The procedure adopted here to compare the asymptotic power of tests is the limiting power at *local alternatives* of the form $H_n : \boldsymbol{\theta} = (\boldsymbol{\theta}_0 + n^{-1/2}\boldsymbol{\delta})$ where $\boldsymbol{\delta}$ is fixed and the level of the test is also fixed. The sequence of hypotheses $H_n : \boldsymbol{\theta} = \boldsymbol{\theta}_o + n^{-1/2}\boldsymbol{\delta}$ are also known as Pitman-type local alternatives; for a thorough discussion of these see Serfling (1980) or Hájek et al. (1999). Here, we shall indicate the main arguments and results.

Let $\boldsymbol{\theta}_n = (\boldsymbol{\theta}_0 + n^{-1/2}\boldsymbol{\delta})$ where $\boldsymbol{\delta}$ is fixed. Let a sequence of local hypotheses be defined by $H_n : \boldsymbol{\theta} = \boldsymbol{\theta}_n$. Then, by a one-term Taylor expansion of $\boldsymbol{S}(\boldsymbol{\theta}_0)$ about $\boldsymbol{\theta}_n$, we have that

$$n^{-1/2}\boldsymbol{S}(\boldsymbol{\theta}_0) = n^{-1/2}\boldsymbol{S}(\boldsymbol{\theta}_n) + \mathcal{I}(\boldsymbol{\theta}_n)\boldsymbol{\delta} + o_p(1) \quad \text{under } H_n : \boldsymbol{\theta} = \boldsymbol{\theta}_n \tag{4.53}$$

Since $\mathcal{I}(\boldsymbol{\theta}_n)\boldsymbol{\delta} \xrightarrow{p} \mathcal{I}(\boldsymbol{\theta}_0)\boldsymbol{\delta}$ and $\boldsymbol{S}(\boldsymbol{\theta}_n)$ is a sum of *iid* random variables, we have that

$$n^{-1/2}\boldsymbol{S}(\boldsymbol{\theta}_0) \xrightarrow{d} N\{\mathcal{I}(\boldsymbol{\theta}_0)\boldsymbol{\delta}, \mathcal{I}(\boldsymbol{\theta}_0)\} \quad \text{and} \quad S_L \xrightarrow{d} \chi^2_p(\Delta) \quad \text{under } H_n, \tag{4.54}$$

where $\Delta = \boldsymbol{\delta}^T\mathcal{I}(\boldsymbol{\theta}_0)\boldsymbol{\delta}$ is the noncentrality parameter. Further, by modifying the arguments to accommodate that $\boldsymbol{\theta}_n$ is the true value that may not be a fixed point, it can be shown that

$$n^{1/2}(\hat{\boldsymbol{\theta}} - \boldsymbol{\theta}_0) \xrightarrow{d} N\{\boldsymbol{\delta}, \mathcal{I}(\boldsymbol{\theta}_0)^{-1}\} \text{ and } T_W \xrightarrow{d} \chi^2_p(\Delta) \quad \text{under } H_n \tag{4.55}$$

where $\Delta = \boldsymbol{\delta}^T\mathcal{I}(\boldsymbol{\theta}_0)\boldsymbol{\delta}$. Therefore, by (4.54) and (4.55), the score and Wald tests have the same asymptotic distribution under $H_n : \boldsymbol{\theta} = \boldsymbol{\theta}_n$ and hence the same limiting power against the sequences of local alternatives, H_n.

Now, *assume that $\boldsymbol{\theta}_0$ is the true value.* Then expansion of $n^{-1/2}\boldsymbol{S}(\hat{\boldsymbol{\theta}})$ about $\boldsymbol{\theta}_0$ leads to

$$n^{-1/2}\mathcal{I}(\boldsymbol{\theta}_0)^{-1}\boldsymbol{S}(\boldsymbol{\theta}_0) = n^{1/2}(\hat{\boldsymbol{\theta}} - \boldsymbol{\theta}_0) + o_p(1). \tag{4.56}$$

Consequently, we can interchange $n^{-1/2}\mathcal{I}(\boldsymbol{\theta}_0)^{-1}\boldsymbol{S}(\boldsymbol{\theta}_0)$ and $n^{1/2}(\hat{\boldsymbol{\theta}} - \boldsymbol{\theta}_0)$ in many cases without affecting the first-oder asymptotic properties of the resulting test. Thus, we have that $T_W = S_L + o_p(1)$. An expansion of $\ell(\boldsymbol{\theta}_0)$ about $\hat{\boldsymbol{\theta}}$ leads to (see (4.4))

$$\ell(\boldsymbol{\theta}_0) = \ell(\hat{\boldsymbol{\theta}}) - (1/2)n(\boldsymbol{\theta}_0 - \hat{\boldsymbol{\theta}})^T\mathcal{I}(\hat{\boldsymbol{\theta}})(\boldsymbol{\theta}_0 - \hat{\boldsymbol{\theta}}) + o_p(1). \tag{4.57}$$

It follows from (4.51), (4.52), (4.56), and (4.57) that

$$LRT = T_W + o_p(1) = S_L + o_p(1) \qquad \text{under } H_0. \tag{4.58}$$

Consequently, LRT, T_W, and S_L, have the same asymptotic null distribution. It can also be shown that (for example, see Basawa (1985))

$$LRT = T_W + o_p(1) = S_L + o_p(1) \qquad \text{under } H_n. \tag{4.59}$$

This follows easily if H_n is contiguous to H_0; otherwise quadratic approximations can be used. Consequently, we have that

$$LRT, T_W, S_L \xrightarrow{d} \chi^2(\Delta) \qquad \text{under } H_n$$

where $\Delta = \boldsymbol{\delta}^T \mathcal{I}_{\boldsymbol{\theta}_0} \boldsymbol{\delta}$, and hence the three tests have the same local power. Usually, once we have shown that the difference between two test statistics is $o_p(1)$ at $\boldsymbol{\theta}_0$ in the null parameter space, then the same would also hold under $H_n : \boldsymbol{\theta} = \boldsymbol{\theta}_0 + n^{-1/2}\boldsymbol{\delta}$ as well. The proof of this may or may not be simple. For the *iid* setting, we can adopt an approach based on a quadratic approximation of $\ell(\boldsymbol{\theta})$ in $n^{-1/2}$-neighborhoods. A different approach would be to use methods based on contiguity of hypotheses.

Essentially the same types of arguments are used when the null hypothesis is not simple. These are discussed in the next subsection.

4.5.2 Test of $H_0 : \boldsymbol{\psi} = \boldsymbol{\psi}_0$ Against $H_2 : \boldsymbol{\psi} \neq \boldsymbol{\psi}_0$

The discussions in the previous subsection are particularly useful because the simplicity of the null hypothesis with no nuisance parameters enabled us to bring out some important features without delicate technical details. In many practical situations, tests of hypotheses usually involve only some components of $\boldsymbol{\theta}$. Let $\boldsymbol{\theta}$ be partitioned as $(\boldsymbol{\lambda} : \boldsymbol{\psi})$; recall that when $\boldsymbol{\theta}$ is partitioned as $(\boldsymbol{\lambda}^T, \boldsymbol{\psi}^T)^T$ it would be denoted by $(\boldsymbol{\lambda} : \boldsymbol{\psi})$. Let the null and alternative hypotheses be

$$H_0 : \boldsymbol{\psi} = \boldsymbol{\psi}_0 \text{ and } H_2 : \boldsymbol{\psi} \neq \boldsymbol{\psi}_0. \tag{4.60}$$

For this testing problem, *LRT*, Wald, and score tests are known to be asymptotically equivalent.

Let $\boldsymbol{\lambda}_0$ be the true value of $\boldsymbol{\lambda}$ and $\boldsymbol{\theta}_0 = (\boldsymbol{\lambda}_0 : \boldsymbol{\psi}_0)$ be the true value of $\boldsymbol{\theta}$ when $H_0 : \boldsymbol{\psi} = \boldsymbol{\psi}_0$ holds. Different extensions of the score statistic in the previous section are available for this testing problem. Let the score function $\boldsymbol{S}(\boldsymbol{\theta})$ in (4.48) be partitioned into $(\boldsymbol{S}_\lambda : \boldsymbol{S}_\psi)$ to conform with $(\boldsymbol{\lambda} : \boldsymbol{\psi})$. Let $\{\mathcal{I}_{\lambda\lambda}, \mathcal{I}_{\lambda\psi} \mid \mathcal{I}_{\psi\lambda}, \mathcal{I}_{\psi\psi}\}$ be obtained by partitioning the information matrix $\mathcal{I}$ to conform with $(\boldsymbol{\lambda} : \boldsymbol{\psi})$; similarly, let $\{\mathcal{I}^{\lambda\lambda}, \mathcal{I}^{\lambda\psi} \mid \mathcal{I}^{\psi\lambda}, \mathcal{I}^{\psi\psi}\}$ be obtained by partitioning $\mathcal{I}^{-1}$. Let

$$\mathcal{I}_{\psi\psi.\lambda} = (\mathcal{I}_{\psi\psi} - \mathcal{I}_{\psi\lambda}\mathcal{I}_{\lambda\lambda}^{-1}\mathcal{I}_{\lambda\psi});$$

then $\mathcal{I}^{\psi\psi} = \mathcal{I}_{\psi\psi.\lambda}^{-1}$.

Let $\bar{\boldsymbol{\lambda}}$ denote the *mle* of $\boldsymbol{\lambda}$ under H_0, defined by $\boldsymbol{S}_\lambda(\bar{\boldsymbol{\theta}}) = \mathbf{0}$ where $\bar{\boldsymbol{\theta}} = (\bar{\boldsymbol{\lambda}} : \boldsymbol{\psi}_0)$; thus $\bar{\boldsymbol{\theta}}$ is the *mle* of $\boldsymbol{\theta}$ under H_0. Let a sequence of local hypotheses be defined by

$$H_n : \boldsymbol{\psi} = \boldsymbol{\psi}_0 + n^{-1/2}\boldsymbol{\delta},$$

where $\boldsymbol{\delta} \in \mathbb{R}^k$ is fixed. By employing Taylor series expansions similar to those leading to (4.53), we have

$$n^{-1/2}\boldsymbol{S}_\psi(\bar{\boldsymbol{\theta}}) \xrightarrow{d} N\{\mathcal{I}_{\psi\psi.\lambda}(\boldsymbol{\theta}_0)\boldsymbol{\delta}, \mathcal{I}_{\psi\psi.\lambda}(\boldsymbol{\theta}_0)\} \quad \text{under } H_n. \tag{4.61}$$

Therefore, a score statistic for testing $H_0 : \boldsymbol{\psi} = \boldsymbol{\psi}_0$ vs $H_2 : \boldsymbol{\psi} \neq \boldsymbol{\psi}_0$ is

$$S_L = n^{-1}[\boldsymbol{S}_\psi^T \mathcal{I}_{\psi\psi.\lambda}^{-1} \boldsymbol{S}_\psi]_{\boldsymbol{\theta}=\bar{\boldsymbol{\theta}}} \tag{4.62}$$

where the suffix $\boldsymbol{\theta} = \bar{\boldsymbol{\theta}}$ for [.] indicates that [.] is evaluated at $\boldsymbol{\theta} = \bar{\boldsymbol{\theta}}$ (see Rao (1973), p 418). It follows from (4.61) that

$$S_L \xrightarrow{d} \chi_k^2(\Delta) \text{ under } H_n, \tag{4.63}$$

where $\Delta = \boldsymbol{\delta}^T \mathcal{I}_{\psi\psi.\lambda}(\boldsymbol{\theta}_0)\boldsymbol{\delta}$. Because $\mathcal{I}(\boldsymbol{\theta}_0)$ is positive definite, it follows that $\mathcal{I}(\boldsymbol{\theta}_0)^{-1}$ and $\mathcal{I}_{\psi\psi.\lambda}(\boldsymbol{\theta}_0)$ are positive definite, and hence $\Delta > 0$ for $\boldsymbol{\delta} \neq \mathbf{0}$. Therefore, a test of $H_0 : \boldsymbol{\psi} = \boldsymbol{\psi}_0$ against $H_2 : \boldsymbol{\psi} \neq \boldsymbol{\psi}_0$ based on S_L in (4.62) rejects H_0 when it is large. This is the simple form of the score test of $\boldsymbol{\psi} = \boldsymbol{\psi}_0$ against $\boldsymbol{\psi} \neq \boldsymbol{\psi}_0$.

Because $n^{1/2}(\hat{\boldsymbol{\psi}} - \boldsymbol{\psi}_0) \xrightarrow{d} N\{\mathbf{0}, \mathcal{I}_{\psi\psi.\lambda}^{-1}(\boldsymbol{\theta}_0)\}$, the Wald statistic for testing $H_0 : \boldsymbol{\psi} = \boldsymbol{\psi}_0$ vs $H_2 : \boldsymbol{\psi} \neq \boldsymbol{\psi}_0$ is defined as

$$T_W = n(\hat{\boldsymbol{\psi}} - \boldsymbol{\psi}_0)^T \mathcal{I}_{\psi\psi.\lambda}(\hat{\boldsymbol{\theta}})(\hat{\boldsymbol{\psi}} - \boldsymbol{\psi}_0). \tag{4.64}$$

Since $n^{1/2}(\hat{\boldsymbol{\psi}} - \boldsymbol{\psi}_0) \xrightarrow{d} N\{\boldsymbol{\delta}, \mathcal{I}_{\psi\psi.\lambda}^{-1}(\boldsymbol{\theta}_0)\}$ under H_n, it follows that

$$T_W \xrightarrow{d} \chi_k^2(\Delta) \text{ under } H_n.$$

Therefore, T_W in (4.64) and S_L in (4.62) have the same asymptotic power against H_n. In fact, by arguments essentially similar to those in the previous subsection, we have

$$LRT = S_L + o_p(1) = T_W + o_p(1) \text{ under } H_0 \text{ and under } H_n.$$

Thus, the three tests have the same asymptotic distribution under the null hypothesis, and also under the sequence of local alternatives, H_n; consequently, they have the same local power as well.

4.5.3 Tests Based on Estimating Equations

As in the previous subsection, let $\boldsymbol{\theta}$ be partitioned as $(\boldsymbol{\lambda} : \boldsymbol{\psi})$, and let the null and alternative hypotheses be the same as in the previous subsection, namely

$$H_0 : \boldsymbol{\psi} = \boldsymbol{\psi}_0 \quad \text{and} \quad H_2 : \boldsymbol{\psi} \neq \boldsymbol{\psi}_0,$$

respectively. Instead of defining $\boldsymbol{S}(\boldsymbol{\theta}) = (\partial/\partial\boldsymbol{\theta})\ell(\boldsymbol{\theta})$, where $\ell(\boldsymbol{\theta})$ is the loglikelihood, a much larger class of tests can be developed by allowing $\boldsymbol{S}(\boldsymbol{\theta})$ to be a suitable estimating function. For example, for large sample inference on $\boldsymbol{\beta}$ in the linear regression model, $y = \boldsymbol{x}^T\boldsymbol{\beta} + \epsilon$, there is no need to assume that the errors are normal; the estimator of $\boldsymbol{\beta}$ defined as the solution of the normal equations can be used for inference on $\boldsymbol{\beta}$ without assuming normal error distribution. Other important special cases of the estimating equation approach includes (i) the class of quasi-likelihood equations, which are particularly useful in generalized linear models (see McCullagh and Nelder (1998, Chapter 9), and (ii) the Generalized Estimating Equation (GEE) approach, which has found use in many areas of applications. The essential results are mentioned below; they are based on Basawa (1985); for a brief summary of these see Silvapulle and Silvapulle (1995) and for an in-depth and detailed coverage of quasi-likelihood see Heyde (1997).

Let $\boldsymbol{S}(\boldsymbol{\theta}) = \boldsymbol{0}$ be a $p \times 1$ *estimating equation* for $\boldsymbol{\theta}$; $\boldsymbol{S}(\boldsymbol{\theta})$ is called an *estimating function*. First, we shall state a set of conditions that are used in the literature to ensure that the choice of the estimating function is appropriate. However, later we shall relax this and state a simpler set of high-level sufficient conditions.

Assume that the following condition is satisfied

Condition A:
There exist nonsingular matrices $\boldsymbol{G}(\boldsymbol{\theta})$ and $\boldsymbol{V}(\boldsymbol{\theta})$ such that, for $a > 0$, as $n \to \infty$,

$$n^{-1/2}\boldsymbol{S}(\boldsymbol{\theta}_0) \xrightarrow{d} N\{\boldsymbol{0}, \boldsymbol{V}(\boldsymbol{\theta}_0)\}, \tag{4.65}$$

$$\sup_{\|\boldsymbol{u}\| \le a} \|n^{-1/2}\{\boldsymbol{S}(\boldsymbol{\theta}_0 + n^{-1/2}\boldsymbol{u}) - \boldsymbol{S}(\boldsymbol{\theta}_0)\} + \boldsymbol{G}(\boldsymbol{\theta}_0)\boldsymbol{u}\| = o_p(1) \tag{4.66}$$

and the convergences in (4.65) and (4.66) are uniform in $\boldsymbol{\theta}_0$. ■

As in the previous subsection, let $\boldsymbol{\theta}_0 = (\boldsymbol{\lambda}_0 : \boldsymbol{\psi}_0)$ denote the true value of $\boldsymbol{\theta}$ when $H_0 : \boldsymbol{\psi} = \boldsymbol{\psi}_0$ holds. Let $\hat{\boldsymbol{\theta}}$ be defined by $\boldsymbol{S}(\hat{\boldsymbol{\theta}}) = \boldsymbol{0}$. Then

$$\sqrt{n}(\hat{\boldsymbol{\theta}} - \boldsymbol{\theta}_0) \xrightarrow{d} N\{\boldsymbol{0}, \boldsymbol{A}(\boldsymbol{\theta}_0)\}, \quad \text{where } \boldsymbol{A} = \boldsymbol{G}^{-1}\boldsymbol{V}\boldsymbol{G}^{-T} \text{ and } \boldsymbol{G}^{-T} = (\boldsymbol{G}^{-1})^T.$$

We can use $\hat{\boldsymbol{\theta}}$ or $\boldsymbol{S}(\boldsymbol{\theta})$ for testing hypotheses about $\boldsymbol{\theta}$. Let $(\boldsymbol{S}_\lambda : \boldsymbol{S}_\psi)$ be the partition of $\boldsymbol{S}$ to conform with $\boldsymbol{\theta} = (\boldsymbol{\lambda} : \boldsymbol{\psi})$. Similarly, let the partition of $\boldsymbol{G}$ and $\boldsymbol{V}$ be $\{\boldsymbol{G}_{\lambda\lambda}, \boldsymbol{G}_{\lambda\psi} \mid \boldsymbol{G}_{\psi\lambda}, \boldsymbol{G}_{\psi\psi}\}$ and $\{\boldsymbol{V}_{\lambda\lambda}, \boldsymbol{V}_{\lambda\psi}, \mid \boldsymbol{V}_{\psi\lambda}, \boldsymbol{V}_{\psi\psi}\}$, respectively. Define

$$\boldsymbol{C}(\boldsymbol{\theta}) = [(\boldsymbol{V}_{\psi\psi} - \boldsymbol{G}_{\psi\lambda}\boldsymbol{G}_{\lambda\lambda}^{-1}\boldsymbol{V}_{\lambda\psi}) - (\boldsymbol{V}_{\lambda\psi}^T - \boldsymbol{G}_{\psi\lambda}\boldsymbol{G}_{\lambda\lambda}^{-1}\boldsymbol{V}_{\lambda\lambda})\boldsymbol{G}_{\lambda\lambda}^{-T}\boldsymbol{G}_{\psi\lambda}^T]_{\boldsymbol{\theta}}. \tag{4.67}$$

$$\text{and} \qquad \boldsymbol{Z}(\boldsymbol{\theta}) = n^{-1/2}[\boldsymbol{S}_\psi - \boldsymbol{G}_{\psi\lambda}\boldsymbol{G}_{\lambda\lambda}^{-1}\boldsymbol{S}_\lambda]_{\boldsymbol{\theta}}. \tag{4.68}$$

The crucial result that enables us to construct local score-type tests is the following:

Proposition 4.5.1 *Let $\bar{\boldsymbol{\lambda}}$ be the estimator of $\boldsymbol{\lambda}$ under H_0 defined by $\boldsymbol{S}_{\boldsymbol{\lambda}}(\bar{\boldsymbol{\lambda}} : \boldsymbol{\psi}_0) = \boldsymbol{0}$. Further, let $\bar{\boldsymbol{\theta}} = (\bar{\boldsymbol{\lambda}} : \boldsymbol{\psi}_0)$ and let the sequence of local hypotheses be defined by $H_n : \boldsymbol{\psi} = \boldsymbol{\psi}_0 + n^{-1/2}\boldsymbol{\delta}$ where $\boldsymbol{\delta}$ is fixed. Then,*

1. $\boldsymbol{Z}(\bar{\boldsymbol{\theta}}) = \boldsymbol{Z}(\boldsymbol{\theta}_0) + o_p(1)$ *under* H_0.

2. *$\boldsymbol{Z}(\boldsymbol{\theta}_0)$ and $\boldsymbol{Z}(\bar{\boldsymbol{\theta}}) \xrightarrow{d} N\{\boldsymbol{G}_{\psi\psi.\lambda}(\boldsymbol{\theta}_0)\boldsymbol{\delta}, \boldsymbol{C}(\boldsymbol{\theta}_0)\}$ under H_n.*

3. *For testing $\boldsymbol{\psi} = \boldsymbol{\psi}_0$ against $\boldsymbol{\psi} \neq \boldsymbol{\psi}_0$ a (local) score-type statistic is*

$$S_L = [\boldsymbol{Z}^T \boldsymbol{C}^{-1} \boldsymbol{Z}]_{\bar{\boldsymbol{\theta}}}. \tag{4.69}$$

The asymptotic distribution of S_L in (4.69), under H_n, is $\chi^2_k(\Delta)$, where $\Delta = \boldsymbol{\delta}^T [\boldsymbol{G}^T_{\psi\psi.\lambda} \boldsymbol{C}^{-1} \boldsymbol{G}_{\psi\psi.\lambda}]_{\boldsymbol{\theta}_0} \boldsymbol{\delta}$.

Proof: Let $\boldsymbol{B}(\boldsymbol{\theta}) = [-\boldsymbol{G}_{\psi\lambda}\boldsymbol{G}^{-1}_{\lambda\lambda}, \boldsymbol{I}]_{\boldsymbol{\theta}}$. Then $\boldsymbol{Z}(\boldsymbol{\theta}) = n^{-1/2}\boldsymbol{B}(\boldsymbol{\theta})\boldsymbol{S}(\boldsymbol{\theta})$ is a linear function of the statistic $\boldsymbol{S}(\boldsymbol{\theta})$. It follows from (4.65) that $\boldsymbol{Z}(\boldsymbol{\theta}_0) \xrightarrow{d} N\{\boldsymbol{0}, \boldsymbol{C}(\boldsymbol{\theta}_0)\}$ under H_0. One term Taylor expansion of $\boldsymbol{Z}(\bar{\boldsymbol{\theta}})$ about $\boldsymbol{\theta}_0$ leads to

$$\boldsymbol{Z}(\bar{\boldsymbol{\theta}}) - \boldsymbol{Z}(\boldsymbol{\theta}_0) \xrightarrow{p} \boldsymbol{0} \quad \text{under } H_0.$$

It follows from (4.65) and (4.66) that $n^{-1/2}\boldsymbol{S}(\boldsymbol{\theta}_0) \xrightarrow{d} N\{\boldsymbol{G}(\boldsymbol{\theta}_0)\boldsymbol{\Delta}, \boldsymbol{V}(\boldsymbol{\theta}_0)\}$ under $H_n : \boldsymbol{\psi} = \boldsymbol{\psi}_0 + n^{-1/2}\boldsymbol{\delta}$, where $\boldsymbol{\Delta} = (\boldsymbol{0} : \boldsymbol{\delta})$ and therefore,

$$\boldsymbol{Z}(\bar{\boldsymbol{\theta}}) \xrightarrow{d} N\{\boldsymbol{G}_{\psi\psi.\lambda}(\boldsymbol{\theta}_0)\boldsymbol{\delta}, \boldsymbol{C}(\boldsymbol{\theta}_0)\} \text{ under } H_n : \boldsymbol{\psi} = \boldsymbol{\psi}_0 + n^{-1/2}\boldsymbol{\delta}. \tag{4.70}$$

■

Now, consider the special case when $\boldsymbol{S}(\boldsymbol{\theta})$ is the score function corresponding to loglikelihood. Let $\bar{\boldsymbol{S}} = \boldsymbol{S}(\bar{\boldsymbol{\theta}})$ and $\bar{\mathcal{I}} = \mathcal{I}(\bar{\boldsymbol{\theta}})$. Then $\boldsymbol{G}(\boldsymbol{\theta}) = \boldsymbol{V}(\boldsymbol{\theta}) = \mathcal{I}(\boldsymbol{\theta})$, and

$$\boldsymbol{Z}(\bar{\boldsymbol{\theta}}) = n^{-1/2}(\bar{\boldsymbol{S}}_\psi - \bar{\mathcal{I}}_{\psi\lambda}\bar{\mathcal{I}}^{-1}_{\lambda\lambda}\bar{\boldsymbol{S}}_\lambda) = n^{-1/2}\bar{\boldsymbol{S}}_{\boldsymbol{\psi}}.$$

In this case, the S_L in (4.69) reduces to Rao's score statistic (4.62) (see Rao (1973), p 418),

$$n^{-1}\bar{\boldsymbol{S}}^T_{\boldsymbol{\psi}}\bar{\mathcal{I}}^{-1}_{\psi\psi.\lambda}\bar{\boldsymbol{S}}_\psi.$$

It follows from *Condition A* given at the beginning of this subsection that

$$n^{-1/2}\boldsymbol{G}^{-1}(\boldsymbol{\theta}_0)\boldsymbol{S}(\boldsymbol{\theta}_0) = n^{1/2}(\hat{\boldsymbol{\theta}} - \boldsymbol{\theta}_0) + o_p(1).$$

Because of this asymptotic linear relationship between $\boldsymbol{S}(\boldsymbol{\theta}_0)$ and $(\hat{\boldsymbol{\theta}} - \boldsymbol{\theta}_0)$ one would conjecture that a test based on $\boldsymbol{S}(\boldsymbol{\theta}_0)$ and one based on $\hat{\boldsymbol{\theta}}$ are likely to be equivalent; this in fact is the case, as will be seen. It follows from *Condition A* that

$$n^{1/2}(\hat{\boldsymbol{\theta}} - \boldsymbol{\theta}_0) \xrightarrow{d} N\{\boldsymbol{0}, \boldsymbol{A}(\boldsymbol{\theta}_0)\}, \quad \text{where } \boldsymbol{A} = [\boldsymbol{G}^{-1}\boldsymbol{V}\boldsymbol{G}^{-T}].$$

Therefore,

$$\sqrt{n}(\hat{\boldsymbol{\psi}} - \boldsymbol{\psi}_0) \xrightarrow{d} N\{\boldsymbol{0}, \boldsymbol{A}_{\psi\psi}(\boldsymbol{\theta}_0)\},$$

and hence a Wald-type statistic for testing $H_0 : \boldsymbol{\psi} = \boldsymbol{\psi}_0$ against $H_1 : \boldsymbol{\psi} \neq \boldsymbol{\psi}_0$ is

$$T_W = n(\hat{\boldsymbol{\psi}} - \boldsymbol{\psi}_0)^T(\hat{\boldsymbol{A}}_{\psi\psi})^{-1}(\hat{\boldsymbol{\psi}} - \boldsymbol{\psi}_0). \tag{4.71}$$

where $\hat{A}_{\psi\psi}$ is a consistent estimator of $A_{\psi\psi}$, for example, $A_{\psi\psi}(\hat{\theta})$. Now we have the following:

Proposition 4.5.2 *Under $H_n : \psi = \psi_0 + n^{-1/2}\delta$, S_L in (4.69) and T_W in (4.71) converge to $\chi^2_k(\Delta)$ where $\Delta = \delta^T G^T_{\psi\psi.\lambda} C^{-1} G_{\psi\psi.\lambda}\delta$.* ■

The definition of S_L in (4.69) depends on $\bar{\theta}$, which needs to be obtained by solving $S_\lambda(\theta) = 0$ under H_0. Similarly, T_W depends on $\hat{\theta}$, which is defined by $S(\hat{\theta}) = 0$. In large samples, the $\bar{\theta}$ and $\hat{\theta}$ in these definitions can be replaced by one-step estimators, which may be significantly easier to compute in complicated models. These one-step estimators are obtained by applying Newton-Raphson iteration once to a suitably chosen starting value. Usually they are asymptotically as efficient as the fully iterated solution and the difference between them is of order $o_p(n^{-1/2})$. The main steps are indicated below.

Let $t_{n\lambda}$ be a preliminary estimator of λ such that $n^{1/2}(t_{n\lambda} - \lambda_0) = O_p(1)$ and let $\bar{\lambda}^* = t_{n\lambda} + n^{-1}G^{-1}_{\lambda\lambda}(\theta^\dagger)S_\lambda(\theta^\dagger)$, where $\theta^\dagger = (t_{n\lambda} : \psi_0)$, and $\bar{\theta}^* = (\bar{\lambda}^* : \psi_0)$. Then, under H_0, $Z(\bar{\theta}^*) - Z(\bar{\theta}) = o_p(1)$. Consequently, the $\bar{\theta}$ in (4.69) can be replaced by $\bar{\theta}^*$ without affecting the large sample distribution of S_L under H_0. More details and references to the results of this subsection may be found in Basawa (1985).

4.6 LOCAL SCORE-TYPE TESTS OF $H_0 : \psi = 0$ AGAINST $H_1 : \psi \in \Psi$

In this section, the main results mentioned in section 4.5 will be used to develop a local score-type test of $\psi = 0$ against $\psi \in \Psi$ where Ψ is a given subset of $\mathbb{R}^k$ and $\theta = (\lambda : \psi)$. Initially, we shall consider the case when $\Psi = \mathcal{C}$, where $\mathcal{C}$ is a closed convex cone. In the last subsection, we shall consider the case when Ψ is not necessarily a cone. We shall consider the settings in subsections 4.5.1, 4.5.2, and 4.5.3 in turn. To introduce the ideas, we will first consider the special case of testing $\theta = 0$ against $\theta \in \mathcal{C}$ where $\mathcal{C}$ is a cone, and provide sufficient motivation for the definition of a suitable statistic. This is then extended in subsection 4.6.2 to the case when nuisance parameters are present. Then in subsection 4.6.3 we extend these and define a statistic based on estimating equations. In all these, the score-type statistic will be denoted by S_L, but the setting will become progressively more general. In the final subsection, it will be general enough to incorporate quasi-likelihood, partial likelihood, and estimating equations. This section is based on Silvapulle and Silvapulle (1995) with particular attention to the local score type tests introduced there.

4.6.1 Local Score Test of $H_0 : \theta = 0$ Against $H_1 : \theta \in \mathcal{C}$

Let $f(y; \theta)$ be a density function and let $Y_1, \ldots, Y_n$ be independently and identically distributed with common density function $f(y; \theta)$ where $\theta \in \Theta \subset \mathbb{R}^p$. Assume that $\mathbf{0}$ is an interior point of Θ. Let $\mathcal{I}_0$ denote $\mathcal{I}(\mathbf{0})$. Assume that Condition Q in subsection

4.2 is satisfied. Let the null and alternative hypotheses be

$$H_0 : \boldsymbol{\theta} = \mathbf{0} \text{ and } H_1 : \boldsymbol{\theta} \in \mathcal{C}, \tag{4.72}$$

respectively, where $\mathcal{C}$ is a closed convex cone. Let $\boldsymbol{S}(\boldsymbol{\theta})$ denote the score function, $(\partial/\partial\boldsymbol{\theta})\ell(\boldsymbol{\theta})$, where $\ell(\boldsymbol{\theta})$ is the loglikelihood, and let

$$\boldsymbol{U} = n^{-1/2}\mathcal{I}_0^{-1}\boldsymbol{S}(\mathbf{0}). \tag{4.73}$$

Let $\boldsymbol{\delta} \in \mathbb{R}^p$ be a fixed point and $\boldsymbol{\theta}_n = n^{-1/2}\boldsymbol{\delta}$. It follows from (4.53) that

$$\boldsymbol{U} \xrightarrow{d} N(\boldsymbol{\delta}, \mathcal{I}_0^{-1}) \text{ under } H_n : \boldsymbol{\theta} = \boldsymbol{\theta}_n. \tag{4.74}$$

This can be used as a guide to constructing a test of $\boldsymbol{\theta} = \mathbf{0}$ against $H_1 : \boldsymbol{\theta} \in \mathcal{C}$. To this end, suppose that the true value of $\boldsymbol{\theta}$ is $n^{-1/2}\boldsymbol{\delta}$. Because $\boldsymbol{\theta} \in \mathcal{C}$ is equivalent to $\boldsymbol{\delta} \in \mathcal{C}$, it may be adequate to construct a test of $\boldsymbol{\delta} = \mathbf{0}$ against $\boldsymbol{\delta} \in \mathcal{C}$ for testing $\boldsymbol{\theta} = \mathbf{0}$ against $\boldsymbol{\theta} \in \mathcal{C}$. In view of (4.74), $\boldsymbol{U}$ can be regarded as an estimator of $\boldsymbol{\delta}$. If the limiting distribution of $\boldsymbol{U}$ in (4.74) were the exact distribution of $\boldsymbol{U}$ for every n under H_n, then the likelihood ratio statistic for testing $\boldsymbol{\delta} = \mathbf{0}$ against $\boldsymbol{\delta} \in \mathcal{C}$ based on a single observation of $\boldsymbol{U}$ would be

$$S_L = \boldsymbol{U}^T\mathcal{I}_0\boldsymbol{U} - \min\{(\boldsymbol{U} - \boldsymbol{a})^T\mathcal{I}_0(\boldsymbol{U} - \boldsymbol{a}) : \boldsymbol{a} \in \mathcal{C}\}. \tag{4.75}$$

The arguments leading to this definition of S_L are based on the local properties of $\boldsymbol{U}$ and $\boldsymbol{S}(\mathbf{0})$. Therefore, these motivations are closely related to those leading to the classical Rao's score statistic. We shall refer to S_L as the *local score statistic*.

The foregoing local arguments were used only as a motivation and guide to constructing S_L. Whether or not it is a good statistic would depend on its properties and performance. Therefore, it would be of interest to compare the asymptotic performance of S_L with other statistics. Because $n^{1/2}(\hat{\boldsymbol{\theta}} - \boldsymbol{\theta}_0) \xrightarrow{d} N\{\mathbf{0}, \mathcal{I}(\boldsymbol{\theta}_0)^{-1}\}$, where $\boldsymbol{\theta}_0$ is the true value, a distance statistic for testing $\boldsymbol{\theta} = \mathbf{0}$ against $\boldsymbol{\theta} \in \mathcal{C}$ is

$$D = n[\hat{\boldsymbol{\theta}}^T\hat{\mathcal{I}}\hat{\boldsymbol{\theta}} - \min\{(\hat{\boldsymbol{\theta}} - \boldsymbol{a})^T\hat{\mathcal{I}}(\hat{\boldsymbol{\theta}} - \boldsymbol{a}) : \boldsymbol{a} \in \mathcal{C}\}], \tag{4.76}$$

where $\hat{\mathcal{I}}$ is a consistent estimator of $\mathcal{I}_0$. It follows from (4.56), (4.75), and (4.76) that $S_L = D + o_p(1)$ under H_0. Therefore, S_L and D are asymptotically equivalent for testing $\boldsymbol{\theta} = \mathbf{0}$ against $\boldsymbol{\theta} \in \mathcal{C}$. These are summarized in the following proposition.

Proposition 4.6.1 *Assume that Condition Q is satisfied. Let $\boldsymbol{U}, S_L, D$, and the hypotheses be as in (4.73), (4.75), (4.76), and (4.72), respectively. Then, under H_0,*

$$LRT = S_L + o_p(1) = D + o_p(1). \tag{4.77}$$

Further, the common asymptotic null distribution of LRT, S_L and D is $\bar{\chi}^2(\mathcal{I}_0^{-1}, \mathcal{C})$.

A consequence of this result is that the local score statistic, the likelihood ratio statistic, and the distance statistic have the same asymptotic null distribution. Further, they also have the same local power if (4.77) holds under $H_n : \boldsymbol{\theta} = n^{-1/2}\boldsymbol{\delta}$, which would be the case under fairly general conditions. In this sense all three tests are asymptotically equivalent. Since there are no nuisance parameters under H_0, S_L may be computed without estimating the model. By contrast, the *LRT* requires the model to be estimated subject to the inequality constraints of H_1, and D requires estimation of the full model without the constraints.

Since the Rao score statistic is based on the score $\boldsymbol{S}(\mathbf{0})$ as the basic building block and thinking of it as the slope of $\ell(\boldsymbol{\theta})$ at the null value, one might be tempted to define a score statistic for testing $\boldsymbol{\theta} = \mathbf{0}$ vs $\boldsymbol{\theta} \in \mathcal{C}$ by replacing the role of $\boldsymbol{U}$ in (4.75) by $\boldsymbol{S}(\mathbf{0})$. This leads to the statistic

$$R^* = \boldsymbol{S}_0^T \mathcal{I}_0^{-1} \boldsymbol{S}_0 - \min\{(\boldsymbol{S}_0 - \boldsymbol{a})^T \mathcal{I}_0^{-1} (\boldsymbol{S}_0 - \boldsymbol{a}) : \boldsymbol{a} \in \mathcal{C}\}.$$

Since $n^{-1/2}\boldsymbol{S}(\mathbf{0}) \xrightarrow{d} N(\mathcal{I}_0 \boldsymbol{\delta}, \mathcal{I}_0)$ under H_n, this essentially assumes that a test of $\boldsymbol{\theta} = \mathbf{0}$ vs $\mathcal{I}_0 \boldsymbol{\theta} \in \mathcal{C}$ would also be suitable for testing $\boldsymbol{\theta} = \mathbf{0}$ vs $\boldsymbol{\theta} \in \mathcal{C}$. Since this is not a reasonable assumption, S_L appears to be better than R^*.

4.6.2 Local Score Test of $H_0 : \psi = 0$ Against $H_1 : \psi \in \mathcal{C}$

It is very rarely that a null hypothesis would specify a value for every component of $\boldsymbol{\theta}$. Therefore, the testing problem considered in the previous subsection is very rare, although the discussions therein were meant to be instructive. In this subsection, we consider the more realistic setting in which the hypotheses impose constraints on some components only. The main ideas leading to the foregoing local score statistic, S_L, can also be applied when there are nuisance parameters.

Let $\boldsymbol{\theta} = (\boldsymbol{\lambda} : \boldsymbol{\psi})$, where $\boldsymbol{\psi}$ is $k \times 1$, and let the null and alternative hypotheses be

$$H_0 : \boldsymbol{\psi} = \mathbf{0} \quad \text{and} \quad H_1 : \boldsymbol{\psi} \in \mathcal{C}, \tag{4.78}$$

respectively, where $\mathcal{C}$ is a closed convex cone in $\mathbb{R}^k$. It is convenient to adopt the notation in subsection 4.5.2. Let $\boldsymbol{S}(\boldsymbol{\theta})$ denote the score function $(\partial/\partial\boldsymbol{\theta})\ell(\boldsymbol{\theta})$ and let it also be partitioned as $(\boldsymbol{S}_\lambda : \boldsymbol{S}_\psi)$. Let $\bar{\boldsymbol{\lambda}}$ denote the *mle* of $\boldsymbol{\lambda}$ under H_0 defined by $\boldsymbol{S}_\lambda(\bar{\boldsymbol{\lambda}} : \mathbf{0}) = \mathbf{0}$ and let $\bar{\boldsymbol{\theta}} = (\bar{\boldsymbol{\lambda}} : \mathbf{0})$ be the *mle* of $\boldsymbol{\theta}$ under H_0. Define

$$\boldsymbol{U} = n^{-1/2} \mathcal{I}^{\psi\psi}(\bar{\boldsymbol{\theta}}) \boldsymbol{S}_\psi(\bar{\boldsymbol{\theta}}). \tag{4.79}$$

Let $\boldsymbol{\delta}$ be a fixed point in $\mathbb{R}^k$. Then, since $\mathcal{I}_{\psi\psi.\lambda} = (\mathcal{I}_{\psi\psi} - \mathcal{I}_{\psi\lambda}\mathcal{I}_{\lambda\lambda}^{-1}\mathcal{I}_{\lambda\psi})$ and $(\mathcal{I}_{\psi\psi})^{-1} = \mathcal{I}_{\psi\psi.\lambda}$, we have that (see (4.61))

$$\boldsymbol{U} \xrightarrow{d} N\{\boldsymbol{\delta}, \mathcal{I}_{\psi\psi.\lambda}^{-1}(\boldsymbol{\theta}_0)\}, \text{ under } H_n : \boldsymbol{\psi} = n^{-1/2}\boldsymbol{\delta}, \tag{4.80}$$

where $\boldsymbol{\theta}_0 = (\boldsymbol{\lambda}_0 : \mathbf{0})$, the true value of $\boldsymbol{\theta}$ under H_0. Now, by arguments similar to those leading to (4.75), we define the *local score statistic* as

$$S_L = \boldsymbol{U}^T \bar{\mathcal{I}}_{\psi\psi.\lambda} \boldsymbol{U} - \inf\{(\boldsymbol{U} - \boldsymbol{a})^T \bar{\mathcal{I}}_{\psi\psi.\lambda} (\boldsymbol{U} - \boldsymbol{a}) : \boldsymbol{a} \in \mathcal{C}\} \tag{4.81}$$

for testing (4.78), where $\bar{\mathcal{I}} = \mathcal{I}(\bar{\boldsymbol{\theta}})$.

Because $n^{1/2}(\hat{\boldsymbol{\psi}} - \boldsymbol{\psi}_0) \xrightarrow{d} N\{\mathbf{0}, \mathcal{I}^{-1}_{\psi\psi\cdot\boldsymbol{\lambda}}(\boldsymbol{\theta}_0)\}$, a distance statistic for testing $H_0 : \boldsymbol{\psi} = \mathbf{0}$ vs $H_1 : \boldsymbol{\psi} \in \mathcal{C}$ is

$$D = n[\hat{\boldsymbol{\psi}}^T \hat{\mathcal{I}}_{\psi\psi.\lambda} \hat{\boldsymbol{\psi}} - \min\{(\hat{\boldsymbol{\psi}} - \boldsymbol{a})^T \hat{\mathcal{I}}_{\psi\psi.\lambda} (\hat{\boldsymbol{\psi}} - \boldsymbol{a}) : \boldsymbol{a} \in \mathcal{C}\}] \tag{4.82}$$

where $\hat{\mathcal{I}}_{\psi\psi\cdot\boldsymbol{\lambda}} = \mathcal{I}_{\psi\psi\cdot\boldsymbol{\lambda}}(\hat{\boldsymbol{\theta}})$ which is a consistent estimator of $\mathcal{I}_{\psi\psi\cdot\boldsymbol{\lambda}}(\boldsymbol{\theta}_0)$. Now, we have the following result, which says that the local score statistic S_L in (4.81) is asymptotically equivalent to the likelihood ratio and distance statistics; see Silvapulle and Silvapulle (1995) for a proof.

Proposition 4.6.2 *Let $\boldsymbol{U}, S_L, D$, and the hypotheses be as in (4.79), (4.81), (4.82), and (4.78), respectively. Then, under H_0,*

$$LRT = S_L + o_p(1) = D + o_p(1). \tag{4.83}$$

Further, the common asymptotic null distribution of LRT, S_L, and D is $\bar{\chi}^2(\mathcal{I}^{-1}_{\psi\psi.\lambda}(\boldsymbol{\theta}_0), \mathcal{C})$.

If the alternative hypothesis in (4.78) is $H_1 : \boldsymbol{\psi} \neq \mathbf{0}$ then S_L in (4.81) reduces to the Rao score statistic; further, since S_L is asymptotically equivalent to the *LRT* and they both are based on the same building block, the S_L in (4.81) can be considered as a natural generalization of Rao's score statistic.

4.6.3 Local Sore-Type Test Based on Estimating Equations

By employing arguments similar to those in the previous subsection, the main results in subsection 4.5.3 can be extended to introduce a test based on estimating equations. Let $\boldsymbol{\theta} = (\boldsymbol{\lambda} : \boldsymbol{\psi})$ where $\boldsymbol{\psi}$ is $k \times 1$ and $\boldsymbol{S}(\boldsymbol{\theta}) = \mathbf{0}$ be an estimating equation for $\boldsymbol{\theta}$. Initially, we shall assume that Condition A on page 173 is satisfied. In the next subsection, we will relax this further to develop a large class of tests. Consequently, the procedure becomes more flexible; for example, quasi-likelihood, partial likelihood, and Generalized Estimating Equations (GEE) are incorporated.

Let the null and alternative hypotheses be

$$H_0 : \boldsymbol{\psi} = \mathbf{0} \text{ and } H_1 : \boldsymbol{\psi} \in \mathcal{C}, \tag{4.84}$$

respectively, where $\mathcal{C}$ is a closed convex cone and $\boldsymbol{\theta}_0 = (\boldsymbol{\lambda}_0 : \mathbf{0})$, the true value of $\boldsymbol{\theta}$ when H_0 is true. Let $\bar{\boldsymbol{\theta}} = (\bar{\boldsymbol{\lambda}} : \mathbf{0})$ denote the estimator obtained by solving $\boldsymbol{S}_{\boldsymbol{\lambda}}(\boldsymbol{\lambda} : \mathbf{0}) = \mathbf{0}$; i.e., solve $\boldsymbol{S}_{\boldsymbol{\lambda}}(\boldsymbol{\theta}) = \mathbf{0}$ under H_0. Let

$$\boldsymbol{U} = \bar{\boldsymbol{G}}^{-1}_{\psi\psi.\lambda} \bar{\boldsymbol{Z}}, \tag{4.85}$$

where $\bar{\boldsymbol{G}} = \boldsymbol{G}(\bar{\boldsymbol{\theta}})$, $\bar{\boldsymbol{Z}} = \boldsymbol{Z}(\bar{\boldsymbol{\theta}})$, and $\boldsymbol{G}(.)$ and $\boldsymbol{Z}(.)$ are as in subsection 4.5.3. Then, by Proposition 4.5.1,

$$\boldsymbol{U} \xrightarrow{d} N\{\boldsymbol{\delta}, \boldsymbol{A}_{\psi\psi}(\boldsymbol{\theta}_0)\} \quad \text{under } H_n : \boldsymbol{\psi} = n^{-1/2}\boldsymbol{\delta}, \tag{4.86}$$

where $\boldsymbol{A}_{\psi\psi} = \boldsymbol{G}^{-1}_{\psi\psi.\lambda}\boldsymbol{C}(\boldsymbol{G}^{-1}_{\psi\psi.\lambda})^T$. Therefore, we define the local score-type test statistic for testing $H_0 : \psi = \mathbf{0}$ against $H_1 : \psi \in \mathcal{C}$ as

$$S_L = \boldsymbol{U}^T\bar{\boldsymbol{A}}^{-1}_{\psi\psi}\boldsymbol{U} - \min\{(\boldsymbol{U} - \boldsymbol{a})^T\bar{\boldsymbol{A}}^{-1}_{\psi\psi}(\boldsymbol{U} - \boldsymbol{a}) : \boldsymbol{a} \in \mathcal{C}\}, \tag{4.87}$$

where $\bar{\boldsymbol{A}}_{\psi\psi} = \boldsymbol{A}_{\psi\psi}(\bar{\boldsymbol{\theta}})$. Further, the limiting distribution of S_L is $\bar{\chi}^2\{\boldsymbol{A}_{\psi\psi}(\boldsymbol{\theta}_0), \mathcal{C}\}$ under H_0.

If $\boldsymbol{S}(\boldsymbol{\theta})$ is the likelihood score function $(\partial/\partial\boldsymbol{\theta})\ell(\boldsymbol{\theta})$, then $\boldsymbol{G}(\boldsymbol{\theta}) = \boldsymbol{V}(\boldsymbol{\theta}) = \mathcal{I}(\boldsymbol{\theta})$ and

$$\boldsymbol{U} = n^{1/2}\bar{\mathcal{I}}^{\psi\psi}(\bar{\boldsymbol{S}}_\psi - \bar{\mathcal{I}}_{\psi\lambda}\bar{\mathcal{I}}^{-1}_{\lambda\lambda}\bar{\boldsymbol{S}}_\lambda),$$

where $\bar{\boldsymbol{S}} = \boldsymbol{S}(\bar{\boldsymbol{\theta}})$ and $\bar{\mathcal{I}} = \mathcal{I}(\bar{\boldsymbol{\theta}})$; consequently, the S_L in (4.87) reduces to the local score statistic in the previous subsection (see 4.81). Further, if the alternative hypothesis is $\psi \neq \mathbf{0}$ then S_L reduces to

$$n^{-1}\{\bar{\mathcal{I}}^{\psi\psi}(\bar{\boldsymbol{S}}_\psi - \bar{\mathcal{I}}_{\psi\lambda}\bar{\mathcal{I}}^{-1}_{\lambda\lambda}\bar{\boldsymbol{S}}_\lambda)\}^T\bar{\mathcal{I}}^{-1}_{\psi\psi.\lambda}\{\bar{\mathcal{I}}^{\psi\psi}(\bar{\boldsymbol{S}}_\psi - \bar{\mathcal{I}}_{\psi\lambda}\bar{\mathcal{I}}^{-1}_{\lambda\lambda}\bar{\boldsymbol{S}}_\lambda)\},$$

which is Neyman's $C(\alpha)$-statistic. In view of these observations, the local score-type statistic, S_L, in (4.87) may be seen as a generalization of the $C(\alpha)$-statistic. Since $\boldsymbol{S}(\boldsymbol{\theta})$ is not necessarily the likelihood score function, the S_L in (4.87) and the *LRT* may not be equivalent.

4.6.4 A General Local Score-Type Test of $H_0 : \psi = 0$

This section assumes familiarity with the definition and properties of *approximating cone* introduced later in this chapter. However, the topic of this subsection fits in better in this section, because it is a direct extension of the results in the previous subsections. An approximating cone is purely a mathematical object. The introduction to this topic in Section 4.7 can be read independently; readers who are not familiar with it (i.e., approximating cone) may wish to read about it before reading the rest of this section.

So far we have assumed that the alternative hypothesis is of the form $\psi \in \mathcal{C}$ where $\mathcal{C}$ is a closed convex cone. The results in the previous subsection extend in a natural way to provide a large class of tests of $\psi = \mathbf{0}$ against quite general alternatives based on quantities that are not necessarily the likelihood score function. Let $\mathbf{0} \in \Psi \subset \mathbb{R}^k$, and let the testing problem be

$$H_0 : \psi = \mathbf{0} \quad \text{vs} \quad H_1 : \psi \in \Psi.$$

Although this formulation allows $\mathbf{0}$ to be an interior point of Ψ, our main focus will be on the case when it is on the boundary of Ψ. Let $\mathcal{A}$ denote the *approximating cone* of Ψ at the null value $\mathbf{0}$. If the null hypothesis is formulated as $H_0 : \psi = \psi_0$, then $\mathcal{A}$ should be the approximating cone of Ψ at ψ_0. Although we will continue to use the notation $\boldsymbol{\theta} = (\boldsymbol{\lambda} : \psi)$, here we allow $\boldsymbol{\lambda}$ to be an infinite dimensional nuisance parameter. Thus, *semiparametric* models are also incorporated in this subsection. Let $\boldsymbol{\theta}_0 = (\boldsymbol{\lambda}_0 : \mathbf{0})$ and let $\boldsymbol{\delta} \in \mathcal{A}$ be arbitrary but fixed. Now, let us introduce the following condition:

Condition LS:

1. There exists a statistic $\boldsymbol{U}$ (i.e., a function of the data only) and a positive definite matrix $\boldsymbol{D}(\boldsymbol{\theta}_0)$ such that

$$\boldsymbol{U} \xrightarrow{d} N\{\boldsymbol{\delta}, \boldsymbol{D}(\boldsymbol{\theta}_0)\} \text{ under } H_n : \boldsymbol{\psi} = n^{-1/2}\boldsymbol{\delta}.$$

2. There exists a consistent estimator, $\bar{\boldsymbol{D}}$ of $\boldsymbol{D}(\boldsymbol{\theta}_0)$. ■

Assume that this condition is satisfied. Then, a local score-type statistic for testing $H_0 : \boldsymbol{\psi} = \mathbf{0}$ against $H_1 : \boldsymbol{\psi} \in \Psi$ is

$$S_L = \boldsymbol{U}^T \bar{\boldsymbol{D}}^{-1} \boldsymbol{U} - \inf_{\boldsymbol{b} \in \mathcal{A}} (\boldsymbol{U} - \boldsymbol{b})^T \bar{\boldsymbol{D}}^{-1} (\boldsymbol{U} - \boldsymbol{b}). \tag{4.88}$$

This is the general form of the local score-type statistic introduced in Silvapulle and Silvapulle (1995) and further developed in Silvapulle et al. (2002) for the semiparametric model. If $\mathcal{A}$ is a closed convex cone then the limiting distribution of S_L is given by

$$S_L \xrightarrow{d} \bar{\chi}^2\{\boldsymbol{D}(\boldsymbol{\theta}_0), \mathcal{A}\} \text{ under } H_0.$$

The asymptotic distribution of S_L under H_n is equal to the expression on the right-hand side (4.88) with $\boldsymbol{U} \sim N\{\boldsymbol{\delta}, \boldsymbol{D}(\boldsymbol{\theta}_0)\}$ and $\bar{\boldsymbol{D}}$ replaced by $\boldsymbol{D}(\boldsymbol{\theta}_0)$. If $\boldsymbol{\lambda}$ is finite dimensional and $\boldsymbol{U}$ is the efficient score then S_L is asymptotically equivalent to the likelihood ratio test [see Theorem 1 in Silvapulle and Silvapulle (1995)].

The test statistic in (4.88) provides a very flexible class of simple tests because its implementation requires only the null model to be estimated and it is based on essentially the same building blocks as those used for the traditional score tests. The applicability of the foregoing local score test in several semiparametric models (including linear regression, AR(p), ARCH, and nonlinear regression models) in which the error distribution is treated as an unknown infinite dimensional nuisance parameter, is discussed in Silvapulle et al. (2002) where *adaptive tests* and *efficient semiparametric tests* are developed. The results in Silvapulle et al. (2002) suggest that once an adaptive test is obtained for testing against an unconstrained hypothesis, it is likely that the same test can be extended to incorporate restricted alternatives as well. The reason for this is that a semiparametric test against an unrestricted alternative is likely to be based on a $\boldsymbol{U}$ satisfying Condition LS.

4.6.5 A Simple One-Dimensional Test of $\boldsymbol{\psi} = \mathbf{0}$ Against $\boldsymbol{\psi} \in \mathcal{C}$

The local score-type statistic, S_L, in (4.87) is based on the distribution of the score vector, $\boldsymbol{S}$, and its asymptotic null distribution would typically be a chi-bar-square. Although the computation of this may be considerably easier than that of the *LRT*, it is not as easy as the usual t-ratio for a single parameter. There is a need for a test that is simpler than S_L even if it is less powerful. To this end, we introduce the following simple test.

Let the testing problem be $H_0 : \psi = \mathbf{0}$ vs $H_1 : \psi \in \mathcal{C}$, where $\mathcal{C}$ is a closed convex cone. Assume that there exists a vector $\boldsymbol{c}$ in $\mathcal{C}$ such that $\boldsymbol{c}^T\boldsymbol{x} > 0$ for every $\boldsymbol{x} \in \mathcal{C} \setminus \{\mathbf{0}\}$. For example, if $\mathcal{C}$ is the positive orthant, then $\boldsymbol{c}$ could be the center direction $(1, \ldots, 1)^T$; if $\mathcal{C}$ is a polyhedral generated by $\boldsymbol{a}_1, \ldots, \boldsymbol{a}_k$ then it suffices to find a $\boldsymbol{c}$ such that $\boldsymbol{a}_i^T\boldsymbol{c} > 0$ for $i = 1, \ldots, k$. Now, the basic idea is to adopt the working assumption that $\psi = \eta\boldsymbol{c}$, and test $\eta = 0$ against $\eta > 0$. It appears reasonable to expect that a test that can detect a departure in the direction of $\boldsymbol{c}$ would also be able to detect departures in other directions of $\mathcal{C}$ that are close to $\boldsymbol{c}$ as well.

Let $\boldsymbol{U}$ be as in (4.85). Then, it follows from (4.86) that

$$\boldsymbol{c}^T\boldsymbol{U} \xrightarrow{d} N\{\boldsymbol{c}^T\boldsymbol{\delta}, \boldsymbol{c}^T\boldsymbol{A}_{\psi\psi}(\boldsymbol{\theta}_0)\boldsymbol{c}\} \quad \text{under } H_n : \psi = n^{-1/2}\boldsymbol{\delta}. \tag{4.89}$$

Because $\boldsymbol{c}$ was chosen such that $\boldsymbol{c}^T\boldsymbol{\delta} \geqslant 0$ for $\boldsymbol{\delta} \in \mathcal{C}$, a simple test of $H_0 : \psi = \mathbf{0}$ against $H_1 : \psi \in \mathcal{C}$ rejects H_0 for large values of $T_{\boldsymbol{c}}$ where

$$T_{\boldsymbol{c}} = \boldsymbol{c}^T\boldsymbol{U}/\{\boldsymbol{c}^T\bar{\boldsymbol{A}}_{\psi\psi}\boldsymbol{c}\}^{1/2}, \tag{4.90}$$

and $\bar{\boldsymbol{A}}_{\psi\psi}$ is a consistent estimator of $\boldsymbol{A}_{\psi\psi}(\boldsymbol{\theta}_0)$. Clearly, this test is considerably simpler than the ones in the previous subsections. The cost of simplicity is that this test is unlikely to be as powerful as the local score test, S_L, on average, although $T_{\boldsymbol{c}}$ is likely to be more powerful than S_L in the direction of $\boldsymbol{c}$.

A closely related procedure was suggested by King and Wu (1997); this test has some locally optimal properties. For this test, let $\boldsymbol{S}(\boldsymbol{\theta})$ be the likelihood score function. It follows from the foregoing results that

$$n^{-1/2}\boldsymbol{c}^T\boldsymbol{S}_\psi(\bar{\boldsymbol{\theta}}) \xrightarrow{d} N\{\boldsymbol{c}^T\mathcal{I}_{\psi\psi.\boldsymbol{\lambda}}(\boldsymbol{\theta}_0)\boldsymbol{\delta}, \boldsymbol{c}^T\mathcal{I}_{\psi\psi.\boldsymbol{\lambda}}(\boldsymbol{\theta}_0)\boldsymbol{c}\} \quad \text{under } H_n : \psi = n^{-1/2}\boldsymbol{\delta}.$$

Now, suppose that the following is satisfied: $\boldsymbol{c}^T\mathcal{I}_{\psi\psi.\boldsymbol{\lambda}}(\boldsymbol{\theta}_0)\boldsymbol{\delta} > 0$ for$\boldsymbol{\delta} \in \mathcal{C} \setminus \{\mathbf{0}\}$. This condition would be satisfied if the elements of $\mathcal{I}_{\psi\psi.\boldsymbol{\lambda}}(\boldsymbol{\theta}_0)$ are nonnegative and $\boldsymbol{c} = (1, \ldots, 1)^T$. Then a test of H_0 against H_1 rejects H_0 if

$$n^{-1/2}\boldsymbol{c}^T\boldsymbol{S}_\psi(\bar{\boldsymbol{\theta}})/\{\boldsymbol{c}^T\bar{\mathcal{I}}_{\psi\psi.\boldsymbol{\lambda}}\boldsymbol{c}\}^{1/2}$$

is large compared with the critical value from the standard normal distribution. For an example to illustrate the application of this to ARCH/GARCH models in time series, see Lee and King (1993); see also King and Wu (1997).

4.6.6 A Data Example: Test Against ARCH Effect in an ARCH Model.

A common feature of many financial time series such as commodity prices, exchange rates, and stock returns is that periods of large changes tend to cluster together, and similarly periods of small changes tend to cluster together. Because high variability is associated with high risk, studying the variability of such time series is important. The AutoRegressive Conditional Heteroscedasticity (ARCH) model and its various generalizations/modifications are now widely used in modeling such volatility clustering patterns. For a comprehensive account of ARCH modeling see Gourieroux (1997).

Let $\mathcal{F}_t$ denote the information available up to time t, Y_t denote the dependent variable, and $\boldsymbol{x}_t$ be a column vector of explanatory variables. Typically, $\boldsymbol{x}_t$ includes lagged values of the dependent variable. The pth order ARCH model of Engle (1982) can be stated as

$$Y_t = \boldsymbol{x}_t^T\boldsymbol{\beta} + \epsilon_t, E(\epsilon_t|\mathcal{F}_{t-1}) = 0, \text{ and } h_t = \alpha_0 + \psi_1\epsilon_{t-1}^2 + \ldots + \psi_p\epsilon_{t-p}^2, \quad (4.91)$$

where $h_t = var(Y_t|\mathcal{F}_{t-1})$. Note that (4.91) provides a specification not only for the mean but also for the variance. If ψ_j in (4.91) is negative for some j ($j = 1, \ldots, p$), then a large value for ϵ_{t-j} would result in a negative variance for $Y_t|\mathcal{F}_{t-1}$. Therefore, the admissible range for $\psi_1, \ldots, \psi_p$ is $\{\psi_1 \geqslant 0, \ldots, \psi_p \geqslant 0\}$. In the context of this example, the following question is important: "is the variance, h_t, constant over time?" As a general rule, this testing problem is formulated as

$$H_0 : \psi_1 = \ldots = \psi_p = 0 \quad \text{against} \quad H_1 : \psi_1 \geqslant 0, \ldots, \psi_p \geqslant 0. \quad (4.92)$$

In this data example, we consider an ARCH model for an overall rate of return from stocks. Let $Y_t = \log(I_t/I_{t-1})$ where I_t is the *All Ordinaries Index (Australia)* on day t. This index is a weighted average of the prices of selected shares in Australia; it corresponds to the *Dow Jones Index* in the United States. The data that we used are for the period 5/January/1984 to 29/November/1985. This provides a total of 484 observations; see Silvapulle and Silvapulle (1995) for the data. For illustrative purposes, we chose $p = 3$; the method would be applicable for other values of p. The specific model that we consider is

$$\boldsymbol{x}_t^T\boldsymbol{\beta} = \beta_0 + \beta_1 Y_{t-1} + \ldots + \beta_4 Y_{t-4} \text{ and } h_t = \alpha_0 + \psi_1\epsilon_{t-1}^2 + \psi_2\epsilon_{t-2}^2 + \psi_3\epsilon_{t-3}^2. \quad (4.93)$$

Let $\boldsymbol{\theta} = (\boldsymbol{\lambda} : \boldsymbol{\psi})$ where $\boldsymbol{\lambda} = (\beta_0, \ldots, \beta_4, \alpha_0)^T$ and $\boldsymbol{\psi} = (\psi_1, \psi_2, \psi_3)^T$. The null and alternative hypotheses are

$$H_0 : \boldsymbol{\psi} = \mathbf{0} \text{ and } H_1 : \boldsymbol{\psi} \geqslant \mathbf{0}.$$

Let $\boldsymbol{S}(\boldsymbol{\theta}) = (\partial/\partial\boldsymbol{\theta})L(\boldsymbol{\theta})$ where $L(\boldsymbol{\theta}) = -(1/2)\sum\{\log(2\pi h_t) + \epsilon_t^2/h_t\}$; thus $L(\boldsymbol{\theta})$ is the loglikelihood if we were to assume that $Y_t|\mathcal{F}_{t-1} \sim N(\boldsymbol{x}_t^T\boldsymbol{\beta}, h_t)$. Now, let us compute the local score-type test statistic S_L in (4.87). First, we need to estimate the nuisance parameter under the null hypothesis. Therefore, we solve the equation $\boldsymbol{S}_{n\lambda}(\bar{\boldsymbol{\theta}}) = \mathbf{0}$ where $\bar{\boldsymbol{\theta}} = (\bar{\boldsymbol{\lambda}} : \mathbf{0})$. Under the null hypothesis, the error variance is constant, and therefore the null model is estimated by ordinary least squares. Thus we have that $\bar{\beta} = (\boldsymbol{X}^T\boldsymbol{X})^{-1}\boldsymbol{X}^T\boldsymbol{Y}$ and $\bar{\alpha}_0 = n^{-1}\sum(Y_t - \boldsymbol{x}_t^T\bar{\boldsymbol{\beta}})^2$. Now, $\bar{\lambda} = (\bar{\boldsymbol{\beta}} : \bar{\alpha}_0)$. Note that to apply the local score-type test statistic S_L, this is the only estimation required despite the fact that the ARCH model itself is rather involved. The next step is to compute $\bar{\boldsymbol{G}}$, $\bar{\boldsymbol{V}}$, $\bar{\boldsymbol{Z}}$, $\boldsymbol{U}$, and $\bar{\boldsymbol{A}}_{\psi\psi}$. Closed form expressions for these quantities are obtained directly from the closed-form available for $\boldsymbol{S}(\boldsymbol{\theta})$; see Silvapulle and Silvapulle (1995) for these formulas.

The computed value of $\boldsymbol{U}$ is $(3.56, 1.30, 2.20)^T$. Because every component of $\boldsymbol{U}$ is positive, $\min\{(\boldsymbol{U} - \boldsymbol{a})^T\bar{\boldsymbol{A}}_{\psi\psi}^{-1}(\boldsymbol{U} - \boldsymbol{a}) : \boldsymbol{a} \geqslant \mathbf{0}\} = 0$ and hence

$$S_L = \boldsymbol{U}^T\bar{\boldsymbol{A}}_{\psi\psi}^{-1}\boldsymbol{U} - \min\{(\boldsymbol{U} - \boldsymbol{a})^T\bar{\boldsymbol{A}}_{\psi\psi}^{-1}(\boldsymbol{U} - \boldsymbol{a}) : \boldsymbol{a} \geqslant \mathbf{0}\} = \boldsymbol{U}^T\bar{\boldsymbol{A}}_{\psi\psi}^{-1}\boldsymbol{U} = 4.36.$$

Let

$$\zeta(\boldsymbol{\lambda}) = [\sum_{i=1}^{3} w_i\{3, \boldsymbol{A}_{\psi\psi}(\boldsymbol{\theta})\}\text{pr}(\chi_j^2 \geqslant 4.36)]_{\boldsymbol{\theta}=(\boldsymbol{\lambda}:\mathbf{0})}.$$

Using the closed forms given in the last chapter for the chi-bar square weights in this expression (see (3.25)), we can compute $sup_{\lambda}\zeta(\boldsymbol{\lambda})$. However, before we do so, let us apply a bounds test. It follows from (3.23) that

$$\zeta(\boldsymbol{\lambda}) \leq 0.5\{\text{pr}(\chi_2^2 \geqslant 4.36) + \text{pr}(\chi_3^2 \geqslant 4.36)\} = 0.17.$$

This upper bound is too high to reject H_0. Therefore, we computed $p_{sup} = sup_{\lambda}\zeta(\boldsymbol{\lambda})$ and $\hat{p} = \zeta(\bar{\boldsymbol{\lambda}})$. The computed values are 0.10 and 0.085, respectively. Thus, it appears that there may be some evidence in favor of ARCH effect, but it is weak.

For testing $H_0 : \boldsymbol{\psi} = \mathbf{0}$ against $H_1 : \boldsymbol{\psi} \geqslant \mathbf{0}$, it is also valid to apply an unrestricted test ignoring the inequality constraints in H_1. To this end, let the testing problem be

$$H_0 : \boldsymbol{\psi} = \mathbf{0} \quad \text{vs} \quad H_2 : \boldsymbol{\psi} \neq \mathbf{0}.$$

Even though the ARCH model would not be well-defined with a negative value for any of the components of $\boldsymbol{\psi}$, it is valid to apply a score-type test of $H_0 : \boldsymbol{\psi} = \mathbf{0}$ against $H_2 : \boldsymbol{\psi} \neq \mathbf{0}$; in fact, this is the procedure that is found in much of the econometrics literature. The only information the score-type statistic $\boldsymbol{U}^T\bar{\boldsymbol{A}}_{\psi\psi}^{-1}\boldsymbol{U}$ uses is that it is approximately $N(0, 1)$ under H_0; if H_0 is not true then it is expected to be large. The sample value of this is

$$\boldsymbol{U}^T\bar{\boldsymbol{A}}_{\psi\psi}^{-1}\boldsymbol{U} = 4.36;$$

and its large sample p-value is $\text{pr}(\chi_3^2 \geqslant 4.36) = 0.22$. As expected, because the components of $\boldsymbol{U}$ turned out to be all positive, the statistics for the one-sided and two-sided tests have the same numerical value and the latter has a larger p-value.

4.7 APPROXIMATING CONES AND TANGENT CONES

Most of the results in the previous sections have generalizations to the case when the parameter space takes more general forms. Typically, such parameter spaces arise because the constraints are nonlinear in the parameter. It turns out that the main first-order results, namely the asymptotic distribution of an estimator, the asymptotic null distribution of a test statistic, and the asymptotic local power of a test are unchanged when the null and alternative parameter spaces are replaced by cones that approximate them at the true value of the parameter provided that the boundaries of the parameter spaces are sufficiently smooth. The objective of this section is to introduce the basics that are required to study the asymptotics when the parameter spaces are neither linear spaces nor cones; for example, the parameter space may have boundaries that are curved or the slope of the boundary may be discontinuous (i.e., kinked).

As an example, let $\Omega = \{\boldsymbol{\theta} \in \mathbb{R}^2 : \theta_2 \geqslant g_1(\theta_1) \text{ and } \theta_2 \geqslant g_2(\theta_1)\}$ where $g_1(\theta_1) = \theta_1^2 + \theta_1$ and $g_2(\theta_1) = \theta_1^2 - \theta_1$. In Fig. 4.1, O is the origin; the two

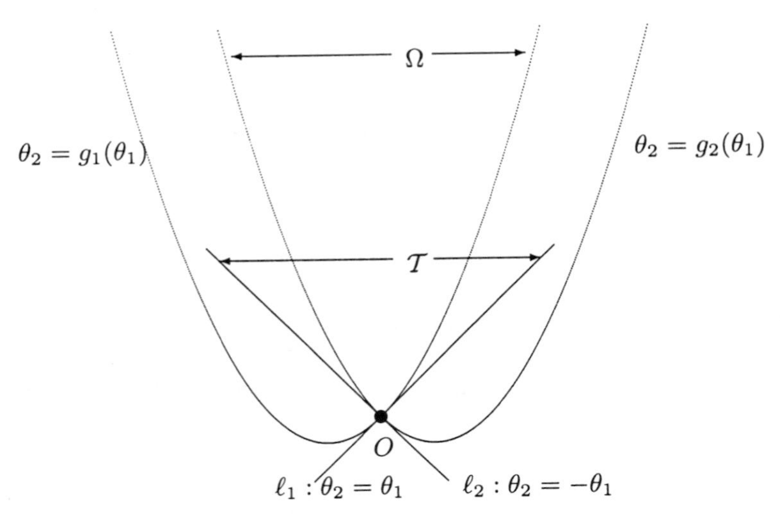

Fig. 4.1 The *cone of tangents,* $\mathcal{T}$, of Ω at O.

parabolas, $\theta_2 = g_1(\theta_1)$ and $\theta_2 = g_2(\theta_1)$, are also shown. Further, Ω is the region bounded by these two parabolas, as shown.

Let ℓ_1 and ℓ_2 be the tangents at O to the two smooth curves that form the boundary of Ω; thus the equations for ℓ_1 and ℓ_2 are $\theta_2 = \theta_1$ and $\theta_2 = -\theta_1$, respectively. Let $\mathcal{T} = \{\boldsymbol{\theta} : \theta_2 \geqslant \theta_1 \text{ and } \theta_2 \geqslant -\theta_1\}$. Locally at the origin, Ω is close to the cone $\mathcal{T}$. In this example, $\mathcal{T}$ is known as the *linearizing cone* of Ω at O, because it is obtained by linearizing the boundaries defining the set Ω at O. The cone is also called the *contingent cone* and *cone of tangents* of Ω at O. The formal definitions of these terms will be introduced in the next subsection.

Let $\boldsymbol{\theta}_0 (\neq \mathbf{0})$ be a given fixed point in $\mathbb{R}^p$. Now, let us move $\{\Omega, \ell_1, \ell_2, \mathcal{T}\}$ by $\boldsymbol{\theta}_0$, and let the new objects be denoted by $\{\Theta, \ell_1', \ell_2', \mathcal{A}\}$ (see Fig. 4.2). Therefore, $\Theta = \boldsymbol{\theta}_0 + \Omega, \mathcal{A} = \boldsymbol{\theta}_0 + \mathcal{T}$, ℓ_1' and ℓ_2' are the tangents to the boundaries of Θ at $\boldsymbol{\theta}_0$, and $\mathcal{A}$ is the cone formed by ℓ_1' and ℓ_2'. Then $\mathcal{A}$ is close to Θ at $\boldsymbol{\theta}_0$, in the same way as $\mathcal{T}$ is close to Ω at O. The cone $\mathcal{A}$ with vertex at $\boldsymbol{\theta}_0$ is called the *approximating cone* of Θ at $\boldsymbol{\theta}_0$. An important result relating to the approximating cone is that if Θ is the parameter space and θ_0 is the true value then Θ can be replaced by its approximating cone $\mathcal{A}$ at $\boldsymbol{\theta}_0$ for obtaining the asymptotic null distribution of *LRT* and other first order asymptotic results when $\boldsymbol{\theta}_0$ is the true value.

Example 4.7.1 *Inference on normal mean when the parameter space is not a cone.*

Let $\boldsymbol{Y}_1, \ldots, \boldsymbol{Y}_n$ be *iid* as the bivariate normal distribution, $N(\boldsymbol{\theta}, \boldsymbol{I})$, where $\boldsymbol{\theta} \in \mathbb{R}^2$ and let $\ell(\boldsymbol{\theta})$ denote the kernel of the loglikelihood $-(n/2)\|\bar{\boldsymbol{Y}} - \boldsymbol{\theta}\|^2$. Let the null and alternative hypotheses be

$$H_0 : \boldsymbol{\theta} = \boldsymbol{\theta}_0 \qquad \text{and} \qquad H_1 : \boldsymbol{\theta} \in \Theta,$$

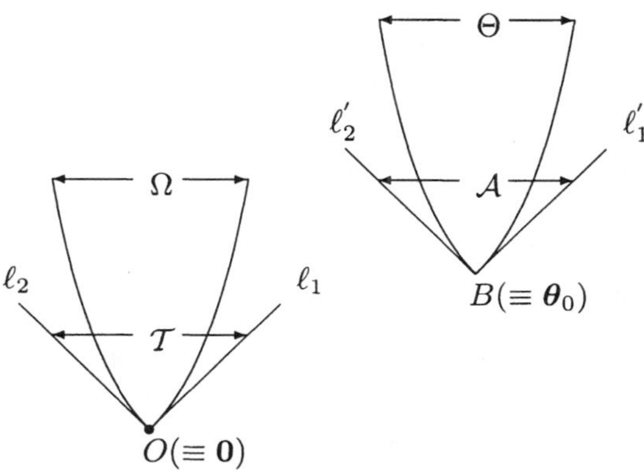

Fig. 4.2 The *cone of tangents*, $\mathcal{T}$, and the *approximating cone*, $\mathcal{A}$, of Θ at B.

respectively, where $\boldsymbol{\theta}_0 = (1,1)^T$, $\Theta = \boldsymbol{\theta}_0 + \Omega$ and Ω is as in Fig. 4.2. Let $\hat{\boldsymbol{\theta}}_\Theta$, $\hat{\boldsymbol{\theta}}_\Omega$ and $\hat{\boldsymbol{\theta}}_\mathcal{T}$ denote the *mle* over Θ, Ω, and $\mathcal{T}$, respectively. Then

$$LRT = 2\{\ell(\hat{\boldsymbol{\theta}}_\Theta) - \ell(\boldsymbol{\theta}_0)\} = n\{\|\bar{\boldsymbol{Y}} - \boldsymbol{\theta}_0\|^2 - \|\bar{\boldsymbol{Y}} - \Theta\|^2\}$$

where $\|\bar{\boldsymbol{Y}} - \Theta\|^2 = \inf_{\boldsymbol{\theta}\in\Theta}(\bar{\boldsymbol{Y}} - \boldsymbol{\theta})^T(\bar{\boldsymbol{Y}} - \boldsymbol{\theta})$ is the squared distance between $\bar{\boldsymbol{Y}}$ and Θ. Suppose that the true value of $\boldsymbol{\theta}$ is $\boldsymbol{\theta}_0$. Now, the results to be presented later in this section show that the asymptotic distribution of the constrained *mle* $\hat{\boldsymbol{\theta}}_\Theta$ and that of the *LRT* of $\boldsymbol{\theta} = \boldsymbol{\theta}_0$ against $\boldsymbol{\theta} \in \Theta$ remain the same even if the constrained parameter space Θ is replaced by $\mathcal{A}$, the approximating cone of Θ at $\boldsymbol{\theta}_0$. More specifically,

$$\begin{aligned} LRT &= n\{\|\bar{\boldsymbol{Y}} - \boldsymbol{\theta}_0\|^2 - \|\bar{\boldsymbol{Y}} - \Theta\|^2\} \\ &= n\{\|\bar{\boldsymbol{Y}} - \boldsymbol{\theta}_0\|^2 - \|\bar{\boldsymbol{Y}} - \mathcal{A}\|^2\} + o_p(1), \qquad (4.94) \\ \text{and} \qquad & n^{1/2}(\hat{\boldsymbol{\theta}}_\Theta - \hat{\boldsymbol{\theta}}_\mathcal{A}) = o_p(1). \end{aligned}$$

Further, the constrained estimator $\hat{\boldsymbol{\theta}}_\Theta$ and the projection of the unconstrained estimator $\hat{\boldsymbol{\theta}}$ onto the approximating cone have the same asymptotic distributions in the following sense:

$$\sqrt{n}(\hat{\boldsymbol{\theta}}_\Theta - \boldsymbol{\theta}_0) \xrightarrow{d} \Pi(\boldsymbol{Z}, \mathcal{T}) \quad \text{and} \quad \sqrt{n}\{\Pi(\hat{\boldsymbol{\theta}}, \mathcal{A}) - \boldsymbol{\theta}_0\} \xrightarrow{d} \Pi(\boldsymbol{Z}, \mathcal{T}),$$

where $\boldsymbol{Z} \sim N(\boldsymbol{0}, \boldsymbol{I})$. ■

It is worth noting that the process of replacing Θ by $\mathcal{A}$ is equivalent to replacing the nonlinear constraints defining Θ by their linear approximations at the null value. In the following subsections, we will extend these ideas to more general settings. To do so, one important requirement is that the parameter space must satisfy certain regularity conditions. These conditions must be satisfied irrespective of the properties of the likelihood function and other stochastic conditions. These are discussed in the next subsection.

Before we introduce regularity conditions relating to parameter spaces, let us make a remark regarding our terminology. Recall that a cone is defined to have vertex at the origin unless the contrary is clear (see Appendix 1). According to our definition, the approximating cone does not necessarily have its vertex at the origin. Some authors refer to the $\mathcal{T}$ in Fig. 4.2 as the approximating cone of Θ at $\boldsymbol{\theta}_0$. We shall deviate from this terminology because it would be helpful to identify $\mathcal{T}$ and $\mathcal{A}$ separately using different terms; we shall refer to them as tangent cone of Θ at $\boldsymbol{\theta}_0$ and approximating cone of Θ at $\boldsymbol{\theta}_0$, respectively.

4.7.1 Chernoff Regularity

Let $\Theta \subset \mathbb{R}^p$ and $\boldsymbol{\theta}_0 \in \Theta$. A vector $\boldsymbol{w} \in \mathbb{R}^p$ is said to be a *tangent* to Θ at θ_o if there exists a sequence $\{\boldsymbol{\theta}_n\}$ in Θ and a sequence of positive numbers $\{t_n\}$ converging to zero such that $t_n^{-1}(\boldsymbol{\theta}_n - \boldsymbol{\theta}_0) \to \boldsymbol{w}$ as $n \to \infty$. This essentially means that $\boldsymbol{w} \in \mathbb{R}^p$ is a tangent to Θ at $\boldsymbol{\theta}_0$ if either $\boldsymbol{w} = \mathbf{0}$ or there exists a sequence of points in Θ converging to $\boldsymbol{\theta}_0$ such that the direction of the vector from $\boldsymbol{\theta}_0$ to $\boldsymbol{\theta}_n$ converges to the direction of $\boldsymbol{w}$. Therefore, a tangent at $\boldsymbol{\theta}_o$ is a vector rather than the set of points on a line passing through $\boldsymbol{\theta}_0$. This definition of a tangent is more general than the usual one where it is seen as a line that touches a smooth curve at a point. The following examples illustrate this generalized notion of a tangent.

Example 4.7.2 *Illustration of a generalized notion of tangents*

1. For $\Theta = \{\boldsymbol{\theta} \in \mathbb{R}^2 : \theta_2 = \theta_1^2\}$, the tangents at $\mathbf{0}$ are $k(1,0)$ and $k(-1,0)$ where $k \geqslant 0$ (see Fig. 4.3). We do not view the line $\theta_2 = 0$ as a tangent.

2. For $\Theta = \{\boldsymbol{\theta} \in \mathbb{R}^2 : \theta_2 = \theta_1^2 \text{ and } \theta_1 \geqslant 0\}$ the only tangent at $\mathbf{0}$ is $k(1,0)$ where $k \geqslant 0$; note that $(-1,0)$ is not a tangent at $\mathbf{0}$.

3. For $\Theta = \{\boldsymbol{\theta} \in \mathbb{R}^2 : \theta_2 \geqslant \theta_1^2\}$, the tangents at $\mathbf{0}$ are (θ_1, θ_2) where $\theta_2 \geqslant 0$ and $\theta_1 \in \mathbb{R}$ (see Fig. 4.4). Thus, the tangents to the curve $\theta_2 = \theta_1^2$, and those to the area bounded by the curve are different.

4. Let $\Theta = \{\boldsymbol{\theta} \in \mathbb{R}^2 : \theta_2 = \sin(1/\theta_1) \text{ and } \theta_1 > 0\} \cup \{\mathbf{0}\}$. The tangents to Θ at $\mathbf{0}$ are $k\boldsymbol{\theta}$ where $\boldsymbol{\theta} = (\theta_1, \theta_2), \theta_1 > 0$ and $k \geqslant 0$. Thus, even though $(d\theta_2/d\theta_1)$ does not exist at the origin, the set $\{\boldsymbol{\theta} : \theta_2 = \sin(1/\theta_1)\}$ has an infinite number of tangents at the same point. ■

The *cone of tangents,* denoted $\mathcal{T}(\Theta; \boldsymbol{\theta}_0)$, of Θ at $\boldsymbol{\theta}_0$ is defined as the set of all tangents to Θ at $\boldsymbol{\theta}_0$. This cone is also called the *contingent cone, Boulingard tangent cone,* and the *ordinary tangent cone.* Note that $\mathcal{T}(\Theta; \boldsymbol{\theta}_0)$ has vertex at the origin irrespective of whether or not $\boldsymbol{\theta}_0$ is the origin. A tangent $\boldsymbol{v}$ is said to be *derivable* if there exists a function $\boldsymbol{f} : [0, \epsilon] \to \Theta$ for some $\epsilon > 0$ such that $\boldsymbol{f}(0) = \boldsymbol{\theta}_0$ and $(d/dt+)\boldsymbol{f}(0) = \boldsymbol{v}$. Intuitively, this means that there is a smooth curve in Θ with one end at $\boldsymbol{\theta}_0$ such that the slope of the curve at $\boldsymbol{\theta}_0$ is parallel to $\boldsymbol{v}$. The set of all derivable tangents to Θ at $\boldsymbol{\theta}_0$ is called the *derived tangent cone* of Θ at $\boldsymbol{\theta}_0$. It is also called the *cone of attainable directions*; see Bazaraa et al (1993).

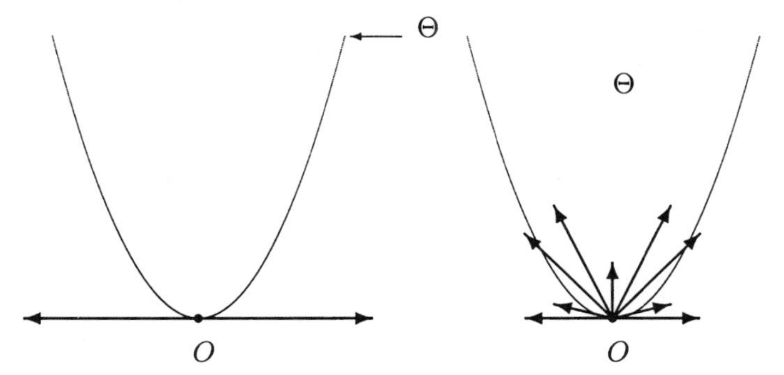

Fig. 4.3 The tangents to $\{\theta_2 = \theta_1^2\}$ at O.

Fig. 4.4 Tangents to $\{\theta_2 \geqslant \theta_1^2\}$ at O.

While it is not essential for us at this stage, let us also state equivalent definitions of the cones just introduced. The derived tangent cone is equal to

$$\{\boldsymbol{v} : \forall t_n \downarrow 0, \exists \boldsymbol{\theta}_n \in \Theta \text{ such that } \boldsymbol{\theta}_n \to \boldsymbol{\theta}_0 \text{ and } t_n^{-1}(\boldsymbol{\theta}_n - \boldsymbol{\theta}_0) \to \boldsymbol{v}\},$$

and the cone of tangents is equal to

$$\{\boldsymbol{v} : \exists t_n \downarrow 0, \exists \boldsymbol{\theta}_n \in \Theta \text{ such that } \boldsymbol{\theta}_n \to \boldsymbol{\theta}_0 \text{ and } t_n^{-1}(\boldsymbol{\theta}_n - \boldsymbol{\theta}_0) \to \boldsymbol{v}\}.$$

It is clear from these definitions that the derived tangent cone is contained in the cone of tangents, because the former must satisfy a condition for all $t_n \downarrow 0$ while the cone of tangents must satisfy the same condition for some $t_n \downarrow 0$.

The following example of (Rockafellar and Wets, 1998, p 199) is helpful to illustrate the difference between the cone of tangents and the derived tangent cone.

Example 4.7.3 *Illustration of a difference between cone of tangents and the derived tangent cone*

Let $\Theta = \{\boldsymbol{\theta} \in \mathbb{R}^2 : \theta_2 = \theta_1 \sin(1/\theta_1), \theta_1 > 0\} \cup \{\mathbf{0}\}$. Note that there is no curve that lies in Θ and is smooth at $\mathbf{0}$. Therefore, the only member of the derived tangent cone is $\{\mathbf{0}\}$. By contrast, the straight line $\theta_2 = k\theta_1$ where $|k| \leq 1$, intersects with Θ at a sequence of points $\{\boldsymbol{\theta}_n\}$ and $\|\boldsymbol{\theta}_n\|^{-1}\boldsymbol{\theta}_n \to (1, k)$. Therefore, the cone of tangents is $\{\boldsymbol{\theta} \in \mathbb{R}^2 : |\theta_2| \leq \theta_1, \theta_1 > 0\}$. ■

The foregoing definitions and the example suggest that if the cone of tangents and the derived tangent cone of Θ at $\boldsymbol{\theta}_0$ are not equal, then Θ is likely to be quite irregular near θ_0. A set $\Theta \subset \mathbb{R}^p$ is said to be *Chernoff regular* at $\boldsymbol{\theta}_0$ if the cone of tangents and the derived tangent cone of Θ at $\boldsymbol{\theta}_0$ are equal. If Θ is Chernoff regular at $\boldsymbol{\theta}_0$ then we shall refer to the *cone of tangents* and the *derived tangent cone* as *tangent cone.*

An intuitively simple way of verifying Chernoff regularity of Θ at $\boldsymbol{\theta}_0$ is the following (this is due to Rockafellar and Wets (1998), page 198). Consider the set $\mathcal{A}_t = t(\Theta - \boldsymbol{\theta}_0)$ where $t > 0$. Now take the limit as t increases to ∞ (see Fig. 4.5). Note that $\mathcal{A}_t$ provides a magnified view of Θ near $\boldsymbol{\theta}_0$ because it is like viewing Θ

through a magnifying glass focused at $\boldsymbol{\theta}_0$. As t increases to ∞, the magnification increases to ∞ and if the magnified view of Θ at $\boldsymbol{\theta}_0$ converges to a limit (i.e., if $\mathcal{A}_t$ has a limit) then Θ is Chernoff regular at $\boldsymbol{\theta}_0$, and the limit is the tangent cone of Θ at $\boldsymbol{\theta}_0$.

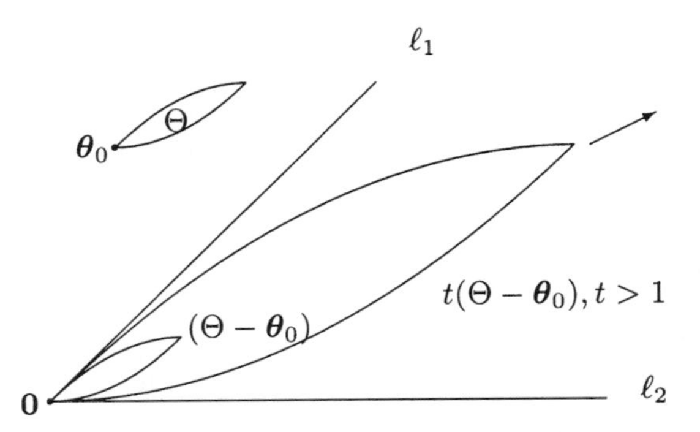

Fig. 4.5 The limit of $t(\Theta - \boldsymbol{\theta}_0)$ as $t \to \infty$ is the cone between the lines ℓ_1 and ℓ_2, and it is equal to the tangent cone, $\mathcal{T}(\Theta; \boldsymbol{\theta}_0)$, of Θ at $\boldsymbol{\theta}_0$.
Reproduced from: *Variational Analysis*, Rockafellar and Wets, Copyright (1998), with permission from Springer, New York.

Let $\Theta \subset \mathbb{R}^p$ and $\boldsymbol{\theta}_0 \in \Theta$. The set Θ is said to be *approximated by a cone $\mathcal{A}$ at $\boldsymbol{\theta}_0$* if

$$\begin{aligned} (a) \quad & d(\boldsymbol{\theta}, \mathcal{A}) = o(\|\boldsymbol{\theta} - \boldsymbol{\theta}_0\|) \text{ for } \boldsymbol{\theta} \in \Theta, \\ \text{and } (b) \quad & d(\boldsymbol{x}, \Theta) = o(\|\boldsymbol{x} - \boldsymbol{\theta}_0\|) \text{ for } \boldsymbol{x} \in \mathcal{A}, \end{aligned} \tag{4.95}$$

where d is the distance between a point and a set defined by

$$d(\boldsymbol{\theta}, A) = \inf_{\boldsymbol{x} \in A} \|\boldsymbol{\theta} - \boldsymbol{x}\|.$$

In this case, $\mathcal{A}$ is called an *approximating cone of* Θ *at* $\boldsymbol{\theta}_0$ and will be denoted by $\mathcal{A}(\Theta_0; \boldsymbol{\theta}_0)$.

This definition was introduced by Chernoff (1954) and it is illustrated in Fig. 4.6 using two different shapes for Θ. In each of the two cases, let Θ be the region bounded by the two curves as shown and O be the point corresponding to $\boldsymbol{\theta}_0$. In the diagram on the right-hand side, A is a point in Θ. Let $\boldsymbol{\theta} = OA$ and B be the projection of A onto the approximating cone $\mathcal{A}$; therefore, $\|AB\| = d(\boldsymbol{\theta}, \mathcal{A})$. Condition (a) of the definition (4.95) says that $\|AB\|/\|OA\| \to 0$ as A converges to O while staying in Θ. In the diagram on the left-hand side, A is a point in $\mathcal{A}$ and let $\boldsymbol{x} = OA$. Let B be the projection of A on to Θ. Condition (b) of the definition (4.95) says that $\|AB\|/\|OA\| \to 0$ as A converges to O while staying in $\mathcal{A}$.

Andrews (1999, page 1358) noted that the foregoing definition of approximating cone can be stated using Hausdorff distance; see also Le Cam (1970, p 819). For two sets A and B, define the *Hausdorff distance* between A and B by

$$h(A, B) = \max\{\sup_{b \in B} d(b, A), \sup_{a \in A} d(a, B)\}.$$

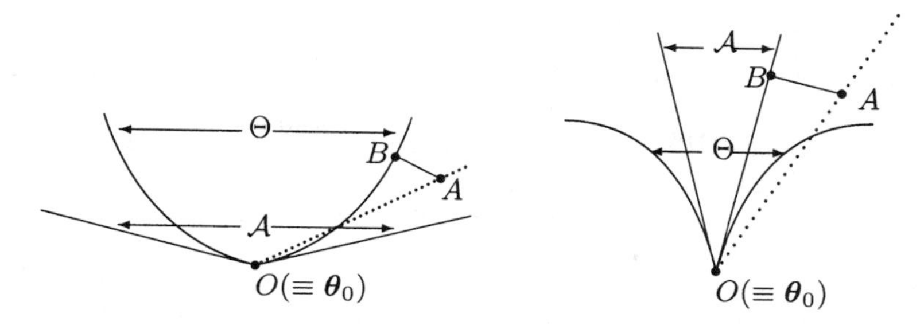

Fig. 4.6 Chernoff's definition of approximating cone $\mathcal{A}$ of Θ at $\boldsymbol{\theta}_0$: $\|AB\| = o(\|OA\|$

Now, the next result provides an equivalent form of definition (4.95).

Proposition 4.7.1 *Let the setting be as in* (4.95) *and* $B_\epsilon = \{\boldsymbol{\theta} : \|\boldsymbol{\theta} - \boldsymbol{\theta}_0\| < \epsilon\}$, *where* $\epsilon > 0$. *Let* $\mathcal{A}$ *be a cone with vertex at* $\boldsymbol{\theta}_0$. *Then* $\mathcal{A}$ *is an approximating cone of* Θ *at* $\boldsymbol{\theta}_0$ *as defined by* (4.95) *if and only if*

$$h\{\Theta \cap B_\epsilon, \mathcal{A} \cap B_\epsilon\} = o(\epsilon) \quad \textit{as } \epsilon \searrow 0. \tag{4.96}$$

Proof: For a given $\epsilon > 0$ let $\Theta_\epsilon = \Theta \cap B_\epsilon, \mathcal{A}_\epsilon = \mathcal{A} \cap B_\epsilon, \boldsymbol{\theta}_\epsilon = \arg\max_{\boldsymbol{\theta} \in \Theta_\epsilon} d(\boldsymbol{\theta}, \mathcal{A}_\epsilon)$, and $\boldsymbol{x}_\epsilon = \arg\max_{\boldsymbol{x} \in \mathcal{A}_\epsilon} d(\boldsymbol{x}, \Theta_\epsilon)$. Then, the Hausdorff distance between Θ_ϵ and $\mathcal{A}_\epsilon$ takes the form

$$h(\Theta_\epsilon, \mathcal{A}_\epsilon) = \max\{d(\boldsymbol{\theta}_\epsilon, \mathcal{A}_\epsilon), d(\boldsymbol{x}_\epsilon, \Theta_\epsilon)\}.$$

Now suppose that $\mathcal{A}$ is an approximating cone according to Chernoff's definition (4.95). Then

$$d(\boldsymbol{\theta}_\epsilon, \mathcal{A}_\epsilon) = o(\|\boldsymbol{\theta}_\epsilon - \boldsymbol{\theta}_0\|) = o(\epsilon);$$

the first equality follows by (4.95) and the second follows since $\|\boldsymbol{\theta}_\epsilon - \boldsymbol{\theta}_0\| \le \epsilon$. Similarly, $d(\boldsymbol{x}_\epsilon, \Theta_\epsilon) = o(\epsilon)$. Therefore, $h(\Theta_\epsilon, \mathcal{A}_\epsilon) = o(\epsilon)$.

Now to prove the converse, suppose that (4.96) is satisfied. Let $\boldsymbol{\theta} \in \Theta$ and $\delta = \|\boldsymbol{\theta} - \boldsymbol{\theta}_0\|$. Then,

$$d(\boldsymbol{\theta}, \mathcal{A}) = d(\boldsymbol{\theta}, \mathcal{A}_\delta) \le d(\boldsymbol{\theta}_\delta, \mathcal{A}_\delta) = o(\delta);$$

the first equality follows because $\|\Pi(\boldsymbol{\theta}, \mathcal{A}) - \boldsymbol{\theta}_0\| \le \|\boldsymbol{\theta} - \boldsymbol{\theta}_0\|$ and hence $\Pi(\boldsymbol{\theta}, \mathcal{A}) = \Pi(\boldsymbol{\theta}, \mathcal{A})$, the middle inequality follows by the definition of $\boldsymbol{\theta}_\delta$, and the third equality follows by (4.96). ■

For a discussion about Hausdorff distance in statistics, see van der Vaart and Wellner (1996, p 162). Andrews (1999, p 1358) extended the definition of Chernoff (1954) to define a sequence of sets being approximated by a cone; such a generalization may be required/helpful in more general settings. We say that a sequence of sets $\{\Theta_n\}$ is locally approximated at $\boldsymbol{\theta}_0$ by a cone $\mathcal{A}$ if

$$\begin{aligned} &(a) \quad d(\boldsymbol{\theta}_n, \mathcal{A}) = o(\|\boldsymbol{\theta}_n - \boldsymbol{\theta}_0\|) \text{ for } \boldsymbol{\theta}_n \in \Theta_n \text{ and } \|\boldsymbol{\theta}_n - \boldsymbol{\theta}_0\| \to 0 \\ \text{and } &(b) \quad d(\boldsymbol{x}_n, \Theta_n) = o(\|\boldsymbol{x}_n - \boldsymbol{\theta}_0\|) \text{ for } \boldsymbol{x}_n \in \mathcal{A} \text{ and } \|\boldsymbol{x}_n - \boldsymbol{\theta}_0\| \to 0. \end{aligned} \tag{4.97}$$

Now the foregoing results for Θ also hold for Θ_n with appropriate modifications.

Throughout this book, parameter spaces will be assumed to be Chernoff regular unless the contrary is made clear; this is a very mild condition as was indicated earlier. The following important result, due to Geyer (1994, Theorem 2.1), establishes the link between the notion of approximating cone introduced by Chernoff and several tangent cones that have been known in the mathematics literature.

Proposition 4.7.2 *(Geyer 1994). The cone $\mathcal{K}$ is an approximating cone of Θ at $\boldsymbol{\theta}_0$ if and only if the derived tangent cone and the cone of tangents are equal and*

$$\text{Closure of } \mathcal{K} = \boldsymbol{\theta}_0 + \mathcal{T}(\Theta; \boldsymbol{\theta}_0) = \boldsymbol{\theta}_0 + \text{ Derived tangent cone of } \Theta \text{ at } \boldsymbol{\theta}_0.$$

■

This essentially says that Θ has an approximating cone at $\boldsymbol{\theta}_0$ if an only if Θ is Chernoff regular at $\boldsymbol{\theta}_0$.

The following result provides a convenient way of obtaining the approximating cone of Θ at $\boldsymbol{\theta}_0$ when Θ is defined by a set of nonlinear equality and inequality constraints, in many cases; the proof is given in the appendix to this chapter.

Proposition 4.7.3 *Suppose that Ω is open and let*

$$\Theta = \{\boldsymbol{\theta} \in \Omega : h_1(\boldsymbol{\theta}) = \ldots = h_\ell(\boldsymbol{\theta}) = 0, h_{\ell+1}(\boldsymbol{\theta}) \geqslant 0, \ldots, h_k(\boldsymbol{\theta}) \geqslant 0\},$$

where $h_1, \ldots, h_k$ are continuously differentiable. Let $\boldsymbol{\theta}_0$ be a point in Θ, and let $\boldsymbol{a}_i = (\partial/\partial\boldsymbol{\theta})h_i(\boldsymbol{\theta}_0)$ for $i = 1, \ldots, k$, and $J(\boldsymbol{\theta}_0) = \{i : h_i(\boldsymbol{\theta}_0) = 0 \text{ and } \ell + 1 \leq i \leq k\}$. Assume that the following condition, known as the Mangasarian-Fromowitz constraint qualification (MF-CQ), is satisfied at $\boldsymbol{\theta}_0$: There exists a nonzero $\boldsymbol{b} \in \mathbb{R}^p$ such that $\boldsymbol{a}_1^T\boldsymbol{b} = \ldots = \boldsymbol{a}_\ell^T\boldsymbol{b} = 0$, $\boldsymbol{a}_1, \ldots, \boldsymbol{a}_\ell$ are linearly independent and $\boldsymbol{a}_i^T\boldsymbol{b} > 0$ for $i \in J(\boldsymbol{\theta}_0)$. Then $\mathcal{A}(\Theta; \boldsymbol{\theta}_0)$ is equal to

$$\{\boldsymbol{\theta} \in \mathbb{R}^p : \boldsymbol{a}_i^T(\boldsymbol{\theta} - \boldsymbol{\theta}_0) = 0 \text{ for } i = 1, \ldots, \ell; \boldsymbol{a}_i^T(\boldsymbol{\theta} - \boldsymbol{\theta}_0) \geqslant 0 \text{ for } i \in J(\boldsymbol{\theta}_0)\}.$$

The foregoing result shows that if MF-CQ is satisfied then the approximating cone at $\boldsymbol{\theta}_0$ is obtained by substituting the first order approximation (i.e., linear approximation at $\boldsymbol{\theta}_0$), $\{h_i(\boldsymbol{\theta}_0) + (\boldsymbol{\theta} - \boldsymbol{\theta}_0)^T(\partial/\partial\boldsymbol{\theta})h_i(\boldsymbol{\theta}_0)\}$, of $h_i(\boldsymbol{\theta})$ at $\boldsymbol{\theta}_0$ if $h_i(\boldsymbol{\theta}_0) = 0$; if $h_s(\boldsymbol{\theta}_0) > 0$ (i.e., $h_s(\boldsymbol{\theta}_0) \geq 0$ is inactive) then the constraint $h_s(\boldsymbol{\theta}_0) \geqslant 0$ does not play a part in determining the approximating cone at $\boldsymbol{\theta}_0$ ($s = \ell + 1, \ldots, k$).

The approximating cone obtained by applying the result in the previous proposition is a polyhedral. Thus, if the approximating cone is not a polyhedral then this method of obtaining the approximating cone would not be applicable in its current form. For example, in Fig. 4.1 if $\Theta = A \cup B$ where A is the region bounded by $\theta_2 = g_2(\theta_1)$ and $\theta_2 = \theta_1$, and B is the region bounded by $\theta_2 = g_1(\theta_1)$ and $\theta_2 = -\theta_1$, then the approximating cone of Θ at O is not convex. Therefore, for this Θ, the method in Proposition 4.7.3 is not directly applicable; in any case, this would be obvious because Θ is not of the form required by Proposition 4.7.3. It can be seen that A and B are of the form required by Proposition 4.7.3 and $\mathcal{A}(A \cup B; \mathbf{0}) = \mathcal{A}(A; \mathbf{0}) \cup \mathcal{A}(B; \mathbf{0})$; now the method based on MF-CQ can be applied to A and B separately to obtain $\mathcal{A}(A \cup B; \mathbf{0})$. More generally, the foregoing method can be applied to a parameter space that is equal to the union of sets each of which satisfies the conditions of Proposition 4.7.3.

Example 4.7.4 *Tangent Cones*

(1) Let $\Theta = \{\boldsymbol{\theta} \in \mathbb{R}^2 : \theta_2 \geqslant (1/2)\theta_1, \theta_2 \leq 2\theta_1\}$ (see Fig. 4.7). Then, we have the following:
$\mathcal{T}(\Theta_0; \boldsymbol{\theta}_0) = \Theta$, where $\boldsymbol{\theta}_0 = (0,0)^T$.
$\mathcal{T}(\Theta; \boldsymbol{\theta}_a) = \{\boldsymbol{x} \in \mathbb{R}^2 : x_2 \geqslant (1/2)x_1\}$, where $\boldsymbol{\theta}_a = (2,1)^T$.
$\mathcal{T}(\Theta; \boldsymbol{\theta}_b) = \mathbb{R}^2$, where $\boldsymbol{\theta}_b = (2,2)^T$.
$\mathcal{T}(\Theta; \boldsymbol{\theta}_c) = \{\boldsymbol{x} \in \mathbb{R}^2 : x_2 \leq 2x_1\}$, where $\boldsymbol{\theta}_c = (1,2)^T$.

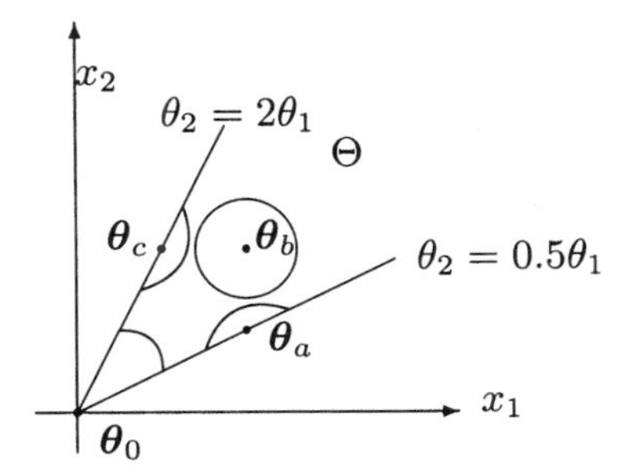

Fig. 4.7 The tangent cones of $\Theta = \{\boldsymbol{\theta} : \theta_2 \leq 2\theta_1, \theta_2 \geqslant 0.5\theta_1\}$ at $\boldsymbol{\theta}_0, \boldsymbol{\theta}_a, \boldsymbol{\theta}_b$ and $\boldsymbol{\theta}_c$

Note that at $\boldsymbol{\theta}_a$, the inequality $\theta_2 \leq 2\theta_1$ holds strictly (i.e., it is not active) and therefore this inequality constraint can be ignored for determining the tangent cone at $\boldsymbol{\theta}_a$. At $\boldsymbol{\theta}_b$, both inequalities are satisfied strictly, and therefore both constraints can be ignored for determining the tangent cone. Since there are no other active constraints, the tangent cone is the full space $\mathbb{R}^2$. At $\boldsymbol{\theta}_c$, the inequality $\theta_2 \geqslant (1/2)\theta_1$ holds strictly and hence it can be ignored for determining the tangent cone; the rest of the arguments are similar to those for $\boldsymbol{\theta}_a$. ■

(2) Let the matrices $\boldsymbol{R}_1$ and $\boldsymbol{R}_2$ be given and let $\Theta = \{\boldsymbol{\theta} \in \mathbb{R}^p : \boldsymbol{R}_1\boldsymbol{\theta} \geqslant \mathbf{0}, \boldsymbol{R}_2\boldsymbol{\theta} \geqslant \mathbf{0}\}$. Then, by Proposition 4.7.3 we have the following:

$$\begin{aligned}
&\boldsymbol{R}_1\boldsymbol{\theta}_a > \mathbf{0}, \boldsymbol{R}_2\boldsymbol{\theta}_a = \mathbf{0} \Rightarrow \mathcal{T}(\Theta; \boldsymbol{\theta}_a) = \{\boldsymbol{x} \in \mathbb{R}^p : \boldsymbol{R}_2\boldsymbol{x} \geqslant \mathbf{0}\}.\\
&\boldsymbol{R}_1\boldsymbol{\theta}_0 = \mathbf{0}, \boldsymbol{R}_2\boldsymbol{\theta}_0 = \mathbf{0} \Rightarrow \mathcal{T}(\Theta; \boldsymbol{\theta}_0) = \{\boldsymbol{x} \in \mathbb{R}^p : \boldsymbol{R}_1\boldsymbol{x} \geqslant \mathbf{0}, \boldsymbol{R}_2\boldsymbol{x} \geqslant \mathbf{0}\}. \quad (4.98)\\
&\boldsymbol{R}_1\boldsymbol{\theta}_b > \mathbf{0}, \boldsymbol{R}_2\boldsymbol{\theta}_b > \mathbf{0} \Rightarrow \mathcal{T}(\Theta; \boldsymbol{\theta}_b) = \mathbb{R}^p.
\end{aligned}$$

(3) Let $\Theta = \{\boldsymbol{\theta} \in \mathbb{R}^2 : \theta_2 \geqslant -\theta_1^2 + 2\theta_1, \theta_2 \geqslant -\theta_1^2 - 2\theta_1\}$ and $\boldsymbol{\theta}_0 = \mathbf{0}$.

It is easily verified that the MF-CQ is satisfied at $\boldsymbol{\theta}_0$ and therefore the tangent cone is determined by linearizing the constraints at $\boldsymbol{\theta}_0$ as follows. Let the two constraints be expressed as $g_1(\boldsymbol{\theta}) \geqslant 0$ and $g_2(\boldsymbol{\theta}) \geqslant 0$. Then the linearized versions of these constraints at $\boldsymbol{\theta}_0$ are

$$g_1(\boldsymbol{\theta}_0) + (\boldsymbol{\theta} - \boldsymbol{\theta}_0)^T(\partial/\partial\boldsymbol{\theta})g_1(\boldsymbol{\theta}_0) \geqslant 0 \text{ and } g_2(\boldsymbol{\theta}_0) + (\boldsymbol{\theta} - \boldsymbol{\theta}_0)^T(\partial/\partial\boldsymbol{\theta})g_2(\boldsymbol{\theta}_0) \geqslant 0,$$

respectively. These linearized versions are precisely $\theta_2 \geqslant 2\theta_1$ and $\theta_2 \geqslant -2\theta_1$, respectively. Since the two constraints defining Θ are active at $\boldsymbol{\theta}_0$, we have $\mathcal{T}(\Theta; \boldsymbol{\theta}_0) = \{\boldsymbol{\theta} : \theta_2 \geqslant 2\theta_1, \theta_2 \geqslant -2\theta_1\}$. In this particular example, we can sketch the shape of Θ easily and the tangent cone can be written down quickly; the above analytical approach is meant to be instructive.

(4) Let $\Theta = \{\boldsymbol{\theta} \in \mathbb{R}^2 : \theta_2 \geqslant \theta_1^2, \theta_2 \leq \theta_1\}$. Let $g_1(\boldsymbol{\theta}) = \theta_2 - \theta_1^2$ and $g_2(\boldsymbol{\theta}) = \theta_1 - \theta_2$. Then $\nabla g_1(\boldsymbol{\theta}) = (-2\theta_1, 1)^T, \nabla g_2(\boldsymbol{\theta}) = (1, -1)^T$. Now, applying the same technique as in the previous example, we have $\mathcal{T}(\Theta; \boldsymbol{\theta}_0) = \{\boldsymbol{\theta} \in \mathbb{R}^2 : \theta_1 \geqslant \theta_2 \geqslant 0\}$, and Θ is Chernoff regular at $\mathbf{0}$ (see Fig. 4.8). ■

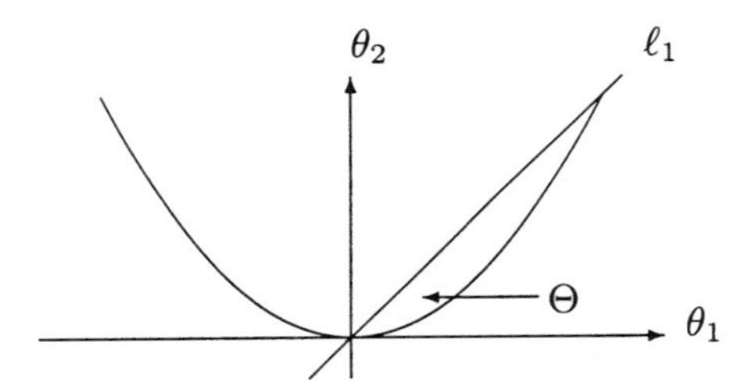

Fig. 4.8 The tangent cone of $\Theta = \{\theta_2 \geqslant \theta_1^2, \theta_2 \leq \theta_1\}$ at the origin is the cone between ℓ_1 and the θ_1 axis.

The next result is useful in asymptotic derivations because it enables us to substitute the approximating cone for the parameter space at the true value even if it (i.e., the true parameter value) is on the boundary which may or may not be smooth.

Proposition 4.7.4 *Let $\Theta \subset \mathbb{R}^p$, $\boldsymbol{V}$ be a positive definite matrix of order $p \times p$ and $\boldsymbol{\theta}_0 \in \Theta$ and $\mathcal{A}$ be an approximating cone of Θ at $\boldsymbol{\theta}_0$. Then*

$$\|\boldsymbol{y} - \Theta\|_{\boldsymbol{V}}^2 - \|\boldsymbol{y} - \mathcal{A}(\Theta; \boldsymbol{\theta}_0\|_{\boldsymbol{V}}^2 = o(\|\boldsymbol{y} - \boldsymbol{\theta}_0\|_{\boldsymbol{V}}^2) \quad as\ \boldsymbol{y} \to \boldsymbol{\theta}_0.$$

The proof is given in the Appendix; in fact, two different proofs are given, one uses Chernoff's definition of an approximating cone and simple geometry and the other uses Hausdorff distance. Fig. 4.9 illustrates this result and it shows two possible shapes of Θ bounded by two curves intersecting at O. Let A be an arbitrary point, and C and D be the points in $\mathcal{A}(\Theta; \boldsymbol{\theta}_0)$ and Θ, respectively, that are closest to A. Now Proposition 4.7.4 says that $|AD^2 - AC^2| = o(OA^2)$.

A useful corollary to the foregoing proposition is the following.

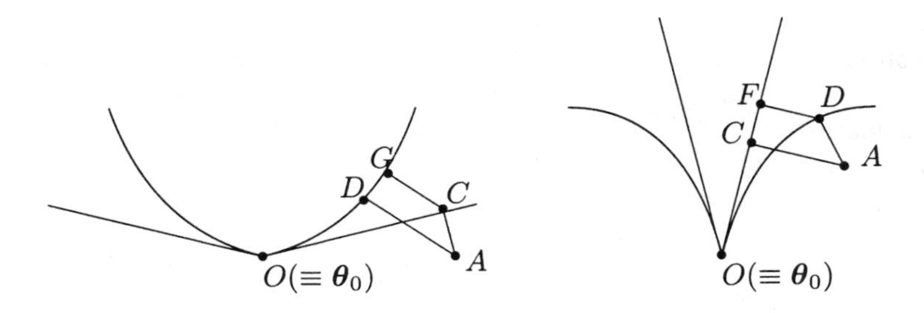

Fig. 4.9 $\|\boldsymbol{y} - \Theta\|^2 - \|\boldsymbol{y} - \mathcal{A}(\Theta; \boldsymbol{\theta}_0)\|^2 = o(\|\boldsymbol{y} - \boldsymbol{\theta}_0\|^2)$ [i.e., $|AD^2 - AC^2| = o(OA^2)$]

Corollary 4.7.5 *Let $\Theta \subset \mathbb{R}^p$, $\boldsymbol{\theta}_0 \in \Theta$, $\boldsymbol{V}$ be a $p \times p$ positive definite matrix and Θ be Chernoff regular at $\boldsymbol{\theta}_0$. Then we have the following:*

1. *If $n^{1/2}(\boldsymbol{T}_n - \boldsymbol{\theta}_0) = O_p(1)$ then*

$$\|\boldsymbol{T}_n - \Theta\|_{\boldsymbol{V}}^2 - \|\boldsymbol{T}_n - \mathcal{A}(\Theta; \boldsymbol{\theta}_0)\|_{\boldsymbol{V}}^2 = o_p(n^{-1}).$$

2. *If $n^{1/2}(\boldsymbol{T}_n - \boldsymbol{\theta}_0) \xrightarrow{d} \boldsymbol{Z}$ then $n\|\boldsymbol{T}_n - \Theta\|_{\boldsymbol{V}}^2 \xrightarrow{d} \|\boldsymbol{Z} - \mathcal{T}(\Theta; \boldsymbol{\theta}_0)\|_{\boldsymbol{V}}^2$*

Proof: Let us first state a lemma.

Lemma 4.7.6 *Let $g : \mathbb{R}^p \to \mathbb{R}$ be such that $g(\boldsymbol{y}) = o(\|\boldsymbol{y} - \boldsymbol{\theta}_0\|^\beta)$ as $\|\boldsymbol{y} - \boldsymbol{\theta}_0\| \to 0$, where $\beta > 0$. Let $\boldsymbol{T}_n$ be a sequence of random p-vectors such that $\boldsymbol{T}_n - \boldsymbol{\theta}_0 = O_p(n^{-\gamma})$ where $\gamma > 0$. Then $g(\boldsymbol{T}_n) = o_p(n^{-\beta\gamma})$.*

Proof: Given $\epsilon > 0$, there exists a δ such that $|g(\boldsymbol{y})/\|\boldsymbol{y} - \boldsymbol{\theta}_0\|^\beta| < \epsilon$ for $\|\boldsymbol{y} - \boldsymbol{\theta}_0\| < \delta$. Therefore,

$$\text{pr}\{|g(\boldsymbol{T}_n)/\|\boldsymbol{T}_n - \boldsymbol{\theta}_0\|^\beta < \epsilon) \geqslant \text{pr}(\|\boldsymbol{T}_n - \boldsymbol{\theta}_0\| < \delta) \to 1 \text{ as } n \to \infty.$$

Thus, $g(\boldsymbol{T}_n)/\|\boldsymbol{T}_n - \boldsymbol{\theta}_0\|^\beta = o_p(1)$. Now,

$$g(\boldsymbol{T}_n) = \|\boldsymbol{T}_n - \boldsymbol{\theta}_0\|^\beta o_p(1) = |n^{-\gamma}O_p(1)|^\beta o_p(1) = o_p(n^{-\beta\gamma}).$$

This completes the proof of the lemma.

Now, the proof of the first part of the corollary follows from the foregoing Lemma and Proposition 4.7.4 by substituting $g(\boldsymbol{y}) = \|\boldsymbol{y} - \Theta\|_{\boldsymbol{V}}^2 - \|\boldsymbol{y} - \mathcal{A}(\Theta; \boldsymbol{\theta}_0)\|_{\boldsymbol{V}}^2$, $\beta = 2$ and $\gamma = 1/2$.

To prove the second part, let $\boldsymbol{Z}_n = n^{1/2}(\boldsymbol{T}_n - \boldsymbol{\theta}_0)$ and note that

$$\begin{aligned} n\|\boldsymbol{T}_n - \Theta\|_{\boldsymbol{V}}^2 &= n\|\boldsymbol{T}_n - \mathcal{A}(\Theta; \boldsymbol{\theta}_0)\|_{\boldsymbol{V}}^2 + o_p(n^{-1}) \\ &= \|\boldsymbol{Z}_n - \mathcal{T}(\Theta; \boldsymbol{\theta}_0)\|_{\boldsymbol{V}}^2 + o_p(1) \xrightarrow{d} \|\boldsymbol{Z} - \mathcal{T}(\Theta; \boldsymbol{\theta}_0)\|_{\boldsymbol{V}}^2 \end{aligned}$$

■

The corollary essentially says that $\mathcal{A}(\Theta; \boldsymbol{\theta}_0)$ can be substituted for Θ in most of the expressions; in particular, if $\boldsymbol{T}_n$ is a $\sqrt{n}$-consistent estimator, $\boldsymbol{\theta}_0$ is the true value and $\boldsymbol{\theta}_0 \in \Theta$ then the squared distance between $\boldsymbol{T}_n$ and Θ is equal to that between $\boldsymbol{T}_n$ and the approximating cone of Θ at $\boldsymbol{\theta}_0$ except for a $o_p(n^{-1})$ term. The corollary can also be extended to the case when Θ_n is a sequence of sets such that $\boldsymbol{\theta}_0 \in \Theta_n$ and $\mathcal{A}(\Theta_n; \boldsymbol{\theta}_0)$ converges to a cone, or Θ_n converges to a cone.

4.8 GENERAL TESTING PROBLEMS

Let $f(y; \boldsymbol{\theta})$ be a density function where $\boldsymbol{\theta} \in \Theta \subset \mathbb{R}^p$ and $\Theta_0 \subset \Theta_1 \subset \Theta$. Let Y be a random variable with density function $f(y; \boldsymbol{\theta})$. In this section, test of

$$H_0 : \boldsymbol{\theta} \in \Theta_0 \text{ against } H_1 : \boldsymbol{\theta} \in \Theta_1 \tag{4.99}$$

is studied when Θ_0, Θ_1 and Θ may have boundary points and the boundaries of these sets may or may not be smooth; for example, the boundary of a cone at its vertex is not smooth. The special case when Θ_0 is a linear space and Θ_1 is a polyhedral was studied in Section 4.3, where the special structure on the parameter spaces enabled us to obtain explicit expressions for the asymptotic null distributions of the test statistics. Now, we consider more general cases and obtain general results. It will be assumed that Θ_0 and Θ_1 have approximating cones at every point in Θ_0; in the terminology introduced in Section 4.7, this is same as saying that Θ_0 and Θ_1 are Chernoff regular at every $\boldsymbol{\theta} \in \Theta_0$. For simplicity, it is assumed that the observations on Y are independently and identically distributed; however, the main results of this section hold, under appropriate regularity conditions, for much more general stochastic processes although such generalizations are not considered.

Let $Y_1, \ldots, Y_n$ be independently and identically distributed as Y. As in the earlier sections, let

$$\ell(\boldsymbol{\theta}) = \sum \log f(Y_i; \boldsymbol{\theta}), \qquad \boldsymbol{S}(\boldsymbol{\theta}) = (\partial/\partial\boldsymbol{\theta})\ell(\boldsymbol{\theta}), \tag{4.100}$$

$$\text{and} \quad \mathcal{I}(\boldsymbol{\theta}) = E\{(\partial/\partial\boldsymbol{\theta}) \log f(Y; \boldsymbol{\theta})(\partial/\partial\boldsymbol{\theta}^T) \log f(Y; \boldsymbol{\theta})\} \tag{4.101}$$

denote the loglikelihood, the score function, and the information matrix, respectively. Some important fundamental results on the *LRT* for testing problems of the form (4.99) were obtained by Chernoff (1954). The technical details in Chernoff (1954) assumed that Θ was open and that inference was based on likelihood. Chernoff's (1954) results have natural extensions to more general cases although modifications would be required to the regularity conditions and technical details; for example, see Self and Liang (1987), Andrews (1998), Vu and Zhou (1997), Shapiro (1985, 1989, 2000a), Geyer (1994), Silvapulle (1992a, 1992b, 1992c, 1994), and El Barmi (1996) among others. The areas covered include the cases when (i) Θ is not necessarily open (ii) observations are not necessarily independently and identically distributed, and (iii) inference is based on quasi-likelihood, partial likelihood, or general objective functions such as those used in, for example, M-estimation, minimum distance methods, and empirical likelihood.

In the next subsection, we consider the simple case of likelihood ratio test when Θ is open. The simplicity of this case makes it easier to convey the essentials. On the other hand, this case is also of independent interest because many practical problems fall into this category. Then, in subsection 4.8.2, we consider likelihood inference when Θ is not necessarily open and the true value of $\boldsymbol{\theta}$ can be a boundary point of Θ, Θ_0, and Θ_1 simultaneously. The distribution and implementation of likelihood ratio test are virtually the same whether or not $\boldsymbol{\theta}_0$ is a boundary point of Θ provided some conditions are satisfied; essentially all that we require is that it should be possible to approximate the objective function by a quadratic similar to that in (4.2) and (4.3). In this sense, the results in subsection 4.8.1 are not very different from those in subsection 4.8.2. In subsection 4.8.3, we consider inference based on general objective functions rather than just likelihood. Some results obtained in subsections 4.8.1 and 4.8.2 do not necessarily remain the same in this general setting. In subsection 4.8.4, some simple examples in two dimensions are considered to illustrate the theory. In subsection 4.8.5, a general form of the Type B testing problem, test of $\boldsymbol{h}(\boldsymbol{\theta}) \geqslant \mathbf{0}$ against $\boldsymbol{h}(\boldsymbol{\theta}) \ngeqslant \mathbf{0}$, is studied; this particular type of problems arises in economics, for example, the interest may be to test whether or not a regression function deviates from concavity.

As usual, we adopt the notation $\hat{\boldsymbol{\theta}}, \tilde{\boldsymbol{\theta}}$, and $\bar{\boldsymbol{\theta}}$ for the estimators over Θ, Θ_1, and Θ_0 respectively.

4.8.1 Likelihood Approach When Θ is Open

Assume that Θ is open and that $H_0 : \boldsymbol{\theta} \in \Theta_0$ holds. Let $\boldsymbol{\theta}_0$ denote the true value of $\boldsymbol{\theta}$ in Θ_0. The set of regularity conditions required for this setting are virtually the same as those usually required for establishing that $(\hat{\boldsymbol{\theta}} - \boldsymbol{\theta}_0)$ is asymptotically normal and the asymptotic null distribution of the *LRT* of $\boldsymbol{R\theta} = \mathbf{0}$ against $\boldsymbol{R\theta} \neq \mathbf{0}$ is chi-squared. There are different sets of conditions available for these results; some involve third-order partial derivatives of the log density, some involve only the second-order partial derivatives, and some replace the differentiability by other special types of differentiability (quadratic mean, stochastic, etc.). For simplicity, we shall work with conditions on the third-order derivatives of the log density, but we recognize that our technical conditions can be improved. Let

$$\|\boldsymbol{x}\|^2 = \boldsymbol{x}^T \mathcal{I}_{\boldsymbol{\theta}_0} \boldsymbol{x}. \tag{4.102}$$

Define,

$$W_D = \min_{\boldsymbol{\theta} \in \Theta_0} n(\hat{\boldsymbol{\theta}} - \boldsymbol{\theta})^T \hat{\mathcal{I}} (\hat{\boldsymbol{\theta}} - \boldsymbol{\theta}) - \min_{\boldsymbol{\theta} \in \Theta_1} n(\hat{\boldsymbol{\theta}} - \boldsymbol{\theta})^T \hat{\mathcal{I}} (\hat{\boldsymbol{\theta}} - \boldsymbol{\theta}), \tag{4.103}$$

where $\hat{\mathcal{I}} = \mathcal{I}(\hat{\boldsymbol{\theta}})$; however, in general $\hat{\mathcal{I}}$ can be any consistent estimator of $\mathcal{I}(\boldsymbol{\theta}_0)$. Note that W_D is a measure of $\{dist(\hat{\boldsymbol{\theta}}, H_0) - dist(\hat{\boldsymbol{\theta}}, H_1)\}$, where $dist(\hat{\boldsymbol{\theta}}, H)$ is a measure of distance between $\hat{\boldsymbol{\theta}}$ and the parameter space defined by H. Hence it is a suitable statistic for testing $H_0 : \boldsymbol{\theta} \in \Theta_0$ against $H_1 : \boldsymbol{\theta} \in \Theta_1$. As in the case of test against the unrestricted alternative, *LRT* and W_D are asymptotically equivalent. A simple method of obtaining their common asymptotic distribution is given in the next result; essentially, it says that the value of *LRT* changes by a term of order $o_p(1)$ if $\mathcal{A}(\Theta_i;\boldsymbol{\theta}_0)$ is substituted for Θ_i $(i=0,1)$.

Proposition 4.8.1 *Assume that Θ is open, Condition Q is satisfied, and $H_0:\boldsymbol{\theta}\in\Theta_0$ holds. Then, with the notation in (4.102), we have the following:*

1. *$LRT = W_D + o_p(1)$, where W_D is given in (4.103).*

2. *The common asymptotic null distribution of LRT and W_D is equal to the distribution of*

$$\|\boldsymbol{X}-\mathcal{A}(\Theta_0;\boldsymbol{\theta}_0)\|^2-\|\boldsymbol{X}-\mathcal{A}(\Theta_1;\boldsymbol{\theta}_0)\|^2, \tag{4.104}$$

which is also equal to the distribution of

$$\|\boldsymbol{Z}-\mathcal{T}(\Theta_0;\boldsymbol{\theta}_0)\|^2-\|\boldsymbol{Z}-\mathcal{T}(\Theta_1;\boldsymbol{\theta}_0)\|^2 \tag{4.105}$$

where $\boldsymbol{X}\sim N(\boldsymbol{\theta}_0,\mathcal{I}(\boldsymbol{\theta}_0)^{-1})$, $\boldsymbol{Z}\sim N(\boldsymbol{0},\mathcal{I}(\boldsymbol{\theta}_0)^{-1})$ and $\|\cdot\|$ is as in (4.102).

Proof: Let $\mathcal{A}_i=\mathcal{A}(\Theta_i;\boldsymbol{\theta}_0)$ and $\mathcal{T}_i=\mathcal{T}(\Theta_i;\boldsymbol{\theta}_0)$ for $i=0,1$. Now, by using the quadratic approximation (4.4), we have that

$$\begin{aligned} LRT &= 2[\sup_{\boldsymbol{\theta}\in\Theta_1}\ell(\boldsymbol{\theta})-\sup_{\boldsymbol{\theta}\in\Theta_0}\ell(\boldsymbol{\theta})] \\ &= n\|\hat{\boldsymbol{\theta}}-\Theta_0\|^2-n\|\hat{\boldsymbol{\theta}}-\Theta_1\|^2+o_p(1) \\ &= W_D+o_p(1); \end{aligned} \tag{4.106}$$

the last step, which says that $\|\cdot\|_{\mathcal{I}(\boldsymbol{\theta}_0)}$ can be replaced by $\|\cdot\|_{\hat{\mathcal{I}}}$, follows from Lemma 4.10.2 in the Appendix.

Now, let us prove the second part. By Corollary 4.7.5, Θ_i in (4.106) can be replaced by its approximating cone $(i=0,1)$. Let $\boldsymbol{Z}_n=\sqrt{n}(\hat{\boldsymbol{\theta}}-\boldsymbol{\theta}_0)$. Now, by applying essentially the same arguments as those used in the proof of the first part, we have

$$\begin{aligned} LRT &= n\{\|\hat{\boldsymbol{\theta}}-\mathcal{A}_0\|^2-\|\hat{\boldsymbol{\theta}}-\mathcal{A}_1\|^2\}+o_p(1), \\ &= \|\boldsymbol{Z}_n-\mathcal{T}_0\|^2-\|\boldsymbol{Z}_n-\mathcal{T}_1\|^2+o_p(1), \\ &\xrightarrow{d} \|\boldsymbol{Z}-\mathcal{T}_0\|^2-\|\boldsymbol{Z}-\mathcal{T}_1\|^2; \end{aligned} \tag{4.107}$$

the last step follows since $\|\boldsymbol{z}-\mathcal{T}_i\|^2$ is continuous in $\boldsymbol{z}$ and $\boldsymbol{Z}_n\xrightarrow{d}\boldsymbol{Z}$. ■

The foregoing results show that, if the null hypothesis is true and $\boldsymbol{\theta}_0$ denotes the true value, then the asymptotic distribution of the *LRT* of $\boldsymbol{\theta}\in\Theta_0$ against $\boldsymbol{\theta}\in\Theta_1$ is equal to the distribution of

1. The *LRT* of $\boldsymbol{\theta}\in\mathcal{A}(\Theta_0;\boldsymbol{\theta}_0)$ against $\boldsymbol{\theta}\in\mathcal{A}(\Theta_1;\boldsymbol{\theta}_0)$ based on a single observation of $\boldsymbol{X}$, where $\boldsymbol{X}\sim N\{\boldsymbol{\theta},\mathcal{I}(\boldsymbol{\theta}_0)^{-1}\}$ and the true value of $\boldsymbol{\theta}$ is $\boldsymbol{\theta}_0$.

2. The *LRT* of $\boldsymbol{\alpha}\in\mathcal{T}(\Theta_0;\boldsymbol{\theta}_0)$ against $\boldsymbol{\alpha}\in\mathcal{T}(\Theta_1;\boldsymbol{\theta}_0)$ based on a single observation of $\boldsymbol{Z}$ where $\boldsymbol{Z}\sim N\{\boldsymbol{\alpha},\mathcal{I}(\boldsymbol{\theta}_0)^{-1}\}$ and the true value of $\boldsymbol{\alpha}$ is $\boldsymbol{0}$.

Thus, as in classical inference and in Section 4.3, the asymptotic distribution of the test statistics can be expressed in terms of a single observation from the normal distribution. However, this apparently simpler problem may be still quite complicated, as was noted in Section 4.3, mainly due to the presence of the nuisance parameter in the null distribution; the nuisance parameter appears because the asymptotic null distribution of the *LRT* depends on $\mathcal{I}(\boldsymbol{\theta}_0)$ where $\boldsymbol{\theta}_0$ is the assumed true value in the null parameter space. This is in sharp contrast to the case where the asymptotic null distribution of *LRT* for testing $\boldsymbol{R\theta} = \mathbf{0}$ against $\boldsymbol{R\theta} \neq \mathbf{0}$ is a chi-squared and hence does not depend on $\boldsymbol{\theta}_0$ in the null parameter space.

If $\mathcal{T}(\Theta_0; \boldsymbol{\theta}_0)$ is a linear space and $\mathcal{T}(\Theta_1; \boldsymbol{\theta}_0)$ is a closed convex cone, then it follows from the foregoing proposition that the asymptotic null distribution of the *LRT* is a chi-bar-square. For example, if the null and alternative hypotheses are $\boldsymbol{R\theta} = \mathbf{0}$ and $\boldsymbol{R\theta} \geqslant \mathbf{0}$, respectively, then the asymptotic null distribution of the *LRT* is chi-bar-square; this was also obtained in Proposition 4.3.1. In fact, Proposition 4.3.1 on the asymptotic null distribution of *LRT* is a special case of the foregoing proposition. However, in general, the asymptotic null distribution of the *LRT* is not a chi-bar-square.

The results in the foregoing proposition provide elegant asymptotic representations of the *LRT* and they lead to its asymptotic distribution, which has a simpler structure because the original parameter spaces have been replaced by their approximating cones and the limiting distribution is a function of $\boldsymbol{Z}$ where $\boldsymbol{Z} \sim N(\mathbf{0}, \mathcal{I}_{\boldsymbol{\theta}_0}^{-1})$. Although Proposition 4.8.1 was established for the case when Θ is open, similar results are available even when Θ is not open and $\boldsymbol{\theta}_0$ is a boundary point of Θ.

4.8.2 Likelihood Approach When Θ is not Open

Now assume that Θ is not necessarily open and that $\boldsymbol{\theta}_0$ may not be an interior point of Θ. Then $(\partial/\partial\boldsymbol{\theta})\ell(\hat{\boldsymbol{\theta}})$ may not be zero and $n^{1/2}(\hat{\boldsymbol{\theta}} - \boldsymbol{\theta}_0)$ may not be asymptotically normal. Consequently, the arguments in the previous subsection that depend on the asymptotic normality of $n^{1/2}(\hat{\boldsymbol{\theta}} - \boldsymbol{\theta}_0)$ are not valid; also, arguments based on Taylor series may not be valid. However, most of the results still hold. For example, the asymptotic null distribution of *LRT* remains unchanged even if $\boldsymbol{\theta}_0$ is a boundary point Θ, but the technical details leading to it are different from those in the previous subsection. *Assume that the null hypothesis holds.* In most practical situations, $\ell(\boldsymbol{\theta})$ would have directional derivatives at $\boldsymbol{\theta}_0$, and hence would admit a quadratic approximation in a neighborhood of $\boldsymbol{\theta}_0$. Further, the maximizers of $\ell(\boldsymbol{\theta})$ over Θ_0 and over Θ_1 are likely to be $n^{1/2}$-consistent. This is because consistency of the estimator is not affected by the true parameter being on the boundary, and $\sqrt{n}$-consistency follows once we have a quadratic approximation of $\ell(\boldsymbol{\theta})$ similar to (4.2) by arguments indicated at the beginning of this chapter. If the objective function $\ell(\boldsymbol{\theta})$ does not have a quadratic approximation similar to (4.2) or the estimators are not $\sqrt{n}$-consistent, then the general approach of this section will need modifications to obtain the asymptotic distribution of *LRT*.

The Condition $Q2$ defined below is a slight modification of Condition Q; the modification is made in view of the fact that $\boldsymbol{\theta}_0$ may be a boundary point of Θ.

Condition Q2. There exist $\boldsymbol{S}_*(\boldsymbol{\theta})$ and $\mathcal{I}_*(\boldsymbol{\theta})$ such that (4.3) holds with $\boldsymbol{S}(\boldsymbol{\theta})$ and $\mathcal{I}(\boldsymbol{\theta})$ replaced by $\boldsymbol{S}_*(\boldsymbol{\theta})$ and $\mathcal{I}_*(\boldsymbol{\theta})$ respectively. Further,

1. $n^{-1/2}\boldsymbol{S}_*(\boldsymbol{\theta}_0) \xrightarrow{d} N\{\mathbf{0}, \mathcal{I}_*(\boldsymbol{\theta}_0)\}$.
2. The maximizers of $\ell(\boldsymbol{\theta})$ over Θ_0 and over Θ_1 are $\sqrt{n}$-consistent for $\boldsymbol{\theta}_0$. ■

There is room for fine tuning this condition; for example, $\sqrt{n}$-consistency would follow from (4.3) and consistency. However, it is convenient for us to state them explicitly.

If $\boldsymbol{\theta}_0$ is an interior point of Θ, then $\boldsymbol{S}_*$ and $\mathcal{I}_*$ would be the $\boldsymbol{S}$ and $\mathcal{I}$ in the previous subsection. If Θ is a rectangle in the Euclidean space, $\boldsymbol{\theta}_0$ is a boundary point of Θ, and $\ell(\boldsymbol{\theta})$ has partial derivatives at $\boldsymbol{\theta}_0$ from the appropriate directions then Condition $Q2$ could be established using Taylor series expansions (for example, see Self and Liang (1987) and Andrews (1999)).

If Θ does not have the shape of a rectangle and $\boldsymbol{\theta}_0$ is a boundary point of Θ, then it is likely that Condition $Q2$ can be established using directional derivatives (see Shapiro (2000b)). Directional derivatives of $\ell(\boldsymbol{\theta})$ at $\boldsymbol{\theta}_0$ would be required only in directions that take us into the approximating cone of Θ at $\boldsymbol{\theta}_0$. The proof of the next result is essentially the same as that for the previous one, hence omitted.

Proposition 4.8.2 *Assume that Condition $Q2$ is satisfied and that $H_0 : \boldsymbol{\theta} \in \Theta_0$ holds. Let $\boldsymbol{\theta}_0$ denote the true value of $\boldsymbol{\theta}$ in Θ_0, $\|\boldsymbol{x}\|^2 = \boldsymbol{x}^T\mathcal{I}_*(\boldsymbol{\theta}_0)\boldsymbol{x}$, $\boldsymbol{Z} \sim N\{\mathbf{0}, \mathcal{I}_*(\boldsymbol{\theta}_0)^{-1}\}$, $\boldsymbol{X} = \boldsymbol{\theta}_0 + \boldsymbol{Z}$, and $\boldsymbol{U}_n = n^{-1/2}\mathcal{I}_*(\boldsymbol{\theta}_0)^{-1}\boldsymbol{S}_*(\boldsymbol{\theta}_0)$. Then,*

1. *$LRT = \|\boldsymbol{U}_n - \mathcal{T}(\Theta_0; \boldsymbol{\theta}_0)\|^2 - \|\boldsymbol{U}_n - \mathcal{T}(\Theta_1; \boldsymbol{\theta}_0)\|^2 + o_p(1)$.*
2. *The asymptotic null distribution of LRT is equal to the distribution of*

$$\|\boldsymbol{X} - \mathcal{A}(\Theta_0; \boldsymbol{\theta}_0)\|^2 - \|\boldsymbol{X} - \mathcal{A}(\Theta_1; \boldsymbol{\theta}_0)\|^2, \tag{4.108}$$

which is also equal to the distribution of

$$\|\boldsymbol{Z} - \mathcal{T}(\Theta_0; \boldsymbol{\theta}_0)\|^2 - \|\boldsymbol{Z} - \mathcal{T}(\Theta_1; \boldsymbol{\theta}_0)\|^2. \tag{4.109}$$

■

The conditions under which Condition $Q2$ holds when $\boldsymbol{\theta}_0$ is a boundary point of Θ, appear to give the impression that the likelihood must behave at $\boldsymbol{\theta}_0$ almost as if it is an interior point of an extended parameter space over which Condition $Q2$ is satisfied, and $E\{\ell(\boldsymbol{\theta})\}$ has a maximum at $\boldsymbol{\theta}_0$ that is also a stationary point of $E\{\ell(\boldsymbol{\theta})\}$. Thus, although Condition Q for the case when $\boldsymbol{\theta}_0$ is an interior point of Θ and Condition $Q2$ when $\boldsymbol{\theta}_0$ is a boundary point of Θ appear to be different, the requirements on the statistical model appear to be not very different. A question of interest is, if Condition $Q2$ is satisfied, perhaps with $\boldsymbol{S}_*$ and $\mathcal{I}_*$ defined using directional derivatives, then is it possible to extend the parameter space such that it includes an open neighborhood of $\boldsymbol{\theta}_0$ and the derivations in the previous subsection still hold? Although we believe that the answer is "yes," the technical details do not appear to be available yet.

If $\boldsymbol{\theta}_0$ is an interior point of Θ, then it follows from

$$n^{1/2}(\hat{\boldsymbol{\theta}} - \boldsymbol{\theta}_0) = n^{-1/2}\mathcal{I}_{\boldsymbol{\theta}_0}^{-1}\boldsymbol{S}(\boldsymbol{\theta}_0) + O_p(n^{-1/2})$$

that $n^{1/2}(\hat{\boldsymbol{\theta}} - \boldsymbol{\theta}_0)$ and $n^{-1/2}\mathcal{I}(\boldsymbol{\theta}_0)^{-1}\boldsymbol{S}(\boldsymbol{\theta}_0)$ are essentially interchangeable in the proofs of the asymptotic results (the $\boldsymbol{Z}_n$ in the proof of Proposition 4.8.1 and $\boldsymbol{U}_n$ in the proof Proposition 4.8.2 are equivalent, and hence the two derivations differ by an $o_p(1)$ term only).

4.8.3 Tests with General Objective Functions

For the case when there are no inequality constraints, the classical inference procedures include M-estimation, minimum distance, partial likelihood, quasi-likelihood, and empirical likelihood and more generally, pseudo-likelihoods (see Pace and Salvan (1997)). The principles underlying these also lend themselves for inference when there are inequality constraints on parameters. In this subsection, we will discuss some of the results that are available in these directions.

Let the objective function be denoted by $R_n(\boldsymbol{\theta})$; this replaces the role of loglikelihood. The estimator of $\boldsymbol{\theta}$ is the maximizer of $R_n(\boldsymbol{\theta})$. Assume that the following condition, which is a more generalized version of those used in the previous subsections, is satisfied.

Condition $Q3$:
There exist a vector function $\boldsymbol{S}_*(\boldsymbol{\theta})$ and symmetric and positive definite matrices, $\boldsymbol{G}_*(\boldsymbol{\theta})$ and $\mathcal{I}_*(\boldsymbol{\theta})$, such that the following are satisfied:

1. (4.3) holds with $\{\ell(\boldsymbol{\theta}), \boldsymbol{S}(\boldsymbol{\theta}), \mathcal{I}(\boldsymbol{\theta})\}$ replaced by $\{R_n(\boldsymbol{\theta}), \boldsymbol{S}_*(\boldsymbol{\theta}), \boldsymbol{G}_*(\boldsymbol{\theta})\}$.
2. $n^{-1/2}\boldsymbol{S}_*(\boldsymbol{\theta}_0) \xrightarrow{d} N\{\boldsymbol{0}, \mathcal{I}_*(\boldsymbol{\theta}_0)\}$.
3. The maximizer of $R_n(\boldsymbol{\theta})$ over Θ_i is $n^{1/2}$-consistent for $\boldsymbol{\theta}_0$ when $\boldsymbol{\theta}_0 \in \Theta_i$ $(i = 0, 1)$. ■

Now, a likelihood ratio type statistic for testing $\boldsymbol{\theta} \in \Theta_0$ against $\boldsymbol{\theta} \in \Theta_1$ is

$$T_n = 2\{\sup_{\boldsymbol{\theta}\in\Theta_1} R_n(\boldsymbol{\theta}) - \sup_{\boldsymbol{\theta}\in\Theta_0} R_n(\boldsymbol{\theta})\}. \tag{4.110}$$

The next proposition shows that the asymptotic null distribution of T_n is obtained by an argument similar to those in the previous subsection.

Proposition 4.8.3 *Assume that Condition $Q3$ is satisfied and that $H_0 : \boldsymbol{\theta} \in \Theta_0$ holds. Let $\boldsymbol{\theta}_0$ denote the true value of $\boldsymbol{\theta}$ in Θ_0, $\|\boldsymbol{x}\|^2 = \boldsymbol{x}^T\boldsymbol{G}_*(\boldsymbol{\theta}_0)\boldsymbol{x}$, and $\boldsymbol{U}_n = n^{-1/2}\boldsymbol{G}_*(\boldsymbol{\theta}_0)^{-1}\boldsymbol{S}_*(\boldsymbol{\theta}_0)$. Then*

1. $T_n = \|\boldsymbol{U}_n - \mathcal{T}(\Theta_0; \boldsymbol{\theta}_0)\|^2 - \|\boldsymbol{U}_n - \mathcal{T}(\Theta_1; \boldsymbol{\theta}_0)\|^2 + o_p(1)$.
2. *The asymptotic null distribution of T_n is equal to that of*

$$\|\boldsymbol{U} - \mathcal{T}(\Theta_0; \boldsymbol{\theta}_0)\|^2 - \|\boldsymbol{U} - \mathcal{T}(\Theta_1; \boldsymbol{\theta}_0)\|^2 \tag{4.111}$$

where $\boldsymbol{U} \sim N\{\mathbf{0}, \boldsymbol{G}_*(\boldsymbol{\theta}_0)^{-1}\mathcal{I}_*(\boldsymbol{\theta}_0)\boldsymbol{G}_*(\boldsymbol{\theta}_0)^{-1}\}$.

Proof : It follows from Condition $Q3(1)$ that, for $\boldsymbol{\theta} - \boldsymbol{\theta}_0 = O(n^{-1/2})$,

$$R_n(\boldsymbol{\theta}) = R_n(\boldsymbol{\theta}_0) - (n/2)\|n^{-1/2}\boldsymbol{U}_n - (\boldsymbol{\theta} - \boldsymbol{\theta}_0)\|^2 + (1/2)n^{-1}\|\boldsymbol{U}(\boldsymbol{\theta}_0)\|^2 + o_p(1).$$

Now,

$$\begin{aligned} T_n &= 2[\sup\{R_n(\boldsymbol{\theta}) : \boldsymbol{\theta} \in \Theta_1\} - \sup\{R_n(\boldsymbol{\theta}) : \boldsymbol{\theta} \in \Theta_0\}] \\ &= n\|n^{-1/2}\boldsymbol{U}_n - (\Theta_0 - \boldsymbol{\theta}_0)\|^2 - n\|n^{-1/2}\boldsymbol{U}_n - (\Theta_1 - \boldsymbol{\theta}_0)\|^2 + o_p(1) \\ &= \|\boldsymbol{U}_n - \mathcal{T}(\Theta_0; \boldsymbol{\theta}_0)\|^2 - \|\boldsymbol{U}_n - \mathcal{T}(\Theta_1; \boldsymbol{\theta}_0)\|^2 + o_p(1) \\ &\xrightarrow{d} \|\boldsymbol{U} - \mathcal{T}(\Theta_0; \boldsymbol{\theta}_0)\|^2 - \|\boldsymbol{U} - \mathcal{T}(\Theta_1; \boldsymbol{\theta}_0)\|^2. \end{aligned}$$ ■

Note that the quadratic form $\|\boldsymbol{U} - \boldsymbol{\alpha}\|^2$ in (4.111), which is equal to $(\boldsymbol{U} - \boldsymbol{\alpha})^T\boldsymbol{G}_*(\boldsymbol{\theta}_0)(\boldsymbol{U} - \boldsymbol{\alpha})$, is with respect to the matrix $\boldsymbol{G}_*(\boldsymbol{\theta}_0)^{-1}$ but the covariance matrix of $\boldsymbol{U}$ is $\boldsymbol{G}_*(\boldsymbol{\theta}_0)^{-1}\mathcal{I}_*(\boldsymbol{\theta}_0)\boldsymbol{G}_*(\boldsymbol{\theta}_0)^{-1}$. Therefore, if $\mathcal{I}_*(\boldsymbol{\theta}_0) \neq \boldsymbol{G}_*(\boldsymbol{\theta}_0)$, then the quadratic form $\|\boldsymbol{U} - \boldsymbol{\alpha}\|^2$ may not be with respect to the covariance matrix of $\boldsymbol{U}$ and hence the results in Chapter 2 for normal distributions are not applicable for obtaining the asymptotic distribution of T_n. In particular, (i) for testing $\boldsymbol{\theta} = \mathbf{0}$ against $\boldsymbol{\theta} \neq \mathbf{0}$, the asymptotic null distribution of T_n may not be a chi-square, and (ii) for testing $\boldsymbol{R\theta} = \mathbf{0}$ against $\boldsymbol{R\theta} \geqslant \mathbf{0}$, the asymptotic null distribution of T_n in (4.111) may not be a chi-bar-square.

Suppose that Θ is open and that $H_0 : \boldsymbol{\theta} \in \Theta_0$ holds. As usual, let $\hat{\boldsymbol{\theta}}$ denote the unrestricted estimator of $\boldsymbol{\theta}$. Then by standard arguments, $n^{1/2}(\hat{\boldsymbol{\theta}} - \boldsymbol{\theta}_0) \xrightarrow{d} N\{\mathbf{0}, \boldsymbol{V}(\boldsymbol{\theta}_0\}$, where $\boldsymbol{V}(\boldsymbol{\theta}_0) = \boldsymbol{G}_*(\boldsymbol{\theta}_0)^{-1}\mathcal{I}_*(\boldsymbol{\theta}_0)\boldsymbol{G}_*(\boldsymbol{\theta}_0)^{-1}$. Therefore, a W_D-type statistic for testing $\boldsymbol{\theta} \in \Theta_0$ against $\boldsymbol{\theta} \in \Theta_1$ is

$$W_D = \min_{\boldsymbol{\theta} \in \Theta_0} n(\hat{\boldsymbol{\theta}} - \boldsymbol{\theta})^T\hat{\boldsymbol{V}}^{-1}(\hat{\boldsymbol{\theta}} - \boldsymbol{\theta}) - \min_{\boldsymbol{\theta} \in \Theta_1} n(\hat{\boldsymbol{\theta}} - \boldsymbol{\theta})^T\hat{\boldsymbol{V}}^{-1}(\hat{\boldsymbol{\theta}} - \boldsymbol{\theta}), \quad (4.112)$$

where $\hat{\boldsymbol{V}}$ is a consistent estimator of $\boldsymbol{V}(\boldsymbol{\theta}_0)$, for example, $\boldsymbol{V}(\hat{\boldsymbol{\theta}})$. The asymptotic distribution of the foregoing W_D is equal to that of

$$\min_{\boldsymbol{\theta} \in \mathcal{T}(\Theta_0; \boldsymbol{\theta}_0)} (\boldsymbol{Z} - \boldsymbol{\theta})^T\boldsymbol{V}(\boldsymbol{\theta}_0)^{-1}(\boldsymbol{Z} - \boldsymbol{\theta}) - \min_{\boldsymbol{\theta} \in \mathcal{T}(\Theta_1; \boldsymbol{\theta}_0)} (\boldsymbol{Z} - \boldsymbol{\theta})^T\boldsymbol{V}(\boldsymbol{\theta}_0)^{-1}(\boldsymbol{Z} - \boldsymbol{\theta})$$

where $\boldsymbol{Z} \sim N\{\mathbf{0}, \boldsymbol{V}(\boldsymbol{\theta}_0)\}$.

Note that the quadratic form $(\boldsymbol{Z} - \boldsymbol{\theta})^T\boldsymbol{V}(\boldsymbol{\theta}_0)^{-1}(\boldsymbol{Z} - \boldsymbol{\theta})$, which appears in the asymptotic null distribution of W_D is with respect to the covariance matrix $\boldsymbol{V}(\boldsymbol{\theta}_0)$ of $\boldsymbol{Z}$. Consequently, it follows that for testing $\boldsymbol{R\theta} = \mathbf{0}$ against $\boldsymbol{R}_1\boldsymbol{\theta} \geqslant \mathbf{0}$, the asymptotic null distribution of W_D is a chi-bar-square. Therefore, the asymptotic null distribution of W_D in (4.112) and that of the likelihood ratio type statistic T_n in (4.110) are not necessarily the same, and hence are unlikely to be asymptotically equivalent in this type of setting.

For some partial-likelihood and quasi-likelihood, $\boldsymbol{G}_*(\boldsymbol{\theta})$ and $\mathcal{I}_*(\boldsymbol{\theta})$ are equal except for a scalar multiplicative constant. Since the multiplicative constant factors out from matrix products, it (i.e., multiplicative constant) can be replaced by a consistent estimator of it and the required results can be obtained (see quasi-likelihood method on page 159 and Silvapulle (1994)).

4.8.4 Examples

Example 4.8.1 *Asymptotic distribution of the LRT of* $\boldsymbol{g}(\boldsymbol{\theta}) = \mathbf{0}$ *vs* $\boldsymbol{g}(\boldsymbol{\theta}) \geqslant \mathbf{0}$.

Let $\boldsymbol{Y} = (Y_1, Y_2)^T \sim N(\boldsymbol{\theta}, \boldsymbol{V})$ where $\boldsymbol{\theta} \in \Theta = \mathbb{R}^2$ and $\boldsymbol{V}$ is known and does not depend on $\boldsymbol{\theta}$. Let the null and alternative hypotheses be

$$H_0 : \boldsymbol{g}(\boldsymbol{\theta}) = \mathbf{0} \text{ and } H_1 : \boldsymbol{g}(\boldsymbol{\theta}) \geqslant \mathbf{0},$$

respectively, where $\boldsymbol{g} = (g_1, g_2)^T, g_1(\boldsymbol{\theta}) = \theta_2 - \theta_1^2 - \theta_1$ and $g_2(\boldsymbol{\theta}) = \theta_2 - \theta_1^2 + \theta_1$; see Fig. 4.1. Let us obtain the asymptotic null distribution of the *LRT* given by Proposition 4.8.1. Let $\bar{\boldsymbol{Y}}$ denote the mean of n independently and identically distributed observations on $\boldsymbol{Y}$; then $\bar{\boldsymbol{Y}} \sim N(\boldsymbol{\theta}, n^{-1}\boldsymbol{V})$. Now, $\Theta_0 = \{\boldsymbol{\theta} \in \mathbb{R}^2 : \boldsymbol{g}(\boldsymbol{\theta}) = \mathbf{0}\}$ and $\Theta_1 = \{\boldsymbol{\theta} \in \mathbb{R}^2 : \boldsymbol{g}(\boldsymbol{\theta}) \geqslant 0\}$, Θ_1 is shown as Ω in Fig. 4.1 on page 184. Clearly, $\Theta_0 = \{\mathbf{0}\}$. Assume that H_0 is true; then the true value $\boldsymbol{\theta}_0$ is $\mathbf{0}, \mathcal{T}(\Theta_0; \boldsymbol{\theta}_0) = \{\mathbf{0}\}$, and $\mathcal{T}(\Theta_1; \boldsymbol{\theta}_0) = \{\boldsymbol{\theta} : \theta_2 \geqslant \theta_1, \theta_2 \geqslant -\theta_1\} = \{\boldsymbol{\theta} : \boldsymbol{R}\boldsymbol{\theta} \geqslant \mathbf{0}\}$ where $\boldsymbol{R}$ is the 2×2 matrix $[-1, 1|1, 1]$; $\mathcal{T}(\Theta_1; \mathbf{0})$ is shown as $\mathcal{T}$ in Fig. 4.1. Let $\boldsymbol{Z} \sim N(\mathbf{0}, \boldsymbol{V})$. Now, since $\ell(\boldsymbol{\theta}) = (-1/2)n(\bar{\boldsymbol{Y}} - \boldsymbol{\theta})^T\boldsymbol{V}^{-1}(\bar{\boldsymbol{Y}} - \boldsymbol{\theta}) + const$ and $\mathcal{I}(\boldsymbol{\theta}) = \boldsymbol{V}^{-1}$, we have the following:

$$\begin{aligned} LRT &= n\|\bar{\boldsymbol{Y}}\|_{\boldsymbol{V}}^2 - n\|\bar{\boldsymbol{Y}} - \Theta_1\|_{\boldsymbol{V}}^2 \\ &= n\|\bar{\boldsymbol{Y}}\|_{\boldsymbol{V}}^2 - n\|\bar{\boldsymbol{Y}} - \mathcal{A}(\Theta_1; \mathbf{0})\|_{\boldsymbol{V}}^2 + o_p(1) \\ &\xrightarrow{d} \|\boldsymbol{Z}\|_{\boldsymbol{V}}^2 - \|\boldsymbol{Z} - \mathcal{T}(\Theta_1; \mathbf{0})\|_{\boldsymbol{V}}^2 \sim \bar{\chi}^2\{\boldsymbol{V}, \mathcal{T}(\Theta_1; \mathbf{0})\}. \end{aligned}$$

Therefore, by (3.8)

$$\text{pr}(LRT \geqslant c|H_0) \to 0.5\text{pr}(\chi_1^2 \geqslant c) + (0.5 - q)\text{pr}(\chi_2^2 \geqslant c) \quad c > 0$$

where $(0.5 - q)$ is as in (3.8). ∎

Example 4.8.2 *Asymptotic null distribution of the LRT of* $\boldsymbol{g}(\boldsymbol{\theta}) \geqslant \mathbf{0}$ *vs* $\boldsymbol{g}(\boldsymbol{\theta}) \not\geqslant \mathbf{0}$.

Let $\boldsymbol{Y}, \bar{\boldsymbol{Y}}, \boldsymbol{V}$, and $\boldsymbol{g}$ be as in the previous example. Let us obtain the asymptotic null distribution of the *LRT* for testing

$$H_0 : \boldsymbol{g}(\boldsymbol{\theta}) \geqslant \mathbf{0} \text{ against } H_1 : \boldsymbol{g}(\boldsymbol{\theta}) \not\geqslant \mathbf{0}.$$

The relevant asymptotic result is given in Proposition 4.8.1. By definition, $\Theta_0 = \{\boldsymbol{\theta} \in \mathbb{R}^2 : \boldsymbol{g}(\boldsymbol{\theta}) \geqslant 0\}$ and $\Theta_1 = \mathbb{R}^2$. Assume that the null hypothesis holds and let $\boldsymbol{\theta}_0$ denote the true value of $\boldsymbol{\theta}$. Clearly, $\mathcal{T}(\Theta_1; \boldsymbol{\theta}_0) = \mathbb{R}^2$. However, $\mathcal{T}(\Theta_0; \boldsymbol{\theta}_0)$ depends on exactly which of the inequalities defining Θ_0 are active at $\boldsymbol{\theta}_0$ (recall that $g_i(\boldsymbol{\theta}) \geqslant 0$ is said to be *active* at $\boldsymbol{\theta}_0$ if $g_i(\boldsymbol{\theta}_0) = 0$). We shall consider four different cases corresponding to different tangent cones of Θ_0 at $\boldsymbol{\theta}_0$.

Case 1: $g_1(\boldsymbol{\theta}_0) > 0$ and $g_2(\boldsymbol{\theta}_0) > 0$.

The condition "$g_1(\boldsymbol{\theta}_0) > 0$ and $g_2(\boldsymbol{\theta}_0) > 0$" is equivalent to saying that $\boldsymbol{\theta}_0$ is an interior point of Θ_0. Therefore, $\mathcal{T}(\Theta_0; \boldsymbol{\theta}_0) = \mathbb{R}^2$ and hence $\mathcal{T}(\Theta_0; \boldsymbol{\theta}_0) = \mathcal{T}(\Theta_1; \boldsymbol{\theta}_0)$.

Now, it follows from (4.105) that $LRT \xrightarrow{p} 0$. Therefore, the probability of rejecting H_0 when the true value is an interior point of Θ_0 converges to zero. Intuitively, this is to be expected, because it follows from $\hat{\boldsymbol{\theta}} \xrightarrow{p} \boldsymbol{\theta}_0$ that $LRT = 0$ and hence the *LRT* will not reject the null hypothesis with probability going to one.

Case 2: $g_1(\boldsymbol{\theta}_0) = 0, g_2(\boldsymbol{\theta}_0) > 0$.

The tangent cone at $\boldsymbol{\theta}_0$ is obtained by replacing the curved boundaries at $\boldsymbol{\theta}_0$ by their linear approximations. Equivalently, by Proposition 4.7.3, we have that $\mathcal{A}(\Theta_0;\boldsymbol{\theta}_0) = \{\boldsymbol{\theta} : (\boldsymbol{\theta}-\boldsymbol{\theta}_0)^T(\partial/\partial\boldsymbol{\theta})g_1(\boldsymbol{\theta}_0) \geqslant 0\}$ and $\mathcal{T}(\Theta_0;\boldsymbol{\theta}_0) = \mathcal{A}(\Theta_0;\boldsymbol{\theta}_0) - \boldsymbol{\theta}_0$. Therefore, $\mathcal{T}(\Theta_0;\boldsymbol{\theta}_0) = \{\boldsymbol{x} : \boldsymbol{x}^T(\partial/\partial\boldsymbol{\theta})g_1(\boldsymbol{\theta}_0) \geqslant 0\} = \{\boldsymbol{x} : x_2 \geqslant x_1(1+2\theta_{01})\}$. Now, with $\boldsymbol{Z} \sim N(\boldsymbol{0}, \boldsymbol{V})$ we have that

$$\text{pr}_{\boldsymbol{\theta}_0}(LRT \geqslant c \mid H_0) \to \text{pr}_{\boldsymbol{\theta}_0}[\|\boldsymbol{Z} - \mathcal{T}(\Theta_0;\boldsymbol{\theta}_0)\|^2_{\boldsymbol{V}} \geqslant c] = (1/2)\text{pr}(\chi^2_1 \geqslant c).$$

Case 3: $g_1(\boldsymbol{\theta}_0) = g_2(\boldsymbol{\theta}_0) = 0$.

As was shown in the previous example, $\boldsymbol{\theta}_0 = \boldsymbol{0}$ and $\mathcal{T}(\Theta_0;\boldsymbol{\theta}_0) = \{\boldsymbol{d} \in \mathbb{R}^2 : d_2 \geqslant d_1, d_2 \geqslant -d_1\} = \{\boldsymbol{d} : \boldsymbol{R}\boldsymbol{d} \geqslant \boldsymbol{0}\}$, where $\boldsymbol{R}$ is the 2×2 nonsingular matrix $[-1,1|1,1]$. Therefore, by (4.105), $LRT \xrightarrow{d} \|\boldsymbol{Z} - \mathcal{T}(\Theta_0;\boldsymbol{\theta}_0)\|^2_{\boldsymbol{V}}$ where $\boldsymbol{Z} \sim N(\boldsymbol{0}, \boldsymbol{V})$. Now by (3.39) it follows that

$$\text{pr}_{\boldsymbol{\theta}_0}(LRT \geqslant c) \to 0.5\text{pr}(\chi^2_1 \geqslant c) + q\text{pr}(\chi^2_2 \geqslant c), \quad c > 0$$

where

$$q = (2\pi)^{-1}\cos^{-1}[(\boldsymbol{R}_1\boldsymbol{V}\boldsymbol{R}_2^T)/\{(\boldsymbol{R}_1\boldsymbol{V}\boldsymbol{R}_1^T)(\boldsymbol{R}_2\boldsymbol{V}\boldsymbol{R}_2^T)\}^{1/2}].$$

Case 4: $g_1(\boldsymbol{\theta}_0) > 0$ and $g_2(\boldsymbol{\theta}_0) = 0$.

This is similar to Case 2 and hence $\text{pr}(LRT \geqslant c|H_0) \to (1/2)\text{pr}(\chi^2_1 \geqslant c)$.

It follows from the different cases considered above that $\text{pr}(LRT \geqslant c \mid H_0)$ is a maximum when $g_1(\boldsymbol{\theta}_0) = g_2(\boldsymbol{\theta}_0) = 0$. Therefore, Case 3 is the least favorable null configuration, and hence $\boldsymbol{\theta}_0 = \boldsymbol{0}$ is the least favorable null value, and

$$\lim \text{pr}(LRT \geqslant c|H_0) \leq 0.5\text{pr}(\chi^2_1 \geqslant c) + q\text{pr}(\chi^2_2 \geqslant c).$$

Since q is known, we can apply the *LRT* easily in this case. It will be seen later that if the null hypothesis involves several inequality constraints, for example, say $g_1(\boldsymbol{\theta}) \geqslant 0, \ldots, g_k(\boldsymbol{\theta}) \geqslant 0$, $(k \geq 3)$, then a least favorable null value may or may not be a point in $\{\boldsymbol{\theta} : g_1(\boldsymbol{\theta}) = \ldots = g_k(\boldsymbol{\theta}) = \boldsymbol{0}\}$. The foregoing example illustrates how the asymptotic null distribution of the LRT depends on the shape of the parameter space at the null value. ■

4.8.5 Test of $h(\theta) \geqslant 0$ Against $h(\theta) \not\geqslant 0$

Let $\boldsymbol{h}(\boldsymbol{\theta})$ be a continuously differentiable $k \times 1$ vector function. Let the null and alternative hypotheses be

$$H_0 : \boldsymbol{h}(\boldsymbol{\theta}) \geqslant \boldsymbol{0} \text{ and } H_1 : \boldsymbol{h}(\boldsymbol{\theta}) \not\geqslant \boldsymbol{0}, \tag{4.113}$$

respectively. The formulation allows for the number of inequalities in H_0 to be larger than the number of components of $\boldsymbol{\theta}$. Consequently, the rank of $\nabla \boldsymbol{h}(\boldsymbol{\theta})$ may be less than k. Therefore, even if $\sqrt{n}(\hat{\boldsymbol{\theta}} - \boldsymbol{\theta}_0)$ converges to $N(\mathbf{0}, \boldsymbol{V})$ where $\boldsymbol{V}$ is positive definite, the asymptotic covariance of $\boldsymbol{h}(\hat{\boldsymbol{\theta}})$ may not be positive definite. The foregoing null and alternative hypotheses may also be stated as

$$H_0 : \boldsymbol{\theta} \in \Theta_0 \qquad \text{and} \qquad H_0 : \boldsymbol{\theta} \in \Theta_1,$$

respectively, where $\Theta_0 = \{\boldsymbol{\theta} : \boldsymbol{h}(\boldsymbol{\theta}) \geqslant \mathbf{0}\}$ and $\Theta_1 = \mathbb{R}^p$. Therefore, $LRT = \sup_{\boldsymbol{h}(\boldsymbol{\theta})\geqslant \mathbf{0}} \ell(\boldsymbol{\theta})$. As usual, assume that Θ_0 is Chernoff regular. Further, assume that Condition Q2 in Section 4.8.2 is satisfied.

To study the asymptotic null distribution of the *LRT*, assume that the null hypothesis holds and let $\boldsymbol{\theta}_0$ denote the true value. Wolak (1987, 1989, 1991) studied this testing problem and obtained important results; he also highlighted some of the difficulties that arise in this type of problems. Wolak (1988) pointed out that there is no guarantee that $\boldsymbol{h}(\boldsymbol{\theta}) = \mathbf{0}$ would correspond to the least favorable null value. It follows from Section 4.8.2 that $LRT \xrightarrow{d} \|\boldsymbol{Z} - \mathcal{T}(\Theta_0; \theta_0)\|^2$ where $\boldsymbol{Z} \sim N\{\mathbf{0}, \mathcal{I}(\boldsymbol{\theta}_0)^{-1}\}$ and $\|\boldsymbol{x}\|^2 = \boldsymbol{x}^T \mathcal{I}(\boldsymbol{\theta}_0)\boldsymbol{x}$.

For simplicity, let us consider the case when $\mathcal{T}(\Theta_0; \theta_0)$ is a closed convex cone. Now, from Theorem 3.4.2 and the fact that

$$\|\boldsymbol{Z} - \mathcal{T}(\Theta_0; \boldsymbol{\theta}_0)\|^2 = \|\boldsymbol{Z}\|^2 - \|\boldsymbol{Z} - \mathcal{T}^o(\Theta_0; \boldsymbol{\theta}_0)\|^2,$$

it follows that

$$\text{pr}_{\boldsymbol{\theta}_0}(LRT \geqslant c \mid H_0) \to \sum_{i=0}^{p} w_i \text{pr}(\chi_i^2 \geqslant c) \tag{4.114}$$

where $w_i = w_i\{p, \mathcal{I}(\boldsymbol{\theta}_0)^{-1}, \mathcal{T}^o(\Theta_0; \boldsymbol{\theta}_0)\}$ is the usual chi-bar square weight and $\mathcal{T}^o$ denotes the polar cone of $\mathcal{T}$ with respect to $\mathcal{I}(\boldsymbol{\theta}_0)^{-1}$. In general, the $\bar{\chi}^2$-distribution in (4.114) depends on $\boldsymbol{\theta}_0$ through the weights, $w_i\{p, \mathcal{I}(\boldsymbol{\theta}_0)^{-1}, \mathcal{T}^o(\Theta_0; \boldsymbol{\theta}_0)\}$. Therefore, to implement a large sample test, we need to find

$$\sup_{\boldsymbol{\theta}\in\Theta_0} \sum_{i=0}^{p} w_i\{p, \mathcal{I}(\boldsymbol{\theta}_0)^{-1}, \mathcal{T}^o(\Theta_0; \boldsymbol{\theta}_0)\}\text{pr}(\chi_i^2 \geqslant c), \quad c > 0. \tag{4.115}$$

Computation of this supremum can be a difficult task; however, we can always obtain bounds for it.

If $\boldsymbol{\theta}_0$ is an interior point of Θ_0 then $\mathcal{T}(\Theta_0; \boldsymbol{\theta}_0) = \mathbb{R}^p$ and hence $\text{pr}(LRT \geqslant c | \boldsymbol{\theta} = \boldsymbol{\theta}_0) \to 0$ for $c > 0$. Therefore, the supremum in (4.115) cannot be achieved at an interior point of Θ_0, and hence for the purposes of computing (4.115), we may restrict the attention to values of $\boldsymbol{\theta}_0$ on the boundary of Θ_0.

It can be shown that at the least favorable null value, at least two of the inequality constraints in $\boldsymbol{h}(\boldsymbol{\theta}_0) \geqslant \mathbf{0}$ must be active; this was shown by Wolak (1991). Thus, it follows that if there are only two inequality constraints, say $h_1(\boldsymbol{\theta}) \geqslant 0$ and $h_2(\boldsymbol{\theta}) \geqslant 0$, then they both must be active at the least favorable null value.

In what follows, we shall assume that $\boldsymbol{\theta}_0$ is a boundary point of Θ_0. Let us consider a setting that is simpler, yet general enough for many practical situations. Assume that $\boldsymbol{h}(\boldsymbol{\theta})$ is continuously differentiable and that the Mangasarian-Fromowitz condition (see Proposition 4.7.3) is satisfied. Because $\boldsymbol{\theta}_0$ is a boundary point of Θ_0, it follows that $h_s(\boldsymbol{\theta}_0) = 0$ for at least one s in $\{1, \ldots, k\}$. Let $\boldsymbol{a}_i$ denote $(\partial/\partial\boldsymbol{\theta})h_i(\boldsymbol{\theta}_0)$ and let $\{j_1, \ldots, j_m\}$ denote $\{i : h_i(\boldsymbol{\theta}_0) = 0\}$, the set of all indices of the active constraints at $\boldsymbol{\theta}_0$; note that m, the number of active constraints at $\boldsymbol{\theta}_0$, is itself a function of $\boldsymbol{\theta}_0$. Then

$$\mathcal{T}(\Theta_0; \boldsymbol{\theta}_0) = \{\boldsymbol{\alpha} : \boldsymbol{a}_{j_1}^T \boldsymbol{\alpha} \geqslant 0, \ldots, \boldsymbol{a}_{j_m}^T \boldsymbol{\alpha} \geqslant 0\}.$$

Because $\mathcal{T}(\Theta_0; \boldsymbol{\theta}_0)$ is a polyhedral, it is a closed convex cone, and hence it follows that $\|\boldsymbol{Z} - \mathcal{T}(\Theta_0; \theta_0)\|^2$, the limiting distribution of *LRT*, is a chi-bar-square (see Theorem 3.8.2).

Now, let $\boldsymbol{H}(\boldsymbol{\theta}_0) = [\boldsymbol{a}_{j_1}, \ldots, \boldsymbol{a}_{j_m}]^T$. Let us consider the case when $rank\{\boldsymbol{H}(\boldsymbol{\theta}_0)\} = m(\boldsymbol{\theta}_0)$. Since $LRT \xrightarrow{d} \|\boldsymbol{Z} - \mathcal{T}(\Theta_0; \theta_0)\|^2$, we have that

$$\text{pr}(LRT \geqslant c | \boldsymbol{\theta} = \boldsymbol{\theta}_0) \to \sum_{i=0}^{m} w_{m-i}\{m, \boldsymbol{V}(\boldsymbol{\theta}_0)\}\text{pr}(\chi_i^2 \geqslant c), \tag{4.116}$$

where $\boldsymbol{V}(\boldsymbol{\theta}_0) = \boldsymbol{H}(\boldsymbol{\theta}_0)\mathcal{I}(\boldsymbol{\theta}_0)^{-1}\boldsymbol{H}(\boldsymbol{\theta}_0)^T$. With t denoting the sample value of *LRT*, the large sample p-value, p_{sup}, is

$$\sup_{\boldsymbol{\theta}_0 \in \Theta_0} \sum_{i=0}^{m} w_{m-i}\{m, \boldsymbol{V}(\boldsymbol{\theta}_0)\}\text{pr}(\chi_i^2 \geqslant t).$$

In general, the value of $\boldsymbol{\theta}_0$ at which this supremum is achieved may also depend on t. Consequently, there may not be a single least favorable null value for all possible values of t. It is not easy to maximize the tail probability of a chi-bar square distribution numerically using Newton-type iteration, because analytical expressions are not available for the derivatives of the tail probability of (4.115). Recent developments using simulation approaches may be helpful (see Shapiro, 2000); the utility of this approach remains to be investigated. In general, there is no easy way to compute the supremum in (4.115).

If computation of the supremum in (4.115) does not appear feasible, we can apply a bounds test using,

$$\sup_{\boldsymbol{\theta}_0 \in \Theta_0} \sum_{i=0}^{p} w_i\{p, \mathcal{I}(\boldsymbol{\theta}_0)^{-1}, \mathcal{T}^o(\Theta_0; \boldsymbol{\theta}_0)\}\text{pr}(\chi_i^2 \geqslant c)$$
$$\leq 0.5\{\text{pr}(\chi_{p-1}^2 \geqslant t) + \text{pr}(\chi_p^2 \geqslant t)\}.$$

If the maximum number of inequalities in the null hypothesis that can be satisfied simultaneously as equalities is, say q, and $q \leq p$, then the foregoing upper bound could be sharpened to $0.5\{\text{pr}(\chi_{q-1}^2 \geqslant t) + \text{pr}(\chi_q^2 \geqslant t)\}$. Thus, while we can write down the asymptotic distribution of statistics for testing $\boldsymbol{h}(\boldsymbol{\theta}) \geqslant \boldsymbol{0}$ vs $\boldsymbol{h}(\boldsymbol{\theta}) \not\geqslant \boldsymbol{0}$, there is scope for developing practically feasible computational procedures to implement them.

The foregoing result extends in a natural way for testing

$$H_0 : \boldsymbol{h}_1(\boldsymbol{\theta}) = \boldsymbol{0}, \boldsymbol{h}_2(\boldsymbol{\theta}) \geqslant \boldsymbol{0} \quad \text{vs} \quad H_1 : \text{ No restrictions on } \boldsymbol{\theta}$$

where $\boldsymbol{h} = (\boldsymbol{h}_1 : \boldsymbol{h}_2)$ is a partition of $\boldsymbol{h}$ into two subvectors; only minor changes are required for this extension (see Kodde and Palm (1986) and Wolak (1989a)).

Example 1. Let $\boldsymbol{\theta} \in \mathbb{R}^2$. Consider a statistical model satisfying Condition Q2 in Section 4.8.2. Let the null and alternative hypotheses be

$$H_0 : \boldsymbol{\theta} \geqslant \mathbf{0} \text{ and } H_1 : \boldsymbol{\theta} \not\geqslant \mathbf{0},$$

respectively, where $\boldsymbol{\theta} \in \mathbb{R}^2$. Then $LRT = \sup\{\ell(\boldsymbol{\theta}) : \boldsymbol{\theta} \geqslant \mathbf{0}\}$. Suppose that the null hypothesis holds and let $\boldsymbol{\theta}_0$ denote the true null value. Then, the asymptotic null distribution of *LRT* is equal to that of $\|\boldsymbol{Z} - \mathcal{T}(\Theta_0; \boldsymbol{\theta}_0)\|^2$ where Θ_0 is the positive orthant, $\boldsymbol{Z} \sim N\{\mathbf{0}, \mathcal{I}(\boldsymbol{\theta}_0)^{-1}\}$ and $\|\boldsymbol{x}\|^2 = \boldsymbol{x}^T\mathcal{I}(\boldsymbol{\theta}_0)\boldsymbol{x}$. Clearly, $\mathcal{T}(\Theta_0; \boldsymbol{\theta}_0)$, and hence the foregoing null asymptotic distribution of *LRT*, depend on $\boldsymbol{\theta}_0$. We shall consider four different cases.

Case 1: $\boldsymbol{\theta}_0 = (0, 0)^T$. It is easily seen that $\mathcal{T}(\Theta_0; \boldsymbol{\theta}_0)$ is the positive orthant and therefore,

$$\text{pr}(LRT \geqslant c \mid \boldsymbol{\theta} = \mathbf{0}) \rightarrow (1/2)\text{pr}(\chi_1^2 \geqslant c) + q\text{pr}(\chi_2^2 \geqslant c),$$

where $q = (2\pi)^{-1}\cos^{-1}\{\mathcal{I}^{12}/(\mathcal{I}^{11}\mathcal{I}^{22})^{1/2}\}$ and $\mathcal{I}(\mathbf{0})^{-1} = (\mathcal{I}^{ij})_{2\times 2}$.

Case 2: $\boldsymbol{\theta}_0 = (0, a)^T$, where $a > 0$. Because $\mathcal{T}(\Theta_0; \boldsymbol{\theta}_0)$ is the half-space $\{\boldsymbol{\alpha} \in \mathbb{R}^2 : \alpha_1 \geqslant 0\}$,

$$\text{pr}\{LRT \geqslant c \mid \boldsymbol{\theta} = (0, a)\} \rightarrow (1/2)\text{pr}(\chi_1^2 \geqslant c).$$

Case 3: $\boldsymbol{\theta}_0 = (a, 0)^T$, where $a > 0$. The asymptotic null distribution is the same as that in Case 2.

Case 4: $\boldsymbol{\theta}_0 = (a, b^T)$, where $a > 0$ and $b > 0$. Now, $\mathcal{T}(\Theta_0; \boldsymbol{\theta}_0) = \mathbb{R}^2$, and therefore, $\|\boldsymbol{Z} - \mathcal{T}(\Theta_0; \boldsymbol{\theta}_0)\| = 0$ and $\text{pr}\{LRT = 0 \mid \boldsymbol{\theta} = (a, b)\} \rightarrow 1$.

It follows from the foregoing four cases that $\boldsymbol{\theta} = \mathbf{0}$ is the least favorable null value. Therefore, the large sample *p*-value corresponding to $LRT = t$ is $(1/2)\text{pr}(\chi_1^2 \geqslant t) + q\text{pr}(\chi_2^2 \geqslant t)$. ■

Example 2. Let the testing problem be

$$H_0 : \boldsymbol{h}(\boldsymbol{\theta}) \geqslant \mathbf{0} \text{ vs } H_1 : \boldsymbol{h}(\boldsymbol{\theta}) \not\geqslant \mathbf{0}$$

where $\boldsymbol{h} = (h_1 : h_2)$. Suppose that $h_1(\boldsymbol{\theta}) = h_2(\boldsymbol{\theta}) = 0$ has a solution, say at $\boldsymbol{\theta}_0$ and $rank[\boldsymbol{H}(\boldsymbol{\theta}_0)] = 2$. As in the previous example, it would be convenient to consider four different cases.

Case 1: $h_1(\boldsymbol{\theta}_0) = 0$ and $h_2(\boldsymbol{\theta}_0) = 0$

$$\text{pr}\{LRT \geqslant c \mid h_1(\boldsymbol{\theta}_0) = 0, h_2(\boldsymbol{\theta}_0) = 0\}$$
$$\rightarrow (1/2)\text{pr}(\chi_1^2 \geqslant c) + q(\boldsymbol{\theta}_0)\text{pr}(\chi_2^2 \geqslant c),$$

where

$$q(\boldsymbol{\theta}) = (2\pi)^{-1}\cos^{-1}\{V_{12}/(V_{11}V_{22})^{1/2}\}, \text{ and } \boldsymbol{V} = \boldsymbol{V}(\boldsymbol{\theta}_0).$$

Case 2: $h_1(\boldsymbol{\theta}_0) > 0$ and $h_2(\boldsymbol{\theta}_0) = 0$

$$\text{pr}\{LRT \geqslant c \mid h_1(\boldsymbol{\theta}_0) > 0, h_2(\boldsymbol{\theta}_0) = 0\} \to (1/2)\text{pr}(\chi_1^2 \geqslant c).$$

Case 3: $h_1(\boldsymbol{\theta}_0) = 0$ and $h_2(\boldsymbol{\theta}_0) > 0$

$$\text{pr}\{LRT \geqslant c \mid h_1(\boldsymbol{\theta}_0) = 0, h_2(\boldsymbol{\theta}_0) > 0\} \to (1/2)\text{pr}(\chi_1^2 \geqslant c);$$

Case 4: $h_1(\boldsymbol{\theta}_0) > 0$ and $h_2(\boldsymbol{\theta}_0) > 0$

$$\text{pr}\{LRT = 0 \mid h_1(\boldsymbol{\theta}_0) > 0, h_2(\boldsymbol{\theta}_0) > 0\} \to 1.$$

Therefore, the least favorable null value is a point at which $h_1(\boldsymbol{\theta}) = h_2(\boldsymbol{\theta}) = 0$. If $h_1(\boldsymbol{\theta}) = h_2(\boldsymbol{\theta}) = 0$ has more than one solution, then the solution at which $q(\boldsymbol{\theta})$ is the maximum is the least favorable null value. ■

Example 3 (Wolak (1991)). Let $\boldsymbol{\theta} = (\theta_1, \theta_2)$ and

$$H_0 : h_1(\boldsymbol{\theta}) \geqslant 0 \text{ and } h_2(\boldsymbol{\theta}) \geqslant 0,$$

where $h_1(\boldsymbol{\theta}) = \theta_2 - \theta_1^2$ and $h_2(\boldsymbol{\theta}) = \theta_1 - \theta_2$; the parameter space is shown in Fig. 4.8. As in the previous example, there are four cases to be considered. It follows that the least favorable null value is a $\boldsymbol{\theta}_0$ satisfying $h_1(\boldsymbol{\theta}_0) = 0$ and $h_2(\boldsymbol{\theta}_0) = 0$. Let t denote the sample value of *LRT*. Because there are two values of $\boldsymbol{\theta}_0$, namely $(0, 0)$ and $(1, 1)$, satisfying these equations, and

$$\text{pr}(LRT \geqslant t | \boldsymbol{h}(\boldsymbol{\theta}_0) = \boldsymbol{0}) \to (1/2)\text{pr}(\chi_1^2 \geqslant t) + q(\boldsymbol{\theta}_0)\text{pr}(\chi_2^2 \geqslant t),$$

the least favorable null value is the $\boldsymbol{\theta}_0$ in $\{(0, 0), (1, 1)\}$ for which

$$(1/2)\text{pr}(\chi_1^2 \geqslant t) + q(\boldsymbol{\theta}_0)\text{pr}(\chi_2^2 \geqslant t)$$

is a maximum. If $\mathcal{I}(\boldsymbol{\theta}_0)$ is known for $\boldsymbol{\theta}_0 = (0, 0)$ and $(1, 1)$, then we can compute the foregoing tail probability for each of the two values of $\boldsymbol{\theta}_0$ and choose the largest as the p-value. ■

In the foregoing examples with only two inequality constraints in the null hypothesis, the least favorable null value turned out to be a point at which all the inequalities in the null hypothesis are active. We also observed in Chapter 2 that for testing $H_0 : \boldsymbol{\theta} \geqslant \boldsymbol{0}$ against $H_1 : \boldsymbol{\theta} \not\geqslant \boldsymbol{0}$, based on observations from $N(\boldsymbol{\theta}, \boldsymbol{V})$ where $\boldsymbol{V}$ does not depend on $\boldsymbol{\theta}$, the least favorable null value, namely $\boldsymbol{0}$, is the point at which every inequality constraint in the null hypothesis is active. However, in general, if the null hypothesis is $H_0 : h_1(\boldsymbol{\theta}) \geqslant 0, \ldots, h_k(\boldsymbol{\theta}) \geqslant 0$, where $k > 2$, then the least favorable null value may not be a point at which all the inequalities are active. The following example from Wolak (1991) illustrates this.

Example 4 (Wolak (1991)). Let $\boldsymbol{X} = (X_1, X_2)^T \sim N(\boldsymbol{0}, \boldsymbol{V})$ where

$$\boldsymbol{V} = \begin{pmatrix} \sigma^2 & \rho\sigma\tau \\ \rho\sigma\tau & \tau^2 \end{pmatrix}.$$

Let $\boldsymbol{\theta} = (\sigma^2, \tau^2, \rho)^T$,

$$\Theta = \{\boldsymbol{\theta} : \theta_1 > 0, \theta_2 > 0, -1 < \theta_3 < 1\},$$

$\boldsymbol{\theta}^A = (1, 1, 0.95)^T$ and $\boldsymbol{\theta}^B = (1, 1, 0)^T$. Let the null hypothesis be $H_0 : \boldsymbol{\theta} \leq \boldsymbol{\theta}^A$. To express this testing problem in the notation of (4.113), let $\boldsymbol{h}(\boldsymbol{\theta}) = (1 - \theta_1, 1 - \theta_2, 0.95 - \theta_3)$; then $\boldsymbol{H}(\boldsymbol{\theta}_0) = -\boldsymbol{I}$. The inverse of the information matrix and the corresponding correlation matrix are given below (see Lehmann (1983), p 441).

$$\mathcal{I}(\boldsymbol{\theta})^{-1} = \begin{bmatrix} 2\sigma^4 & 2\rho^2\sigma^2\tau^2 & \rho(1-\rho^2)\sigma^2 \\ 2\rho^2\sigma^2\tau^2 & 2\tau^4 & \rho(1-\rho^2)\tau^2 \\ \rho(1-\rho^2)\sigma^2 & \rho(1-\rho^2)\tau^2 & (1-\rho^2)^2 \end{bmatrix}$$

$$corr\{\mathcal{I}(\boldsymbol{\theta})^{-1}\} = \begin{bmatrix} 1 & \rho^2 & \rho/\sqrt{2} \\ \rho^2 & 1 & \rho/\sqrt{2} \\ \rho/\sqrt{2} & \rho/\sqrt{2} & 1 \end{bmatrix}.$$

Note that Θ is open, $\sqrt{n}(\hat{\boldsymbol{\theta}} - \boldsymbol{\theta}_0) \xrightarrow{d} N\{\mathbf{0}, \mathcal{I}(\boldsymbol{\theta}_0)^{-1}\}$, and Condition $Q2$ in subsection 4.8.2 is satisfied. Now,

$$\lim_{n\to\infty} \text{pr}(LRT \geqslant c \mid \boldsymbol{\theta} = \boldsymbol{\theta}^A) = w_1\{3, \mathcal{I}(\boldsymbol{\theta}_0)\}\text{pr}(\chi_3^2 \geqslant c) + w_2\{3, \mathcal{I}(\boldsymbol{\theta}_0)\}\text{pr}(\chi_2^2 \geqslant c) + w_3\{3, \mathcal{I}(\boldsymbol{\theta}_0)\}\text{pr}(\chi_1^2 \geqslant c), \tag{4.117}$$

where w_1, w_2, and w_3 are determined by the formulas in (3.26). By direct substitution into these formulas, we have $w_1 = 0.0153, w_2 = 0.1682, w_3 = 0.4847$,

$$\lim_{n\to\infty} \text{pr}\{LRT \geqslant c \mid \boldsymbol{\theta} = \boldsymbol{\theta}^B\} = 0.5\{\text{pr}(\chi_1^2 \geqslant c) + 0.5\text{pr}(\chi_2^2 \geqslant c)\}, \tag{4.118}$$

and $$\lim_{n\to\infty} \text{pr}(LRT \geqslant c \mid \boldsymbol{\theta} = \boldsymbol{\theta}^A) < \text{pr}\{LRT \geqslant c \mid \boldsymbol{\theta} = \boldsymbol{\theta}^B\}.$$

Therefore, the least favorable null value is not $\boldsymbol{\theta}^A$, the only point at which all the inequalities in the null hypothesis are active.

Let $\boldsymbol{\theta}^*$ be a point at which exactly two of those inequalities are active. Then,

$$\lim \text{pr}\{LRT \geqslant c \mid \boldsymbol{\theta} = \boldsymbol{\theta}^*\} = 0.5\text{pr}(\chi_1^2 \geqslant c) + q\text{pr}(\chi_2^2 \geqslant c),$$

for some q in the range, $0 \leq q \leq 0.5$. Let $\boldsymbol{\theta}^+$ be a point at which exactly one of the inequalities in H_0 is active. Then,

$$\lim \text{pr}\{LRT \geqslant c \mid \boldsymbol{\theta} = \boldsymbol{\theta}^+\} = 0.5\text{pr}(\chi_1^2 \geqslant c).$$

Since this limit is smaller than the RHS of (4.118), it follows that $\boldsymbol{\theta} = \boldsymbol{\theta}^B$ is the least favorable null value. ∎

Example 4.8.3 (Stram and Lee, 1994) *Testing for the presence of random effects.*

Consider the growth curve model in Example 1.2.10 (p 14). The data consist of measurements of the distance (in *mm*) from the center pituitary to the pteryomaxillary fissure for 27 children (11 girls and 16 boys). Every subject has four measurements, taken at ages 8, 10, 12 and 14 years. The object is to model the growth in the distance as a function of age and sex. To allow for individual variability in the growth function, the model being considered is

$$\boldsymbol{y}_i = (1, Age, Sex, Age.Sex)\boldsymbol{\alpha} + (1, Age)\boldsymbol{\beta}_i + \boldsymbol{\epsilon}_i$$

where $\boldsymbol{y}_i$ is 4×1 for the four measurements of the ith subject, $Age = (8, 10, 12, 14)^T$, $Sex = (a, a, a, a)^T$ where a is 0 for boys and 1 for girls. The parameter $\boldsymbol{\alpha}(4 \times 1)$ captures the fixed effects and $\boldsymbol{\beta}_i(2 \times 1)$ captures the random effect of subject i. Assume that $\boldsymbol{\epsilon}_i \sim N(0, \sigma^2 \boldsymbol{I})$, $\boldsymbol{\beta}_i \sim N(\boldsymbol{0}, \Psi)$ for some 2×2 positive semi-definite matrix Ψ and that ϵ_i and $\boldsymbol{\beta}_i$ are independent. The following models are of interest:

1. Model $M1 : \Psi = \boldsymbol{0}$. There are no random effects.
2. Model $M2 : \Psi = (\psi_{11}, 0 \mid 0, 0)$. The individual variability is captured by a random intercept; therefore, the regression planes are parallel for different subjects.
3. Model $M_3 : \Psi = (\psi_{11}, \psi_{12} \mid \psi_{21}, \psi_{22})$. Individual variability results in not only a random intercept but also a different slope parameter for age.

To simplify notation, let us write $\Psi = (\psi_1, \psi_2 \mid \psi_2, \psi_3)$. In the process of modeling these data, it is possible that we may be interested to know whether or not the fit for Model M3 is significantly better than that for the simpler Model M2. Therefore, we formulate the testing problem as

$$H_0 : \psi_1 \geqslant 0, \psi_2 = \psi_3 = 0 \quad \text{vs} \quad H_1 : \psi_1 \geqslant 0, \psi_3 \geqslant 0, \psi_1\psi_3 - \psi_2^2 \geqslant 0.$$

The regression model can be written as

$$\boldsymbol{y}_i = \boldsymbol{X}_i\boldsymbol{\alpha} + \boldsymbol{Z}_i\boldsymbol{\beta}_i + \boldsymbol{\epsilon}_i.$$

With the assumptions as introduced earlier for the distributions of $\boldsymbol{\epsilon}_i$ and $\boldsymbol{\beta}_i$, we have

$$\ell(\boldsymbol{\theta}) = \sum_{i=1}^{n}(-n_i/2)\log(2\pi) - (1/2)\log\{\det(\boldsymbol{\Sigma}_i^{-1})\} - (1/2)\text{trace}(\boldsymbol{\Sigma}_i^{-1}\boldsymbol{S}_i)$$

where $\boldsymbol{\Sigma}_i = \sigma^2\boldsymbol{I} + \boldsymbol{Z}_i\Psi\boldsymbol{Z}_i^T$ and $\boldsymbol{S}_i = (\boldsymbol{y}_i - \boldsymbol{X}_i\boldsymbol{\alpha})^T(\boldsymbol{y}_i - \boldsymbol{X}_i\boldsymbol{\alpha})$. Now, under the null,

$$LRT \xrightarrow{d} \|\boldsymbol{U} - \mathcal{T}(\Theta_0; \boldsymbol{\theta}_0)\|^2 - \|\boldsymbol{U} - \mathcal{T}(\Theta_1; \boldsymbol{\theta}_0)\|^2$$

where $\boldsymbol{U} \sim N(\boldsymbol{0}, \mathcal{I}_{\boldsymbol{\theta}_0}^{-1})$,

$$\Theta_0 = \{\boldsymbol{\psi} \in \mathbb{R}^3 : \psi_2 = \psi_3 = 0, \psi_1 \geqslant 0\},$$
$$\text{and } \Theta_1 = \{\boldsymbol{\psi} \in \mathbb{R}^3 : \psi_1 \geqslant 0, \psi_3 \geqslant 0, \psi_1\psi_3 - \psi_2^2 \geqslant 0\}.$$

There are two possible values of $\boldsymbol{\theta}_0$ in Θ_0 that we need to consider separately: $(0,0,0)^T$ and $(a,0,0)^T$ where $a > 0$.

Case 1: $\boldsymbol{\theta}_0 = (0,0,0)^T$.

It may be verified that

$$\mathcal{T}(\Theta_0; \boldsymbol{\theta}_0) = \{\boldsymbol{\psi} \in \mathbb{R}^3 : \psi_2 = \psi_3 = 0, \psi_1 \geqslant 0\}.$$

To determine the tangent cone of Θ_1, note that

$$h_3(\boldsymbol{\theta}) = \psi_1\psi_3 - \psi_2^2, \qquad \text{and} \qquad (\partial/\partial\boldsymbol{\theta})h_3(\boldsymbol{\theta}) = (\psi_3, -2\psi_2, \psi_1)^T.$$

At $\boldsymbol{\theta}_0 = (0,0,0)^T$ we have $h_3(\boldsymbol{\theta}_0) = 0$ and $(\partial/\partial\boldsymbol{\theta})h_3(\boldsymbol{\theta}_0) = \mathbf{0}$. Therefore, MF-CQ is not satisfied at $\boldsymbol{\theta}_0$, and hence we cannot use Proposition 4.7.3 to write down the tangent cone; thus we should not simply replace the nonlinear constraints by their linear approximations at $\boldsymbol{\theta}_0$.

Since Θ_1 is a closed convex cone with vertex at the origin it follows that $\mathcal{T}(\Theta_1; \boldsymbol{\theta}_0) = \Theta_1$. Since neither $\mathcal{T}(\Theta_1; \boldsymbol{\theta}_0)$ nor $\mathcal{T}(\Theta_0; \boldsymbol{\theta}_0)$ is a linear space, the asymptotic null distribution of *LRT* may not be a chi-bar square. Exact distribution of *LRT* for this case is nonstandard. A bound can be obtained by replacing $\mathcal{T}(\Theta_1; \boldsymbol{\theta}_0)$ by the larger set $\{\psi_1 \geqslant 0, \psi_3 \geqslant 0\}$. In this case, the distribution is a mixture of chi-square distributions.

Case 2: $\boldsymbol{\theta}_0 = (a,0,0)^T$ where $a > 0$.

It may be verified that

$$\begin{aligned} \mathcal{T}(\Theta_0; \boldsymbol{\theta}_0) &= \{\boldsymbol{\psi} \in \mathbb{R}^3 : \psi_2 = \psi_3 = 0\}, \\ \text{and} \quad \mathcal{T}(\Theta_1; \boldsymbol{\theta}_0) &= \{\boldsymbol{\psi} \in \mathbb{R}^3 : \psi_3 \geqslant 0\}, \end{aligned}$$

the Mangasarian-Fromowitz constraint qualification is satisfied at $\boldsymbol{\theta}_0$ and therefore we can use Proposition 4.7.3 to write down the tangent cones. Since $\mathcal{T}(\Theta_0; \boldsymbol{\theta}_0)$ is a linear space and $\mathcal{T}(\Theta_1; \boldsymbol{\theta}_0)$ is a closed convex cone it follows that the asymptotic null distribution of *LRT* is a chi-bar square. ∎

4.9 PROPERTIES OF THE MLE WHEN THE TRUE VALUE IS ON THE BOUNDARY

We continue to consider the *iid* setting for simplicity although most of the results in this section would hold in more general settings as will be indicated later. Let $\boldsymbol{X}_1, \ldots, \boldsymbol{X}_n$ denote *iid* observations from a population with density function $f(\boldsymbol{x}; \boldsymbol{\theta})$. Let $\ell(\boldsymbol{\theta})$ denote the loglikelihood, where $\boldsymbol{\theta} \in \Theta \subset \mathbb{R}^p$. Let $\Omega \subset \Theta$ where Ω is not necessarily open. In this section we study large sample properties of local and global *mle*'s of $\boldsymbol{\theta}$. Let us denote an *mle*, global or local, by $\tilde{\boldsymbol{\theta}}$. The true value, denoted by $\boldsymbol{\theta}_0$, will be assumed to be a boundary point of Ω unless the contrary is made clear. *Assume that Condition Q is satisfied.* For simplicity, we shall restrict to maximum likelihood. However, most of the results of this section hold for M-estimators as well; for details of this case see Geyer (1994) and Shapiro (2000a).

Distribution of the *mle* over a restricted parameter space that is not open has not attracted as much attention as the corresponding hypothesis testing problems. If $\boldsymbol{\theta}_0$ were an interior point then $\sqrt{n}\hat{\mathcal{I}}^{-1/2}(\hat{\boldsymbol{\theta}} - \boldsymbol{\theta}_0) \xrightarrow{d} N(\mathbf{0}, \boldsymbol{I})$. Since this limiting distribution does not depend on any unknown parameters (i.e., $\sqrt{n}\hat{\mathcal{I}}^{-1/2}(\hat{\boldsymbol{\theta}} - \boldsymbol{\theta}_0)$ is a pivotal quantity), we can use this result to construct a confidence region for $\boldsymbol{\theta}$. Unfortunately, the asymptotic distribution of $\sqrt{n}\hat{\mathcal{I}}^{-1/2}(\tilde{\boldsymbol{\theta}} - \boldsymbol{\theta}_0)$ is a discontinuous function of $\boldsymbol{\theta}_0$. Further, it is unknown whether or not there is a a pivotal quantity based on $\tilde{\boldsymbol{\theta}}$ that would lend itself for constructing a confidence region. Consequently, statistical inference based on the constrained estimator $\tilde{\boldsymbol{\theta}}$ has not attracted much attention. Nevertheless, it is of interest to study properties of the constrained *mle* because it has an important role to play in the theory of statistical inference. In this section, we shall consider some examples to illustrate the main results. Then we shall state the main results and provide references to sources where detailed statements of the regularity conditions and proofs may be found.

At this stage it would be helpful to reconsider Example 3.3.1.

Example 4.9.1 *Asymptotic distribution of mle when* $\Omega = \mathbb{R}^{+2}$ *and* $\boldsymbol{X} \sim N(\boldsymbol{\theta}, \boldsymbol{I})$.

Let $\boldsymbol{X} \sim N(\boldsymbol{\theta}, \boldsymbol{I})$ and $\Omega = \mathbb{R}^{+2}$. Let $\boldsymbol{X}_1, \ldots, \boldsymbol{X}_n$ be *iid* as $\boldsymbol{X}$. Then,

$$\ell(\boldsymbol{\theta}) = (-n/2)\|\bar{\boldsymbol{X}} - \boldsymbol{\theta}\|^2$$

and $\tilde{\boldsymbol{\theta}}$ is the point in $\mathbb{R}^{+2}$ that is closest to $\bar{\boldsymbol{X}}$. Thus,

$$\tilde{\boldsymbol{\theta}} = \Pi(\bar{\boldsymbol{X}}, \mathbb{R}^{+2}) \text{ and } \tilde{\boldsymbol{\theta}} \xrightarrow{p} \boldsymbol{\theta}_0.$$

If $\boldsymbol{\theta}_0$ is an interior point of $\mathbb{R}^{+2}$ then $\text{pr}(\bar{\boldsymbol{X}} = \tilde{\boldsymbol{\theta}}) \to 1$ and hence $\sqrt{n}(\tilde{\boldsymbol{\theta}} - \boldsymbol{\theta}_0) \xrightarrow{d} N(\mathbf{0}, \boldsymbol{I})$. Let $\mathcal{T} = \mathcal{T}(\mathbb{R}^{+2}; \boldsymbol{\theta}_0)$, $\boldsymbol{Z}_n = \sqrt{n}(\bar{\boldsymbol{X}} - \boldsymbol{\theta}_0)$ and $\boldsymbol{Z} \sim N(\mathbf{0}, \boldsymbol{I})$. Suppose that $\boldsymbol{\theta}_0$ lies on the boundary of $\mathbb{R}^{+2}$. Then it is easily seen that

$$(\tilde{\boldsymbol{\theta}} - \boldsymbol{\theta}_0) = \Pi(\bar{\boldsymbol{X}} - \boldsymbol{\theta}_0, \mathcal{T}) \text{ with probability approaching 1.}$$

Therefore,

$$\sqrt{n}(\tilde{\boldsymbol{\theta}} - \boldsymbol{\theta}_0) = \Pi(\boldsymbol{Z}_n, \mathcal{T}) \text{ with prob approaching 1.}$$

Since $\Pi(\boldsymbol{z}, \mathcal{T})$ is continuous in $\boldsymbol{z}$ and $\boldsymbol{Z}_n \xrightarrow{d} \boldsymbol{Z}$ it follows that $\Pi(\boldsymbol{Z}_n, \mathcal{T}) \xrightarrow{d} \Pi(\boldsymbol{Z}, \mathcal{T})$. Therefore, we have

$$\sqrt{n}(\tilde{\boldsymbol{\theta}} - \boldsymbol{\theta}_0) \xrightarrow{d} \Pi(\boldsymbol{Z}, \mathcal{T}).$$

This may be verified directly (in fact, it would be instructive to do so) by considering the following three cases separately: (i) $\boldsymbol{\theta}_0$ is on the positive θ_1-axis (this is the case shown in Fig. 4.10), (ii) $\boldsymbol{\theta}_0$ is on the positive θ_2-axis and (iii) $\boldsymbol{\theta}_0$ is at the origin. It is perhaps easier to visualize this result as

$$(\tilde{\boldsymbol{\theta}} - \boldsymbol{\theta}_0) \approx \Pi(\boldsymbol{Z}/\sqrt{n}; \mathcal{T}),$$

for large n as shown in Fig. 4.10. ■

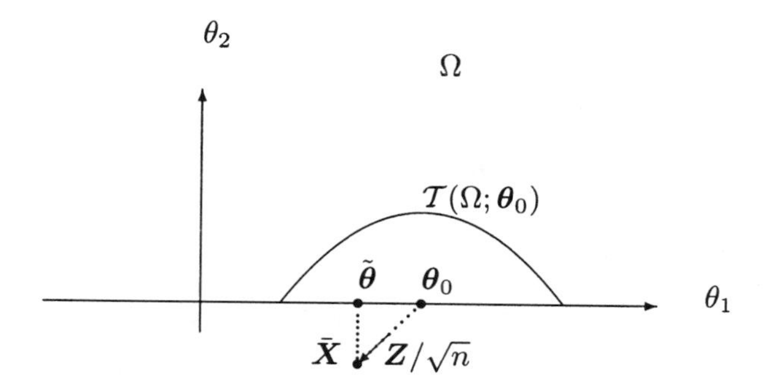

Fig. 4.10 $(\tilde{\boldsymbol{\theta}} - \boldsymbol{\theta}_0) \approx \Pi\{\boldsymbol{Z}/\sqrt{n}; \mathcal{T}(\Omega;\boldsymbol{\theta}_0)\}$,where $\boldsymbol{Z} \sim N(\boldsymbol{0}, \boldsymbol{I})$.

When $\boldsymbol{\theta}_0$ is an interior point, a consistent *mle* $\hat{\boldsymbol{\theta}}$, whether it is a global or local *mle*, satisfies $n^{-1/2}\hat{\mathcal{I}}^{-1/2}\boldsymbol{S}(\boldsymbol{\theta}_0) = \sqrt{n}(\hat{\boldsymbol{\theta}} - \boldsymbol{\theta}_0) + o_p(1)$. This is a neat representation of $\hat{\boldsymbol{\theta}}$ as a sum of independently and identically distributed variables ($\boldsymbol{S}(\boldsymbol{\theta}_0)$ is a sum of *iid* variables). This is a direct result of applying a one-term Taylor expansion on the first-order condition $\nabla\ell(\hat{\boldsymbol{\theta}}) = \boldsymbol{0}$. If $\boldsymbol{\theta}_0$ is not an interior point, then the first-order conditions, such as the Kuhn-Tucker conditions, involve inequalities. Consequently, $\sqrt{n}(\tilde{\boldsymbol{\theta}} - \boldsymbol{\theta}_0)$ does not have a simple representation as for $\sqrt{n}(\hat{\boldsymbol{\theta}} - \boldsymbol{\theta}_0)$. In fact, the properties of local *mle*'s are very subtle. Let us consider a simple example to illustrate this.

Example 4.9.2 *Nonconvex parameter space*

Let $\boldsymbol{X}$ have the bivariate normal distribution $N(\boldsymbol{0}, \boldsymbol{I})$ and let $\boldsymbol{X}_1, \ldots, \boldsymbol{X}_n$ be *iid* as $\boldsymbol{X}$. Then $\ell(\boldsymbol{\theta}) = (-n/2)\|\bar{\boldsymbol{X}} - \boldsymbol{\theta}\|^2$. Let Ω be the union of the θ_1-axis and the θ_2-axis; thus $\Omega = \{(\theta_1, \theta_2) : \theta_1 = 0 \text{ or } \theta_2 = 0\}$. In this case, there are two local *mle*'s: $\tilde{\boldsymbol{\theta}}^a = (\bar{X}_1, 0)$ and $\tilde{\boldsymbol{\theta}}^b = (0, \bar{X}_2)$. It is clear that their large sample distributions are different. Further, when $\boldsymbol{\theta}_0$ is on the θ_1-axis, we have the following with probability $\to 1$:

$$\tilde{\boldsymbol{\theta}}^a - \boldsymbol{\theta}_0 = \text{ projection of } (\bar{\boldsymbol{X}} - \boldsymbol{\theta}_0) \text{ onto the } \theta_1\text{-axis},$$

and $\sqrt{n}(\tilde{\boldsymbol{\theta}}^a - \boldsymbol{\theta}_0)$ converges to $\Pi(\boldsymbol{Z}, \mathcal{P})$ where $\boldsymbol{Z} \sim N(\boldsymbol{0}, \boldsymbol{I})$ and $\mathcal{P} = \{\boldsymbol{\theta} \in \mathbb{R}^2 : \theta_2 = 0\}$. In fact, we have

$$\sqrt{n}(\tilde{\boldsymbol{\theta}}^a - \boldsymbol{\theta}_0) \begin{cases} \xrightarrow{d} \Pi(\boldsymbol{Z}, \mathcal{P}) & \text{if } \boldsymbol{\theta}_0 \in \theta_1\text{-axis} \\ = \begin{bmatrix} \sqrt{n}\bar{X}_1 \\ -\sqrt{n}\theta_{02} \end{bmatrix} & \text{if } \boldsymbol{\theta}_0 \notin \theta_1\text{-axis.} \end{cases}$$

A similar comment applies to $\tilde{\boldsymbol{\theta}}^b$ by symmetry.

By contrast, with $\tilde{\boldsymbol{\theta}}$ denoting the global *mle*, it is easily seen that

$$(\tilde{\boldsymbol{\theta}} - \boldsymbol{\theta}_0) = \Pi(\bar{\boldsymbol{X}} - \boldsymbol{\theta}_0, \mathcal{T}(\Omega;\boldsymbol{\theta}_0)) \text{ with prob approaching 1.}$$

Therefore,

$$\sqrt{n}(\tilde{\boldsymbol{\theta}} - \boldsymbol{\theta}_0) \xrightarrow{d} \Pi(\boldsymbol{Z}, \mathcal{T}(\Omega; \boldsymbol{\theta}_0)), \text{ where } \boldsymbol{Z} \sim N(\boldsymbol{0}, \boldsymbol{I}).$$

Thus, local and global *mle*'s have different asymptotic distributions. If the parameter space is $\mathbb{R}^{+2}$ then there is only one maximum, hence local and global *mle*'s are the same. This example illustrates that if the parameter space is not convex, or at least nearly convex in a way to be explained later, the global and local *mle*'s may have different properties. A consequence of such a difference between local *mle*'s is that if an iterative algorithm with Newton-Raphson type steps were to be used to find the global maximum, it could be trapped around a local maximum. ■

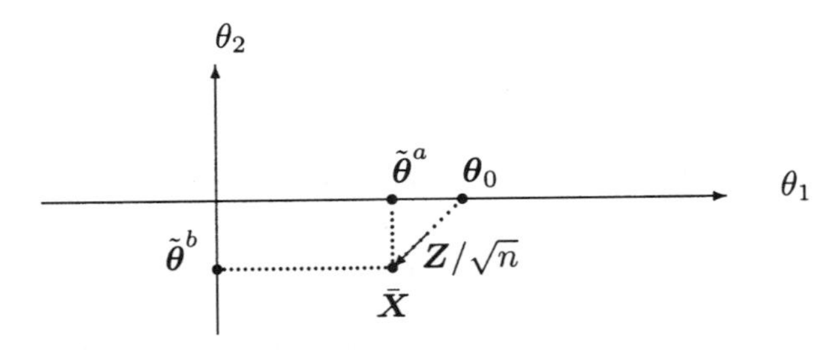

Fig. 4.11 Two local *mles* when Ω is the union of θ_1-axis and the θ_2-axis, and $\boldsymbol{\theta}_0$ is not on the θ_2-axis; $\sqrt{n}(\tilde{\boldsymbol{\theta}}^a - \boldsymbol{\theta}_0)$ converges in distribution but $\|\sqrt{n}(\tilde{\boldsymbol{\theta}}^b - \boldsymbol{\theta}_0)\| \to \infty$.

The standard method that is adopted in the literature to study large sample properties of consistent *mle*'s is very similar to that was adopted so far to study the properties of *LRT*. To illustrate the main idea, let us first consider the case when the parameter space Ω is a closed convex cone with its vertex at $\boldsymbol{\theta}_0$. Let $\mathcal{T} = \Omega - \boldsymbol{\theta}_0$; thus $\mathcal{T}$ is the tangent cone of Ω at $\boldsymbol{\theta}_0$. Suppose that Condition Q is satisfied. Let us write the quadratic approximation of $\ell(\boldsymbol{\theta})$ as (see (4.3))

$$\ell(\boldsymbol{\theta}) = K_n - 2^{-1}(\boldsymbol{Z}_n - \boldsymbol{u})^T \boldsymbol{I}_{\boldsymbol{\theta}_0}(\boldsymbol{Z}_n - \boldsymbol{u}) + \delta_n(\boldsymbol{u}),$$

where $\boldsymbol{u} = \sqrt{n}(\boldsymbol{\theta} - \boldsymbol{\theta}_0)$ and K_n does not depend on $\boldsymbol{\theta}$. Let

$$\boldsymbol{\theta}^\dagger = \text{argmin}_{\boldsymbol{\theta} \in \Omega}(\boldsymbol{Z}_n - \sqrt{n}(\boldsymbol{\theta} - \boldsymbol{\theta}_0))^T \boldsymbol{I}_{\boldsymbol{\theta}_0}(\boldsymbol{Z}_n - \sqrt{n}(\boldsymbol{\theta} - \boldsymbol{\theta}_0)).$$

A consequence of $\delta_n(\boldsymbol{u}) = n^{-1/2}\|\boldsymbol{u}\|^3 O_p(1)$ is that

$$n^{1/2}(\tilde{\boldsymbol{\theta}} - \boldsymbol{\theta}^\dagger) = o_p(1)$$

(see Self and Liang (1987, Lemma 2)). Therefore, $n^{1/2}(\tilde{\boldsymbol{\theta}} - \boldsymbol{\theta}_0)$ and $n^{1/2}(\boldsymbol{\theta}^\dagger - \boldsymbol{\theta}_0)$ have the same asymptotic distribution. Now, recall the following notation for projection:

$$\Pi_{\boldsymbol{W}}(\boldsymbol{z}, \mathcal{T}) = \text{argmin}_{\boldsymbol{u} \in \mathcal{T}}(\boldsymbol{z} - \boldsymbol{u})^T \boldsymbol{W}^{-1}(\boldsymbol{z} - \boldsymbol{u}).$$

Since $\mathcal{T}$ is convex and projection onto a convex set is distance reducing, it follows that $\Pi_{\boldsymbol{W}}(\boldsymbol{z}, \mathcal{T})$ is continuous in $\boldsymbol{z}$ for every fixed $\boldsymbol{W}$; in fact, it follows from

$$\|\boldsymbol{z}^t(\boldsymbol{W}_1 - \boldsymbol{W}_2)\boldsymbol{z}\| \leq \|\boldsymbol{W}_1 - \boldsymbol{W}_2\|\|\boldsymbol{z}\|^2$$

that

$$\Pi_{\boldsymbol{W}}(\boldsymbol{z}, \mathcal{T}) \text{ is continuous in } (\boldsymbol{z}, \boldsymbol{W}).$$

Since $\boldsymbol{Z}_n \xrightarrow{d} \boldsymbol{Z} \sim N(\boldsymbol{0}, \boldsymbol{I}_{\boldsymbol{\theta}_0}^{-1})$ and $\sqrt{n}(\boldsymbol{\theta}^\dagger - \boldsymbol{\theta}_0) = \Pi(\boldsymbol{Z}_n, \mathcal{T})$, it follows that

$$\sqrt{n}(\boldsymbol{\theta}^\dagger - \boldsymbol{\theta}_0) \xrightarrow{d} \Pi(\boldsymbol{Z}, \mathcal{T}).$$

Therefore, the asymptotic distribution of the global *mle* is given by

$$\sqrt{n}(\tilde{\boldsymbol{\theta}} - \boldsymbol{\theta}_0) \xrightarrow{d} \Pi(\boldsymbol{Z}, \mathcal{T}).$$

These arguments can also be modified when Ω is not necessarily a cone, but provided $\mathcal{T}(\Omega, \boldsymbol{\theta}_0)$ is convex; see Andrews (1999) and Le Cam (1970, p 820) for details. This result has also been extended to the case when $\mathcal{T}(\Omega, \boldsymbol{\theta}_0)$ is not convex, but Ω is Chernoff regular at $\boldsymbol{\theta}_0$. The next result states the most general form known for the *iid* setting.

Proposition 4.9.1 *Suppose that Ω is Chernoff regular at $\boldsymbol{\theta}_0$, and let $\tilde{\boldsymbol{\theta}}$ denote the global mle. Suppose that Condition Q is also satisfied. Then*

$$\sqrt{n}(\tilde{\boldsymbol{\theta}} - \boldsymbol{\theta}_0) \xrightarrow{d} \Pi_{\mathcal{I}(\boldsymbol{\theta}_0)^{-1}}\{\boldsymbol{Z}, \mathcal{T}(\Omega; \boldsymbol{\theta}_0)\}, \quad \textit{where } \boldsymbol{Z} \sim N(\boldsymbol{0}, \mathcal{I}(\boldsymbol{\theta}_0)^{-1}).$$

Proof: See Self and Liang (1987, Theorem 2) and Geyer (1994, Theorem 4.4); the proof in Geyer (1994) is applicable for M-estimators as well. ■

To state the foregoing result differently, let $\boldsymbol{Y} \sim N(\boldsymbol{\theta}, \mathcal{I}(\boldsymbol{\theta}_0)^{-1})$ where $\boldsymbol{\theta} \in \Omega$. Let G denote the distribution of the *mle* of $\boldsymbol{\theta}$ based on a single observation of $\boldsymbol{Y}$ when $\boldsymbol{\theta} = \boldsymbol{0}$. Then the asymptotic distribution of $\sqrt{n}(\tilde{\boldsymbol{\theta}} - \boldsymbol{\theta}_0)$ is G. This is illustrated in Fig. 4.12. It shows that the first order asymptotic behavior of $(\tilde{\boldsymbol{\theta}} - \boldsymbol{\theta}_0)$ is the same as that of $\Pi\{\boldsymbol{Z}/\sqrt{n}, \mathcal{T}(\Omega; \boldsymbol{\theta}_0)\}$,where $\boldsymbol{Z} \sim N(\boldsymbol{0}, \mathcal{I}(\boldsymbol{\theta}_0)^{-1})$.

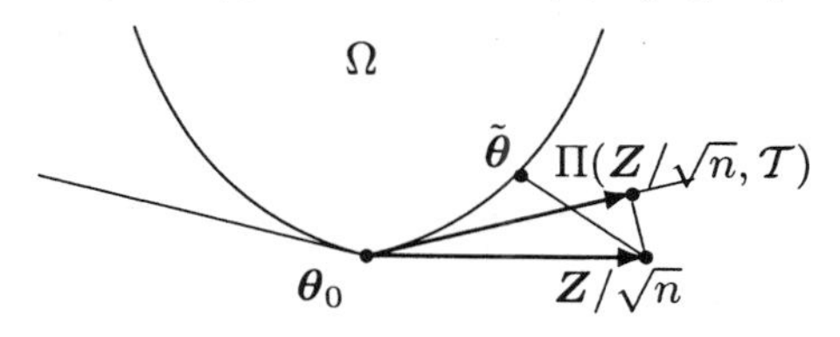

Fig. 4.12 $(\tilde{\boldsymbol{\theta}} - \boldsymbol{\theta}_0) \approx \Pi\{\boldsymbol{Z}/\sqrt{n}, \mathcal{T})\}$,where $\boldsymbol{Z} \sim N(\boldsymbol{0}, \mathcal{I}(\boldsymbol{\theta}_0))$ and $\mathcal{T} = \mathcal{T}(\Omega; \boldsymbol{\theta}_0)$.

To study properties of local *mle*'s, let us introduce the terms *near convexity of a set* and *pro-regular*.

Definition: We say that the set Ω is *nearly convex* at $\boldsymbol{\theta}_0$ if there exists a neighborhood V of $\boldsymbol{\theta}_0$ and a function $k(\boldsymbol{\theta}, \boldsymbol{\theta}^*)$ tending to zero as $\boldsymbol{\theta} \to \boldsymbol{\theta}_0, \boldsymbol{\theta}^* \to \boldsymbol{\theta}_0$, such that

$$dist\{\boldsymbol{\theta}^* - \boldsymbol{\theta}, \mathcal{T}(\Omega; \boldsymbol{\theta})\} \leq k(\boldsymbol{\theta}, \boldsymbol{\theta}^*)\|\boldsymbol{\theta}^* - \boldsymbol{\theta}\|, \quad \forall \boldsymbol{\theta}, \boldsymbol{\theta}^* \in \Omega \cap V.$$

We say that the set Ω is *pro-regular* at $\boldsymbol{\theta}_0$ if there exists a neighborhood V of $\boldsymbol{\theta}_0$ and a constant K such that

$$dist\{\boldsymbol{\theta}^* - \boldsymbol{\theta}, \mathcal{T}(\Omega; \boldsymbol{\theta})\} \leq K\|\boldsymbol{\theta}^* - \boldsymbol{\theta}\|^2, \quad \forall \boldsymbol{\theta}, \boldsymbol{\theta}^* \in \Omega \cap V.$$

If Ω is convex at $\boldsymbol{\theta}_0$ then it is nearly convex and pro-regular at $\boldsymbol{\theta}_0$. The concept of *near convexity* was introduced by Shapiro and Al-Khayyal (1993). *Pro-regularity* appears in Shapiro (1994) under the name "O(2)-Convexity"; for related details on this topic see Rockafellar and Wets (1998).

Another concept that is related to this is *monotonicity of normals*. This is closely related to what is known as *Clark regularity*, convexity of tangent cones and near convexity. See Shapiro (2000a) for details about these concepts. Geyer (1994) provides an excellent discussion of *Clark regularity*; for more detailed discussions see Rockafellar and Wets (1998).

If Ω satisfies the Mangasarian and Fromowitz constraint qualification at $\boldsymbol{\theta}_0$, then Ω is nearly convex at $\boldsymbol{\theta}_0$. This is particularly relevant because many of the constrained inference problems in statistics are likely to be specified by equality and inequality constraints on functions of $\boldsymbol{\theta}$, and the Mangasarian and Fromowitz condition is likely to be satisfied in many cases. Further, if the constraining functions $h_i(\boldsymbol{\theta})$ are Lipchitz continuous in a neighborhood of $\boldsymbol{\theta}_0$ then Ω is pro-regular at $\boldsymbol{\theta}_0$.

Proposition 4.9.2 *(Shapiro (2000a)). Suppose that Ω is nearly convex at $\boldsymbol{\theta}_0$ and Condition Q is satisfied. Let $\tilde{\boldsymbol{\theta}}^a$ and $\tilde{\boldsymbol{\theta}}^b$ be two local mle's. Then,*

$$\sqrt{n}(\tilde{\boldsymbol{\theta}}^a - \tilde{\boldsymbol{\theta}}^b) = o_p(1). \quad \blacksquare$$

Shapiro (2000a) showed, by constructing a counter example, that even if the tangent cone of Ω at $\boldsymbol{\theta}_0$ is convex, two local *mle*'s may fail to be asymptotically equivalent [i.e., $\sqrt{n}(\tilde{\boldsymbol{\theta}}^a - \tilde{\boldsymbol{\theta}}^b)$ may not be $o_p(1)$]. In fact, Shapiro (2000a) showed that even if Ω is *Clark regular*, $\sqrt{n}(\tilde{\boldsymbol{\theta}}^a - \tilde{\boldsymbol{\theta}}^b)$ may not be $o_p(1)$. Clark regularity is weaker than near convexity, but it is a considerably stronger than Chernoff regularity; for example, it ensures that the tangent cone is convex.

Suppose that Ω is pro-regular at $\boldsymbol{\theta}_0$ and $\tilde{\boldsymbol{\theta}}^a$ and $\tilde{\boldsymbol{\theta}}^b$ are strongly consistent. Then, under some regularity conditions on the smoothness of and uniformity of convergence of $n^{-1}\ell(\boldsymbol{\theta})$ we have $\tilde{\boldsymbol{\theta}}^a = \tilde{\boldsymbol{\theta}}^b$ with probability approaching one (see Shapiro (2000a) for details).

Throughout this section we restricted our discussions to the simple *iid* setting. However, these results hold under much more general conditions. The developments in Geyer (1994) and Shapiro (2000a) include M-estimation; Andrews (1999) studied properties of the global *mle* under conditions that are more general than the usual *iid* setting. For example, his framework includes many time series models used in econometric modeling. He considered several important econometric models and illustrated the relevance of the theoretical results. In these papers quadratic approximations are used even if the objective function is not differentiable; see also Shapiro (1989).

4.10 APPENDIX: PROOFS

Proof of Proposition 4.3.2: Assume that the null hypothesis is true. The test statistic $T_{\boldsymbol{\lambda}}$ is a function of the sample. Let $T_{\boldsymbol{\lambda}}(\mathcal{X})$ denote the value of the test statistic $T_{\boldsymbol{\lambda}}$ for the sample $\mathcal{X}$. Let $\mathcal{Y}$ be *iid* as $\mathcal{X}$. Then

$$p^*(\mathcal{X}) = \alpha_1 + \sup_{\boldsymbol{\lambda} \in \mathcal{A}(\mathcal{X})} \mathrm{pr}_{\mathcal{Y}|\mathcal{X}}\{T_{\boldsymbol{\lambda}}(\mathcal{Y}) \geqslant T_{\boldsymbol{\lambda}}(\mathcal{X}) \mid \boldsymbol{\theta} = (\mathbf{0} : \boldsymbol{\lambda})\}.$$

Now,

$$\begin{aligned} \mathrm{pr}_{\mathcal{X}}\{p^*(\mathcal{X}) \le \alpha\} &= \mathrm{pr}_{\mathcal{X}}\{p^*(\mathcal{X}) \le \alpha \text{ and } \boldsymbol{\lambda}_0 \in \mathcal{A}(\mathcal{X})\} + \\ &\quad \mathrm{pr}\{p^*(\mathcal{X}) \le \alpha \text{ and } \boldsymbol{\lambda}_0 \notin \mathcal{A}(\mathcal{X})\} \\ &\le \mathrm{pr}\{p^*(\mathcal{X}) \le \alpha \text{ and } \boldsymbol{\lambda}_0 \in \mathcal{A}(\mathcal{X})\} + \mathrm{pr}(\boldsymbol{\lambda}_0 \notin \mathcal{A}(\mathcal{X})\}. \end{aligned} \tag{4.119}$$

Let F denote the cumulative distribution function of $T_{\boldsymbol{\lambda}_0}(\mathcal{X})$. Then $F\{T_{\boldsymbol{\lambda}_0}(\mathcal{X})\}$ has the uniform distribution on $(0, 1)$, and hence $\mathrm{pr}_{\mathcal{Y}|\mathcal{X}}\{T_{\boldsymbol{\lambda}_0}(\mathcal{Y}) \geqslant T_{\boldsymbol{\lambda}_0}(\mathcal{X})\}$, which is equal to $[1 - F\{T_{\boldsymbol{\lambda}_0}(\mathcal{X})\}]$, has the uniform distribution on $(0, 1)$. Now,

$$\begin{aligned} &\mathrm{pr}_{\mathcal{X}}\{p^*(\mathcal{X}) \le \alpha \text{ and } \boldsymbol{\lambda}_0 \in \mathcal{A}(\mathcal{X})\} \\ &\quad = \mathrm{pr}_{\mathcal{X}}[\alpha_1 + \\ &\qquad \sup_{\boldsymbol{\lambda} \in \mathcal{A}(\mathcal{X})} \mathrm{pr}_{\mathcal{Y}|\mathcal{X}}\{T_{\boldsymbol{\lambda}}(\mathcal{Y}) \geqslant T_{\boldsymbol{\lambda}}(\mathcal{X}) | \boldsymbol{\theta} = (\mathbf{0} : \boldsymbol{\lambda})\} \le \alpha \text{ and } \boldsymbol{\lambda}_0 \in \mathcal{A}(\mathcal{X})] \\ &\quad \le \mathrm{pr}_{\mathcal{X}}[\alpha_1 + \mathrm{pr}_{\mathcal{Y}|\mathcal{X}}\{T_{\boldsymbol{\lambda}_0}(\mathcal{Y}) \geqslant T_{\boldsymbol{\lambda}_0}(\mathcal{X}) | \boldsymbol{\theta} = (\mathbf{0} : \boldsymbol{\lambda}_0)\} \le \alpha] \\ &\quad = \alpha - \alpha_1. \end{aligned}$$

Now pr(Type I error) $= \mathrm{pr}\{p^*(\mathcal{X}) \le \alpha\} \le \alpha$, by (4.119). ■

In the proof of Proposition 4.8.1 and that of other similar asymptotic results, it is often sufficient to restrict attention to neighborhoods of the true value that shrink at the rate of $n^{-1/2}$, more precisely, neighborhoods of the form $\{\boldsymbol{\theta} : \sqrt{n}\|\boldsymbol{\theta} - \boldsymbol{\theta}_0\| \le K\}$ for large K. The next lemma provides a justification.

Lemma 4.10.1 *(1) Let* $\Theta \subset \mathbb{R}^p$, $\boldsymbol{\theta}_0 \in \Theta$, $n^{1/2}(\boldsymbol{T}_n - \boldsymbol{\theta}_0) = O_p(1)$ *and* $\Theta_{n,K} = \Theta \cap \{\boldsymbol{\theta} : n^{1/2}\|\boldsymbol{\theta} - \boldsymbol{\theta}_0\| < K\}$. *Then* $\|\boldsymbol{T}_n - \Theta\| = \|\boldsymbol{T}_n - \Theta_{n,K}\|$ *with arbitrarily large probability for sufficiently large* K *and* n. *More precisely, given* ϵ *there exist* $n_0(\epsilon)$ *and* $K_0(\epsilon)$ *such that pr*$\{\Pi(\boldsymbol{T}_n, \Theta) = \Pi(\boldsymbol{T}_n, \Theta_{n,K})\} > 1 - \epsilon$ *for* $n > n_0$ *and* $K > K_0$.

(2) Suppose that $\tilde{\boldsymbol{\theta}}$ is the maximizer of the objective function $R_n(\boldsymbol{\theta})$ over Θ. Suppose also that $\sqrt{n}(\tilde{\boldsymbol{\theta}} - \boldsymbol{\theta}_0) = O_p(1)$. Then, given $\epsilon > 0$ there exists $n_0 > 0$ and $K_0 > 0$ such that

$$\sup_{\boldsymbol{\theta}\in\Theta} R_n(\boldsymbol{\theta}) = \sup_{\boldsymbol{\theta}\in\Theta_{n,K}} R_n(\boldsymbol{\theta})$$

with arbitrarily large probability for $n > n_0$ and $K > K_0$.

Proof: Given $\epsilon > 0$ there exist $n_0(\epsilon)$ and $K_0(\epsilon)$ such that

$$\text{pr}(n^{1/2}\|\boldsymbol{T}_n - \boldsymbol{\theta}_0\| < K/2) > 1 - \epsilon, \text{ for } n > n_0 \text{ and } K > K_0.$$

Since $\boldsymbol{\theta}_0 \in \Theta$, we have that

$$\|\boldsymbol{T}_n - \boldsymbol{\theta}_0\| \geqslant \|\boldsymbol{T}_n - \Pi(\boldsymbol{T}_n, \Theta)\|.$$

Therefore,

$$\|\Pi(\boldsymbol{T}_n, \Theta) - \boldsymbol{\theta}_0\| \leq \|\Pi(\boldsymbol{T}_n, \Theta) - \boldsymbol{T}_n\| + \|\boldsymbol{T}_n - \boldsymbol{\theta}_0\| \leq 2\|\boldsymbol{T}_n - \boldsymbol{\theta}_0\|.$$

Now, if $n^{1/2}\|\boldsymbol{T}_n - \boldsymbol{\theta}_0\| < K/2$ then $n^{1/2}\|\Pi(\boldsymbol{T}_n, \Theta) - \boldsymbol{\theta}_0\| < K$ and hence $\Pi(\boldsymbol{T}_n, \Theta) = \Pi(\boldsymbol{T}_n, \Theta_{n,K})$. Therefore,

$$\text{pr}\{\Pi(\boldsymbol{T}_n, \Theta) = \Pi(\boldsymbol{T}_n, \Theta_{n,K})\} \geqslant \text{pr}(n^{1/2}\|\boldsymbol{T}_n - \boldsymbol{\theta}_0\| < K/2) > 1 - \epsilon$$

for $n > n_0$ and $K > K_0$.

The proof of the second part is similar. ■

Proof of Proposition 4.8.1: First let us first establish a lemma that would be of independent interest.

Lemma 4.10.2 *Suppose that $\boldsymbol{Y}_n = O_p(1)$, $A \subset \mathbb{R}^p$, $\boldsymbol{V}$ is a positive definite matrix of order $p \times p$ and $\hat{\boldsymbol{V}}$ is a consistent estimator of $\boldsymbol{V}$. Then*

1. $\Pi_{\boldsymbol{V}}(\boldsymbol{Y}_n, A) = O_p(1)$.
2. $\|\boldsymbol{Y}_n - A\|_{\hat{\boldsymbol{V}}} = \|\boldsymbol{Y}_n - A\|_{\boldsymbol{V}} + o_p(1)$.
3. $\|\Pi_{\hat{\boldsymbol{V}}}(\boldsymbol{Y}_n, A)\|_{\hat{\boldsymbol{V}}} = O_p(1)$.

Proof:
(1) The first part follows by triangular inequality. Let P be a fixed point in A, O be the origin, $\boldsymbol{Y}_n = OQ$, and R be the point in A that is $\boldsymbol{V}$-closest to Q. Then

$$\|\boldsymbol{Y}_n - A\|_{\boldsymbol{V}} = QR \leq QP \leq QO + OP = O_p(1)$$

$$\|\Pi(\boldsymbol{Y}_n, A)\|_{\boldsymbol{V}} = OR \leq OQ + QR = O_p(1).$$

(2) For a given matrix $\boldsymbol{A}$, let $\|\boldsymbol{A}\|$ denote the matrix norm that is equal to the largest

eigenvalue of $\boldsymbol{A}\boldsymbol{A}^T$; then an upper bound for $\|\boldsymbol{A}\|$ is $\sum\sum|a_{ij}|$. Here we shall use the inequality

$$\|A\boldsymbol{x}\|_{\boldsymbol{W}} \leq \|\boldsymbol{A}\|\|\boldsymbol{x}\|_{\boldsymbol{W}} \leq \sum\sum|a_{ij}|\|\boldsymbol{x}\|_{\boldsymbol{W}}$$

for any $\boldsymbol{x}$ and any positive definite matrix $\boldsymbol{W}$ (for example, see Pryce (1973, p 101)). From this and the Cauchy-Schwartz inequality, we have $\boldsymbol{x}^T\boldsymbol{A}\boldsymbol{x} \leq \sum\sum|a_{ij}|\|\boldsymbol{x}\|^2$. Since $\hat{\boldsymbol{V}}$ is consistent, it follows that $\|\boldsymbol{V}-\hat{\boldsymbol{V}}\| = o_p(1)$. Let

$$f_n(\boldsymbol{\theta}) = (\boldsymbol{Y}_n - \boldsymbol{\theta})^T\hat{\boldsymbol{V}}^{-1}(\boldsymbol{Y}_n - \boldsymbol{\theta}) \quad \text{and} \quad g_n(\boldsymbol{\theta}) = (\boldsymbol{Y}_n - \boldsymbol{\theta})^T\boldsymbol{V}^{-1}(\boldsymbol{Y}_n - \boldsymbol{\theta}).$$

Then

$$\begin{aligned}\sup_{\boldsymbol{\theta}\in A}|f_n(\boldsymbol{\theta}) - g_n(\boldsymbol{\theta})| &\leq \sup_{\boldsymbol{\theta}\in A}(\|\boldsymbol{Y}_n - \boldsymbol{\theta}\|^2)(\|\hat{\boldsymbol{V}}^{-1} - \boldsymbol{V}^{-1}\|\,) \\ &\leq (\|\boldsymbol{Y}_n - A\|_{\boldsymbol{I}}^2)(\|\hat{\boldsymbol{V}}^{-1} - \boldsymbol{V}^{-1}\|) = O_p(1)o_p(1) = o_p(1)\end{aligned}$$

Note that, since f_n and g_n are nonnegative, we have

$$\inf f_n(\boldsymbol{\theta}) \leq |f_n(\boldsymbol{\theta})| \leq |f_n(\boldsymbol{\theta}) - g_n(\boldsymbol{\theta}) + g_n(\boldsymbol{\theta})| \leq |f_n(\boldsymbol{\theta}) - g_n(\boldsymbol{\theta})| + g_n(\boldsymbol{\theta}).$$

Therefore, $\inf f_n(\boldsymbol{\theta}) \leq \sup|f_n(\boldsymbol{\theta}) - g_n(\boldsymbol{\theta})| + g_n(\boldsymbol{\theta})$, and hence

$$\inf f_n(\boldsymbol{\theta}) \leq \sup|f_n(\boldsymbol{\theta}) - g_n(\boldsymbol{\theta})| + \inf g_n(\boldsymbol{\theta}).$$

Thus, $\inf f_n(\boldsymbol{\theta}) - \inf g_n(\boldsymbol{\theta}) \leq \sup|f_n(\boldsymbol{\theta}) - g_n(\boldsymbol{\theta})|$. By symmetry

$$\inf g_n(\boldsymbol{\theta}) - \inf f_n(\boldsymbol{\theta}) \leq \sup|f_n(\boldsymbol{\theta}) - g_n(\boldsymbol{\theta})|.$$

Therefore, we have

$$|\inf_{\boldsymbol{\theta}\in A} f_n(\boldsymbol{\theta}) - \inf_{\boldsymbol{\theta}\in A} g_n(\boldsymbol{\theta})| \leq \sup_{\boldsymbol{\theta}\in A}|f_n(\boldsymbol{\theta}) - g_n(\boldsymbol{\theta})| = o_p(1).$$

This completes the proof of part (2). Part (3) follows from the first two. ■

Proof of Proposition 4.7.3: The proof of this uses several results concerning Constraint Qualification (CQ) in nonlinear programming and optimization. The main steps are indicated here without attempting to define terms and concepts in nonlinear optimization that are used in this proof. We shall use the same notation as in Bazaraa et al. (1993, section 5.3); for convenience, the relevant notations therein are stated below:

$$\begin{aligned}T &= \{\boldsymbol{d} : \boldsymbol{d} = \lim_{k\to\infty}\lambda_k(\boldsymbol{\theta}_k - \boldsymbol{\theta}_0), \boldsymbol{\theta}_k \in \Theta_1, \boldsymbol{\theta}_k \to \boldsymbol{\theta}_0\}; \\ G' &= \{\boldsymbol{d} : \boldsymbol{d}^T\nabla\boldsymbol{h}_i(\boldsymbol{\theta}_0) \geqslant 0 \quad \text{for } i \in J(\boldsymbol{\theta}_0)\}, \\ H_0 &= \{\boldsymbol{d} : \boldsymbol{d}^T\nabla\boldsymbol{h}_i(\boldsymbol{\theta}_0) = 0 \quad \text{for } i = 1, \ldots, \ell\}, \\ A &= \{\boldsymbol{d} : \exists\delta > 0, \exists\boldsymbol{f} : \mathbb{R} \to \mathbb{R}^p \text{ such that } \boldsymbol{f}(\lambda) \in \Theta_1 \text{ for } \lambda \in (0,\delta), \\ &\qquad \boldsymbol{f}(\boldsymbol{0}) = \boldsymbol{\theta}_0, \text{ and } \lim_{\lambda\to 0}\lambda^{-1}[\boldsymbol{f}(\lambda) - \boldsymbol{f}(\boldsymbol{0})] = \boldsymbol{d}.\}\end{aligned}$$

Thus, T is the cone of tangents and A is the derived tangent cone. For these cones the following holds (Bazarra et al. (1993, p 193):

1. Closure of $A \subset T \subset G' \cap H_0$.
2. Kuhn-Tucker CQ : Closure of $A = G' \cap H_0$.
3. Cottle CQ $\Rightarrow$ Kuhn-Tucker (KT) CQ
4. Cottle CQ is equivalent to Mangasarian-Fromowitz (MF) CQ provided Θ is open.

Since Θ is open and the MF-CQ is satisfied, it follows that Cottle-CQ is satisfied, and hence KT-CQ is satisfied. Therefore, it follows that Closure of $A = T = G' \cap H_0$, and hence Θ_1 is Chernoff regular and the tangent cone is $G' \cap H_0$. ■

Proof of Proposition 4.7.4 using Hausdorff distance definition

Let $\epsilon > 0$ be given and $\|\boldsymbol{y} - \boldsymbol{\theta}_0\| < \epsilon/2$. We shall refer to Figure 4.9 for this proof.

Let $\boldsymbol{y} = OA$ and $\Pi(A, \Theta)$ denote the point in Θ that is $\boldsymbol{V}$-closest (i.e., closest with respect to $\|.\|_{\boldsymbol{V}}$) to A. Let $C = \Pi(A, \mathcal{A}), D = \Pi(A, \Theta)$, $G = \Pi(C, \Theta)$ and $F = \Pi(D, \mathcal{A})$. First, note that since $C \in \mathcal{A}$ and $D \in \Theta$,

$$CD \leq h\{\Theta \cap B_\epsilon, \mathcal{A} \cap B_\epsilon\} = o(\epsilon).$$

By triangle inequality, $AC \leq AD + CD$; squaring both sides and subtracting AD^2, we have

$$AC^2 - AD^2 \leq CD^2 + 2AD.CD \leq o(\epsilon^2) + 2.OA.CD = o(\epsilon^2) + \epsilon o(\epsilon) = o(\epsilon^2).$$

Similarly, by considering $AD \leq AC + CD$ we have $AD^2 - AC^2 \leq o(\epsilon^2)$. Therefore, $AD^2 - AC^2 = o(\epsilon^2)$. ■

Proof of Proposition 4.7.4 without using Hausdorff distance

Let $\boldsymbol{y} = OA$ and $\Pi(A, \Theta)$ denote the point in Θ that is $\boldsymbol{V}$-closest (i.e., closest with respect to $\|.\|_{\boldsymbol{V}}$) to A. Let $C = \Pi(A, \mathcal{A}), D = \Pi(A, \Theta)$, $G = \Pi(C, \Theta)$ and $F = \Pi(D, \mathcal{A})$.

Case 1: $AC > AD$; see the diagram on the right in Fig.4.9. Since $AD \leq OA$, we have that $OD/OA \leq (OA + AD)/OA \leq (OA + OA)/OA = 2$. Now by making use of the definition of "closest point", we have that $0 \leq (AC - AD)/OA \leq (AF - AD)/OA \leq (AD + DF - AD)/OA = DF/OA = (DF/OD)(OD/OA) \leq 2(DF/OD)$.

Case 2: $AC < AD$; see the diagram on the left in Fig.4.9. Since OAC is a right-angle triangle, we have $OC \leq OA$. Now, $0 \leq (AD - AC)/OA \leq (AG - AC)/OA \leq (AC + CG - AC)/OA \leq CG/OA \leq CG/OC$.

From cases (1) and (2), we have that

$$|(AC - AD)|/OA \leq \max\{2(DF/OD), CG/OC\}.$$

Since $AC \leq OA$ and $AD \leq OA$, it follows that $(AC + AD)/OA \leq 2$. Now, $|(AC^2 - AD^2)|/OA^2 = \{|(AC - AD)|/OA\}\{(AC + AD)/OA\} \leq 2|(AC - AD)|/OA \leq 2\max\{2DF/OD, CG/OC\}$.

Now, as $OA \to 0$, we have $OD \to 0$ and $OC \to 0$ because $OD \leq OA + AD \leq 2OA$. As $OD \to 0$ we have $DF/OD \to 0$ by condition (a) of the definition of approximating cone; as $OC \to 0$ we have $CG/OC \to 0$ by condition (b) of the definition of approximating cone. Therefore, as $OA \to 0$, we have that $\max\{2DF/OD, CG/OC\} \to 0$, $(|AC^2 - AD^2)|/OA^2 \to 0$ and this establishes the claim of the proposition. ■

5

Likelihood and Alternatives

5.1 INTRODUCTION

Likelihood, sufficiency, and invariance principles have fortified the infrastructure of statistical inference covering point as well as confidence set estimation theory, hypothesis testing, and statistical decision theory as a whole. Without sufficiency, the likelihood principle may not lead to finite sample optimal resolutions; even when sufficiency prevails, but the parameter space (and or the sample space) may be *non-regular* in a certain sense, such exact optimality properties may vanish. Yet in the kind of problems encountered, for example, in the preceding two chapters there could be various constraints on the parameters, often, in nonlinear forms or in terms of inequalities, and as a result, invariance structures on the parameter (and/or the sample) space would be distorted to an extent so as to raise concerns on the optimality or desirable properties of conventional likelihood based statistical inference procedures in those setup. Such irregularities may become more perceptible when the parameter space is large (the Neyman-Scott problem is a classical example). Ramifications of likelihood formulations have therefore been advocated to restore some workable solution and so seek (pseudo-) optimality (or at least some desirable) properties of procedures within such a class. In a relatively more general decision theoretic setup the likelihood principle has been amalgamated with the Bayesian principle to yield interesting theoretical results, much of which have recently been directed towards computational Bayes methodology with varied applications in diverse fields. Yet there are roadblocks to such adoptions in inequality constrained statistical inference problems. There is also a basic query: how much importance to be attached to the likelihood based procedures when the form of the underlying distributions may not be taken for granted? Or, in other words, how statistical inference under inequality constraints can be made in a nonparametric (or even semiparametric) setup?

In this chapter, some of these methodological issues are discussed and alternative approaches are presented. It is in this spirit, in Section 5.2, the union-intersection principle (UIP) is introduced, its genesis outlined, and its role in statistical inference problems under inequality constraints is appraised. Section 5.3 deals with a variant form, known as the intersection-union test (IUT) that is especially useful in bioequivalence models. Section 5.4 is devoted to models that are not entirely parametric, and alternative formulations based on nonparametric methods are presented there. Some decision theoretic approaches, along with an appraisal of variants of *LRT* by Perlman and Wu (1999), Wang and McDermott (1998a), and Sen and Tsai (1999), among others, would be outlined in the last chapter.

5.2 THE UNION-INTERSECTION PRINCIPLE

The UIP plays a significant role in *simultaneous confidence set* estimation theory, *multiple comparisons,* and *multiparameter hypothesis testing* problems. It is relatively more flexible in many nonregular or nonstandard problems where the parameter space (and or the sample space) may not have suitable group-theoretic invariance structures that pertain to the optimality properties of LP-based approaches. For example, consider an inference problem for a multivariate normal distribution with mean vector $\boldsymbol{\theta}$ and dispersion matrix $\boldsymbol{\Sigma}$, both unspecified. For testing the null hypothesis $H_0 : \boldsymbol{\theta} = \mathbf{0}$, or some specified vector, it may be intended to consider (one-sided) alternatives that $\boldsymbol{\theta}$ belongs to the nonnegative orthant space $\mathbb{R}^{+p} = \{\boldsymbol{x} \in \mathbb{R}^p : \boldsymbol{x} \geq \mathbf{0}\}$, or to some other subspace of $\mathbb{R}^p$ for which the null hypothesis space is a vertex or an edge; even we might have a cone which is defined by suitable inequality restraints. In the univariate case, an optimal ((uniformly most powerful (UMP))) exists for such one-sided alternative. However, in the genuine multivariate case, such UMP one-sided tests may not exist. Yet, in many problems, such one-sided alternatives crop-up in a natural manner. Similarly, an estimation problem can be posed with the inequality constraint that $\boldsymbol{\theta} \in R^{+p}$. Maximization of the likelihood function over such a restricted parameter space may not always yield a closed expression for the so-called (restricted) RMLE, and even if a closed expression is available (as is in this case), verifying its optimality properties can often be cumbersome, if not difficult. For the same model, for the hypothesis testing problem against the global alternative $\boldsymbol{\theta} \neq \mathbf{0}$, a UMP test may not exist. One may take recourse to the LP for test construction, albeit in the presence of nuisance parameters, such tests may not be similar or optimal (i.e., UMP) in a sense. The Hotelling T^2 test is functionally related to the likelihood ratio test (*LRT*) and is optimal in the sense that among tests that are invariant under affine transformation on the sample observations as well as on the parameters, it has the maximum power so that T^2-test is *uniformly most powerful invariant* (UMPI) test. Of course, there are other tests that are not affine-invariant and might have better power for alternatives belonging to some part of the parameter space, but cannot dominate the Hotelling T^2 test over the entire parameter space. If nonnegative orthant alternatives are considered then the parameter space Θ coincides with $\mathbb{R}^{+p}$, which is not affine invariant. Hence, the Hotelling T^2 test though valid and consistent for such restricted alternatives may not have the best property in terms of power. In Chapter 8, this aspect of *LRT* and its ramifications would be considered in detail. For a (simultaneous) confidence set for the mean vector $\boldsymbol{\theta}$ when $\boldsymbol{\Sigma}$ is not known, one makes use of the Hotelling T^2-statistics along with its null distribution; again, in the restricted case (i.e., for multivariate one-sided confidence sets) the use of the Hotelling T^2-statistics will result in a larger set, and hence, will entail some loss of efficiency. It is therefore of interest to appraise statistical inference procedures under such restricted set-ups, and explore other alternatives that might have some natural appeal on their own.

As a second illustration, consider a $(p_1 + p_2)$-variate multinormal distribution with mean vector $\boldsymbol{\theta}$ and dispersion matrix $\boldsymbol{\Sigma}$; we partition the latter into $\boldsymbol{\Sigma}_{ij}$ of order $p_i \times p_j, i, j = 1, 2$. The null hypothesis (of independence of the two subsets of variates) that $\boldsymbol{\Sigma}_{12} = 0$ is to be tested against possible dependence that relates to $\boldsymbol{\Sigma}_{12} \neq 0$, treating the mean vector as a nuisance parameter. Whereas there are $p_1 \times p_2$ elements of $\boldsymbol{\Sigma}_{12}$ that are under testing, by an appeal to the Hotelling (1936) *canonical correlation* coefficients, we may reduce the hypothesis testing problem to a $p^* = p_1 \wedge p_2$ parameter problem; nevertheless, the resolution of the likelihood ratio test may not be very simple. The UIP can as well be applied in this context, leading to the (S.N.) Roy (1953) largest root criterion that differs from the *LRT*, and may have some advantages from the simultaneous confidence sets perspectives. We shall illustrate this later in detail including the case of restricted canonical correlations, treated in Das and Sen (1995, 1996).

In a multiparameter setting, especially in the presence of nuisance parameter(s), it might not be possible to obtain a test that performs optimally for the entire class of envisaged alternative hypotheses. Moreover, with possible nuisance parameters and a lack of a Neyman-structure, a test may not qualify for a similar test in the sense that it may not have the exact size (say α) for some preassigned $0 < \alpha < 1$. However, even in such a complex setup, it might be possible to express the null hypothesis H_0 as a (finite or not) intersection of a number of more primitive hypotheses, and, similarly, the alternative H_1 as the union (or even intersection) of a number of hypotheses that are more manageable from a statistical point of view. Symbolically, writing the component null and alternative hypotheses as $H_{0i}, H_{1i}, i \in \mathcal{I}$ where $\mathcal{I}$ refers to an index set, this formulation may be expressed as

$$H_0 = \bigcap_{i \in \mathcal{I}} H_{0i} \qquad \text{and} \qquad H_1 = \bigcup_{i \in \mathcal{I}} H_{1i}, \tag{5.1}$$

where for each $i \in \mathcal{I}, (H_{0i}, H_{1i})$ form a natural pair for which an optimal test may exist. The beauty of this UIP formulation is that such a decomposition, though not necessarily unique in all cases, could be well tailored for statistical inference under inequality restraints. Further, in a general multiparameter set-up, the UIP, having good alliance with the Roy's (1953) largest root criterion, may often provide relatively simpler simultaneous confidence sets, even in a contemplated restricted parameter space. As an illustration, let us go back to the multinormal mean testing problem against the nonnegative orthant alternative model. In this mode, the basic formulation of the classical UIT is outlined along with two of its variants, namely, *finite union-intersection tests* (FUIT) and *step-down procedure* (SDP).

5.2.1 UIT on Multinormal Mean

Let $\boldsymbol{X}_1, \ldots, \boldsymbol{X}_n$ be n *iid* observation from the p-variate multinormal $N(\boldsymbol{\theta}, \boldsymbol{\Sigma})$ where $\boldsymbol{\Sigma}$ is unknown. For a p-vector $\boldsymbol{a}$, let us define $H_{0\boldsymbol{a}} : \boldsymbol{a}^T\boldsymbol{\theta} = \boldsymbol{0}$, and $H_{1\boldsymbol{a}} : \boldsymbol{a}^T\boldsymbol{\theta} \neq \boldsymbol{0}$. Then for testing

$$H_0 : \boldsymbol{\theta} = \boldsymbol{0} \quad \text{against} \quad H_1 : \boldsymbol{\theta} \neq \boldsymbol{0},$$

one may write

$$H_0 = \bigcap_{\boldsymbol{a} \in \mathbb{R}^p} H_{0\boldsymbol{a}} \quad \text{and} \quad H_1 = \bigcup_{\boldsymbol{a} \in \mathbb{R}^p} H_{1\boldsymbol{a}}. \tag{5.2}$$

For testing $H_{0\boldsymbol{a}}$ against $H_{1\boldsymbol{a}}$, for a given $\boldsymbol{a}$, an optimal test is based on the student t-statistic constructed from the scalar random variables $\boldsymbol{a}^T\boldsymbol{X}_i, i = 1, \ldots, n$. The UIP prescribes that the null hypothesis H_0 is accepted only when all the component null hypotheses are acceptable, while the alternative hypothesis H_1 is accepted if at least for some $\boldsymbol{a}$, $H_{1\boldsymbol{a}}$ is accepted. Marginally, for each $\boldsymbol{a}$, the classical test based on the student t-statistic, denoted by $t(\boldsymbol{a})$, is a similar (and optimal) test, and further, has the same distribution under H_0, this testing procedure amounts to finding a critical value say, $C_{n,\alpha}$, such that H_0 is accepted when

$$t^2(\boldsymbol{a}) \leq C_{n,\alpha}, \forall \boldsymbol{a} \in \mathbb{R}^p, \tag{5.3}$$

and reject H_0 whenever the opposite inequality holds for at least some $\boldsymbol{a} \in \mathbb{R}^p$; the crux of the problem is therefore to choose $C_{n,\alpha}$ in such a way that test procedure has the specified level of significance $\alpha(0 < \alpha < 1)$.

For the original sample of observations $\boldsymbol{X}_1, \ldots, \boldsymbol{X}_n$, let us denote the mean vector and the sum of product matrix by $\bar{\boldsymbol{X}}_n$ and $\boldsymbol{S}_n$ respectively. Then we have

$$t^2(\boldsymbol{a}) = n(n-1)(\boldsymbol{a}^T\bar{\boldsymbol{X}}_n)^2/(\boldsymbol{a}^T\boldsymbol{S}_n\boldsymbol{a}), \forall \boldsymbol{a} \in \mathbb{R}^p. \tag{5.4}$$

As such, by the celebrated Courant theorem on the ratio of two nonnegative definite quadratic forms,

$$\sup_{\boldsymbol{a} \in \mathbb{R}^p} t^2(\boldsymbol{a}) = n(n-1)\text{ch}_{\max}(\bar{\boldsymbol{X}}_n\bar{\boldsymbol{X}}^T\boldsymbol{S}_n^{-1}) = n(n-1)\bar{\boldsymbol{X}}_n^T\boldsymbol{S}_n^{-1}\bar{\boldsymbol{X}}_n, \tag{5.5}$$

which is the classical Hotelling T^2-statistic. Therefore, for testing $H_0 : \boldsymbol{\theta} = \boldsymbol{0}$ against the global alternative $H_1 : \boldsymbol{\theta} \neq \boldsymbol{0}$, the UIP yields the same procedure as the *LRT*.

The foregoing isomorphism between *LRT* and UIP when the alternative is unrestricted does not hold in general for our contemplated restricted alternatives. To exhibit this lack of isomorphism let us consider test of

$$H_0 : \boldsymbol{\theta} = \boldsymbol{0} \quad \text{against} \quad H_1 : \boldsymbol{\theta} \in \mathbb{R}^{+p},$$

where $\mathbb{R}^{+p}$ is the nonnegative orthant. Recall that $H_1 : \boldsymbol{\theta} \in \mathbb{R}^{+p}$ can be equivalently expressed as $\cup_{\boldsymbol{a}\in\mathbb{R}^{+p}} \boldsymbol{a}^T\boldsymbol{\theta} > 0$. Hence, here the set of possible component hypotheses is governed by the nonnegative orthant $\mathbb{R}^{+p}$ itself; it is therefore taken

$$H_0 : \bigcap_{\boldsymbol{a}\in\mathbb{R}^{+p}} H_{0\boldsymbol{a}}, \text{ and } H_1 : \bigcup_{\boldsymbol{a}\in\mathbb{R}^{+p}} H_{1\boldsymbol{a}}^{+}, \tag{5.6}$$

where $H_{0\boldsymbol{a}} : \boldsymbol{a}^T\boldsymbol{\theta} = 0$ as in (5.2) and $H_{1\boldsymbol{a}}^{+} : \boldsymbol{a}^T\boldsymbol{\theta} > 0$, which is the same as $H_{1\boldsymbol{a}}$ in (5.2) except that $\boldsymbol{a}^T\boldsymbol{\theta} \neq 0$ is replaced by $\boldsymbol{a}^T\boldsymbol{\theta} > 0$.

Unlike in (5.3), we need to work with the left-hand side acceptance region $t(\boldsymbol{a}) \leq C_{n,\alpha}^{+}$, for all $\boldsymbol{a} \in \mathbb{R}^{+p}$, and as a result, unlike in (5.5), we need to consider here the following test statistic

$$T_n^* = \sup_{\boldsymbol{a}\in\mathbb{R}^{+p}} t(\boldsymbol{a}) \tag{5.7}$$

and the corresponding acceptance region:

$$T_n^* \leq C_{n,\alpha}^{+*}, \tag{5.8}$$

with a Type I error rate of $\alpha (= P\{T^* \geq C_{n,\alpha}^{+*} \mid H_0\})$.

To obtain a closed expression for T^*, we proceed as follows. Let $P = \{1, \ldots, p\}$ and for every $a : \emptyset \subseteq a \subseteq P$, let a' be its complement, and $\mid a \mid$ its cardinality. Thus, there are 2^p such possible subsets $a : \emptyset \subseteq a \subseteq P$, and $0 \leq \mid a \mid \leq p$. For each a, we partition (following possible rearrangements) $\bar{\boldsymbol{X}}_n$ and $\boldsymbol{S}_n$ as

$$\bar{\boldsymbol{X}}_n = \begin{pmatrix} \bar{\boldsymbol{X}}_{na} \\ \bar{\boldsymbol{X}}_{na'} \end{pmatrix} \quad \boldsymbol{S}_n = \begin{pmatrix} \boldsymbol{S}_{naa} & \boldsymbol{S}_{naa'} \\ \boldsymbol{S}_{na'a} & \boldsymbol{S}_{na'a'} \end{pmatrix} \tag{5.9}$$

and write

$$\bar{\boldsymbol{X}}_{na:a'} = \bar{\boldsymbol{X}}_{na} - \boldsymbol{S}_{naa'}\boldsymbol{S}_{na'a'}^{-1}\bar{\boldsymbol{X}}_{na'}, \tag{5.10}$$

$$\boldsymbol{S}_{naa:a'} = \boldsymbol{S}_{naa} - \boldsymbol{S}_{naa'}\boldsymbol{S}_{na'a'}^{-1}\boldsymbol{S}_{na'a}. \tag{5.11}$$

Further, let

$$I_{na} = I(\bar{\boldsymbol{X}}_{na:a'} > 0, \boldsymbol{S}_{na'a'}^{-1}\bar{\boldsymbol{X}}_{na'} \leq 0), \tag{5.12}$$

for $\emptyset \subseteq a \subseteq P$, where $I(A)$ stands for the indicator function of the set A. Then only one of the 2^p subsets has the indicator function equal to 1 while the rest are all equal to 0 (though this nonzero indicator function relates to a random a). Further, as in the preceding chapter, we use the Kuhn-Tucker-Lagrange (KTL) point formula theorem, and obtain that

$$T_n^{*2} = \sum_{\emptyset\subseteq a\subseteq P} [n\bar{\boldsymbol{X}}_{na:a'}^T\boldsymbol{S}_{naa:a'}^{-1}\bar{\boldsymbol{X}}_{na:a'}]I_{na}. \tag{5.13}$$

Thus, basically, for a given sample, we need to look into the 2^p possible subsets $\emptyset \subseteq a \subseteq P$ and identify the (random) a^* for which $I_{na*} = 1$. Then

$$T_n^{*2} = n\{\bar{\boldsymbol{X}}_{na*:a*'}^T\boldsymbol{S}_{na*a*:a*'}^{-1}\bar{\boldsymbol{X}}_{na*:a*'}\}. \tag{5.14}$$

Thus, for testing $H_0 : \boldsymbol{\theta} = \boldsymbol{0}$ against $H_1 : \boldsymbol{\theta} \in \mathbb{R}^p$, the UIT statistic T_n^{*2} has the closed-form (5.14)

In this set-up, $\boldsymbol{a}$ is allowed to vary over $\mathbb{R}^{+p}$, so that we have an infinite UIT. A finite version of UIT will be formulated later, but before that we compare the above UIT with the *LRT* developed in Section 3.9. We may express the *LRT* test statistic in the current setup and under the present notation as (see also Proposition 3.10.1)

$$L_n^* = \sum_{\emptyset \subseteq a \subseteq P} I_{na}(n\bar{\boldsymbol{X}}_{na;a'}^T \boldsymbol{S}_{naa:a'}^{-1} \bar{\boldsymbol{X}}_{na:a'})(1 + n\bar{\boldsymbol{X}}_{na'}^T \boldsymbol{S}_{na'a'}^{-1} \bar{\boldsymbol{X}}_{na'})^{-1}. \quad (5.15)$$

It follows from the above two equations that

$$L_n^* \leq T_n^{*2}, \quad (5.16)$$

where the equality sign holds only when $a = P$, i.e., the vector $\bar{\boldsymbol{X}}_n$ is itself in the positive orthant. Since the event $a = P$ occurs with a probability less than one (unless n is either very large and the true parameter point is in the interior of the positive orthant), the performance of the two tests may not be isomorphic for the entire parameter space $\mathbb{R}^{+p}$. This feature also calls for an appraisal of the distribution theory of the UIT and *LRT* statistics, which even under the null hypothesis may not be very simple. We will discuss this aspect in a later section.

Example 5.2.1 *Simultaneous confidence interval based on* T_n^*

Let $\boldsymbol{X}_1, \ldots, \boldsymbol{X}_n$ be *iid* r.v.'s having a p-variate normal distribution with mean vector $\boldsymbol{\theta}$ and dispersion matrix $\boldsymbol{\Sigma}$, both unknown. For every $\boldsymbol{a} \in \mathbb{R}^{+p}$, consider the test statistic

$$t_n(\boldsymbol{a}) = \boldsymbol{a}^T[\bar{\boldsymbol{X}}_n - \boldsymbol{\theta}]/(\boldsymbol{a}^T \boldsymbol{S}_n \boldsymbol{a})^{1/2}.$$

Let then

$$t_n^* = \sup_{\boldsymbol{a} \in \mathbb{R}^{+p}} t_n(\boldsymbol{a}).$$

Note that under $\boldsymbol{\theta}$, t_n^* has the same distribution as T_n^*, defined by (5.14), under the null hypothesis $H_0 : \boldsymbol{\theta} = \boldsymbol{0}$. Thus, if we have a means to compute the $(1-\alpha)$-percentile point of the null distribution of T_n^*, denoted by $T_{n,1-\alpha}^*$, then we could pose

$$\text{pr}\{\boldsymbol{a}^T\boldsymbol{\theta} \leqslant T_{n,1-\alpha}^*(\boldsymbol{a}^T \boldsymbol{S}_n \boldsymbol{a})^{1/2}, \forall \boldsymbol{a} \in \mathbb{R}^{+p} \mid \boldsymbol{\theta})\} = 1 - \alpha.$$

As a result, we obtain that

$$\text{pr}\{\boldsymbol{a}^T\boldsymbol{\theta} \geq \boldsymbol{a}^T\bar{\boldsymbol{X}}_n - T_{n,1-\alpha}^*(\boldsymbol{a}^T \boldsymbol{S}_n \boldsymbol{a})^{1/2}, \forall \boldsymbol{a} \in \mathbb{R}^{+p}\} = 1 - \alpha.$$

Moreover, for any finite set $\{\boldsymbol{a}_j, j = 1, \ldots, m\}$ of linearly independent vectors, all defined on $\mathbb{R}^{+p}$, we obtain that

$$\text{pr}\{\boldsymbol{a}_j^T\boldsymbol{\theta} \geqslant \boldsymbol{a}_j^T\bar{\boldsymbol{X}}_n - T_{n,1-\alpha}^*(\boldsymbol{a}^T \boldsymbol{S}_n \boldsymbol{a})^{1/2}, \forall 1 \leqslant j \leqslant m\} \geqslant 1 - \alpha.$$

In particular, if we take $\boldsymbol{a}_1^T = (1, 0, 0, \ldots, 0)$, $\boldsymbol{a}_2^T = (0, 1, 0, \ldots, 0)$, $\ldots$, $\boldsymbol{a}_p^T = (0, 0, 0, \ldots, 0, 1)$, then we have from the above

$$\text{pr}\{\theta_j \geqslant \bar{X}_{nj} - T_{n,1-\alpha}^*\sqrt{s_{njj}}, \ \forall\, 1 \leqslant j \leqslant p\} \geqslant 1 - \alpha.$$

Side by side, if we use the Bonferroni inequality, we obtain that

$$\text{pr}\{\theta_j \geqslant \bar{X}_{nj} - t_{n-1,1-\alpha/p}\sqrt{s_{njj}}, \forall 1 \leqslant j \leqslant p\} \geqslant 1-\alpha,$$

where $t_{m,1-\epsilon}$ is the upper $(1-\epsilon)$-quantile of the t-distribution with m degrees of freedom. If p is not too small, and n is not small either, then $T^*_{n,1-\alpha}$ is smaller than $t_{n-1,(1-\alpha/p)}$, and as a result, the UIP provides a better bound than the Bonferroni bound. Of course the success of the UIP hinges on the knowledge of $T^*_{n,1-\alpha}$, which may not be the universal case. Further, by (5.16), we may note that the LRP-based procedure may have the same shortcoming. ■

Let us consider now finite UIT procedures among which the step-down procedure (Roy (1958)) deserves a special mention. In the UIT considered above, we have allowed the coefficient vector $\boldsymbol{a}$ to range over the entire $\mathbb{R}^{+p}$. In some testing problems there may be *a priori* an ordering of the importance of the p variates. For example, in a multiple end point clinical trial, the primary variate may be the failure time, and there may be other variates that are of some interest too (even a surrogate end point might be potentially available whenever the primary end point is costly to observe). As such, it might be more appealing to attach differential importance to the variates under consideration. Let us first formulate such a procedure for a general class of alternatives, Roy et al. (1971), and then review the case of restricted ones.

The overall hypothesis is assumed to be decomposable into a finite number of subhypotheses, based on a specification of a sequence of responses in an *a priori* order. For simplicity, we assume that the responses $1, \ldots, p$ are set in that order of consideration. For the first variate, based on the set of observations $X_{11}, \ldots, X_{1n}$ that are independently and identically distributed according to a normal distribution with mean θ_1, and variance σ_{11}, we may consider the null hypothesis $H_{10} : \theta_1 = 0$ against $H_{11} : \theta_1 \neq 0$ (or > 0) for which an optimal test based on the Student t-statistic exists. In the second step, conditioned on $X_{1i}, i = 1, \ldots, n$ we consider the set of observations $X_{21}, \ldots, X_{2n}$ and exploit the fact that the conditional mean and variance of X_{2i}, given X_{1i}, are

$$\theta_2 + \beta_{21}(X_{1i} - \theta_1) \quad \text{and} \quad \sigma_{22.1} = \sigma_{22} - \sigma_{21}^2/\sigma_{11}, \tag{5.17}$$

respectively, where $\beta_{21} = \sigma_{21}/\sigma_{11}$. Thus, if we let $\theta_2^* = \theta_2 - \beta_{21}\theta_1$, then the null hypothesis $H_{02}^* : \theta_1 = \theta_2 = 0$ can be equivalently expressed as the intersection of H_{10} and $H_{20} : \theta_2^* = 0$. The alternative hypothesis $H_{12}^* : (\theta_1, \theta_2) \neq 0$ can be similarly expressed as the union of H_{11} and $H_{12} : \theta_2^* \neq 0$. Proceeding this way, we could express the overall null and alternative hypotheses as the intersection and union of stepwise hypotheses, where at the jth step, we consider a conditional set-up, given the responses of the previous steps, for $j = 1, \ldots, p$.

For a random vector $\boldsymbol{X}$, having a multinormal distribution with mean vector $\boldsymbol{\theta}$ and dispersion matrix $\boldsymbol{\Sigma}$, let $Y_1 = X_1, Y_2 = X_2 - \beta_{21}X_1, Y_3 = X_3 - \beta_{31}X_1 - \beta_{32}X_2$, and so on. Then $Y_1, \ldots, Y_p$ are independent with variances $\sigma_{11}, \sigma_{22.1}, \ldots, \sigma_{pp\cdot 1\ldots(p-1)}$, respectively. Their means are $\theta_1, \theta_2 - \beta_{21}\theta_1, \ldots, \theta_p - \beta_{p1}\theta_1 - \ldots - \beta_{p(p-1)}\theta_{p-1}$. Moreover $\sigma_{jj\cdot 1\ldots(j-1)}$ and β_{ij} involve only $\sigma_{rs}, r, s \leqslant j$. Therefore, given $X_{i1}, \ldots, X_{i(j-1)}, i = 1, \ldots, n, X_{ij}, i = 1, \ldots, n$ are independent normal with means given as above and variance equal to $\sigma_{jj\cdot 1\ldots(j-1)}$. Thus, under the null hypothesis H_{0j}, t_j, given $X^{(j-1)}$, the preceding $j-1$ columns of $\boldsymbol{X}$, has the t-distribution with $n-j$ degrees of freedom, and as this conditional distribution does not depend on $X^{(j-1)}$, they are unconditionally independent too (under H_0). More details may be found in Roy et al. (1971) and Krishnaiah (1981).

At each step, we could use the Student t-test (with degrees of freedom adjusted for the estimation of $\sigma_{jj\cdot 1\ldots(j-1)}$). Under the null hypothesis, these test statistics, denoted by $t_j, j = 1, \ldots, p$ are independently distributed, so that it is possible to choose a set of critical values $c_j, j = 1, \ldots, p$, along with a set of significance levels $\alpha_j, j = 1, \ldots, p$ such that

$$P\{\mid t_j \mid \leq c_j \mid H_{0j}\} = 1 - \alpha_j, j = 1, \ldots, p, \tag{5.18}$$

and

$$P\{\mid t_j \mid \leq c_j, \forall j = 1, \ldots, p \mid H_0\} = \Pi_{j=1}^{p} P\{\mid t_j \mid \leq c_j \mid H_{0j}\} \tag{5.19}$$

$$= \Pi_{j=1}^{p}(1 - \alpha_j) \quad = 1 - \alpha, \tag{5.20}$$

for some preassigned $\alpha (0 < \alpha < 1)$, the overall significance level.

In this formulation, the acceptance region is given by the intersection of the $[\mid t_j \mid \leq c_j]$ and the critical region is the union of the complementary sets of $[\mid t_j \mid > c_j]$. The computation of the c_j is facilitated by existing tables on the t-distribution, and the choice of the $\alpha_j, j = 1, \ldots, p$ (or equivalently the c_j) can be tailored to the degree of importance to be attached to the p responses. In the absence of such a strong preference, we may even prescribe the simple solution

$$\alpha_1 = \cdots = \alpha_p = 1 - (1 - \alpha)^{1/p}, \tag{5.21}$$

though the ordering of the responses is to be done judiciously. The modification for one-sided alternatives are quite simple.

The discouraging feature is the fact that the ordering of the responses in which the component hypotheses are set may have a bearing on the statistical conclusions. For example, even if the α_j are all the same, but instead of taking the variates in the order $1, \ldots, p$, we take in some other order, the data set at hand might lead to opposite conclusion regarding the acceptance or rejection of the overall hypotheses. This can be avoided in a different way, and under the Simes procedure, we shall discuss that in a later section.

It is also possible to consider the p marginal t-statistics corresponding to the p responses, and then formulate a finite UIT based on them without the step-down part in the previous formulation. Again, here also, one may consider one-sided alternatives for each response (to suit the orthant alternative problem). Although marginally these t-statistics will have the same t-distribution with $(n-1)$ degrees of freedom, they are not stochastically independent. Hence, using the UIP to obtain a critical level for the maximum of these statistics would require the knowledge of the dependence pattern of these statistics. In general, it is difficult to obtain explicit expressions for the percentile points of correlated t or F or even chi-square statistics. We pose a few exercises at the end of this chapter to illustrate these points. Based on all these considerations, it seems that the infinite UIT framed before might have some advantages over the step-down procedure or the finite UIT versions in many statistical uses. We shall comment more on it in a later section.

We consider some additional interesting problems to illustrate the role of UIP in the current context.

(I) Multivariate paired-sample models

Let $(\boldsymbol{X}_i, \boldsymbol{Y}_i), i = 1, \ldots, n$ be n independent and identically distributed random vectors having a 2p-variate multinormal distribution, $N(\boldsymbol{\mu}, \boldsymbol{\Sigma})$, where $\boldsymbol{\mu}$ and $\boldsymbol{\Sigma}$ are partitioned as

$$\boldsymbol{\mu} = \begin{pmatrix} \boldsymbol{\mu}_1 \\ \boldsymbol{\mu}_2 \end{pmatrix} \text{ and } \boldsymbol{\Sigma} = \begin{pmatrix} \boldsymbol{\Sigma}_{11} & \boldsymbol{\Sigma}_{12} \\ \boldsymbol{\Sigma}_{21} & \boldsymbol{\Sigma}_{22} \end{pmatrix}, \tag{5.22}$$

each $\boldsymbol{\mu}_i$ being of order $p \times 1$ and each $\boldsymbol{\Sigma}_{ij}$ being of the order $p \times p$; *as in the previous chapters, population covariance matrices will assumed to be positive definite unless the contrary in made clear.* We wish to test

$$H_0 : \boldsymbol{\mu}_1 = \boldsymbol{\mu}_2 (= \boldsymbol{\mu}_0, \text{ unknown}) \tag{5.23}$$

against the set of ordered alternatives

$$H_1 : \boldsymbol{\mu}_1 \geqslant \boldsymbol{\mu}_2, \boldsymbol{\mu}_1 \neq \boldsymbol{\mu}_2, \tag{5.24}$$

and treating $\boldsymbol{\mu}_0$ and $\boldsymbol{\Sigma}$ as nuisance parameters.

In this context, it may not be necessary to assume that $\boldsymbol{\Sigma}_{11} = \boldsymbol{\Sigma}_{22}$ and/or that $\boldsymbol{\Sigma}$ has some specified pattern. Such additional restraints on $\boldsymbol{\Sigma}$ may lead to a more complex and general hypothesis testing problem. For example, for a hypertensive patient, let $\boldsymbol{X}$ stand for the vector of diastolic and systolic blood pressures before a treatment is administered and $\boldsymbol{Y}$ for the same measures after the treatment is administered. If the treatment is effective then the mean vector of $\boldsymbol{Y}$ should be smaller than that of $\boldsymbol{X}$. It is also likely that the covariance matrix of $\boldsymbol{Y}$ should be smaller than that of $\boldsymbol{X}$ (in the matrix sense). Thus, we may like to test

$$H_0^* : \boldsymbol{\mu}_1 = \boldsymbol{\mu}_2 (= \boldsymbol{\mu}_0 \text{ unknown}) \text{ and } \boldsymbol{\Sigma}_{11} = \boldsymbol{\Sigma}_{22}, \tag{5.25}$$

against an alternative of the form

$$H_{11}^* \cup H_{12}^* : [\boldsymbol{\mu}_1 \geqslant \boldsymbol{\mu}_2] \cup [\boldsymbol{\Sigma}_{11} \succcurlyeq \boldsymbol{\Sigma}_{22}] \tag{5.26}$$

where $\boldsymbol{\Sigma}_{11} \succcurlyeq \boldsymbol{\Sigma}_{22}$, equivalently $\boldsymbol{\Sigma}_{22} \preccurlyeq \boldsymbol{\Sigma}_{11}$, means that $\boldsymbol{\Sigma}_{11} - \boldsymbol{\Sigma}_{22}$ is positive semi-definite.

It is also possible to formulate such hypotheses in terms of *stochastic order;* this will be considered in a later section. As is common in *bioassays*, if one is willing to accept $\boldsymbol{\mu}_1 = \boldsymbol{\Lambda}\boldsymbol{\mu}_2$ where $\boldsymbol{\Lambda} = \text{Diag}\,(\lambda_1, \ldots, \lambda_p)$ has nonnegative diagonal elements, $\boldsymbol{\Sigma}_{12} = \boldsymbol{\Lambda}\boldsymbol{\Sigma}_{11}, \boldsymbol{\Sigma}_{21} = \boldsymbol{\Sigma}_{11}\boldsymbol{\Lambda}$ and $\boldsymbol{\Sigma}_{22} = \boldsymbol{\Lambda}\boldsymbol{\Sigma}_{11}\boldsymbol{\Lambda}$ (in this case each component has the same coefficient of variation before and after the treatment), then the null and alternative hypotheses can be stated as $H_0 : \boldsymbol{\Lambda} = \boldsymbol{I}$ and $H_1 : \boldsymbol{\Lambda} \preccurlyeq \boldsymbol{I}$, respectively.

In the foregoing example on blood pressure, if $(\boldsymbol{X}, \boldsymbol{Y})$ denotes log-blood pressure, the difference between the two treatments is likely to be only in the mean vectors. In this case, the hypothesis testing problem in (5.23)-(5.24) is of natural interest.

Let us proceed to consider these hypothesis testing problems in the order they are formulated. Consider the (linear) transformation

$$\boldsymbol{U}_i = \boldsymbol{X}_i - \boldsymbol{Y}_i \quad \text{and} \quad \boldsymbol{V}_i = \boldsymbol{X}_i + \boldsymbol{Y}_i \quad (i = 1, \ldots, n) \tag{5.27}$$

and let $\boldsymbol{\delta} = \boldsymbol{\mu}_1 - \boldsymbol{\mu}_2, \boldsymbol{\nu} = \boldsymbol{\mu}_1 + \boldsymbol{\mu}_2$,

$$\begin{aligned}
\boldsymbol{\Gamma}_{11} &= \boldsymbol{\Sigma}_{11} + \boldsymbol{\Sigma}_{22} - \boldsymbol{\Sigma}_{12} - \boldsymbol{\Sigma}_{21}, \quad \boldsymbol{\Gamma}_{12} = \boldsymbol{\Sigma}_{11} + \boldsymbol{\Sigma}_{12} - \boldsymbol{\Sigma}_{21} - \boldsymbol{\Sigma}_{22}, \\
\boldsymbol{\Gamma}_{21} &= \boldsymbol{\Sigma}_{11} - \boldsymbol{\Sigma}_{12} + \boldsymbol{\Sigma}_{21} - \boldsymbol{\Sigma}_{22}, \quad \boldsymbol{\Gamma}_{22} = \boldsymbol{\Sigma}_{11} + \boldsymbol{\Sigma}_{12} + \boldsymbol{\Sigma}_{21} + \boldsymbol{\Sigma}_{22}.
\end{aligned}$$

Then $(\boldsymbol{U}_i, \boldsymbol{V}_i), i = 1, \ldots, n$, are *iid* random vectors having a 2p-variate normal distribution with mean vector $(\boldsymbol{\delta}, \boldsymbol{\nu})$ and dispersion matrix $\boldsymbol{\Gamma}$, partitioned into $\boldsymbol{\Gamma}_{ij}, i, j = 1, 2$. We then rewrite (5.23) vs (5.24) as $H_0 : \boldsymbol{\delta} = \boldsymbol{0}$ vs $H_1 : \boldsymbol{\delta} \geqslant \boldsymbol{0}$. Thus, if we use the marginal likelihood based on $\boldsymbol{U}_1, \ldots, \boldsymbol{U}_n$ only, we have the one-sample orthant alternative model, wherein the dispersion matrix $\boldsymbol{\Gamma}_{11}$ is unknown. For this problem, both the *LRT* and (more conveniently) the UIT worked out in (5.1) through (5.16) can be adapted without any major changes. This is essentially a reduction, by the marginal likelihood principle, to a one-sample model. The details are therefore omitted.

Consider next the hypothesis testing problem in (5.25) - (5.26). We write $H_0^* = H_{01}^* \cap H_{02}^*$ where $H_{01}^* : \boldsymbol{\mu}_1 = \boldsymbol{\mu}_2$ and $H_{02}^* : \boldsymbol{\Sigma}_{11} = \boldsymbol{\Sigma}_{22}$. Likewise, we write

$$H_{11}^* : \boldsymbol{\mu}_1 \geqslant \boldsymbol{\mu}_2 \equiv \bigcup_{\boldsymbol{a} \in \mathbb{R}^{+p}} H_{11\boldsymbol{a}}^*, \tag{5.28}$$

where

$$H_{11\boldsymbol{a}}^* : \boldsymbol{a}^T(\boldsymbol{\mu}_1 - \boldsymbol{\mu}_2) \geqslant 0, \boldsymbol{a} \in \mathbb{R}^{+p}.$$

Also, we write $H_{12}^* = \cup_{\boldsymbol{b} \in \mathbb{R}^p} H_{12\boldsymbol{b}}^*$, i.e.,

$$H_{12}^* : \boldsymbol{\Sigma}_{11} \not\succeq \boldsymbol{\Sigma}_{22} \equiv \bigcup_{\boldsymbol{b} \in \mathbb{R}^p} H_{12\boldsymbol{b}}^*, \tag{5.29}$$

where

$$H_{12\boldsymbol{b}}^* : \boldsymbol{b}^T \boldsymbol{\Sigma}_{11} \boldsymbol{b} \geqslant \boldsymbol{b}^T \boldsymbol{\Sigma}_{22} \boldsymbol{b}.$$

Let us denote the sample sum of product matrix by

$$\boldsymbol{S}_n = \sum_{i=1}^{n} \begin{pmatrix} \boldsymbol{X}_i - \bar{\boldsymbol{X}}_n \\ \boldsymbol{Y}_i - \bar{\boldsymbol{Y}}_n \end{pmatrix} (\boldsymbol{X}_i^T - \bar{\boldsymbol{X}}_n^T, \boldsymbol{Y}_i^T - \bar{\boldsymbol{Y}}_n^T)$$

and note that

$$\boldsymbol{S}_n \sim \text{Wishart } (n - 1, 2p, \boldsymbol{\Sigma}). \tag{5.30}$$

However, if we partition $\boldsymbol{S}_n$ into $\boldsymbol{S}_{nij}, i, j = 1, 2$, then marginally, $\boldsymbol{S}_{nii} \sim$ Wishart $(n - 1, p, \boldsymbol{\Sigma}_{ii}), i = 1, 2$, but $\boldsymbol{S}_{n11}$ and $\boldsymbol{S}_{n22}$ are not independent. In that way, the characteristic roots of $\boldsymbol{S}_{n11}^{-1}\boldsymbol{S}_{n22}$, as will crop up in the sample counterpart of (5.29), will depend on two dependent Wishart matrices, rendering difficulties for their distribution theory. Nevertheless, in principle, the UIT can be formulated by invoking steps similar to those in (5.1)–(5.16), but working with the double maximization process.

(II) Some multivariate two-sample models

Let $\boldsymbol{X}_i, i = 1, \ldots, n_1$ be i.i.d.r. vectors distributed according to a p-variate normal distribution with mean vector $\boldsymbol{\mu}_1$ and dispersion matrix $\boldsymbol{\Sigma}_1$; both $\boldsymbol{\mu}_1$ and $\boldsymbol{\Sigma}_1$ are unspecified. Let $\boldsymbol{Y}_i, i = 1, \ldots, n_2$, be an independent set of n_2 i.i.d.r.v's distributed according to a p-variate normal distribution with mean vector $\boldsymbol{\mu}_2$ and dispersion matrix $\boldsymbol{\Sigma}_2$. First consider the restricted alternative Behrens-Fisher problem where

$$H_0 : \boldsymbol{\mu}_1 = \boldsymbol{\mu}_2 (= \boldsymbol{\mu}_0, \text{ unknown}) \tag{5.31}$$

against the one-sided alternatives

$$H_1 : \boldsymbol{\mu}_1 \geqslant \boldsymbol{\mu}_2 \text{ and } \boldsymbol{\mu}_1 \neq \boldsymbol{\mu}_2, \tag{5.32}$$

treating $(\boldsymbol{\mu}_0, \boldsymbol{\Sigma}_1, \boldsymbol{\Sigma}_2)$ as nuisance parameters. If, however, we assume that $\boldsymbol{\Sigma}_1 = \boldsymbol{\Sigma}_2 = \boldsymbol{\Sigma}$, then the problem is isomorphic to an orthant alternative problem that has been treated earlier in this section.

Treating $\boldsymbol{\Sigma}_1$ and $\boldsymbol{\Sigma}_2$ as arbitrary, a formulation of the *LRT* for (5.31) against (5.32), in this generality may be highly complex, and its distribution theory, even under the H_0 in (5.31), remains intractable for finite sample sizes (n_1, n_2). The situation with UIT is comparatively simpler from formulation and computation perspectives, though it has similar distributional impasses.

Without any loss of generality we take $n_1 \leqslant n_2$. One simple coupling method would be to pair the $(\boldsymbol{X}_i, \boldsymbol{Y}_i)$, $i = 1, \ldots, n_1$, and then appeal to the paired-sample formulation and solution worked out in (5.22)–(5.24) and after that. However, if n_1 and n_2 differ considerably, then this pairing amounts to sacrificing information on $n_2 - n_1$ observations $(\boldsymbol{Y}_j, j > n_1)$ and hence, will be less efficient. We therefore prescribe a *coupling method* due to Scheffé (1943) and extended to the multivariate (global alternative) case by Anderson (1958); incorporating UIT, this procedure is extended here to CSI. Chatterjee (1962) contains a detailed account of the global alternative problem, including the associated simultaneous confidence intervals. The advantage of the Scheffe method is that it is more efficient than the conventional two-sample method based on random pairing of observations based on the minimum of n_1, n_2. However, it still suffers from some loss of information, and, moreover, because of the arbitrariness of the choice of the linear functions, the procedure may not yield unique conclusions from the given data set.

Let $\boldsymbol{A} = ((a_{ij}))$ be a $n_1 \times n_2$ matrix of real numbers, such that

$$\sum_{j=1}^{n_2} a_{ij} = 1 \text{ and } \sum_{j=1}^{n_2} a_{ij} a_{i'j} = \delta_{ii'} \quad (i, i' = 1, \ldots, n_1) \tag{5.33}$$

where δ_{ab} is the usual Kronecker delta (i.e., equal to 1 or 0 according as $a = b$ or not). Let then

$$\boldsymbol{Z}_i = \sum_{j=1}^{n_2} a_{ij} \boldsymbol{Y}_j \quad (i = 1, \ldots, n_1).$$

Then $E\boldsymbol{Z}_i = \boldsymbol{\mu}_2$ and $E(\boldsymbol{Z}_i - \boldsymbol{\mu}_2)(\boldsymbol{Z}_i - \boldsymbol{\mu}_2)^T = \boldsymbol{\Sigma}_2, \forall i \geqslant 1$, and the $\boldsymbol{Z}_i$ have multinormal distributions, too. Let us then couple the vectors ($\boldsymbol{X}_i$ and $\boldsymbol{Z}_i$) for $i = 1, \ldots, n_1$. Having done this coupling, we reduce the hypothesis testing problem in (5.31)–(5.32) to the one in (5.22)–(5.24) with a simplification that $\boldsymbol{\Sigma}_{12} = 0$. Hence the discussion made in (I), after (5.27), pertains to the contended Behrens-Fisher restricted alternative problem in (5.31)–(5.32). In passing, we may remark that we may choose $\boldsymbol{A}$ to maximize the information from the $\boldsymbol{Y}$-sample (subject to (5.33)); nevertheless, the choice of $\boldsymbol{A}$ or the coupling of $\boldsymbol{X}_i$ with $\boldsymbol{Z}_i$ may not be unique.

We now proceed to consider the hypothesis testing problem in (5.25)–(5.26) but relating to the two-sample model under consideration. Here also, we make use of the coupling $(\boldsymbol{X}_i, \boldsymbol{Z}_i)$ as above for $i = 1, \ldots, n_1$, and proceed as in (5.28)–(5.30). For the component hypothesis testing in (5.29), in the current two-sample context, we may define the individual sample sum of product matrices by $\boldsymbol{S}_X$ and $\boldsymbol{S}_Y$, respectively, so that

$$\boldsymbol{S}_X \sim \text{Wishart}(n_1 - 1, p, \boldsymbol{\Sigma}_1), \qquad \boldsymbol{S}_Y \sim \text{Wishart}\,(n_2 - 1, p, \boldsymbol{\Sigma}_2),$$

and $\boldsymbol{S}_X$ and $\boldsymbol{S}_Y$ are independent. Consider then

$$\boldsymbol{U} = \boldsymbol{S}_X \boldsymbol{S}_Y^{-1}[(n_2 - 1)/(n_1 - 1)] \tag{5.34}$$

and note that, under $\boldsymbol{\Sigma}_1 = \boldsymbol{\Sigma}_2, \boldsymbol{U}$ has the same distribution as $\boldsymbol{U}^0 = \boldsymbol{S}_1 \boldsymbol{S}_2^{-1}[(n_2 - 1)/(n_1 - 1)]$ where $\boldsymbol{S}_j \sim$ Wishart $(n_j - 1, p, \boldsymbol{I}), j = 1, 2$. Under the restricted alternative $H : \boldsymbol{\Sigma}_{11} \succcurlyeq \boldsymbol{\Sigma}_{22}$, note that $\boldsymbol{\Sigma}_{11}\boldsymbol{\Sigma}_{22}^{-1} \succcurlyeq \boldsymbol{I}$ so that

$$Ch_j(\boldsymbol{\Sigma}_{11}\boldsymbol{\Sigma}_{22}^{-1}) \geqslant 1, \;\; \forall j = 1, \ldots, p, \tag{5.35}$$

with at least one strict inequality being true; here $Ch_j(\boldsymbol{A})$ stands for the jth eigenvalue (characteristic root) of $\boldsymbol{A}$, and we take them as ordered, i.e., $Ch_1(\boldsymbol{A}) \leqslant \cdots \leqslant Ch_p(\boldsymbol{A})$, so that $Ch_1(\boldsymbol{A})$ and $Ch_p(\boldsymbol{A})$ stand for the minimum and maximum eigenvalues, respectively. The UIP, as applied to the matrix $\boldsymbol{U}$ in (5.34), leads to the UIT statistic

$$\mathcal{L} = Ch_1(\boldsymbol{U}) = \left(\frac{n_2 - 1}{n_1 - 1}\right) Ch_1(\boldsymbol{S}_X \boldsymbol{S}_Y^{-1}). \tag{5.36}$$

(Simultaneous diagonalization of two p.d. matrices is employed here.) Under $H_0 : \boldsymbol{\Sigma}_1 = \boldsymbol{\Sigma}_2$, the distribution of $\mathcal{L}$ is independent of the common $\boldsymbol{\Sigma}$, and there are existing tables for the critical level of $\mathcal{L}$ that can be used; we refer to Anderson (1984), Table 4, pp. 634–637.

(III) Multi-sample restricted dispersion alternative problem

Let $\boldsymbol{X}_{i1}, \ldots, \boldsymbol{X}_{in_i}$ be iid r.v.'s having the multinormal distribution with mean vector $\boldsymbol{\mu}_i$ and dispersion matrix $\boldsymbol{\Sigma}_i$, for $i = 1, \ldots, k(\geqslant 2)$; the k samples are themselves assumed to be independent of each other. We are interested in testing

$$H_0 : \boldsymbol{\Sigma}_1 = \boldsymbol{\Sigma}_2 = \cdots = \boldsymbol{\Sigma}_k \; (= \boldsymbol{\Sigma}, \text{ unknown}) \tag{5.37}$$

against the ordered alternative

$$H_1 : \boldsymbol{\Sigma}_1 \succcurlyeq \cdots \succcurlyeq \boldsymbol{\Sigma}_k, \tag{5.38}$$

treating the $\boldsymbol{\mu}_i$ and $\boldsymbol{\Sigma}$ as nuisance parameters.

Let $\boldsymbol{S}_i = \sum_{j=1}^{n_i} (\boldsymbol{X}_{ij} - \bar{\boldsymbol{X}}_i)(\boldsymbol{X}_{ij} - \bar{\boldsymbol{X}}_i)^T, i = 1, \ldots, k$ be the sample sum of product matrices, so that by the sufficiency principle, we reduce the hypothesis testing problem in (5.37)–(5.38) to that based solely on $\boldsymbol{S}_1, \ldots, \boldsymbol{S}_k$. Even so, constructing a *LRT* statistic based on the likelihood of $\boldsymbol{S}_1, \ldots, \boldsymbol{S}_k$ (product of independent Wishart densities) and incorporating the ordering in (5.38) may be quite difficult, when $k \geqslant 3$. On the other hand, we may write

$$H_1 = \bigcup_{j=1}^{k-1} H_{1j}, \quad \text{where} \quad H_{1j} : \boldsymbol{\Sigma}_j \succcurlyeq \boldsymbol{\Sigma}_{j+1}, \; 1 \leqslant j \leqslant k-1.$$

Similarly, we set $H_0 = \bigcap_{j=1}^{k-1} H_{0j}$ where $H_{0j} : \boldsymbol{\Sigma}_j = \boldsymbol{\Sigma}_{j+1}, 1 \leqslant j \leqslant k-1$. Thus, we are tempted to using (a finite) UIT wherein we incorporate the componentwise tests for H_{0j} vs $H_{1j}, 1 \leqslant j \leqslant k-1$. Fortunately, for testing H_{0j} vs H_{1j}, we may proceed as in (5.36) and consider the test statistic

$$\mathcal{L}_j = \left(\frac{n_{j+1} - 1}{n_j - 1} \right) Ch_1(\boldsymbol{S}_j \boldsymbol{S}_{j+1}^{-1}), \tag{5.39}$$

for $j = 1, \ldots, k-1$. Unfortunately, $\mathcal{L}_1, \ldots, \mathcal{L}_{k-1}$ are not stochastically independent. Moreover, if $n_1, \ldots, n_k$ are not equal, the marginal distribution of $\mathcal{L}_j$ (under H_0) may depend on j. Therefore, we might not be able to set the overall test statistic as $\max_{1 \leqslant j \leqslant k-1} \mathcal{L} = \mathcal{L}^*$, say. Rather, we take recourse to a (finite) step-down procedure. We decompose the significance level $\alpha(0 < \alpha < 1)$ into $\alpha_1 + \cdots + \alpha_{k+1} (= \alpha)$ and test for H_{0j} vs H_{1j}, using $\mathcal{L}_j$, at the significance level α_j, $j = 1, \ldots, k-1$. It is possible to set $\alpha_1 = \cdots = \alpha_{k-1} = \alpha/(k-1) = \alpha^0$, say, and use the α^0 level critical values for the $\mathcal{L}_j$ in (5.39). Though this UIT is intuitively simpler and appealing, because of possible dependence among $\{\mathcal{L}_1, \ldots, \mathcal{L}_{k-1}\}$, the above decomposition leads to a somewhat conservative procedure. For other related tests and tables see Chapter 21 in Krishnaiah (1981); the tables do not depend on $\boldsymbol{\Sigma}$ because, under the null hypothesis, $\boldsymbol{\Sigma}_{11} \boldsymbol{\Sigma}_{22}^{-1} = \boldsymbol{I}$.

Let us look into the scenario from the simultaneous confidence set perspectives, where also we may have similar inequality constraints on the parameters. First, let us illustrate the statistical aspects with the multinormal mean problem when the dispersion matrix is unknown. For testing a null hypothesis $H_0 : \boldsymbol{\theta} = \boldsymbol{\theta}_0$, the *LRT* statistic is a monotone function of the Hotelling T^2-statistic:

$$T_n^2(\boldsymbol{\theta}_0) = n(\bar{\boldsymbol{X}}_n - \boldsymbol{\theta}_0)^T \boldsymbol{S}_n^{-1}(\bar{\boldsymbol{X}}_n - \boldsymbol{\theta}_0), \tag{5.40}$$

and the acceptance region for this null hypothesis corresponds to

$$T_n^2(\boldsymbol{\theta}_0) \leq T_{n,\alpha}^2, \tag{5.41}$$

where the cutoff point $T_{n,\alpha}^2$ is the $100(1-\alpha)$ percentile point of the null distribution of the Hotelling $T_n^2(\boldsymbol{\theta}_0)$ and does not depend on $\boldsymbol{\theta}_0$. For the given sample, both $\bar{\boldsymbol{X}}_n$ and $\boldsymbol{S}_n$ are given, and the above acceptance region represents an ellipsoid in $\boldsymbol{\theta}_0$ with center at $\bar{\boldsymbol{X}}_n$ and orientation depending on $\boldsymbol{S}_n$. This provides the following confidence set for $\boldsymbol{\theta}$ with a confidence coefficient (coverage probability) $1-\alpha$:

$$C_n(\boldsymbol{\theta}) = \{\boldsymbol{\theta} : T_n^2(\boldsymbol{\theta}) \leq T_{n,\alpha}^2\}. \tag{5.42}$$

This confidence set can be interpreted as a simultaneous confidence interval (SCI) in the following sense (Scheffé (1951)). Consider an arbitrary $\boldsymbol{a}$ and define $Y_i = \boldsymbol{a}^T \boldsymbol{X}_i, i = 1, \ldots, n$. Then based on the Y_i, and using the usual student t-statistic, we obtain a confidence interval for $\boldsymbol{a}^T\boldsymbol{\theta}$ in terms of the average $\bar{Y}_n = \boldsymbol{a}^T\bar{\boldsymbol{X}}_n$ and $\boldsymbol{a}^T\boldsymbol{S}_n\boldsymbol{a}$. If we now allow $\boldsymbol{a}$ to vary over the entire $\mathbb{R}^p$, then the locus of all such parallel tangents constitutes the confidence set $C(\boldsymbol{\theta})$. The same SCI can be obtained by incorporating the UIP along with the directional wise Student t-statistics treated earlier in this section. As in the case of the testing problem, for SCI, the LRP and UIP lead to the same solution for the unrestricted case. The situation could be quite different in inequality constrained inference problems, and we illustrate this aspect with the multinormal mean problem when the mean vector is restricted to the positive orthant.

At the very start, we can mention that the confidence set in (5.42) has exact coverage probability $1-\alpha$ when $\boldsymbol{\theta}$ is allowed to vary over the entire $\mathbb{R}^p$. On the other hand, if we restrict ourselves to the domain $\mathbb{R}^{+p}$, a part of the ellipsoid in (5.42) will be in $\mathbb{R}^{+p}$, while the complementary part would be outside this positive orthant; this effect is more perceptible if $\boldsymbol{\theta}$ lies near the lower boundary (edges) of the restricted parameter space $\mathbb{R}^{+p}$. Whereas the unrestricted confidence set estimation problem is invariant under affine transformations, in the restricted parameter case, this invariance property may not generally hold. Thus, the coverage probability may well depend on $\boldsymbol{\theta}$, and in that way, it could be quite conservative in nature. Problem 5.2 is set to illustrate this point with respect to the bivariate normal mean estimation problem.

The UIP, possibly in a finite intersection version, may have some advantage in such restricted parameter space confidence set estimation problems. For each coordinate of $\boldsymbol{\theta}$ we may set an $1-\alpha^*$ confidence interval that takes into account the nonnegative nature of the parameter (where we set $\alpha^* = \alpha/p$), and in that way, though we would have a conservative confidence set, it could be much less conservative than the one based on (5.42). The primary limitation of such a confidence set would be that it will not be affine invariant—a property that we would not value much when the basic problem itself may not have that property. It is also possible to incorporate the step-down procedure wherein in each step a range-preserving confidence interval may be set. However, the presence of the nuisance regression parameters that could be both positive or negative may create some impasses in this respect. Problem 5.3 is set to verify this in the bivariate normal mean problem. In the rest of this section, we consider some simpler confidence set problems, and illustrate the various alternative procedures more vividly.

The ordered alternative problem in a univariate setup: Consider k independent samples of sizes $n_1, \ldots, n_k$, respectively, drawn from k normal populations with means $\theta_1, \ldots, \theta_k$ and a common (unknown) variance σ^2. The first problem is to test for the null hypothesis $H_0 : \theta_1 = \cdots = \theta_k = \theta$ (unknown) against ordered alternatives $H_1^> : \theta_1 \leq \theta_2 \cdots \leq \theta_k$, with at least one strict inequality sign being true. This hypothesis testing problem can be reduced to the orthant alternative problem by writing $\theta_j = \theta_{j-1} + \delta_j, j = 2, \ldots, k$ wherein the δ_j all belong to $\mathbb{R}^+$. Moreover, in this case, the covariance matrix has a simple form $\sigma^2 V$ where V is a known matrix. Thus, the results of Chapter 2 can be incorporated to consider an *LRT* for this ordered alternative testing problem. In the same way, one can use the UIP to construct an UIT for the same testing problem, and the two tests are isomorphic. Consider next the confidence set and SCI problem under the inequality restraints on the θ_j. For this purpose, we need to provide a SCI for the δ_j under the inequality restraints that they are all nonnegative, and treating θ_1 as a nuisance parameter. The usual procedure of inverting a test statistic to obtain the SCI (as done in (5.42)) may not work here, as the hypothesis testing problem is not invariant to a general group of transformations. On an ad hoc basis, one can consider the set of $k - 1$ paired difference $\theta_j - \theta_{j-1}, j = 2, \ldots, k$, and for each pair consider a t-statistic to obtain a confidence interval. But these t-statistics are not all independent, and, hence, one needs to work with the joint density of a set of correlated t-statistics; some of these details are put in Problem 5.5. Step-down procedure and other variants of the UIP can also be posed along the same line. Some of these problems are relegated to the end of the chapter as exercises.

5.3 INTERSECTION UNION TESTS (IUT)

For most hypotheses testing problems, the *LRT* is the preferred first option. This is particularly so for testing $\theta = \theta_0$ against $\theta \neq \theta_0$ in regular cases with θ_0 being an interior point of the parameter space. When the hypotheses are more complicated, there may be simper alternatives to *LRT*. In this section, we shall consider intersection union tests for a particular class of testing problems.

To introduce the basic idea of intersection union test, let us consider a simple example. Let X_1 and X_2 be two random variables with means θ_1 and θ_2, respectively. Suppose that we have random samples on X_1 and on X_2; the samples may be dependent. We are interested to test

$$H_0 : \theta_1 \leqslant 0 \text{ or } \theta_2 \leqslant 0, \text{ against } H_1 : \theta_1 > 0 \text{ and } \theta_2 > 0.$$

The results to be introduced later in this section say that a 5% level intersection union test (IUT) of H_0 against H_1 is, reject H_0 and accept H_1 if a 5% level test rejects $\theta_1 \leqslant 0$ and accepts $\theta_1 > 0$ and another 5% level test rejects $\theta_2 \leqslant 0$ and accepts $\theta_2 > 0$. Therefore, to construct a 5% level IUT for the foregoing testing problem all we need is a 5% level test of $\theta_i \leqslant 0$ against $\theta_i > 0$, for every i. A remarkable feature of this IUT is that it does not require the extent or the form of dependence between the statistic for testing against $\theta_1 > 0$ and that for testing against $\theta_2 > 0$; further, there is no need to make Bonferroni-type corrections to control the overall probability of type I error. By contrast, to develop a likelihood ratio test, we need to know the joint likelihood of (X_1, X_2), which implicitly requires the exact form of dependence between the two test statistics. Just because the likelihood approach uses more information than the IUT, it does not follow that the former is always better than the latter; an example will illustrate this later.

To provide a general formulation of the intersection union test let J be an indexing set and $\{H_0^{(j)} : j \in J\}$ and $\{H_1^{(j)} : j \in J\}$ be two collections of hypotheses. For each $j \in J$, suppose that a level-α_j test of $H_0^{(j)}$ against $H_1^{(j)}$ is available, where $0 < \alpha_j < 1$ is given $(j \in J)$. Let the null and alternative hypotheses be

$$H_0 : H_0^{(j)} \text{is true for some } j \in J, \quad \text{and} \quad H_1 : H_1^{(j)} \text{is true for every } j \in J, \quad (5.43)$$

respectively. A level-α IUT of H_0 against H_1 is the following:

> Reject H_0 and accept H_1 if a level-α test of $H_0^{(j)}$ against $H_1^{(j)}$ rejects $H_0^{(j)}$ and accepts $H_1^{(j)}$ for every $j \in J$.

The rationale for this is simple: since H_1 says that $H_1^{(j)}$ is true for every $j \in J$, H_1 would be accepted if there is evidence to support $H_1^{(j)}$ for every $j \in J$. The fact that the foregoing IUT is level-α is stated and proved in Proposition 5.3.1 below.

In most cases, the hypothesis testing problem in (5.43) takes the following more convenient form. Let $\boldsymbol{\theta} \in \mathbb{R}^k$ denote the parameter of interest and $\Theta \subseteq \mathbb{R}^k$ denote the parameter space. Assume that, for every $j \in J$, there exists a subset Θ_j of Θ such that the component hypotheses, $H_0^{(j)}$ and $H_1^{(j)}$, are

$$H_0^{(j)} : \boldsymbol{\theta} \in \Theta_j \quad \text{and} \quad H_1^{(j)} : \boldsymbol{\theta} \notin \Theta_j, \quad (5.44)$$

respectively. Then, the null and alternative hypotheses in (5.43) may be stated as

$$H_0 : \boldsymbol{\theta} \in \bigcup_{j \in J} \Theta_j \quad \text{and} \quad H_1 : \boldsymbol{\theta} \notin \bigcup_{j \in J} \Theta_j, \quad (5.45)$$

respectively. Note that there is a natural pairing of $\boldsymbol{\theta} \in \Theta_j$ in H_0 with $\boldsymbol{\theta} \notin \Theta_j$ in H_1, for each $j \in J$. For example, when $J = \{1, \ldots, k\}$, it is helpful to note that

$$H_0 : \boldsymbol{\theta} \in \Theta_1 \quad \text{or} \quad \boldsymbol{\theta} \in \Theta_2 \quad \text{or} \ldots \text{or} \quad \boldsymbol{\theta} \in \Theta_k,$$

and

$$H_1 : \boldsymbol{\theta} \notin \Theta_1 \quad \text{and} \quad \boldsymbol{\theta} \notin \Theta_2 \quad \text{and} \ldots \text{and} \quad \boldsymbol{\theta} \notin \Theta_k.$$

Now, for (5.45), a level-α IUT of H_0 against H_1 is, reject H_0 and accept H_1 if a level-α test rejects $\boldsymbol{\theta} \in \Theta_j$ and accepts $\boldsymbol{\theta} \notin \Theta_j$, for every $j \in J$.

Note that the null and the alternative hypotheses can also be stated as

$$H_0 : \boldsymbol{\theta} \in \bigcup_{j \in J} \Theta_j \qquad H_1 : \boldsymbol{\theta} \in \bigcap_{j \in J} \Theta_j^c,$$

respectively, where $\Theta_j^c = \Theta \backslash \Theta_j$, the complement of Θ_j in Θ. The reason for the term "intersection union test" is related to the fact that the null and alternative parameter spaces can be expressed as an union and an intersection, respectively; for the closely related "union intersection test," the roles of union and intersection are interchanged in the formulation of the problem.

The basic idea underlying IUT is not new. It appeared in Lehmann (1952); see also Gleser (1973). Important contributions to this topic have been made by R.L. Berger in a series of publications [see Berger (1982, 1984a, 1997), Berger and Hsu (1996), and Liu and Berger (1995)]. Other related publications include Cohen, Gatsonis and Marden (1983a, 1983b), Davis (1989), Laska and Meisner (1989), Laska et al. (1992), Sasabuchi (1980, 1988a, 1988b), Nomakuchi and Sakata (1987), Iwasa (1991), and Gutmann (1987). The topic is also discussed in Perlman and Wu (1999).

Let $\boldsymbol{X}$ denote the data vector, and R_j denote the critical region for testing $\boldsymbol{\theta} \in \Theta_j$ against $\boldsymbol{\theta} \notin \Theta_j$, $j \in J$. Then, by definition, the critical region for the IUT is

$$R = \bigcap_{j \in J} R_j.$$

The next result provides a formal statement of the fact that the IUT as introduced earlier is valid.

Proposition 5.3.1 *Let the test of $H_0^{(j)}$ against $H_1^{(j)}$ be level-α_j, $j \in J$. Then the corresponding IUT of H_0 against H_1 in (5.43) is level-α, where $\alpha = \sup\{\alpha_j : j \in J\}$. Consequently, the p-value for the IUT does not exceed* $\sup\{p_j : j \in J\}$ *where p_j is the p-value for testing $H_0^{(j)}$ against $H_1^{(j)}$, $j \in J$.*

Proof: The size of the IUT is equal to

$$\sup_{\boldsymbol{\theta}} pr_{\boldsymbol{\theta}} \left(\boldsymbol{X} \in \cap_{i \in J} R_i \middle| \boldsymbol{\theta} \in H_0\right) = \sup_{j \in J} \sup_{\boldsymbol{\theta}} pr_{\boldsymbol{\theta}} \left(\boldsymbol{X} \in \cap_{i \in J} R_i \middle| \boldsymbol{\theta} \in \Theta_j\right)$$
$$\leqslant \sup_{j \in J} \sup_{\boldsymbol{\theta}} pr_{\boldsymbol{\theta}} (\boldsymbol{X} \in R_j \middle| \boldsymbol{\theta} \in \Theta_j) = \sup_{j \in J} \alpha_j. \qquad \blacksquare$$

It follows from this result that in order to maximize the power of IUT, it is better to choose α_j to have a common value for different $j \in J$.

Let us first consider an example to illustrate the basics of an IUT.

Example 5.3.1 *An illustration of IUT.*

Let $X_1 \sim N(\theta_1, 1)$ and $X_2 \sim N(\theta_2, 1)$ where X_1 and X_2 are univariate and independent. Let the null and alternative hypotheses be

$$H_0 : \theta_1 \leqslant 0 \text{ or } \theta_2 \leqslant 0, \text{ and } H_1 : \theta_1 > 0 \text{ and } \theta_2 > 0,$$

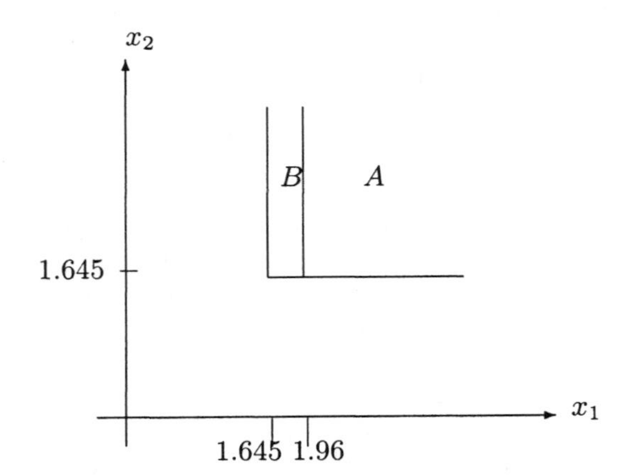

Fig. 5.1 Intersection union test.

respectively.

(a) For illustrative purposes, let us choose $\alpha_1 = 0.025$ and $\alpha_2 = 0.05$. Consider the following standard tests.

A size-α_1 test on θ_1: reject $\theta_1 \leqslant 0$ and accept $\theta_1 > 0$ if $X_1 \geqslant 1.96$.

A size-α_2 test on θ_2: reject $\theta_2 \leqslant 0$ and accept $\theta_2 > 0$ if $X_2 \geqslant 1.645$.

Now, the IUT corresponding to these two tests is,

$$\text{reject } H_0 \text{ and accept } H_1 \text{ if } X_1 \geqslant 1.96 \text{ and } X_2 \geqslant 1.645.$$

It follows from Proposition 5.3.1, that this IUT has level 0.05, its critical region is $A = \{\boldsymbol{x} \in \mathbb{R}^2 : x_1 \geqslant 1.96, x_2 \geqslant 1.645\}$ (see Fig. 5.1).

(b) Although the foregoing IUT is a 5% level test, its size cannot be deduced from Proposition 5.3.1. However, in this particular example, the size of the IUT is also $\max\{\alpha_1, \alpha_2\} = 0.05$. This can be verified as follows. Let $\boldsymbol{\theta} = (\theta_1, \theta_2)$ be a point in H_0, and let $f(\boldsymbol{\theta}) = pr_{\boldsymbol{\theta}}$ *(IUT rejects H_0)*. Then $f(\boldsymbol{\theta}) = pr_{\boldsymbol{\theta}}(X_1 \geqslant 1.96) pr_{\boldsymbol{\theta}}(X_2 \geqslant 1.645)$. If $\theta_1 \leqslant 0$ then $f(\theta_1, \theta_2) \leqslant f(0, \theta_2)$; if $\theta_2 \leqslant 0$ then $f(\theta_1, \theta_2) \leqslant f(\theta_1, 0)$. Since $f(\theta_1, 0)$ and $f(0, \theta_2)$ are nondecreasing functions of θ_1 and θ_2, respectively, it follows that $f(\theta_1, \theta_2) \leqslant \max\{f(0, \infty), f(\infty, 0)\} = \max\{0.025, 0.05\} = 0.05$. Therefore, the least favorable null value of $\boldsymbol{\theta}$ is $(\infty, 0)$. At this least favorable null value, the probability of type I error is 0.05; hence the size of the corresponding IUT is 0.05.

(c) Now, let us consider the case when the test on θ_1 and that on θ_2 each has size-0.05. Then, the IUT of H_0 against H_1 is

$$\text{reject } H_0 \text{ and accept } H_1 \text{ if } X_1 \geqslant 1.645 \text{ and } X_2 \geqslant 1.645. \tag{5.46}$$

The critical region for this IUT is $A \cup B$ in Figure 5.1; thus, the critical region has now expanded by B compared to that in part (a). Consequently, the test would be more powerful, and the probability of Type I error would be higher at every point in

the null parameter space. However, by Proposition 5.3.1, the level of the test is still 0.05. At the point $(0, \infty)$, the probability of Type I error increased from 0.025 to 0.05. Now, $(0, \infty)$ and $(\infty, 0)$ are both least favorable null values; at each of these points, the probability of type I error is 0.05 and hence the size of the test is also 0.05.
(d) Suppose that the sample value of (X_1, X_2) is (1.645, 1.96). Then an upper bound for the p-value of IUT is $\max\{0.05, 0.025\} = 0.05$. ■

It would be helpful to note some similarities and differences between the IUT and a corresponding multiple testing problem; the latter may involve Bonferroni-type adjustments. The following example illustrates the main points.

Example 5.3.2 *Relationship between IUT and multiple testing.*

As in Example 5.3.1, let $X_1 \sim N(\theta_1, 1)$ and $X_2 \sim N(\theta_2, 1)$ where X_1 and X_2 are independent, and let the null and alternative hypotheses be $H_0 : \theta_1 \leqslant 0$ or $\theta_2 \leqslant 0$, and $H_1 : \theta_1 > 0$ and $\theta_2 > 0$, respectively.

It is possible to apply a set of multiple tests for this problem. As an example, suppose that a test of H_0 against H_1 is desired at 5% level. A multiple testing procedure with Bonferroni adjustment would be to test each of $\theta_1 \leqslant 0$ against $\theta_1 > 0$, and $\theta_2 \leqslant 0$ against $\theta_2 > 0$ at 2.5% level; if they both reject their null hypotheses then H_1 would be accepted at 5% level. An IUT also follows the same procedure except that the same two component tests are carried out at 5% level each instead of 2.5% level. Thus, it follows that the IUT is more powerful than the multiple test.

To further illustrate an important difference between the multiple test and the IUT, suppose that the sample value of (X_1, X_2) is (2.4, 1.4). Since $X_2 = 1.4 \ngeq 1.645$, the foregoing IUT would not reject H_0. Now, consider the following multiple decision rule at 5% level: "reject $\theta_1 \leqslant 0$ and accept $\theta_1 > 0$ if $X_1 \geqslant 1.96$, and reject $\theta_2 \leqslant 0$ and accept $\theta_2 > 0$ if $X_2 \geqslant 1.96$." The result of this multiple decision rule is "accept $\theta_1 > 0$, and do not reject $\theta_2 \leqslant 0$." Note that the multiple decision rule makes a decision about each θ_i separately. By contrast, an IUT would either accept "$\theta_1 > 0$ and $\theta_2 > 0$" in its entirety or not; it does not attempt to make one decision about θ_1 and then another decision about θ_2, as a multiple testing problem does. Therefore, we conclude that an IUT is preferable to a multiple test for testing against $H_1 : \theta_1 > 0$ and $\theta_2 > 0$. In fact, the two hypotheses are not isomorphic and hence their resolutions are different too. ■

In Example 5.3.1, we considered the two cases, "$\alpha_1 = 0.025$ and $\alpha_2 = 0.05$" and "$\alpha_1 = \alpha_2 = 0.05$." For each of these cases, it follows from Proposition 5.3.1 that the IUT also has level, $\max\{\alpha_1, \alpha_2\} = 0.05$. The calculations in Example 5.3.1 show that the size of the IUT is also $\max\{\alpha_1, \alpha_2\} = 0.05$ in each of the two cases and that the least favorable null value is at infinity. Proposition 5.3.2 generalizes these and gives conditions under which the size of IUT is $\sup\{\alpha_j : j \in J\}$; it is a slight modification of a result in Berger (1997).

Proposition 5.3.2 *For simplicity of notation, assume that $1 \in J$. Let α denote the size of R_1. Suppose that R_j is a level-α critical region for $j \neq 1$ and that there exists a sequence $\{\boldsymbol{\theta}_\ell\}$ in Θ_1 such that as $\ell \to \infty$, pr$(\boldsymbol{X} \in R_1 | \boldsymbol{\theta} = \boldsymbol{\theta}_\ell) \to \alpha$ and pr$(\boldsymbol{X} \in \cap_{j \neq 1} R_j | \boldsymbol{\theta} = \boldsymbol{\theta}_\ell) \to 1$. Then, the IUT with rejection region $\cap_{j \in J} R_j$ is size-α and every limit point of $\{\boldsymbol{\theta}_\ell\}$ is a least favorable null value.*

Proof: The size of the IUT is equal to

$$\begin{aligned}
&\sup_{\boldsymbol{\theta}\in H_0} pr\{\text{reject } H_0|\boldsymbol{\theta}\} \geqslant \lim_{\ell\to\infty} pr(\text{reject } H_0|\boldsymbol{\theta}_\ell) \\
&= \lim_{\ell\to\infty} \{1 - pr\left(\boldsymbol{X} \in \cup_{j\in J}\mathbb{R}_j^c\middle|\boldsymbol{\theta} = \boldsymbol{\theta}_\ell\right)\} \\
&\geqslant 1 - \lim_{\ell\to\infty} (pr(\boldsymbol{X} \in \mathbb{R}_1^c|\boldsymbol{\theta} = \boldsymbol{\theta}_\ell) - \lim_{\ell\to\infty} pr\left(\boldsymbol{X} \in (\cap_{j\neq 1}R_j)^c\middle|\boldsymbol{\theta} = \boldsymbol{\theta}_\ell\right) \\
&= 1 - (1-\alpha) - 0 = \alpha.
\end{aligned}$$

The fact that $\sup_{\boldsymbol{\theta}} pr$ (reject $H_0\big|\boldsymbol{\theta} \in H_0) \leqslant \alpha$ follows from Proposition 5.3.1; this completes the proof. ■

This result can also be re-stated in terms of p-values as follows: Let $\boldsymbol{x}$ denote the sample value of $\boldsymbol{X}$, p_j denote the p-value for testing $H_0^{(j)}$ against $H_1^{(j)}$, Q_j denote the event that the outcome is at least as extreme as $\boldsymbol{x}$ for testing $H_0^{(j)}$ against $H_1^{(j)}$ so that $p_j = \sup_{\boldsymbol{\theta}} \text{pr}\{\boldsymbol{X} \in Q_j|\boldsymbol{\theta} \in \Theta_j\}$. Suppose that $p_j \leqslant p_1$, for $j \in J$, and that there exists a sequence $\{\boldsymbol{\theta}_\ell\}$ such that $\text{pr}\{\boldsymbol{X} \in Q_1|\boldsymbol{\theta} = \boldsymbol{\theta}_\ell\} \to p_1$ and $\text{pr}\{\boldsymbol{X} \in \cap_{j\neq 1}Q_j|\boldsymbol{\theta} = \boldsymbol{\theta}_\ell\} \to 1$ as $\ell \to \infty$. Then the p-value for the IUT is p_1.

The result takes the following simpler form when J is finite (see Berger (1997)).

Corollary 5.3.3 *Suppose that $J = \{1,\ldots,k\}, R_1$ is size-α, R_j is level-α for $j = 2,\ldots,k$, there exists a sequence $\{\boldsymbol{\theta}_\ell\}$ in Θ_1 such that, as $\ell \to \infty$, $pr(\boldsymbol{X} \in R_1\big|\boldsymbol{\theta} = \boldsymbol{\theta}_\ell) \to \alpha$ and $pr(\boldsymbol{X} \in R_j\big|\boldsymbol{\theta} = \boldsymbol{\theta}_\ell) \to 1$ for $j = 2,\ldots,k$. Then, the IUT is size-α and every limit point of $\{\boldsymbol{\theta}_\ell\}$ is a least favorable null value.* ■

To provide a numerical illustration of this result, consider Example 5.3.1 with $\alpha_1 = 0.025$ and $\alpha_2 = 0.05$. Then $\Theta_1 = \{(\theta_1,\theta_2) : \theta_1 \leqslant 0\}$, $\Theta_2 = \{(\theta_1,\theta_2) : \theta_2 \leqslant 0\}$, $R_1 = \{\boldsymbol{x} \in \mathbb{R}^2 : x_1 \geqslant 1.96\}$, $R_2 = \{\boldsymbol{x} \in \mathbb{R}^2 : x_2 \geqslant 1.645\}$; note that Θ_1 is a half-space, not the negative part of the θ_1-axis. Let $\boldsymbol{\theta}_\ell = (\ell, 0), \ell = 1, 2, \ldots$. Then, $\text{pr}(\boldsymbol{X} \in R_2|\boldsymbol{\theta} = \boldsymbol{\theta}_\ell) \to 0.05$ and $\text{pr}(\boldsymbol{X} \in R_1|\boldsymbol{\theta} = \boldsymbol{\theta}_\ell) \to 1$ as $\ell \to \infty$. Therefore, it follows from Corollary 5.3.3 that the size of the IUT is 0.05 and $(\infty, 0)$ is a least favorable null value. ■

Now, we consider two examples to illustrate practical applications of IUT.

Example 5.3.3 *Acceptance sampling (Berger (1982)).*

Consider Example 1.2.15 on acceptance sampling. Let θ_1 denote the parameter characterizing the *breaking strength*; for example, it can be the mean, median, or a quantile. Similarly, define θ_2 for *tong tear strength*, θ_3 for *dimensional change (increase as a %)*, etc. According to Table 1.11, the minimum standard required of these parameters are $\theta_1 > 50, \theta_2 > 6$, $\theta_3 > -5, \theta_3 < 2$, etc. Each of these can be tested using t-ratios, for example. The parameter relating to colorfastness and flammability may be characterized using binomial parameters. For example, let θ_5 denote the proportion of the batch having a grade of four or above in colorfastness. Further, suppose that the minimum standard of "class 4" for water colorfastness is that $\theta_5 > 0.9$. The problem may be formulated similarly for the other parameters. Consequently, the null and alternative hypotheses are

$H_0 : \theta_1 \leqslant 50$ *or* $\theta_2 \leqslant 6$ *or* $\theta_3 \leqslant -5$ *or* $\theta_3 \geqslant 2$, *etc.*, and
$H_1 : \theta_1 > 50,\ \theta_2 > 6,\ \theta_3 > -5,\ \theta_3 < 2$, *etc.*,
respectively. Now an IUT of H_0 against H_1 at 5% level is, reject H_0 and accept H_1 at 5% level if $\theta_1 \leqslant 50$ is rejected and $\theta_1 > 50$ is accepted at 5% level, and $\theta_2 \leqslant 6$ is rejected and $\theta_2 > 6$ is accepted at 5% level, and $\theta_3 \leqslant -5$ is rejected and $\theta_3 > -5$ is accepted at 5% level, etc. ■

Example 5.3.4 *Analysis of* $2 \times 2 \times 2$ *tables (Cohen et al. (1983a)).*

Consider Example 1.2.16. First suppose that we wish to test the claim that vaccination is beneficial in reducing infection. This claim involves only the relationship between vaccination and Infection; it does not involve the immune Level. Therefore, the claim can be tested using the data in Table 5.1 obtained by collapsing Table 1.12 over immune Level. For this testing problem, the null and alternative hypotheses may be stated as
H_0^*: *Vaccination does not reduce the risk of infection,* and
H_1^*: *Vaccination reduces the risk of infection.*
An exact conditional p-value for testing H_0^* against H_1^* is

$$\binom{30}{15} \sum_{i=0}^{5} \binom{17}{10+i} \binom{13}{5-i} = 0.231.$$

Since this is large, we conclude that there is no evidence to support the claim that vaccination reduces infection.

Now, let us make the following reasonable assumption.
Conditional Independence: Vaccination affects the risk of infection only through its effect on the level of immunity.

We would like to establish that *vaccination improves immunity and improved immunity reduces the risk of infection.* To apply an IUT for this, let $H_1^{(1)}$ and $H_1^{(2)}$ be defined as $H_1^{(1)}$: *Vaccination improves immune level,* and $H_1^{(2)}$: *Improved immune level reduces the risk of infection.* Now, the null and alternative hypotheses are

$$H_0 : \text{ not } H_1^{(1)} \text{ or not } H_1^{(2)} \qquad \text{and} \qquad H_1 : H_1^{(1)} \text{ and } H_1^{(2)}.$$

Table 5.1 Effect of vaccination on infection

Infection	Vaccination No	Yes	Total
Well	7	10	17
Sick	8	5	13
Total	15	15	30

Source: See Table 1.12 and the footnote therein.

Table 5.2 Collapsed tables on effect of vaccination

Infection	Immune level Low	High	Total	Immune level	Vaccinated No	Yes	Total
Well	7	10	17	Low	11	6	17
Sick	10	3	13	High	4	9	13
Total	17	13	30	Total	15	15	30

Source: Cohen et al. (1983a); see footnote to Table 1.12.

Using Table 5.2, exact conditional p-values for testing against $H_1^{(1)}$ and against $H_1^{(2)}$ are

$$\binom{30}{15}^{-1}\sum_{i=0}^{4}\binom{13}{4-i}\binom{17}{11+i} \text{ and } \binom{30}{13}^{-1}\sum_{i=0}^{3}\binom{13}{3-i}\binom{17}{10+i},$$

respectively; these simplify to 0.0697 and 0.0552, respectively. Therefore, the p-value for testing H_0 against H_1 is less than or equal to $\max\{0.0552, 0.0697\} = 0.0697$. Thus, there appears to be some weak evidence in favor of H_1.

This p-value is marginally larger than the traditional nominal level 0.05. However, since the IUT is not similar and every least favorable null configuration turns out to be an extreme scenario, it may be reasonable to consider nominal levels slightly higher than 0.05; Cohen et al. (1983a) also made the same comment. A formal procedure of this would be useful; for example, it may be appropriate to consider a weighted probability of type I error with respect to some prior distribution of the nuisance parameter. For a related procedure, see Silvapulle (1996b) and Berger and Boos (1994).

It is of interest to note that a test of H_0^* against H_1^* based on Table 5.1 failed to support (p-value = 0.231) the claim that vaccination reduces the risk of infection. Perhaps this is not unexpected. There are two points worthy of note in this regard. First, if the conditional independence assumption is true then an IUT of H_0 against H_1 is likely to be more powerful than a test of H_0^* against H_1^*. This is because H_1 is tighter than H_1^*; both of them claim that vaccination reduces infection but the former also says that the improvement is achieved only through the beneficial effects of improved immune level resulting from vaccination.

The second point is that the data are showing trends in the same direction as that suggested by H_1 which is a special case of H_1^*. Consequently, we would expect the p-value for a test of H_0 against H_1 to be smaller than that for a test of H_0^* against H_1^*; however, there is no rigorous result that says that this should be the case.

In this example, exact conditional tests were used because the frequencies are small. In addition to illustrating the foregoing exact conditional procedure, Cohen, Gatsonis and Marden (1983) also developed a large sample procedure for testing H_0 against H_1, in which the alternative hypothesis takes the form, $H_1 : \theta_1 > 0$ and $\theta_2 > 0$ where θ_1 and θ_2 are some known functions of cell probabilities; their approach is an IUT using $\{\hat{\theta}_1/se(\hat{\theta}_1)\}$ and $\{\hat{\theta}_2/se(\hat{\theta}_2)\}$ as test statistics for computing large sample

p-values for the tests on θ_1 and θ_2, respectively. Some of these problems will be considered in detail in Chapter 6 dealing with categorical data models. ■

5.4 NONPARAMETRICS

Nonparametrics relating to statistical inference under order restrictions (or inequality constraints) can be organized into two major categories:

(i) *Analogues of parametrics* wherein the likelihood is based on unspecified densities or probability laws, but having some parametric structures that are under such inequality constraints, and

(ii) *Beyond parametrics* where more general statistical functionals (or measures) are incorporated in the formulation of suitable inequality constrained models.

In the first category, in some cases, one has a *distribution free test* or *confidence set* while in the majority of other cases, nonparametrics provide globally robust statistical inference procedures that are asymptotically distribution free. Formulations of nonparametric functional may be a more delicate task, albeit they encompass a wider class of models. We shall deal mostly with (i) in this section, relegating the more complex developments in (ii) to Chapter 7.

As in nonparametrics there is less emphasis on precise likelihoods, it is not surprising the UIP has a relatively bigger appeal in nonparametrics. Motivated by this stand, we consider first some simple illustrations.

5.4.1 Two-Sample Problem Revisited

Let $X_1, \ldots, X_m$ be m *iid* random variables from a continuous d.f. F, defined on $\mathbb{R}$, and let $Y_1, \ldots, Y_n$ be an independent sample of n *iid* random variables from a continuous d.f. G, also defined on $\mathbb{R}$. In this set-up, the simplest case relates to the so called *shift model*:

$$G(x) = F(x - \Delta), \qquad x, \Delta \in \mathbb{R},$$

and the d.f. F is allowed to be arbitrary to a certain extent (viz., $F \in \mathcal{F}$, the class of all continuous d.f.'s on $\mathbb{R}$ or $F \in \mathcal{F}_0$, the class of all d.f.'s symmetric around median). We are interested in testing a null hypothesis say $H_0 : \Delta = 0$, against a one-sided alternatives $H_1 : \Delta > 0$. Then, F is treated as a nuisance (functional) parameter. For normal F, we have the classical student t-test that is uniformly most powerful (UMP) test against such one-sided alternatives. Without the normality of F, the t-test is no longer UMP and it may not be powerful (or even valid in some cases), while an optimal test may not exist for finite sample sizes. *Asymptotically optimal* tests deem the knowledge (and square integrability) of the Fisher score function $-f'(x)/f(x)$, where f is the density function and f' is its first derivative (both assumed to exist). The situation is vastly different in a nonparametric walk. For example, one may use a Wilcoxon-Mann-Whitney (WMW) test based on the two-sample rank statistic

$$W_{mn} = \frac{1}{mn}\sum_{i=1}^{m}\sum_{j=1}^{n}\phi(X_i, Y_j), \tag{5.47}$$

where the (generalized) kernel is defined as

$$\phi(x, y) = \text{sign}(x - y), \qquad x, y \in \mathbb{R}. \tag{5.48}$$

Since F is assumed to be continuous, the event $X = Y$ has null probability, so that *sign* (x, y) is $+1$ or -1 according as $x - y$ is positive or not. Under the null hypothesis, the two d.f.'s are the same, and hence, W_{mn} has a distribution independent of the underlying F, and has mean 0. Thus, if we use the critical level from this null distribution of W_{mn} we will have an Exactly Distribution Free (EDF) test for H_0; here EDF means that the null distribution of W_{mn} does not depend on the underlying distribution. This test is quite efficient when actually F is normal, and may be even more efficient than the t-test when F departs from normality. For example, for logistic or Laplace F, it is more efficient than the t-test, and for the Cauchy d.f., the t-test is inconsistent while the WMW-test is very good. In this sense, it is a globally robust test. Actually, the scenario is broader in this perspective. We say that Y is *stochastically larger (smaller)* than X if $P\{X < Y\}$ is $>$ or < 0.5. Note that

$$P\{X < Y\} = \int_{\mathbb{R}} F(y)dG(y) = \theta(F, G), \tag{5.49}$$

so that we have the null hypothesis $H_0 : F \equiv G$ against the one-sided alternative $\theta(F, G) > 1/2$. Going back to the shift model in (5.47), we may note that

$$\theta(F, G) = \int_{\mathbb{R}} F(y - \Delta)dF(y) \lesseqqgtr 0.5 \text{ according as } \Delta \text{ is } \gtreqqless 0. \tag{5.50}$$

Therefore, for the shift model (5.47), the alternative of stochastically larger (smaller) is equivalent to directed deviation of Δ from 0. However, note that $\theta(F, G)$ may be equal to $1/2$, without requiring that $F = G$ (for example, if both F and G are symmetric about a common median but otherwise of possibly different scale factors or even functional form, $\theta(F, G)$ will still be equal to $1/2$). Thus, if one wants to test for the null hypothesis that $\theta(F, G) = 1/2$ vs $\theta(F, G) > 1/2$, that would be more general than testing the null hypothesis of $\Delta = 0$ vs $\Delta > 0$, under the shift model. For testing the hypothesis of stochastically larger or smaller r.v.'s we may still use W_{mn} as a valid test statistic, although if F and G are not the same, the test would not be EDF. However, it will be asymptotically distribution free (ADF) under very general regularity conditions, and, in any case, will be robust in a nonparametric setup. This feature of the WMW statistic is shared by a general class of rank statistics which are EDF under $F = G$, and robust for a general class of (constrained or not) alternatives. The basic point is that instead of seeking *optimality* at a given F, the emphasis here is on validity and robustness perspectives. We shall incorporate *local optimality* properties to illustrate these aspects in a more general setup. However, this example motivates us to proceed to a general class of restrained models that we consider in the next sub-section.

5.4.2 Ordered Alternative Problem Revisited

Consider $C(\geq 2)$ independent samples of sizes $n_1, \ldots, n_C$, drawn from continuous distributions $F_1, \ldots, F_C$, respectively. In a parametric (ANOVA) set-up, we set

$$F_j(x) = F(x - \theta_j), \qquad j = 1, \ldots, C, x \in \mathbb{R},$$

where the θ_j are all real valued parameters, and the d.f. F is taken to be normal. In a nonparametric formulation, though retaining the shift model flavor, we allow the d.f. $F \in \mathcal{F}$, the class of all continuous d.f.'s on the real line $\mathbb{R}$. Consider then the null hypothesis of equality of locations, i.e., $H_0 : \theta_1 = \cdots = \theta_C (= \theta$, unknown), against ordered alternatives of the form:

$$H_1^{<} : \theta_1 \leq \theta_2 \leq \cdots \leq \theta_C,$$

with at least one strict inequality sign being true. In Chapter 3, for normal F, we have considered suitable (LRP driven) tests for this ordered alternative testing problem. Consider now a nonparametric analogue wherein we allow $F \in \mathcal{F}$. Note that under the shift model, we have

$$H_1^{<} : F_1(x) \geq F_2(x) \geq \cdots \geq F_C(x), \qquad \forall x \in \mathbb{R},$$

with at least one strict inequality holding on a set of measure nonzero. Note that there is an implied stochastic ordering of the d.f.'s namely,

$$\theta(F_i, F_j) \geq 1/2, \qquad \forall 1 \leq i < j \leq C,$$

with at least one strict inequality being true. Here also, we may note that $\theta(F_i, F_j)$ may be equal to 1/2, without necessarily requiring that $F_i = F_j$, although under the shift model, that is an equivalence relation. For testing the hypothesis of homogeneity of the F_j against the ordered alternatives in (5.51), one requires the RMLE of the θ_j, under the inequality constraints in (5.51). Though for normal F we have some efficient algorithms to carry out this estimation problem, the task is not very simple. The problem becomes much more complex when F is of unknown form, and even more when we consider ordered alternatives of the form (5.51), instead of (5.51). As such, a LRP-based test for homogeneity of the F_j against ordered alternatives of the form (5.51) or (5.51) could be computationally more difficult.

The UIP works out well in such a nonparametric setup wherein we may consider (5.51) or (5.51). We consider a set of component hypotheses:

$$H_{0ij} : F_i(x) = F_j(x), \quad i < j = 1, \ldots, C;$$
$$H_{1ij} : \theta(F_i, F_j) \geqslant 1/2, \quad i < j = 1, \ldots, C.$$

Note that the null hypothesis of homogeneity of the C d.f.'s, $F_1, \ldots, F_C$, can be expressed as

$$H_0 = \bigcap_{1 \leqslant i < j \leqslant C} H_{0ij},$$

and the ordered alternative hypothesis as

$$H_1 = \bigcup_{1 \leqslant i < j \leqslant C} H_{1ij}.$$

Note that under (5.51), we have for any pair $(i,j) : j \geqslant i+1$,

$$\begin{aligned} \theta(F_i, F_j) &= \int_{\mathbb{R}} F_i(x) dF_j(x) \\ &= \int_{\mathbb{R}} F_i(x) dF_{i+1}(x) + \int_{\mathbb{R}} [F_{i+1}(x) - F_j(x)] dF_i(x) \geqslant \theta(F_i, F_{i+1}), \end{aligned}$$

so that under the ordered alternative (5.51), we have

$$H_1 = H_{112} \bigcup H_{123} \bigcup \ldots H_{1(C-1)C}. \tag{5.51}$$

Based on this consideration, under the stochastic ordering in (5.51), we may formulate

$$H_0 = \bigcap_{1 \leqslant i \leqslant C-1} H_{0i,i+1}; \qquad H_1 = \bigcup_{1 \leqslant i \leqslant C-1} H_{1i,i+1}$$

For testing $H_{0i,i+1}$ against $H_{1i,i+1}$, we may appeal to the two-sample problem treated earlier, and consider a WMW-statistics which we denote as $W_{ni}, i = 1, \ldots, C-1$. If the n_j are all equal, then under H_0, all the W_{ni} are marginally identically distributed, though they are not independent of each other. Even if the n_j are not equal, under H_0, all the $n(= n_1 + \cdots + n_C)$ observations in the pooled sample are *iid* random variables, so that noting that the $\theta(F_i, F_j)$ are all rank-statistics, the joint (discrete) distribution of $W_{n1}, \ldots, W_{n(C-1)}$ can be obtained by enumeration of all possible $n!$ permutations of the n observations among themselves. Thus, for small sample sizes, $n_1, \ldots, n_C$, the exact null distribution of the UIT based on the W_{ni} can be enumerated by all possible $n!/\{\Pi_{i=1}^{C} n_i!\}$ partitioning of n observations into C subsets of sizes $n_1, \ldots, n_C$. This task becomes prohibitively laborious as the n_i increase. However, the W_{ni} are all generalized U-statistics, and their joint distribution (under H_0) can be approximated by the classical *permutational central limit theorem* (PCLT) (viz., Sen (1981) and Hájek et al. (1999) where other references are cited) for large n, the permutation distribution of

$$n^{1/2}(W_{n1}, \ldots, W_{n(C-1)})$$

can be well approximated by a multinormal distribution with null mean vector and a completely known covariance matrix $\frac{1}{3}[\xi_{ij}]$, where $\xi_{i,i} = (n/n_i + n/n_{i+1}), \xi_{i,i+1} = -n/n_{i+1}, i = 1, \ldots, C-1$ and $\xi_{i,j} = 0$ for $| i - j | \geqslant 2$. As such, we may appeal to the multinormal one-sided mean testing problem (with known dispersion matrix), worked out in Chapter 4. As such, we leave the details as Problem 5.6 at the end of this chapter.

Let us look into the same problem without imposing (5.51) in the alternative. That is, we want to test for the homogeneity of the d.f.'s against the ordered alternatives in (5.51). For testing H_0 in (5.51) against one-sided alternatives (H_1) in (5.51), the UI-characterization in (5.52) may not generally hold. Jonkheere (1954a), apparently unaware of the UIP, proposed a linear nonparametric test statistic. Consider all possible $\binom{C}{2}$ WMW-statistics $W_{nij}, 1 \leqslant i < j \leqslant C$. His ad hoc test statistic for testing the null hypothesis of homogeneity of the $F_j(1 \leqslant j \leqslant C)$ against the ordered alternatives in (5.51) is

$$J_n = \sum_{1 \leqslant i < j \leqslant C} n_i n_j W_{nij} = \sum_{1 \leqslant i < j \leqslant C} \sum_{r=1}^{n_i} \sum_{s=1}^{n_j} \text{sign}\,(X_{js} - X_{ir}) \qquad (5.52)$$

where $X_{i1}, \ldots, X_{in_i}$ denote the sample observation in the ith sample, for $i = 1, \ldots, C$. The test rejects the null hypothesis in favor of H_1 when J_n exceeds a (positive) critical level $J_{n\alpha}$. As under H_0, all the X_{ir}, $1 \leqslant r \leqslant n_i, 1 \leqslant i \leqslant C$ are *iid* random variables with a continuous F, for small values of $n_1, \ldots, n_C$, one may use the permutation distribution of J_n, generated by all possible $\{n!/(n_1! \ldots n_c!)\}$ partitioning of the combined sample observation into C subsets of sizes $n_1, \ldots, n_C$, respectively. This task becomes prohibitively laborious as the n_j increase. However, noting that under H_0, J_n has zero expectation and variance

$$\sigma^2_{J,n} = \{n^2(2n+3) - \sum_{j=1}^{C} n_j^2(2n_j + 3)/72,$$

one can incorporate the PCLT on the standardized form $J_n/\sigma_{J,n}$, and thereby approximate $J_{n\alpha}/\sigma_{J,n}$ by τ_α, the upper α-quantile of the standard normal distribution function; we leave the details as Problem 5.7.

Note that for the nonparametric UIT discussed earlier, typically, we have asymptotically chi-bar square distribution (under H_0) whereas $J_n/\sigma_{J,n}$ is asymptotically normal. Thus, their null distributions are not of comparable form. For contiguous alternatives their asymptotic distributions can be obtained by the usual asymptotic statistical methods (viz, Problem 5.8 with hints provided). However, as they are of different forms, their asymptotic power functions may not be comparable to a single summary measure, such as the Pitman (1948) asymptotic relative efficiency (PARE). In this respect, the situation is comparable to the Tukey-Abelson linear test vs the likelihood ratio type test for normal mean problem, described in the previous chapter. If we want to formulate a UIT for testing H_0 vs H_1 in (5.52), based on all the $\binom{C}{2}$ W_{nij}'s we may note that, under H_0 (the homogeneity of all the F_j), these $\binom{C}{2}$ statistics have asymptotically a $\binom{C}{2}$-dimensional multinormal distribution of rank $(C-1)$. This singularity of the asymptotic distribution can be avoided by taking a subset of $(C-1)$ of these statistics, as we did earlier; a natural choice is the set $W_{i,(i=1)}, i = 1, \ldots, C-1$. For finite sample sizes, the situation is quite involved.

Sparked by the developments on nonparametrics in the 1960s and 1970s, the use of asymptotically normal test statistics became quite popular. Basically, like the WMW-statistics, two-sample linear rank statistics (based on monotone rank scores) can be incorporated as in (5.52) to test for the homogeneity of the F_j against ordered (shift) alternatives. Chacko (1963) and Hogg (1965) considered models and hypotheses with restricted alternatives, and various extensions are due to Conover (1967), Shorack (1967), Johnson and Mehrota (1971), Tryon and Hettmansperger (1973), and others. Puri (1965) used normal scores statistics (instead of WMW statistics) and formulated an analogue of the Jonkheere statistic. For normal densities, his test has asymptotic power superiority for ordered location alternatives though based on validity considerations, it may not be as robust as the test based on J_n in (5.52). In terms of power, the normal scores procedure is asymptotically better than the normal theory based test for ordered location alternatives only, while for other ordered alternatives or nonnormal populations, the Wilcoxon-scores procedure may perform better. For a nice treatise of these tests, albeit at an application level, we refer to Chakraborti and Gibbons (1992).

5.4.3 Ordered Alternatives: Two-Way Layouts

Consider $n(\geqslant 2)$ blocks of $p(\geqslant 2)$ plots where p different treatments have been used. Let X_{ij} stand for the yield of the plot in the block receiving the jth treatment, for $j = 1, \ldots, p,\ i = 1, \ldots, n$. In the conventional ANOVA model, we let

$$X_{ij} = \mu + \beta_i + \tau_j + \varepsilon_{ij},\ 1 \leqslant j \leqslant p,\ 1 \leqslant i \leqslant n, \tag{5.53}$$

where μ stand for the mean effect, β_i for the ith block effect, τ_j for the jth treatment effect, and ε_{ij} are *iid* random variables, assumed to be normally distributed with zero mean and variance $\sigma^2(< \infty)$. In the so called fixed-effects model, μ, β_i and τ_j are all regarded as unknown parameters, and without loss of generality, we set

$$\sum_{i=1}^{n} \beta_i = 0 \quad \text{and} \quad \sum_{j=1}^{p} \tau_j = 0. \tag{5.54}$$

The null hypothesis of no treatment effect relates to

$$H_0 : \tau_1 = \cdots = \tau_p = 0 \tag{5.55}$$

and the global alternatives to $\boldsymbol{\tau} = (\tau_1, \ldots, \tau_p)^T \neq \mathbf{0}$. As in (5.51), we may conceive of ordered alternatives

$$H_1^< : \tau_1 \leqslant \cdots \leqslant \tau_p, \quad \text{with at least one " } <\text{"}. \tag{5.56}$$

The classical LRP can be incorporated to formulate a test for H_0 against $H_1^<$, and this will have the usual $\bar{E}^2$-distribution discussed in Chapter 4. The UIP also relates to the same test statistic (as here the covariance matrix is known up to a scalar constant, σ^2). The situation is slightly more complex for a mixed-effects model wherein the τ_j are regarded as fixed parameters, while the β_i are treated as random variables. In a nonparametric formulation, we have a gradation of models which include fixed-effects and mixed effects models.

We assume that $\{X_{ij}\}$ are independent and X_{ij} has a continuous distribution function $F_{ij}, i = 1, \ldots, n, j = 1, \ldots, p$. The null hypothesis is then formulated as

$$H_0 : F_{i1} = \cdots = F_{ip} = F_i \ \forall i = 1, \ldots, n \tag{5.57}$$

where $F_1, \ldots, F_n$ need not be equal (nor related to each other by shifts (block effects)). Thus, the additivity of the block effects is not necessarily imposed in (5.57). Then, in a shift-alternative model, we set

$$F_{ij}(x) = F_i(x - \tau_j), \; i = 1, \ldots, n, j = 1, \ldots, p$$

treating the F_i as nuisance parameters (functions). Then, we may set the ordered (location/treatment effect) alternatives as in (5.56), but treating $F_1, \ldots, F_n$ to be all nuisance parameters. Parallel to (5.51), we may set

$$H_1^{<*} : F_{i1}(x) \geqslant \cdots \geqslant F_{ip}(x), \; \forall i = 1, \ldots, n, x \in \mathbb{R} \tag{5.58}$$

with at least one strict ">" being true on a set of nonzero measure. The null hypothesis of interest (that all the τ_j are equal to zero) can be framed in a more general way as follows: observations within a block are interchangeable (or exchangeable). This means that the joint distribution $F_i(x_{i1}, \ldots, x_{ip})$ is a symmetric function of its p arguments for each $i = 1, \ldots, n$. This definition implies that all possible ($p!$) permutations of $(X_{i1}, \ldots, X_{ip})$ have the same distribution (which need not be the same for all i). This relates to the mixed-model (β_i random) case as well. Moreover, in a more general situation where there are n_i judges each ranking (independently of each other) p objects (e.g. players), so that for each judge, we have a set of rankings, the numbers $1, \ldots, p$ (if ties can be neglected). For the ith judge, this (intra-block) rank vector is denoted by $(r_{i1}, \ldots, r_{ip})$ for $i = 1, \ldots, n$. This is called the method of *n-ranking*. Under H_0, all these rank vectors are independent and each has $p!$ realizations with equal probability $1/(p!)$. Suppose now that we want to test for this null hypothesis against an ordered alternative that the objects have a different ranking. Since the r_{ij} are monotone functions of the underlying (trait) variable (say X_{ij}), this problem can be reduced to (5.58), even in a more geneal set up, allowing interchangeability to a greater extent. Keeping this in mind, we formulate the null hypothesis of interchangeability as in above, against ordered alternatives of the form:

$$H_1^{<} : X_{ij} \text{ is stochastically larger than } X_{i\ell}, \forall 1 \leqslant \ell < j \leqslant p, \; i \geqslant 1. \tag{5.59}$$

For testing H_0 in (5.55) against H_1 in (5.56), but allowing the distribution functions $F_1, \ldots, F_n$ to be completely arbitrary (but continuous a.e.), we may consider nonparametric tests that are natural analogues of the ones in the one-way layout model. However the *intrablock rankings* are to be used (so that interblock comparisons are to be de-emphasized). Let Q_i be a *measure of agreement* of $r_{i1}, \ldots, r_{ip}$ with the natural integers $1, \ldots, p$ (for $i = 1, \ldots, n$); such measures are specifically tailored for ordered alternatives. Then, typically a test statistic is based on

$$T_Q = \sum_{i=1}^{n} Q_i. \tag{5.60}$$

For example, we may set for every $j < \ell (= 1, \ldots, p)$,

$$\begin{aligned} Q_{ij\ell} &= \{\text{sign}(r_{i\ell} - r_{ij}) + 1\}/2 = I(r_{ij} < r_{i\ell}), \quad i = 1, \ldots, n; \\ Q_i &= \sum_{1 \leqslant j < \ell < p} Q_{ij\ell}, \quad i = 1, \ldots, n. \end{aligned} \tag{5.61}$$

Note that $\binom{p}{2}^{-1}Q_i$ is essentially the Kendall's tau statistic based on $((1, r_{i1}), \ldots, (p, r_{ip}))$, $(i = 1, \ldots, n)$. Motivated by this, Jonkheere (1954b) considered the test statistic

$$J_n = \binom{p}{2}^{-1} n^{-1} \sum_{i=1}^{n} \sum_{1 \leqslant j < \ell \leqslant p} \text{sign}(r_{i\ell} - r_{ij}) \tag{5.62}$$

rejecting the null hypothesis H_0 in favor of $H_1^<$, whenever J_n exceeds (the right-hand side) critical level $J_{n\alpha}$.

Page (1963) considered the case of

$$Q_i = \frac{12}{p(p^2-1)} \sum_{i=1}^{p} \left(j - \frac{p+1}{2}\right)\left(r_{ij} - \frac{p+1}{2}\right), i = 1, \ldots, n, \tag{5.63}$$

which are the intra-block Spearman rank correlations, and proposed the test statistic

$$P_n(1) = \frac{12}{np(p^2-1)} \sum_{i=1}^{n} \sum_{j=1}^{p} \left(j - \frac{p+1}{2}\right)\left(r_{ij} - \frac{p+1}{2}\right) \tag{5.64}$$

rejecting the null hypothesis H_0 in favor of $H_1^<$ for large (positive) values of P_n.

Both J_n and P_n lie in the interval $(-1, 1)$. Moreover, under H_0, the $\boldsymbol{r}_i (= r_{i1}, \ldots, r_{ip})^T)$ are *iid* random variables each assuming $p!$ possible equally likely realizations, so that the null distributions of J_n and P_n can be obtained by enumeration of all the $(p!)^n$ equally likely permutations of the intra-block rankings–a procedure that becomes exceedingly laborious for large values of n (or p). However, the classical central limit theorem can be used to show that for large n, under H_0,

$$n^{1/2} J_n / \sigma_J \xrightarrow{d} N(0,1) \text{ and } n^{1/2} P_n / \sigma_P \xrightarrow{d} N(0,1) \tag{5.65}$$

where

$$\sigma_J^2 = 2(2p-5)/\{9p(p-1)\} \text{ and } \sigma_P^2 = 1/(p-1); \tag{5.66}$$

(Problem 5.9 is set to verify these details). As such $J_{n\alpha}$ and $P_{n\alpha}$ may be approximated by $n^{-1/2}\sigma_J \tau_\alpha$ and $n^{-1/2}\sigma_P \tau_\alpha$, respectively, τ_α being the upper α-percentile of the standard normal distribution.

Since the r_{ij} are translation-invariant and their null distribution depends only on the $p!$ possible equally likely permutations, the (exact as well as asymptotic) tests considered above remain valid for the mixed-effects model. Moreover, they do not presume the additivity of the block-effects, though in that way, they do not utilize fully inter-block information, and thereby may be less efficient, particularly when the block effects are additive. This is better achieved by aligned ranking methods that we describe below.

We consider the model (5.53) (where the β_i may or may not be stochastic), and consider an alignment procedure that essentially eliminates the block-effect β_i, and induces meaningful inter-block comparison. Although, we can consider an arbitrary translation invariant symmetric function of the within block observations, for simplicity, we let $\bar{X}_i(= p^{-1}(X_{i1} + \ldots + X_{ip})), i = 1, \ldots, n$ as estimates of $\mu + \beta_i$, and define the aligned observations as

$$\begin{aligned} Y_{ij} &= X_{ij} - \bar{X}_i = \tau_j + (\varepsilon_{ij} - \bar{\varepsilon}_i) \\ &= \tau_j + e_{ij},\ 1 \leqslant j \leqslant p,\ 1 \leqslant i \leqslant n, \end{aligned} \tag{5.67}$$

which are free from the β_i (though the e_{ij}, within each block are not independent any more). Nevertheless, for each $i\ (= 1, \ldots, n), (e_{i1}, \ldots, e_{ip})$ are exchangeable random variables, and they are identically distributed $i\ (= 1, \ldots, n)$. Thus, under $H_0 : \boldsymbol{\tau} = \mathbf{0}$, $(Y_{i1}, \ldots, Y_{ip})$ are interchangeable random variables. We pool the Y_{ij} into a combined set of $N(= np)$ observations and define the aligned ranks as

$$R_{ij} = \sum_{r=1}^{n} \sum_{\ell=1}^{p} I(Y_{r\ell} \leqslant Y_{ij}),\ 1 \leqslant i \leqslant n,\ 1 \leqslant j \leqslant p. \tag{5.68}$$

Then a natural extension of J_n in (5.62) would be

$$J_n^* = \binom{p}{2}^{-1} \sum_{1 \leqslant j < \ell \leqslant p} W_{nj\ell}, \tag{5.69}$$

where $W_{nj\ell} = \binom{n+1}{2}^{-1} \sum_{1 \leqslant i \leqslant i' \leqslant n} \text{sign}(Y_{ij\ell} + Y_{i'j\ell})$, (with $Y_{ij\ell} = Y_{i\ell} - Y_{ij}$, for $1 \leqslant j < \ell \leqslant p, i = 1, \ldots, n$) are the Wilcoxon signed-rank statistics. This incorporates signed-ranks instead of signs only, and accounts for a higher PARE in general. Similarly, analogous to $P_n(1)$ in (5.64) for the one-way layout, we have the following aligned-rank statistic for the two-way layout:

$$P_n(2) = \frac{12}{p(p^2 - 1)} \sum_{j=1}^{p} \left(j - \frac{p+1}{2}\right) \left(\bar{R}_{\cdot j} - \frac{N+1}{2}\right), \tag{5.70}$$

where

$$\bar{R}_{\cdot j} = \frac{1}{n} \sum_{i=1}^{n} R_{ij},\ \text{for } j = 1, \ldots, p. \tag{5.71}$$

Note that unlike the $\mathbf{r}_i$, the R_{ij} are not permutable with respect to the same group of $(p!)^n$ intra-block permutations (as R_{ij} and $R_{i'\ell}$ are not independent for $i \neq i'$). For this reason, a conditionally distribution-free approach for aligned rank statistics in two-way layouts has been developed systematically in Sen (1968). Based on the R_{ij} and incorporating a monotone score function $\phi(u), u \in (0, 1)$, Sen (1968) considered general aligned linear rank statistics which we denote by $T_{N,j}, j = 1, \ldots, p$. We choose, without loss of generality, the scores $a_N(r)$ such that their average value $\bar{a}_N = 0$.

In a comparatively general set-up, we consider a nondecreasing, square-integrable score generating function $\phi = \phi(u), u \in (0, 1)$ and set $\bar{\phi} = \int_0^1 \phi(u)du$ (= 0, without loss of gerenality) and $A_\phi^2 = \int_0^1 \phi^2(u)du$. Also let $U_{N:1} < ... < U_{N:N}$ be the order statistics in a random sample of size N from the Uniform[0,1] distribution. For every $N \geqslant 1$, we define a set of scores $a_N(r) = E\phi(U_{N:r}), r = 1, ..., N$. It is also possible to define $a_N(r) = \phi(r/(N+1)), r = 1, \ldots, N$, or in some other convenient (and asymptotically equivalent) way; without loss of generality, we set $\bar{a}_N = N^{-1}\sum_{r=1}^N a_N(r) = 0$. Let then $T_{N,j} = n^{-1}\sum_{i=1}^n a_N(R_{ij}), j = 1, \ldots, p$. Further, define

$$\bar{a}_{NR_{i\cdot}} = p^{-1}\sum_{j=1}^{p} a_N(R_{ij}),\ 1 \leqslant i \leqslant n, \tag{5.72}$$

and let

$$\sigma_{P_n}^2 = \frac{1}{n(p-1)}\sum_{i=1}^{n}\sum_{j=1}^{p}\{a_N(R_{ij}) - \bar{a}_{NR_{i\cdot}}\}^2 \tag{5.73}$$

Then, a general aligned rank test statistic for testing H_0 vs $H_1^<$ is of the form

$$S_N^0 = \left(\frac{12n}{p(p^2-1)\sigma_{P_n}^2}\right)^{1/2}\sum_{j=1}^{p}\left(j - \frac{p+1}{2}\right)T_{N,j}. \tag{5.74}$$

Note that under H_0, the residuals within each block are interchangeable. As such, given the aligned rank vectors $\boldsymbol{R}_{i\cdot} = (R_{i1}, ..., R_{ip})', i = 1, \ldots, n$, we can generate $(p!)^n$ permutations of the intra-block aligned ranks, and these permutations are conditionally equally likely, under H_0. Hence, under H_0, S_N^0 is conditionally (or permutationally) distribution-free (CDF). Generally, if the block-effects are additive (albeit the errors may not be normally distributed), as the aligned ranks pertain to inter-block comparisons (which the intra-block ranks may not), S_N^0 is more powerful than P_{n1} and J_n. Moreover, for nonnormal errors, S_N^0 is generally more powerful than the Tukey-Abelson type of tests. Even when the block-effects are not additive in the sense that the errors for different blocks are heteroscedastic, the aligned rank tests seem to perform better than the intra-block rank tests (Sen (1968)). Further, it might be remarked in passing that the UIP may as well be applied to the vector $\boldsymbol{T}_N$ (taking into consideration that $\sum_{j=1}^p T_{N,j} = 0$), leading to the UIT statistic

$$S_N^* = (12n/\{p(p^2-1)\sigma_{P_n}^2\})^{1/2}\sup\{\boldsymbol{a}'\boldsymbol{T}_n : a_1 \leq a_2 \leq ... \leq a_p, \boldsymbol{a}'\boldsymbol{a} = 1\}.$$

The KTL-point formula employed in Chapters 3 and 4 can then be incorporated to express S_N^* in the familiar way, incorporating the pooled-adjacent algorithm as in the normal theory (known variance) case. However, such tests are only CDF, and, hence, tables are not available for their critical levels.

Since S_N^0 is CDF, its conditional (permutational) distribution, though tedious, can be obtained by enumeration of intra-block permutation of the Y_{ij} (which are exchangeable under H_0). For large n, this conditional null distribution can be well approximated (though, in probability) by a standard normal law. Generally, S_N^0 is more powerful than P_n or J_n. (Problems 5.10 and 5.11 are set to verify these details.)

5.4.4 UIP and LMPR Tests for Restricted Alternatives.

As has been discussed in Section 5.2, the UIP is essentially based on a (finite or infinite) decomposition of the null hypothesis H_0 as the intersection of component null hypothesis $H_{0i}, i \in I$, and the alternative H_1 (may or may not be restricted ones) as the union of $H_{1i}, i \in I$, under the presumption that for testing H_{0i} vs H_{1i}, there is a test, preferably optimal, that can be incorporated in formulating the UIT for H_0 vs H_1. In a general framework, especially in nonparametric inference, as we have seen in the preceding two subsections, such as an optimal test statistic (for H_{0i} vs $H_{1i}, i \in I$) may not exist, or be difficult to construct. However, in a variety of nonparametric testing problems (even involving some restricted alternatives), if the null hypothesis H_0 relates to an hypothesis of invariance (under suitable groups of transformations that map the sample space onto itself), then it may be possible to have some test for H_{0i} vs H_{1i} which is a locally most powerful rank (LMPR) test (for $i \in I$). A test is LMPR if among the class of rank tests, it is uniformly most powerful (UMP) (at some level $\alpha(0 < \alpha < 1)$) for H_0 against a class $H_1(\varepsilon)$ of alternatives that are indexed by a parameter Δ, such that $0 < \Delta < \varepsilon$, $\varepsilon > 0$. There are many important cases in which LMPR exists in the classical nonparametric hypothesis testing problems, the null hypothesis relates to an hypothesis of invariance under a specific group of transformations that maps the sample space onto itself. For example, for the null hypothesis of symmetry about a given origin (which can be taken as 0, without loss of generality), this group is the finite group of all possible sign-inversions; for testing the hypothesis of randomness of a set of observations, this group is the group of all possible permutations of the observations. For testing bivariate independence, a similar group is the permutation of the second coordinates in the context of pairing with the first one; for the randomized block design, for the hypothesis of no treatment effect, the within block permutations form the group. For details, we may refer to Hájek, Šidák, and Sen (1999, ch. 3); particularly, their Theorem 1 on p.71 is the most general result. In such a case, it is possible to have an exact distribution-free (EDF) test that is also *locally most powerful rank test* (LMPRT) against a specific parametric alternative. For example, for the hypothesis of symmetry (about 0) against a shift alternative, tests based on signed-rank statistics are all EDF. Within this class of rank tests, the Wilcoxon signed-rank test is LMPR when the underlying density is logistic, the normal-scores test is LMPR for underlying normal density, and the sign test is LMPR when the underlying density is Laplace. Thus, it may be possible to have a test for H_{0i} vs. H_{1i} that is LMPR. For clarification of ideas, we relegate some of these details as Problem 5.12 at the end of this chapter.

Such LMPR tests are, by construction, rank tests. Hence, whenever a null hypothesis H_0 pertains to some invariance (under appropriate groups of transformations that map the sample space into itself), such LMPR tests are EDF. In general, multivariate models, as we shall see later on, an invariance structure yields conditional (permutational) distribution-freeness, and hence such LMPR tests may be conditionally distribution free (CDF), and as a result, asymptotically distribution free. In this respect, the situation is quite comparable to the Jonckheere's (1954a,b) tests for ordered alternatives, discussed in the earlier subsections. Note that the Jonckheere tests are LMPR if the underlying density is logistic and there is some linear ordering of the location parameters. Without the linear ordering or the logistic density, this LMPRT property may not be tenable for the Jonckheere test. However, there is a general theory of LMPRT outlined in Hájek, Šidák and Sen (1999, p.71), which may be incorporated to characterize other rank tests as locally optimal for general restricted alternatives.

Of course, such LMPRT properties are tenable only for certain specific type of hypotheses. Even if this LMPRT characterization is not tenable, in principle, the underlying UIP can be incorporated to formulate suitable rank tests for restricted alternatives in a way that lends to nice interpretations and generally better power properties. With this characterization, the rest of the formulation of UIP-based LMPR tests for restricted alternatives can be handled in a way quite similar to that in Section 4.2. We therefore motivate this type of test with some additional nonparametric problems (not treated in the earlier subsection) and illustrate with specific cases.

In some of these problems to be considered rank tests may not be EDF; they are sometimes conditionally distribution-free (CDF) and mostly only asymptotically distribution-free (ADF). The main difficulty in enforcing the EDF characterization stems from the fact that even if an invariant group of transformations exist the associated orbits may not have a distribution (under the null hypothesis) independent of the underlying densities. However, as we shall see, the UIP works out quite well.

(I) Nonparametric multivariate location: orthant alternative problem

Let $\mathbf{X}_1, \ldots, \mathbf{X}_n$ be n *iid* random vectors with a continuous distribution function, $F_{\boldsymbol{\theta}}$ defined on $\mathbb{R}^p$, for some $p \geqslant 1$. It is assumed that

$$F_{\boldsymbol{\theta}}(\mathbf{x}) = F_0(\mathbf{x} - \boldsymbol{\theta}), \qquad \boldsymbol{\theta} \in \mathbb{R}^p, \ \mathbf{x} \in \mathbb{R}^p \tag{5.75}$$

where F_0 in symmetric about $\mathbf{0}$ in a well-defined manner (and its form does not involve $\boldsymbol{\theta}$). For $p = 1$, F_0 continuous, the symmetry is defined by $F_0(-\mathbf{y}) = 1 - F_0(\mathbf{y}), \ \forall \mathbf{y} \in \mathbb{R}$. For $p \geqslant 2$, there are some definitions and interpretations of symmetry that are not isomorphic.

(1) *Elliptical Symmetry:* If $F_0(\mathbf{x})$ has a pdf $f_0(\mathbf{x})$, such that $f_0(\mathbf{x}) = h_0(\mathbf{x}^T\mathbf{A}\mathbf{x})$ for some positive definite $\mathbf{A}$, where $h_0(t), t \in \mathbb{R}$, is a scalar function. (2) *Spherical Symmetry:* The *spherical symmetry* is a special case where $\mathbf{A} = a^2\mathbf{I}_p$, for $a^2 > 0$. (3) *Total Symmetry:* An r-vector $\mathbf{X}$ is said to have a *totally symmetric* (about $\mathbf{0}$) distribution function F_0, if $\mathbf{X}$ and $\big((-1)^{i_1}X_1, \ldots, (-1)^{i_p}X_p\big)$ both have the same distribution function F_0 $\forall i_j = 0, 1, 1 \leqslant j \leqslant p$. (4) *Rotational Symmetry:* A more specialized case is the *rotational symmetry* where for any orthogonal $\mathbf{P}, \mathbf{X}$ and $\mathbf{PX}$ both have the same distribution function F_0. A more general case is the *diagonal symmetry* of F_0: If both $\mathbf{X}$ and $(-1)\mathbf{X}$ have the common distribution function F_0, then F_0 is diagonally symmetric about 0. Note that F_0 may be diagonally symmetric without being elliptically symmetric; in the multinormal case, we have elliptical symmetry and, hence, diagonal symmetry too. However, if we have a (common mean) mixture of two or more multinormal distribution functions (whose dispersion matrices are not the same) then diagonal symmetry holds, but not necessarily the elliptic symmetry. In the discussion below, we take F_0 to be diagonally symmetric about $\mathbf{0}$ (Sen and Puri (1967)).

Consider now the hypothesis testing problem

$$H_0 : \boldsymbol{\theta} = \mathbf{0} \ \text{vs} \ H_1 : \boldsymbol{\theta} \geqslant \mathbf{0}, \ \boldsymbol{\theta} \neq \mathbf{0}. \tag{5.76}$$

For $p = 1$, we have $H_0 : \theta = 0$ vs $H_1 : \theta > 0$, and the entire clan of signed rank statistics provide EDF tests. Within this class, using Theorem 1 of Hájek, Šidák and Sen (1999, p.71) one has a specific signed-rank statistic that is LMPR for a specific $F_{\mathbf{0}}$ (and at the same time is EDF, for all symmetric (about $\mathbf{0}$) $F_{\mathbf{0}}$). The basic idea of UIP is to incorporate this finding for a suitable test for H_0 vs H_1 (orthant alternatives) in (5.76).

As in Section 5.2, we write (finite UI):

$$H_0 = \bigcap_{j=1}^{p} H_{0j}, \quad \text{where } H_{0j} : \theta_j = 0, \ 1 \leqslant j \leqslant p,$$

$$H_1 = \bigcup_{j=1}^{p} H_{1j}, \quad \text{where } H_{1j} : \theta_j \geqslant 0, \ 1 \leqslant j \leqslant p, \tag{5.77}$$

with at least one strict inequality being true. For testing H_{0j} vs H_{1j}, we choose a suitable signed-rank statistic T_{nj}; this could be a LMPR test statistic for suitable F_{0j}, the jth marginal distribution function of $F_{\mathbf{0}}$, for $j = 1, \dots, p$.

If we define $T_{nj} = n^{-1/2} \sum_{i=1}^{n} s_{ij} a_{nj}(R_{ij}^{+}), forj = 1, \dots, p$, where $S_{ij} = sign(X_{ij}) R_{ij} = rank|X_{ij}|$ among $|X_{1j}|, \dots, |X_{nj}|, a_{nj}(k), k = 1, \dots, n$ are the scores for the jth coordinate, then $\boldsymbol{V}_n$ has the elements $v_{njj'}$ defined by $v_{njj'} = n^{-1} \sum_{i=1}^{n} s_{ij} S_{ij'} a_{nj}(R_{ij}^{+}) a_{nj'}(R_{ij'}^{+}), j, j' = 1, \dots, p$. This matrix $\boldsymbol{V}_n$ is derived from the conditional distribution of the matrices of signs and absolute ranks over the permutation group of sign inversions and rank permutations. Note that $\mathbf{V}_n$ is a stochastic matrix, but its diagonal elements are all nonstochastic and nonnegative. By normalizing the score functions, we can take all these diagonal elements $v_{nii} = 1, 1 \leqslant i \leqslant p$, without loss of generality. In that sense, $\mathbf{V}_n$ is interpreted as a rank-score correlation or association matrix. The characteristic property of $\mathbf{V}_n$ is that whereas the (conditional) null distribution of $\mathbf{T}_n = (T_{n1}, \dots, T_{np})^T$ is generated by the $2^n(n!)$ possible (conditionally) equally likely signs-inversions and rank column permutations (Sen and Puri (1967)), $\mathbf{V}_n$ is invariant with respect to this group of transformations. Or, in other words, if the test statistic is formulated as a function of $(\mathbf{T}_n, \mathbf{V}_n)$, then its (conditional) null distribution is completely determined by the totality of $2^n(n!)$ (conditionally) equally likely realization of the elements of this group of transformations. Moreover, for monotone nondecreasing score functions, T_{nj} is stochastically tilted to the right or left according as θ_j is positive or not; for $\theta_j = 0, T_{nj}$ has distribution symmetric about 0. Note that a distribution function G is tilted to the left of F if $G(x) \geq F(x)$, for all x; it includes the shift in the mean as a very notable case. This can be easily verified as follows. Let $T_{nj}(b)$ be the same signed-rank statistic based on $X_{ij} - b, \ 1 \leqslant i \leqslant n$, for $b \in \mathbb{R}$. Then (see Problem 5.13) it can be shown that

$$T_{nj}(b) \ \text{is nonincreasing in} \ \ b \in \mathbb{R} \tag{5.78}$$

Secondly, $T_{nj}(\theta_j)$ has the same distribution as T_{nj}, under $\theta_j = 0$, the latter being symmetric about 0. These two results imply the stochastic tiltedness of the T_{nj}, $1 \leqslant j \leqslant p$. Further (see Problem 5.14), using PCLT, it can be shown that under H_0,

$$n^{1/2}(\mathbf{T}_n - \mathbf{0}) \xrightarrow{d} N(0, \boldsymbol{\nu}) \tag{5.79}$$

where $\boldsymbol{\nu}$ (positive definite) is the stochastic limit of $\mathbf{V}_n$ (positive definite in probability). An expression for $\boldsymbol{\nu}$ may be obtained as follows. Assume that the score-generating functions $\phi_j^*(u), u \in (0,1)$ are all square integrable and nondecreasing. Also, let $F_{[j]}(x)$ be the d.f. of X_{ij} (under H_0), and $F_{[jj']}$ be the joint d.f. of $(X_{ij}, X_{ij'})$, for $j, j' = 1, \ldots, p$. Further, let $\phi_j^*(u) = \phi_j((1+u)/2), u \in (0,1), j = 1, \ldots, p$, where the $\phi_j(.)$ are all skew-symmetric about $u = 1/2$, i.e., $\phi_j(u) + \phi_j(1-u) = 0$, for all $u \in (0,1)$. Then, $\nu = ((\nu_{jj'})$ has the elements $\nu_{jj'} = \int\int \phi_j(F_{[j]}(x))\phi_j'(F_{[j']}(y))dF_{[jj']}(x,y)$, for $j, j' = 1, \ldots, p$.

In passing, we may remark that $T_{nj}, j = 1, \ldots, p$, for an arbitrary $\boldsymbol{\theta}$, having the stochastic tiltedness property mentioned before, have a centering vector $\boldsymbol{\tau}$ such that $n^{1/2}(\boldsymbol{T}_n - \boldsymbol{\tau})$ converges in law to a multinormal distribution with null mean vector and dispersion matrix $\boldsymbol{\nu}$ mentioned before; see Puri and Sen (1971, ch. 6) where these results are given in detail. Here $\boldsymbol{\tau} = (\tau_1, \ldots, \tau_p)^T$ where τ_j and θ_j have the same sign, for $1 \leqslant j \leqslant p$. Moreover, if θ_j is close to 0, then we may write $\tau_j = \gamma_j\theta_j + o(\|\theta_j\|), 1 \leqslant j \leqslant p$ where the γ_j are suitable nonnegative functionals of the jth marginal pdf (f_{0j}, say) and the underlying score function $\phi_j(u), 0 < u < 1$ for T_{nj}, $1 \leqslant j \leqslant p$. In particular, if f_{0j} is absolutely continuous, with a finite Fisher information,

$$I(f_{0g}) = \int_0^1 \left[-f_{0g}'\{F_{0j}^{-1}(u)/f_{0j}\{F_{0j}^{-1})u)\}\right]^2 du, \; j = 1, \ldots, p, \tag{5.80}$$

and $\phi_j(u)$ is square integrable inside $(0,1)$, then

$$\gamma_j = \int_0^1 \left[-f_{0j}'\{F_{0j}^{-1}(u)\}/f_{0j}\{F_{0j}^{-1}\}\right] \phi_j(u)du \;\; j = 1, \ldots, p. \tag{5.81}$$

Recall further that for a LMPR test statistic T_{nj}, $\phi_j(\cdot)$ is isomorphic to $-f_{0j}'\{F_{0j}^{-1}(\cdot)\}/f_{0j}\{F_{0j}^{-1}(\cdot)\}$, and, hence, γ_j is a scalar multiple of $I(f_{0j})$. If $\phi_j(u)$ is not isomorphic to the Fisher score function but is monotone and square integrable then it can be shown that (Puri and Sen (1971), Ch. 4)

$$n^{1/2}(\mathbf{T}_n - \boldsymbol{\tau}) \xrightarrow{d} N(\mathbf{0}, \boldsymbol{\nu}^*) \tag{5.82}$$

where $\boldsymbol{\tau}$ and $\boldsymbol{\nu}^*$ may both depend on $\boldsymbol{\theta}$ and are defined in terms of the score functions and the density $f_0(\mathbf{x})$. In fact, for local (Pitman-type) alternatives ($\boldsymbol{\theta} = n^{-1/2}\boldsymbol{\delta}$), (5.82) simplifies to

$$n^{1/2}(\mathbf{T}_n - \boldsymbol{\Gamma\theta}) \xrightarrow{d} N(\mathbf{0}, \boldsymbol{\nu}) \tag{5.83}$$

where $\boldsymbol{\nu}$ appears in (5.79) and $\boldsymbol{\Gamma} = \text{diag}(\gamma_1, \ldots, \gamma_p)$. The situation becomes quite comparable to the multinormal mean orthant alternative problem, treated in Section

5.2. As $\boldsymbol{\nu}$ is unknown, we use $\mathbf{V}_n$ as an estimator of $\boldsymbol{\nu}$ and motivate the UIT as follows. This test is CDF and is ADF, too. The asymptotic result (5.83) can be used for constructing a local test of $\boldsymbol{\theta} = \mathbf{0}$ vs $\boldsymbol{\theta} \in \mathcal{C}$, where $\mathcal{C}$ is a closed convex cone, by arguments similar to those in Silvapulle and Silvapulle (1996). Let $\hat{\mathbf{\Gamma}}$ denote a consistent estimator of $\mathbf{\Gamma}$. Then $n^{1/2}\hat{\mathbf{\Gamma}}^{-1}\mathbf{T}_n - \boldsymbol{\delta} \xrightarrow{d} N(\mathbf{0}, \mathbf{\Gamma}^{-1}\boldsymbol{\nu})$. Now, a local test of $\boldsymbol{\theta} = \mathbf{0}$ against $\boldsymbol{\theta} \in \mathcal{C}$ is a test of $\boldsymbol{\delta} = \mathbf{0}$ against $\boldsymbol{\delta} \in \mathcal{C}$ based on a single observation of $n^{1/2}\hat{\mathbf{\Gamma}}^{-1}\mathbf{T}_n$. To apply this a consistent estimator of $\hat{\mathbf{\Gamma}}$ may be constructed as follows. For the jth coordinate, consider a distribution-free confidence interval for θ_j (based on aligned $T_{nj}(b)$ where each X_{ij} is shifted by the amount b before sign and absolute ranks are computed); corresponding to a coverage probability $1 - \alpha$, let $\hat{\theta}_{nj,L}$ and $\hat{\theta}_{nj,U}$ be the lower and upper confidence limits so obtained. These are obtained by equating $T_{nj}(b)$ to $+C_{n,\alpha/2}$ and $-C_{nj\alpha/2}$, respectively. By virtue of the distribution-freeness of T_{nj}, under H_{0j}, $C_{n,\alpha/2}$ does not depend on the underlying df F as long as the later is continuous and symmetric. Then, the ratio $(\hat{\theta}_{nj,U} - \hat{\theta}_{nj,l})/(2C_{n,\alpha/2})$ provides a consistent estimator of γ_j, for $j = 1, \ldots, p$ (see Sen (1966)).

Given the invariance of $\mathbf{V}_n$ under $\mathcal{P}_n$, and the weak convergence of the permutation distribution, we are tempted to adapt the UIP to formulate a rank test for $H_0 : \boldsymbol{\theta} = \mathbf{0}$ vs $H_1 : \boldsymbol{\theta} \geqslant \mathbf{0}$, $\boldsymbol{\theta} \neq \mathbf{0}$; here $\mathcal{P}_n$ refers to the conditional probability law given the vectors of signs and the absolute rank collection matrix (Puri and Sen (1971), Ch. 4). We proceed as in Section 5.2, and consider the set $\mathbb{P}$ of all 2^p vectors $\mathbf{a} = (a_1, \ldots, a_p)^T$, where each a_j can be either 0 or 1, and partition $\mathbf{T}_n$ and $\mathbf{V}_n$ into $(\mathbf{T}_{na}, \mathbf{T}_{na'})$ and

$$\begin{pmatrix} \mathbf{V}_{naa} & \mathbf{V}_{naa'} \\ \mathbf{V}_{na'a} & \mathbf{V}_{na'a'} \end{pmatrix}$$

$\boldsymbol{a}'$ being the complement of $\mathbf{a}$, $\phi \subset \mathbf{a} \subseteq \mathbb{P}$. Further, we define $T_{na:\boldsymbol{a}'}$ and $V_{naa:\boldsymbol{a}'}$ as in there (i.e., $\mathbf{T}_{n\mathbf{a}:\mathbf{a}'} = \mathbf{T}_{na} - \mathbf{V}_{n\mathbf{aa}'}\mathbf{V}^{-1}_{n\mathbf{aa}'}\mathbf{T}_{na'}$ and $\mathbf{V}_{n\mathbf{aa}:\mathbf{a}'} = \mathbf{V}_{n\mathbf{aa}} - \mathbf{V}_{n\mathbf{aa}'}\mathbf{V}^{-1}_{n\mathbf{a}'\mathbf{a}'}$ $\mathbf{V}_{n\mathbf{a}'\mathbf{a}}$, $\forall \emptyset \subseteq \mathbf{a} \subseteq \mathbb{P}$. Let then [1]

$$\mathcal{L}_n = \sum_{\emptyset \subseteq \mathbf{a} \subseteq \mathbb{P}} I(\mathbf{T}_{n\mathbf{a}:\mathbf{a}'} > \mathbf{0}, \mathbf{V}^{-1}_{n\mathbf{a}'\mathbf{a}'}\mathbf{T}_{n\mathbf{a}'} \leqslant \mathbf{0})(n\mathbf{T}^T_{n\mathbf{a}:\mathbf{a}'}\mathbf{V}^{-1}_{n\mathbf{aa}:\mathbf{a}'}\mathbf{T}_{n\mathbf{a}:\mathbf{a}'}) \qquad (5.84)$$

where we note that $I(\cdot)$ is different from 0 only for one specific (random) $\mathbf{a}$.

As $\mathbf{V}_n$ is $\mathcal{P}_n$ invariant, the partitioning of $\boldsymbol{V}_n$ are also so. Therefore, the (conditional) permutational distribution of $\mathcal{L}_n$ (over the set of $2^n(n!)$ conditionally equally likely realizations) can be obtained by direct enumeration of the $\boldsymbol{T}_n$ when n is not large. For large n, by virtue of the weak convergence of $n^{1/2}(\mathbf{T}_n - \mathbf{0})$ (under H_0) to a p-variate normal law (and the stochastic convergence of $\mathbf{V}_n$ to $\boldsymbol{\nu}$), we claim that

[1] The statistic $\mathcal{L}_n$ is numerically equal to the likelihood ratio statistic if $\boldsymbol{T}_n$ were exactly $N(\boldsymbol{\theta}, \boldsymbol{V}_n)$ and $\boldsymbol{V}_n$ is treated as nonstochastic. Therefore, $\mathcal{L}_n = [n\boldsymbol{T}_n^T\boldsymbol{V}_n^{-1}\boldsymbol{T}_n - \inf\{(\boldsymbol{T}_n - \boldsymbol{b})^T\boldsymbol{V}_n^{-1}(\boldsymbol{T}_n - \boldsymbol{b}) : \boldsymbol{b} \geq \mathbf{0}\}]$, and hence it can be computed easily using a quadratic program. However, its small sample distribution under P_n has to be enumerated by considering all possible $2^n n!$ conditionally equally likely realizations. For large samples, the same computation and asymptotic distribution as in Chapter 3 hold.

under $H_0 : \boldsymbol{\theta} = \mathbf{0},\ F_0 \in \mathcal{F}_0$,

$$\mathcal{L}_n \xrightarrow{d} \sum_{k=0}^{p} \omega_k \chi_k^2 \tag{5.85}$$

where the χ_k^2 are independent chi-square random variables with degrees of freedom $k(= 0, 1, \ldots, p)$ and the normal orthant probabilities (with respect to $\mathbf{V}_n$) lead to the approximation for the ω_k (when sorted by the cardinality of the element $\mathbf{a} : \emptyset \subseteq \mathbf{a} \subseteq \mathbb{P}$.)

II. Nonparametric multivariate two-sample problem: Layer alternatives

Let $\mathbf{X}_1, \ldots, \mathbf{X}_{n_1}$ be n_1*iid* random vectors with a continuous distribution function $F(\mathbf{x})$ defined on R^p for some $p \geqslant 1$, and let $\mathbf{Y}_1, \ldots, \mathbf{Y}_{n_2}$ be an independent sample of n_2*iid* random vectors with a continuous distribution function $G(\mathbf{x})$, also defined on R^p. It is assumed that

$$G(\mathbf{x}) = F(\mathbf{x} - \boldsymbol{\Delta}),\ \mathbf{x} \in R^p,\ F \in \mathcal{F} \tag{5.86}$$

where $\boldsymbol{\Delta}$ is a p-vector of real (unknown) elements, and $\mathcal{F}$ is the class of all continuous (not necessarily symmetric) distribution functions on R^p. (Thus, $\mathcal{F}_0$ in (5.75) is a subclass of $\mathcal{F}$). Consider now the following hypothesis testing problem:

$$H_0 : \boldsymbol{\Delta} = \mathbf{0} \ \text{ vs } \ H_1 : \boldsymbol{\Delta} \geqslant \mathbf{0},\ \boldsymbol{\Delta} \neq \mathbf{0}. \tag{5.87}$$

This is an extension of the hypothesis testing problem in (5.76) to the two-sample model, without the symmetry of F. H_1 is referred to, in the literature, as *layer alternative* (Bhattacharya and Johnson (1970)). Writing $\boldsymbol{\Delta} = (\Delta_1, \ldots, \Delta_p)^T$ we may write as in (5.77),

$$H_0 = \bigcap_{j=1}^{p} H_{0j} \quad \text{and} \quad H_1 = \bigcup_{j=1}^{p} H_{1j},$$

where $H_{0j} : \Delta_j = 0$ and $H_{1j} : \Delta_j > 0,\ j = 1, \ldots, p$.

For F_j, the jth marginal distribution function for F (and G_j for G), we have by (5.86) $G_j(x) = F_j(x - \Delta_j),\ x \in \mathbb{R}$, so that when F_j admits an absolutely continuous pdf $f_j(x)$ with a finite Fisher information $I(f_j)$, we can construct a LMPR test statistic, for H_{0j} vs H_{1j}, based on an appropriate two-sample linear rank statistic, for $j = 1, \ldots, p$. In this sense, the WMW-statistic described in (5.47) is LMPR when F_j is logistic; the log-rank test is LMPR when F_j is exponential, the median test is LMPR when F_j is Laplace, and the normal scores test is LMPR when F_j is normal (Problem 5.15). If F_j is not of assumed form (apart from unknown location and scale parameters) we may choose a suitable linear rank statistic, albeit without being able to label it as LMPR. In this way we choose a set of p two-sample linear rank statistics $\mathbf{T}_n = (T_{n1}, \ldots, T_{np})^T$, where $n = n_1 + n_2$. Under H_0, the two samples are from a common population, so that the joint distribution of all the n-vectors

$(\mathbf{X}_1, \ldots, \mathbf{X}_{n_1}, \mathbf{Y}_1, \ldots, \mathbf{Y}_{n_2})$ remains invariant under any permutation of them, and hence the columns of the vectors are *iid* random vectors, and, hence, the columns of the rank-collection matrix $\boldsymbol{R}_n$ are exchangeable vectors. Therefore, if we rearrange the columns in such a way that the top row is in the natural order $1, \ldots, n$, the resulting matrix , denoted by $\boldsymbol{R}_n^o$, is called the reduced rank-collection matrix. Consider then the finite group of $n!$ possible matrices, which can be obtained by column permutations of the reduced rank collection matrix. A discrete uniform probability measure on this generated group is termed the rank permutation measure in the multivariate case (Chatterjee and Sen (1964)). Under this *rank collection principle*, we can generate a set of $n!$ possible rank-collection matrices by assigning a (row-wise) coordinatewise ranks and performing all possible $n!$ column permutations. These $n!$ rank-collection matrices are conditionally equally likely (under H_0), and they therefore lead to a permutational (conditional) probability law which we denote by $\mathcal{P}_n$.

Let R_{ij} be the rank of X_{ij} among the n observations $X_{1j}, \ldots, X_{n_1j}, Y_{1j}, \ldots, Y_{n_2j}$ for $i = 1, \ldots, n_1$ and $j = 1, \ldots, p$, and let $R_{n_1+i,j}$ be the rank of Y_{ij} among the same set for $i = 1, \ldots, n_2$ and $j = 1, \ldots, p$. Then the rank collection matrix is

$$\mathbf{R}_n = \begin{pmatrix} R_{11}, & \ldots & R_{1n} \\ \vdots & \vdots & \vdots \\ R_{p1}, & \ldots & R_{pn} \end{pmatrix} \tag{5.88}$$

while the collection $\boldsymbol{R}_n$ of $n!$ matrices is obtained by permuting the columns of $\boldsymbol{R}_n$ in all possible $(n!)$ ways. If we denote the scores for the jth row by $a_{nj}(r), r = 1, \ldots, n$, then typically T_{nj} can be expressed as

$$T_{nj} = \frac{1}{n_2} \sum_{i=n_1+1}^{n} a_{nj}(R_{ij}) - \frac{1}{n_1} \sum_{i=1}^{n_1} a_{nj}(R_{ij}) \; 1 \leqslant j \leqslant p. \tag{5.89}$$

In the same way, the permutation rank scores covariance matrix can be expressed as

$$\mathbf{V}_n = \frac{1}{n} \left[\sum_{i=1}^{n} \{a_{nj}(R_{ij}) - \bar{a}_{nj}\}\{a_{n\ell}(R_{i\ell}) - \bar{a}_{n\ell}\} \right]_{j,\ell=1,\ldots,p} \tag{5.90}$$

where $\bar{a}_{nj} = \{a_{nj}(1) + \cdots + a_{nj}(n)\}/n$, $1 \leqslant j \leqslant p$. Here also, note that $\mathbf{V}_n$ is a stochastic matrix, although its diagonal elements $v_{njj}, 1 \leqslant j \leqslant p$ are nonstochastic. $\boldsymbol{V}_n$ converges in probability to a positive definite $\boldsymbol{\nu}$ under quite general conditions (Problem 5.16). Results similar to (5.82)–(5.83) hold for two-sample and general regression model as well (Puri and Sen (1985), Ch. 7). These results can be used for constructing local tests of $\boldsymbol{\theta} = \mathbf{0}$ against $\boldsymbol{\theta} \in \mathcal{C}$, where $\mathcal{C}$ is a closed convex cone.

We can then proceed as in (5.78)–(5.83), where we incorporate $\mathcal{P}_n$ (the distribution of $\boldsymbol{R}_n$ over $\mathcal{R}_n$), and replace n by $n_0 = n_1 n_2/n$. We partition $\mathbf{T}_n$ and $\boldsymbol{V}_n$ as in (5.84) and consider the following UIT statistic (noting that $\mathcal{L}_{n_1 n_2}$ is structurally similar to the *LRT*): the likelihood ratio statistic if $\boldsymbol{T}_n$ were exactly $N(\boldsymbol{\theta}, \boldsymbol{V}_n)$ and $\boldsymbol{V}_n$ is nonstochastic. Therefore, it can also be expressed as a projection and hence can be computed using a quadratic program.

$$\mathcal{L}_{n_1 n_2} = \sum_{\phi \subseteq \mathbf{a} \subseteq \mathbb{P}} I(\mathbf{T}_{na:a'} > 0) I(\mathbf{V}_{na'a'}^{-1} \mathbf{T}_{na'} \leqslant 0) \; \{n_0 \mathbf{T}_{na:a'}^T \mathbf{V}_{naa:a'} \mathbf{T}_{na:a'}\}. \tag{5.91}$$

Problems 5.17–5.19 are set to verify some of these results. $\mathcal{L}_{n_1 n_2}$ provides a conditionally (permutationally) distribution-free test; its permutation distribution can be obtained by enumerating the possible realizations of $\mathbf{T}_n$ (under $\mathcal{P}_n$), noting that $\mathbf{V}_n$ is $\mathcal{P}_n$-invariant. The asymptotic null distribution in (5.85) also pertains to $\mathcal{L}_{n_1 n_2}$. Here the asymptotic distribution theory follows from general results on multivariate multisample rank statistics presented in Puri and Sen (1971, Ch.5).

III. Nonparametric multivariate regression problem: orthant alternative

Let $\mathbf{X}_1, \ldots, \mathbf{X}_n$ be n independent p-vectors with continuous distribution functions $F_1, \ldots, F_n$, all defined on $\mathbb{R}^p$, for some $p \geqslant 1$. It is assumed that

$$F_i(\mathbf{x}) = F(\mathbf{x} - \boldsymbol{\theta} - \boldsymbol{\beta} c_i), \quad 1 \leqslant i \leqslant n, \; F \in \mathcal{F} \tag{5.92}$$

where $c_1, \ldots, c_n$ are known (regression) constants, not all equal, $\boldsymbol{\theta}$ and $\boldsymbol{\beta}$ are unknown parameter (vectors), and $\mathcal{F}$ is the class of all continuous p-variate distribution functions. Consider then the following hypothesis testing problem:

$$H_0 : \boldsymbol{\beta} = \mathbf{0} \text{ vs } H_1 : \boldsymbol{\beta} \geqslant \mathbf{0}, \boldsymbol{\beta} \neq \mathbf{0} \tag{5.93}$$

treating $\boldsymbol{\theta}$ as a nuisance parameter and allowing $F \in \mathcal{F}$. This is a direct extension of the two-sample model in (5.86) which corresponds to $c_1 = \cdots = c_{n_1} = 0, c_{n_1+1} = \cdots = c_n = 1$. Here also, we write

$$H_0 = \bigcap_{j=1}^{p} H_{0j} \quad \text{and} \quad H_1 = \bigcup_{j=1}^{p} H_{1j},$$

where $H_{0j} : \beta_j = 0$ and $H_{1j} : \beta_j > 0, \; j = 1, \ldots, p$.

If F_{j0}, the jth marginal distribution function for F in (5.92) admits an absolutely continuous pdf f_{j0} with finite Fisher information $I(f_{j0})$, then for testing H_{0j} vs H_{1j}, there is a LMPR-test based on the linear rank statistic

$$T_{nj} = \sum_{i=1}^{n} (c_i - \bar{c}_n) a_{nj}(R_{ij}) \; j = 1, \ldots, p \tag{5.94}$$

where $\bar{c}_n = n^{-1} \sum_{i=1}^{n} c_i$ and the scores $a_{nj}(r), 1 \leqslant r \leqslant n$ and $R_{ij} (1 \leqslant i \leqslant n, j = 1, \ldots, p$, are all defined as in the two-sample model. The characterization of LMPR tests for H_{0j} vs H_{1j}, made in Problem 5.15, remains in tact for this regression alternative problem as well. We define the rank collection matrix as in (5.88) and formulate the permutational (conditional) probability measure $\mathcal{P}_n$ as in the two-sample model. Also, we define $\boldsymbol{V}_n$ as in (5.90). Further, we let $C_n^2 = \sum_{i=1}^{n} (c_i - \bar{c}_n)^2$ and assume that the *Noether Condition* holds, i.e.,

$$\max_{1 \leqslant i \leqslant n} (c_i - \bar{c}_n)^2 / C_n^2 \to 0 \text{ as } n \to \infty. \tag{5.95}$$

Further, we partition $\boldsymbol{T}_n$ and $\boldsymbol{V}_n$ as in (5.91), and consider the following UIT statistic:

$$\mathcal{L}_n^* = \sum_{\phi \subset a \subset \mathbb{P}} I(\mathbf{T}_{n\mathbf{a}:\mathbf{a}'} > 0, \mathbf{V}_{n\mathbf{a}'\mathbf{a}'}^{-1}\mathbf{T}_{n\mathbf{a}'} \leqslant 0)\,\{C_n^{-2}\mathbf{T}_{n\mathbf{a}:\mathbf{a}'}^T\mathbf{V}_{\mathbf{aa}:\mathbf{a}'}^{-1}\mathbf{T}_{n\mathbf{a}:\mathbf{a}'}\}; \quad (5.96)$$

this $\mathcal{L}_n^*$ is also the likelihood ratio statistic if $\boldsymbol{T}_n$ were $N(\boldsymbol{\theta}, C_n^2\boldsymbol{V}_n)/$ and $C_n^2\boldsymbol{V}_n$ is not stochastic. Like the two-sample model, here also we can generate the permutational (conditional) distribution of $\mathcal{L}_n^*$ (under H_0) by enumerating all possible $(n!)$ conditionally equally likely (under $\mathcal{P}_n$) realizations of $\mathbb{R}_n$ over $\mathcal{R}_n$ (noting that $\boldsymbol{V}_n$ is $\mathcal{P}_n$-invariant). The asymptotic null distribution of $\mathcal{L}_n^*$ will be the same as in (5.85). The asymptotic distribution theory for $\mathcal{L}_n^*$ under H_0, and some local (orthant) alternatives follow from general results on multivariate rank statistics for general linear models, presented in Chapter 7 of Puri and Sen (1985), and we leave some of these as exercises at the end of the Chapter (Problems 5.20, 5.21).

Note that, if we drop the LMPR criterion, $\mathcal{L}_n^*$ can be used for general linear rank statistics–without requiring finite Fisher information (but more stringent regularity conditions on the scores $a_{nj}(\cdot)$. We pose some of these technicalities as Problem 5.24. In the next subsection, we proceed to consider more general nonparametric UITs.

5.4.5 Nonparametric UIT for Restricted Alternatives

Even when we may not have LMPR tests, UIT can be based on suitable nonparametric statistics. We consider some illustrative examples.

(I) Multiple regression model: restricted alternatives
Consider the classical linear model

$$\boldsymbol{Y}_n = \theta \mathbf{1}_n + \boldsymbol{X}_n\boldsymbol{\beta} + \boldsymbol{e}_n; \quad \boldsymbol{e}_n = (e_1 \ldots, e_n)^T, \quad (5.97)$$

where $\boldsymbol{X}_n$ is a matrix of known regression constants, $\theta, \boldsymbol{\beta}$ are unknown parameters, and $e_1, \ldots, e_n$ are *iid* random variables with a d.f. F, defined on $\mathbb{R}$. Consider the hypothesis testing problem:

$$H_0 : \boldsymbol{\beta} = \mathbf{0}, \text{ vs. } H_1^> : \boldsymbol{\beta} \geq \mathbf{0}, \|\boldsymbol{\beta}\| > 0. \quad (5.98)$$

We may also consider the hypothesis testing problem:

$$H_0 : \beta_1 = \cdots = \beta_p = \beta \text{ (unknown) vs } H_1^< : \beta_1 \leq \cdots \leq \beta_p, \quad (5.99)$$

where at least one "$<$" holds. (5.99) can be reduced to (5.98) by a reparametrization in (5.97). Hence, without loss of generality, we consider (5.98).

When F is normal with variance $\sigma^2(< \infty)$, the MLE of $\boldsymbol{\beta}$ is $\hat{\boldsymbol{\beta}}_n = (\boldsymbol{X}^T\boldsymbol{X})^{-1}$ $\boldsymbol{X}^T\boldsymbol{Y}_n, \hat{\boldsymbol{\beta}}_n$ is unbiased for $\boldsymbol{\beta}$ and

$$\hat{\boldsymbol{\beta}}_n - \boldsymbol{\beta} \stackrel{\mathcal{D}}{=} \mathcal{N}_p(\mathbf{0}, \sigma^2(\boldsymbol{X}^T\boldsymbol{X})^{-1}). \quad (5.100)$$

Since $(\boldsymbol{X}^T\boldsymbol{X})$ is known, and $(n-p-1)^{-1}(\boldsymbol{Y}_n - \hat{\theta}_n\mathbf{1}_n - \boldsymbol{X}_n\hat{\boldsymbol{\beta}}_n)^T(\boldsymbol{Y}_n - \hat{\theta}_n\mathbf{1}_n - \boldsymbol{X}_n\hat{\boldsymbol{\beta}}_n) = s_e^2$ is unbiased for σ^2, irrespective of H_0 being true or not, we may reduce the hypothesis testing problem in (5.98) by sufficiency, and based on the joint distribution of $\hat{\boldsymbol{\beta}}_n$ and s_e^2 (which are independent), we can use the *LRT* when the covariance matrix is known up to a scalar constant (σ^2) (Problem 5.23). This prescription may not work out when F is nonnormal, or more noticeably in the nonparametric case, which we consider here. We allow $F \in \mathfrak{F}$, the class of all continuous d.f.'s on $\mathbb{R}$.

We denote by R_{ni} the rank of Y_i among $Y_1, \ldots, Y_n$, for $i = 1, \ldots, n$ (ties neglected, with probability 1 as $F \in \mathfrak{F}$), and consider suitable scores $a_n(1), \ldots, a_n(n)$, assumed to be monotone. Then, define $\boldsymbol{T}_n = (T_{n1}, \ldots, T_{np})^T$ by

$$\boldsymbol{T}_n = \sum_{i=1}^{n} (\boldsymbol{x}_i - \bar{\boldsymbol{x}}_n)^T a_n(R_{ni}), \tag{5.101}$$

where $\boldsymbol{X}_n^T = (\boldsymbol{x}_1^T, \ldots, \boldsymbol{x}_n^T)$ and $\bar{\boldsymbol{x}}_n = n^{-1}\sum_{i=1}^{n} \boldsymbol{x}_i$. Let us also write

$$A_n^2 = (n-1)^{-1}\sum_{i=1}^{n}[a_n(i) - \bar{a}_n] \text{ and } \boldsymbol{C}_n = (\boldsymbol{X}_n - \mathbf{1}_n\bar{\boldsymbol{x}}_n^T)^T(\boldsymbol{X}_n - \mathbf{1}\boldsymbol{x}_n^T) \tag{5.102}$$

where $\bar{a}_n = n^{-1}\sum_{i=1}^{n} a_n(i)$. Note that under $H_0 : \boldsymbol{\beta} = \mathbf{0}$, $(R_{n1}, \ldots, R_{nn})$ takes on all possible $n!$ permutations of $(1, \ldots, n)$ each with the common probability $(n!)^{-1}$. As such, $\boldsymbol{T}_n$ is EDF with null mean vector and covariance matrix $A_n^2\boldsymbol{C}_n$. Without loss of generality, we use the standardized scores, so that $\bar{a}_n = 0$ and $A_n^2 = 1$. Then, by permutational central limit theorems, we claim that under H_0, as $n \to \infty$,

$$\boldsymbol{C}_n^{-1/2}[\boldsymbol{T}_n - \mathbf{0}] \to \mathcal{N}_p(\mathbf{0}, \boldsymbol{I}_p), \tag{5.103}$$

where C_n is a known (p.d.) matrix; here, we assume that the generalized Noether condition holds:

$$\max_{1 \le i \le n} \boldsymbol{x}_i(\boldsymbol{X}_n^T\boldsymbol{X}_n)^{-1}\boldsymbol{x}_i^T \to 0 \text{ as } n \to \infty. \tag{5.104}$$

We introduce the same set $\mathcal{P}$ of all possible (2^p) subsets of $\{1, \ldots, p\}$ (as in (5.84)): $\emptyset \subseteq \boldsymbol{a} \subseteq \mathbb{P}$, and partition $\boldsymbol{T}_n$ and $\boldsymbol{C}_n$ in the same manner as in there (replacing $\boldsymbol{V}_n$ by $\boldsymbol{C}_n$). Then, we consider the following UIT based on $(\boldsymbol{T}_n, \boldsymbol{C}_n)$:

$$\mathcal{L}_n^o = \sum_{\emptyset \subseteq \boldsymbol{a} \subseteq \mathbb{P}} I(\boldsymbol{T}_{na:a'} > \mathbf{0}; \boldsymbol{C}_{na'a'}^{-1}\boldsymbol{T}_{na'} \le \mathbf{0})\{\boldsymbol{T}_{naa'}^T : \boldsymbol{C}_{naa:a'}^{-1}\boldsymbol{T}_{na:a'}\}. \tag{5.105}$$

Here, also, $\boldsymbol{C}_n$ (and its partitioning) remain invariant under the set of $n!$ permutations of $(R_{n1}, \ldots, R_{nn})$ over $(1, \ldots, n)$ while the $\boldsymbol{T}_n$ (and its partitioning) are affected by these rank permutations. Therefore, the (null hypothesis) distribution of $\mathcal{L}_n^o$ can be enumerated by considering the $n!$ equally likely permutations of $R_{n1}, \ldots, R_{nn}$ over $1, \ldots, n$. This shows that $\mathcal{L}_n^o$ is EDF. However, this enumeration task becomes prohibitively laborious as n increases, though by virtue of (5.103) and (5.104), we may claim that the asymptotic null distribution in (5.85) also pertains to the present situation.

We may consider contiguous alternatives $\{H_{1(n)}^{>}\}$ where

$$H_{1(n)}^{>} : \boldsymbol{\beta} = \boldsymbol{\beta}_{(n)} = n^{-1/2}\boldsymbol{\lambda}, \boldsymbol{\lambda} \in \mathbb{R}^{p+} \tag{5.106}$$

where we assume that $n^{-1}\boldsymbol{C}_n \rightarrow \boldsymbol{C}$ (p.d.). Then, invoking Le Cam's third lemma [Hájek, Šidák and Sen (1999, p.259)] it follows that under $H_{1(n)}^{>}$, as $n \rightarrow \infty$

$$\boldsymbol{C}_n^{-1/2}\boldsymbol{T}_n \rightarrow \mathcal{N}(\boldsymbol{C}^{1/2}\boldsymbol{\lambda}\gamma, \boldsymbol{I}_p); \tag{5.107}$$

where $\gamma(> 0)$ is a scalar constant, depending on the score function. Problem 5.26 is set to verify the above result. This makes it comparable to the multinormal mean problem (for orthant alternatives) when the covariance matrix is known. Hence, in this asymptotic set-up, $\mathcal{L}_n^o$ in (5.105) is comparable to the normal theory *LRT*, though $\mathcal{L}_n^o$ is EDF and more motivated by the UIP.

(II) Multivariate linear model: restricted alternatives

As a natural extension of (5.97), consider the following:

$$\boldsymbol{Y}_n = \boldsymbol{1}_n\boldsymbol{\theta}^T + \boldsymbol{X}_n\boldsymbol{\beta} + \boldsymbol{\epsilon}_n; \boldsymbol{\epsilon}_n^T = (\boldsymbol{e}_1^T, \ldots, \boldsymbol{e}_n^T) \tag{5.108}$$

where the $\boldsymbol{e}_i$ are *iid* random vectors having a continuous (p-variate) d.f. F, defined on $\mathbb{R}^p$; $\boldsymbol{\beta}$ is a $m \times p$ matrix of unknown (regression) parameters, $\boldsymbol{\theta}$ is the vector of regression parameters, and the (design) matrix $\boldsymbol{X}_n$ of order $m \times n$, is known and assumed to be of full rank $m(\leq n)$. We may then frame a null hypothesis H_0 and alternatives $H_1^{>}$ as in (5.98); i.e., for each column of $\boldsymbol{\beta}$, we have the orthant alternative in (5.98). Keeping this in mind, we write $\boldsymbol{\beta} = (\boldsymbol{\beta}_1, \ldots, \boldsymbol{\beta}_p)$ where each $\boldsymbol{\beta}_j$ is an m-vector. Then we write $H_0 = \bigcap_{j=1}^{p} H_{0j}$ and $H_1^{>} = \bigcup_{j=1}^{p} H_{ij}^{>}$ where $H_{0j} : \boldsymbol{\beta}_j = \boldsymbol{0}$ and $H_{1j}^{>} : \boldsymbol{\beta}_j \geq 0, \|\boldsymbol{\beta}_j\| > 0$, for $j = 1, \ldots, p$. Without loss of generality, we assume that $\boldsymbol{1}_n^T\boldsymbol{X}_n = \boldsymbol{0}$. For normal F, null mean vector ($\boldsymbol{0}$) and (unknown) dispersion matrix $\boldsymbol{\Sigma}$ (p.d.), the MLE of β is

$$\hat{\boldsymbol{\beta}}_n = (\boldsymbol{X}_n^T\boldsymbol{X}_n)^{-1}\boldsymbol{X}_n^T\boldsymbol{Y}_n, \tag{5.109}$$

and

$$\hat{\boldsymbol{\beta}}_n - \boldsymbol{\beta} = \mathcal{N}(\boldsymbol{0}, (\boldsymbol{X}_n^T\boldsymbol{X}_n)^{-1} \otimes \boldsymbol{\Sigma})) \tag{5.110}$$

where $\otimes$ stands for the Kronecker product. In this set-up, $\boldsymbol{\Sigma}$ being unknown, we encounter the same difficulties (with the distribution theory) as in the multinormal mean, orthant alternative problem, when $\boldsymbol{\Sigma}$ is unspecified (Problem 5.26). The situation becomes more complex for nonnormal F.

For the univariate case (i.e., $p = 1$), (5.110) simplifies to a known dispersion matrix $(\boldsymbol{X}_n^T\boldsymbol{X}_n)^{-1}\sigma^2$, apart from a scalar σ^2, so that the methodology developed in the previous chapter applies, and we have seen how nonparametrics could be placed side by side without requiring the underlying distribution to be normal. However, in the multivariate case, though the rank statistics can be defined as in (5.94) (for each coordinate), they are not genuinely distribution-free, so that the treatment of (5.103) through (5.107) fails to produce an EDF procedure. Asymptotically, for this matrix case, (5.103) extends to (under H_0)

$$n^{-1/2}\boldsymbol{T}_n \xrightarrow{\mathcal{L}} \mathcal{N}_{mp}(\boldsymbol{0}, n^{-1}\boldsymbol{C}_n \otimes \boldsymbol{\nu}) \tag{5.111}$$

where the $(p \times p)$ dispersion matrix $\boldsymbol{\nu}$ can be estimated (consistently and nonparametrically) by the permutation covariance matrix of the scores (Problem 5.27), and $\boldsymbol{C}_n$ (is now an $m \times m$ matrix) is defined as in (5.102). In the same way, for local alternatives as in (5.106) (but for $\boldsymbol{\Lambda} \in \mathbb{R}^{m \times p}$), (5.107) extends to (under $\{H_{(n)}\}$)

$$n^{-1/2}\boldsymbol{T}_n \xrightarrow{\mathcal{D}} \mathcal{N}_{mp}(\boldsymbol{\Lambda\Gamma}, \boldsymbol{C} \otimes \boldsymbol{\nu}) \tag{5.112}$$

where $\boldsymbol{\Gamma} = \text{diag}(\gamma_1, \ldots, \gamma_p)$, and for each $j(= 1, \ldots, p)$, γ_j is a scalar constant (> 0) depending on the score function used for the jth coordinate and the Fisher score function for the same; we leave the details of verification as Problem 5.28.

Motivated by (5.111) and (5.82), Chinchilli and Sen (1981a, b) appealed to the UIP and derived some robust (asymptotic) tests for H_0 vs H_1 in the multivariate set-up. We refer to their development in the form of Problem 5.29 and 5.30.

5.5 RESTRICTED ALTERNATIVES AND SIMES-TYPE PROCEDURES

In many environmental, epidemiological, and biomedical studies, as well as multicenter clinical trials, there is a genuine need to pool the *statistical information* from related (or compartmental) studies to "boost" the statistical evidence in the (main) study under consideration. This constitutes the main theme of meta-analysis and small area estimation is annexed to this research area. In many of these studies, conducted in diverse set-ups, the experimental (or observational) patterns may not conform to a common design, and in some cases, even the response variables may not be the same. Of course, pooling of statistical information in such a diverse set-up may need some sort of concordance of the statistical patterns (e.g. the null hypotheses and alternatives for different studies are conformable in a well defined sense), and in many applications this conformity can be taken for granted on various other considerations. We explain this aspect with the help of some pertinent examples.

Example 5.5.1. Consider a multicenter clinical trial for comparing a *treatment* (say, a drug for lowering the blood cholesterol level) and *placebo* group, where for different centers, the subjects (patients) under the study are possibly quite different (with respect to age, diet, physical exercise, smoking and drinking practice, and other demographic criteria). This lack of homogeneity across the centers makes the conventional linear model with additive center-effect (even allowing it to be random) much less appealing, if not inappropriate. Yet, it is conceivable that the drug (treatment) has an effect (although not equally) across the centers, so that it is plausible to frame, for each center, a null hypotheses of equality of treatment and placebo effects, against the one-sided alternative that the treatment is better. Without a linear model set-up or homogeneity of experimental units across the centers, conventional methods for pooling of statistical information may not be appropriate. It is therefore, often, advocated to formulate statistical procedures based on the individual center observed significant level (OSL) or the so-called p-values.

Incidentally, if the treatment has an interaction effect with the centers, in the sense that the treatment effect is not concordant across the centers then in pooling statistical information, one needs to take into account this treatment $\times$ center interaction, and that can lead to a different appraisal of the problem. Such a problem may arise, for example, if the prescribed drug has some adverse effect if taken with alcoholic drinks or some other food/drink/smoking items—a factor that can not be ruled out in double blind trials unless a pretrial appraisal of this effect is properly made. However, in such a case, there might be much less appeal for a restricted (viz., one-sided) alternatives, and, hence, for our interest, we shall by-pass some of these complications at this stage. ■

Example 5.5.2. Drug researchers and drug developers, especially the multinational groups, try to incorporate the pooling of statistical information from various national/regional studies to boost the statistical picture of a prospective drug, for easier entry into the drug market of a specific country (which may need certain agency approval). Here also, the basic approach is to appeal to the similarity of the drug-effect in all the reported studies so that statistical pooling makes sense. However, there may be considerable inter-study variation, so that conventional (mixed or fixed-effects) linear (or generalized linear) models may not be that appealing. Therefore, as in the previous example, OSL or p-value based pooling procedures may sense better.

It is not uncommon that the drug developers may not report some studies that were not completed (on the ground that they were not heading for any conclusive decisions or resulted in no-significant effect of the drug) in the present pooling effort. This results in the so-called *publication bias*. In meta analysis, this is a genuine concern. ■

Example 5.5.3. In many epidemiologic studies covering a broad region, *stratification* into smaller (or relatively homogeneous) areas is often advocated. In this way, the *between-stratum* variation may not be insignificant, though this variation may often be accounted in terms of a number of *auxiliary variables* that may vary (considerably) from stratum to stratum. Thus, conventional *analysis of variance* (or even *analysis of covariance* procedures may not be suitable. Thus, pooling of statistical information from the *within-stratum* sources may be subjected to the same difficulties as in the previous examples. OSL or p-value based procedures are appealing especially when we have restricted alternatives in mind. This is because in the global case, often the pooling is done in an additive way with access to simple null hypothesis distribution. For example, in ANOVA, the sum of square due to regression can be pooled from several homogeneous studies, and so the sum of square due to error, so that a combined test statistic having the usual variance ratio distribution can be constructed. For restricted alternatives, this pooling may not lead to a simple additive form of individual test statistics. Moreover, with heterogeneous studies, a combination or pooling will be even more cumbersome. On the other hand, for individual data sets, restricted alternative test statistics can be obtained by a variety of methods discussed in this and earlier chapters, so that even for heterogeneous studies, pooling of the p-values would be more easy to handle. ■

Example 5.5.4. In environmental pollution and toxicity studies, though intended for a region or area, data are to be recorded or collected at a number of *grid points* or *sites* (which are much smaller in dimension), and pooled then to get a comprehensive picture for the entire area. Here also the basic idea is that the statistical pictures at the grid points should be concordant, though they may differ (often considerably) with respect to the toxicity/pollution levels, depending on the location of the grid points, their proximity to the emitting sources, methods of recording data, and many other demographic and socioeconomic factors. Without the homogeneity of the within-site (grid-point) statistics and lacking the usual *spatial homogeneity*, the usual *variogram* based (Kriging and other) methods may not work out properly. Nevertheless, a case for restricted (one-sided) alternatives can be made (given the basic fact that toxicants are harmful), and hence, we may as well proceed with OSL or p-value based procedures. We may need to take into account the spatial dependence of these statistics. ■

Example 5.5.5. With environmental chemical toxics, *human experimentation* is not permissible, *subhuman primates* are used in relatively controlled laboratory set-ups, with the intention that the statistical information acquired from such *animal studies* may then be pooled for human exposure experience. A dose-response model exhibits the mathematical relationship between an amount of exposure or treatment and the degree of a biological or health effect, generally a measure of an adverse outcome. Dosimetry models intend to provide a general description of the uptake and distribution of inhaled/ingested/absorbed toxics (or compounds having adverse health effects) on the entire body system. for judgement on human population, such dosimetric models for animal studies need to be extrapolated with a good understanding of the interspecies differences. which is now widely used. However, the animal studies are more of the laboratory type but the human exposure is not. The human metabolism is different from subhuman primates. The human exposure to environmental toxicity is rather complex (as it varies across demographic strata) and highly heterogeneous (compared to animal studies in laboratories). Actually, the human exposure variation with respect to socioeconomic, racial, outdoor/indoor activity, diet, physical exercise (and many other factors) is much more prominent than subhuman primates. Yet dosimetry aims to provide the key to the human stochastics! Clearly, conventional pooling procedures may not make much sense in this context as well. On the other hand, a case for restricted alternatives can be made relatively easily, and we may find it more appealing to incorporate OSL or p-value based procedures. ■

Other important examples include the evolving field of *bioinformatics* and large biological systems in general. The basic characteristic of this field is an enormously large dimension of acquired data sets. This aspect brings us more into the topic, curse of (high)-dimensionality, where one might have (too much) *nonsummarizable information.* The relevance of restricted alternatives has to be properly assessed in this interdisciplinary field, and as such, we shall by-pass this area in our treatise of OSL based methods.

Even in conventional statistical analysis, OSL or p-value based analysis may arise in a number of problems (Folks (1984)):
(1) t-tests for the equality of two-treatment effects versus the one-sided alternative that one is better than another; among these multiple t-tests, one may be in a *completely randomized* (one-way) design, the second one may be in a *randomized block design*, a third one in a *Latin square design*, etc. Even if they were all based on a common design, the error variances within each plan may not be all the same. Thus, the inherent heterogeneity of the model merits attention .
(2) F-tests for equality of two variances versus one-sided alternatives when the designs are not necessarily the same; even the F-tests need not all involve the same treatments. Such a case may arise in agricultural studies conducted in different area.
(3) Chi-squared tests for homogeneity of variances against one-sided (ordered) alternatives.
(4) Chi-squared tests for independence in contingency tables, against positive or negative (i.e., one-sided) alternatives.
(5) Hotelling T^2-tests when the within-group covariance matrices may not all be the same (Monti and Sen (1976)).

In addition, the UIT and the step-down procedures, discussed in previous sections, may need to be appraised from this pooling perspective. The usual OSL or p-value based methods work out well when under the null hypotheses, the p-values (to be defined precisely) are distributed as uniform random variables on $[0, 1]$. This, in turn, may require that the individual test-statistics have all continuous distribution, under the null hypotheses. This may not always hold, especially for the contingency table or other discrete distributions, though in large samples, it is quite close to a simple continuous distribution (e.g., normal, chi-square, etc.)

5.5.1 Probability Integral Transformation (PIT) and OSL or p-Values

Let T be a statistic having a continuous distribution function $F(\cdot)$, defined on $\mathbb{R}$. Then, note that

$$F(T) \text{ has the uniform } [0,1] \text{ distribution.} \tag{5.113}$$

Note that sans continuity of F, (5.113) is not true. The Probability Integral Transformation (PIT) is then defined by $T \to F(T)$, which is a monotone function bounded by 0 and 1 on the left and right, respectively.

If T is a test statistic whose distribution under a null hypothesis (H_0) is $F_T^0(\cdot)$, and $F_T^0(\cdot)$ is completely known (and continuous), led by (5.113), we define the OSL for T by

$$L_T(t) = 1 - F_T^0(t) = \boldsymbol{P}\{T \geqslant t | H_0\}. \tag{5.114}$$

Note that under H_0, L_T has the uniform $(0, 1)$ distribution. Thus, OSL is complementary to the PIT and has the same uniform $(0, 1)$ distribution under H_0. For an observed value of T having the null distribution $F_T^0(\cdot)$, $L_T(T)$ is termed the OSL or the p-value. There is no difficulty in extending this definition to restricted alternatives, as the only relevant part is the known F_T^0 (though the formulation of T may depend on such restricted alternatives).

Suppose now that there are $k(\geqslant 2)$ different (and independent) test statistics $T_1, \ldots, T_k$, based on null hypotheses $H_{01}, \ldots, H_{0k}$ against (possibly restricted) alternatives $H_{11}, \ldots, H_{1k}$, respectively, whose null distributions are denoted by $F_1^0, \ldots, F_k^0$, respectively. Then their OSL or p-values are

$$L_j = 1 - F_j^0(T_j), \quad j = 1, \ldots, k. \tag{5.115}$$

Under $H_0 = H_{01} \cap \ldots \cap H_{0k}$, the L_j are independent and identically distributed uniform $(0, 1)$ random variables, and a test for H_0 against $H_1 = H_{11} \cup \ldots \cup H_{1k}$ is then intended to be based solely on $L_1, \ldots, L_k$.

I. Tippett (1931) and Wilkinson (1951) method

Let $L_{k:1} < \cdots < L_{k:k}$ be the ordered values of $L_1, \ldots, L_k$; we neglect the ties in probability, as the underlying d.f. is continuous. Tippett proposed the statistic $L_{k:1}$, the smallest order statistic, for testing H_0 vs H_1. Since the L_j are *iid* uniform $(0, 1)$ r.v.'s, it is easy to show (Problem 5.31) that under H_0,

$$P_0\{L_{k:1} \leqslant u\} = 1 - (1 - u)^k, \ 0 \leqslant u \leqslant 1,$$

so that it has the Beta $(1, k)$ distribution. Wilkinson proposed the test statistic $L_{k:m}$, for some $m : 1 \leqslant m < k$. Thus, under H_0 [Problem 5.32],

$$P_0\{L_{k:m} \leqslant u\} = \frac{k!}{(m-1)!(k-m)!} \int_0^u v^{m-1}(1 - v)^{k-m} dv$$

i.e., $L_{k:m}$ has the Beta $(m, k - m + 1)$ distribution (under H_0). Equivalently, under H_0 [Problem 5.33]:

$$\{1 - L_{k:m}\}/L_{k:m} \stackrel{\mathcal{D}}{=} F_{2(m-k+1),2m}$$

where $F_{2(m-k+1),2m}$ stands for the variance-ratio statistic with degrees of freedom (DF) $(2(m - k + 1), 2m)$. In a sense, the Tippett-Wilkinson method is a precursor of the Simes (1986) method.

II. Fisher (1932) and Pearson (1933) method

Fisher noted that under $H_0, -2 \log L_j, j = 1, \ldots, k$ are *iid* random variables having the central chi squared d.f. with 2 DF (Problem 5.34), and proposed the test statistic

$$\sum_{j=1}^{k} \{-2 \log L_j\} = \mathcal{L}_F, \text{ say.}$$

so that under $H_0, \mathcal{L}_F$ has the central chi-square d.f. with $2k$ degrees of freedom. Pearson advocated the same statistic but suggested that the L_j should be the minimum of the right- and left-hand tail areas (not the OSL defined earlier). David (1934) corrected the error in Pearson's derivation, and this is termed the P_λ-test, which is essentially the Fisher test amended to include the two-sided alternatives.

III. **Liptak (1958) and Lancaster (1961) method**

Let $\Phi(x)$ stand for the standard normal d.f. Liptak proposed the test statistic

$$\sum_{j=1}^{k} \Phi^{-1}(L_j) = \mathcal{L}_p, \text{ say}$$

so that under H_0, $\mathcal{L}_p$ has the normal distribution with 0 mean and variance k (Problem 5.35). Lancaster considered the statistic $\sum_{j=1}^{k} \Gamma_{\alpha_j}^{-1}(L_j)$ where $\Gamma_{\alpha_j}(\cdot)$ stands for the gamma $(\alpha_j, 1/2)$ distribution function, which has the density function $\Gamma(\alpha_j)^{-1} 2^{\alpha_j} x^{\alpha_j - 1} \exp(-2x)$. Both these statistics are analogous to $\mathcal{L}_F$ but based on other transformations on the L_j. Other proposals are due to Good (1955), Zelen (1957), George and Mudholkar (1977), and others. With this class, the Fisher statistic has emerged as a desirable one, based on certain considerations, and we briefly summarize this here.

An OSL (or p-value) based (combined) test has a test statistic $\mathcal{L}_0^*$ of the form

$$\mathcal{L}_0^* = \psi(L_1, \ldots, L_k), \quad \psi \in \Psi$$

where Ψ is the class of real valued functions. In all the specific cases, mentioned before, $\psi(\cdot)$ is a symmetric function of its k arguments, and hence, it can also be expressed as $\psi^*(L_{k:1}, \ldots, L_{k:k})$, though ψ^* may not be symmetric in the order statistics. Birnbaum (1954) showed that if $\psi(\cdot)$ is a *monotone increasing* function of each of its k arguments $(L_1, \ldots, L_k)$, then the test based on $\mathcal{L}_0^*$ is *admissible*. Thus, whenever $\mathcal{L}_0^*$ is $\nearrow$ in each $L_{k:j} (1 \leqslant j \leqslant k)$, $\mathcal{L}_0^*$ is admissible. In that way the tests of the Fisher type are all admissible. However, among them there may not be an optimal one (in the UMP sense). But, the Fisher (1932) test has the optimality property (for fixed k and large individual sample sizes on which the T_j are based) in the *Bahadur-efficiency sense* (Littell and Folks (1977)). Berk and Cohen (1979) showed that there are some other methods (like the Lancaster and George Mudholkar ones) that are also Bahadur-efficient. Hence too much significance should not be attached to such an asymptotic optimality property. Primarily driven by such motivations, Mudholkar and Subbaiah (1980) and Sen (1983, 1988), considered some MANOVA tests in the light of the Fisher OSL method, and we shall briefly present this in a more general context later on.

5.5.2 Simes-Type Procedures

We start with the Simes (1986) theorem. As in (5.114)–(5.115), consider k independent test statistics $T_1, \ldots, T_k$, a null hypothesis $H_0 = \cap_{j=1}^{k} H_{0j}$, and alternative $H_1 = \cup_{j=1}^{k} H_{1j}$ (which, in our case, could be relating to restricted ones). Also, let $L_1, \ldots, L_k$ be defined as i (5.115); thus L_j is the p-value for testing H_{0j} against H_{1j}. Then, we have the following

Theorem 5.5.1 *(Simes). Under H_0 and for every $\alpha : 0 < \alpha < 1$,*

$$P\{L_{k:j} > \frac{j}{k}\alpha, \text{ for all } j = 1, \ldots, k\} = 1 - \alpha, \tag{5.116}$$

where the $L_{k:j}$ are the ordered values of $L_1, \ldots, L_k$.

It follows from this theorem that a level α test of H_0 against H_1 may be carried out as follows: reject H_0 if the jth smallest p-value, $L_{k:j}$, is smaller than $j\alpha/k$ for some $j = 1, \ldots, k$.

We shall consider some comparatively more general results here, and, hence, the proof of the theorem is omitted. However, it is interesting to note that the Simes theorem is essentially the Ballot theorem (Karlin (1969), Takâcs (1967)), so that we sketch the latter first.

Theorem 5.5.2 *(Ballot theorem). Let $U_1, \ldots, U_k$ be iid random variables having the $U(0,1)$ distribution, and let $G_k(u) = k^{-1}\sum_{j=1}^{k} I(U_j < u), u \in (0,1)$ be the associated empirical d.f.. Then, for every $\gamma \geqslant 1$ and every $k \geqslant 1$,*

$$P\{G_k(u) \leqslant \gamma u, \ \forall u \in (0,1)\} = 1 - \gamma^{-1}. \tag{5.117}$$

Note that $[G_k(u) \leqslant \gamma u, \forall u \in (0,1)] \Leftrightarrow [U_{k:i} \geqslant i/k\gamma, \ \forall i = 1, \ldots, k]$ where the $U_{k:i}$ are the ordered values of the U_j, so that letting $\gamma^{-1} = \alpha$, (5.116) follows from (5.117). Possibly, it is an oversight of Simes (1986) for not noticing this equivalence.

The Simes (-Ballot) theorem provides only a test for the overall null hypothesis $H_0(= \cap_{j=1}^{k} H_{0j})$ against $H_1(= \cup_{j=1}^{k} H_{1j})$. In *multiple comparison procedures* (MCP), the context in which this theorem was posed, it is not uncommon to go for componentwise hypothesis testing (as in UIT), so that there is more of *multiple decision theory* in MCP. In this spirit, Hommel (1988, 1989) considered a *step-wise rejective* (SWR) *multiple testing procedure* (MTP); Hochberg (1988) showed that

$$P\{L_{k:j} \geqslant \frac{\alpha}{k - j + 1}, j = 1, \ldots, k | H_0\} = 1 - \alpha, \tag{5.118}$$

and incorporated this interesting result in a MTP set-up. It follows from (5.118) that a level α test of H_0 against H_1 may be carried out as follows: reject H_0 if the jth smallest p-value, $L_{k:j}$, is smaller than $\alpha/(k - j + 1)$ for some $j = 1, \ldots, k$. Some comments on the relationship between this and the test based on Simes theorem would be helpful.

Hochberg and Rom (1995) utilized (5.118) for some *logically related hypothesis testing* (LRHT) problems. Benjamini and Hochberg (1995) developed some further results related to the Simes theorem; their main contention was to develop the measure of *false discovery rate* (FDR) and use it in MTP. We sketch these developments in the following context.

In order to control the multiplicity effect when testing a family of hypotheses $\{H_1, \ldots, H_k\}$ simultaneously, MCP's have been designed to control the Type I family-wise error (FWE) rate defined as the probability of one or more false rejection of true hypotheses, irrespective of how many hypotheses are true and what values the parameters of the false hypotheses take (Hochberg and Tamhane (1987)) The control of the FWE rate is, however, at the expense of substantially lower power in detecting false hypotheses. Benjamini and Hochberg (1995), in order to adjust for this conservativeness property of multiple testing, introduced the concept of FDR.

Let $H_1, \ldots, H_k$ be the component hypotheses with H_{0j} and H_{1j} standing for the null and alternative ones, for $j = 1, \ldots, k$. Let L_j be the OSL value for testing H_{0j} versus H_{1j}, for $j = 1, \ldots, k$. In this set-up, there is no loss of generality in accommodating one-sided or other restricted alternative, as long as, the L_j has $U[0, 1]$ distribution under the null hypothesis H_{0j}, for $j = 1, \ldots, k$. As in (5.116), we denote the ordered values of $L_1, \ldots, L_k$ by $L_{k:1} < \cdots < L_{k:k}$ (ties neglected in probability). Borrowing the notion of *ranks* we let

$$L_j = L_{k:R_j}, j = 1, \ldots, k,$$

where $(R_1, \ldots, r_k)$ is some random permutation of $(1, \ldots, k)$. The induced ordering of the hypotheses may then be formulated by letting $H_{(j)}$ as the hypothesis corresponding to $L_{k:j}$, for $j = 1, \ldots, k$. For every $k(\geqslant 1)$, let $0 < a_{k1} < \cdots < a_{kk} < 1$ be a set of real numbers, such that whenever $L_{k:j}$ is $< a_{kj}$, $H_{0(j)}$ is rejected in favor of $H_{1(j)}$, for $j = 1, \ldots, k$. Thus, the probability of rejecting H_0 when it is actually true is equal to

$$\begin{aligned} &P\{L_{k:j} < a_{kj}, \text{ for at least one } j(= 1, \ldots, k)|H_0\} \\ &\quad = 1 - P\{L_{k:j} \geqslant a_{kj}, \text{ for all } j = 1, \ldots, k|H_0\}, \end{aligned} \tag{5.119}$$

so that Simes' type result can be used to control (5.119) to any preassigned level $\alpha(0 < \alpha < 1)$. We shall formulate this a bit more generally. But before that, let us define FDR more formally.

Let $H_1, \ldots, H_k$ be the $k(> 2)$ null hypotheses under consideration, and let $L_1, \ldots, L_k$ be the corresponding OSL values; these are assumed to be independent random variables, and under H_j, L_j has the uniform (0,1) distribution. As has been stated earlier, the Type I FWE rate is defined as the probability of one or more rejection of true hypotheses, irrespective of how many hypotheses are true and what values the parameters of the false hypotheses take. α, the Type I error rate is often replaced by the FWE rate. Let R denote the number of hypotheses rejected, and let V denote the number of true hypotheses erroneously rejected. Let then

$$Q = \begin{cases} V/R, & \text{if } R > 0; \\ 0, & \text{if } R = 0. \end{cases}$$

(The interpretation of $Q = 0$ for $R = 0$ is that no error of false rejection is committed in this case.) The FDR, τ_{FDR}, is then defined as

$$\tau_{FDR} = E\{Q\} = E\{V/R\},$$

while the FWE rate, τ_{FWE}, is simply

$$\tau_{FWE} = P\{V \geqslant 1\}. \tag{5.120}$$

Note that τ_{FDR} is intended to be bounded by some β and

$$\begin{aligned}\tau_{FDR} &= E\{Q\} = 0.P\{R=0\} + E\{(V/R)I(R \geqslant 1)\}\\ &= E\{V/R)I(R \geqslant 1)\} = E\{(V/R)I(V \geqslant 1, R \geqslant 1)\}\\ &\leqslant E\{I(V \geqslant 1, R \geqslant 1)\} \leqslant E\{I(V \geqslant 1)\} = P\{V \geqslant 1\} = \tau_{FWE}.\end{aligned}$$

Motivated by (5.121) and Theorem 5.5.1, Benjamini and Hochberg (1995) considered the following MTP: Let K be the largest $j(\leqslant k)$ for which $L_{k:j} \leqslant \frac{j}{k}\beta$; take $K = 0$ if $L_{k:j} > (j/k)\beta \ \forall j \geqslant 1$. Then reject $H_{(1)}, \ldots, H_{(K)}$, where $H_{(j)}$ is formulated (as before) the hypothesis corresponding to the index (anti-ranks) r_j for which $L_{r_j} = L_{k:j}, j = 1, \ldots, k$, and $\beta(0 < \beta < 1)$ is a preassigned level ($\beta < \alpha$, by (5.121)). Though this procedure controls the FDR at level β, by (5.121), its FWE rate may be greater than β, and hence, we need to assign $\beta < \alpha$.

Benjamini and Hochberg (1995) appraised some *step-up* (and step-down) procedures wherein one starts with the largest (and smallest) OSL values to smaller (and larger) values and examine the relative values of the $L_{k:j}$ and $\frac{\beta j}{k}, 1 \leqslant j < k$. For some other modification we may refer to Benjamini and Liu (1999).

A central problem in this set-up is to incorporate (5.119) in a more general and flexible manner. Sen (1999a) posed the following result that covers a general set-up pertaining to all tests of the Simes'-type.

Theorem 5.5.3 *For every $k(\geqslant 1)$ and $\boldsymbol{a}_k = (a_{k1}, \ldots, a_{kk}) : 0 < a_{k1} \leqslant \cdots < a_{kk} < 1$,*

$$P\{\boldsymbol{a}_k\} = P\{U_{k:j} \geqslant a_{kj}, 1 \leqslant j < k\} = k!\mathcal{H}_{kk}(\boldsymbol{a}_k, 1),$$

where for each $j(= 1, \ldots, k)$ and $u \in (a_{kj}, 1)$,

$$\begin{aligned}\mathcal{H}_{kj}(\boldsymbol{a}_k, u) &= \int_{a_{kj}}^{u} \int_{a_{k(j-1)}}^{u_j} \cdots \int_{a_{k1}}^{u_1} du_1, \ldots du_j\\ &= \int_{a_{kj}}^{u} \mathcal{H}_{k(j-1)}(g_k, u_i) du_j, 1 \leqslant j \leqslant k, \end{aligned} \tag{5.121}$$

and $U_{k:1} < \cdots < U_{k:k}$ are the uniform order statistics. [Conventionally, we let $\mathcal{H}_{k0}(\boldsymbol{a}_k, u) = I(u > a_{k1})$.]

Proof: Note that for each $k(\geqslant 1), U_{k:1} < \cdots < U_{k:k}$ are the ordered $r : v$'s corresponding to the *iid* random variables $U_1, \ldots, U_k$, having the uniform distribution on [0,1]. Therefore the joint density of $(U_{k:1}, \ldots, U_{k:k})$ is given by

$$k!I(0 < u < \cdots < u_k < 1). \tag{5.122}$$

Further, note that $\{u_j \geqslant a_{kj}, j = 1, \ldots, k\}$ is a subset of the cone $\{0 < u_1 < u_2 < \cdots < u_k < 1\}$. If we let

$$S_{kj} = \{u_l \geqslant a_{kl},\ l \leqslant j\}, j = 1, \ldots, k, \tag{5.123}$$

then we may note that $S_{kj} \subseteq S_{k(j-1)}, \forall j \geq 1$. Then (5.121) follows from (5.122) and (5.123); this in turn implies (5.118). ∎

It may be noted that generally the a_{kj} depend on k as well as j, so that the $\mathcal{H}_{kj}$ even though dependent only on the $a_{kr}, r \leqslant j$, may depend on k. These recursive relations are helpful for actual numerical evaluation of $P\{\boldsymbol{a}_k\}$ for specific $\boldsymbol{a}_k$. Theorem 5.5.3 can be incorporated in developing a general Simes'-type multiple testing procedure involving a *spending function* approach which pertains to an exact overall significance level and, at the same time, allows more flexibility in the choice of the a_{kj} (Sen 1999b). For example, for $k = 2$, we have

$$
\begin{aligned}
P\{a_{21}, a_{22}\} &= 2\mathcal{H}_{22}(\boldsymbol{a}_2, 1) = 2\int_{a_{22}}^{1} \mathcal{H}_{21}(\boldsymbol{a}_2, u)du \\
&= 2\{(1-a_{21})(1-a_{22}) - \frac{1}{2}(1-a_{22})^2\} = (1-a_{22})(1-2a_{21}+a_{22}).
\end{aligned}
$$

Thus, if we let $a_{21} = 2^{-1}a_{22} = 2^{-1}\alpha$, then (5.120) is exactly equal to $1-\alpha$, so that a size α test can be based on the OSL values $L_{2:1} < L_{2:2}$. This is in agreement with both Simes (1986) and Hochberg (1988). The attainment of the exact level $1-\alpha$ may also be accomplished with some other combination of a_{21}, a_{22} for which the right-hand side of (5.120) is equal to $1-\alpha$. Such procedures can be adapted better to multiple hypotheses testing problems.

For $k \geqslant 2$, if we take

$$a_{kj} = a.m(j,k), j = 1, \ldots, k, a > 0,$$

where the $m(j,k)$ are assumed to be nonnegative constants (and nondecreasing in j), then by Theorem 5.5.3,

$$k!\mathcal{H}_{kk}(\boldsymbol{a}_k, 1) = \text{a polynomial of degree} k \text{ in } \boldsymbol{a},$$

with coefficients that are functions of the preassumed $m(j,k)$. Therefore, we are at liberty to choose a spending function $m(j,k), j \leqslant k$ in such a way that (5.118) is appropriate in a particular context, and it leads to an attainment of the exact significance level α. Of course, such a spending function needs to be so chosen that the closure principle (Hochberg 1988, Hochberg and Rom 1995) holds for the subset hypothesis testing problem. Hochberg and Rom (1995) considered the LRH family (Shaffer 1986) and Theorem 5.5.3 remain pertinent for that as well. There is, however, a basic theoretical query: Can some of these multiple testing procedures (even for unrestricted alternatives) be justified on the ground of admissibility or minimaxity? As of now, the primary emphasis has been laid down on the attainment of overall significance level(or FWE rate) α and FDR for subfamilies of hypotheses; when it comes to the question of power properties, the alternative hypotheses may not allow a simple resolution. This is particularly true for restricted alternative hypothesis testing problems, and hence, we should appraise these concepts (of FWE, EDR, etc.) properly before routinely using them.

There is another aspect of MTP that deserves our attention. In many problems of practical interest, we encounter multiple tests of significance where k, the number of components, may not be small, and the alternatives may be restricted in the sense of our main treatise. For example, in a *clinical trial*, as *interim analysis* scheme may involve a *group-sequential methodology* relating to a treatment group versus a placebo group; the null hypothesis relates to the homogeneity of the two groups while the alternative relate to *treatment better than control*. Accumulating data sets at regular time intervals, in a follow-up scheme are incorporated, in a time-sequential set-up, to suit the interim analysis, and this *multiple look* needs a careful statistical appraisal (Sen 1999a). The test statistics at these time points are no longer independent and no optimality property of conventional tests may be easily established (or even be true). Although Theorem 5.5.3 can be extended to cover such set-ups, there remains the open question: Is there any optimality or desirable property of the Simes'-type tests in this context?

As has been explained in Sen (1999a), in a multiparameter setting, even in most simple parametric models, universally optimal MTP may not generally exist. The situation is worse with restrictive alternative hypotheses, as has been explained in earlier sections. As such, the choice of a MTP is largely guided by practical considerations and robustness cum validity perspectives (which tend to deemphasize FDR or other measures). In that way, the flexibility of choice of the $a_{kj} (1 \leqslant j \leqslant k)$, granted by Theorem 5.5.3, offers a variety of OSL-based tests. In this context, the Fisher method of combining independent tests has some (asymptotic) optimality properties, though in finite sample cases and for broader families of alternatives, this asymptotic optimality criterion might not be very appealing. It is well known that the (approximate) Bahadur slope that is usually employed (Hájek et al. (1999), Chapter 8) is insensitive to certain deficiency aspects, and may not lead to a finite sample coherence of the optimality property. For example, for two tests having asymptotic chi-square distributions with p_1 and p_2 degrees of freedom ($p_1 < p_2$) but having a common noncentrality will have the same Bahadur efficacy, although for the smaller degrees of freedom test, the power would be better than the other one.

Guided by robustness and power considerations, for large k, it might be appealing to choose a comparatively smaller value of $r (\leqslant k)$ and let either

$$a_{k1} < \cdots < a_{kr} = a_{k(r+1)} = \cdots = a_{kk}, \tag{5.124}$$

or

$$a_{k1} = \cdots = a_{kr} < a_{k(r+1)} < \cdots < a_{kk}, \tag{5.125}$$

depending on the type of alternatives. Sen (1999a,b) illustrated this aspect in the context of interim analysis. Under (5.124) or (5.125), the expressions for $\mathcal{H}_{kj}(\boldsymbol{a}_k m, u)$ simplify to a certain extent; we leave the details as Problem 5.36.

It may be noted that the above development relates to the case where under the null hypothesis, the L_j have all uniform (0,1) distribution and they are independent. The impact of discreteness of the null hypothesis distribution of the L_j as well as their plausible (stochastic) dependence need to be appraised. When L_j has a distribution admitting jump discontinuities (under H_0), these jump points as well as the jumps may depend on the underlying distributions, though the distribution is still defined on [0,1]. Thus, the uniform [0,1] distribution of the L_j is not the case, and moreover, the L_j may not even be identically distributed (under H_0). [In the case of test statistics having a null distribution with jump discontinuities, the L_j would have a discrete distribution on (0,1) with jump discontinuities depending on the underlying null distributions. Hence, Simes type of results resting on a continuous uniform distribution assumption may not be tenable.] As a result, Theorems 5.5.1–5.5.3 may not hold. For this reason, often randomized test statistics are used to render the resulting OSL having a continuous uniform distribution (on [0,1]); this enables Theorems 5.5.1–5.5.3 to be applicable. We leave Problem 5.37 to illustrate this point.

Let us make some comments on the dependent case where the test statistics (and hence, the OSL L_j) for the component hypotheses are not stochastically independent (even under H_0). Sans the stochastic independence of the L_j, Theorems 5.5.1 - 5.5.3 may not be generally true. If the L_j are independent and identically distributed then the $L_{k:j} (1 \leqslant j \leqslant k)$ are (pair-wise) positively associated. This result may not be generally true (as may be verified by considering the case where $L_1, \ldots, L_k$ are exchangeable (symmetric dependent) with a *negative dependence* pattern. We refer to Sarkar and Chang (1997) where the Simes method has been extended to positively dependent test statistics; Sarkar (1998) incorporated an MTP_2 property to provide a proof of the Simes' conjecture for such positively dependent test statistics.

5.6 CONCLUDING REMARKS

This chapter has been specifically devoted to an appraisal of likelihood principle and union-intersection principle, along with other variants, for order-restricted statistical inference in a multiparameter set-up. In this context, attention has mainly been paid to continuous distributions relating to suitable test statistics or estimators. *Qualitative data models* are also often encountered in live applications. Somewhere in between purely qualitative and continuous response models are the so called *ordered categorical data* models. We shall discuss such models in a latter chapter, and in that context too, the LRP and UIP will be appraised.

Even for the most simple multiparameter (parametric) models, for restricted alternative (including inequality constraints and ordered restrictions), the *LRT* or its variants may not be *admissible* while their modified versions may be too conservative. Conditional likelihoods and marginal likelihoods are often used to capture the LRP, albeit in a partial way, with some loss of information. In such a situation, the UIT may have some advantages. This feature deserves an in depth appraisal (in a more general set-up), and some of these theoretical foundations are relegated to the Appendix.

In the context of bioinformatics, genomics and environmetrics, typically an enormously large dimensional data model is encountered. This leads to the curse of dimensionality problem in statistical inference. For order (or inequality)-restricted statistical inference, this curse of dimensionality may be even much more complex, specially when linear statistical inference may not be that appropriate; this is particularly the case in genomics, computational biology, and bioinformatics, in general. In this set-up, we not only have an exceedingly high-dimensional data models but also the response variables are (often purely) qualitative in nature. Although *generalized linear models* (GLM) and *generalized additive models* (GAM) have been commonly used for such models, the usual (linear) dimension-reduction tools (such as *canonical analysis, principal component models, projection pursuit, neural network* and *classification* and *regression trees*) are of very limited help. As a result, the classical LRP (or its variants) may not be appropriate from conceptual as well as applicational point of view. *Data mining* (or more fashionably, *knowledge discovery* and *data mining* (KDDM)) tools, though quite appealing from data analysis perspectives, may lack sound statistical reasoning (and interpretations too). The UIP may be based on some other considerations (without putting too much emphasis on the likelihood), and thereby may have some advantages. We intend to cover some aspects of these recent developments in a later chapter.

Problems

5.1

(a) If $\boldsymbol{X} \sim N_p(\boldsymbol{\mu}, \boldsymbol{\Sigma})$, show that given $\boldsymbol{X}_{(j)} = (X_1, \ldots, X_j)^T = \boldsymbol{x}, X_{j+1}$ is normally distributed with mean $\mu_j = \boldsymbol{\beta}_{j+1}^T(\boldsymbol{x} - \boldsymbol{\mu}_{(j)})$ and variance $\sigma_{(j+1)(j+1)\cdot 12..j}$ where $\boldsymbol{\mu}_{(j)} = (\mu_1, \ldots, \mu_j)^T$, $\boldsymbol{\beta}_{j+1} = \boldsymbol{\Sigma}_{(j)}^{-1}\boldsymbol{\sigma}_{j+1}, \boldsymbol{\Sigma}_{(j)} = (\sigma_{rs})_{r,s=1,\ldots,j}, \boldsymbol{\sigma}_{j+1} = (\sigma_{1(j+1)}, \ldots, \sigma_{j(j+1)})^T$ and $\sigma_{(j+1)(j+1)\cdot 12..j} = \sigma_{(j+1)(j+1)} - \boldsymbol{\sigma}_{j+1}^T\boldsymbol{\Sigma}_{(j)}^{-1}\boldsymbol{\sigma}_{j+1}$.

(b) Hence, or otherwise show that $Y_{j+1} = X_{j+1} - \beta_{j+1}^T\boldsymbol{X}_{(j)}, 1 \leq j \leq k-1, Y_1 = X_1$ are stochastically independent.

(c) Consider now the null hypothesis $H_{0(j+1)}^* = \cap_{i=1}^{j+1} H_{0i} (0 \leq j \leq p-1)$. Also, let $\boldsymbol{X}_n$ be the $n \times p$ observation matrix whose rows are *iid* $N(\boldsymbol{\mu}, \boldsymbol{\Sigma})$ vector. Then, by (a), under $H_{0(j+1)}^*$, given $\boldsymbol{X}_n^{(j)}$, the first j columns of $\boldsymbol{X}_n$, (noting that $\mu_{j+1} - \boldsymbol{\beta}_{j+1}^T\boldsymbol{\mu}_{(j)} = 0$; μ_{j+1} the MLE of $\boldsymbol{\beta}_{j+1}$ and $\sigma_{(j+1)(j+1)\cdot 12..j}$ (conditional on $X_n^{(j)}$) are the sample counterparts involving $\boldsymbol{S}_{n(j)}^{-1}\boldsymbol{s}_{nj+1} = \boldsymbol{b}_{j+1}$ and $\boldsymbol{s}_{(nj+1)(j+1)\cdot 12..j}$, and further with these $(j+1)$ parameters being estimated and $s_{(nj+1)(j+1).12..j}$, having the distribution $\sigma^2\chi_{n-j-1}^2/(n-j-1)$, the statistic

$$t_{j+1} = n(\bar{X}_{nj+1} - \boldsymbol{b}_{j+1}^T\bar{\boldsymbol{X}}_{n(j)})/\sqrt{s_{(nj+1)(j+1).12..j}}$$

has the student's t-distribution with $n - j - 1$ degrees of freedom.

(d) Show that this conditional law does not depend on $\boldsymbol{X}_n^{(j)}$, and, hence, unconditionally, under the null hypothesis, t_{j+1} has the student t-distribution with $n - j - 1$ DF.

(e) Invoke the orthogonality result in (b) to show that under $H_{0p}^* (\equiv H_0), t_1, \ldots, t_p$ are independently distributed as student's t-distribution with DF = $n - 1, \ldots, n - p$, respectively.

(f) Appraise the difficulties of testing H_0 against one-sided alternatives ($\boldsymbol{\mu} \geq \mathbf{0}$ based on $t_1, \ldots, t_p$).

5.2 Let $\boldsymbol{X}_1, \ldots, \boldsymbol{X}_n$ be *iid* $N_2(\boldsymbol{\mu}, \boldsymbol{\Sigma})$ r-vectors. We want to set a confidence interval for $\boldsymbol{\mu}$, subject to the constraint that $\boldsymbol{\mu} \geq 0$. Let $\bar{\boldsymbol{X}}_n$ and $\boldsymbol{S}_n$ be the sample mean vector and covariance matrix, and define the Hotelling T^2-statistic by $T_n^2 = n(\bar{\boldsymbol{X}}_n - \boldsymbol{\mu})^T \boldsymbol{S}_n^{-1} (\bar{\boldsymbol{X}}_n - \boldsymbol{\mu})$. Since $P\{T_n^2 \leq T_{n,\alpha}^2\} = 1 - \alpha$, we have

$$P_{\boldsymbol{\mu}}\{(\boldsymbol{\mu} - \bar{\boldsymbol{X}}_n)^T \boldsymbol{S}_n^{-1} (\boldsymbol{\mu} - \bar{\boldsymbol{X}}_n) \leq n^{-1} T_{n,\alpha}^2\} = 1 - \alpha.$$

This specifies an ellipse with origin $\bar{X}_n$. Whenever $\boldsymbol{\mu}$ is close to the vertex $\mathbf{0}$ or an edge $(\mu_1 = 0, \mu_2 = 0)$, $\bar{\boldsymbol{X}}_n$ will have concentration around $\boldsymbol{\mu}$, and, hence, part of the confidence ellipse as obtained above will lie outside the positive orthant. As a result, the coverage probability would be less than the prefixed level $1 - \alpha$. This deficiency depends on the realized value of $\bar{\boldsymbol{X}}_n$ and $\boldsymbol{S}_n$, and it is also possible that $\bar{\boldsymbol{X}}_n$ itself lies outside the positive orthant (with a positive probability), so that the deficiency could be even more. Illustrate this deficiency picture with various position of $\bar{\boldsymbol{X}}_n$ (through the underlying values of $\boldsymbol{\mu}$) and comment on the sensitivity of the coverage probability on $\boldsymbol{\mu}$ itself. For the unrestricted case, the coverage probability is equal to $1 - \alpha$, for all $\boldsymbol{\mu} \in \mathbb{R}^2$.

5.3 Make use of Problem 5.1 and verify that in the bivariate case, for a step-down procedure, the coverage probability for $\boldsymbol{\mu}$ may depend on β_{21} being positive or negative.

5.4 Note that for the paired differences $\theta_j - \theta_{j-1}, 2 \leq j \leq k$, the sample estimates $\bar{X}_j - \bar{X}_{j-1}, 2 \leq j \leq k$, are negatively dependent. As a result, though marginally $t_2, \ldots, t_k$ have student's t-distributions, they are negatively dependent. For a multi $(k - 1)$-variate normal distribution with mean $\mathbf{0}$ and covariance matrix $\boldsymbol{\Gamma}$ having all off-diagonal elements nonpositive, show that the positive orthant probability is less than $2^{-(k-1)}$, and an opposite inequality holds when all the elements of Γ are nonnegative. [Hint: Start with $k = 3$, and prove by induction.] Define a multivariate t-distribution as in Johnson and Kotz (1972) Show that for $k = 3$, a similar result holds for nonpositively correlated t-statistics. Appraise the general case of $k \geq 3$.

5.5 Define a multivariate $\boldsymbol{t}$-pdf as

$$p_t(\boldsymbol{y}) = \frac{\Gamma((\nu + p)/2)}{(\pi\nu)^{(p/2)}\Gamma(p/2)|\boldsymbol{R}|^{1/2}}(1 + \nu^{-1}\boldsymbol{y}^T \boldsymbol{R}^{-1} \boldsymbol{y})^{-(p+\nu)/2},$$

where $\boldsymbol{R}$ is the correlation matrix for the p t-statistics. Side by side, place the p marginal densities and hence, verify that if $\boldsymbol{R}$ is not equal to $\boldsymbol{I}$, then the t-statistics are not independently distributed. The difficulties for tabulating the critical values of the multivariate t-statistics have been thoroughly discussed in chapter 5 of Johnson and Kotz (1972). Verify that as ν increases, the multivariate t-distribution converges to a multivariate normal one with dispersion matrix $\boldsymbol{R}$.

5.6 Use the results in Problem 5.4 and appraise the exact testing problem with nonpositive dependence in mind. Asymptotically, however, this problem reduces to that of a multinormal orthant mean problem, and hence, outline the asymptotic solution. Use the asymptotic test statistic, and for small sample sizes, discuss how the permutation principle leads to a EDF test.

5.7 Show that the permutation distribution of J_n depends on the combined sample observations only through their ranks, and hence, it is also the (unconditional) null distribution. Apply then the permutational central limit theorem and verify that if the relative sample sizes (i.e., $n_i/n; n = \sum_{j=1}^{C} n_j$) are all bounded away from zero (and one), the conditions for the PCLT hold, and thereby the asymptotic normality holds.

[Hint: Consider a linear rank statistic $L_n = \sum_{i=1}^{n} a_{ni} b_{nR_i}$ where the a_{ni}, b_{ni} are real numbers, and $R_1, \ldots, R_n$ takes on all permutations of $1, \ldots, n$ with the common probability $(n!)^{-1}$. Compute the mean and variance of L_n under this permutation law. Then rewrite the standardized form as $L_n^* = (L_n - EL_n)/\sqrt{V(L_n)} = \sum_{i=1}^{n} c_{ni} d_{nR_i}$ where $c_{ni} = (a_{ni} - \bar{a}_n)/A_n, d_{ni} = (b_{ni} - \bar{b}_n)/B_n,\ i = 1, \ldots, n$ and $\bar{a}_n$ and $\bar{b}_n$ are the averages of the a_{ni} and b_{ni}, respectively, $A_n^2 = \sum_{i=1}^{n} (a_{ni} - \bar{a}_n)^2$ and $B_n^2 = (n-1)^{-1} \sum_{i=1}^{n} (b_{ni} - \bar{b}_n)^2$. A primitive way to establish the PCLT is to show that $E(L_n^*)^{2k} \to (2k)!/\{2^k \Gamma(k)\}$ and $E(L_n^*)^{2k+1} \to 0$, as $n \to \infty$, for every $k \geq 1$. In the present case, this treatise works out well, as both the sequences a_{ni}, b_{ni} have bounded elements. For a more general version of this PCLT, due to Hájek (1961), we refer to Hájek et al. (1999).]

5.8 Show that if a statistic L_n^* is asymptotically normal with zero mean and unit variance, under a hypothesis H_0, and if we have a sequence of alternatives $H_{(n)}$ that are contiguous to H_0, then L_n^* is asymptotically normal with unit variance and mean that can be directly computed by incorporating the contiguity.

[Hints: Consider two sequences of densities $\{p_n\}, \{q_n\}$, defined on measure spaces $(\mathcal{X}_n, \mathcal{A}_n, \mu_n)$, $n \geq 1$. If for any sequence of events $A_n : A_n \in \mathcal{A}_n$, $[P(A_n \mid p_n) \to 0]$ implies that $[P(A_n| \mid q_n) \to 0]$, we say that the densities q_n are contiguous to p_n. The celebrated LeCam's (second) lemma asserts that if under p_n-measure, the likelihood ratio L_n is asymptotically normal with mean $\mu = -\frac{1}{2}\sigma^2$ and variance σ^2, then the densities q_n are contiguous to p_n. Further, if for another statistic S_n, under H_0, $(S_n, \log L_n)$ is asymptotically bivariate normal $(\mu_1, \mu_2; \sigma_1^2, \sigma_2^2, \sigma_{12})$ with $\mu_2 = -\frac{1}{2}\sigma_2^2$, then under $H_{(n)}$, $S_n - \xi_n$ is asymptotically normal $(\mu_1 + \sigma_{12}, \sigma_1^2)$ (Hájek et al. (1999), Ch. 7). Use the above contiguity result to establish the asymptotic normality of J_n under contiguous alternatives.]

5.9 Show that the intra-block rank vectors are independent for different blocks. Express both J_n and $P_n(1)$ as average over independent within block rank statistics, and hence, apply the classical central limit theorem to establish their asymptotic normality. Show that by virtue of the aforesaid independence of the intrablock ranks, the asymptotic normality result holds with different normalizing factors even if the null hypothesis is not true, and even under noncontiguous alternatives.

5.10 Show that the aligned ranks, given the intra-block order statistics (of the aligned observations), are conditionally independent for different blocks. Exploit this conditional independence and establish the asymptotic normality result for S_N^0 in (5.74) under the null hypothesis.

[Hints: You may need to show that the conditional variance asymptotically converges to a positive constant, in probability, and then proceed from the conditional to the unconditional null distribution, under H_0.] [Sen (1968)]

5.11 For contiguous (shift) alternatives, use the results in Problem 5.8 and verify that the asymptotic normality of S_N^0 holds. Hence, or otherwise, obtain the asymptotic mean of this normal distribution, and find out the efficacy of the test best on S_N^0. Compare the efficacy results of all the three statistics P_n, J_n, and S_N^0, and show that for normal, logistic, double-exponential and some other densities, S_N^0 has higher asymptotic relative efficiency. [Sen 1968]

5.12 Consider the two-sample location model, write down the likelihood function, based on the ranks, instead of the observations themselves. For local alternatives, expand the loglikelihood (rank) statistic, and verify that a LMPR test statistic is a linear rank statistic based on the Fisher score generating function. Next, consider all possible pairs of samples from the C samples, and use the LMPR test statistics, from each pair, along with the UPI to construct a suitable overall test for the ordered alternative problem. [Hints: You may need to extend the results in Theorem 1 (p.71) in Hájek et al. (1999) wherein the pair-sample ranks are to be replaced by the overall ranks.]

5.13 Consider a signed-rank statistic $T_n(b) = \sum_{i=1}^{n} sign(X_i - b)a_n(R_i^+(b))$ where the scores $a_n(k)$ are monotone (in $k : 1 \leq k \leq n$) and $R_i^+(b)$ is the rank of $|X_i - b|$ among the $|X_j - b|, j = 1, \ldots, n$. For any given b, divide the set of observations into two subgroups for which the $X_i - b$ are positive or negative, so that $T_n(b)$ can be expressed as the difference of two rank statistics, say, $T_{n1}(b)$ and $T_{n2}(b)$, where $T_{n1}(b)$ is nonincreasing in b while $T_{n2}(b)$ is nondecreasing. Hence, show that $T_n(b)$ is monotone nonincreasing in b.

5.14 Invoke the stochastic independence of the vector of signs and the vector of absolute ranks, and first, conditioning on the absolute ranks, show that the multivariate central limit applies to $\boldsymbol{T}_N$ (i.e., to any arbitrary linear combination of the elements of $\boldsymbol{T}_n$). Further, use the stochastic convergence of $\boldsymbol{V}_n$ and verify that the result of the previous problem extends to its unconditional null distribution too.

5.15 (Continuation of Problem 5.12). For normal, logistic, Laplace and exponential densities, obtain the LMPR test statistics under the two-sample location model. What can you say about the multivariate case?

5.16 Define the rank-covariance matrix $\boldsymbol{V}_n$ as in (5.90). Express the R_{ij} in terms of the empirical distribution function for the jth marginal, and the scores $a_{nj}(k)$ in terms of some square integrable score generating functions $\phi_j(.)$. Invoke the a.s. convergence of the empirical d.f. in the multivariate case (in the sup-norm) and show that $\boldsymbol{V}_n$ stochastically converges to a positive definite matrix $\boldsymbol{\nu}$, where the positive definiteness holds when the score functions for a vector are linearly independent, in probability.

5.17 Show that the permutation covariance matrix $\boldsymbol{V}_n$ in the previous problem is invariant under any column-permutation of the rank-collection matrix. Thus, it is invariant under $\mathcal{P}_n$. [Puri and Sen (1971, Chapter 5)]

5.18 Generate a set of $n!$ equally likely column-permutations of the reduced rank-collection matrix, and with respect to this conditional (discrete uniform) measure, show that $\boldsymbol{T}_n$ has asymptotically, in probability, a multinormal distribution, with dispersion matrix $\boldsymbol{V}_n$.

5.19 Use the results in the previous two problems, and the KTL-point formula to show that $\mathcal{L}_{n_1,n_2}$ in (5.91) has a permutation distribution (under $\mathcal{P}_n$) that converges, in probability, to the chi-bar distribution, derived in Chapter 3.

5.20 Consider the multivariate simple linear regression model, so that under H_0 : $\boldsymbol{\beta} = 0$, the vector of linear rank statistics has a permutation distribution that converges (in probability) to a multinormal law with null mean vector and dispersion matrix $\boldsymbol{V}_n C_n^2$. Hence use the result in Problem 5.17, and conclude on the unconditional null distribution.

5.21 Use the KTL-point formula and the results of the previous problem and verify that the asymptotic null distribution of $\mathcal{L}_n^*$ in 5.96 is the chi-bar distribution defined in Chapter 3.

5.22 For densities having finite Fisher information, invoking contiguity and the asymptotic under the null hypothesis (previous Problems), derive the asymptotic distribution under contiguous alternatives.

5.23 If the score functions are monotone and square integrable, show that the existence of the finite Fisher information clause can be avoided for the derivation of the asymptotic distribution theory of linear rank statistics. As such, for the Wilcoxon scores statistics, obtain the asymptotic distribution of $\mathcal{L}_n^*$, for local Pitman-type alternatives, sans contiguity.

5.24 For the univariate linear model, when σ^2 is unknown, show that the results in Chapters 3 and 4 extend for the regression parameter vector for orthant and other restricted alternatives. What happens if we drop the assumption of underlying normality (of errors)? Can we still claim that asymptotically we have the chi-bar distribution? Examine its connection with the multivariate one-sample orthant alternative problem, sans the multinormality assumption.

5.25 Invoke contiguity to extend the asymptotic distribution of rank statistics for the multiple regression model to contiguous alternatives, and hence, discuss the asymptotic distribution of $\mathcal{L}_n^o$ in (5.105) under contiguity. [Puri and Sen (1985, Ch.7)]

5.26 Extend the results in Problem 5.4.19 to the multivariate linear model, and show that if $\boldsymbol{\Sigma}$ is unknown, as is usually the case, the normal theory *LRT* has the same difficulties as in the case of the orthant alternative mean problem.

5.27 Show that the matrix of linear rank statistics, in the multivariate linear model, under the null hypothesis of no regression, has a permutation (conditional) distribution (given the reduced rank-collection matrix), which converges, in probability, to a multinormal law. [You may need to use the convergence result in Problem 5.17 in this context.]

5.28 Invoke contiguity, and extend the results in the previous problem to asymptotic distribution theory for contiguous alternatives. Provide expressions for the elements of $\mathbf{\Gamma}$ in (5.105).

5.29 Consider the multivariate linear model in (5.108), roll out $\boldsymbol{\beta}$ into a pm vector, and partition that into two subvectors $\boldsymbol{\beta}_{(1)}$ and $\boldsymbol{\beta}_{(2)}$ of a and $pm - a$ components, respectively; a being a positive integer less than pm. The orthant problem consists of testing for $H_0 : \boldsymbol{\beta} = \mathbf{0}$ vs $H_1^+ : \boldsymbol{\beta} \neq \mathbf{0}, \boldsymbol{\beta}_{(1)} \geq \mathbf{0}$. Show that H_1^+ can be equivalently posed as $\boldsymbol{\beta} : \boldsymbol{A\beta} \geq \mathbf{0}$ where $\boldsymbol{A} = (\boldsymbol{I}_a, \mathbf{0}_{pm-a})$. Use (5.111), (5.112) and the UIP to obtain a UIT, and show that it has the same asymptotic null distribution as the corresponding restricted likelihood ratio test. Comment on the robustness perspectives of the two tests. Show that the ordered alternative problem can be reduced to the above orthant alternative problem, and a similar rank test applies.

[Chinchilli and Sen 1981 a, b].

5.30 In the set-up of the Problem 5.29, consider a profile analysis model where the alternative hypothesis is specified by some equality constraints (instead of the inequalities that define the orthant /ordered alternatives), then show that the UIT statistic has under the null hypothesis asymptotically chi square distribution with degrees of freedom equal to mp - number of equality restraints. Also, show that under local alternatives, it has a noncentral chi square distribution with the same d.f. and appropriate noncentrality parameter.

[Chinchilli and Sen 1982]

5.31 Let $U_1, \ldots, U_k$ be k *iid* r.v.'s having the uniform(0,1) distribution function. Then show that $P\{U_{k:1} > u\} = [P\{U_1 > u\}]^k = (1-u)^k$.

5.32 Show that for every $m : 1 \leq m \leq k, P\{U_{k:m} \leq u\} = P\{m$ or more of the $U_j, 1 \leq j \leq k$, are $\leq u\}$ is equal to $\sum_{r=m}^{k}(k!/\{r!(k-r)!\}u^r(1-u)^{k-r}$. Hence, use the relationship of binomial tail and incomplete beta functions, and verify Wilkinson's result.

5.33 Use the result in the previous problem, and verify the F-distribution result for the odds-ratio $(1 - L_{k:m})/L_{k:m}$.

5.34 Show that for U, a uniform (0,1) r.v., $-2\log U$ has the exponential distribution, which is the same as a chi square with two DF. Hence, use the reproductive property of chi-squared distribution, and show that the Fisher statistic has under the null hypothesis a chi squared law with $2k$ DF.

5.35 Using the inverse normal transformation on U in the previous problem, show that Liptak statistic has a normal law under the null hypothesis. In the same way, use the gamma distributional reproductive property to show that the Lancaster statistic has a gamma distribution under the null hypothesis.

5.36 Define LRH family as in Hochberg and Rom (1995), and show that Theorem 5.5.3 pertains to this family as well.

5.37 Use (5.121) in Theorem 5.5.3, and use the two special cases considered in (5.124) and (5.124), and obtain by induction the expression for $H_{kk}(\boldsymbol{a}_k, 1)$, in both the special cases. You may start with $k = 2, 3$, and then proceed by induction.

5.38 Consider some nonparametric test statistics where under the null hypothesis, they have exact but discrete distributions generated by the equally likely permutations of the observations. Hence, show that the p-values do not have the uniform $(0, 1)$ distribution (under the null hypothesis), as postulated. If the sample size is large, then the PCLT eliminates this problem. On the other hand, for moderate to small sample sizes, examine the impact of this discreteness on all the results considered in Section 5.5. It is possible to render continuity by randomization, but that results in loss of power. On the other hand, the discrete distribution has less variability, and hence, by resampling methods, such as the bootstrap, a better result can be obtained for the test statistics considered in Section 5.5. Nevertheless, the Simes type of theorems may not stand valid for small sample sizes. Show that the problem reduces to that of discrete order statistics where ties cannot be neglected.

5.39 Consider the special case in (5.124). Then show that $P\{\boldsymbol{a}_k\}$ reduces to $k^{[r]} \int_{a_{kr}}^{1} H_{k(r-1)}(u)(1-u)^{k-r} du$. Show that a similar case holds for the configuration (5.125).

5.40 If the L_j have discrete distribution on $[0, 1]$ with mass points at a denumerable set of points, the uniform $[0, 1]$ distribution fails to capture the distortion caused by these mass points, and hence, Theorem (5.5.1)–(5.5.3) may not generally hold. Verify this with the simple case of $k = 2$ with L_1 and L_2 independent but not having the continuous uniform $[0, 1]$ distribution.

5.41 If the L_j are not independent, the simple law for the order statistics may not hold. If the L_j are exchangeable then there are some simplification. However, if for example, the L_j are pairwise positively dependent then $P\{L_j > x, L_l > y\}$ is $\geq P\{L_j > x\}P\{L_l > y\}, \forall x, y \in [0, 1]$. This inequality sign in general distorts the identity in Theorem (5.5.3). Again, verify this with $k = 2$ and L_1, L_2 positively dependent. Examine the case where they are negatively dependent.

6

Analysis of Categorical Data

6.1 INTRODUCTION

Categorical data arise in may areas of applications such as biology, economics, engineering, medicine, and social sciences. There are two distinct types of categorical data: nominal and ordinal. In this chapter our main interest would be on studies in which at least one variable is ordinal. For example, one may be interested to compare a treatment with a control where the response variable is recorded as {Excellent, Very good, Good, Poor, Very Poor}. In such a study, the objective may be to establish that treatment is better than the control. In general this is formulated as a hypothesis testing problem involving inequality constraints on several parameters; such constraints may appear in the null or alternative or both hypotheses. The categorical structure may arise through, for example, (a) a multinomial distribution characterized by the vector of cell probabilities $(\pi_1, \ldots, \pi_c)$ where $\pi_1 + \ldots + \pi_c = 1$, or (b) a dose response model in which the probability of success π_i, where i is the dose level, may be a nondecreasing function of dose, or (c) a contingency table arising from a comparison of r treatments in which the response variable is a c-category ordinal variable. In the foregoing dose-response model, the index i may correspond to a cell in a factorial design. This chapter develops the theory in a unified manner for statistical inference problems arising in the analysis of categorical data when there are order restrictions/inequality constraints on parameters. As in the previous chapter, we shall deal with mainly hypothesis testing problems.

Let us write the null and alternative hypotheses as

$$H_0 : \boldsymbol{\theta} \in \Theta_0 \text{ and } H_1 : \boldsymbol{\theta} \in \Theta_1,$$

respectively, where $\boldsymbol{\theta}$ is a vector parameter. For every context envisaged in this chapter, the likelihood is the product multinomial or Poisson probability functions. Further, the loglikelihood is smooth and concave in $\boldsymbol{\theta}$. Consequently, it follows from the results in Chapter 4 that the asymptotic null distributions of *LRT*, score and Wald-type test statistics belong to the chi-bar-square family. This simplifies matters considerably because we do not have to derive the asymptotic null distributions of the test statistics separately for each problem. Another important consequence of the loglikelihood being a concave function of the parameter is that it is fairly easy to compute the maximum of the loglikelihood subject to linear or nonlinear inequality constraints. For example, software packages/libraries such as IMSL, GAUSS, and MATLAB have efficient computer programs. For most practical applications, the programs in these are likely to be more than adequate. We shall not discuss details relating to efficient computational algorithms.

Once we have derived the asymptotic null distributions, we encounter some practical difficulties in applying them. First of all, there is the usual issue of whether or not the sample size is large enough to rely on asymptotic results. Another difficulty that we encounter is that the distribution of the test statistic usually depends on the null parameter value. The usual method in such cases is to compute the p-value with respect to the least favorable null distribution. However, it is not always easy to find the least favorable null distribution. Fortunately, for most of the situations encountered in this chapter on the analysis of categorical data, the least favorable null value/distribution is known. However, a typical feature of such distributions that has been a concern is the fact the least favorable null values are usually located at the boundary of the parameter space. Consequently the p-value computed at the least favorable null value tends to be conservative. One possible way of addressing this problem, at least theoretically, is to consider the least favorable distribution within a confidence region of the nuisance parameter (see Chapter 3). Although this has a sound theoretical basis, the computational aspects need further developments. Another approach to overcoming the nuisance parameter problem is to eliminate the nuisance parameters by conditioning on sufficient statistics, and apply *exact conditional tests.* Several such tests have been developed recently. For these tests, the suggested critical and p-values are valid for any sample size, and they do not depend on the nuisance parameter. It is possible that this is achieved at the cost of power, but at this stage nothing much is known about this. However, exact conditional tests are known to perform well in unconstrained inference. We would conjecture that the performance of the corresponding conditional procedure in constrained inference is likely to be good. In those situations for which exact conditional tests are available, it turns out that it is considerably easy to compute the p-value by simulation than to compute its exact value. Consequently, the computing time is unlikely to be a concern in most practical applications.

In Section 6.2 several data examples are presented to motivate and to provide an overview of the topics. In Section 6.3, inference from several independent binomial samples is discussed. Odds ratios and their relevance in formulating hypotheses in constrained inference are discussed in Section 6.4. Section 6.5 provides a discussion on the comparison of two treatments and the analysis of $2 \times c$ contingency tables.

These results are extended to incorporate more than two rows in Section 6.7. Marginal homogeneity is studied in Section 6.8. Exact conditional tests are discussed in Section 6.9. It will be noted that such exact conditional methods appear to be quite suitable irrespective of the sample size.

The likelihood functions that arise in the analysis of categorical data are smooth and concave, and hence are known to satisfy the regularity conditions such as Condition Q stated at the beginning of Chapter 4 (see page 146). Therefore, Proposition 4.3.1 is applicable, and hence we have a general result of the following form: Let $\boldsymbol{\theta}$ denote the model parameter; it may be a loglinear model or a more elaborate generalized linear model. Let the testing problem be

$$H_0 : \boldsymbol{\theta} \in \Theta_0 \quad \text{vs} \quad H_1 : \boldsymbol{\theta} \in \Theta_1$$

where $\Theta_0 \subset \Theta_1$. For every inference problem that we will encounter in this chapter, Θ_0 and Θ_1 are Chernoff regular at every point in Θ_0. Let T denote the likelihood ratio statistic, local score statistic, or any of their asymptotic equivalents such as W_D or W_H; see (4.103) and page 166. Then

$$T \xrightarrow{d} \|\boldsymbol{Z} - \mathcal{T}_0\|^2 - \|\boldsymbol{Z} - \mathcal{T}_1\|^2, \text{ under } H_0 \tag{6.1}$$

where

$$\mathcal{T}_0 = \mathcal{T}(\Theta_0; \boldsymbol{\theta}_0), \quad \mathcal{T}_1 = \mathcal{T}(\Theta_1; \boldsymbol{\theta}_0), \quad \boldsymbol{Z} \sim N(\boldsymbol{0}, \mathcal{I}_{\boldsymbol{\theta}_0}^{-1}), \quad \|\boldsymbol{x}\|^2 = \boldsymbol{x}^T \mathcal{I}_{\boldsymbol{\theta}_0} \boldsymbol{x},$$

and $\boldsymbol{\theta}_0$ is the assumed true value of $\boldsymbol{\theta}$ in Θ_0. Therefore, results such as the *LRT* of $\boldsymbol{R\theta} = \boldsymbol{0}$ against $\boldsymbol{R\theta} \geqslant \boldsymbol{0}$ is asymptotically chi-bar-square, follow directly from general results, and hence need not be established separately for each case; see Silvapulle (1991, 1994) for these results covering a large class of models including loglinear models for contingency tables, binomial response models including linear logistic regression models, limited dependent variable models, generalized linear models, Cox's regression model, proportional hazards models, and more generally quasi-likelihood and partial likelihood models. Therefore, we can by pass the asymptotics most of the time by appealing to such general results. Some of our tasks are made easier by the fact that the regularity conditions required for those results are the same as those required for the case when there are no inequality constraints, except that the parameter spaces must be Chernoff regular. Finally, note that the quantiles of the limiting distribution in (6.1) can be computed by simulation at any $\boldsymbol{\theta}$ in H_0.

6.2 MOTIVATING EXAMPLES

In this section, we consider several data examples, describe the context in which they arise, state a range of constrained inference problems that may arise in each of these examples, and discuss formulation of the inference problems in such a way that they can be answered using the theory developed in Chapters 3 and 4. The calculations to carry out statistical inference in some of these examples will be provided in later sections.

Example 6.2.1 *Independent binomial samples*

The results of a prospective study relating the occurrence of congenital sex organ malformation to maternal alcohol consumption are given in Table 6.1. Women completed a questionnaire early in their pregnancy concerning alcohol use in the first trimester. Later, the outcome of the pregnancy was recorded; the response variable is congenital sex organ malformation, recorded as present or absent.

Let π_i denote the probability of malformation corresponding to the ith level of alcohol consumption where $i = 1, \ldots, 5$ (see Table 6.1). Let us consider some statistical inference questions that may arise in this example and in similar ones with binomial probabilities.

(a) Is there any evidence of maternal alcohol consumption being related to malformation of sex organ? To answer this question, the null and alternative hypotheses may be formulated as

$$H_0 : \pi_1 = \cdots = \pi_5 \quad \text{and} \quad H_1 : \pi_1, \ldots, \pi_5 \text{ are not all equal,}$$

respectively. However, this formulation is unlikely to be appropriate because the main issue of interest is the possible increase in the probability of malformation with the increase in alcohol consumption.

(b) Is there any evidence that an increase in maternal alcohol consumption is associated with an increase in the probability of malformation? This question, as it stands, is quite broad to give a precise formulation of the null and the alternative hypotheses. Two possible formulations are given below.

$$(i) \qquad H_0 : \pi_1 = \cdots = \pi_5 \quad \text{and} \quad H_1 : \pi_1 \leqslant \cdots \leqslant \pi_5; \tag{6.2}$$

$$(ii) \qquad H_0 : \text{ not } H_1 \quad \text{and} \quad H_1 : \pi_1 < \cdots < \pi_5, \tag{6.3}$$

There is an important difference between the formulations in (6.2) and (6.3). Let us consider the formulation in (6.3), and suppose that a test rejects the H_0. Then we can conclude that there is sufficient statistical evidence to reject any value of $(\pi_1, \ldots, \pi_5)$ that does not satisfy $\pi_1 < \cdots < \pi_5$ and to accept the claim, $\pi_1 < \cdots < \pi_5$. Thus,

Table 6.1 Congenital sex-organ malformation relating to maternal alcohol consumption

	Alcohol consumption (Average number of drinks per day)				
	0	< 1	1-2	3-5	$\geqslant 6$
Number of malformations	48	38	5	1	1
Total number of births	17114	14502	793	127	38
Probability of malformation	π_1	π_2	π_3	π_4	π_5

there is statistical evidence to establish that maternal alcohol consumption is positively associated with the probability of malformation, and the data are inconsistent with any other possibility.

By contrast the same or a similar conclusion cannot be reached from the hypothesis testing problem (6.2), unless we make the *a priori* assumption that $\pi_1 \leqslant \cdots \leqslant \pi_5$ holds. To illustrate this, consider the testing problem (6.2), and suppose that a test rejects the null hypothesis. Then we can conclude that there is sufficient statistical evidence to support the claim that $\pi_1, \ldots, \pi_5$ are not all equal, but not to claim that $\pi_1 \leqslant \cdots \leqslant \pi_5$ holds with at least one strict inequality, even though this order restriction was the alternative hypothesis. However, if we make the *a priori* assumption that $\pi_1 \leqslant \cdots \leqslant \pi_5$ holds, then this assumption together with the statistical evidence to reject $H_0 : \pi_1 = \cdots = \pi_5$, leads to the conclusion that $\pi_1 \leqslant \cdots \leqslant \pi_5$ holds with at least one strict inequality and hence we would be able to accept that the probability of malformation increases with maternal alcohol consumption. It is important to note that this conclusion is based on the statistical evidence against $H_0 : \pi_1 = \cdots = \pi_5$ and the *a priori* assumption $\pi_1 \leqslant \cdots \leqslant \pi_5$. Such an assumption is not necessary if the problem is formulated as in (6.3).

(c) For illustrative purposes, let us state another type of questions that may arise, although it may not be relevant for the particular context of this example. Suppose that a scientific hypothesis that has been advanced is that the probability of malformation must be a nondecreasing function of the amount of maternal alcohol consumption. The question of interest is whether or not there is any evidence against this scientific hypothesis. In this case, the null and alternative hypothesis may be formulated as

$$H_0 : \pi_1 \leqslant \cdots \leqslant \pi_5 \quad \text{and} \quad H_1 : \text{Not } H_0,$$

respectively.

Example 6.2.2 *Comparison of two treatments when the response is ordinal*

The results of an experiment to compare two treatments for ulcer are given in Table 6.2. Let Y denote the response variable and X denote the explanatory variable; Y takes the ordinal values $1, \ldots, 4$ and $X = 1$ or 2 according as the Treatment is 1 or 2. For $i = 1, 2$ and $j = 1, \ldots, 4$, let

$$\pi_{ij} = pr(Y = j \mid X = i) \quad \text{and} \quad \boldsymbol{\pi}_i = (\pi_{i1}, \ldots, \pi_{i4})^T. \tag{6.4}$$

Table 6.2 Responses of two treatments for ulcer

	Change in the size of ulcer crater			
	Larger	<1/3 Healed	>2/3 Healed	Healed
Treatment 1 (Control)	12	10	4	6
Treatment 2 (Treatment)	5	8	8	11

The main interest concerns the possibility that the treatment is better than the control. There are several ways of formulating the statement "the treatment is better than the control." For the time being, let us adopt the following. Treatment 2 is *at least as good as* Treatment 1 if

$$\text{pr}(Y \geqslant j \mid \text{Treatment} = 2) \quad \geqslant \quad \text{pr}(Y \geqslant j \mid \text{Treatment} = 1) \qquad \text{for every } j. \tag{6.5}$$

This is equivalent to

$$\pi_{2j} + \cdots + \pi_{24} \geqslant \pi_{1j} + \cdots + \pi_{14} \quad \textit{for every } j.$$

Further, Treatment 2 is *better* than Treatment 1 if (6.5) holds with at least one strict inequality. Now let us consider some questions that would be of interest.

(a) Is there any evidence of differences between the two treatments? The null and alternative hypotheses for this may be formulated as $H_0 : \boldsymbol{\pi}_1 = \boldsymbol{\pi}_2$, and $H_1 : \boldsymbol{\pi}_1 \neq \boldsymbol{\pi}_2$, respectively. In this formulation, the ordinal nature of the response does not play a part. As was indicated earlier, this type of unconstrained testing problems will not be discussed here; excellent discussions may be found in Agresti (2002) and Lloyd (1999).

(b) Assuming that Treatment 2 is at least as good as Treatment 1 (i.e., (6.5) holds), is there any evidence to support the claim that Treatment 2 is better? In this case the null and the alternative hypotheses may be

$$H_0 : \boldsymbol{\pi}_1 = \boldsymbol{\pi}_2, \quad \text{and} \quad H_1 : \pi_{2j} + \cdots + \pi_{24} \geqslant \pi_{1j} + \cdots + \pi_{14}, j = 2, 3, 4,$$

respectively. If H_0 is rejected and H_1 is accepted then it follows that the inequalities in H_1 hold with at least one strict inequality, and hence we can conclude that Treatment 2 is better than Treatment 1.

(c) Is there any evidence to support the claim that Treatment 2 is better than Treatment 1, without making the *a priori* assumption that the former is at least as good as the latter? In this case, the null and the alternative hypotheses may be

$$H_0 : \boldsymbol{\pi} \notin A \quad \text{and} \quad H_1 : \boldsymbol{\pi} \in A$$

where $\boldsymbol{\pi} = (\boldsymbol{\pi}_1 : \boldsymbol{\pi}_2)$ $(= (\boldsymbol{\pi}_1^T, \boldsymbol{\pi}_2^T)^T$ and A is the set of all values of $(\boldsymbol{\pi}_1 : \boldsymbol{\pi}_2)$ for which Treatment 2 is considered to be better than Treatment 1; one possible choice is

$$A = \{(\boldsymbol{\pi}_1 : \boldsymbol{\pi}_2) : \pi_{2j} + \cdots + \pi_{24} > \pi_{1j} + \cdots + \pi_{14}, \ \text{for } j = 2, 3, 4\}.$$

Example 6.2.3 *Comparison of several treatments when the response is ordinal*

The result of a clinical trial regarding the outcome for patients who experienced trauma due to subarachnoid hemorrhage are given in Table 6.3. This set of data, appeared in Chuang-Stein and Agresti (1997); they were also analyzed in Agresti and Coull (2002) to illustrate constrained inference in the analysis of categorical data. For $i = 1, \ldots, 4$ and $j = 1, \ldots, 5$, let

$$\pi_{ij} = \text{pr}(\text{Outcome} = j \mid \text{Treatment} = i)$$

Table 6.3 Response from a clinical trial comparing four treatments on the extent of trauma due to subarachnoid hemorrhage

	Outcome				
Treatment	Death	Vegetative state	Major disability	Minor disability	Good recovery
Placebo	59	25	46	48	32
Low dose	48	21	44	47	30
Medium dose	41	14	54	64	31
High dose	43	4	49	58	41

Source: Dr. Christy Chuang-Stein, Pharmacia and Upjohn Company.

and $\boldsymbol{\pi}_i = (\pi_{i1}, \ldots, \pi_{i5})^T$. Note that the rows and columns in Table 6.3 are ordinal variables, and the main interest is on some form of positive dependence between rows and columns. More specifically, we are interested in the possibility of higher dose being more effective. Let us write

$$\boldsymbol{\pi}_l \preceq_s \boldsymbol{\pi}_m \text{ if } \operatorname{pr}(Y \geqslant j \mid X = l) \leqslant \operatorname{pr}(Y \geqslant j \mid X = m), \forall j,$$

where Y = Outcome and X = Treatment. The notation $\preceq$ may be read as "stochastically at least as small as." Later in this chapter, we shall refer to this as *simple stochastic order*, hence the suffix 's'. Some questions of interest relating to Table 6.3 are mentioned below.

(a) Assuming that pr(Outcome $\geqslant j \mid$ Dose $= l$) is a nondecreasing function of dose, is there any evidence that dose has an effect on the outcome? For this testing problem, the null and alternative may be

$$H_0 : \boldsymbol{\pi}_1 = \boldsymbol{\pi}_2 = \boldsymbol{\pi}_3 = \boldsymbol{\pi}_4, \quad \text{and} \quad H_1 : \boldsymbol{\pi}_1 \preceq_s \boldsymbol{\pi}_2 \preceq_s \boldsymbol{\pi}_3 \preceq_s \boldsymbol{\pi}_4, \tag{6.6}$$

respectively.

(b) Suppose that the objective is to establish that an increase in dose is associated with an increase in the probability of better outcomes. Then the null and alternative hypotheses may be

$$H_0 : \boldsymbol{\pi} \notin A \quad \text{and} \quad H_1 : \boldsymbol{\pi} \in A$$

where

$$\boldsymbol{\pi} = (\boldsymbol{\pi}_1 : \boldsymbol{\pi}_2 : \boldsymbol{\pi}_3 : \boldsymbol{\pi}_4) \quad (= (\boldsymbol{\pi}_1^T, \boldsymbol{\pi}_2^T, \boldsymbol{\pi}_3^T, \boldsymbol{\pi}_4^T)^T)$$

and A is a suitably chosen set; two possible choices of A are A_1 and A_2 where

$$\begin{aligned} A_1 = \{\boldsymbol{\pi} :& \pi_{1j} + \cdots + \pi_{15} < \pi_{2j} + \cdots + \pi_{25} < \pi_{3j} + \cdots + \pi_{35} < \\ & \pi_{4j} + \cdots + \pi_{45}, \text{ for } j = 2, \ldots, 5\}, \\ \text{and } A_2 = \{\boldsymbol{\pi} :& \pi_{1j} + \cdots + \pi_{15} < \pi_{ij} + \cdots + \pi_{i5} \text{ for } j = 2, \ldots, 5 \\ & \text{and } i = 2, 3, 4\}. \end{aligned}$$

If $\pi \in A_2$ then a non-zero dose is more effective than the placebo but it does not say that the treatments become more effective as the dose level increases. By contrast, if $\pi \in A_1$ then the treatments do become more effective as the dose increases.

Example 6.2.4 *Comparison of treatments with stratified data*

The results of a survey on job satisfaction are given in Table 6.4. Let $\pi_{ij(k)}$ denote pr(Response = j | Income = i) for Sex = k ($i = 1, 2, 3; j = 1, 2, 3; k = 1, 2$). The type of dependence that may be of interest includes increasing levels of job satisfaction as income increases, for males and females. Two testing problems that are of interest in such examples are given below.
(a) The null hypothesis says that the level of satisfaction does not depend on the level of income for males and females. The alternative hypothesis says that income has a positive association with the level of job satisfaction for males and females. More precisely,

$$\begin{aligned} H_0 &: \pi_{1j(k)} = \pi_{2j(k)} = \pi_{3j(k)} \text{ for any } (j,k), \\ H_1 &: \pi_{3j(k)} + \cdots + \pi_{33(k)} \geqslant \pi_{2j(k)} + \cdots + \pi_{23(k)} \geqslant \\ &\qquad \pi_{1j(k)} + \cdots + \pi_{13(k)} \text{ for any}(j,k). \end{aligned}$$

Table 6.4 Relationship between job satisfaction and income

	Sex					
	Female			Male		
	Job satisfaction			Job satisfaction		
Income	Low	Medium	High	Low	Medium	High
< 5000	4	11	2	2	2	1
$5000 - 25000$	6	25	8	3	12	4
> 25000	2	4	2	1	9	6

Source: 1991 General Social Survey, National Opinion Research Center, University of Chicago. This table appeared in this form in the *Journal of Statistical Planning and Inference*, Volume 107, A. Agresti and B. A. Coull, The analysis of contingency tables under inequality constraints, Pages 45-73.

(b) The hypotheses that correspond to (6.3) are the following:

$$H_0 : \pi \notin A \text{ and } H_1 : \pi \in A$$

where A may be $\{\pi : \pi_{3j(k)} + \cdots + \pi_{33(k)} > \pi_{2j(k)} + \cdots + \pi_{23(k)} > \pi_{1j(k)} + \cdots + \pi_{13(k)}$, for $j = 2, 3$, and $k = 1, 2\}$, and π is the vector formed by all the $\pi_{ij(k)}$ s.

In this example, there are only two strata, *Male* and *Female*. If there are a large number of strata corresponding to a nominal variable, a usual assumption is that the association is the same within different strata. In this case, the testing problem can be formulated in terms of association parameters that are common for different strata; this would be in the spirit of the Cochran-Mantel-Haenzel test.

Example 6.2.5 *Marginal homogeneity in matched pair designs*

Table 6.5 shows the results of a case-control study to evaluate the relationship between cataracts and the use of head coverings during summer; the response variable is head cover; this example is discussed in detail in Agresti and Coull (2002).

Table 6.5 Case-control study on cataracts and head coverings

Head cover for case	Head cover for control Almost always	Freq	Occ	Never	Total
Almost always	29	3	3	4	39
Frequently	5	0	1	1	7
Occasionally	9	0	2	0	11
Never	7	3	1	0	11
Total	50	9	7	5	68

Reprinted from: *American Journal of Epidemiology*, Volume 129, J.M. Dolezal, E.S. Perkins, and R.B. Wallace, Sunlight, skin sensitivity, and senile cataracts, pages 559–568, Copyright(1989), with permission from Oxford University Press.

For a randomly chosen matched pair consisting of a case and a control, let $\pi_{ij} =$ pr(Head cover $= i$ and j for case and control, respectively), $\pi_{i+} = \sum_j \pi_{ij}$, and $\pi_{+j} = \sum_i \pi_{ij}$; see Table 6.6. The main question that arises here relates to the protective effect of head cover. The questions could be formulated in terms of the marginal probabilities, $\{\pi_{i+}\}$ and $\{\pi_{+i}\}$. The following hypotheses are of interest.

$$H_0 : (\pi_{1+}, \pi_{2+}, \pi_{3+}, \pi_{4+}) = (\pi_{+1}, \pi_{+2}, \pi_{+3}, \pi_{+4});$$
$$H_1 : \pi_{1+} + \ldots + \pi_{j+} \leq \pi_{+1} + \ldots + \pi_{+j}, \quad j = 1, 2, 3;$$
$$H_2 : (\pi_{1+}, \pi_{2+}, \pi_{3+}, \pi_{4+}) \neq (\pi_{+1}, \pi_{+2}, \pi_{+3}, \pi_{+4});$$
$$H_4 : \pi_{1+} + \ldots + \pi_{j+} < \pi_{+1} + \ldots + \pi_{+j}, \quad j = 1, 2, 3.$$

In this chapter we will discuss statistical methods that are applicable to the forego-ing type of problems. This includes asymptotic as well as exact finite sample methods.

Table 6.6 Probabilities corresponding to Table 6.5

Head cover for case	Head cover for control Almost always	Freq	Occ	Never	Total
Almost always	π_{11}	π_{12}	π_{13}	π_{14}	π_{1+}
Frequently	π_{21}	π_{22}	π_{23}	π_{24}	π_{2+}
Occasionally	π_{31}	π_{32}	π_{33}	π_{34}	π_{3+}
Never	π_{41}	π_{42}	π_{43}	π_{44}	π_{4+}
Total	π_{+1}	π_{+2}	π_{+3}	π_{+4}	1.0

The general asymptotic results obtained in Chapter 4 are applicable for the foregoing type of settings.

6.3 INDEPENDENT BINOMIAL SAMPLES

Let $Y_1, \ldots, Y_k$ be k independent binomial random variables with $Y_i \sim B(n_i, \pi_i)$ for $i = 1, \ldots, k$. Suppose that we wish to test a hypothesis that involves inequality constraints on $\{\pi_1, \ldots, \pi_k\}$. For convenience, assume that $(n_i/n) \to a_i$ as $n \to \infty$ where $n = (n_1 + \cdots + n_k)$ and $0 < a_i < 1, i = 1, \ldots, k$. This essentially says that $n_1, \ldots, n_k$ are similar order of magnitude. Apart from that, it is not really a regularity condition, but it simplifies statement of asymptotic results; even if a_i is 0 or 1, asymptotic results can be obtained. In the rest of this section, whenever a_i appears in the limiting distribution, we may replace it by n_i/n for every i to obtain a large sample approximation $(i = 1, \ldots, k)$.

Let $\boldsymbol{\pi} = (\pi_1, \ldots, \pi_k)^T, \boldsymbol{Y} = (Y_1, \ldots, Y_k)^T$ and let $\ell(\boldsymbol{\pi})$ denote the loglikelihood. Since $Y_1, \ldots, Y_k$ are independent, the likelihood function is the product of binomial probabilities, and hence we have that

$$\ell(\boldsymbol{\pi}) = f_1(\pi_1) + \ldots + f_k(\pi_k)$$

where $f_i(\pi_i)$ is the log of the binomial probability for Y_i and is given by

$$f_i(\pi_i) = Y_i \log \pi_i + (n_i - Y_i) \log(1 - \pi_i) + \log[n_i!\{Y_i!(n_i - Y_i)!\}^{-1}] \quad (6.7)$$

for $i = 1, \ldots, k$. It is easily seen that $f_i(\pi_i)$ is a concave function in π_i, $(i = 1, \ldots, k)$, and hence $\ell(\boldsymbol{\pi})$ is also concave in $\boldsymbol{\pi}$. Further, $\ell(\boldsymbol{\pi})$ is also quite smooth in $\boldsymbol{\pi}$. Such functions are well behaved and would typically satisfy the regularity conditions in Proposition 4.8.1. Thus, the asymptotic null distribution of *LRT*, Wald-type, and score statistics for a range of hypotheses test can be deduced from the general results in Chapter 4.

Let us first consider a general form of the testing problem, and then consider important special cases. Let

$$\Omega = \{(\pi_1, \ldots, \pi_k)^T : 0 < \pi_i < 1 \text{ for } i = 1, \ldots, k\},$$

Ω_0 be a closed subset of Ω, and Ω_1 be a subset of Ω such that $\Omega_0 \subset \Omega_1$. Let the null and alternative hypotheses be

$$H_0 : \boldsymbol{\pi} \in \Omega_0 \quad \text{and} \quad H_1 : \boldsymbol{\pi} \in \Omega_1, \quad (6.8)$$

respectively. Assume that Ω_0 and Ω_1 are Chernoff regular at every point in Ω_0. Let T denote the *LRT* or any of its equivalent forms considered in Proposition 4.8.1. Assume that H_0 holds and let $\boldsymbol{\pi}_0$ denote the assumed true null value of $\boldsymbol{\pi}$. Now, by Proposition 4.8.1, the asymptotic null distribution of T is given by

$$T \xrightarrow{d} \|\boldsymbol{Z} - \mathcal{T}(\Omega_0; \boldsymbol{\pi}_0)\|^2 - \|\boldsymbol{Z} - \mathcal{T}(\Omega_1; \boldsymbol{\pi}_0)\|^2 \quad (6.9)$$

where

$$\|\boldsymbol{x}\|^2 = \boldsymbol{x}^T \boldsymbol{i}_0 \boldsymbol{x}, \quad \boldsymbol{i}_0 = \boldsymbol{i}(\boldsymbol{\pi}_0),$$

$\boldsymbol{i}(\boldsymbol{\pi})$ is the information matrix given by

$$\boldsymbol{i}(\boldsymbol{\pi}) = \text{diag}\{a_i\{\pi_i(1-\pi_i)\}^{-1}\}, \tag{6.10}$$

$\boldsymbol{Z} \sim N(\boldsymbol{0}, \boldsymbol{i}_0^{-1})$, $\mathcal{T}(A; \boldsymbol{a})$ is the tangent cone of A at $\boldsymbol{a}$ and

$$\|\boldsymbol{Z} - \mathcal{T}(\Omega_i; \boldsymbol{\pi}_0)\|^2 = \inf\{(\boldsymbol{Z} - \boldsymbol{t})^T \boldsymbol{i}_0 (\boldsymbol{Z} - \boldsymbol{t}) : \boldsymbol{t} \in \mathcal{T}(\Omega_i; \boldsymbol{\pi}_0)\},$$

which is the squared distance between $\boldsymbol{Z}$ and the tangent cone $\mathcal{T}(\Omega_i; \boldsymbol{\pi}_0)$, $i = 1, 2$. The asymptotic null distributions of *LRT* for many testing problems can be deduced from (6.9) in three stages:

Main steps for obtaining the asymptotic null distribution of the LRT

1. Write down the form of the tangent cones, $\mathcal{T}(\Omega_0; \boldsymbol{\pi}_0)$ and $\mathcal{T}(\Omega_1; \boldsymbol{\pi}_0)$.

2. Use the fact that the asymptotic null distribution of *LRT* in (6.9) is also the distribution of the *LRT* for testing $\boldsymbol{\mu} \in \mathcal{T}(\Omega_0; \boldsymbol{\pi}_0)$ against $\boldsymbol{\mu} \in \mathcal{T}(\Omega_1; \boldsymbol{\pi}_0)$ based on a single observation of $\boldsymbol{X}$ where $\boldsymbol{X} \sim N(\boldsymbol{\mu}, \boldsymbol{i}_0^{-1})$ and $\boldsymbol{i}_0$ is treated as known.

3. Now, appeal to exact results, including those in Chapter 4, for tests on the mean of a multivariate normal mean with known covariance matrix.

This general approach is applied in the subsections to follow.

In practically every problem involving several binomial parameters encountered in practice, the constraints are pairwise contrasts. In this case, the constraints may be expressed as $\boldsymbol{A\pi} \geqslant \boldsymbol{0}$ where every row of $\boldsymbol{A}$ is a permutation of $(1, -1, 0, \ldots, 0)$. It follows from Proposition 2.4.3 that the *mle* $\tilde{\boldsymbol{\pi}}$ of $\boldsymbol{\pi}$ under $\boldsymbol{A\pi} \geqslant \boldsymbol{0}$ is also the solution of

$$\min_{\boldsymbol{A\pi} \geqslant \boldsymbol{0}} \sum (\hat{\pi}_i - \pi_i)^2 n_i.$$

In other words, $\tilde{\boldsymbol{\pi}}$ is the isotonic regression of $\hat{\boldsymbol{\pi}}$ with respect to the weights $\{n_i\}$. Consequently, if the constraint is $\pi_1 \leq \ldots \leq \pi_k$ then $\tilde{\boldsymbol{\pi}}$ can be computed by PAVA.

6.3.1 Test of $H_0 : \boldsymbol{R}_1\boldsymbol{\pi} = \boldsymbol{0}$ and $\boldsymbol{R}_2\boldsymbol{\pi} = \boldsymbol{0}$ Against $H_1 : \boldsymbol{R}_2\boldsymbol{\pi} \geqslant \boldsymbol{0}$

Let $\boldsymbol{R}$ be a $q \times k$ matrix of full row rank $(q \leqslant k)$ and let it be partitioned as

$$\boldsymbol{R} = \begin{bmatrix} \boldsymbol{R}_1 \\ \boldsymbol{R}_2 \end{bmatrix}$$

where $\boldsymbol{R}_1$ and $\boldsymbol{R}_2$ are $m \times k$ and $(q-m) \times k$, respectively, $(q \geqslant m \geqslant 0, q \geqslant 1)$. Let the testing problem be

$$H_0 : \boldsymbol{R\pi} = \boldsymbol{0} \quad \text{vs} \quad H_1 : \boldsymbol{R}_2\boldsymbol{\pi} \geqslant \boldsymbol{0}. \tag{6.11}$$

Assume that the null hypothesis holds. Let T denote the *LRT* or any of its asymptotically equivalent forms, $\boldsymbol{\pi}_0$ denote the assumed true null value of $\boldsymbol{\pi}$, $\boldsymbol{i}(\boldsymbol{\pi})$ be as in (6.10), and $\boldsymbol{i}_0 = \boldsymbol{i}(\boldsymbol{\pi}_0)$. Now, $\Omega_0 = \{\boldsymbol{\pi} \in \Omega : \boldsymbol{R}\boldsymbol{\pi} = \boldsymbol{0}\}$, $\Omega_1 = \{\boldsymbol{\pi} \in \Omega : \boldsymbol{R}_2\boldsymbol{\pi} \geqslant \boldsymbol{0}\}$, $\mathcal{T}(\Omega_0; \boldsymbol{\pi}_0) = \{\boldsymbol{x} \in \mathbb{R}^k : \boldsymbol{R}\boldsymbol{x} = \boldsymbol{0}\}$, and $\mathcal{T}(\Omega_1; \boldsymbol{\pi}_0) = \{\boldsymbol{x} \in \mathbb{R}^k : \boldsymbol{R}_2\boldsymbol{x} \geqslant \boldsymbol{0}\}$. Then it follows from (6.9) that the asymptotic null distribution of T is equal to the null distribution of the *LRT* for the testing problem (6.11) based on a single observation of $\boldsymbol{X}$ where $\boldsymbol{X} \sim N(\boldsymbol{\pi}, \boldsymbol{i}_0^{-1})$, and $\boldsymbol{i}_0$ is assumed known. Therefore, it follows from Corollary 3.7.2 that

$$\mathrm{pr}(T \geqslant t \mid \boldsymbol{\pi} = \boldsymbol{\pi}_0) \to \sum_{i=0}^{q-m} w_i\{q - m, \boldsymbol{A}(\boldsymbol{\pi}_0)\} pr(\chi^2_{m+i} \geqslant t) \tag{6.12}$$

as $n \to \infty$, where $\boldsymbol{A}(\boldsymbol{\pi}_0) = \boldsymbol{R}_2\boldsymbol{i}_0^{-1}\boldsymbol{R}_2^T$. If t is the sample value of T then the asymptotic p-value is the supremum of the limit in (6.12) where the supremum is taken over $\boldsymbol{\pi}_0 \in H_0$. A closed-form is not available for this p-value, for an arbitrary $\boldsymbol{R}$. However, if the null hypothesis is $\pi_1 = \ldots = \pi_k$ then the $\bar{\chi}^2$-weights do not depend on any unknown nuisance parameters. To show this, suppose that the testing problem is

$$H_0 : \pi_1 = \cdots = \pi_k \quad \text{and} \quad H_1 : \boldsymbol{B}\boldsymbol{\pi} \geqslant \boldsymbol{0} \tag{6.13}$$

where $\boldsymbol{B}$ is $r \times k$, $\mathrm{rank}(\boldsymbol{B}) = r$ and $H_0 \subset H_1$. This is a special case of (6.11) with $\boldsymbol{R}$ being a $(k-1) \times k$ matrix of full row-rank and $q - m = r$. Let

$$\boldsymbol{V} = \boldsymbol{B}\,\mathrm{diag}\{a_i\}^{-1}\boldsymbol{B}^T.$$

Since

$$\boldsymbol{A}(\boldsymbol{\pi}_0) = \{\pi_*(1 - \pi_*)\}\boldsymbol{V},$$

where π_* is the common value of $\pi_1, \ldots, \pi_k$ under H_0, it follows that

$$\mathrm{pr}(T \geqslant t \mid H_0) \to \sum_{i=0}^{r} w_i(r, \boldsymbol{V})\mathrm{pr}(\chi^2_{k-1-r+i} \geqslant t); \tag{6.14}$$

where $\boldsymbol{V}$ does not depend on any unknown nuisance parameters (see Problem 6.4).

It is worth noting that the $\bar{\chi}^2$-distribution in (6.14) is the same as that arises for the testing problem (6.13) based on independent $X_1, \ldots, X_k$ where $X_i \sim N(\pi_i, n_i^{-1}), i = 1, \ldots, k$. For example, if the alternative hypothesis is the simple order $\pi_1 \leq \cdots \leq \pi_k$ then the $\bar{\chi}^2$-distribution in (6.14) is the one that corresponds to testing against simple order of normal means.

A natural question that arises in the above case is how large the sample size needs to be for the chi-bar square distribution in (6.14) to provide a reasonable approximation? The nature of the approximations involved in the proof of (6.14) suggest that if the sample size is large enough for χ^2_q to provide a good approximation for the test of $\boldsymbol{B}\boldsymbol{\pi} = \boldsymbol{0}$ against $\boldsymbol{B}\boldsymbol{\pi} \neq \boldsymbol{0}$, then it is likely that the sample size is large enough for the chi-bar square approximation in (6.14). At this stage, this is only a conjecture.

6.3.2 Test of $H_0 : A\pi \geqslant 0$ Against $H_1 : A\pi \ngeq 0$

The general result, (6.9), can be applied when there are inequality constraints in the null hypothesis. Let us consider test of

$$H_0 : \boldsymbol{A\pi} \geqslant \mathbf{0} \text{ against } H_1 : \boldsymbol{A\pi} \ngeq \mathbf{0} \tag{6.15}$$

where $\boldsymbol{A}$ is a given matrix of constants. In the notation of (6.8), we have $\Omega_0 = \{\boldsymbol{\pi} \in \Omega : \boldsymbol{A\pi} \geqslant \mathbf{0}\}$ and $\Omega_1 = \Omega$; hence $\mathcal{T}(\Omega_0; \boldsymbol{\pi}_0)$ is the polyhedral $\{\boldsymbol{\pi} \in \mathbb{R}^k : \boldsymbol{A\pi} \geqslant \mathbf{0}\}$ and $\mathcal{T}(\Omega_1; \boldsymbol{\pi}_0) = \mathbb{R}^k$ for any $\boldsymbol{\pi}_0$ in the null parameter space Ω_0. It follows from (6.9) that the asymptotic null distribution of *LRT* is a chi-bar -square at every point in Ω_0. In general, this distribution depends on the particular value of $\boldsymbol{\pi}$ in the null parameter space, and the least favorable null value is unknown. However, if every row of $\boldsymbol{A}$ is a pairwise contrast, in other words every row of $\boldsymbol{A}$ is a permutation of $(1, -1, 0 \ldots, 0)$, then the least favorable null value is a point in $\{\boldsymbol{\pi} : \boldsymbol{A\pi} = \mathbf{0}\}$, and the least favorable null distribution is a chi-bar square and it does not depend on any unknown nuisance parameters; this is stated in the next proposition and the proof is given in the Appendix to this chapter (see Lemma 6.11.1 on page 336).

Proposition 6.3.1 *Let the testing problem be (6.15) where $\boldsymbol{A}$ is a matrix in which every row is a permutation of $(1, -1, 0, \ldots, 0)$. Then for the asymptotic distribution of LRT, the least favorable null value is a point in $\{\boldsymbol{\pi} : \boldsymbol{A\pi} = \mathbf{0}\}$, and the least favorable null distribution is equal to that of*

$$\inf\{\boldsymbol{Z} - \boldsymbol{\pi})^T \boldsymbol{W}(\boldsymbol{Z} - \boldsymbol{\pi}) : \boldsymbol{A\pi} \geqslant \mathbf{0}\}$$

where $\boldsymbol{Z} \sim N(\mathbf{0}, \boldsymbol{W}^{-1})$ and $\boldsymbol{W}$ is the diagonal matrix whose ith diagonal is $n_i, i = 1, \ldots, k$; this least favorable null distribution is also equal to $\bar{\chi}^2(\boldsymbol{W}^{-1}, \mathcal{C}^0)$ where $\mathcal{C}^0 = \{\boldsymbol{y} : \boldsymbol{y}^T \boldsymbol{W x} \leq 0 \;\forall \boldsymbol{x}$ satisfying $\boldsymbol{Ax} \geqslant \mathbf{0}\}$.

6.3.3 Test of $H_0 : A\pi \ngtr 0$ Against $H_1 : A\pi > 0$

In Example 6.2.1, a question of particular interest is whether or not there is sufficient statistical evidence to establish that an increase in alcohol consumption is associated with an increase in the probability of malformation of sex organs, without making any assumption about $\boldsymbol{\pi}$. In this regard, one possible formulation of the hypothesis testing problem is the following:

$$H_0 : \mathit{not}\ H_1 \quad \text{and} \quad H_1 : \pi_1 < \cdots < \pi_k$$

where $\boldsymbol{\pi} = (\pi_1, \ldots, \pi_k)^T$. These hypotheses may be expressed in the following forms suitable for IUT (i.e., Intersection Union Test):

$$H_0 : \pi_1 \geqslant \pi_2, \text{ or } \ldots, \text{ or } \pi_{k-1} \geqslant \pi_k \quad \text{and} \quad H_1 : \pi_1 < \pi_2, \ldots, \pi_{k-1} < \pi_k.$$

Now, for $i = 1, \ldots, k - 1$, we can apply a procedure for the comparison of two independent binomial samples to test $\pi_i \geqslant \pi_{i+1}$ against $\pi_i < \pi_{i+1}$, and then combine these $k-1$ tests using an IUT. While there are no theoretical difficulties in applying an IUT for this, it may need a large number of observations to have a reasonable level of power because the test is not similar and the null parameter space is large; intuitively, the reason for low power is clear once we note that, to reject the null hypothesis, the IUT demands the p-value for testing $\pi_i \geqslant \pi_{i+1}$ against $\pi_i < \pi_{i+1}$ to be small for every $i = 1, \ldots, k-1$.

Now let us consider test of $H_0 : \boldsymbol{A}\boldsymbol{\pi} \not> \boldsymbol{0}$ against $H_1 : \boldsymbol{A}\boldsymbol{\pi} > \boldsymbol{0}$ for some given matrix $\boldsymbol{A}$ of order $r \times k$. Let $\boldsymbol{a}_i^T$ denote the ith row of $\boldsymbol{A}$ $(i = 1, \ldots, r)$. Then the foregoing hypotheses can be stated as

$$H_0 : \boldsymbol{a}_1^T \boldsymbol{\pi} \leq 0, \text{ or }, \ldots, \text{ or } \boldsymbol{a}_r^T \boldsymbol{\pi} \leq 0, \text{ and } H_1 : \boldsymbol{a}_1^T \boldsymbol{\pi} > 0, \ldots, \boldsymbol{a}_r^T \boldsymbol{\pi} > 0.$$

Now, we can use any of the standard procedures for testing $\boldsymbol{a}_i^T \boldsymbol{\pi} \leq 0$ against $\boldsymbol{a}_i^T \boldsymbol{\pi} > 0$, and then combine these r tests using an IUT.

6.3.4 Pairwise Contrasts of Binomial Parameters

The foregoing methods can be applied easily when $\{\pi_1, \ldots, \pi_k\}$ are the cell probabilities of a factorial design. Let us consider some special cases to illustrate the flexibility of this approach. Consider a binomial response model with two explanatory variables which we denote by X and Y. Suppose that X and Y are ordinal with I and J levels, respectively. As an example, the response variable could be whether or not a person suffers from bronchitis, $X =$ *level of exposure to cigarette smoke* and $Y =$ *the air-pollution level.* To give another example, the response variable could be whether or not a university student obtains at least a B-grade, $X =$ *aptitude-score* and $Y=$ *high school GPA*. For such a two-way design, let π_{ij} denote the probability of success when $(X, Y) = (i, j)$. In such examples, a hypothesis of interest is that π_{ij} is nondecreasing in i for fixed j, and in j for fixed i. Thus, the following hypotheses are of interest.

$$\begin{aligned}
&H_0 : \pi_{ij} = \pi_{kl} \text{ for every } (i, j) \text{ and } (k, l) \\
&H_1 : \pi_{ij} \leq \pi_{kl} \text{ for } i \leq k \text{ and } j \leq l \\
&H_2 : \pi_{ij}\ s \text{ are unrestricted} \\
&H_3 : \pi_{ij} < \pi_{kl} \text{ for } i \leq k, j \leq l \text{ and } (i, j) \neq (k, l).
\end{aligned}$$

The hypothesis H_1 is also known as the *matrix hypothesis*. Let $\boldsymbol{\pi}$ denote the vector formed by all the $\pi_{ij}\ s$. Now, the foregoing hypotheses can be stated as

$$H_0 : \boldsymbol{A}\boldsymbol{\pi} = \boldsymbol{0}, \quad H_1 : \boldsymbol{A}\boldsymbol{\pi} \geqslant \boldsymbol{0}, \quad H_2 : \boldsymbol{\pi} \text{ is unrestricted}, \quad H_3 : \boldsymbol{A}\boldsymbol{\pi} > \boldsymbol{0},$$

where $\boldsymbol{A}$ is of order $(2IJ - I - J) \times (IJ)$, and every row of $\boldsymbol{A}$ is a permutation of $(1, -1, 0, \ldots, 0)$.

Similar hypotheses may be formulated even if the design has more than two factors. Let us now revert to our original notation and write $\pi_1, \ldots, \pi_k$ for the cell probabilities for the k cells, and let $\boldsymbol{\pi} = (\pi_1, \ldots, \pi_k)^T$, as introduced at the beginning of this section. Let the null and the alternative hypotheses be

$$H_0 : \pi_1 = \ldots = \pi_k \text{ and } H_1 : \boldsymbol{A\pi} \geqslant \mathbf{0},$$

respectively, where each row of $\boldsymbol{A}$ is a permutation of $(1, -1, 0, \ldots, 0)$, and let $a_i = lim(n_i/n)$. When the null hypotheses is true, let π_0 denote the common unknown value of the π_i s. It follows from (6.9) and Theorem 3.7.1 (on page 84) that the asymptotic null distribution of *LRT* is given by

$$LRT \xrightarrow{d} \bar{\chi}^2(\text{diag}\{a_i\}^{-1}, \mathcal{C}) \quad \text{where } \mathcal{C} = \{\boldsymbol{\pi} : \boldsymbol{A\pi} \geqslant \mathbf{0}, \sum a_i\pi_i = 0\}. \quad (6.16)$$

For the purposes of computing approximate critical values and p-values, we may simulate the *LRT* directly or its limiting distribution in (6.16) with a_i replaced by n_i for every i. Since $\mathcal{C}$ is a polyhedral, the large sample p-value of *LRT* can be computed easily.

Finally, let us note that since $\boldsymbol{A}$ is a matrix of pairwise contrasts, Proposition 6.3.1 provides a result for testing $\boldsymbol{A\pi} \geqslant \mathbf{0}$ against $\boldsymbol{A\pi} \not\geqslant \mathbf{0}$, and Section 6.3.3 provides a method of testing $\boldsymbol{A\pi} \not> \mathbf{0}$ against $\boldsymbol{A\pi} > \mathbf{0}$.

6.3.5 Discussion

Apart from the large sample results discussed in this chapter for constrained inference on binomial parameters, there have been several related studies. For example, see Piegorsch (1990), Peddada et al. (2001), Morris (1988), Gelfand and Kuo (1991), and Lin (1998).

Peddada et al. (2001) developed a bootstrap procedure for testing the hypothesis that the binomial probabilities are equal against an alternative of the form $\boldsymbol{A\pi} \geqslant 0$ where every row of $\boldsymbol{A}$ is a permutation of $(1, -1, 0 \ldots \mathbf{0})$. They proposed a method based on the estimation procedure of Hwang and Peddada (1994). Their discussions also cover the setting when some observations may be missing not at random. For example, consider a toxicological study in which tumor incidence is assessed over a 2-year period. In such a study, animals may die before the end of the study. In this case it is not justified to assume that the observations are missing at random. They also showed that their bootstrap test against order restrictions on binomial parameters is asymptotically correct.

In some studies the interest may be to test $H_0 : \pi_1 = \ldots = \pi_k$ against an alternative which says that $\pi_1, \ldots, \pi_k$ follow an umbrella order with known or unknown turning point. Peddada et al. (2001) considered an example in toxicology in which the π_i are monotonically nondecreasing in i up to a certain point and then a down turn occurs. They noted " ...in some toxicological experiments the dose levels of the chemical of interest are positively correlated with weight loss and, because tumor incidence is also positively correlated with body weight, animals in higher-dose groups may experience a reduced incidence of certain types of tumors, such as those occurring in the liver and mammary gland." In such experiments, test of $H_0 : \pi_1 = \ldots = \pi_k$ against $H_1 : \cup_{j=2}^{k}\{(\pi_1, \ldots, \pi_k) : \pi_1 \leq \ldots \leq \pi_j \geqslant \ldots \geqslant \pi_k\}$ may be of interest. The test developed in Peddada et al. (2001) is based on a bootstrap approach. Mack and Wolfe (1981) developed a test based on ranks for similar testing problems; see also Simpson and Margolin (1986).

Morris (1988) developed a finite sample method of constructing a confidence interval for a scalar parameter θ_i for a fixed i in $\{1, \ldots, k\}$ under the constraint $\theta_1 \leq \cdots \leq \theta_k$. Assume that there are independent samples from k populations and θ_i is the unknown population parameter from the ith population ($i = 1, \ldots, k$). The method is based on inverting a test, and it is applicable when θ_i is the binomial parameter. In a simulation study, Morris (1988) observed that competing asymptotic methods can produce large discrepancies between nominal and actual coverage probabilities.

Bayesian estimation of $\pi_1, \ldots, \pi_k$ subject to $\pi_1 \leq \cdots \leq \pi_k$ is studied in Gelfand and Kuo (1991). The traditional Bayesian approaches have had only limited success due to computational difficulties. Gelfand and Kuo (1991) use Gibbs sampling to evaluate objects of interest from the posterior density. Use of Gibbs sampling overcomes some of the traditional computational difficulties. The authors also show that their method extends to ordinal response data. For example, consider k doses of a treatment. Suppose that the response is an ordinal variable, which we assume to have only three categories for simplicity. For the ith treatment let π_{ij} denote the probability that the response is category $j (j = 1, 2, 3)$. Gelfand and Kuo (1991) illustrated the application of their method for inference under the constraint that the treatments are stochastically ordered [i.e., π_{i1} is nondecreasing in i, and $(\pi_{i1} + \pi_{i2})$ is nondecreasing in i]. Sedransk et al. (1985) also applied a Bayesian approach for estimating $\pi_1, \ldots, \pi_k$ subject to $\pi_1 \leq \cdots \leq \pi_k$. They used importance sampling for computing the quantities of interest.

There have been several simulation studies to compare the performance of different methods of testing hypotheses involving order-restrictions on binomial parameters. For example, consider test of $H_0 : \pi_1 = \cdots = \pi_k$ against $H_1 : \pi_1 \leq \cdots \leq \pi_k$ based on k independent binomial observations. For this testing problem, it is valid to apply a test of H_0 against $H_2 : \pi_1, \ldots, \pi_k$ *are not all equal.* Agresti and Coull (1996) investigated this and concluded that the order-restricted approach is better than an unrestricted test of H_0 against H_2. Piegorsch (1990) compared several order-restricted tests including a Bonferroni corrected one-sided Wald test. He concluded that the latter is conservative when the sample sizes are small. Piegorsch (1990) also provide a detailed discussion to compare the performances of other asymptotic methods, and noted that the asymptotic theory, when applied to small samples, did not perform well in terms of maintaining the size of the test.

6.4 ODDS RATIOS AND MONOTONE DEPENDENCE

When the variables defining a contingency table are ordinal, one would typically expect a form of monotone or nonnegative dependence. In this section, we consider ways of quantifying different types of monotone dependence. This would help to formulate the statistical inference problem.

To introduce the notation to be adopted throughout this chapter, let us consider an $r \times c$ table that corresponds to the comparison of r treatments and a response variable with c ordinal categories. Let Y denote the response variable; the range of Y is $\{1, \ldots, c\}$ where the values 1, 2,... etc are interpreted as first, second, ..., etc. Let X denote the variable corresponding to the rows of the $r \times c$ table. The range of X is $\{1, \ldots, r\}$; $X = i$ for Treatment i, $(i = 1, \ldots, r)$. The variable Y refers to the response variable without referring to any particular row, while Y_i refers to the response variable when $X = i$, $(i = 1, \ldots, r)$. For $i = 1, \ldots, r$ and $j = 1, \ldots, c$, let

$$\pi_{ij} = \text{pr}(Y = j \mid X = i) \quad \text{and} \quad \boldsymbol{\pi}_i = (\pi_{i1}, \ldots, \pi_{i,c-1}).$$

Thus, $\pi_{i1}, \ldots, \pi_{ic}$ are the conditional probabilities given that Treatment = i, and $\pi_{i1} + \ldots + \pi_{ic} = 1$. Note that, in the notation just introduced, we also have

$$\pi_{ij} = \text{pr}(Y_i = j).$$

Let

$$\gamma_{ij} = \text{pr}(Y \leq j \mid X = i) \quad \text{and} \quad \boldsymbol{\gamma}_i = (\gamma_{i1}, \ldots, \gamma_{i,c-1}). \tag{6.17}$$

Then

$$\gamma_{ij} = \pi_{i1} + \ldots + \pi_{ij} = \text{pr}(Y_i \leq j); \tag{6.18}$$

it is the cumulative probability for ith row. These cumulative probabilities play a significant role in the analysis of ordinal response data.

6.4.1 2×2 Tables

Consider an experiment to compare Treatment 1 with Treatment 2 in which the response variable, Y, is binary and hence $c = 2$; assume that "2" corresponds to "success," the preferred outcome. Let n_{ij} denote the frequency in cell, (i, j); these are shown in Table 6.7. For the 2×2 table in Table 6.7, the *odds* of failure for Treatments 1 and 2 are defined as π_{11}/π_{12} and π_{21}/π_{22}, respectively, and the *odds ratio*, θ, is defined as

$$\theta = (\pi_{11}/\pi_{12})/(\pi_{21}/\pi_{22});$$

it is also called the *cross-product ratio* because $\theta = (\pi_{11}\pi_{22})/(\pi_{21}\pi_{12})$. As an example, suppose that $\theta = 2$. Then the odds of failure,(π_{11}/π_{12}), for Treatment1 is twice of that for Treatment 2. Equivalently, the odds of success, (π_{22}/π_{21}), for Treatment 2 is twice of that for Treatment 1. It is easily verified that

$$\theta > 1 \Leftrightarrow \pi_{22} > \pi_{12} \Leftrightarrow \pi_{11} > \pi_{21}; \quad \text{and} \quad \theta = 1 \Leftrightarrow \boldsymbol{\pi}_1 = \boldsymbol{\pi}_2. \tag{6.19}$$

Table 6.7 Notation for comparison of two treatments

	Probability			Frequency		
	Y			Y		
	1	2	Total	1	2	Total
Treatment 1	π_{11}	π_{12}	1	n_{11}	n_{12}	n_{1+}
Treatment 2	π_{21}	π_{22}	1	n_{21}	n_{22}	n_{2+}

Since $Y = 2$ corresponds to "success", the preferred outcome, Treatment 2 is better than Treatment 1 if and only if $\theta > 1$. If we wish to test the hypothesis of no difference between the treatments against the alternative that Treatment 2 is better, then the null and the alternative hypotheses could be $H_0 : \pi_{22} = \pi_{12}$ and $H_1 : \pi_{22} > \pi_{12}$, respectively, or equivalently $H_0 : \theta = 1$ and $H_1 : \theta > 1$, respectively.

The usual estimator $\hat{\theta}$ of θ is the sample odds ratio, $(n_{11}n_{22})/(n_{21}n_{12})$. For large sample inference on $\boldsymbol{\theta}$, we may use (for example, see Agresti (2002))

$$\{\log\hat{\theta} - \log\theta\}/(n_{11}^{-1} + n_{12}^{-1} + n_{21}^{-1} + n_{22}^{-1})^{1/2} \approx N(0,1). \qquad (6.20)$$

The reason for using $\log\hat{\theta}$ instead of $\hat{\theta}$ is that the former provides a better approximation to normal distribution.

6.4.2 $2 \times c$ Tables

Now, let us suppose that the response variable Y is ordinal with c categories where $c > 2$; let us also assume that larger values of Y are preferred. For illustrative purposes, the notations for the special case when $c = 4$ are given in Table 6.8. Now, for $j = 1, \ldots, c-1$, define

$$\textit{Local odds ratio} : \theta_j^L = (\pi_{1j}/\pi_{1,j+1})/(\pi_{2j}/\pi_{2,j+1});$$

$$\textit{Cumulative odds ratio} : \theta_j^C = \{\gamma_{1j}/(1-\gamma_{1j}\}/\{\gamma_{2j}/(1-\gamma_{2j})\},$$

where γ_{ij} is as defined in (6.17). These two odds ratios have been used extensively in the literature for the comparison two treatments when the response variable is ordinal. The local odds ratio is the basic quantity of interest in the standard log-linear models for the analysis of nominal categorical data. By contrast, the cumulative odds ratio appears to be better suited than the local odds ratio for the purposes of formulating an inference problem when the response is ordinal.

Although, there is no unique way of formulating what is meant by *Treatment 2 is at least as good as Treatment 1*, one that is likely to be quite suitable is,

$$\text{pr}(Y \geqslant j \mid X = 2) \geqslant \text{pr}(Y \geqslant j \mid X = 1) \quad \text{for every } j. \qquad (6.21)$$

In terms of γ_{ij}, (6.21) is equivalent to $\gamma_{1j} \geqslant \gamma_{2j}$, for every j; and it is also equivalent to $1 - \gamma_{2j} \geqslant 1 - \gamma_{1j}$ for every j. If the constraints in (6.21) hold with at least one

Table 6.8 Multinomial probabilities for two treatments

	Response $(Y)^{(1)}$				
	1	2	3	4	Total
Treatment 1	π_{11}	π_{12}	π_{13}	π_{14}	1
Treatment 2	π_{21}	π_{22}	π_{23}	π_{24}	1

Table 6.9 All possible collapsed 2 × 2 tables of Table 6.8

	Y: 1	Y: 2,3,4	Y: 1,2	Y: 3,4	Y: 1,2,3	Y: 4
Tr 1	π_{11}	$\pi_{12}+\pi_{13}+\pi_{14}$	$\pi_{11}+\pi_{12}$	$\pi_{13}+\pi_{14}$	$\pi_{11}+\pi_{12}+\pi_{13}$	π_{14}
Tr 2	π_{21}	$\pi_{22}+\pi_{23}+\pi_{24}$	$\pi_{21}+\pi_{22}$	$\pi_{23}+\pi_{24}$	$\pi_{21}+\pi_{22}+\pi_{23}$	π_{24}
O.R.	θ_1^C		θ_2^C		θ_3^C	

strict inequality, then we would consider Treatment 2 to be *better* than Treatment 1. To provide another interpretation of (6.21), let us consider all possible 2 × 2 tables that can be formed by merging adjacent columns; they are shown in Table 6.9 for the particular case when $c = 4$. Note that $\theta_2^C > 1$ is equivalent to saying that Treatment 2 is better than Treatment 1 when the response variable is binary and success is defined as $\{3, 4\}$ and Failure as $\{1, 2\}$. Similar interpretations can be given to $\theta_j^C > 1$ for other values of j. Let $\boldsymbol{\theta}^C = (\theta_1^C, \ldots, \theta_{c-1}^C)^T$. Now, we have the following.

Proposition 6.4.1

$$(i)\quad \boldsymbol{\theta}^C \geqslant \mathbf{1} \quad\Leftrightarrow\quad \gamma_1 \geqslant \gamma_2 \quad\Leftrightarrow\quad (6.21)\ \textit{holds.}$$

$$(ii)\quad \boldsymbol{\theta}^C = \mathbf{1} \quad\Leftrightarrow\quad \gamma_1 = \gamma_2 \quad\Leftrightarrow\quad \boldsymbol{\pi}_1 = \boldsymbol{\pi}_2.$$

$$(iii)\quad \boldsymbol{\theta}^C \gneqq \mathbf{1} \quad\Leftrightarrow\quad \gamma_1 \gneqq \gamma_2$$

$$\Leftrightarrow\quad (6.21)\ \textit{holds with at least one strict inequality.}$$

Proof: The proof of this result follows easily by applying (6.19) to each of the 2 × 2 tables similar to those in Table 6.9.
(i) $\boldsymbol{\theta}^C \geqslant \mathbf{1} \Leftrightarrow \gamma_{1j}(1-\gamma_{2j})/\{\gamma_{2j}(1-\gamma_{1j})\} \geqslant 1 \forall j \Leftrightarrow \gamma_{1j} \geqslant \gamma_{2j} \forall j \Leftrightarrow$ (6.21) holds.
(ii) $\boldsymbol{\theta}^C = \mathbf{1} \Leftrightarrow \gamma_{1j}(1-\gamma_{2j})/\{\gamma_{2j}(1-\gamma_{1j})\} = 1 \forall j \Leftrightarrow \gamma_{1j} = \gamma_{2j} \forall j \Leftrightarrow \boldsymbol{\pi}_1 - \boldsymbol{\pi}_2 = \mathbf{0}$.
(iii) Follows from (i) and (ii). ■

The null and alternative hypotheses for establishing that Treatment 2 is better than Treatment 1 can be formulated in several ways; two are given below:

$$(a)\qquad H_0 : \gamma_1 = \gamma_2 \quad\text{and}\quad H_1 : \gamma_1 \geqslant \gamma_2; \tag{6.22}$$

$$(b)\qquad H_0^* : \gamma_1 \not> \gamma_2 \quad\text{and}\quad H_1^* : \gamma_1 > \gamma_2. \tag{6.23}$$

In view of Proposition 6.4.1, these can also be stated directly in terms of cumulative odds ratios as follows:

$$(a)\qquad H_0 : \boldsymbol{\theta}^C = \mathbf{1} \quad\text{and}\quad H_1 : \boldsymbol{\theta}^C \geqslant \mathbf{1}; \tag{6.24}$$

$$(b)\qquad H_0^* : \boldsymbol{\theta}^C \not> \mathbf{1} \quad\text{and}\quad H_1^* : \boldsymbol{\theta}^C > \mathbf{1}. \tag{6.25}$$

The local odds ratios have also been used for formulating hypotheses testing problems. It can be shown that (see Proposition 6.4.2)

$$\boldsymbol{\theta}^L = \mathbf{1} \Leftrightarrow \boldsymbol{\pi}_1 = \boldsymbol{\pi}_2; \quad \boldsymbol{\theta}^L \geqslant \mathbf{1} \Rightarrow \boldsymbol{\theta}^C \geqslant \mathbf{1}; \text{ and } \boldsymbol{\theta}^C \geqslant \mathbf{1} \not\Rightarrow \boldsymbol{\theta}^L \geqslant \mathbf{1}.$$

Table 6.10 Probabilities of response (hypothetical values[1])

	Response			
	Worse	Good	Very good	Excellent
Treatment 1	0.6	0.01	0.02	0.37
Treatment 2	0.5	0.01	0.01	0.48

[1]These values show that stochastic order does not imply likelihood ratio order.

Thus, $\boldsymbol{\theta}^L = \mathbf{1}$ is equivalent to no difference between the two treatments. Since $\{\boldsymbol{\theta}^L \geqslant \mathbf{1}\}$ is a strict subset of $\{\boldsymbol{\theta}^C \geqslant \mathbf{1}\}$, it follows that $\boldsymbol{\theta}^L \geqslant \mathbf{1}$ also says that Treatment 2 is at least as good as Treatment 1, but in a more strict sense than $\boldsymbol{\theta}^C \geqslant \mathbf{1}$. The following hypothetical example illustrates this. Suppose that the multinomial probabilities are as in Table 6.10. For these values note that $\boldsymbol{\gamma}_1 \gtrapprox \boldsymbol{\gamma}_2$. Therefore, Treatment 2 is better than Treatment 1. However, $\theta_2^L (= (0.01 \times 0.01)/(0.01 \times 0.02) = 1/2)$ is less than one. An interpretation of this is that conditional on the response being "Good" or "Very Good," Treatment 2 has lower probability of better response than Treatment 1. However, this should not change our original opinion based on $\boldsymbol{\gamma}_1 \gtrapprox \boldsymbol{\gamma}_2$ that Treatment 2 is better than Treatment 1. This phenomenon is likely if some of the ordered response categories in the middle range have small probabilities.

Another odds ratio that has attracted some attention is the *continuation odds ratio*; it is defined as

$$\theta_j^{CO} = \pi_{1j}(\pi_{2,j+1} + \ldots + \pi_{2c})/\{\pi_{2j}(\pi_{1,j+1} + \ldots + \pi_{1c})\}.$$

The stochastic order defined by $\boldsymbol{\theta}^{CO} \geqslant \mathbf{1}$ has also been called the *hazard rate order*. At the beginning this subsection, a simple interpretation was given for the cumulative odds ratio. Similarly, to interpret $\boldsymbol{\theta}^{CO} \geqslant \mathbf{1}$, let us note that

$$\theta_j^{CO} \geqslant 1 \text{ if and only if } \mathrm{pr}(Y_2 \geqslant j+1 \mid Y_2 \geqslant j) \geqslant \mathrm{pr}(Y_1 \geqslant j+1 \mid Y_1 \geqslant j).$$

Thus, if Y is thought of as the survival time then $\theta_j^{CO} \geqslant 1$ if and only if the probability of survival beyond time $j+1$ given that the individual has survived up to j, is higher for Treatment 2 than for Treatment 1.

The following results summarize the main relationships among the foregoing odds ratios; see Douglas et al. (1991) for more details.

Proposition 6.4.2 *1.* $\boldsymbol{\theta}^L \geqslant \mathbf{1} \Rightarrow \boldsymbol{\theta}^{CO} \geqslant \mathbf{1} \Rightarrow \boldsymbol{\theta}^C \geqslant \mathbf{1}$.

2. $\boldsymbol{\theta}^L = 1 \Leftrightarrow \boldsymbol{\theta}^{CO} = \mathbf{1} \Leftrightarrow \boldsymbol{\theta}^C = \mathbf{1}$.

3. $\boldsymbol{\theta}^L \gtrapprox \mathbf{1} \Rightarrow \boldsymbol{\theta}^{CO} \gtrapprox \mathbf{1} \Rightarrow \boldsymbol{\theta}^C \gtrapprox \mathbf{1}$.

4. $\boldsymbol{\theta}^C \geqslant \mathbf{1} \not\Rightarrow \boldsymbol{\theta}^{CO} \geqslant \mathbf{1} \not\Rightarrow \boldsymbol{\theta}^L \geqslant \mathbf{1}$.

Proof: First, let us prove $\boldsymbol{\theta}^L \geqslant \mathbf{1} \Rightarrow \boldsymbol{\theta}^{CO} \geqslant \mathbf{1}$. Let $p_j = \pi_{1j}$ and $q_j = \pi_{2j}$ ($j = 1, \ldots, c$) for simplicity. It follows from $\boldsymbol{\theta}^L \geqslant \mathbf{1}$ that $p_j q_{j+1} \geqslant q_j p_{j+1}$ for every

j, and hence (p_j/q_j) decreases as j increases; thus

$$p_1 q_1^{-1} \geqslant p_2 q_2^{-1} \geqslant \ldots \geqslant p_c q_c^{-1}. \tag{6.26}$$

To prove $\boldsymbol{\theta}^{CO} \geqslant 1$, we need to show that

$$p_j(q_{j+1} + \ldots + q_c) \geqslant q_j(p_{j+1} + \ldots p_c)$$

for every j; this is equivalent to

$$p_j q_{j+1} + \ldots + p_j q_c \geqslant q_j p_{j+1} + \ldots + q_j p_c \text{ for every } j. \tag{6.27}$$

It follows from (6.26), by cross-multiplication, that

$$p_j q_{j+1} \geqslant q_j p_{j+1}, \quad p_j q_{j+2} \geqslant q_j p_{j+2}, \quad \ldots, \quad p_j q_c \geqslant q_j p_c. \tag{6.28}$$

Now (6.26) follows by adding the inequalities in (6.28). The proof of the other parts are omitted because the arguments are very similar. ■

It is fairly easy to construct a table similar to Table 6.10 to show that the reverse implications of (1) in the foregoing Proposition do not hold.

Since (6.21) is equivalent to $\boldsymbol{\theta}^C \geqslant \mathbf{1}$, a form of *stochastic order* is captured by $\boldsymbol{\theta}^C \geqslant \mathbf{1}$; more precisely, it says that Y_2 is stochastically larger than Y_1. Similarly, each of $\boldsymbol{\theta}^L \geqslant \mathbf{1}$ and $\boldsymbol{\theta}^{CO} \geqslant \mathbf{1}$ also captures a form of stochastic order. The following terms have been used in the literature for these stochastic orders.

$\boldsymbol{\theta}^L \geqslant \mathbf{1}$: *Likelihood ratio order,* which we denote by $\boldsymbol{\pi}_1 \preceq_\ell \boldsymbol{\pi}_2$

$\boldsymbol{\theta}^{CO} \geqslant \mathbf{1}$: *Uniform stochastic order,* which we denote by $\boldsymbol{\pi}_1 \preceq_u \boldsymbol{\pi}_2$

$\boldsymbol{\theta}^C \geqslant \mathbf{1}$: *Simple stochastic order,* which we denote by $\boldsymbol{\pi}_1 \preceq_s \boldsymbol{\pi}_2$

6.4.3 $r \times c$ Tables

Now, let us consider an $r \times c$ table in which the rows and columns are ordered. There are at least 16 different odds ratios; they are referred to as *generalized odds ratios.* For illustrative purposes, four of them are defined below. Let $i = 1, \ldots, r-1$, and $j = 1, \ldots, c-1$. Then

$$\textit{Local odds ratio:} \quad \theta_{ij}^L = \frac{\mathrm{pr}(Y{=}j|X{=}i)\ \mathrm{pr}(Y{=}j{+}1|X{=}i{+}1)}{\mathrm{pr}(Y{=}j|X{=}i{+}1)\ \mathrm{pr}(Y{=}j{+}1|X{=}i)} \tag{6.29}$$

$$\textit{Cumulative odds ratio:} \quad \theta_{ij}^C = \frac{\mathrm{pr}(Y{\le}j|X{=}i)\ \mathrm{pr}(Y{>}j|X{=}i{+}1)}{\mathrm{pr}(Y{\le}j|X{=}i{+}1)\ \mathrm{pr}(Y{>}j|X{=}i)} \tag{6.30}$$

$$\textit{Global odds ratio:} \quad \theta_{ij}^G = \frac{\mathrm{pr}(Y{\le}j|X{\le}i)\ \mathrm{pr}(Y{>}j|X{>}i)}{\mathrm{pr}(Y{\le}j|X{\geqslant}i)\ \mathrm{pr}(Y{>}j|X{<}i)} \tag{6.31}$$

$$\textit{Continuation odds ratio:} \quad \theta_{ij}^{CO} = \frac{\mathrm{pr}(Y{=}j|X{=}i)\ \mathrm{pr}(Y{>}j|X{=}i{+}1)}{\mathrm{pr}(Y{=}j|X{=}i{+}1)\ \mathrm{pr}(Y{>}j|X{=}i)}. \tag{6.32}$$

These definitions show that generalized odds ratios are odds ratios for 2×2 tables obtained from the $r \times c$ table, by collapsing adjacent categories if necessary. Fig. 6.1

shows several 2×2 tables obtained by merging adjacent categories. For the purposes of interpreting different generalized odds ratios, it would be helpful to view them as odds ratios of the corresponding 2×2 tables, which have simple interpretations.

If $\boldsymbol{\theta}^C \geqslant \mathbf{1}$ then clearly the rows satisfy the simple stochastic order $\boldsymbol{\pi}_1 \preceq_s \dots \preceq_s \boldsymbol{\pi}_r$. Similarly, if $\boldsymbol{\theta}^L \geqslant \mathbf{1}$ then the rows satisfy the likelihood ratio order $\boldsymbol{\pi}_1 \preceq_l \dots \preceq_l \boldsymbol{\pi}_r$, and if $\boldsymbol{\theta}^{CO} \geqslant \mathbf{1}$ then the rows satisfy the uniform stochastic order $\boldsymbol{\pi}_1 \preceq_u \dots \preceq_u \boldsymbol{\pi}_r$. Each of these constraints define a form of monotone order of the rows involving only two rows at a time. By contrast, the constraint $\boldsymbol{\theta}^G \geqslant \mathbf{1}$, defines a monotone relationship which involves more than two rows at a time; this constraint is equivalent to

$$\text{pr}(Y \geqslant j \mid X > i) \geqslant \text{pr}(Y \geqslant j \mid X \leq i).$$

There are important connections among the generalized odds ratios. An excellent discussion of generalized odds ratios is provided in Douglas et al. (1991). Since the use of different terms may cause ambiguity, we shall use the hieroglyphic-like symbolic notation in Douglas et al. (1991). A single cell is denoted by $\bullet$; a horizontal strip of the table formed by several adjacent columns in one row is denoted by $-$; a vertical strip of the table formed by several adjacent rows in one column is denoted by $|$; a rectangular section consisting of the cells in the bottom-right, top-left, top-right, and bottom-left are denoted by $\lceil$, $\rfloor$, $\lfloor$ and $\rceil$, respectively. In this notation,

$$(\pi_{ij}\pi_{i+1,j+1})/(\pi_{i+1,j}\pi_{i,j+1}) \geqslant 1, \ \forall(i,j) \text{ is represented by } \bullet\bullet \geqslant \mathbf{1};$$

the first $\bullet$ says to construct a 2×2 table in which the $(1,1)$ cell is the single (i,j) cell of the original $r \times c$ table, and the second $\bullet$ says that the $(2,2)$ cell is the single $(i+1, j+1)$ cell of the original $r \times c$ table. Similarly, $--\ \geqslant \mathbf{1}$ means that for the relevant 2×2 table, the $(1,1)$ cell is formed by merging $(i,1), \dots, (i,j)$ and the $(2,2)$ cell is formed by merging $(i+1, j+1), \dots, (i+1, k)$.

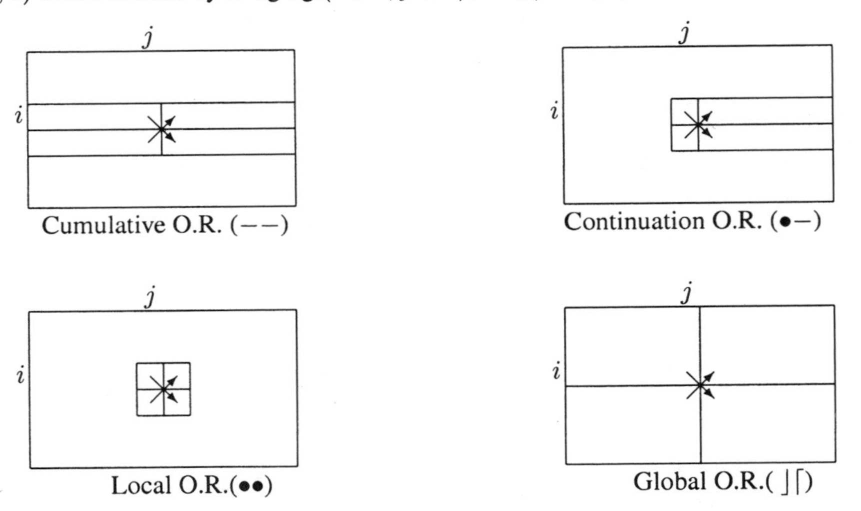

Fig. 6.1 Four generalized Odds ratios; continuous lines define a typical 2×2 table for the odds ratio

Let $\boldsymbol{\theta}$ denote a vector of generalized odds ratios of a particular type; for example it could be $\boldsymbol{\theta}^L$ or $\boldsymbol{\theta}^C$. Then, $\boldsymbol{\theta} \geqslant \mathbf{1}$ defines a corresponding stochastic order among the rows, and also a form of monotone dependence between X and Y. In this regard, the following result is useful; see Douglas et al. (1991) for proof.

Proposition 6.4.3 *Every implication in Fig. 6.2 is strict; in other words, for any implication, the reverse implication does not hold. If $\boldsymbol{\theta}$ denotes the vector of anyone of the generalized odds ratios in Figure 6.1, then $\boldsymbol{\theta} = \mathbf{1}$ if and only if the row and columns are independent. Consequently, $\boldsymbol{\theta}^L \gneqq \mathbf{1} \Rightarrow \boldsymbol{\theta} \gneqq \mathbf{1} \Rightarrow \boldsymbol{\theta}^G \gneqq \mathbf{1}$; further $\boldsymbol{\theta}^A = \mathbf{1}$ if and only if $\boldsymbol{\theta}^B = \mathbf{1}$ where $\boldsymbol{\theta}^A$ and $\boldsymbol{\theta}^B$ are any two generalized odds ratios.*

Thus, $\boldsymbol{\theta}^L \geqslant \mathbf{1}$ defines the most stringent form of stochastic order, $\boldsymbol{\theta}^G \geqslant \mathbf{1}$ defines the most liberal form of stochastic order, and $\boldsymbol{\theta} \geqslant \mathbf{1}$ defines a stochastic order that is in between these two, where $\boldsymbol{\theta}$ is any of the other generalized odds ratios.

The hierarchy among the bivariate dependence properties is given in Fig. 6.2. Every implication in this figure is strict; in other words, the reverse implication does not hold. Further, the rows and columns are independent if and only if the vector of any one of the generalized odds ratios is equal to $\mathbf{1}$.

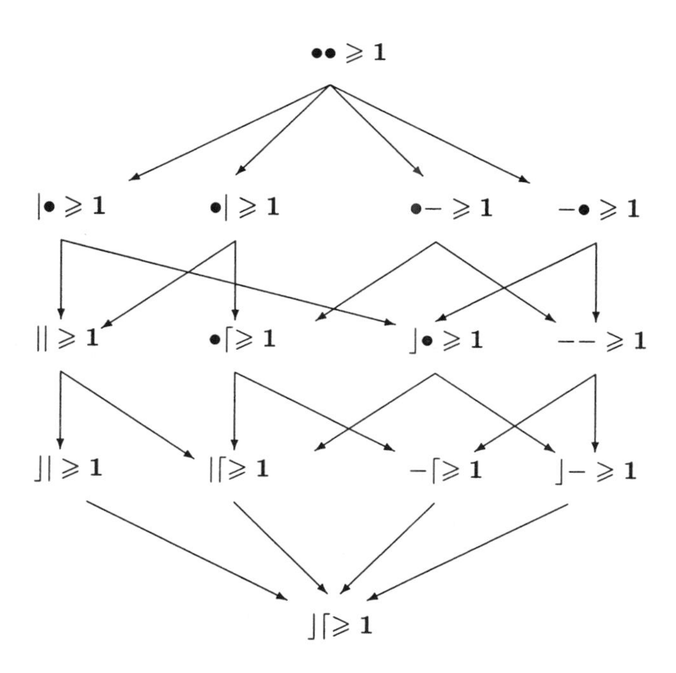

Fig. 6.2 Hierarchy among bivariate dependence properties. [Reprinted from: Topics in Statistical Dependence, R. Douglas, S. E. Fienberg, M. -L. Lee, A. R. Sampson, and L. R. Whitaker, Positive dependence concepts for ordinal contingency tables, Pages 189–202, Copyright (1991), with permission from the Institute of Mathematical Statistics.]

6.5 ANALYSIS OF $2 \times C$ CONTINGENCY TABLES

In this section, we consider approaches to statistical inference for the comparison of two treatments when the response variable is ordinal. The traditional method of inference for this problem is based on likelihood ratio and its equivalents. However, exact conditional methods also have very attractive features including simplicity and validity for any sample size. Once the approaches to inference in $2 \times c$ tables have been developed, they can be extended to more than two rows where ordering of the rows may be partial or total. Such extensions will be considered in the next section.

Let us consider two treatments for which the response variable is ordinal with c categories. The two treatments can be thought of as two ordinal categories of a variable X. Let the response variable be denoted by Y. Let us recall

$$\pi_{ij} = \text{pr}(Y = j \mid X = i) \text{ and } \boldsymbol{\pi}_i = (\pi_{i1}, \ldots, \pi_{i,c-1})^T$$

for $i = 1, 2$ and $j = 1, \ldots, c$. Let

$$\boldsymbol{\pi} = (\boldsymbol{\pi}_1 : \boldsymbol{\pi}_2).$$

Note that $\boldsymbol{\pi}$ does not include π_{1c} and π_{2c}; this was done to avoid the redundancy due to $\pi_{i1} + \cdots + \pi_{ic} = 1$, for $i = 1, 2$. Let n_{ij} denote the frequency for cell (i, j), $n_{i+} = n_{i1} + \cdots + n_{ic}$ and $n_{+j} = n_{1j} + n_{2j}$, for $i = 1, 2$ and $j = 1, \ldots, c$ (see Table 6.11). Let

$$\ell(\boldsymbol{\pi}) = \sum_{i=1}^{2}\sum_{j=1}^{c} n_{ij} \log \pi_{ij} \tag{6.33}$$

$$= \sum_{i=1}^{2}\left\{\left(\sum_{j=1}^{c-1} n_{ij} \log \pi_{ij}\right) + n_{ic} \log(1 - \pi_{i1} \cdots - \pi_{i,c-1})\right\}; \tag{6.34}$$

this is the kernel of the log likelihood. Let $\hat{\boldsymbol{\pi}}$ denote the unconstrained *mle* of $\boldsymbol{\pi}$. Then,

$$\sqrt{n}(\hat{\boldsymbol{\pi}} - \boldsymbol{\pi}) \xrightarrow{d} N\{\mathbf{0}, \boldsymbol{i}(\boldsymbol{\pi})^{-1}\}$$

Table 6.11 Notation for comparison of two treatments

	Probability				Frequency			
	Y				Y			
	1	...	c	Total	1	...	c	Total
Treatment 1	π_{11}	...	π_{1c}	1	n_{11}	...	n_{1c}	n_{1+}
Treatment 2	π_{21}	...	π_{2c}	1	n_{21}	...	n_{2c}	n_{2+}
Total	π_{+1}	...	π_{+c}	1	n_{+1}	...	n_{+c}	n

where $\boldsymbol{i}(\boldsymbol{\pi})$ is the information matrix. It may be verified that

$$\boldsymbol{i}(\boldsymbol{\pi}) = \text{diag}\{a_1\boldsymbol{W}(\boldsymbol{\pi}_1), a_2\boldsymbol{W}(\boldsymbol{\pi}_2)\} \tag{6.35}$$

where

$$a_i = \lim(n_i/n) \quad \text{and} \quad \boldsymbol{W}(\boldsymbol{\pi}_i) = \text{diag}(\boldsymbol{\pi}_i)^{-1} + \mathbf{1}\mathbf{1}^T/\pi_{i,c}, i = 1, 2. \tag{6.36}$$

For a direct derivation of the asymptotic covariance matrix of $\sqrt{n}(\hat{\boldsymbol{\pi}} - \boldsymbol{\pi})$, see Agresti (1990, p 423). It is assumed that $0 < a_1, a_2 < 1$, and that a_1 and a_2 are known; in applications, we would simply substitute (n_i/n) for $a_i (i = 1, 2)$. For computations, it is helpful to note that

$$\boldsymbol{W}(\boldsymbol{\pi}_i)^{-1} = \text{diag}(\boldsymbol{\pi}_i) - \boldsymbol{\pi}_i\boldsymbol{\pi}_i^T, \; i = 1, 2. \tag{6.37}$$

Let Ω denote the unconstrained parameter space for $\boldsymbol{\pi}$; thus, we have that

$$\Omega = \{\boldsymbol{\pi} : 0 < \pi_{ij} \text{ for every } (i, j) \text{ and } \pi_{i1} + \cdots + \pi_{i,c-1} < 1 \text{ for every } i\}.$$

Since this is open, subsection 4.8.1 (page 195) is relevant. Let us write the testing problem as

$$H_0 : \boldsymbol{\pi} \in \Omega_0 \quad \text{vs} \quad H_1 : \boldsymbol{\pi} \in \Omega_1, \tag{6.38}$$

respectively, where $\Omega_0 \subset \Omega_1 \subseteq \Omega$. We shall assume that Ω_0 and Ω_1 are Chernoff regular; this mild requirement is satisfied in every example considered here and it is difficult to think of a practical scenario where this is not satisfied. As was mentioned in Chapter 4, Chernoff regularity of parameter spaces is assumed throughout unless the contrary is made clear. Let T denote the likelihood ratio or any of its equivalents, such as W_D, in Proposition 4.8.1. Then, the asymptotic null distribution of T at $\boldsymbol{\pi}_0$ in Ω_0 is equal to that of

$$\|\boldsymbol{Z} - \mathcal{T}(\Omega_0; \boldsymbol{\pi}_0)\|^2 - \|\boldsymbol{Z} - \mathcal{T}(\Omega_1; \boldsymbol{\pi}_0)\|^2 \tag{6.39}$$

where

$$\|\boldsymbol{x}\|^2 = \boldsymbol{x}^T\boldsymbol{i}_0\boldsymbol{x}, \quad \boldsymbol{i}_0 = \boldsymbol{i}(\boldsymbol{\pi}_0), \quad \boldsymbol{Z} \sim N(\mathbf{0}, \boldsymbol{i}_0^{-1}),$$

and $\mathcal{T}$ denote the tangent cone. This general result will be used throughout this section for deducing the asymptotic null distribution of T. As a special case of the foregoing general results, let us consider the testing problem,

$$H_0 : \boldsymbol{\pi}_2 = \boldsymbol{\pi}_1 \quad \text{vs} \quad H_1 : \boldsymbol{h}(\boldsymbol{\pi}_2) \geqslant \boldsymbol{h}(\boldsymbol{\pi}_1) \tag{6.40}$$

where $\boldsymbol{h}$ is a smooth vector function. Several important testing problems are of this form. It is easily seen that if $\boldsymbol{\theta}$ denotes one of the generalized odds ratios in Fig. 6.2, then "$H_0 : \boldsymbol{\theta} = \mathbf{1}$ vs $H_1 : \boldsymbol{\theta} \geqslant \mathbf{1}$" is of the form (6.40) for some smooth function $\boldsymbol{h}$. Thus, the likelihood ratio order (i.e., $\boldsymbol{\theta}^L \geqslant \mathbf{1}$), simple stochastic order (i.e., $\boldsymbol{\theta}^C \geqslant \mathbf{1}$), uniform stochastic order (i.e., $\boldsymbol{\theta}^{CO} \geqslant \mathbf{1}$) are of the form $\boldsymbol{h}(\boldsymbol{\pi}_2) \geqslant \boldsymbol{h}(\boldsymbol{\pi}_1)$ for some smooth vector function $\boldsymbol{h}$. Hence (6.40) encompasses a very large class of testing problems of interest. Suppose that $H_0 : \boldsymbol{\pi}_1 = \boldsymbol{\pi}_2$ is true, and let $\boldsymbol{\pi}_0$ denote the true value of π; then we have that $\pi_0 = (\pi_{01} : \pi_{01})$ for some π_{01}. Now, it is easily seen that

$$\mathcal{T}(\Omega_0; \pi_0) = \{\pi : \pi_1 = \pi_2\} \quad \text{and} \quad \mathcal{T}(\Omega_1; \pi_0) = \{\pi : A_0\pi \geqslant 0\}$$

where $A_0 = (-B_0 : B_0)$ and $B_0 = (\partial h/\partial \pi^T)$ evaluated at π_{01} (see Proposition 4.7.3). Since $\mathcal{T}(\Omega_0; \pi_0)$ is a linear space, $\mathcal{T}(\Omega_1; \pi_0)$ is a closed convex cone and $\mathcal{T}(\Omega_0; \pi_0) \subset \mathcal{T}(\Omega_1; \pi_0)$, it follows from Corollary 3.7.1 that

$$\text{pr}(T \geqslant t \mid \pi = \pi_0) \to \text{pr}\{\bar{\chi}^2(i_0^{-1}, \mathcal{C}_0) \geqslant t\},$$

where $i_0 = i(\pi_0)$, $i(\pi)$ is as in (6.35) and

$$\mathcal{C}_0 = \mathcal{T}(\Omega_0; \pi_0)^{\perp} \cap \mathcal{T}(\Omega_1; \pi_0)$$

with the orthogonal complement of $\mathcal{T}(\Omega_0; \pi_0)$ being with respect to i_0^{-1}. In general, $\bar{\chi}^2(i_0^{-1}, \mathcal{C}_0)$ depends on the assumed true null value π_0, and therefore asymptotic critical values and p-values need to be computed at a least favorable null value, which would depend on the form of h. Now, let us consider some important special forms of h.

6.5.1 LRT of $\pi_1 = \pi_2$ Against Simple Stochastic Order

In this section we shall develop the results to apply *LRT* for testing the null hypothesis of no difference between the two treatments against the alternative that Treatment 2 is better than Treatment 1 with respect to simple stochastic order, which by definition means that $\gamma_1 \geqslant \gamma_2$; see (6.17) (page 299) for definition of γ parameters.

Without loss of generality, assume that the ordinal categories of the response variable Y are in an increasing order so that higher values of Y are preferred. Let the null and the alternative hypotheses be

$$H_0 : \pi_1 = \pi_2 \text{ and } H_1 : \pi_{11} + \ldots + \pi_{1j} \geqslant \pi_{21} + \ldots + \pi_{2j}, \text{ for every } j, \quad (6.41)$$

respectively. For the rest of this subsection, we shall interpret the foregoing H_1 as equivalent to "Treatment 2 is at least as good as Treatment 1." Further, if H_1 holds and one of the inequalities in it holds strictly then we shall interpret it as equivalent to "Treatment 2 is better than Treatment 1." To express (6.41) in a more compact form, let A be the $(c-1) \times (c-1)$ lower triangular matrix in which every element on the diagonal and below is one, and every element above the diagonal is a zero. Let $R = [A, -A]$; thus R is of order $(c-1) \times (2c-2)$. Now (6.41) may be written as

$$H_0 : R\pi = 0 \text{ and } H_1 : R\pi \geqslant 0. \quad (6.42)$$

Let $\tilde{\pi}$ and $\bar{\pi}$ denote the *mle*'s under H_1 and H_0, respectively. Then, by (6.34),

$$LRT = 2\sum_{i=1}^{2}\sum_{j=1}^{c} n_{ij} \log(\tilde{\pi}_{ij}/\bar{\pi}_{ij}). \quad (6.43)$$

Let $\boldsymbol{\pi}_0 = (\boldsymbol{\pi}_{01} : \boldsymbol{\pi}_{01})$ be an arbitrary point in the null parameter space. It follows from (6.39) and Corollary 4.3.1 that the asymptotic null distribution of the *LRT* at $\boldsymbol{\pi}_0$ is given by

$$\text{pr}(\text{LRT } \geqslant t \mid \boldsymbol{\pi} = \boldsymbol{\pi}_0) \to \sum_{i=0}^{c-1} w_i(c-1, \boldsymbol{V}_0)\text{pr}(\chi_i^2 \geqslant t), \tag{6.44}$$

where $\boldsymbol{V}_0 = \boldsymbol{R}\boldsymbol{i}_0^{-1}\boldsymbol{R}^T$ and $\boldsymbol{i}(\boldsymbol{\pi})$ is defined in (6.35). Let $\boldsymbol{W}_0 = \boldsymbol{W}(\boldsymbol{\pi}_{01})$ where $\boldsymbol{W}(\cdot)$ is as in (6.36).

It is easily verified that

$$\boldsymbol{V}_0 = c_0 \boldsymbol{A}\boldsymbol{W}_0^{-1}\boldsymbol{A}^T, \tag{6.45}$$

where $c_0 = (a_1 + a_2)^{-1}$ with $a_i = lim(n_i/n)$ (see (6.36)) and hence the limit in (6.44) does not depend on it (i.e., c_0). Since the $\boldsymbol{V}_0$ in (6.44) is a function of $\boldsymbol{\pi}_0$ in the null parameter space, the asymptotic p-value corresponding to $LRT = t$ is the supremum over $\boldsymbol{\pi}_0$ of the $\bar{\chi}^2$ tail probability in (6.44). Thus,

$$\text{asymptotic } p\text{-value} = \sup_{\boldsymbol{\pi}_{01}} \sum_{i=0}^{c-1} w_i(c-1, \boldsymbol{A}\boldsymbol{W}_0^{-1}\boldsymbol{A}^T)\text{pr}(\chi_i^2 \geqslant t). \tag{6.46}$$

The next result provides a simple closed-form for this asymptotic p-value and the least favorable null value for the asymptotic distribution of *LRT*.

Proposition 6.5.1 *Let the testing problem be as in (6.41). Then (i) the LRT is given by (6.43), (ii) a least favorable null value for the asymptotic distribution of LRT is* $\boldsymbol{\pi}_*$ *where* $\boldsymbol{\pi}_* = (\boldsymbol{\pi}_{*1} : \boldsymbol{\pi}_{*1})$ *and* $\boldsymbol{\pi}_{*1} = (\frac{1}{2}, 0, \ldots, 0)^T$, *and (iii)*

$$\sup_{H_0} \lim_{n \to \infty} pr(LRT \geqslant t \mid H_0) = (1/2)\{pr(\chi_{c-2}^2 \geqslant t) + pr(\chi_{c-1}^2 \geqslant t)\}. \tag{6.47}$$

The proof is given in the Appendix. It follows from the foregoing result that if the sample value of *LRT* is t, then

$$\text{the asymptotic } p\text{-value } = (1/2)\{\text{pr}(\chi_{c-2}^2 \geqslant t) + \text{pr}(\chi_{c-1}^2 \geqslant t)\}. \tag{6.48}$$

In view of part (ii) of the foregoing proposition, the asymptotic p-value in (6.48) is computed at the limiting null value $\boldsymbol{\pi}_* = (\boldsymbol{\pi}_{*1} : \boldsymbol{\pi}_{*1})$ where $\boldsymbol{\pi}_{*1} = (1/2, 0, \ldots, 0)^T$; thus, the least favorable null value assigns probability (1/2) for the first and the last ordinal categories, and hence $(\pi_{*11}, \ldots, \pi_{*1c}) = (1/2, 0, \ldots, 0, 1/2)$. In many practical situations, a common multinomial probability vector close to this least favorable null value would be considered unrealistic, although it would be subjective to partition the null parameter space, $\{(\boldsymbol{\pi}_1 : \boldsymbol{\pi}_2) : \boldsymbol{\pi}_1 = \boldsymbol{\pi}_2\}$, into a set of realistic and another set of unrealistic values. This difficulty can be overcome to a great extent by rejecting the null hypothesis for small values of the supremum in (6.46) where the supremum is taken over a confidence region for $\boldsymbol{\pi}_{01}$, rather than over all possible values of $\boldsymbol{\pi}_{01}$. For details and discussion of this approach see Section 4.3.2; although this approach is easy to understand, the implementation of the procedure is not easy and more work remains to be done.

Another approach that may be useful to present the evidence against H_0 is the following: Let $\pi_0(\epsilon) = (\pi_{01}(\epsilon) : \pi_{01}(\epsilon))$ where $\pi_{01}(\epsilon) = (1/2)(1-(c-2)\epsilon, \epsilon, \ldots, \epsilon)^T$, where $\epsilon > 0$. Then, as $\epsilon \to 0, \pi_0(\epsilon)$ converges to the least favorable null value and the tail probability of the chi-bar square distribution in (6.44) at $\pi_0(\epsilon)$ converges to the asymptotic p-value in (6.48). Therefore, it would be helpful to tabulate values of the tail probability of the foregoing Chi-bar-square distribution at $\pi_0(\epsilon)$ against ϵ; this is illustrated in the next example.

Example 6.5.1

Let us consider Example 6.2.2, which involves the comparison of two treatments. Let the null and the alternative hypotheses be

$$H_0 : \pi_1 = \pi_2 \text{ and } H_1; \pi_{11} + \ldots + \pi_{1j} \geqslant \pi_{21} + \ldots + \pi_{2j}, j = 1, 2, 3,$$

respectively. This is the same as (6.40) with $A_{3\times 3} = [1,0,0 \mid 1,1,0 \mid 1,1,1]$ and $\boldsymbol{R} = [\boldsymbol{A}, -\boldsymbol{A}]$. For cell (i,j), the unconstrained estimator $\hat{\pi}_{ij}$ is the observed sample proportion for that cell. Therefore, $\hat{\pi} = 32^{-1}(12, 10, 4, 5, 8, 8^T)$. Since $\boldsymbol{R}\hat{\pi} = 32^{-1}(7, 9, 5)^T > \mathbf{0}$, the unconstrained estimator of π satisfies the constraints in H_1 and hence $\tilde{\pi} = \hat{\pi}$. Since H_0 is equivalent to $\pi_1 = \pi_2$, the *mle* of π_{ij} under H_0 is (n_{+j}/n) for every $(i,j), i = 1, 2,\ j = 1, \ldots, 4$. Therefore, $\bar{\pi} = 64^{-1}(17, 18, 12, 17, 18, 12)^T$. Now, LRT $= 2\sum_{i=1}^{2}\sum_{j=1}^{4} n_{ij}\log(\tilde{\pi}_{ij}/\bar{\pi}_{ij}) = 6.0$. Let π_0 be an arbitrary point in the null parameter space. Then a large sample approximation of $\text{pr}(LRT \geqslant 6.0 \mid \pi = \pi_0)$ is

$$\sum_{i=0}^{3} w_i\{3, \boldsymbol{V}(\pi_0)\}\text{pr}(\chi_i^2 \geqslant 6.0) \tag{6.49}$$

where $\boldsymbol{V}(\pi_0) = \boldsymbol{A}\{\text{diag}(\pi_{01}) - \pi_{01}\pi_{01}^T\}\boldsymbol{A}^T$. Since this tail probability depends on π_0, the asymptotic p-value is its the supremum over $\pi_0 \in H_0$. Thus, the asymptotic p-value $= (1/2)\{\text{pr}(\chi_2^2 \geqslant 6.0) + \text{pr}(\chi_3^2 \geqslant 6.0)\} = 0.083$.

The least favorable null value for the asymptotic distribution of LRT is $\pi_* = (\pi_{*1} : \pi_{*1})$ where $\pi_{*1} = (1/2, 0, 0)^T$. Let $\pi_{01}(\epsilon) = (1/2)(1-2\epsilon, \epsilon, \epsilon)^T, \pi_0(\epsilon) = (\pi_{01}(\epsilon), \pi_{01}(\epsilon))$ and $q(\epsilon)$ denote the value of (6.49) at $\pi_0(\epsilon)$. Values of $q(\epsilon)$ for different values of ϵ are given in Table 6.12.

Fig. 6.3 shows that $q(\epsilon)$ approaches its supremum, the asymptotic p-value, at an increasing rate as ϵ decreases to zero. In practice, it is not difficult to rule out very

Table 6.12 Tail probability near the least favorable null value.

ϵ	0.0	0.0001	0.001	0.01	0.05
$q(\epsilon)$	0.083	0.0803	0.0795	0.077	0.072

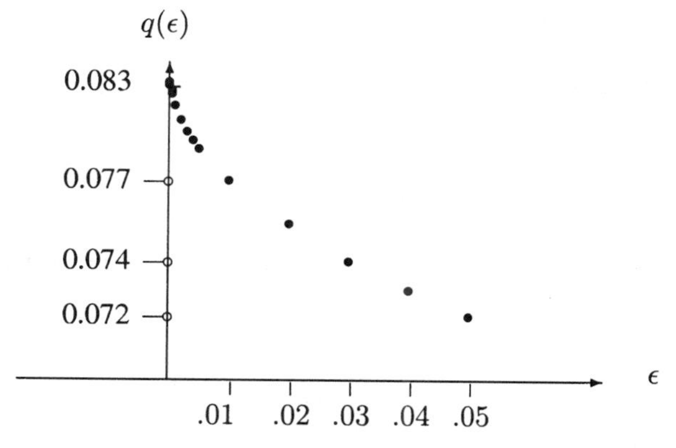

Fig. 6.3 The tail probability near the least favourable null value

small values for any π_{ij}. Thus, it appears that there is some merit in summarizing the statistical evidence as in Fig. 6.3.

If we were to test $H_0 : \boldsymbol{\pi}_1 = \boldsymbol{\pi}_2$ against the unrestricted alternative $H_1 : \boldsymbol{\pi}_1 \neq \boldsymbol{\pi}_2$, then $LRT = 2\sum_{i=1}^{2}\sum_{j=1}^{4} n_{ij}\log(\hat{\pi}_{ij}/\bar{\pi}_{ij}) = 6.0$; in this particular example, this is equal to the *LRT* for testing $H_0 : \boldsymbol{\pi}_1 = \boldsymbol{\pi}_2$ against the restricted alternative $H_1 : \boldsymbol{R\pi} \geqslant \boldsymbol{0}$, because $\hat{\boldsymbol{\pi}} = \tilde{\boldsymbol{\pi}}$. However, the asymptotic p-value for testing against H_2 is $\text{pr}(\chi_3^2 \geqslant 6.0) = 0.11$. Note that the p-value for testing against the unrestricted alternative is substantially larger than that for testing against the restricted alternative $\boldsymbol{R\pi} \geqslant \boldsymbol{0}$ because the $\bar{\chi}^2$ distribution has shorter tails than the chi-square distribution for the test against the corresponding unrestricted alternative. More generally, let us note the following, which has been pointed out earlier. Suppose that the unconstrained estimator satisfies the constraints in H_1. Then the *LRT* for testing against the unrestricted alternative H_2 and that against the restricted alternative H_1 are equal. Since the asymptotic null distribution of *LRT* against the restricted alternative has shorter tails than that against the unrestricted alternative, the latter has larger p-value. Consequently, the test against the restricted alternative would give stronger evidence to reject H_0. ■

6.5.2 LRT for Simple Stochastic Order: $\boldsymbol{R\pi} \geqslant \boldsymbol{0}$ vs $\boldsymbol{R\pi} \not\geqslant \boldsymbol{0}$

Let the null and alternative hypotheses be

$$H_1 : \pi_{11}+\ldots+\pi_{1j} \geqslant \pi_{21}+\ldots+\pi_{2j} \text{ for } j = 1,\ldots,c-1, \text{ and } H_2 : \textit{not } H_1, \quad (6.50)$$

respectively. Let $\boldsymbol{R}$ be as in (6.42). Then (6.50) may be written as

$$H_1 : \boldsymbol{R\pi} \geqslant \boldsymbol{0} \quad \text{and} \quad H_2 : \boldsymbol{R\pi} \not\geqslant \boldsymbol{0}, \quad (6.51)$$

respectively. Let $\tilde{\boldsymbol{\pi}}$ denote the *mle* of $\boldsymbol{\pi}$ under H_1; similarly, let $\hat{\boldsymbol{\pi}}$ denote the unconstrained *mle* of $\boldsymbol{\pi}$. Then the *LRT* for testing H_1 against H_2 in (6.51) is

$$2\sum_{i=1}^{2}\sum_{j=1}^{c} n_{ij}\log(\hat{\pi}_{ij}/\tilde{\pi}_{ij}). \tag{6.52}$$

This test is not similar, but closed forms are available for its least favorable null value and the p-value; these are given in the next result.

Proposition 6.5.2 *Let the testing problem be as in (6.50) or equivalently (6.51). Then the LRT is given by (6.52), the least favorable null value, $\boldsymbol{\pi}_*$, is given by $\boldsymbol{\pi}_* = (\boldsymbol{\pi}_{*1} : \boldsymbol{\pi}_{*1})$ where $\boldsymbol{\pi}_{*1} = (1, 0, \ldots, 0)^T$, and*

$$\sup_{\boldsymbol{R\pi}\geqslant \mathbf{0}} \lim pr_{\boldsymbol{\pi}}(LRT \geqslant t) = \sum_{i=1}^{c-1} 2^{c-1}(c-1)!\{i!(c-1-i)!\}^{-1} pr(\chi_i^2 \geqslant t). \tag{6.53}$$

See Dardanoni and Forcina (1998) or Theorem 5.4.6 in Robertson et al. (1988) for a proof. If t is the sample value of *LRT*, then it follows that the asymptotic p-value for *LRT* is the limit in (6.53).

To provide more insight into the form of the behavior of the asymptotic null distribution of *LRT*, let $\boldsymbol{\pi}_0$ be a point in the null parameter space, $\Omega_0 = \{\boldsymbol{\pi} : \boldsymbol{R\pi} \geqslant \mathbf{0}\}$. Let $\Omega_1 = \{\boldsymbol{\pi} : \pi_{ij} > 0 \text{ for every } (i,j)\}$, $\boldsymbol{i}(\boldsymbol{\pi})$ be the information matrix in (6.35), $\boldsymbol{i}_0 = \boldsymbol{i}(\boldsymbol{\pi}_0)$, and $\boldsymbol{Z} \sim N(\mathbf{0}, \boldsymbol{i}_0^{-1})$. Since $\boldsymbol{\pi}_0$ is an interior point of $\Omega_1 \subseteq \mathbb{R}^{2(c-1)}$, we have that $\mathcal{T}(\Omega_1; \boldsymbol{\pi}_0) = \mathbb{R}^{2(c-1)}$. Consequently, the asymptotic distribution of the *LRT* at $\boldsymbol{\pi}_0$ is equal to that of $\|\boldsymbol{Z} - \mathcal{T}(\Omega_0; \boldsymbol{\pi}_0)\|^2$. Let $\boldsymbol{P}$ be the matrix formed by the rows of $\boldsymbol{R}$ corresponding to the active constraints at $\boldsymbol{\pi}_0$; let $\boldsymbol{Q}$ be the matrix formed by the remaining rows of $\boldsymbol{R}$. Thus, $\boldsymbol{P\pi}_0 = 0$ and $\boldsymbol{Q\pi}_0 > \mathbf{0}$. Then, it is easily seen that (see Example 4.7.4)

$$\mathcal{T}\{\Omega_0; \boldsymbol{\pi}_0\} = \{\boldsymbol{\pi} : \boldsymbol{P\pi} \geqslant \mathbf{0}\}.$$

Therefore, it follows that the asymptotic null distribution of the *LRT* for the testing problem (6.51), when $\boldsymbol{\pi}_0$ is the true value, is equal to that of

$$\inf\{(\boldsymbol{Z} - \boldsymbol{\pi})^T \boldsymbol{i}_0 (\boldsymbol{Z} - \boldsymbol{\pi}) : \boldsymbol{P\pi} \geqslant \mathbf{0}\}$$

where $\boldsymbol{Z} \sim N(\mathbf{0}, \boldsymbol{i}_0^{-1})$ and $\boldsymbol{i}_0 = \boldsymbol{i}(\boldsymbol{\pi}_0)$; this is a chi-bar square distribution.

For simplicity, let us consider the special case when $\boldsymbol{P}$ has full row-rank. Then by Corollary 3.6.2 (on page 83) we have

$$\text{pr}(LRT \geqslant t \mid \boldsymbol{\pi} = \boldsymbol{\pi}_0) \to \sum_{j=0}^{s} w_{s-j}(s, \boldsymbol{P}\boldsymbol{i}_0^{-1}\boldsymbol{P}^T)\text{pr}(\chi_j^2 \geqslant t), \tag{6.54}$$

where s is the number of rows in $\boldsymbol{P}$. The supremum of this over $\boldsymbol{\pi}_0 \in H_0$ is the limit in (6.53). Since s, the number of rows of $\boldsymbol{P}$, depends on $\boldsymbol{\pi}_0$, the number of summands and the dimension of the covariance matrix in the chi-bar square distribution in (6.54) also depends on π_0. By contrast, for testing $R\pi = 0$ vs $R\pi \geq 0$, the asymptotic null distribution of the *LRT* is the chi-bar square in (6.44) and the number of summands and the dimension of the covariance matrix in the chi-bar square distribution remain the same at every point in the null parameter space. This is closely tied to the fact that the relevant tangent cones for testing $R\pi = 0$ vs $R\pi \geq 0$ are the same at every point in the null parameter space, but not for testing $R\pi \geq 0$ vs $R\pi \not\geq 0$.

6.6 TEST TO ESTABLISH THAT TREATMENT IS BETTER THAN CONTROL

Let the matrix R and the parameter π be as in the previous section for $2 \times c$ tables. Therefore, $R\pi = 0$ is equivalent to $\pi_1 = \pi_2$, which is the same as saying no difference between the treatment and the control; similarly, $R\pi \geqslant 0$ is equivalent to the treatment is at least as good as the control. Suppose that a test of $R\pi = 0$ against $R\pi \geqslant 0$ rejects $R\pi = 0$. Then it does follow that there is statistical evidence to establish that there is difference between the treatment and the control, but it does not follow that there is statistical evidence to accept that the treatment is better than the control. However, if we make the prior assumption that the treatment is at least as good as the control, then the rejection of "no difference between the treatment and the control" together with the prior assumption would lead to the conclusion that the treatment is better than the control.

In some studies the objective is to provide statistical evidence to establish that the treatment is better than the control without making the prior assumption that treatment is at least as good as the control. In a clinical trial, where the objective is to establish that a new drug is better than the placebo, it may be desirable not to assume that the drug is at least as good as the placebo. Consequently, it would be inadequate/unacceptable to formulate the null hypothesis as "no difference between the treatment and the control." We need to formulate the null and alternative hypotheses as

$$H_0: \text{ not } H_1, \quad \text{and} \quad H_1: \text{Treatment is better than the control.}$$

With this formulation, if a test rejects H_0, then there would be sufficient statistical evidence to accept that the treatment is better than the control, without making the prior assumption that the treatment is at least as good as the control. In this section, we shall consider some aspects of this problem. Theoretical aspects of this are also discussed in more detail under Intersection Union Test (IUT).

Since any point in $A = \{\pi : R\pi \geqslant 0, R\pi \neq 0\}$ corresponds to "treatment is better than the control," it is tempting to define the null and alternative hypotheses as $H_0 : \pi \notin A$ and $H_1 : \pi \in A$, respectively. However, this would cause difficulties because A is not a closed set. One way of overcoming the problem is to include the boundary of A into the null parameter space. This leads to a formulation for which an IUT can be applied. However, it may not be the most appropriate in all situations because the null parameter space may be very large, which would result in loss of power. We shall first consider a fairly restrictive formulation of the hypotheses; later we shall suggest a more liberal one.

Let us formulate the null and alternative hypotheses as

$$H_0 : \quad \gamma_{1j} \leq \gamma_{2j} \text{ for some } j = 1, \ldots, c-1, \tag{6.55}$$

$$\text{and} \quad H_1 : \quad \gamma_{1j} > \gamma_{2j} \text{ for every } j = 1, \ldots, c-1, \tag{6.56}$$

respectively, where $\gamma_{ij} = \pi_{i1} + \ldots + \pi_{ij}$ for $i = 1, 2$ and $j = 1, \ldots, c-1$. Equivalently, the hypotheses may be stated in terms of cumulative odds ratios, as

$$H_0 : \theta_1^C \leq 1, \text{or} \ldots, \text{ or } \theta_{c-1}^C \leq 1, \text{ and } H_1 : \theta_1^C > 1, \ldots, \theta_{c-1}^C > 1, \tag{6.57}$$

where θ_j^C is the cumulative odds ratio. Now, we can apply an IUT for testing H_0 against H_1. It requires $c-1$ separate tests, namely a test of $\theta_j^C \leq 1$ against $\theta_j^C > 1$ for each $j = 1, \ldots, c-1$. Recall that θ_j^C is the odds ratio for the 2×2 table obtained by merging columns $1, \ldots, j$ to form the first column and the remaining columns $j+1, \ldots, c$ to form the second column. Therefore, to test $\theta_j^C \leq 1$ against $\theta_j^C > 1$, we can apply any of the test of independence in 2×2 tables; these include (a) an asymptotic test based on $\{\log \hat{\theta}_j - \log \theta_j\}/se(\log \hat{\theta}_j)\} \approx N(0, 1)$, (b) Fisher's exact test, and (c) a test for the comparison of two independent binomial samples with success probabilities γ_{1j} and γ_{2j}. The next example illustrates these ideas.

Example 6.6.1

Let us consider Example 6.2.2. Suppose that we wish to test

H_0 : Treatment 2 is not better than Treatment 1,

against H_1 : Treatment 2 is better than Treatment 1.

Let us formulate this as in (6.57):

$$H_0 : \quad \log \theta_1^C \leq 0, \text{ or } \log \theta_2^C \leq 0 \text{ or } \log \theta_3^C \leq 0, \tag{6.58}$$

$$\text{against} \quad H_1 : \quad \log \theta_1^C > 0, \log \theta_2^C > 0 \text{ and } \log \theta_3^C > 0. \tag{6.59}$$

We have chosen to work with log odds ratios rather than odds ratios because the normal approximation is better for estimates of the former. The three 2×2 tables obtained by merging adjacent columns are shown in Table 6.13. For these three tables, we have $\log \hat{\theta}_1 = \log\{12 \times 27/(5 \times 20)\} = 1.176$; similarly, $\log \hat{\theta}_2 = 1.166$ and $\log \hat{\theta}_3 = 0.815$. For a 2×2 table with cell counts $\{n_{ij}\}$, a large sample estimate of the standard error of the sample log odds ratio is $\{n_{11}^{-1} + n_{12}^{-1} + n_{21}^{-1} + n_{22}^{-1}\}^{1/2}$ (see Agresti (1990, p 54)). This leads to $se(\log \hat{\theta}_1) = (12^{-1} + 5^{-1} + 20^{-1} + 27^{-1}\}^{1/2} = 0.61$; similarly, $se(\log \hat{\theta}_2) = 0.524$ and $se(\log \hat{\theta}_3) = 0.586$. Let $z_i = \log \hat{\theta}_i / se(\log \hat{\theta}_i)$, for $i = 1, 2, 3$; then $z_1 = 1.176/0.61 = 1.93$, $z_2 = 2.24$ and $z_3 = 1.39$. Now, an upper bound for the asymptotic p-value for an IUT based on $\{z_1, z_2, z_3\}$ is

$$\max_i [1 - pr\{N(0,1) \geqslant z_i\}] = 1 - pr\{N(0,1) \geqslant 1.39\} = 0.08.$$

The foregoing asymptotic test assumes that the cell counts in each of the three 2×2 tables formed by collapsing the adjacent columns have large cell counts so that $\log \hat{\theta}_1 / se(\log \hat{\theta}_i)$ is approximately $N(0, 1)$ under H_0 for $i = 1, 2$, and 3. If this assumption is not reasonable, an IUT may be carried out by applying Fisher's exact test for each of the three 2×2 tables. The p-values for Fisher's exact test are $0.0439, 0.0218$, and 0.1287. Therefore, for the corresponding IUT, an upper bound for the p-value is $\max\{0.0439, 0.0218, 0.1287\} = 0.1287$. Although Fisher's test is exact for each 2×2 table, it does not follow that the p-value 0.1278 is exact for the corresponding IUT.

Table 6.13 Summary statistics for the comparison two treatments for the data in Table 6.8

	Y		Y		Y	
	1	2,3,4	1,2	3,4	1,2,3	4
Treatment 1	12	20	22	10	26	6
Treatment 2	5	27	13	19	21	11
$\log \hat{\theta}_j$	1.176		1.166		0.815	
$se(\log \hat{\theta}_j)$	0.61		0.524		0.586	
z_j	1.93		2.24		1.39	
p-value (Fisher's exact test)	0.0439		0.0218		0.1287	

Since the p-value for the first two 2×2 tables are small but not for the last one, it follows that had the response variable been binary with "success" = {Response $\geqslant 3$}, the Fisher's exact would have provided sufficient evidence (p-value = 0.0218) to accept that Treatment 2 is better albeit in a weaker sense than the H_1 in (6.59). The alternative hypothesis in (6.59) is stronger and consequently the statistical evidence to establish it is weaker. It is also clear that for the foregoing IUT to have good power, the cell counts in the first and the last columns must be large; small counts in the middle columns do not affect the power as much. This example illustrates some of the advantages and disadvantages of formulating the testing problem as in (6.57). ■

6.7 ANALYSIS OF $r \times c$ TABLES

The results in the previous section on the comparison of two treatments in $2 \times c$ tables with ordinal response variables can be extended to the comparison of more than two treatments. Suppose that there are r treatments and that the response variable Y is ordinal with c categories. The r treatments will be considered as r levels of a categorical variable X. The observations corresponding to different treatments are assumed to be independent; thus, the likelihood is a product of the multinomial probability functions. Let $\boldsymbol{\theta}$ denote a vector of generalized odds ratios. We are interested in the following two hypothesis testing problems:

$$(i) \quad H_0 : \boldsymbol{\theta} = \mathbf{1} \quad \text{against} \quad H_1 : \boldsymbol{\theta} \geqslant \mathbf{1}, \tag{6.60}$$

$$\text{and} \quad (ii) \quad H_1 : \boldsymbol{\theta} \geqslant \mathbf{1} \quad \text{against} \quad H_2 : \boldsymbol{\theta} \not\geqslant \mathbf{1} \tag{6.61}$$

It will be seen that the loglikelihood is a smooth concave function, and that closed forms are available for its slope and hessian. Therefore, a general purpose nonlinear optimization program can be used to compute the constrained *mle;* such programs are available in IMSL, FSQP, and MATLAB. Once the *mle*'s have been obtained under the null and under the alternative hypotheses, it is a simple matter to compute the *LRT.* For each of the foregoing two testing problems, the asymptotic null distribution of *LRT* is a chi-bar square; this follows from the general result, Proposition 4.8.1. For the particular odds ratios that are of interest, the least favorable null distributions are known and they turn out to be functions of chi-square and chi-bar square random variables. Therefore, the p-value for *LRT* can be computed sufficiently precisely by simulation. Consequently, it is fairly easy to apply the *LRT* for the foregoing two testing problems in large samples.

Let us recall the following notation from the previous sections:

$$\pi_{ij} = \text{pr}(Y = j \mid X = i), \quad \boldsymbol{\pi}_i = (\pi_{i1}, \dots, \pi_{i,c-1})^T, \quad \boldsymbol{\pi} = (\boldsymbol{\pi}_1 : \dots : \boldsymbol{\pi}_r) \tag{6.62}$$

for $i = 1, \dots, r$ and $j = 1, \dots, c$. Note that $\boldsymbol{\pi}_i$ does not include π_{ic}, $(i = 1, \dots, r)$; this avoids redundant parameters arising from the constraint $\pi_{i1} + \dots + \pi_{ic} = 1$. Since the likelihood is the product of multinomial probability functions corresponding to different rows, the loglikelihood takes the form

$$\sum_{i=1}^{r} \sum_{j=1}^{c} n_{ij} \log(\pi_{ij}) + K \tag{6.63}$$

where K does not depend on unknown parameters. By eliminating $\{\pi_{1c}, \dots, \pi_{rc}\}$ from (6.63), the kernel of the loglikelihood can be expressed as

$$\ell(\boldsymbol{\pi}) = \sum_{i=1}^{r} \left[\left\{ \sum_{j=1}^{c-1} n_{ij} \log \pi_{ij} \right\} + n_{ic} \log(1 - \pi_{i1} - \dots - \pi_{i,c-1}) \right]. \tag{6.64}$$

It is easily verified that $\ell(\boldsymbol{\pi})$ is concave in $\boldsymbol{\pi}$. This helps to verify the regularity conditions of Proposition 4.8.1 are also satisfied.

For $i = 1, \dots, r$, let

$$a_i = \lim(n_i/n), \qquad \boldsymbol{W}(\boldsymbol{\pi}_i) = \text{diag}(\boldsymbol{\pi}_i)^{-1} + \mathbf{1}\mathbf{1}^T/\pi_{i,c}$$
$$\boldsymbol{a} = (a_1, \dots, a_{r-1})^T \text{ and } \quad \boldsymbol{i}(\boldsymbol{\pi}) = \text{diag}\{a_1 \boldsymbol{W}(\boldsymbol{\pi}_1), \dots, a_r \boldsymbol{W}(\boldsymbol{\pi}_r)\}.$$

Then

$$\boldsymbol{i}(\boldsymbol{\pi})^{-1} = \text{diag}\{a_1^{-1} \boldsymbol{W}(\boldsymbol{\pi}_1)^{-1}, \dots, a_r^{-1} \boldsymbol{W}(\boldsymbol{\pi}_r)^{-1}\} \tag{6.65}$$

where

$$\boldsymbol{W}(\boldsymbol{\pi}_i)^{-1} = \text{diag}(\boldsymbol{\pi}_i) - \boldsymbol{\pi}_i \boldsymbol{\pi}_i^T, \; i = 1, \dots, r.$$

Let the null and the alternative hypothesis be

$$H_0 : \boldsymbol{\pi} \in \Theta_0 \quad \text{and} \quad H_1 : \boldsymbol{\pi} \in \Theta_1, \tag{6.66}$$

respectively, where $\Theta_0 \subset \Theta_1 \subseteq \Theta$ and

$$\Theta = \{\pi : \pi = (\pi_1, \ldots, \pi_r)^T, 0 < \pi_{ij} < 1 \quad \text{for all} \quad (i, j)\};$$

as usual Θ_0 and Θ_1 are assumed Chernoff regular at every point in Θ_0.

Now, the asymptotic null distribution of the *LRT* for testing H_0 against H_1 in (6.66) is given by

$$LRT \xrightarrow{d} ||Z - \mathcal{T}(\Theta_0; \pi_0)||^2 - ||Z - \mathcal{T}(\Theta_1; \pi_0)||^2 \tag{6.67}$$

where π_0 is the assumed true null value of π in Θ_0, $Z \sim N(\mathbf{0}, i_0^{-1})$, $i_0 = i(\pi_0)$ and $\|x\|^2 = x^T i_0 x$. For the testing problem (6.60), $\mathcal{T}(\Theta_0; \pi_0)$ is a linear space, $\mathcal{T}(\Theta_1; \pi_0)$ is a closed convex cone and $\mathcal{T}(\Theta_0; \pi_0) \subset \mathcal{T}(\Theta_1; \pi_0)$; consequently, the asymptotic null distribution of *LRT* in (6.67) is a chi-bar square. For the testing problem (6.61), $\mathcal{T}(\Theta_0; \pi_0)$ is a closed convex cone, $\mathcal{T}(\Theta_1; \pi_0)$ is a linear space and $\mathcal{T}(\Theta_0; \pi_0) \subset \mathcal{T}(\Theta_1; \pi_0)$; consequently, the asymptotic null distribution of the *LRT* is again a chi-bar square. For the particular stochastic orders of practical interest, namely likelihood ratio order, uniform stochastic order, and simple stochastic order, the least favorable null distributions of these chi-bar square distributions are known and hence the large sample p-values for *LRT* can be computed reasonably easily. In the next subsection, we consider simple stochastic order. Then, in Subsection 6.7.2, all three stochastic orders are considered.

6.7.1 LRT Against Simple Stochastic Order

In this subsection, we consider test of

$$H_0 : \pi_1 = \ldots = \pi_r, \quad \text{vs} \quad H_1 : \gamma_1 \geqslant \ldots \geqslant \gamma_r, \tag{6.68}$$

where $\gamma_{ij} = \pi_{i1} + \ldots + \pi_{ij}$ and $\gamma_i = (\gamma_{i1}, \ldots, \gamma_{i,c-1})^T$. Let $\bar{\pi}$ and $\tilde{\pi}$ denote the *mle* of π under $H_0 : \pi_1 = \ldots = \pi_r$ and that under $H_1 : \gamma_1 \geqslant \ldots \geqslant \gamma_r$, respectively. Then the *LRT* for testing H_0 against H_1 is $2\sum_{i=1}^{r}\sum_{j=1}^{c} n_{ij}\log(\tilde{\pi}_{ij}/\bar{\pi}_{ij})$.

Suppose that null hypothesis holds, and let q denote the common value of $\pi_1, \ldots, \pi_r$. Let T_m denote the $m \times m$ upper triangular matrix in which every element on the main diagonal and above it is one, and let

$$S_a = T_{r-1}^T\{\text{diag}(a) - aa^T\}T_{r-1} \quad \text{and} \quad S_q = T_{c-1}^T\{\text{diag}(q) - qq^T\}T_{c-1}.$$

It may be verified that

$$S_a^{-1} = T_{r-1}^{-1}\{\text{diag}(a)^{-1} + \mathbf{1}\mathbf{1}^T/a_r\}(T_{r-1}^{-1})^T.$$

Now, we have the following (see Dardanoni and Forcina (1998), for a proof).

Proposition 6.7.1 *(i) The asymptotic null distribution of the LRT for the testing problem in (6.68) is $\bar{\chi}^2(S_a^{-1} \otimes S_q, \mathcal{O}^+)$, where $\mathcal{O}^+$ is the positive orthant.*
(ii) For a given a, the least favorable asymptotic null value for the LRT in part (i) is

$q = (1/2, 0, \dots, 0)$.

(iii)

$$\sup_{q} pr\{\bar{\chi}^2(S_a^{-1} \otimes S_q, \mathcal{O}^+) \geqslant t\} = pr\{\chi_\nu^2 + \bar{\chi}^2(S_a^{-1}, \mathcal{O}^+) \geqslant t\} \tag{6.69}$$

where $\nu = (r-1)(c-2)$, *and* χ_ν^2 *and* $\bar{\chi}^2(S_a^{-1}, \mathcal{O}^+)$ *on the right-hand side are independent.*

It follows form part (ii) of this proposition that the least favorable configuration for the asymptotic null distribution of *LRT* for the testing problem in (6.68) is reached when $(\pi_{i1}, \dots, \pi_{ic}) = (1/2, 0, \dots, 0, 1/2)$ for every $i = 1, \dots, r$; to avoid a possible confusion of notation, let us note that the q in part (ii) of the foregoing proposition did not have 1/2 in its last coordinate because q has only $c-1$ elements and it does not include the c th coordinate in $(\pi_{i1}, \dots, \pi_{ic})$.

It follows from part (iii) of the proposition that, corresponding to a sample value t for *LRT*,

$$\text{asymptotic } p\text{-value} = \text{pr}\{\chi_\nu^2 + \bar{\chi}^2(S_a^{-1}, \mathcal{O}^+) \geqslant t\}.$$

This can be computed to high degree of precision by simulation. It may be verified, for example using moment generating functions, that

$$\text{pr}\{\chi_\nu^2 + \bar{\chi}^2(S_a^{-1}, \mathcal{O}^+) \geqslant t\} = \sum_{i=0}^{r-1} w_i(r-1, S_a^{-1}) \text{pr}(\chi_{\nu+i}^2 \geqslant t). \tag{6.70}$$

Therefore, we can also compute the asymptotic p-value by first computing the chi-bar square weights, $\{w_i(r-1, S_a^{-1}), i = 1, \dots, r-1\}$. It follows from (6.70) that an upper bound for the asymptotic p-value is $(1/2)\{\text{pr}(\chi_{M-1}^2 \geqslant t) + \text{pr}(\chi_M^2 \geqslant t)\}$ where $M = (r-1)(c-1)$. Finally, let us remark that part (ii) of the foregoing proposition is stated for a given vector a and hence rows and columns are not treated symmetrically. Consequently, ν, which is equal to $(r-1)(c-2)$ is not symmetric in r and c.

Example 6.7.1

Table 6.14 provides the data for a sample of British males cross-classified by X and Y, where X = father's occupational status and Y = son's occupational status. These data have been analyzed by several other authors, for example, see Goodman (1991, p 1086); the calculations presented here are based on those obtained by Dardanoni and Forcina (1998). A question of interest is whether or not son's occupational status has a positive relationship with that of the father. We are interested in the conditional distribution of Y given X. Therefore, we can treat fathers's occupational status as Treatment and son's occupational status as response, and hence the results of this section based on the product multinomial likelihood are applicable.

Let us apply Proposition 6.7.1 to test H_0 against H_1 where

$$H_0 : \pi_0 = \dots = \pi_6 \quad \text{and} \quad H_1 : \gamma_1 \geqslant \dots \geqslant \gamma_6,$$

respectively. Let $\hat{\pi}_{ij}$ and $\hat{\gamma}_{ij}$ denote the unconstrained estimates of π_{ij} and γ_{ij}, respectively, for $i, j = 1, \ldots, 6$. Then $\hat{\pi}_{ij} = n_{ij}/n_{i+}$ and $\hat{\gamma}_{ij} = \hat{\pi}_{i1} + \ldots + \hat{\pi}_{ij}$ for every (i, j); these are shown in Table 6.15. Since $\{\hat{\gamma}_{ij}\}$ satisfy the constraints in H_1, we have that $\tilde{\pi}_{ij} = \hat{\pi}_{ij}$ for every (i, j), where $\tilde{\boldsymbol{\pi}}$ is the estimator of $\boldsymbol{\pi}$ under the inequality constraints in H_1.

Table 6.14 Occupational status for a sample of British males

Father's status	Son's occupational status 1	2	3	4	5	6	Total
1	125	60	26	49	14	5	279
2	47	65	66	123	23	21	345
3	31	58	110	223	64	32	518
4	50	114	185	715	258	189	1511
5	6	19	40	179	143	71	458
6	3	14	32	141	91	106	387
Total	262	330	459	1430	593	424	3498

Source: Glass (1954); these data have appeared in several publications, including Goodman (1991). The entries in this table are the same as those in Dardanoni and Forcina (1998).

Let $\bar{\boldsymbol{\pi}}$ denote the estimate of $\boldsymbol{\pi}$ under H_0. Then $\bar{\boldsymbol{\pi}} = (\bar{\boldsymbol{\pi}}_1 : \ldots : \bar{\boldsymbol{\pi}}_6)$, where $\bar{\boldsymbol{\pi}}_1, \ldots, \bar{\boldsymbol{\pi}}_6$ are all equal to the vector of marginal column proportions. Thus, we have $\bar{\boldsymbol{\pi}}_i = 3498^{-1}(262, 330, 459, 1430, 593, 424)$ for every i. Substituting these into $LRT = \sum\sum n_{ij} \log(\tilde{\pi}_{ij}/\bar{\pi}_{ij})$, we have $LRT = 839$. An upper bound for the p-value is $0.5\{\text{pr}(\chi^2_{M-1} \geqslant 839) + \text{pr}(\chi^2_M \geqslant 839)\}$ where $M = (r-1)(c-1) = 25$; this upper bound is less than 0.001. Thus, there is very strong evidence to reject H_0. Let us emphasize that this small p-value provides strong evidence to establish that occupational status of the son and that of the father are associated, but it does not go as far as to claim that the association is positive. Therefore, based on the rejection of H_0 alone, we cannot claim that father's occupational status is positively related to son's occupational status. However, if we make the additional assumption that the association between occupational status of the son and that of the father is nonnegative in the sense that $\boldsymbol{\gamma}_1 \geqslant \ldots \geqslant \boldsymbol{\gamma}_6$ holds, then based on the rejection of the $H_0 : \boldsymbol{\pi}_1 = \ldots = \boldsymbol{\pi}_6$ we can conclude that the association is positive in the sense that $\boldsymbol{\gamma}_1 \geqslant \ldots \geqslant \boldsymbol{\gamma}_6$ holds with at least one strict inequality.

Suppose that it is not assumed *a priori* that $\boldsymbol{\gamma}_1 \geqslant \ldots \geqslant \boldsymbol{\gamma}_6$ holds. Since the estimates, $(\hat{\boldsymbol{\gamma}}_1, \ldots, \hat{\boldsymbol{\gamma}}_6)$, satisfies the constraint $\hat{\boldsymbol{\gamma}}_1 \geqslant \ldots \geqslant \hat{\boldsymbol{\gamma}}_6$ and the foregoing p-value is small, one may be tempted to conclude that there is statistical evidence to accept that $\boldsymbol{\gamma}_1 \geqslant \ldots \geqslant \boldsymbol{\gamma}_6$ holds with at least one strict inequality. This argument is incorrect because there may be a high probability of $\{\hat{\boldsymbol{\gamma}}_1 \geqslant \ldots \geqslant \hat{\boldsymbol{\gamma}}_6\}$ and the *LRT* of $H_0 : \boldsymbol{\pi}_0 = \ldots = \boldsymbol{\pi}_6$ vs $H_1 : \boldsymbol{\gamma}_1 \geqslant \ldots \geqslant \boldsymbol{\gamma}_6$ being large even when $H_0 \cup H_1$ is not true and the true parameter value is close to $H_0 \cup H_1$.

6.7.2 LRT for and Against Stochastic Orders

Let H_0, H_s, H_u, H_ℓ and H_2 be defined as

$$H_0: \quad \pi_1 = \ldots = \pi_r, \tag{6.71}$$

$$H_h: \quad \pi_1 \preceq_h \ldots \preceq_h \pi_r \tag{6.72}$$

$$H_2: \quad \pi_1, \ldots, \pi_r \quad \text{are unconstrained,} \tag{6.73}$$

where h is equal to s (= simple stochastic order), or u (= uniform stochastic order), or ℓ (= likelihood ratio order) (see Subsection 6.4.2). Abusing the notation slightly, we shall also write $Z_1 \preceq_s Z_2$ and $G_1 \preceq_s G_2$ when Z_1 and Z_2 are random variables and $\text{pr}(Z_1 \geqslant z) \leqslant \text{pr}(Z_2 \geqslant z)$ for every z, where $G_1(z) = \text{pr}(Z_1 \leqslant z)$ and $G_2(z) = \text{pr}(Z_2 \leqslant z)$. It follows from Proposition 6.4.2 that $H_0 \subset H_\ell \subset H_u \subset H_s \subset H_2$. For $h = s, u$, or ℓ, let

$$T_{0h} = LRT \text{ of } H_0 \text{ vs } H_h \quad \text{and} \quad T_{h2} = LRT \text{ of } H_h \text{ vs } H_2.$$

The main results concerning the asymptotic distributions of these test statistics are given in the next three theorems; see Dardanoni and Forcina (1998) for proofs. These theorems state that (a) the asymptotic null distribution of each of the test statistic is a chi-bar square, and (b) when the test is not similar, its least favorable null distribution is known. The proof of (a) follows from (6.67); Dardanoni and Forcina (1998) also provided the proofs from first principles. The statement of the following theorems also include, as a special case, the results in Proposition 6.7.1 for testing against simple stochastic order.

Theorem 6.7.2 *Let the H_0 in (6.71) be the null hypothesis and suppose that it is true; let $\boldsymbol{q} = (q_1, \ldots, q_{c-1})^T$ denote the common value of $\pi_1, \ldots, \pi_r$, and let $\mathcal{O}^+$ denote the positive orthant. Let $T_{0h(\infty)}$ denote the limiting distribution of T_{0h} ($h = s, u, \ell$). Let $\bar{\chi}_1^2(\boldsymbol{S}_{\boldsymbol{a}}^{-1}, \mathcal{O}^+), \ldots, \bar{\chi}_{c-1}^2(\boldsymbol{S}_{\boldsymbol{a}}^{-1}, \mathcal{O}^+)$ be independent and identically distributed as $\bar{\chi}^2(\boldsymbol{S}_{\boldsymbol{a}}^{-1}, \mathcal{O}^+)$. Then we have the following:*

Table 6.15 Estimates of γ_{ij} for the data in Table 6.14

Father's status	Son's occupational status 1	2	3	4	5	6	Total
1	0.448	0.663	0.756	0.932	0.982	1	
2	0.136	0.325	0.516	0.872	0.939	1	
3	0.060	0.172	0.384	0.815	0.938	1	
4	0.033	0.109	0.231	0.704	0.875	1	
5	0.013	0.055	0.142	0.533	0.845	1	
6	0.008	0.044	0.127	0.491	0.726	1	

(i) $\bar{\chi}^2(\boldsymbol{S}_{\boldsymbol{a}}^{-1}, \mathcal{O}^+) \preceq_s T_{0\ell(\infty)} \sim \bar{\chi}^2(\boldsymbol{S}_{\boldsymbol{a}}^{-1} \otimes \boldsymbol{S}_{\boldsymbol{q}}^{-1}, \mathcal{O}^+) \preceq_s \sum_{i=1}^{c-1} \bar{\chi}_i^2(\boldsymbol{S}_{\boldsymbol{a}}^{-1}, \mathcal{O}^+)$. *The lower and upper bounds of* $T_{0\ell}$ *given here are reached when* $\boldsymbol{q}$ *is equal to* $(1/2, 0, \ldots, 0)^T$ *and* $(1, 0, \ldots, 0)^T$*, respectively; hence the asymptotic least favorable null configuration for* $T_{0\ell}$ *is obtained when the common multinomial probability vector* $(\pi_{i1}, \ldots, \pi_{ic})$ *is* $(1, 0, \ldots, 0)$.
(ii) $T_{0u(\infty)} \sim \sum_{i=1}^{c-1} \bar{\chi}_i^2(\boldsymbol{S}_{\boldsymbol{a}}^{-1}, \mathcal{O}^+)$.
(iii) $\sum_{i=1}^{c-1} \bar{\chi}_i^2(\boldsymbol{S}_{\boldsymbol{a}}^{-1}, \mathcal{O}^+) \preceq_s T_{0s(\infty)} \sim \bar{\chi}^2(\boldsymbol{S}_{\boldsymbol{a}}^{-1} \otimes \boldsymbol{S}_{\boldsymbol{q}}^{-1}, \mathcal{O}^+) \preceq_s \chi_M^2 + \bar{\chi}^2(\boldsymbol{S}_{\boldsymbol{a}}^{-1}, \mathcal{O}^+)$, *where* χ_M^2 *and* $\bar{\chi}^2(\boldsymbol{S}_{\boldsymbol{a}}^{-1}, \mathcal{O}^+)$ *are independent, and* $M = (r-1)(c-2)$*. The lower and upper bounds of* $T_{0s(\infty)}$ *given here are reached when* $\boldsymbol{q}$ *is equal to* $(1, 0, \ldots, 0)^T$ *and* $(1/2, 0, \ldots, 0)^T$*, respectively; hence the asymptotic least favorable null configuration for* T_{0s} *is obtained when the common multinomial probability vector* $(\pi_{i1}, \ldots, \pi_{ic})$ *is* $(1/2, 0, \ldots, 0, 1/2)$.

The foregoing proposition shows that the least favorable asymptotic null distribution of $T_{0\ell}$ is also a lower bound (i.e., "most favorable") for $T_{0s(\infty)}$. Consequently, if the unconstrained estimator of $\boldsymbol{\pi}$ satisfies the constraint in H_ℓ then $T_{0\ell}$ and T_{0s} have the same numerical value but the former will have a smaller asymptotic p-value.

It follows from part (ii) of Proposition 6.7.2 that the asymptotic null distribution of T_{0u} is a constant over the null parameter space; in other words, the asymptotic test is similar. In Theorem 6.7.2, the $a_i = \lim(n_i/n)$ was assumed to be given and fixed $(i = 1, \ldots, r)$. It is possible to obtain bounds which do not depend on $\boldsymbol{a}$ so that the bounds hold uniformly on $\boldsymbol{a}$; these are given in the next result.

Theorem 6.7.3 *Assume that the null hypothesis,* H_0 *in (6.71) holds. Let* $\bar{\chi}^2(1, \mathcal{O}^+)$ *denote the chi-bar square distribution in one dimension with* $\mathcal{O}^+$ *denoting* $[0, \infty)$*, and let* $\nu = (c-1)(r-1)$*. Let* $T_{0h(\infty)}$ *denote the limiting distribution of* T_{0h} $(h = s, u, \ell)$*. Then we have the following:*
(i) $\bar{\chi}^2(1, \mathcal{O}^+) \preceq_s T_{0\ell(\infty)} \preceq_s \bar{\chi}^2(\boldsymbol{I}_\nu, \mathcal{O}^+)$.
(ii) $\bar{\chi}^2(\boldsymbol{I}_{c-1}, \mathcal{O}^+) \preceq_s T_{0u(\infty)} \preceq_s \bar{\chi}^2(\boldsymbol{I}_\nu, \mathcal{O}^+)$.
(iii) $\bar{\chi}^2(\boldsymbol{I}_{c-1}, \mathcal{O}^+) \preceq_s T_{0s(\infty)} \preceq_s \chi_{\nu-1}^2 + \bar{\chi}^2(1, \mathcal{O}^+)$ *where* $\chi_{\nu-1}^2$ *and* $\bar{\chi}^2(1, \mathcal{O}^+)$ *are independent.*

Now, let us consider test of H_h in (6.72) against the unrestricted alternative H_2 where $h = s, u, \ell$. We have already indicated that it follows from (6.67) that the asymptotic null distribution of *LRT* is a chi-bar square. The relevant least favorable null distributions are given in the next result.

Theorem 6.7.4 *Let the null and alternative hypotheses be* H_h *and* H_2*, respectively, where* h *can be* s, u *or* ℓ*. Assume that* H_h *holds. Let* $\nu = (r-1)(c-1)$ *and let* $\bar{\chi}^2(1, \mathcal{O}^+)$ *be as in Theorem 6.7.3. Let* $T_{h2(\infty)}$ *denote the limiting distribution of* T_{h2} $(h = s, u, \ell)$*. Then, the upper bounds for* T_{h2} *given below are also the asymptotic least favorable null distributions:*
(i) $T_{\ell 2(\infty)} \preceq_s \chi_{(r-1)(c-1)-1}^2 + \bar{\chi}^2(1, \mathcal{O}^+)$;
(ii) $T_{u2(\infty)} \preceq_s \chi_{(r-2)(c-1)}^2 + \bar{\chi}^2(\boldsymbol{I}_{c-1}, \mathcal{O}^+)$;

(iii) $T_{s2(\infty)} \preceq_s \chi^2_{(r-2)(c-1)} + \bar{\chi}^2(\boldsymbol{I}_{c-1}, \mathcal{O}^+)$;
in each of (i), (ii) and (iii) the χ^2 *and* $\bar{\chi}^2$ *random variables appearing on the right are independent.*

The tail probabilities of the distributions in the foregoing theorem may be computed using

$$\text{pr}\{\chi^2_\nu + \bar{\chi}^2(\boldsymbol{I}_k, \mathcal{O}^+) \geq c\} = \sum_{i=0}^{k} w_i(k, \boldsymbol{I}_k)\text{pr}(\chi^2_{i+\nu} \geq c),$$

where $w_i(k, \boldsymbol{I}_k) = 2^{-k}k!\{i!(k-i)!\}^{-1}$. This may be verified using (6.70), (3.27), and the fact that χ^2_ν and $\bar{\chi}^2(\boldsymbol{I}_k, \mathcal{O}^+)$ are independent.

Example 6.7.2

Let us consider the occupational status data example discussed in the previous subsection. The results reported below are based on those of Dardanoni and Forcina (1998). Although the unconstrained estimator satisfies the constraints in H_s, they do not satisfy those in H_ℓ or H_u. This reflects the fact that H_s is weaker than H_u and H_ℓ. The sample values of the test statistics are $T_{0u} = 831$ and $T_{0\ell} = 830$. By parts (i) and (ii) of Theorem 6.7.3, upper bounds for the asymptotic critical values are obtained from $\bar{\chi}^2(\boldsymbol{I}_{(c-1)(r-1)}, \mathcal{O}^+)$. It is easily seen that the asymptotic p-value for $T_{0\ell}$ is smaller than $\text{pr}(\chi^2_{25} \geqslant 830)$, and that for T_{0u} is smaller than $\text{pr}(\chi^2_{25} \geqslant 831)$; both are very small. Therefore, a test of H_0 against H_h would reject H_0 for $h = \ell, u$.

Now let us consider test of H_h against H_2 where $h = u$ or ℓ. The sample values of the test statistics are $T_{u2} = 7.79$ and $T_{\ell 2} = 9.39$. For T_{u2} and $T_{\ell 2}$, the least favorable null distributions are given in Theorem 6.7.4. The 5% critical values for these are 34.72 and 37.07, respectively. Therefore, there is insufficient evidence to reject H_u and H_ℓ. ■

6.8 SQUARE TABLES AND MARGINAL HOMOGENEITY

Tests of hypotheses about the marginal distribution of square tables arise in several areas of applications, particularly with matched pairs. In many such cases, the unconstrained large sample methods can be extended to incorporate inequality constrained problems as well. To indicate the nature of the problems, let us consider an example. Table 6.19 provides data for 80 esophageal cancer patients and 80 matching controls. The response variable is the number of beverages reported drunk at "burning hot" temperatures. This set of data is taken from Breslow (1982), and is also analyzed in Agresti (1990, p 364). The question of interest is whether or not cancer patients consumed more beverages at burning hot temperatures than did the controls.

For a matched pair, let

$$\pi_{ij} = \text{pr(The number of beverages for case and control are } i \text{ and } j\text{, respectively)}.$$

Based on the figures in Table 6.19, the marginal distributions for case and control are $(0.51, 0.68, 0.90, 1.0)$ and $(0.79, 0.89, 0.99, 1.0)$, respectively. Note that the latter is stochastically smaller than the former. Therefore, in this sample, the cancer patients consumed more hot beverages than the controls. It is of interest to formulate this as a hypothesis testing problem and compute a p-value. To this end, we suggest a large sample method in this section; an exact finite sample method will be suggested later.

To formulate the inference problem in a general context, let us consider an $I \times I$ table. Let the null and alternative hypotheses be

$$\begin{aligned} &H_0: \quad \pi_{i+} = \pi_{+i} \text{ for } i = 1, \ldots, I, \\ \text{and} \quad &H_1: \quad \pi_{1+} + \ldots + \pi_{i+} \le \pi_{+1} + \ldots + \pi_{+i} \text{ for } i = 1, \ldots, I, \end{aligned}$$

respectively. Let $\alpha_i = (\pi_{+i} - \pi_{i+})$ for $i = 1, \ldots, I$ and $\boldsymbol{\alpha} = (\alpha_1, \ldots, \alpha_{I-1})^T$. Now the null and alternative hypotheses are

$$H_0: \boldsymbol{\alpha} = \mathbf{0} \quad \text{and} \quad H_1: \boldsymbol{A\alpha} \geqslant \mathbf{0},$$

respectively, where $\boldsymbol{A}$ is the $(I-1) \times (I-1)$ lower triangular matrix in which each element on the diagonal and below is one. Thus, $\boldsymbol{A} = (1, 0, \ldots, 0;\ 1, 1, 0, \ldots, 0;\ \ldots;\ 1, 1, \ldots, 1)$, where ";" separates different rows.

Let $\hat{\boldsymbol{\alpha}}$ be an estimator of $\boldsymbol{\alpha}$ such that $\sqrt{n}(\hat{\boldsymbol{\alpha}} - \boldsymbol{\alpha}) \xrightarrow{d} N(\mathbf{0}, \boldsymbol{V})$ for some $\boldsymbol{V}$ and $\hat{\boldsymbol{V}}$ be a consistent estimator of $\boldsymbol{V}$. Then, a suitable statistic for testing H_0 against H_1 is

$$W_D = n[\hat{\boldsymbol{\alpha}}^T \hat{\boldsymbol{V}}^{-1} \hat{\boldsymbol{\alpha}} - \inf\{(\hat{\boldsymbol{\alpha}} - \boldsymbol{b})^T \hat{\boldsymbol{V}}^{-1} (\hat{\boldsymbol{\alpha}} - \boldsymbol{b}) : \boldsymbol{Ab} \geqslant \mathbf{0}\}],$$

and its asymptotic null distribution is $\bar{\chi}^2(\boldsymbol{AVA}^T, \mathcal{O}^+)$.

A suitable choice of $(\hat{\boldsymbol{\alpha}}, \hat{\boldsymbol{V}})$ is given in Bhapkar (1966); this has the following form: Let $\hat{\pi}_{i+} = n_{i+}/n$, $\hat{\pi}_{+i} = n_{+i}/n$, $\hat{\alpha}_i = \hat{\pi}_{+i} - \hat{\pi}_{i+}$, and $\hat{\boldsymbol{\alpha}} = (\hat{\alpha}_1, \ldots, \hat{\alpha}_{I-1})^T$. Now, $\sqrt{n}(\hat{\boldsymbol{\alpha}} - \boldsymbol{\alpha}) \xrightarrow{d} N(\mathbf{0}, \boldsymbol{V})$ and a consistent estimator of $\boldsymbol{V}$ is $\hat{\boldsymbol{V}} = [\hat{v}_{ij}]$ where

$$\hat{v}_{ij} = \begin{cases} -(\hat{\pi}_{ij} + \hat{\pi}_{ji}) - (\hat{\pi}_{i+} - \hat{\pi}_{+i})(\hat{\pi}_{j+} - \hat{\pi}_{+j}) & \text{if } i \neq j \\ (\hat{\pi}_{i+} + \hat{\pi}_{+i}) - 2\hat{\pi}_{ii} - (\hat{\pi}_{i+} - \hat{\pi}_{+i})^2 & \text{if } i = j; \end{cases}$$

see Bhapkar (1966) for derivations; see also Agresti (1990, p 359).

Now, suppose that the null hypothesis holds, and let $\boldsymbol{\pi}^0$ denote the assumed true value of $\boldsymbol{\pi}$ under H_0. Then

$$v_{ij} = \begin{cases} -(\pi_{ij}^0 + \pi_{ji}^0) & \text{if } i \neq j \\ 2(\pi_{i+}^0 - \pi_{ii}^0) & \text{if } i = j, \end{cases}$$

where $\boldsymbol{V} = [v_{ij}]$. Since $\boldsymbol{V}$ depends on nuisance parameters, an upper bound for the asymptotic p-value of W_D is $(1/2)\{\text{pr}(\chi_{I-2}^2 \geqslant t) + \text{pr}(\chi_{I-1}^2 \geqslant t)\}$ where t is the sample value of W_D.

For testing $\boldsymbol{\alpha} = \mathbf{0}$ against $\boldsymbol{\alpha} \neq \mathbf{0}$, the Wald statistic $n\hat{\boldsymbol{\alpha}}^T \hat{\boldsymbol{V}}^{-1} \hat{\boldsymbol{\alpha}}$ is asymptotically optimal (see Bhapkar (1996)). Thus, it is reasonable to conjecture that the foregoing W_D for testing $\boldsymbol{\alpha} = \mathbf{0}$ against $\boldsymbol{A\alpha} \geqslant \mathbf{0}$ is likely to have good properties in large samples.

6.9 EXACT CONDITIONAL TESTS

When the sample size is not large enough to rely on asymptotic results, exact conditional tests are attractive alternatives. Even if the sample size is large enough to rely on asymptotic results, exact tests are still good competitors. An important feature of the implementation of the exact conditional tests discussed in this section is that the p-value is estimated by simulation. Consequently, it is computed with some error, but it (i.e., error) can be made arbitrarily small by increasing the number of simulations. Let us first consider Fisher's exact test to briefly review conditional test in a 2×2 table.

Consider an experiment to compare two treatments. The response variable takes one of two values, Failure or Success; let the notation be as in Table 6.7 on page 299. Let the null and alternative hypotheses be

$$H_0 : \pi_{11} = \pi_{21}, \quad \text{and} \quad H_1 : \pi_{11} > \pi_{21},$$

respectively. Suppose that H_0 is true, and let the common value of π_{11} and π_{21} be denoted by π_1. The exact finite sample distribution of *LRT* depends on π_1. Conditional inference provides a convenient way of ensuring that the nuisance parameter, π_1, does not appear in the finite sample exact null distribution of the relevant test statistic. This is achieved by the standard method of conditioning on a sufficient statistic for the nuisance parameter, π_1. The column total is a sufficient statistic for π_1 and therefore we condition on $\{n_{+1}\}$. To carry out an exact conditional test of H_0 against H_1, we condition on the marginal totals; the row totals are fixed by design. For now, note that if the marginal totals $\{n_{1+}, n_{2+}, n_{+1}, n_{+2}, n\}$ are fixed then only one of the four cell counts is a free variable; without loss of generality, let us choose it to be n_{11}. Therefore, the joint distribution of $\{n_{11}, n_{12}, n_{21}, n_{22}\}$ given the marginal totals is equivalent to the distribution of n_{11} given the marginal totals. This leads to the hypergeometric distribution,

$$\text{pr}(n_{11} \mid n_{1+}, n_{2+}, n_{+1}, n_{+2}) = \binom{n_{1+}}{n_{11}} \binom{n - n_{1+}}{n_{+1} - n_{11}} \binom{n}{n_{+1}}^{-1}. \quad (6.74)$$

Since large values of n_{11} favor H_1, the exact conditional p-value is the probability that n_{11} is at least as large as the observed sample value, where the probability is computed using the conditional distribution of n_{11} in (6.74). The foregoing test is known as the *Fisher's exact test*. This is the uniformly most powerful similar test. For more details on these aspects see Lehmann (1994, p 80), Cox and Hinkley (1974, example 5.3, p 137), Pace and Salvan (1997, p 205) and Agresti (1990, Section 3.5). The complete derivation of the conditioning argument when the observations are based on a single multinomial sample is given in Pace and Salvan (1997, p 205).

Example 6.9.1

Suppose that the observed cell counts are as in Table 6.16(A). Apart from the observed table, there is only one other possible outcome that is at least as favorable to H_1 as the observed one; this is Table 6.16(B). By definition, the exact conditional p-value

Table 6.16 Comparison of a treatment with a control

	Observed outcome			More extreme outcome		
	Failure	Success	Total	Failure	Success	Total
Control	3	1	4	4	0	4
Treatment	1	3	4	0	4	4
Total	4	4	8	4	4	8
		(A)			(B)	

is the probability that the outcome is either Table 6.16(A) or 6.16(B) conditional on the marginal totals and computed under H_0. By (6.74)

$$p\text{-value} = \binom{4}{3}\binom{4}{1}\binom{8}{4}^{-1} + \binom{4}{4}\binom{4}{0}\binom{8}{4}^{-1} = 0.24.$$

Since the p-value is large there is very little statistical evidence to establish that the treatment is better than the control. The data in Table 6.16 are the same as those for the well-known Fisher's Tea tasting example, but we considered slightly different design of the experiment. ■

6.9.1 Exact Conditional Tests in $2 \times c$ Tables

The ideas underlying the Fisher's exact test can be extended to $r \times c$ tables; let us first consider $2 \times c$ tables. Let the testing problem be

$$\begin{array}{ll} & H_0: \text{ No Difference between the two treatments,} \\ \text{vs} & H_1: \text{ Treatment 2 is better than Treatment 1.} \end{array}$$

Suppose that the null hypothesis is true. Then $\pi_{1j} = \pi_{2j}$ for $j = 1, \ldots, c$; let π_j denote the common value of π_{1j} and $\pi_{2j}, j = 1, \ldots, c$. Now, the probability function of $\{n_{ij}\}$ is

$$\{n_{1+}!(n_{11}! \ldots n_{1c}!)^{-1}\pi_1^{n_{11}} \ldots \pi_c^{n_{1c}}\} \times \{n_{2+}!(n_{21}! \ldots n_{2c}!)^{-1}\pi_1^{n_{21}} \ldots \pi_c^{n_{2c}}\}, \tag{6.75}$$

which is proportional to $\pi_1^{n_{+1}} \ldots \pi_c^{n_{+c}}$. To obtain a test statistic for which the distribution function is free of the nuisance parameter, $(\pi_1, \ldots, \pi_c)$, we condition on the marginal totals. The probability function of the marginal total, $(n_{+1}, \ldots, n_{+c})$, is

$$n!(n_{+1}! \ldots n_{+c}!)^{-1}\pi_1^{n_{+1}} \ldots \pi_c^{n_{+c}}. \tag{6.76}$$

The probability function of $\{n_{ij} : i = 1, 2, j = 1, \ldots, c\}$, conditional on the marginal totals is equal to (6.75) divided by (6.76); this leads to

$$(n_{1+}!n_{2+}!)(n_{+1}! \ldots n_{+c}!)/\{n!(n_{11}! \ldots n_{1c}!)(n_{21}! \ldots n_{2c}!)\}. \tag{6.77}$$

This distribution function does not depend on $(\pi_1, \ldots, \pi_c)$ or any other unknown parameters. Now, the exact conditional p-value is the probability that the outcome is at least as favorable to H_1 as the observed one; the probability is computed with respect to the distribution, (6.77).

We have not yet stated precisely how to determine whether or not a given possible outcome is at least as favorable to H_1 as the observed one. In the 2×2 example discussed in the previous subsection, this meant that n_{11} is at least as large as the observed one. However, when there are more than two response categories, there is no unique way of determining whether or not a possible outcome is at least as favorable to H_1 as the observed one. To illustrate this, let us consider a hypothetical example of three possible outcomes with the same marginal totals (see Table 6.17) Note that, compared with the outcome in Table 6.17A, the treatment performs better and the control performs worse in Table 6.17C. Therefore, the latter is more favorable to the treatment than the former. However, it is not so easy to determine whether or not Table 6.17B is more or less favorable to treatment than Table 6.17A; for example, in Table B, the treatment is worse for fewer subjects but it is also better for fewer subjects. Therefore, when there are more than two response categories, we use test statistics to order tables of possible outcomes.

Table 6.17 Three possible outcomes with the same marginal totals

	W (A)	NC (A)	B (A)	W (B)	NC (B)	B (B)	W (C)	NC (C)	B (C)
Control	10	5	10	11	3	11	12	5	8
Treatment	7	4	8	6	6	7	5	4	10
Total	17	9	18	17	9	18	17	9	18

W: worse; NC: no change; B: better.

Reprinted from: *Journal of Multivariate Analysis*, Volume 72, A. Cohen, J.H.B. Kemperman, and H.B. Sackrowitz, Properties of likelihood inference for order restricted models, Pages 50–77, Copyright (2000), with permission from Elsevier.

Let T denote a statistic for testing H_0 against H_1; without loss of generality assume that large values of T are assumed to favor H_1. For example, T can be $T_{0\ell}, T_{0s}$ or T_{0u}. Given two tables, (A) and (B), of possible outcomes, we shall say that Table (B) is more favorable to H_1 than Table (A), if the value of T for (B) is higher than that for (A). This does not solve the problem completely because different test statistics may order the tables differently; nevertheless, it is a reasonably objective way of ordering tables of possible outcomes.

Let t denote the observed sample value of T, and let $\mathcal{A}$ denote the collection of all tables of possible outcomes with the same marginal totals as the observed one and $T \geqslant t$. Now, the *exact conditional p-value* corresponding to $T = t$ has the following forms:

$$\text{exact conditional } p\text{-value } = \text{pr}(T \geqslant t \mid \text{marginal total}, H_0) \tag{6.78}$$

$$= \quad \text{pr}^*(T \geqslant t) \tag{6.79}$$

$$= \quad \sum_{A \in \mathcal{A}} \text{pr}^*(\text{ outcome is table } A), \tag{6.80}$$

where pr* is the probability computed with respect to the conditional distribution of $\{n_{ij} : i = 1, 2, j = 1, \ldots, c\}$ in (6.77). To implement this exact conditional inference procedure, we need to be able to compute the p-value using at least one of the forms in (6.78)–(6.80). When the sample size is small, the number of elements in $\mathcal{A}$ would not be too large and hence it would be possible to use (6.80) and compute the conditional p-value exactly; the summand in (6.80) would be computed using the hypergeometric distribution in (6.77). However, even with moderate sample sizes, the number of elements in $\mathcal{A}$ can be quite large and it would not be practically feasible to use (6.80) and compute the p-value exactly. An approach that is suitable for practical use, irrespective of the sample size, is estimation of the p-value by simulation; the number of simulations depend on the degree of precision desired, not on the sample size. The steps for the simulation are the following:

1. Generate N tables independently, all having the same marginal totals as the observed table.

2. Compute the test statistic T for each table.

3. Estimate the exact conditional p-value by M/N where M is the number of tables generated for which T turned out to be not less than t.

The error in the estimate of the p-value obtained in the above simulation approach can be made arbitrarily small by increasing N. Since computer programs and algorithms are available for generating tables with given marginal totals, (for example, see Patefield (1981) and the subroutine RNTAB in IMSL), the foregoing simulation approach provides a method that appears to be quite promising. Its simplicity and validity for any sample size makes it quite attractive.

6.9.2 Comparison of Several Binomial Parameters

The foregoing method for the analysis of $2 \times c$ tables can also be used for testing the equality of several binomial parameters against an order restriction. Let $X_1, \ldots, X_k$ be k independent binomial random variables such that $X_i \sim B(n_i, \pi_i), i = 1, \ldots, k$. Let the testing problem be defined by

$$H_0 : \pi_1 = \ldots = \pi_k \quad \text{and} \quad H_1 : \boldsymbol{R\pi} \geqslant \mathbf{0}$$

where $\boldsymbol{\pi} = (\pi_1, \ldots, \pi_k)^T$ and $\boldsymbol{R}$ is an $m \times k$ matrix. The joint probability function of $X_1, \ldots, X_k$, under H_0, is

$$\pi^x(1-\pi)^{n-x} \prod_{i=1}^{k} \binom{n_i}{x_i}$$

where π is the common value of $\pi_1, \ldots, \pi_k$ under H_0, $x = x_1 + \ldots + x_k$ and $n = n_1 + \ldots + n_k$. Let $X = X_1 + \ldots + X_k$. The nuisance parameter π can be eliminated by conditioning on X. The probability function of $(X_1, \ldots, X_k)$ conditional on $X = x$ is

$$\binom{n}{x}^{-1} \prod_{i=1}^{k} \binom{n_i}{x_i}. \tag{6.81}$$

Let $T(X_1, \ldots, X_k)$ be a statistic for testing H_0 against H_1. Then the exact p-value corresponding to a sample value of t for T, conditional on $X = x$, is $\text{pr}(T \geqslant t)$ where the probability is evaluated with respect to the probability function in (6.81). Now, in view of (6.77), the distribution in (6.81) is also the joint distribution for a $2 \times k$ table conditional on the row totals being $\{x, n - x\}$ and the column totals being $\{n_1, \ldots, n_k\}$ with the frequencies in the first row being denoted by $\{x_1, \ldots, x_k\}$. Therefore, the exact conditional p-value can be estimated precisely by the same simulation method as that used in the analysis of $2 \times c$ tables. This problem has also been studied by Cohen, Perlman, and Sackrowitz (1990), Agresti and Coull (1996), and Cohen and Sackrowitz (1998).

6.9.3 Exact Conditional Tests in $r \times c$ Tables

The ideas developed in the previous subsections can be extended to the case when there are more than two rows. Let θ_{ij} denote an odds ratio ($i = 1, \ldots, r-1, j = 1, \ldots, c-1$); it could be the cumulative odds ratio, local odds ratio, or any one of the generalized odds ratios. Let $\boldsymbol{\theta}$ denote the vector formed by $\{\theta_{ij} : i = 1, \ldots, r-1, j = 1, \ldots, c-1\}$. Let the null and alternative hypotheses be

$$H_0 : \boldsymbol{\theta} = \mathbf{1} \quad \text{and} \quad H_1 : \boldsymbol{\theta} \geqslant \mathbf{1}, \tag{6.82}$$

respectively. The null hypothesis is equivalent to $\boldsymbol{\pi}_1 = \ldots = \boldsymbol{\pi}_r$ (see Section 6.4). Under the null hypothesis, the joint distribution of the cell frequencies $\{n_{ij} : i = 1, \ldots, r, j = 1, \ldots, c\}$ conditional on the marginal totals is hypergeometric which does not depend on any unknown parameters. This probability function, which is a direct extension of (6.77), takes the form

$$(\prod_{i=1}^{r} n_{i+}!)(\prod_{j=1}^{c} n_{+j}!)/(n! \prod_{i=1}^{r} \prod_{j=1}^{c} n_{ij}!). \tag{6.83}$$

By conditioning on the marginal totals, we have eliminated the dependence of the distribution on the unknown nuisance parameter, which is the vector of probabilities in the common multinomial distribution. Let T denote a test statistic for testing H_0 against H_1. Now,

$$\begin{aligned} \text{exact conditional } p\text{-value} &= \text{pr}(T \geqslant t \mid \text{marginal totals}, H_0) && (6.84)\\ &= \text{pr}^*(T \geqslant t) && (6.85)\\ &= \sum_{A \in \mathcal{A}} \text{pr}^*(\text{ outcome is table } A), && (6.86) \end{aligned}$$

where t is the sample value of T, $\mathcal{A}$ denotes the collection of all tables of possible outcomes with the same marginal totals as the observed one and $T \geqslant t$, and pr^* is the probability computed with respect to the conditional distribution in (6.83). Exact computation of (6.84) is computing intensive except when the sample size is very small. However, it is easy to estimate it by simulation to any desired degree of precision. The simulation steps are the same as those already given in subsection 6.9.1 for $r = 2$ with the obvious changes to accommodate $r \geqslant 2$.

The exact conditional approach discussed in this subsection is valid and appropriate irrespective of the sample size. Patefield (1981) provided a computer program for generating random tables with a given set of marginal totals. For further details and discussions of the foregoing procedure see Agresti and Coull (1998); for applications see Agresti et al. (1979), Patefield (1981, 1982) , Mehta et al. (1988), Kreiner (1987), Forster et al. (1996), Kim and Agresti (1997), and Agresti and Coull (1996).

Example 6.9.2

Table 6.3 (page 289) provides data from a clinical trial. There are four treatments and the response variable, known as the Glasgow Outcome Scale, is ordinal with five categories. The four treatments are a placebo and three doses of a medication. The data in Table 6.3 are a slight modification of the original data. An objective of the study was to determine whether or not a more favorable outcome tends to occur as the dose increases. Let us formulate the null and the alternative hypotheses as

$$H_0 : \boldsymbol{\theta} = \mathbf{1} \quad \text{and} \quad H_1 : \boldsymbol{\theta} \geqslant \mathbf{1}, \tag{6.87}$$

where $\boldsymbol{\theta}$ is a vector of odds ratios; although it could be any one of the generalized odds ratios, we shall restrict our attention to global, cumulative, and local odds ratios. Agresti and Coull (1998) carried out the computations for the testing problem (6.87) using the exact conditional test; the results reported below are based on their computations.

Let T denote the *LRT* for testing H_0 against H_1 in (6.87). Let $\boldsymbol{\theta}^G$ be the vector of global odds ratios. The sample global odds ratios are all greater than one. Therefore, the estimator of $\boldsymbol{\theta}^G$ subject to the constraint $\boldsymbol{\theta}^G \geqslant \mathbf{1}$ is the same as its unconstrained estimator, and the sample value of the *LRT* for testing $H_0 : \boldsymbol{\theta}^G = \mathbf{1}$ against $H_1 : \boldsymbol{\theta}^G \geqslant \mathbf{1}$ is also equal to the *LRT* for testing $H_0 : \boldsymbol{\theta}^G = \mathbf{1}$ against $H_2 : \boldsymbol{\theta}^G \neq \mathbf{1}$; the latter is the same as the *LRT* for testing independence of rows and columns, and its sample value is 27.80. To compute the exact conditional p-value for this, the simulation approach was used. A 95% confidence interval for the p-value is (0.0017, 0.0022). Thus, there is sufficient evidence to reject $H_0 : \boldsymbol{\theta}^G = \mathbf{1}$ in favor of $H_1 : \boldsymbol{\theta}^G \geqslant \mathbf{1}$.

Let $\boldsymbol{\theta}^C$ and $\boldsymbol{\theta}^L$ denote the vector of cumulative and local odds ratios, respectively. The sample values of these odds ratios for the data in Table 6.3 are given in Table 6.18. Let us consider test of $H_0 : \boldsymbol{\theta}^C = \mathbf{1}$ against $H_1 : \boldsymbol{\theta}^C \geqslant \mathbf{1}$. Note that the sample value of every cumulative odds ratio reported in Table 6.18 is greater than one, except for two. The two that are less than one are only marginally less than one; for example, $\hat{\theta}^C_{31} = 0.95$, and a 95% confidence interval for it is $(0.59, 1.4)$. The sample value of *LRT* for testing $H_0 : \boldsymbol{\theta}^C = \mathbf{1}$ against $H_1 : \boldsymbol{\theta}^C \geqslant \mathbf{1}$ is 27.70. By applying the

Table 6.18 Sample odds ratios

Odds ratio	Treatment	Outcome 1	2	3	4
Cumulative	1	1.16	1.17	1.11	1.04
	2	1.25	1.46	1.24	0.94
	3	0.95	1.23	1.22	1.51
Local	1	1.03	1.14	1.02	0.96
	2	0.73	1.84	1.11	0.76
	3	0.29	3.18	0.999	1.46

simulation approach, a 95% confidence interval for the exact conditional p-value is $(0.0002, 0.0004)$, based on 100000 randomly generated tables with margins equal to the observed values. Thus, there is very strong evidence to reject $H_0 : \boldsymbol{\theta}^C = \mathbf{1}$. Although there is strong evidence to reject $H_0 : \boldsymbol{\theta}^C = \mathbf{1}$, it does not follow that there is statistical evidence to accept $\boldsymbol{\theta}^C \gneq \mathbf{1}$.

Now let us consider test of

$$H_0^* : \boldsymbol{\theta}^C \not> \mathbf{1} \quad \text{against} \quad H_1^* : \boldsymbol{\theta}^C > \mathbf{1}.$$

To apply an IUT at 5% level, we test $\theta_{ij}^C \le 1$ against $\theta_{ij}^C > 1$ at 5% level, for every (i, j). If each of these is rejected, then we would reject $\boldsymbol{\theta}^C \not> \mathbf{1}$ and accept $\boldsymbol{\theta}^C > \mathbf{1}$. Since $\hat{\theta}_{32}^C$ is less than one, we would not reject $\theta_{32}^C \le 1$ against $\theta_{32}^C > 1$. Therefore, irrespective of the values of the other odds ratios we would not reject $H_0^* : \boldsymbol{\theta}^C \not> \mathbf{1}$. These calculations illustrate that it is considerably more difficult (i.e., requires very strong evidence) to reject $H_0^* : \boldsymbol{\theta}^C \not> \mathbf{1}$ than $H_0 : \boldsymbol{\theta}^C = \mathbf{1}$. It is even more difficult to reject $\boldsymbol{\theta}^L \not> \mathbf{1}$ because estimates of $\boldsymbol{\theta}^L$ are based on smaller numbers and rejection of $H_0 : \boldsymbol{\theta}^L \not> \mathbf{1}$ is a stronger conclusion than the rejection of $H_0 : \boldsymbol{\theta}^C \not> \mathbf{1}$ because $\{\boldsymbol{\theta}^C \not> \mathbf{1}\} \subset \{\boldsymbol{\theta}^L \not> \mathbf{1}\}$.

It is possible to formulate the testing problem so that the null parameter space is not as large as $\{\boldsymbol{\theta}^C \not> \mathbf{1}\}$; formulations along this line will be discussed under the comparison of treatment with multivariate responses, in particular, under multiple end points. In such an approach, we start with positive numbers $\epsilon_{ij} (i = 1, \ldots, r, j = 1, \ldots, c)$. These numbers are chosen so that if $\theta_{ij} \geqslant -\epsilon_{ij}$ for every (i, j) then the treatments can be considered as not inferior to the control/placebo. Then, we define $\Theta_1 = \{\boldsymbol{\theta}^C : \theta_{ij}^C > -\epsilon_{ij}$ for every (i, j), and $\theta_{ij}^C > 0$ for some $(i, j)\}$, and then the null and alternative hypotheses as $H_0 : \boldsymbol{\theta} \notin \Theta_1$ and $H_1 : \boldsymbol{\theta}^C \in \Theta_1$, respectively. This type of testing problems can be handled by the methods discussed under multiple end points in a later chapter.

6.9.4 Marginal Homogeneity in Square Tables (Agresti-Coull test)

Exact conditional tests can be applied to test marginal homogeneity against an order restriction on the marginal distribution of a square table. Such questions arise particularly when there are matched pairs. An asymptotic test was provided in Subsection 6.8. To illustrate the nature of the problem, let us consider Table 6.19, which relates to a case/control study on the relationship between esophageal cancer and the consumption of very hot beverages; for an analysis of these data that does not involve multiparameter inequality constraints, see Agresti (1990, p 364). In this context, the question that we wish to answer is the following: Is there evidence to support the claim that the cancer patients consumed very hot beverages more than did the control group? In this section, we shall provide an exact conditional test due to Agresti and Coull (1998) for this type of situations. We shall first indicate the implementation of the test, and then discuss the relevant technical details to provide more insight into and justification of the proposed procedure. Let R denote the ordinal response variable, and I denote the number of ordinal categories in R. For a randomly chosen matched pair, let the ordered pair (R_1, R_2) denote the ordinal responses for the two treatments. Let

$$\pi_{ij} = \text{pr}(R_1 = i, R_2 = j).$$

The marginal distributions of R_1 and R_2 are given by $\text{pr}(R_1 = i) = \pi_{i+}$ and $\text{pr}(R_2 = i) = \pi_{+i}$, respectively ($i = 1, \ldots, I$). Let the null and the alternative hypotheses be

$$H_0 : \pi_{j+} = \pi_{+j} \text{ for every } j, \text{ and } H_1 : \pi_{1+} + \ldots + \pi_{j+} \geqslant \pi_{+1} + \ldots + \pi_{+j}, \forall j, \tag{6.88}$$

respectively. Let T denote a statistic for testing H_0 against H_1; it can be the *LRT* but it does not have to be.

To indicate the procedure for computing the p-value for a test of H_0 against H_1, let us suppose that H_0 holds. Therefore, by assumption, there is no difference between the two treatments, namely case and control. Now, in the procedure to be introduced, all nuisance parameters will be eliminated by conditioning on the marginal totals

Table 6.19 Number of beverages drunk at burning hot temperatures for esophageal cancer patients and matching controls

	Controls			
Case	0	1	2	3
0	31	5	5	0
1	12	1	0	0
2	14	1	2	1
3	6	1	1	0

Table 6.20 Sample space for exact conditional test

Pair ID		Observed response						Possible response[1]					
		1	2	3	4	5	Total	1	2	3	4	5	Total
1	Subject 1	0	0	0	1	0	1	0	1	0	0	0	1
	Subject 2	0	1	0	0	0	1	0	0	0	1	0	1
	Total	0	1	0	1	0		0	1	0	1	0	
⋮		⋮						⋮					
k	Subject 1	0	1	0	0	0	1	0	0	0	1	0	1
	Subject 2	0	0	0	1	0	1	0	1	0	0	0	1
	Total	0	1	0	1	0		0	1	0	1	0	
⋮		⋮						⋮					
m	Subject 1	1	0	0	0	0	1	0	0	1	0	0	1
	Subject 2	0	0	1	0	0	1	1	0	0	0	0	1
	Total	1	0	1	0	0		1	0	1	0	0	

[1]This is obtained by interchanging the observed responses of the two subjects; consequently, the marginal totals are unchanged.

for each matched pair. Consider an arbitrary matched pair. For this matched pair, conditional on the marginal totals of the pair, the observed outcome and the one obtained by interchanging the two responses are equally likely and hence any one of the two can occur with probability (1/2); this is true whether or not the observed responses for the two subjects are equal. For a matched pair in which the observed responses for the treatments are equal, conditional on the marginal totals for that pair, the distribution of the cell frequencies is completely determined because there is only one possible outcome. Thus, only those matched pairs for which the responses for the two treatments are different are relevant for conditional inference. For each matched pair, conditional on the marginal totals, the two possible and equally likely outcomes are shown in Table 6.20. This consists of the observed one on the left; and, on the right, the one obtained by interchanging the observed responses of the two treatments.

Let m denote the number of matched pairs for which the responses are different. It is easily seen that, conditional on the marginal totals for each matched pair, there are 2^m possible ways of choosing one of the two possible outcomes for the m individuals; further, they (i.e., the 2^m possibilities) are all equally likely. Therefore, the exact conditional p-value is the proportion of the 2^m tables for which $T \geqslant t$, where t is the sample value of T. Even for moderate values of m, computing this p-value exactly could be prohibitively expensive. [The total number of distinct tables is $\{2^{-c(c-1)/2}\Pi_{j>i}(n_{ij}+n_{ji})(n_{ij}+n_{ji}+1)\}$; this is a large number even for moderate sample sizes.] A simpler approach is to estimate the p-value by simulation.

Simulation steps for computing the exact conditional p-value:

1. For each of the m pairs in Table 6.20, choose one of the two $2 \times I$ tables with equal probability.

2. Construct the $I \times I$ table using the pseudo observations generated in Step (1), and compute the test statistic T for testing H_0 against H_1.

3. Repeat the previous two steps N times independently.

4. Estimate the exact conditional p-value by M/N where M is the number of times $T \geqslant t$ where t is the sample value of T.

The foregoing implementation of the simulation can be made more efficient, perhaps only slightly. Note that there may well be several, say K_1, matched pairs with the same marginal totals as for the first matched pair. In this case, instead of generating one random binary observation for each of the K_1 pairs, we may generate one binomial observation with K_1 trials. This may be implemented as follows:

An efficient simulation approach for computing the exact conditional p-value:

1. Let $m_{ij} = n_{ij} + n_{ji}$, for $1 \leq i < j \leq I$.

2. Generate n_{ij}^* from $B(m_{ij}, 1/2)$ for every $j > i$.

3. Let $n_{ji}^* = m_{ij} - n_{ij}^*$ for $1 \leq j < i \leq I$.

4. Compute the test statistic T for the table $\boldsymbol{n}^*(= \{n_{ij}^*\})$

5. Repeat the first four steps N times and estimate the p-value by M/N where M is the number of times $T \geqslant t$ where t is the sample value of T.

This is more efficient than the procedure outlined earlier, because we need to generate only $(I-1)(I-2)/2$ binomial random variables irrespective of the sample size. By contrast the previous method requires to generate m binary random numbers where m is equal to the total number of matched pairs for which the responses are different. Typically $(I-1)(I-2)/2$ is much smaller than m.

For the special case when $I = 2$, detailed technical arguments to justify the method are given in Agresti (1990, Section 10.1.4). The same idea can be extended, as indicated below, to the case when $I > 2$. The essentials of the following arguments are based on Section 7.5.1 of McCullagh and Nelder (1989).

For the sth matched pair, let $R_{1(s)}$ and $R_{2(s)}$ denote the response variables for treatments 1 and 2, respectively; each one of them takes a value in $\{1, \ldots, I\}$. Further, let the two multinomial response vectors be

$$\boldsymbol{Y}_{1(s)} = (Y_{11(s)}, \ldots, Y_{1I(s)})^T \quad \text{and} \quad \boldsymbol{Y}_{2(s)} = (Y_{21(s)}, \ldots, Y_{2I(s)})^T,$$

respectively. For example, if $R_{1(s)} = j$ then $Y_{1(s)} = (0, \ldots, 0, 1, 0 \ldots, 0)^T$ where the "1" appears at the jth coordinate. Let the column total vector be

$$\boldsymbol{Y}_{\cdot(s)} = (Y_{11(s)} + Y_{21(s)}, \ldots, Y_{1I(s)} + Y_{2I(s)})^T.$$

Now to develop a test of H_0 against H_1, we make the following assumption: *For the sth matched pair, the marginal probabilities for $R_{1(s)}$ and $R_{2(s)}$ satisfy*

$$\begin{aligned} &\text{pr}(R_{1(s)} = j) \propto \exp(\lambda_{j(s)}) \\ \text{and } &\text{pr}(R_{2(s)} = j) \propto \exp(\lambda_{j(s)} + \delta_j), \text{ for } j = 1, \ldots, I. \end{aligned}$$

Let $\boldsymbol{\lambda}(s) = (\lambda_{1(s)}, \ldots, \lambda_{I(s)})^T$ and $\boldsymbol{\delta} = (\delta_1, \ldots, \delta_I)^T$. The parameter $\boldsymbol{\lambda}_{(s)}$ captures the effect of sth matched pair and $\boldsymbol{\delta}$ captures the effect of the treatments, which is assumed to be the same for different pairs. Let

$$A_{(s)} = \sum_j \exp(\lambda_{j(s)}) \quad \text{and} \quad B_{(s)} = \sum_j \exp(\lambda_{j(s)} + \delta_j).$$

Then

$$\begin{aligned} \text{pr}(R_{1(s)} = j) &= A_{(s)}^{-1} \exp(\lambda_{j(s)}) \\ \text{and } \text{pr}(R_{2(s)} = j) &= B_{(s)}^{-1} \exp(\lambda_{j(s)} + \delta_j), \end{aligned}$$

$j = 1, \ldots, c$. The unconditional probability function for the matched pair s is

$$\text{pr}\{\boldsymbol{Y}_{1(s)} = \boldsymbol{y}_{1(s)}, \boldsymbol{Y}_{2(s)} = \boldsymbol{y}_{2(s)}\} = f(\boldsymbol{\lambda}_{(s)}, \boldsymbol{\delta}) \exp(\boldsymbol{\lambda}_{(s)}^T \boldsymbol{y}_{\cdot(s)} + \boldsymbol{\delta}^T \boldsymbol{y}_{2(s)}) \quad (6.89)$$

where $f(\boldsymbol{\lambda}_{(s)}, \boldsymbol{\delta})$ does not depend on $\{\boldsymbol{y}_{1(s)}, \boldsymbol{y}_{2(s)}\}$. The full likelihood is the product of terms like (6.89) for $s = 1, \ldots, n$. Thus, $(\boldsymbol{Y}_{\cdot(1)}, \ldots, \boldsymbol{Y}_{\cdot(s)})$ is sufficient for $(\lambda_{(1)}, \ldots, \lambda_{(s)})$. Therefore, the nuisance parameter $(\lambda_{(1)}, \ldots, \lambda_{(s)})$ can be eliminated by conditioning on $(\boldsymbol{Y}_{\cdot(1)}, \ldots, \boldsymbol{Y}_{\cdot(n)})$. The only possible values of $\boldsymbol{Y}_{\cdot(s)}$ are of the form $(0, \ldots, 0, 1, 0, \ldots, 0, 1, 0 \ldots, 0)$ and $(0, \ldots, 0, 2, 0, \ldots, 0)$. If we condition on $\boldsymbol{Y}_{\cdot(s)} = (0, \ldots, 0, 2, 0, \ldots, 0)$ then $\boldsymbol{Y}_{1(s)} = \boldsymbol{Y}_{2(s)} = (0, \ldots, 0, 1, 0, \ldots, 0)$. Therefore, the probability function of $(\boldsymbol{Y}_{1(s)}, \boldsymbol{Y}_{2(s)})$, conditional on $\boldsymbol{Y}_{\cdot(s)} = (0, \ldots, 0, 2, 0, \ldots, 0)$ does not depend on $\boldsymbol{\delta}$. Let $\boldsymbol{e}_{ij} = (0, \ldots, 0, 1, 0, \ldots, 0, 1, 0 \ldots, 0)$ where 1 appears at locations i and j $(i \neq j)$. The probability function of $(\boldsymbol{Y}_{1(s)}, \boldsymbol{Y}_{2(s)})$, conditional on $\boldsymbol{Y}_{\cdot(s)} = \boldsymbol{e}_{ij}$ is given by

$$\text{pr}\{R_{1(s)} = a, R_{2(s)} = b \mid \boldsymbol{Y}_{\cdot(s)} = \boldsymbol{e}_{ij}\}$$

$$= \begin{cases} \exp(\delta_j)/\{\exp(\delta_i) + \exp(\delta_j)\} & \text{if } (a, b) = (i, j) \\ \exp(\delta_i)/\{\exp(\delta_i) + \exp(\delta_j)\} & \text{if } (a, b) = (j, i). \end{cases}$$

Under H_0, we have that $\boldsymbol{\delta} = \boldsymbol{0}$ and, therefore, the foregoing two probabilities are equal to 1/2. This provides an alternative justification for the Agresti-Coull approach to implementing an exact conditional test of marginal homogeneity.

Let Z_{ij} denote the number of matched pairs for which $(R_1, R_2) = (i, j)$, and let $m_{ij} = Z_{ij} + Z_{ji}$ $(i \neq j)$; let $m_{ii} = Z_{ii}$ for every i. Thus m_{ij} is the number of matched pairs for which the column total is $\boldsymbol{e}_{ij}$. Conditionally on the column total being equal to $\boldsymbol{e}_{ij}$, we have that

$$Z_{ij} \sim B(m_{ij}, \pi_{ij}), \quad \text{and} \quad \text{logit}(\pi_{ij}) = \delta_j - \delta_i. \quad (6.90)$$

This is the so called quasi-symmetry model suggested by Caussinus (1965). For such a model, marginal homogeneity is equivalent to symmetry (i.e., $\pi_{ij} = \pi_{ji}$). The model (6.90) can be used for exact conditional constrained inference on $\boldsymbol{\delta}$.

For discussions on the relevance of these in unconstrained inference see Agresti (1990, Section 10.3) and McCullagh and Nelder (1989, Section 7.5).

6.10 DISCUSSION

The general asymptotic results on tests of hypotheses given in the previous chapter can be applied to test hypotheses concerning generalized odds ratios. Let $\boldsymbol{\theta}$ denote any one of the generalized odds ratios. Let the null and alternative hypotheses be

$$H_0 : \boldsymbol{\theta} = \mathbf{0} \text{ and } H_1 : \boldsymbol{\theta} \geqslant \mathbf{1},$$

respectively. Clearly these hypotheses may be stated as

$$H_0 : \boldsymbol{h}(\boldsymbol{\pi}) = \mathbf{0} \text{ and } H_1 : \boldsymbol{h}(\boldsymbol{\pi}) \geqslant \mathbf{0}$$

where $\boldsymbol{h}$ is a smooth vector function of $\boldsymbol{\pi}$. Now it follows that *LRT*, Wald-type and score test statistics are asymptotically equivalent, and their common asymptotic null distribution is a chi-bar square. The exact form of the chi-bar square can be written down fairly easily. In general this test is not similar; it is similar when $\boldsymbol{\theta}$ is the continuation odds ratio, $\boldsymbol{\theta}^C$ (see Theorem 6.7.2). The location of the least favorable null value is known for some generalized odds ratios. If we wish to test the forgoing hypotheses, a convenient procedure is to apply an exact conditional test with p-value estimated by simulation. Engen and Lillegärd (1997) provide a general method for implementing Monte Carlo simulations conditional on sufficient statistics. This approach may have potential applications in constrained inference.

Now, consider test of $H_1 : \boldsymbol{\theta} \geqslant \mathbf{1}$ against $H_2 : \boldsymbol{\theta} \ngeqslant \mathbf{1}$. The asymptotic null distribution of the *LRT*, Wald-type, and score test statistics are known to be chi-bar-square at every point in the null parameter space, $\{\boldsymbol{\pi} : \boldsymbol{\theta} \geqslant \mathbf{1}\}$. These tests are not similar for any of the generalized odds ratios, and the least favorable null distributions corresponding to $\boldsymbol{\theta}^C$ and $\boldsymbol{\theta}^{CO}$ and $\boldsymbol{\theta}^L$ are known (Theorem 6.7.4).

6.11 PROOFS:

First, we shall prove a lemma related to the least favorable null value for a distribution that arises in Type B testing problems involving contingency tables.

Lemma 6.11.1 *Let $\theta_1, \ldots, \theta_k$ be k scalar parameters and $\boldsymbol{A}$ be an $m \times k$ matrix in which every row is a permutation of $(-1, 1, 0, \ldots, 0)$. Let the null and alternative hypotheses be*

$$H_0 : \boldsymbol{A\theta} \geqslant \mathbf{0} \text{ and } H_1 : \boldsymbol{A\theta} \ngeqslant \mathbf{0},$$

respectively. Suppose that the null distribution (may be asymptotic) of a statistic for testing H_0 against H_1 is

$$\inf\{(\boldsymbol{X} - \boldsymbol{t})^T \boldsymbol{W}(\boldsymbol{\theta}_0)(\boldsymbol{X} - \boldsymbol{t}) : \boldsymbol{t} \in \mathcal{T}(\boldsymbol{\Omega}_0; \boldsymbol{\theta}_0)\} \tag{6.91}$$

where $\boldsymbol{\theta}_0 = (\theta_{01}, \ldots, \theta_{0k})$ is the assumed true null value of $\boldsymbol{\theta}$, $\boldsymbol{X} \sim N\{\mathbf{0}, \boldsymbol{W}(\boldsymbol{\theta}_0)^{-1}\}$, $\boldsymbol{W}(\boldsymbol{\theta}_0) = diag\ \{W_1(\boldsymbol{\theta}_0), \ldots, W_k(\boldsymbol{\theta}_0)\}, W_i(\boldsymbol{\theta}_0) = c_i\sigma^2 g(\theta_{0i}), \{c_1, \ldots, c_k\}$ are known constants, σ^2 is a parameter that may be unknown, $g(\cdot)$ is a real positive *function, and $\mathcal{T}(\Omega_0;\boldsymbol{\theta}_0)$ is the tangent cone of Ω_0 at $\boldsymbol{\theta}_0$. Then the least favorable null value (for the distribution in* (6.91)*) is any point in $\{\boldsymbol{\theta} : \boldsymbol{A\theta} = \boldsymbol{0}\}$ and the least favorable null distribution is*

$$\inf\{(\boldsymbol{U} - \boldsymbol{t})^T \boldsymbol{C}(\boldsymbol{U} - \boldsymbol{t}) : \boldsymbol{At} \geqslant \boldsymbol{0}\}$$

where $\boldsymbol{C} = diag\ \{c_1, \ldots, c_k\}$ and $\boldsymbol{U} \sim N(\boldsymbol{0}, \boldsymbol{C}^{-1})$.

Proof:

Let $\boldsymbol{D} = \text{diag}\ \{d_1, \ldots, d_k\}$ where $d_i = \sigma\sqrt{g(\theta_{0i})}$ for $i = 1, \ldots, k$ and $\boldsymbol{C} = \text{diag}\ \{c_1, \ldots, c_k\}$. Then $\boldsymbol{W}(\theta_0) = \boldsymbol{D}^T\boldsymbol{C}\boldsymbol{D}$. Let $\boldsymbol{U} = \boldsymbol{DX}$. Then, it is easily seen that (A.1) is equal to

$$\inf\{(\boldsymbol{U} - \boldsymbol{u})^T \boldsymbol{C}(\boldsymbol{U} - \boldsymbol{u}) : \boldsymbol{u} \in \boldsymbol{D}\mathcal{T}(\Omega_0;\boldsymbol{\theta}_0)\}. \tag{6.92}$$

Let $\boldsymbol{A}_0$ be the submatrix of $\boldsymbol{A}$ containing all the active constraints at $\boldsymbol{\theta}_0$; in other words, if $\boldsymbol{a}^T$ is a row of $\boldsymbol{A}$ then $\boldsymbol{a}^T$ is also a row $\boldsymbol{A}_0$ if and only if $\boldsymbol{a}^T\boldsymbol{\theta}_0 = 0$. Therefore, $\boldsymbol{A}_0\boldsymbol{\theta}_0 = \boldsymbol{0}$ and $\mathcal{T}(\Omega_0;\boldsymbol{\theta}_0) = \{\boldsymbol{\theta} : \boldsymbol{A}_0\boldsymbol{\theta} \geqslant \boldsymbol{0}\}$. Now, it may be verified that

$$\boldsymbol{D}\mathcal{T}(\Omega_0;\boldsymbol{\theta}_0) = \mathcal{T}(\Omega_0;\boldsymbol{\theta}_0). \tag{6.93}$$

Let us first verify (6.93). Note that

$$\begin{aligned}
\boldsymbol{u} \in \boldsymbol{D}\mathcal{T}(\Omega_0;\theta_0) &\Longleftrightarrow \boldsymbol{u} = \boldsymbol{Dx}, \text{ for some } \boldsymbol{x} \text{ in } \mathcal{T}(\Omega_0;\boldsymbol{\theta_0})\\
&\Longleftrightarrow \boldsymbol{u} = \boldsymbol{Dx} \text{ for some } \boldsymbol{x} \text{ satisfying } \boldsymbol{A}_0\boldsymbol{x} \geqslant \boldsymbol{0}\\
&\Longleftrightarrow \boldsymbol{A}_0\boldsymbol{D}^{-1}\boldsymbol{u} \geqslant 0.
\end{aligned}$$

Now, let $\boldsymbol{e}^T = (0, \ldots, 0, 1, 0, \ldots, 0 - 1, 0, \ldots, 0)$ be an arbitrary row $\boldsymbol{A}_0$ where 1 and -1 appear at the ith and jth positions; the following arguments would hold even if 1 and -1 are interchanged. Since $\boldsymbol{e^T\theta_0} = 0$ it follows that $\theta_{0i} = \theta_{0j}$, and hence $\boldsymbol{e}^T\boldsymbol{D}^{-1}\boldsymbol{u} = \sigma^{-1}\{g(\theta_{0i})\}^{-\frac{1}{2}}(u_i - u_j)$. Now, $\boldsymbol{e}^T\boldsymbol{D}^{-1}\boldsymbol{u} \geqslant 0 \Leftrightarrow \boldsymbol{e}^T\boldsymbol{u} \geqslant 0$, and hence

$$\boldsymbol{u} \in \boldsymbol{D}\mathcal{T}(\Omega_0;\boldsymbol{\theta}_0) \Leftrightarrow \boldsymbol{A}_0\boldsymbol{D}^{-1}\boldsymbol{u} \geqslant 0 \Leftrightarrow \boldsymbol{A}_0\boldsymbol{u} \geq 0 \Leftrightarrow \boldsymbol{u} \in \mathcal{T}(\Omega_0;\boldsymbol{\theta}_0).$$

Therefore, $\boldsymbol{D}\mathcal{T}(\Omega_0;\boldsymbol{\theta}_0) = \mathcal{T}(\Omega_0;\boldsymbol{\theta}_0)$, this shows that (A.3) is true.

Now, it follows from (6.92) and (6.93) that the distribution of (6.91) is equal to

$$\inf\{\boldsymbol{U} - \boldsymbol{u})^T \boldsymbol{C}(\boldsymbol{U} - \boldsymbol{u}) : \boldsymbol{u} \in \mathcal{T}(\boldsymbol{\Omega}_0;\boldsymbol{\theta}_0)\}. \tag{6.94}$$

Clearly, (6.94) becomes larger as $\mathcal{T}(\Omega_0;\boldsymbol{\theta}_0)$ becomes smaller, and $\mathcal{T}(\Omega_0;\boldsymbol{\theta}_0)$ has its smallest value when $\boldsymbol{A}_0 = \boldsymbol{A}$. Therefore, the least favorable null value is any $\boldsymbol{\theta}_0$ such that $\boldsymbol{A\theta}_0 = \boldsymbol{0}$. At any such $\boldsymbol{\theta}_0$, the distribution of (6.94) is a chi-bar square and it does not depend on the particular vale of $\boldsymbol{\theta}_0$ in $\{\boldsymbol{\theta} : \boldsymbol{A\theta} = \boldsymbol{0}\}$. Therefore, any $\boldsymbol{\theta}$ in $\{\boldsymbol{\theta} : \boldsymbol{A\theta} = \boldsymbol{0}\}$ is a least favorable null value for the distribution (6.91). At such a point, (6.94) is equal to $\inf\{(\boldsymbol{U} - \boldsymbol{u})^T\boldsymbol{C}(\boldsymbol{U} - \boldsymbol{u}) : \boldsymbol{Au} \geqslant \boldsymbol{0}\}$; its distribution does not depend on any unknown nuisance parameters. ■

Proof of Proposition 6.5.1 :

[This is based on the proof given by Dardanoni and Forcina (1998)]. The main idea of the proof relating to the least favorable null distribution of the *LRT* for testing $H_0 : \boldsymbol{R\pi} = \mathbf{0}$ against $H_1 : \boldsymbol{R\pi} \geqslant \mathbf{0}$ is given first. Then the technical details are given.

The asymptotic null distribution of the *LRT* at $\boldsymbol{\pi}_0$ is $\bar{\chi}^2(\boldsymbol{V}_0, \mathcal{O}^+)$ where $\boldsymbol{V}_0 = \boldsymbol{V}(\boldsymbol{\pi}_0) = \boldsymbol{A}\{\text{diag}\,(\boldsymbol{\pi}_{01}) - \boldsymbol{\pi}_{01}\boldsymbol{\pi}_{01}^T\}\boldsymbol{A}^T$. Let $\boldsymbol{Y} \sim N(\mathbf{0}, \boldsymbol{V}_0)$ and for any matrix $\boldsymbol{B}$, let $\mathcal{C}(\boldsymbol{B}) = \{\boldsymbol{x} : \boldsymbol{Bx} \geqslant \mathbf{0}\}$. First, by a suitable transformation, transfer the dependence of the chi-bar square distribution, $\bar{\chi}^2\{\boldsymbol{V}(\boldsymbol{\pi}_0), \mathcal{O}^+\}$, on $\boldsymbol{\pi}_0$ from the covariance matrix to the cone. This is done as follows. Let $\boldsymbol{V}(\boldsymbol{\pi}) = \boldsymbol{L}_\pi \boldsymbol{L}_\pi^T$ for some $\boldsymbol{L}_\pi$. In this proof we will use the Cholesky factorization. Then, we have that

$$\begin{aligned}
&\bar{\chi}^2\{\boldsymbol{V}(\boldsymbol{\pi}), \mathcal{O}^+\} \sim \boldsymbol{Y}^T\boldsymbol{V}(\boldsymbol{\pi})^{-1}\boldsymbol{Y} - \inf\{(\boldsymbol{Y}-\boldsymbol{a})^T\boldsymbol{V}(\boldsymbol{\pi})^{-1}(\boldsymbol{Y}-\boldsymbol{a}) : \boldsymbol{a} \geqslant \mathbf{0}\} \\
&= \boldsymbol{Y}^T(\boldsymbol{L}_\pi\boldsymbol{L}_\pi^T)^{-1}\boldsymbol{Y} - \inf\{(\boldsymbol{Y}-\boldsymbol{a})^T(\boldsymbol{L}_\pi\boldsymbol{L}_\pi^T)^{-1}(\boldsymbol{Y}-\boldsymbol{a}) : \boldsymbol{a} \geqslant \mathbf{0}\} \\
&= \boldsymbol{Z}^T\boldsymbol{Z} - \inf\{(\boldsymbol{Z}-\boldsymbol{b})^T(\boldsymbol{Z}-\boldsymbol{b}) : \boldsymbol{L}_\pi\boldsymbol{b} \geqslant \mathbf{0}\}, \text{ where } \boldsymbol{Z} = \boldsymbol{L}_\pi^{-1}\boldsymbol{Y}, \\
&\sim \bar{\chi}^2\{\boldsymbol{I}, \mathcal{C}(\boldsymbol{L}_\pi)\}.
\end{aligned}$$

Recall that $\mathcal{C}_1 \subseteq \mathcal{C}_2 \Rightarrow \bar{\chi}^2(\boldsymbol{V}, \mathcal{C}_1) \leq_d \bar{\chi}^2(\boldsymbol{V}, \mathcal{C}_2)$ for any covariance matrix, $\boldsymbol{V}$, where $\leq_d$ means stochastically larger. Now, the idea is to choose a sequence of points, $\boldsymbol{\pi}(N)$, such that $\mathcal{C}\{\boldsymbol{L}_{\pi(N)}\}$ converges to an upper limit of $\mathcal{C}(\boldsymbol{L}_\pi)$ over $\boldsymbol{\pi} \in H_0$. Then it follows that $\boldsymbol{\pi}(N)$ converges to the least favorable null value. The details of these are given below.

Let $f_1, \ldots, f_{c-1}$ be positive numbers such that $f_1 + \cdots + f_{c-1} < 1$, and let $\boldsymbol{f} = (f_1, \ldots, f_{c-1})^T$. Thus, $\boldsymbol{f}$ serves as a generic value of $\boldsymbol{\pi}_1$. Let $\boldsymbol{V}_f = \boldsymbol{A}\{$ diag $(\boldsymbol{f}) - \boldsymbol{f}\boldsymbol{f}^T\}\boldsymbol{A}^T$. It may be verified that the Cholesky decomposition of $\boldsymbol{V}_f$ is given by $\boldsymbol{V}_f = \boldsymbol{L}_f\boldsymbol{L}_f^T$, where

$$\boldsymbol{L_f} = \begin{bmatrix} \sqrt{\rho_1}(1-F_1) & 0 & 0\ldots0 \\ \sqrt{\rho_1}(1-F_2) & \sqrt{\rho_2}(1-F_2) & 0\ldots0 \\ \vdots & \vdots & \vdots \\ \sqrt{\rho_1}(1-F_{c-1}) & \sqrt{\rho_2}(1-F_{c-1}) & \ldots\sqrt{\rho_{c-1}}(1-F_{c-1}) \end{bmatrix}$$

$$\boldsymbol{F} = \begin{bmatrix} F_1 \\ F_2 \\ \vdots \\ F_{c-1} \end{bmatrix} = \begin{bmatrix} f_1 \\ f_1 + f_2 \\ \vdots \\ f_1 + \cdots + f_{c-1} \end{bmatrix},$$

$$\boldsymbol{P} = \begin{bmatrix} \rho_1 \\ \rho_2 \\ \vdots \\ \rho_{c-1} \end{bmatrix} = \begin{bmatrix} F_1(1-F_1)^{-1} \\ F_2(1-F_2)^{-1} - F_1(1-F_1)^{-1} \\ \vdots \\ F_{c-1}(1-F_{c-1})^{-1} - F_{c-2}(1-F_{c-2}) \end{bmatrix}$$

Let $\boldsymbol{e} = (1, 0, \ldots, 0)^T$. Since the elements of $\boldsymbol{L_f}$ are all positive, it follows that $\mathcal{O}_{c-1}^+ \subseteq \mathcal{C}(\boldsymbol{L}_f) \subseteq \mathcal{C}(\boldsymbol{e}^T)$.

Let $\boldsymbol{f}_1(\epsilon) = 2^{-1}(1-(c-2)\epsilon, \epsilon, \ldots \epsilon)^T$ where $\epsilon > 0$. Now, as $\epsilon \to 0$,

$$\begin{aligned} \boldsymbol{f}_1(\epsilon) &\to (\frac{1}{2}, 0, \ldots, 0)^T \\ \boldsymbol{P}_{f_1} &\to \boldsymbol{e}, \\ \boldsymbol{L}_{f_1} &\to 2^{-1}(\mathbf{1}, \mathbf{0}, \ldots, \mathbf{0}), \\ \mathcal{C}(\boldsymbol{L}_{f_1}) &\to \mathcal{C}(\boldsymbol{e}^T) = \{\boldsymbol{x} : x_1 \geqslant 0\}. \end{aligned}$$

Since $\mathcal{C}(\boldsymbol{L}_f) \subseteq \mathcal{C}(\boldsymbol{e}^T)$ for any $\boldsymbol{f}$, it follows that a least favorable null value is obtained in the limit $\epsilon \to 0$ when $\boldsymbol{\pi} = (\boldsymbol{f_1}(\epsilon) : \boldsymbol{f}_1(\epsilon))$. Thus, the least favorable null value is $\boldsymbol{\pi}_* = (\boldsymbol{\pi}_{*1} : \boldsymbol{\pi}_{*1})$ where $\boldsymbol{\pi}_{*1} = (2^{-1}, 0, \ldots, 0)^T$ and the least favorable null distribution is $\bar{\chi}^2\{\boldsymbol{I}, \mathcal{C}(\boldsymbol{e}^T)\}$ for which the *cdf* at t is

$$(\frac{1}{2})\{pr(\chi^2_{c-2} \leq t) + \mathrm{pr}(\chi^2_{c-1} \leq t)\}.$$

Problems

6.1 Consider an $r \times c$ contingency table in which the rows and columns are ordinal. Let π_{ij} denote the probability that the response is j for row $i (i = 1, \ldots, r, j = 1, \ldots, c)$. Let $\tilde{y}_{ij}$ denote the estimated frequency for cell (i, j) under an order restriction on the odds ratios. Similarly, let $\bar{y}_{ij}$ denote the estimates under the independence assumption. Show that the *LRT* for testing independence against the particular order restriction is $2\sum_i \sum_j \log(\tilde{y}_{ij}/\bar{y}_{ij})$.

6.2 Let $Y_1, \ldots, Y_c$ be independent binomial random variables distributed as $Y_i \sim B(n_i, \pi_i), i = 1, \ldots, c$. Let $\boldsymbol{A}$ be a $q \times c$ matrix, and let the null and alternative hypotheses be

$$H_0 : \boldsymbol{A\pi} \geqslant \mathbf{0} \text{ and } H_1 : \boldsymbol{A\pi} \ngeq \mathbf{0},$$

respectively. Let $\Omega = \{\boldsymbol{\pi} : 0 < \pi_i < 1, \text{ for } i = 1, \ldots, c\}$, $\Omega_0 = \{\boldsymbol{\pi} \in \Omega : \boldsymbol{A\pi} \geqslant \mathbf{0}\}$, $\boldsymbol{\pi}_0$ be a point in the null parameter space Ω_0. Let $\boldsymbol{A}$ be partitioned as $\boldsymbol{A} = [\boldsymbol{A}_1^T, \boldsymbol{A}_2^T]^T$ such that $\boldsymbol{A}_1\boldsymbol{\pi}_0 = 0$ and $\boldsymbol{A}_2\boldsymbol{\pi}_0 > 0$; thus, the submatrix $\boldsymbol{A}_1$ of $\boldsymbol{A}$ corresponds to all the active constraints at $\boldsymbol{\pi}_0$, and similarly $\boldsymbol{A}_2$ corresponds to all the inactive constraints at $\boldsymbol{\pi}_0$.

(i) Show that $\mathcal{T}(\Omega_0; \boldsymbol{\pi}_0) = \{\boldsymbol{\pi} : \boldsymbol{A}_1\boldsymbol{\pi} \geqslant \mathbf{0}\}$ and $\mathcal{T}(\Omega; \boldsymbol{\pi}_0) = \mathbb{R}^c$.

(ii) Deduce that the asymptotic null distribution of *LRT* at $\boldsymbol{\pi}_0$ is equal to that of $\inf\{(\boldsymbol{Z} - \boldsymbol{\pi})^T \boldsymbol{i}_0 (\boldsymbol{Z} - \boldsymbol{\pi}) : \boldsymbol{A}_1\boldsymbol{\pi} \geqslant \mathbf{0}\}$, where $\boldsymbol{Z} \sim N(\mathbf{0}, \boldsymbol{i}_0^{-1})$ and $\boldsymbol{i}_0$ is the diagonal matrix whose ith diagonal is $n_i\{\pi_{0i}(1-\pi_{0i})\}^{-1}$.

6.3 Verify (6.12).
[Hint: Since $\Omega_0 = \{\boldsymbol{\pi} : \boldsymbol{R\pi} = \mathbf{0}\}$ and $\boldsymbol{R\pi}_0 = \mathbf{0}$, it follows that $\mathcal{T}(\Omega_0, \boldsymbol{\pi}_0) = \Omega_0$. Since $\Omega_1 = \{\boldsymbol{\pi} : \boldsymbol{R}_2\boldsymbol{\pi} \geqslant \mathbf{0}\}$ and $\boldsymbol{R}_2\boldsymbol{\pi}_0 = \mathbf{0}$. It follows that $\mathcal{T}(\Omega_1; \boldsymbol{\pi}_0) = \Omega_1$. Substitute these into (6.9). Now, observe that the limiting distribution of T in (6.9)

is the distribution of *LRT* at π_0 for the testing problem (6.11) based on a single observation of Z where $Z \sim N(\pi, i_0^{-1})$.]

6.4 Consider the testing problem in (6.11). Suppose that the null hypothesis is $H_0 : \pi_1 = \ldots = \pi_k$. Show that the limit in (6.12) does not depend on any unknown nuisance parameters. [Hint: Suppose that H_0 holds. Let π_0 denote the unknown common value of $\pi_1, \ldots, \pi_k$. Then $i(\pi) = c\,\text{diag}\{a_i\}$ where c is a positive real number and $A(\pi) = c^{-1} R_2 \,\text{diag}\{a_i\}^{-1} R_2^T$; in fact $c = \{\pi_0(1 - \pi_0)\}^{-1}$, but its value is not important. Now, use the fact that $w_i(p, cV) = w_i(p, V)$ or $\bar{\chi}^2(cV, \mathcal{C}) \sim \bar{\chi}^2(V, \mathcal{C})$ for any $c > 0$.]

6.5 Prove (6.16). [Hint: It follows from (6.9) that the asymptotic null distribution of *LRT* is equal to that of

$$\|Z - \mathcal{T}_0\|^2 - \|Z - \mathcal{T}_1\|^2 \tag{6.95}$$

where $Z \sim N(0, \text{diag}\{a_i\}^{-1})$, $\|x\|^2 = x^T \text{diag}\{a_i\} x$, $\mathcal{T}_0 = \{c\mathbf{1} : c \in \mathbb{R}\}$ where $\mathbf{1}$ is a vector of ones, and $\mathcal{T}_1 = \{\pi : A\pi \geqslant 0\}$. Since $\mathcal{T}_0$ is a linear space, $\mathcal{T}_1$ is a closed convex cone and $\mathcal{T}_0 \subset \mathcal{T}_1$, it follows that the asymptotic null distribution of *LRT* is a chi-bar square. More specifically, this is $\bar{\chi}^2(\text{diag}\{a_i\}^{-1}, \mathcal{C})$ where $\mathcal{C} = \mathcal{T}_0^{\perp} \cap \mathcal{T}_1$ $\mathcal{T}_0^{\perp} = \{\pi : \mathbf{1}^T \text{diag}\{a_i\}\pi = 0\} = \{\pi : \sum a_i \pi_i = 0\}$.]

6.6 Consider a $2 \times c$ table for the comparison of two treatments. Let H_ℓ and H_s denote the likelihood ratio and simple stochastic order, respectively; further, let $H_0 : \pi_1 = \pi_2$. Show that the chi-bar square weights in the asymptotic null distribution of the *LRT* for testing H_0 against H_ℓ and that for testing H_0 against H_s are the same except that they appear in the reverse order; more precisely, if

$$\lim \text{pr}(T_{0\ell} \geqslant t \mid H_0) = \sum_{i=0}^{c-1} w_i \text{pr}(\chi_i^2 \geqslant t)$$

then

$$\lim \text{pr}(T_{0s} \geqslant t \mid H_0) = \sum_{i=0}^{c-1} w_{c-1-i} \text{pr}(\chi_i^2 \geqslant t).$$

[Hint: Let $\mathcal{C}$ denote $\mathbb{R}^{+(c-1)}$. Use Theorem 6.7.2 and note that $T_{0\ell} \approx \bar{\chi}^2(V, \mathcal{C})$ and $T_{0s} \approx \bar{\chi}^2(V^{-1}, \mathcal{C})$ for some V. Use $\bar{\chi}^2(V, \mathcal{C}) \sim \bar{\chi}^2(V^{-1}, \mathcal{C}^o)$ and $w_i(V, \mathcal{C}) = w_{p-i}(V, \mathcal{C})$.]

6.7 Use the probabilities given in Table 6.21 for a 2×4 table to show that $\theta^C \geqslant 1 \nRightarrow \theta^{CO} \geqslant 1$, where θ^C and θ^{CO} are the cumulative and continuation odds ratios.

6.8 Let θ be a generalized odds ratio that is different from θ^L and θ^G. Thus, θ can be any one of the 14 odds ratios in Fig. 6.2 Then we have the following:

1. $\theta^L \geqslant 1 \Rightarrow \theta \geqslant 1 \Rightarrow \theta^G \geqslant 1$.

2. $\theta^L = 1 \Leftrightarrow \theta = 1 \Leftrightarrow \theta^G = 1$.

3. $\theta^L \gneqq 1 \Rightarrow \theta \gneqq 1 \Rightarrow \theta^G \gneqq 1$.

Table 6.21 Probability of response

	Response				
	1	2	3	4	Total
Treatment 1	0.6	0.001	0.029	0.37	1.0
Treatment 2	0.5	0.01	0.01	0.48	1.0

4. $\boldsymbol{\theta}^G \geqslant \mathbf{1} \not\Leftarrow \boldsymbol{\theta} \geqslant \mathbf{1} \not\Leftarrow \boldsymbol{\theta}^L \geqslant \mathbf{1}$.

6.9 Let $G^2(A \mid B)$ denote the *LRT* for testing model B against model A; they are the models under the null and alternative hypotheses, respectively. Show that

$$G^2(I \mid L) \leq \{G^2(I \mid CU), G^2(I \mid CO)\} \leq G^2(I \mid G) \leq G^2(I)$$

where I denotes independence, and L, CU, CO and G denote stochastic orders with respect to local, cumulative, continuation and global odds ratios. [Hint: Apply Proposition 1.3.3.].

[Agresti and Coull, 1998].

6.10 Prove the followings for an $r \times c$ contingency table:

(i) For the cumulative logit model,

$$\text{logit}(\pi_{ij} + \cdots + \pi_{ij}) = \alpha_j - \beta x_i,$$

$\boldsymbol{\theta}^{CU} \geqslant 1$ if the scores $\{x_i\}$ are nondecreasing in i and $\beta > 0$.

(ii) For the linear-by-linear association model,

$$\log\ pr(X = i, Y = j) = \mu + \lambda_i^X + \lambda_j^Y + \beta x_i y_j,$$

$\boldsymbol{\theta}^L \geqslant 1$ if the scores $\{x_i\}$ and $\{y_j\}$ are nondecreasing and $\beta > 0$.

(iii) For the rc model, $\text{pr}(X = i, Y = j) = \alpha_i \beta_j \exp(\phi \mu_i \nu_j)$, $\boldsymbol{\theta}^G \geqslant 1$ if $\{\mu_i\}$ and $\{\nu_j\}$ are nondecreasing and $\phi > 0$. (For further results concerning constrained inference in rc model, see Ritov and Gilula (1991)).

[Agresti and Coull, 1998)]

6.11 Female salamanders were randomly assigned to one of the following four treatments.

Control :	Saline solution
FSH :	Biweekly injections of follicle-stimulating hormone
LH :	Luteinizing hormone
Sup-Fed :	Twice the amount of feed as for the salamanders in all the other groups

Table 6.22 Response of salamanders to different treatments

Treatment	Control	FSH	LH	Sup-Fed
y_i	9	4	6	13
n_i	13	12	13	14

At the end of the experimental period, each animal was evaluated and declared to be either in reproductive condition or not. The data are given in Table 6.22.

Let the treatments be numbered as 1, 2, 3, and 4, and let π_i denote the probability of success for the ith treatment ($i = 1, 2, 3, 4$). The hypothesis of interest is whether or not the salamanders in the treated groups have a higher probability of being in reproductive condition than the control group. Test the following hypotheses:

(a) $H_0 : \pi_1 = \pi_2 = \pi_3 = \pi_4$ against $H_1 : \pi_1 \leq \pi_2, \pi_1 \leq \pi_3$ and $\pi_1 \leq \pi_4$.

(b) $H_0 :$ *not* H_1 against $H_1 : \pi_1 < \pi_2, \pi_1 < \pi_3$ and $\pi_1 < \pi_4$.

[Peddada et al. (2001) and Crespi (2001)]

6.12 The carcinogenicity of turmeric oleoresin was investigated in a 2-year study. A random sample of female mice was assigned to each of the dose groups, 0 ppm, 2000 ppm, 10000 ppm, and 50000 ppm. At the end of the study period the animals were examined for the presence of tumour. The results are given in Table 6.23.

Table 6.23 Carcinogenicity of turmeric

Dose (ppm)	0	2000	10000	50000
y_i	7	8	19	14
n_i	50	50	51	50
n_i^*	45.62	45.52	43.63	46

 Source: Peddada et al. (2001)

Since some of the animals died before the end of the study period, adjustments were made to compute an effective sample size for each treatment. These effective sample sizes, n_i^*, are also given in the table. For details of the adjustments, see Section 3.2.2 in Peddada et al. (2001). Formulate and test suitable hypotheses.

6.13 A study was carried out at a hospital to investigate the occurrence of *adverse events* (AE) following treatment at the hospital (see Table 6.24). A hypothesis of interest is that adverse events are less likely if the patient is managed by a more experienced medical practitioner. The study design is case-control where the cases and controls are matched. A case is defined as one for which an AE occurred. For each case, a control was selected by matching age, sex and medical diagnostic category. The data, modified for confidentiality reasons, are given in Table 6.24; the "most qualified" medical practitioner has rank 1.

Test H_0 against H_1, where

H_0 : The rank of the managing medical practitioner does not depend on whether or not the patient had an adverse event,

H_1 : The rank of the managing medical practitioner for the control is at least as high as that for the control.

Table 6.24 Occurrence of adverse events

Rank of doctor for case	Rank of doctor for control 1	2	3	4	5
1	0	2	0	3	0
2	5	9	4	5	2
3	1	6	4	2	0
4	2	11	4	7	2
5	15	40	30	35	6

Source: J. Hendrie. The original data have been modified for confidentiality reasons, and hence these data cannot be used to draw any substantive conclusions.

6.14 Consider an $r \times c$ table. Suppose that all the sample local odds ratios are at least one. Show that $T_{0\ell} = T_{0s}$ and that $T_{0\ell}$ has a smaller large sample p-value than T_{0s}.

6.15 Consider Example 6.2.2, which involves the comparison of two treatments for ulcer. Let the log-linear model be expressed as

$$\log m_{ij} = \mu + \lambda_i^X + \lambda_j^Y + \lambda_{ij}^{XY}$$

in the usual notation where m_{ij} is the expected frequency for cell (i, j) and the identifiability constraints are $\lambda_1^X = \lambda_1^Y = \lambda_{i1}^{XY} = \lambda_{1j}^{XY} = 0$ for every (i, j).

1. Show that λ_{2j}^{XY} is the odds ratio for the 2×2 table formed by columns 1 and j for $j = 1, \ldots, c$.

2. Show that $\boldsymbol{\theta}^L \geqslant \mathbf{1}$ is equivalent to $\lambda_{2c} \geqslant \lambda_{2,c-1} \geqslant \ldots \geqslant \lambda_{22} \geqslant 0$, and that $\boldsymbol{\theta}^L = \mathbf{1}$ is equivalent to $\boldsymbol{\lambda}^{XY} = \mathbf{0}$.

3. Let $\hat{\boldsymbol{\lambda}}$ denote the *mle* of $\boldsymbol{\lambda}^{XY}$ and $\hat{\boldsymbol{V}}$ denote a consistent estimate of the covariance matrix of $\hat{\boldsymbol{\lambda}}^{XY}$. Use a standard software to compute $\hat{\boldsymbol{\lambda}}^{XY}$ and $\hat{\boldsymbol{V}}$.

4. Let the null and the alternative hypothesis be $H_0 : \boldsymbol{\lambda}^{XY} = \mathbf{0}$ and $H_1 : \lambda_{2c} \geqslant \ldots \geqslant \lambda_{22} \geqslant 0$, respectively, and let W_D denote the Wald-type statistic defined by

$$W_D = \|\hat{\boldsymbol{\lambda}}^{XY}\|^2 - \|\hat{\boldsymbol{\lambda}}^{XY} - \mathcal{C}\|^2$$

for testing $H_0 : \boldsymbol{\theta}^L = \mathbf{1}$ against $H_1 : \boldsymbol{\theta}^L \geqslant \mathbf{1}$ where $\|\boldsymbol{x}\|^2 = \boldsymbol{x}^T \hat{\boldsymbol{V}}^{-1} \boldsymbol{x}$ and $\mathcal{C} = \{\boldsymbol{x} : \boldsymbol{x} = (x_1, \ldots, x_{c-1})^T, x_{c-1} \geqslant \ldots \geqslant x_1\}$.

5. Explain why W_D is asymptotically equivalent to the *LRT* for testing $\boldsymbol{\theta}^L = \mathbf{1}$ against $\boldsymbol{\theta}^L \geqslant \mathbf{1}$.

6. Explain why

$$\text{pr}(W_D \geqslant t \mid H_0) - \sum_{i=0}^{c-1} w_i(c-1, \hat{\boldsymbol{V}}, \mathcal{C}) \text{pr}(\chi_i^2 \geqslant t) \to 0$$

as $n \to \infty$.

7. Use a standard quadratic program to compute W_D and an upper bound for the asymptotic p-value.

6.16 Consider the following analogue of the paired t-test for ordered categorical data models in a restricted alternative set-up. Consider an $r \times r$ table with ordered categories:

	A_1	A_2	$\ldots$	A_r	Total
B_1	π_{11}	π_{21}	$\ldots$	π_{r1}	$\pi_{.1}$
B_2	π_{12}	π_{22}		π_{r2}	$\pi_{.2}$
$\vdots$	$\vdots$	$\vdots$		$\vdots$	$\vdots$
B_r	π_{1r}	π_{2r}	$\ldots$	π_{rr}	$\pi_{.r}$
Total	$\pi_{1.}$	$\pi_{2.}$		$\pi_{r.}$	1

The traits A and B are not assumed to be independent. We denote the marginal correlative probabilities by

$$\alpha_j = \sum_{s \leqslant j} \pi_{s\cdot} \quad \beta_j = \sum_{s \leqslant j} \pi_{\cdot s}, \quad 1 \leqslant j \leqslant r. \tag{6.96}$$

Then consider the null hypothesis (of the homogeneity of the marginal probabilities):

$$H_0 : \alpha_j = \beta_j \text{ for } j = 1, \ldots, r, \tag{6.97}$$

against the set of one-sided alternatives:

$$H_1 : \alpha_j \geqslant \beta_j, \forall j = 1, \ldots, r. \tag{6.98}$$

Corresponding to the probabilities π_{ij}, we consider the observed two-way (contingency) table, where for the cell (A_i, B_j), the observed frequency is n_{ij}, for $i, j = 1, \ldots, r$. Also, we let

$$n_{j\cdot} = \sum_{s \leqslant j} n_{js} \quad \text{and} \quad n_{\cdot j} = \sum_{s \leqslant j} n_{sj}, 1 \leqslant j \leqslant r, \tag{6.99}$$

and let $n = \sum_{s \leqslant r} \sum_{m \leqslant r} n_{sm}$ be the total number of observations. Then the marginal nonparametric *mle* [MNMLE] of $\pi_{j.}$ and $\pi_{.j}$ are $n_{j.}/n$ and $n_{.j}/n$, respectively. However, if we consider the likelihood function for the $r \times r$ table,

$$n! \left\{ \Pi_{i=1}^{r} \Pi_{j=1}^{r} n_{ij}! \right\}^{-1} \Pi_{i=1}^{r} \Pi_{j=1}^{r} \pi_{ij}^{n_{ij}}, \tag{6.100}$$

computing the *mle* of α_j and β_j, under the inequality restraints $\alpha_j \geqslant \beta_j, 1 \leqslant j \leqslant r-1$ requires a nonlinear constrained optimization program. Thus the conventional (restricted) likelihood ratio test for H_0 vs H_1 would require certain amount of computer programming. On the other hand, starting with the MNMLE $(n_{.j}/n, n_{j.}/n), j = 1, \ldots, r$, one can use the UIT in the following manner.

Let $T_j = \sum_{s \leqslant j} (n_{s.} - n_{.s})/n, 1 \leqslant j \leqslant r-1$. Also, assume that under $H_0, \pi_{ij} = \pi_{ji}, \forall i, j = 1, \ldots, r$ (i.e., the two traits are interchangeable). Then, under H_0, the MNMLE of π_{ij} is $\hat{\pi}_{ij}^0 = 2^{-1}(n_{ij} + n_{ji})/n, \ \forall i, j = 1, \ldots, r$, and as a result, the conditional distribution of $T_j (1 \leqslant j \leqslant r-1)$, given the $\hat{\pi}_{ij}^0$, can be used to generate an UIT for the one-sided alternative in (6.98).

7

Beyond Parametrics

7.1 INTRODUCTION

In the previous chapters, in the context of general constrained statistical inference problems, the fundamental role of the likelihood function (along with its variants) has been thoroughly appraised. There has been a persistent parametric flavor in this respect. In the same vein, alternatives to the classical *likelihood principle* (LP), discussed in the preceding chapter, also include some *nonparametrics*, albeit formulated, mostly, without changing the basic (parametric) structure of the model but relaxing some stringent distributional assumptions, customarily made in genuine parametric approaches. Thus, nonparametrics generally induce *distributional robustness* properties (in a global sense).

During the past three decades, considerable attention has been focused on nonparametrics (along with *semiparametrics*) in a far more general set-up where the LP may not be always conveniently adaptable. For example, in the transition from the classical *linear models* to appropriate *nonlinear models* and *generalized linear models* (GLM), finite-sample (or exact) parametric formulations may, often, stumble into road blocks; even in asymptotic set-ups, there could be a genuine concern regarding their robustness prospects (to plausible model departures). This advocates a genuine need for looking beyond the (traditional) parametrics, there being a greater need for constrained statistical inference.

Beyond parametrics, formulations often relate to *functional parameters*; nonparametric (and semiparametric) *regression* and *smoothing* procedures have addressed the need for incorporating functional parameter spaces when finite dimensional parameter spaces are judged inadequate from all practical considerations. Yet in such contemporary developments, it has often been observed that natural restraints (e.g., inequality constraints) crop up in a natural way, and they ought to be taken into account in drawing statistical conclusions in a valid and efficient way. For example, in a functional regression model, it may be of good interest to test for the *monotonicity of regression* without necessarily assuming that such a regression is linear (or a specific type of nonlinear form), or even a GLM. The classical *biological assay* or bioassay models often pertain to this set-up. Likewise, in non- (or semi-) parametric estimation of a *density function* subject to some *shape-restraints*, such as *monotonicity* or *unimodality*, or a *monotone failure rate*, the underlying inequality constraints are functional, and, hence, more complex in nature. A reduction to a finite dimensional parametric model may not address the underlying complexities adequately.

In spite of the fact that there have been some attempts to extend the LP to *local likelihood*, *profile-likelihood*, *pseudo-likelihood*, *quasi-likelihood*,inxxlikelihood,quasi *nonparametric likelihood*, and other related concepts, the dilemma regarding the supposedly *optimality* property of the LP or the associated *maximum likelihood estimator* (MLE), in such constrained inference contexts, continues to reign. The situation may become more complex if robustness considerations are given their due importance in judging the desirability of a statistical resolution; it is well known that MLE and allied *LRT* are generally much less robust than some other competitors which deemphasize the LP motivation to a certain extent. Indeed, in most nonparametric (and, to a certain extent, semiparametric) formulations, though the LP underlines in a subtle way, greater emphasis is laid down on robustness along with some other considerations. Most of these developments have taken place during the past two decades, and will be briefly appraised in this chapter.

Faced with this broad scenario, we intend to provide a wider coverage, albeit at an intermediate level, of problems arising in statistical inference under *shape constraints* on distributions. To a greater extent this coverage has a dominant nonparametric and semiparametric flavor. While this branch of research has gone through an evolutionary phase, we would like to focus primarily on (i) monotone density estimation, (ii) unimodal density, (iii) monotone failure rate, (iv) decreasing mean residual life, and (v) monotone nonparametric regression functions, touching briefly on some other related problems as well. These developments are outlined in the next five sections, namely, Sections 7.2, 7.3, 7.4, 7.5, and 7.6. The concluding section is devoted to some allied hypothesis testing problems. In view of the paucity of finite-sample resolutions, there is a general emphasis on asymptotics in this depiction. We shall incorporate some of these (without invoking too much abstractions, typically observed in asymptotics).

7.2 INFERENCE ON MONOTONE DENSITY FUNCTION

Let $X_1, \ldots, X_n$ be n *iid* r.v.'s having an absolutely continuous distribution function $F(x)$ defined on the real line $\mathbb{R}$. Further assume that F admits a probability density function $f(x)[= (d/dx)F(x)]$, also defined on $\mathbb{R}$. For the time being we shall assume that f is continuous almost everywhere; other conditions would be improved on f as and when needed. Let us denote the order statistics associated with $X_1, \ldots, X_n$ by $X_{n:1} \leqslant X_{n:2} \leqslant \cdots \leqslant X_{n:n}$; by virtue of the assumed continuity of F, ties among the X_i can be neglected with probability 1, and hence,

$$X_{n:1} < X_{n:2} < \cdots < X_{n:n} \quad \text{wp } 1 \tag{7.1}$$

where "wp 1" is the abbreviation for "with probability 1." We also introduce the *empirical distribution function* F_n as

$$F_n(x) = n^{-1} \sum_{i=1}^{n} I(X_i \leqslant x), \; x \in \mathbb{R}. \tag{7.2}$$

Note that (a) there is a one–one relationship between the order statistics and the empirical d.f. F_n, (b) F_n is a step function with jump-discontinuity of equal magnitude, n^{-1}, at each of the n order statistics, and (c)

$$F_n(x) = k/n \text{ for } X_{n:k} \leqslant x < X_{n:k+1}, 0 \leqslant k \leqslant n \tag{7.3}$$

where $X_{n:0} = \inf\{x : F(x) > 0\}$ and $X_{n:n+1} = \sup\{x : F(x) < 1\}$.

Although $F_n(x)$ is an unbiased estimator of $F(x)$ for any x and has nice statistical properties (see Problem 7.1), simple differentiation of F_n does not yield an estimator of $f(\cdot)$ because F_n is a step function. Thus, various methods, such as kernel smoothing, histogram methods, spline, and orthogonal functionals, have been studied extensively in the literature for nonparametric estimation of density and functionals of distribution functions. We may observe that if the density function f (in the contemplated univariate *iid* case) belongs to a constrained subclass (such as monotonically increasing), a smooth estimator may not automatically inherit this property. There are some technical difficulties with the nonparametric *mle* (NPMLE) of monotone densities, and hence, there is a need to examine the smoothness properties of NPMLE's as well.

The likelihood as a function of f can be expressed as

$$L_n(f) = \prod_{i=1}^{n} f(X_{n:i}). \tag{7.4}$$

Therefore, one possibility of obtaining the NPMLE of $f(\cdot)$ would be to maximize (7.4) with respect to $f(\cdot)$ subject to the constraint that $f(x)$ is nonincreasing in x.

For nonincreasing $f(\cdot)$, without loss of generality, we may assume that the support of f is $\mathbb{R}^+ = [0, \infty)$. Since f is nonincreasing on $\mathbb{R}^+, F$ is concave on $\mathbb{R}^+$. Let $\mathcal{F} = \{F : F \text{ is concave on } \mathbb{R}^+\}$. Then the problem of finding NPMLE of $f(\cdot)$ is to maximize (7.4) over $F \in \mathcal{F}$. This functional optimization problem simplifies considerably because it reduces to the finite dimensional isotonic regression problem studied in Chapter 2.

To indicate the reasons for such a simplification, let f be a nonincreasing density on $[0, \infty)$, $f_i = f(X_{n:i})$ and let g be the density defined by

$$g(t) = \begin{cases} cf_i & \text{for } X_{n:i-1} < x \leqslant X_{n:i}, 1 \leqslant i \leqslant n \\ 0 & \text{for } x < 0 \text{ and } x > X_{n:n}. \end{cases}$$

where c is a constant to ensure that $\int g(t)dt = 1$, which leads to $c\sum_{i=1}^{n} f(X_{n:i})(X_{n:i} - X_{n:i-1}) = 1$. Since f is a nonincreasing density on $[0, \infty)$, we have that $\sum_{i=1}^{n} f(X_{n:i})(X_{n:i} - X_{n:i-1}) \leqslant 1$ and hence $c \geqslant 1$. It follows that $L_n(g) = c^n L_n(f) \geqslant L_n(f)$. Therefore, for any nonincreasing density function f, there is a corresponding density g, a left-continuous step-function such that $L_n(g) \geqslant L_n(f)$. Hence, the nonincreasing density f which maximizes $L_n(f)$ must be a left-continuous step function of the form g with $c = 1$. Another way of visualizing this is the following: Suppose that f is a nonincreasing density on $[0, \infty)$ and it is not a constant on $(X_{n:i-1}, X_{n:i}]$. Then, the value of the likelihood would be increased if the value of f on the same interval is changed to the constant

$$(X_{n:i} - X_{n:i-1})^{-1} \int_{X_{n:i-1}}^{X_{n:i}} f(t)dt.$$

Consequently, the maximization of $L_n(f)$ reduces to the following:

$$\max \prod_{i=1}^{n} f_i$$

subject to

$$f_1 \geqslant f_2 \geqslant \cdots \geqslant f_n \quad \text{and} \quad \sum_{1}^{n} (X_{n:1} - X_{n:i-1}) f_i = 1. \tag{7.5}$$

Now, by Corollary 2.4.4 and Proposition 2.4.2, the solution $(\hat{f}_1, \ldots, \hat{f}_n)$ is

$$\hat{f}_i = \min_{r \leqslant i-1} \max_{s \geqslant i} \left\{ \frac{F_n(X_{n:s}) - F_n(X_{n:r})}{X_{n:s} - X_{n:r}} \right\}, 1 \leqslant i \leqslant n \tag{7.6}$$

$$= \min_{r \leqslant i-1} \max_{s \geqslant i} \frac{(s - r)}{(X_{n:s} - X_{n:r})}. \tag{7.7}$$

Therefore, the nonparametric *mle* $\hat{f}_n(\cdot)$ of $f(\cdot)$ is given by

$$\hat{f}_n(x) = \begin{cases} \hat{f}_i & \text{for } X_{n:i-1} < x \leqslant X_{n:i} \\ 0 & \text{for } x \leqslant 0 \text{ and } x > X_{n:n} \end{cases} \tag{7.8}$$

Let $\hat{F}_n$ denote the least concave majorant [LCM] of F_n, defined by

$$\hat{F}_n = \inf\{g : F_n(t) \leqslant g(t) \text{ for every } t \text{ and } g \text{ is concave on } [0, \infty)\}.$$

Thus, $\hat{F}_n$ is the smallest concave function on $[0, \infty)$ such that $F_n(t) \leqslant \hat{F}_n(t)$ for every t. Intuitively, $\hat{F}_n$ has the shape of a tight string tied to the origin and wraps around F_n from above. Now, $\hat{f}_n(x)$ defined in (7.8) is the left-continuous slope of $\hat{F}_n(x)$. Since the likelihood function on the RHS of (7.4) can be thought of as a functional of F, it follows that $\hat{F}_n$ is the NPMLE of F.

Note that if $\delta_n = \sup_x |F_n(x) - F(x)|$, then $F_n^*(x) = (F(x) + \delta_n) \wedge 1$ is a concave function whenever F is concave. Therefore, noting that $\hat{F}_n$ is the smallest concave function such that $F_n \leqslant \hat{F}_n$ and that $F_n(x) \leqslant F_n^*(x)$, $\forall x$, we obtain that $\hat{F}_n(x) \leq F_n^*(x)$, $\forall x$. Similarly, $\hat{F}_n(x) \geq F_n(x) \geq (F(x) - \delta_n) \vee 0, \forall x$. Thus, we obtain that

$$\sup_x \left|\hat{F}_n(x) - F(x)\right| \leqslant \sup_x |F_n(x) - F(x)|. \tag{7.9}$$

Therefore, the rate of convergence of the LHS of (7.9) to zero is at least as fast as that of the RHS of (7.9). For example, by Glivenko-Cantelli theorem, the right-hand side of (7.9) converges to zero with probability one. Hence

$$\sup_x \left|\hat{F}_n(x) - F(x)\right| \to 0 \quad \text{wp 1}. \tag{7.10}$$

A more precise rate of convergence result for $\sup_x \left|\hat{F}_n(x) - F(x)\right|$ is

$$\overline{\lim}\ (n/\log\log n)^{\frac{1}{2}} \sup_x \left|\hat{F}_n(x) - F(x)\right| = 1/\sqrt{2} \quad \text{wp 1}; \tag{7.11}$$

see Problem 7.2.

We may also remark that the histogram-type estimator of the density $f(\cdot)$ is given by

$$\tilde{f}_n(x) = \begin{cases} \{n(X_{n:i} - X_{n:i-1})\}^{-1}, & \text{for } x \in (X_{n:i-1}, X_{n:i}], \text{ and } i = 1, \dots, n \\ 0, \text{ otherwise}, \end{cases} \tag{7.12}$$

so that $\hat{f}_n(x)$ in (7.6) - (7.8) can also be represented as the *isotonic regression* of $\tilde{f}_n$ (Problem 7.3). Note that neither the NPMLE $\hat{f}_n(\cdot)$ nor the estimator $\tilde{f}_n(\cdot)$ is smooth; they are step functions.

To render smoothness of isotonic density estimates, several proposals have been put forward in the literature. Mukerjee (1986) suggested smoothing the isotonic estimate using a kernel estimator based on a *log-concave* (or *strongly unimodal*) density kernel; this may yield isotonic smooth estimators but there is no guarantee. Chaubey and Sen (2002) proposed another smooth estimator that preserves monotonicity. To introduce their approach, define $\hat{f}_n(x), x \in \mathbb{R}^+$, as in (7.6) - (7.8). Also, let

$$\omega_{j,n}(x) = e^{-x\lambda_n}(x\lambda_n)^j/j!, \quad j = 0, 1, 2, \dots, \infty \tag{7.13}$$

where assume that F has support $\mathbb{R}^+$, and let

$$\lambda_n = n/X_{n:n} (= o_p(n), \text{ as } n \to \infty). \tag{7.14}$$

Define then

$$\check{f}_n(x) = \sum_{i=0}^{\infty} \omega_{i,n}(x)\hat{f}_n(i/\lambda_n), x \in \mathbb{R}^+. \tag{7.15}$$

To establish the nonincreasing property of $\tilde{f}_n(\cdot)$, define

$$W_{k,n}(x) = \sum_{j=k}^{\infty} \omega_{j,n}(x), k \geqslant 0, x \in \mathbb{R}^+. \tag{7.16}$$

Using the relationship between incomplete Gamma function and cumulative Poisson probabilities (Problem 7.4), we have

$$W_{k,n}(x) = \{\Gamma(k)\}^{-1} \int_0^{x\lambda_n} e^{-y} y^{k-1} dy, \quad x \geqslant 0 \tag{7.17}$$

and hence,

$$\frac{\partial}{\partial x} W_{k,n}(x) = \lambda_n \{\Gamma(k)\}^{-1} e^{-x\lambda_n} (x\lambda_n)^{k-1} \geqslant 0, \tag{7.18}$$

so that for every $k(0 \leqslant k \leqslant n), W_{k,n}(x)$ is $\uparrow$ in x. We rewrite $f_n(x)$ in (7.15) as

$$\check{f}_n(x) = \hat{f}_n(0) + \sum_{k \geqslant 1} W_{k,n}(x) \left\{ \hat{f}_n \left(\frac{k}{\lambda_n} \right) - \hat{f}_n \left(\frac{k-1}{\lambda_n} \right) \right\}. \tag{7.19}$$

Since, $\hat{f}_n$ is nonincreasing (albeit a step function),

$$\hat{f}_n \left(\frac{k}{\lambda_n} \right) - \hat{f}_n \left(\frac{k-1}{\lambda_n} \right) \leqslant 0, \forall k \geqslant 1. \tag{7.20}$$

Hence, from (7.19) and (7.20), we conclude that

$$\check{f}_n(x) \text{ is nonincreasing for every } x \geqslant 0. \tag{7.21}$$

Thus, $\check{f}_n$, a smoothed version of $\hat{f}_n$, is isotonic.

We may remark that Chaubey and Sen (1996) considered the following smooth estimator of F, defined on $\mathbb{R}^+$:

$$F_n^*(x) = \sum_{k=0}^{n} \omega_{nk}^*(x\lambda_n) F_n(k/\lambda_n), x \in \mathbb{R}^+ \tag{7.22}$$

where

$$\omega_{nk}^*(t) = (t^k/k!)/(\sum_{i=0}^{n} t^i/i!), 0 \leqslant k \leqslant n. \tag{7.23}$$

Their smooth empirical d.f. F_n^* is nondecreasing and continuously differentiable, so that a derived estimator of the density f is

$$f_n^*(x) = \lambda_n \{ [1 - F_n^*(x)][1 - \omega_{n,n}^*(x\lambda_n)] - \sum_{k=0}^{n} \omega_{nk}^*(x\lambda_n)[1 - F_n(\frac{k+1}{\lambda_n})] \}, x \in \mathbb{R}^+ \tag{7.24}$$

However, f_n^* is not necessarily monotone even if f is so. Thus, there is a need to isotonize f_n^* to render it suitable in the current context. This has been discussed in Chaubey and Sen (2002). Their discussion pertains to general kernel smoothing as well.

The kernel method is one of the oldest smoothing methods. For a *kernel function* $k(y)$, typically a symmetric probability density function with mean 0 and variance 1, a kernel estimator of the density $f(\cdot)$ is given by

$$\tilde{f}_n^{[k]}(x) = \int_{-\infty}^{\infty} \frac{1}{h_n} k\left(\frac{y-x}{h_n}\right) dF_n(y) \tag{7.25}$$

$$= \frac{1}{nh_n} \sum_{i=1}^{n} k\left(\frac{X_i - x}{h_n}\right), x \in \mathbb{R}, \tag{7.26}$$

where h_n, known as the *bandwidth*, is so chosen that $h_n \to 0$ but $nh_n \to \infty$, as $n \to \infty$ (Rosenblatt (1956), Parzen (1956), Nadaraya (1964), and others). Typically, $h_n = O(n^{-1/5})$. For a general $k(\cdot)$, not necessarily nonnegative everywhere (but satisfying the conditions that $\int xk(x)dx = 0$ and $\int x^2k(x)dx = 1$), the estimator in (7.26) may yield negative estimates at some values of x (though generally very small). To eliminate this inadequacy, Bagai and Prakasa Rao (1996) modified $k(\cdot)$ by nonnegative density function $k^*(\cdot)$, defined on $\mathbb{R}^+$, such that $\int_0^\infty x^2k^*(x)dx = 1$ (normalized); however, in this case, for $X_{n:r} < x \leqslant X_{n:n+1}$, the estimator in (7.26) (with $k(\cdot)$ replaced by $k^*(\cdot)$) depends only on $X_{n:i}, i \leqslant r; (r \leqslant n)$. By contrast $f_n^*(\cdot)$ in (7.24) depends on the whole data set. In any case, neither (7.24) nor (7.26) may preserve the isotonic property, and hence, may not be that suitable in the present context.

[There is a persistent asymptotic flavor in the statistical treatment of these isotonic density estimators. Conventional $\sqrt{n}$-consistency rates of convergence may not apply here, and in the univariate case, typically, we have $n^{2/5}$-consistency rates].

For the smooth estimator $\check{f}_n(\cdot)$ in (7.15) and (7.26) (and its counterpart under random censoring) asymptotic point-wise normality results have been derived under, typically, $n^{2/5}$ normalizing factor, wherein the *asymptotic bias* (of the same order) merits a critical appraisal too. These results have also been extended to the entire process $n^{2/5}\{\tilde{f}_n(\cdot) - f(\cdot)\}$ and *optimal bandwidth selection* problems have been addressed from this perspective. This is a more difficult problem and it requires more complex asymptotic tools which are somehow outside the scope of the present monograph. At the end of this chapter, we will present a general result on smoothing that would facilitate comprehending the general asymptotics needed in the current context. We also refer to Chaubey and Sen (2002) for some of these asymptotics, even including the smoothed version; other references are also cited there.

The asymptotics are quite different for the isotonic estimators in (7.7) and the smoothed version in (7.16). The asymptotics may depend on the nature of the derivative $f'(\cdot)$, and the asymptotic normality may not hold. If we look into (7.7) we may observe that the base estimator in (7.12) is not asymptotically normal and that the algorithm underlying (7.7), for isotonizing, is not a linear operator. Hence, even for the smooth isotonic estimator based on (7.24) and the algorithm in (7.7), asymptotic normality results may not hold. For the NPMLE in (7.7), if $f'(x)$ is $< 0, \forall x$, then at a fixed point $x_0, n^{1/3}\{\hat{f}_n(x_0) - f(x_0)\}/\{\frac{1}{2}f(x_0)|f'(x_0)|\}^{1/3} \xrightarrow{d} 2Z$, where $2Z$ is the slope of the GCM of $\{W(t) + t^2, t \in \mathbb{R}\}$ at $t = 0$, where $W(\cdot)$ is a two-sided Brownian motion originating at 0 [Groeneboom et al. (2001a)].

Estimation of a convex decreasing density f on $[0, \infty)$, that is $f(x)$ decreasing but $f'(x)$ increasing on $\mathbb{R}^+$, has received a good deal of interest in the recent past, albeit mostly from asymptotics perspectives. We refer to Groeneboom et al. (2001 a,b), and Banerjee and Wellner (2001) where other references are cited. Most of these asymptotics refer to a fixed point $t_0(< \infty)$ (instead of the entire range $\mathbb{R}^+$ or a compact subset of $\mathbb{R}^+$). Further, the asymptotic distribution does not turn out to be normal (but relates to that of the slope of the GCM of $\{W(t) + t^2, t \in \mathbb{R}\}$ at zero, where W is the two-sided Brownian motion, originating from 0).

Let us consider the estimation of a nonincreasing density f under censoring and truncation. For *Type I censoring* (or *truncation*), for a given time point $T(< \infty)$, the observable random variables are $X_i^* = X_i \wedge T$ and $\delta_i = I(X_i^* = X_i)$, for $i = 1, \ldots, n$; note that $n_T = \sum_{i=1}^n \delta_i$ is the (random) number of untruncated observation, while the rest $(n - n_T)$ are all censored; the only available information being that all these unobserved values are greater than T. Further, note that

$$n^{-1}n_T \to F(T) \text{ almost surely, as } n \to \infty. \tag{7.27}$$

Hence, confined to the range $[0, T]$, we can incorporate the uncensored observations $\{X_i^* : \delta_i = 1, i \leqslant n\}$, and proceeding as in before estimate the monotone density $f(\cdot)$ on $[0, T]$; the only change to (7.7) is that the s and r would be restricted to values less than or equal to n_T. Because of (7.27), the general asymptotic theory will also pertain here.

Consider next the case of *Type II censoring.* For a preassigned positive integer $r(= r_n)$, the observable random variables are $X_{n:1} < \cdots < X_{n:r}$ while the remaining $n - r$ observations are censored at the (random) point $X_{n:r}$. The likelihood function for this Type II censored scheme is

$$[n!/(n-r)!] \prod_{i=1}^{r} f(X_{n:i}[1 - F(X_{n:r})]^{n-r}. \tag{7.28}$$

It is generally assumed that $n^{-1}r_n \to p : 0 < p < 1$, so that if we let $T_p : F(T_p) = p$, then it can be shown (Problem 7.5) that

$$X_{n:r_n} \xrightarrow{a.s.} T_p \text{ as } n \to \infty. \tag{7.29}$$

As such, we can proceed as in the case of Type I censoring and estimate the monotone density on $[0, T_p]$. In either Type I or Type II censoring scheme, one is virtually confronted with right censoring at a fixed point (or a stochastic point converging to a fixed point), and, hence, it is not possible to extend the findings beyond the point of truncation (or censoring) when a nonparametric approach is used (as is the case here).

Random censoring schemes have received considerable attention from researchers from all walks of statistics and probability theory as well. We consider a random variable C, independent of X, such that the observable random elements are $X = X \wedge C$ and $S = I(X = C)$. Thus, we observe the censored lifetimes and their indicator variables $(Z_i, \delta_i), i = 1, \ldots, n$. Let $G(y)$ be the distribution function of C_i, and we denote the corresponding survival function by $S_G(y) = 1 - G(y)$; likewise, we denote by $S_F(x) = 1 - F(x)$. Then, the survival function for Z_i is $S_H(x) = 1 - P\{Z \leqslant x\} = S_F(x)S_G(x)$. Our basic task is to infer on the isotonic density $f(\cdot)$ based on the $(Z_i, \delta_i), i = 1, \ldots, n$. Under the postulated independence assumption, the celebrated Kaplan-Meier (1958) *product-limit estimator* (PLE) of $S_F(\cdot)$ is defined as

$$S_n^0(x) = 1 - F_n^0(x) = \prod_{i=1}^{n} (1 - \delta_{[n:i]}/(n - i + 1))^{I(Z_{n:i} \leqslant x)}, \tag{7.30}$$

where $\delta_{[n:i]} = \delta_j$ if $Z_{n:i} = Z_j$, for $i, j = 1, \ldots, n$, and $Z_{n:1}, \ldots, Z_{n:n}$ are the ordered values of $Z_1, \ldots, Z_n$; conventionally, we let $Z_{n:0} = 0$ and $Z_{n:n+1} = +\infty$.

Following Bhattacharjee and Sen (1997), we denote by m_n the total number of failure points (i.e., $m_n = \sum_{i=1}^{n} \delta_i$), and denote these ordered failure points by $Z_{n:1}^* < \cdots < Z_{n:m_n}^*$; also, conventionally, we let $Z_{n:0}^* = 0$ and $Z_{n:m_n+1}^* = +\infty$. Then the Kaplan-Meier PLE in (7.30) may also be expressed as

$$S_n^0(x) = \prod_{i \leqslant j} \left(\frac{n - k_i}{n - k_i + 1} \right), Z_{n:j}^* \leqslant x < Z_{n:j+1}^*, \tag{7.31}$$

for $j = 1, \ldots, m_n$, where $Z_{n:j}^* = Z_{n:k_j}$ for $j = 0, \ldots, m_n$. As a result, we conclude that $S_n^0(x)$ is also a step function with downward steps only at the failure points $Z_{n:j}^*, j = 1, \ldots, m_n$; conventionally, we let $Z_{n:m_n+1}^* = Z_{n:n} I(m_n < n) + I(m_n \geqslant n) Z_{n:n+1}$. Having defined $S_n^0(\cdot)$ as in (7.31), we take recourse to (7.22)–(7.26) with the only change that $F_n(k/\lambda_n)$ is to be replaced by $F_n^0(k/\lambda_n)$ defined in (7.30), $k \geqslant 0$. This smooth estimator may not be isotonic (even if $f(\cdot)$ is so). Hence, having defined $S_n^0(\cdot)$ as in (7.31) and the derived smooth density $\tilde{f}_n^0(\cdot)$, we proceed to isotonize it as follows [Chaubey and Sen (2002)]:

Let (Ω, S, μ) represent a probability space and $L_2(\Omega, S, \mu)$ represent the collection of μ-measurable, square integrable random variables. The isotonic density estimation problem is to seek a function g^* in the class $\mathcal{A}$ of isotonic functions defined on Ω such that

$$\int_{\Omega} \{g(\omega) - g^*(\omega)\}^2 d\mu(\omega) \text{ is a minimum} \tag{7.32}$$

for a given function g in L_2. The solution is represented by $g^*(\omega) = P_\mu(g(\omega)|\mathcal{A})$, where $P_\mu(\cdot)$ represents the projection operator of the Hilbert space $(L_2, \langle\cdot\rangle)$ on its subspace $\mathcal{A}$. In the contemplated case, $g(\cdot)$ can be taken to be the smooth density estimator $\tilde{f}_n^0$, derived from the smoothed version $\tilde{S}_n^0(\cdot)$ of S_n^0 in (7.31). To find $g^*(\cdot)$, we need to exhibit its orthogonal basis, which does not seem to be an easy problem, in general. Nevertheless, taking $\Omega = \{Z_{n:1}, \ldots, Z_{n:n}\}$ and defining an isotonic density estimator as in (7.7), it may be possible to obtain an isotonic density estimator as the smoothed version of the isotonic estimator itself. However, though we may invoke the sufficiency of $\{Z_{n:1}, \ldots, Z_{n:n}\}$ in the uncensored case, for the censored case, the sufficiency relates to the product law $S_H(\cdot) = S_F(\cdot)S_G(\cdot)$, but not $S_F(\cdot)$. Hence, we may only somewhat intuitively consider the smoothed version of the isotonic estimator of $f(\cdot)$ as a plausible smooth isotonic estimator. We may remark in this context (Laslett (1982), McNichols and Laslett (1982)) that, in contrast to the uncensored case, the NPMLE of $f(\cdot)$ is not the slope of $\tilde{F}_n^0$, the LCM of the Kaplan-Meier estimator of F; in fact, the NPMLE generally may not have a simple closed form. We may also remark that the basic assumption of random censoring (namely, the independence of censoring and failure times) may not match the reality in many real applications, so that censoring may be informative and the Kaplan-Meier PLE may not be that rational. The consequences of negation of this independence assumption may be far reaching, especially in shape-restrained inference problems, and have not been studied in detail.

7.3 INFERENCE ON UNIMODAL DENSITY FUNCTION

The Pareto, exponential, Weibull (with shape parameter $\gamma \leqslant 1$), and Gamma distribution (with shape parameter $\gamma \leqslant 1$) have monotone decreasing densities. However, a majority of other distributions may not have this characteristic. For *life distributions*, defined on $\mathbb{R}^+$, often the hazard function $h(\cdot)$, or its cumulative version $H(\cdot)$, plays a basic role in characterizing the *aging properties* of the distributions, and we shall study them in greater detail in the next section. For a *df* $F(\cdot)$ (survival function $S(\cdot) = 1 - F(\cdot)$) having the density function $f(\cdot)$, let $h(x) = f(x)/S(x), x \geqslant 0$ (where $S(0) = 1$) and $H(x) = \int_0^x h(y)dy, x \geqslant 0$. Therefore, by definition,

$$S(x) = e^{-H(x)} \text{ and } f(x) = h(x)e^{-H(x)}, x \geqslant 0. \tag{7.33}$$

This equivalence relationship plays a basic role in the study of various shape constrained inference problems. Note that a density is *unimodal* with a mode M if $f(x)$ is nondecreasing in x for $x \leqslant M$, and $f(x)$ is nonincreasing in x, for $x \geqslant M$; M is termed the *mode* (where the density is a maximum). Similarly, if there exists a point U, such that $f(x)$ is nonincreasing for $x \leqslant U$, and nondecreasing for $x \geqslant U$, then U is the *anti-mode* and $f(\cdot)$ is *uni-antimodal*. We are confronted here with the estimation of such unimodal (or uniantimodal) densities, where M (or U) may or may not be specified.

First, consider the case where M (or U) is known *a priori*. We have defined the LCM of F_n, after (7.5). In a similar manner, the *greatest convex minorant* (GCM) of F_n on $(-\infty, M)$ (or (U, ∞)) is defined to be the largest function $\hat{F}_n$ such that $\hat{F}_n(x) \leqslant F_n(x)$ for all $x \leqslant M$ (or $x \geqslant v$) and is convex on $(-\infty, M)$ (or (v, ∞)). Both LCM and GCM play a vital role in the estimation of unimodal (or uniantimodal) densities. For known M, defining $L_n(f)$ as in (7.4), we express this as a product of two factors; those $X_i < M$ and those $X_i > M$ (the probability that $X_i = M$ for some $i(=1,\ldots,n)$ being zero for continuous F). For the first factor, our task is to maximize with respect to the $f_i(=f(X_{n:i})) : X_{n:i} < M$, subject to the inequality constraints $f(X_{n:i}) \leqslant f(X_{n:i+1}), \forall i : X_{n:i} < M$ and the largest one begin replaced by M. Thus, for this factor, the NPMLE of f is the density corresponding to the GCM of F_n on $(-\infty, M)$. For the second factor (i.e., on (M,∞)), we maximize with respect to the $f_i(=f(X_{n:i})) : X_{n:i} > M$, subject to the inequality constraints $f_i \geqslant f_{i+1}$; the NPMLE of f on this part is therefore given by the density corresponding to the LCM of F_n on (M,∞). [For the estimation of uniantimodal density, we need to use the LCM and GCM of F_n on $(-\infty, v)$, and (v,∞), respectively.]

Consider the two segments of F_n on $(-\infty, M)$ and (M,∞), respectively. For the second segment the estimator in (7.7) applies with little modifications. For the first segment $(-\infty, M)$, replacing the LCM by the GCM, in (7.7), we would have an analogous estimator wherein the "min" and "max" arguments are to be interchanged and the range of $X_{n:s}$ is confined to $(-\infty, M)$; we leave the details as Problem 7.6. The case of antimodal densities can be worked out analogously. It should be noted that the NPMLE of f, in either case, is not smooth (as was the case in (7.7)). Hence, there is a genuine need to have a smoothed version of the NPMLE. For this, we proceed as follows.

Note that for a unimodal density f with a mode M, $f(x)$ is decreasing for $x \geqslant M$, and if the support of f extends to $+\infty$, then $f(x) \to 0$ as $x \to \infty$. Motivated by this feature, we are, at least intuitively, led to incorporate the same idea of smoothing as has been illustrated for monotone density in the previous section; see (7.13) through (7.21). In the present context, we confine ourselves to $[M,\infty)$, and hence, we translate $x \to x^* = x - M, x \geqslant M$. Let $n^+ = \sum_{i=1}^n I(X_i \geqslant M)$ and $n^- = n - n^+$; n^+ and n^- are both random, adding up to the number n. Further, let $X^*_{n^+:1} < \cdots < X^*_{n^+:n^+}$ be the order statistics associated with the X_i values that are greater than M, and translated to the respective X^* values. Then, in (7.14), we take $\lambda_{n^+} = n^+/X^*_{n^+:n^+}$ and define the $\omega^+_{j,n}(x^*\lambda_{n^+})$ as in (7.13) with λ_n being replaced by λ_{n^+} and x by x^*. As such, parallel to (7.15), we have here

$$\check{f}_n^+(M + x^*) = \sum_{i\geqslant 0} \omega^+_{i:n^+}(x^*\lambda_{n^+})\hat{f}_n^+(M + i/\lambda_{n^+}), x^* \geqslant 0, \tag{7.34}$$

where the isotonic estimator $\hat{f}_n^+(\cdot)$ is defined as in (7.7) along with the modifications considered earlier in this section. As such, using the same argument as in (7.16)–(7.21) with modifications discussed above, we claim that

$$\check{f}_n^+(x) \text{ in (7.34) is } \searrow \text{ in } x \geqslant M. \tag{7.35}$$

For the segment $(-\infty, M)$ where the true $f(x)$ is $\nearrow$ in x, we make the transformation $x \to x^* = M - x$, so that x^* has the support $[0,\infty)$ and $f^-(x^*) = f(M - x^*)$ is $\searrow$ in $x^*(\geqslant 0)$. As such, we may proceed as in (7.34) and (7.35) with n^+, λ_{n^+} etc. replaced by n^-, λ_{n^-} etc, and defining the isotonic estimator $\hat{f}_{n^-}(M - x^*)$ as in (7.7) along with the modifications considered above.

The modifications for the (random) censoring case are very similar to the case of monotone density, treated in the previous section. With the definition of n^+ and n^-

as in above, we need to define the Kaplan-Meier PLE separately for the two segments $(-\infty, M)$ and $(M, +\infty)$ translated to $(0, \infty)$ by $x^* = M - x$ and $x^* = x - M$, respectively. Having defined these two PLE, for each one we find the isotonic density estimates and smooth them as in the uncensored case. Therefore, we leave the details to verify as Problem 7.7.

The case where M is not known a priori, is more realistic and, at the same time, more difficult. Wegman (1970 a,b), Reiss (1973, 1976), and Prakase Rao (1983) addressed this problem, while Lo (1986) considered the problem of estimating a unimodal distribution function rather than a density. It seems more natural to estimate simultaneously M and the density f when M is not known and the density f is assumed to be unimodal (Wegman 1970 a,b). However, to derive the NPMLE of M and f, it becomes a harder problem, especially when n is not small. A second way to have this estimation problem tractable (from a computational view point) is to estimate first the mode M by, say, $\hat{M}_n$, and then estimate the unimodal density f in a way treated before, presuming that $\hat{M}_n$ is the known mode. The second approach allows the scope for iteration by the use of some preliminary estimator of M and NPMLE of f, given M, followed by the usual Fisher-Rao method of scoring (i.e., stochastic Newton-Raphson method).

A different approach was considered by Birge (1997). His approach considers the family of Grenander-type estimators that is indexed by the mode M. Then choose the one, indexed by $\hat{M}$, such that the corresponding unimodal density estimates lead to a version of the derived empirical distribution that is "closest" to the actual empirical d.f. in the sense of the Kolmogorov-Smirnov distance. Of course, other measures of distance may also be considered in this respect. Birge allowed considerable amount of flexibility in the choice of the underlying density: apart from unimodality, no other severe assumption on the density near the true mode is imposed. His method is primarily data-driven and has $n^{-1/3}$ rate of convergence. However, his procedure is adaptive to the Grenander estimate, and in that process, it may not necessarily be a restricted MLE. Since Birge did not assume f to be smooth (near M), his estimator is of the histogram type. If f is smooth, as has been assumed by others, a kernel estimator with a suitable band-width will converge at the rate $n^{-2/5}$, compared with $n^{-1/3}$ in his case. Therefore, there is a small loss of efficiency in his approach, though, in terms of applications, it might have better perspectives. Smoothing of the Birge estimator along the lines in (7.34) and (7.35) would be highly advocated.

We have seen in this section, and earlier ones, that asymptotics play a vital role in the related developments. Much of these asymptotic methods involve more sophisticated arguments that can't be presented at our contemplated level. As such, after we have also discussed the testing problems (in Section 7.6), we will present some broader asymptotic results (outline only) that would help the reader to comprehend better the more complex asymptotics referred to here.

7.4 INFERENCE ON SHAPE-CONSTRAINED HAZARD FUNCTIONALS

The basic equivalence representation [see (7.33)]

$$1 - F(x) = S(x) = e^{-H(x)} = e^{-\int_0^x h(y)dy}, x \geqslant 0 \tag{7.36}$$

has often been exploited in the formulation of various shape constraints wherein the hazard function $h(\cdot)$ plays a basic role. For example, using (7.36), whenever $f(\cdot)$ is differentiable, we have

$$f'(x) = -e^{-H(x)}\{h^2(x) - h'(x)\}, x \geqslant 0, \tag{7.37}$$

so that one may equivalently characterize a monotone decreasing density $f(\cdot)$ in terms of the hazard property:

$$h(x) \geqslant h'(x)/h(x), \forall x \geqslant 0. \tag{7.38}$$

Similarly, a pdf $f(\cdot)$ is unimodal around a mode M if

$$h'(x)/h(x) \gtreqless h(x) \text{ according as } x \text{ is } \lesseqgtr M. \tag{7.39}$$

Along the same line, a density $f(\cdot)$ is said to be *strongly unimodal* if $\log f(x)$ is *concave* on the support of f. This strong unimodality property can also be characterized as $H(x) - \log h(x)$ being convex, or equivalently,

$$h(x) - h'(x)/h(x) \nearrow \text{ in } x(\geqslant 0). \tag{7.40}$$

There is another aspect of a life distribution, namely its *aging property*, where the hazard has a prime characterizing feature. This aging property stems from a basic feature of a negative exponential law, for which $S(x) = e^{-\lambda x}, x \geqslant 0$, for some $\lambda > 0$, so that $h(x) = \lambda, \forall x \geqslant 0$. Thus, the simple exponential model can be characterized by a constant hazard (or failure) rate. In this case, we have

$$S(x+y)/S(x) = e^{-\lambda y} = S(y), \forall x, y \geqslant 0. \tag{7.41}$$

In *reliability theory* for repairable systems, one has the *new better (worse) than used* NBU (NWU) property based on the characterization $s(x+y) \geqslant (\text{or} \geqslant) s(x)s(y)$, for every $x, y \geqslant 0$. On using (7.36), we say that a life distribution is NBU (NWU) if $H(x+y)$ is $\geqslant (\leqslant) H(x) + H(y)$, for every $x, y \geqslant 0$; the distribution is NBUE (in expectation) if $\int_0^\infty S(x+u)du)/S(x) \geqslant \mu = \int_0^\infty S(y)dy, \forall x \geqslant 0$. Similarly, a life distribution belongs to the *increasing failure rate* (IFR) class if

$$h(x) \text{ is monotone nondecreasing in } x \in \mathbb{R}^+, \tag{7.42}$$

or, equivalently, $H(x)$ is convex in x. The distribution belongs to the IFRA (in average) class if

$$x^{-1}H(x) \text{ is } \nearrow \text{ in } x(\in \mathbb{R}^+). \tag{7.43}$$

From the above definitions, it follows that

$$\begin{aligned} &\text{IFR} \Rightarrow \text{IFRA} \Rightarrow \text{NBU} \Rightarrow \text{NBUE};\\ &\text{DFR} \Rightarrow \text{DFRA} \Rightarrow \text{NWU} \Rightarrow \text{NWUE}. \end{aligned} \tag{7.44}$$

There are other characterizations of life distributions based on a *decreasing mean remaining life* (DMRL) property, which will be considered in the next section. As we shall see later, the IFR property implies the DMRL property while the DMRL property implies the NBUE property. Therefore, we will have the following implication flow chart:

$$\begin{aligned} &\text{IFR} \Rightarrow \text{IFRA} \Rightarrow \text{NBU} \Rightarrow \text{NBUE}\\ &\text{IFR} \Rightarrow \text{DMRL} \Rightarrow \text{NBUE}. \end{aligned} \tag{7.45}$$

It is, in this sense, of greater interest to estimate a life distribution under the *shape constraint* that it belongs to the IFR class or DMRL class or NBUE class; the other two, namely IFRA and NBU classes, being intermediate may not be of that importance.

A similar implication chart relates to the DFR, DFRA, NWU, NWUE, and IMRL classes of life distributions. It is clear from the above discussion that such classes of life distributions may conveniently be characterized in terms of the hazard (or failure rate) function $h(\cdot)$. In this section, we emphasize on the IFR class for which the constraint in (7.42) is explicitly isotonicity of $h(\cdot)$; for the IFRA class, by (7.43), the constraint involves the monotonicity of $x^{-1}H(x)$, and as such, a similar formulation can be made. For the NBU class, we would have a constraint of the form $H(x+y) - H(x) - H(y) \geqslant 0, \forall x, y$, and a somewhat different formulation would be necessary. A complete treatise of this entire hierarchy of life distributions is somewhat outside the scope of this monograph; there are other classes of life distributions that are larger, but more specialized to repairable systems in reliability theory. We will omit discussion of them, but may refer to Bhattacharjee and Sen (1998) for detailed treatise; other pertinent references have also been cited there.

For the estimation of $S(\cdot)$ under IFR constraint, a simpler task may be to estimate $h(\cdot)$ under the monotonicity constraint in (7.42) and then use the relation: $h(x) = -(d/dx)\log S(x)$ to proceed $S(x)$ via the integrated hazard $H(x)$ (i.e., using the relation: $S(x) = e^{-H(x)}, \forall x \in \mathbb{R}^+$, so that $-\log S(x)$ is an equivalent representation for $H(x)$. Whereas the empirical survival function $S_n(\cdot)(= 1 - F_n(\cdot))$ is a natural estimator of $S(\cdot)$, so that $-\log S_n(\cdot)$ is a naive estimator of $H(\cdot)$, noting that $S_n(\cdot)$ is a step-function, albeit being bounded by 1 (from the left) and 0 (from the right), $-\log S_n(\cdot)$ is also a step (-up) function, which is bounded from the left by 0 but unbounded from the right. As such, differentiation of $-\log S_n(x)$ (with respect to x), fails to provide any sensitive estimate of $h(x)$. Therefore, some alternative routes are explored here.

Recall that $h(x) = f(x)/S(x), x \in \mathbb{R}^+$, so that it might be intuitive to estimate $f(\cdot)$ and $S(\cdot)$ as in the previous sections and plug-in the estimates to estimate $h(\cdot)$. One may even use $\hat{f}_n(\cdot)$, the NPMLE of $f(\cdot)$, discussed in the previous sections (see (7.12)). Then a naive plug-in estimator of $h(x)$ is given by $\hat{h}_n(x) = \hat{f}_n(x)/S_n(x), 0 \leqslant x < X_{n:n}$. Note that $S_n(x) = 0$ for $x \geqslant X_{n:n}$ (and so is $\hat{f}_n(x), x \geqslant X_{n:n}$). Therefore, $\hat{h}_n(x), x \geqslant X_{n:n}$ is not properly defined. In the IFR model, $h(x)$ being $\nearrow$ in x, for $x \geqslant X_{n:n}$, $h(x)$ is $\geqslant h(X_{n:n}-)$, and it could have a finite or infinite upper asymptote. Therefore, setting $\hat{h}_n(x) = \hat{h}_n(X_{n:n}-) \forall x \geqslant X_{n:n}$ may not give a reliable estimate of $h(\cdot)$ beyond the sample largest observation $X_{n:n}$. Moreover, the naive estimator $\hat{f}_n(\cdot)$ in (7.12) may not generally preserve the isotonic nature of $\hat{h}_n(\cdot)$. Even if we use the NPMLE of $f(\cdot)$, defined by (7.7) [so assuming that $f(x)$ is $\searrow$ in $x \in \mathbb{R}^+$], $\hat{h}_n(x)$, being the ratio of two nondecreasing functions, may not be isotonic itself. Therefore, there is a need to isotonize $\hat{h}_n(\cdot)$ in order to preserve the isotonic nature of $h(\cdot)$. Thus, it might be better to consider some other isotonic estimator of $h(\cdot)$.

The solution in (7.32) for the isotonic density case worked out well as $\tilde{f}_n(\cdot)$ is a linear functional. However, in the present case, $h(x) = -(\partial/\partial x) \log S(x)$ is not a linear functional of $S(\cdot)$, and hence, an algebraic solution (involving the L_2-projection operator), even confined to $\Omega = \{X_{n:1}, \ldots, X_{n:n}\}$ may be difficult. Therefore, from an operational point of view, we may consider the estimator in (7.21) [or (7.15)] and isotonize it as in (7.14). Since such statistical functionals can be well approximated by linear functionals, asymptotically the operating rule could be equivalent to the solution of (6.2.19).

The estimators of $h(\cdot)$ adapt well to the various censored cases discussed in Section 6.2. We consider explicitly the case of random censoring, as the other cases follow on similar arguments. Define the Kaplan-Meier PLE $S_n^0(x)$ as in (7.30). Use the Hille's lemma in (7.53) and denote the corresponding smooth version by $\tilde{S}_n^0(\cdot)$. A naive estimator $\tilde{h}_n^0(\cdot)$ can then be directly obtained as

$$\tilde{h}_n^0(x) = -(d/dx) \log \tilde{S}_n^0(x) = \tilde{f}_n^0(x)/\tilde{S}_n^0(x), x \in \mathbb{R}^+. \tag{7.46}$$

Then, isotonize $\tilde{h}_n^0(\cdot)$ as in (7.24). Based on similar arguments as in the complete sample case, asymptotic equivalence of such smooth isotonic estimators of $h(\cdot)$ and smoothed versions of alternative isotonic (but not necessarily smooth) estimators considered by Huang and Zhang (1994) and Huang and Wellner (1995) can be established. Additional regularity assumptions are, however, needed to cover the general case of $h(\cdot)$ being unbounded (IFR) and the domain being extended to cover $\mathbb{R}^+$. Finally, kernel smoothing, as in Section 7.2, is also possible when the kernel is adapted in a similar manner.

The problem of the NPMLE of an IFR $h(\cdot)$ can also be posed as follows. First, consider the naive estimator of the failure rate function by plugging in the estimator $S_n(\cdot)$ and the naive density estimator $f_n(\cdot)$, defined by (7.12). This is given by

$$h_n(x) = \begin{cases} 0, & x < X_{n:1}, \\ \{(n-i)(X_{n:i+1} - X_{n:i})\}^{-1}, & X_{n:i} \leqslant x < X_{n:i+1}, i \leqslant n-1, \\ \infty, & x \geqslant X_{n:n} \end{cases}. \tag{7.47}$$

Note that $h_n(\cdot)$ is a step function and has no guarantee of being monotone even if $h(\cdot)$ is so. In this set-up, to obtain the NPMLE of $h(\cdot)$, writing $\hat{h}_{n,i} = h_n(X_{n:i}), i = 1, \ldots, n$, one needs to maximize $\sum_{i=1}^{n-1} \{\log \hat{h}_{n,i} - (n-1)\hat{h}_{n,i}(X_{n:i+1} - X_{n:i})\}$

subject to the condition $0 \leqslant \hat{h}_{n,1} \leqslant \cdots \leqslant \hat{h}_{n,n-1} \leqslant M$ where M is arbitrarily large. The solution is given by

$$\hat{h}_n(x) = \begin{cases} 0, & x < X_{n:1}, \\ \hat{h}_{n,i}, & X_{n:i} \leqslant x < X_{n:i+1}, i = 1, \ldots, n-1 \\ \infty, & x \geqslant X_{n:n} \end{cases} \tag{7.48}$$

where

$$\hat{h}_{n-i} = \min_{i \leqslant t \leqslant n-1} \max_{1 \leqslant s \leqslant i} \left\{ \frac{t - s + 1}{\sum_{j=s}^{t} (n-j)(X_{n:j+1} - X_{n:j})} \right\}, \tag{7.49}$$

for $i = 1, \ldots, n-1$. The values of $\hat{h}_n(\cdot)$ are the isotonic regression of the naive estimator of $h(\cdot)$, discussed in (7.47). We leave the details as Problem 7.8.

The NPMLE $\hat{h}_n(\cdot)$ of $h(\cdot)$ has, by construction, the desired isotonic property, albeit it is still a step-function having possible jump discontinuities at the ordered failure points $X_{n:1}, \ldots, X_{n:n}$. However, the isotonic $h(\cdot)$ is typically smooth (as both $S(\cdot)$ and $f(\cdot)$ are), and hence, there is a genuine need to preserve the smoothness property of $h(\cdot)$.

Let us present the smooth isotonic estimators of $h(\cdot)$. First, consider the smooth estimator $F_n^*(\cdot)$ of $F(\cdot)$, defined by (7.22), and denote by $S_n^*(x) = 1 - F_n^*(x), x \geqslant 0$, the smooth estimator of $S(\cdot)$. Next, define the smooth density estimator $f_n^*(x), x \geqslant 0$, as in (7.24). Then a smooth estimator of $h(\cdot)$ is

$$h_n^*(x) = -(d/dx) \log S_n^*(x) = f_n^*(x)/S_n^*(x) \tag{7.50}$$

$$= \lambda_n \left\{1 - w_{n,n}^*(x\lambda_n)\right\} - \tag{7.51}$$

$$\lambda_n \left\{ \sum_{k=0}^{n} w_{n,k}^*(x\lambda_n) S_n\left(\frac{k+1}{\lambda_n}\right) \right\} \left\{ \sum_{k=0}^{n} w_{n,k}^*(x\lambda_n) S_n\left(\frac{k}{\lambda_n}\right) \right\}^{-1}, \tag{7.52}$$

for $x \in \mathbb{R}^+$.

The smooth estimator $h_n^*(\cdot)$, defined above, needs to be isotonized to meet the required objective. For this, we follow the initial lead by Mukerjee (1986) and the prescription made in Chaubey and Sen (2002); in this respect, the following (Hille (1948)) lemma plays a basic role.

Lemma 7.4.1 *Let $u(x), x \in \mathbb{R}^+$ be a bounded function, and let*

$$\tilde{u}_n(x) = e^{-x\lambda_n} \sum_{k=0}^{\infty} \{(x\lambda_n)^k / k!\} u(k/\lambda_n), \tag{7.53}$$

converges uniformly to $u(x)$ in any finite sub-interval of $\mathbb{R}^+$, as $\lambda_n \to \infty$. This convergence extends to the whole interval if the function $u(\cdot)$ is monotone.

[We leave the proof of the lemma as Problem 7.9.] Whereas for a monotone decreasing density, survival function, and decreasing failure rate, the boundedness condition of $u(\cdot)$ can be justified, for the IFR case, the boundedness of $h(\cdot)$ may not be generally true. Toward this, we consider the following illustrations.

Example 7.4.1

Weibull distribution: $S(x) = e^{-\lambda x}, x \geqslant 0$, where $\lambda > 0$ and $\gamma > 0$. For γ (the shape parameter) equal to 1, we have the exponential case (with a constant hazard $h(x) = \lambda, \forall x$), while we have an IFR or DFR case according as γ is greater than 1 or less than 1. For the Weibull case, we have $h(x) = \lambda\gamma x^{\gamma-1}$, so that for $\gamma > 1, h(\cdot)$ is not only IFR but also $h(x) \to \infty$ as $x \to \infty$. For the DFR case, $h(x) \to 0$ as $x \to \infty$.

Example 7.4.2

Gamma density: $f(x) = \{\Gamma(p)\}^{-1}\lambda^p e^{-\lambda x}x^{p-1}, x \geqslant 0$, where $\lambda > 0, p > 0$; here also for p (the shape parameter) equal to 1, we have the exponential density, while we have the IFR or DFR property according as p is $>$ or < 1. However, even for $p > 1$, though $h(x)$ is $\nearrow$ in x, it has the upper asymptote equal to λ. We leave the detail as Problem 7.10.

In order to capture the full impact of Hille's lemma for the smooth estimation of an IFR $h(\cdot)$, we may therefore assume (as in the Gamma case) that $h(x)$ has a finite upper asymptote $\lambda(< \infty)$, or (as in the Weibull case) limit ourselves to a finite subinterval ($[0, M), M < \infty$) of $\mathbb{R}^+$. (This was implicitly done in (7.48) while deriving the (nonsmooth) NPMLE of an IFR. This extra condition (or restraint) does not arise in the case of [$S(\cdot)$ or] a monotone decreasing density and unimodal density, treated in the previous sections).

We need to discuss the monotonicity of $\tilde{u}_n(\cdot)$ when $u(\cdot)$ is so. We define

$$W_{n,k}(x) = \sum_{j=k}^{\infty} e^{-x\lambda_n}(x\lambda_n)^j/j! \quad \text{for } k = 0, 1, \ldots, \infty. \tag{7.54}$$

Then using the well-known identity (see Problem 7.4) that

$$e^{-m}\sum_{j=0}^{k} m^j/j! = (k!)^{-1}\int_m^{\infty} e^{-x}x^k dx, \forall k (= 0, 1, \ldots, \infty)$$

we have

$$W_{n,k}(x) = \frac{1}{\Gamma(k)}\int_0^{x\lambda_n} e^{-y}y^{k-1}dy, \quad k \geqslant 0. \tag{7.55}$$

As a result, $(\partial/\partial x)W_{n,k}(x) = \{\Gamma(k)\}^{-1}e^{-x\lambda_n}(x\lambda_n)^{k-1}\lambda_n \geqslant 0, \forall x \geqslant 0$. Therefore, rewriting $\tilde{u}_n(x)$ in (7.53) as

$$\tilde{u}_n(x) = u(0) + \sum_{k=1}^{\infty} W_{n,k}(x)\left\{u\left(\frac{k}{\lambda_n}\right) - u\left(\frac{k-1}{\lambda_n}\right)\right\}, \tag{7.56}$$

and noting that the $u(k/\lambda_n) - u((k-1)/\lambda_n), k \geqslant 1$ are all of the same sign (i.e., either all are nonnegative or all are nonpositive), the (strict) monotonicity of $W_{n,k}(x)$ (in x) ensures the monotonicity of $\tilde{u}_n(x)$ when $u(x)$ is monotone.

Having the desired monotonicity of the smoother in (7.53) in order, we would like to make use of the approach outlined in (7.32) (for the isotonic density case) and seek to find an isotonic function $g(\cdot)$ such that

$$\int_0^{X_{n:n}} \left\{\tilde{h}_n(x) - g(x)\right\}^2 dx \tag{7.57}$$

is minimized, where $\tilde{h}_n(x)$ is a modified version of $h_n^*(x)$ in (7.52), adjusted to suit the monotonicity property in (7.56). Specifically, we let

$$\tilde{h}_n(x) = \frac{\lambda_n e^{-x\lambda_n} \sum_{k=0}^{\infty} \frac{(x\lambda_n)^k}{k!} S_n\left(\frac{k+1}{\lambda_n}\right)}{e^{-x\lambda_n} \sum_{k=0}^{\infty} \frac{(x\lambda_n)^k}{k!} S_n\left(\frac{k}{\lambda_n}\right)}, x \geqslant 0. \tag{7.58}$$

In passing, we may note that for all $k : \frac{k}{\lambda_n} \geqslant X_{n:n}$, by definition, $S_n\left(\frac{k}{\lambda_n}\right) = 0$, and hence, in (7.58), the infinite summation can be replaced by summation over $0 \leqslant k < k_n$ where $S_n\left(\frac{1}{\lambda_n}k_n\right) = 0, k_n$ begin the smallest integer $\geqslant \lambda_n X_{n:n}$, is itself a random variable, covering $a.s$ to ∞ as $\rightarrow \infty$.

7.5 INFERENCE ON DMRL FUNCTIONS

For life distributions, defined on $\mathbb{R}^+$, a functional of considerable statistical interest, especially in the field of reliability and survival analysis, is the mean remaining (or residual) life (MRL) function. Corresponding to a survival function $S(\cdot)$, let $f(\cdot)$ be the pdf; then the MRL at age x is defined as

$$\mu(x) = \{\int_0^{\infty} uf(x+u)du\}/S(x) \tag{7.59}$$

$$= \{S(x)\}^{-1} \int_0^{\infty} S(x+u)du \tag{7.60}$$

$$= \int_0^{\infty} \exp\{-[H(x+u) - H(x)]\}du, \quad x \in \mathbb{R}^+. \tag{7.61}$$

Note that by definition $\mu(0) = \mu$ is the mean lifetime (for $S(\cdot)$). In this framework, a life distribution is said to have a DMRL *(decreasing MRL)* property if $\mu(x)$ is decreasing in $x \in \mathbb{R}^+$. The DMRL property is intricately related to the aging aspects of a life distribution, as can be seen from the following outline. For an exponential pdf, $f(x) = \lambda e^{-\lambda x}, x \geqslant 0$, for some $\lambda > 0$, so that the *no memory* property in (7.41) holds, and, as such, $\mu(x) = \mu(0) = \mu$, for all $x \geqslant 0$. The exponential density is a vertex member of the class of IFR distributions, for which the hazard (failure rate)

$h(x)$ is nondecreasing in $x(\geqslant 0)$. As such, for the IFR class, $H(x)$ is convex and monotone increasing, and hence, $\{H(x+y) - H(x) - H(y)\}$ is nondecreasing in x, for all $y, x \geqslant 0$. Therefore, from (7.61), we conclude that for the IFR class of life distributions, $\mu(x)$ is decreasing in x, that is, the DMRL property holds. Thus, the IFR class is a subclass of the DMRL class. From this nested nature of the various classes of life distributions, it appears that statistical inference on the MRL function, especially under the DMRL constraint, is of genuine interest. In both reliability and survival analysis, there being generally a number of *auxiliary* (or *explanatory*) *variables*, often termed as *covariates*, statistical modeling and analysis of DMRL functions involve greater complexity. Nevertheless, the MRL function plays a primary role in this context.

Let us consider first the estimation of $\mu(x), x \in \mathbb{R}^+$, under the DMRL constraint (i.e., $\mu(x)$ being $\searrow$ in x). Looking at (7.61) and replacing the unknown $S(\cdot)$ by its natural (empirical) counterpart $S_n(x)(= n^{-1}\sum_{i=1}^{n} I(X_i > x)), x \in \mathbb{R}^+$, it seems that a naive estimator of $\mu(x)$ may be intuitively posed as

$$\breve{\mu}_n(x) = \{S_n(x)\}^{-1}\int_0^\infty S_n(x+u)du. \tag{7.62}$$

(See Yang (1978)). Although when the support for $S(\cdot)$ is $\mathbb{R}^+, S(x) > 0, \forall x < \infty$, by definition $S_n(x) = 0, \forall x \geqslant X_{n:n}$. As a result, $\breve{\mu}_n(x)$ is not defined for $x \geqslant X_{n:n}$. Further, note that $S_n(x) = (n-k)/n$ for $X_{n:k} \leqslant x < X_{n:k+1}, 0 \leqslant k \leqslant n$ (where, conventionally, $X_{n:0} = 0$ and $X_{n:n+1} = +\infty$). Therefore, for $x \in [X_{n:k}, X_{n:k+1})$, we write

$$\begin{aligned}
\int_0^\infty & S_n(x+u)du \\
&= \int_0^{X_{n:k+1}-x} S_n(x+u)du + \sum_{j\geqslant 1}\int_{X_{n:k+j}-x}^{X_{n:k+1+j}-x} S_n(x+u)du \\
&= \frac{1}{n}\{(n-k)(X_{n:k+1}-x) + \sum_{j=1}^{n-k-1}(n-k-j)(X_{n:k+j+1}-X_{n:k+j})\}
\end{aligned} \tag{7.63}$$

$$= \frac{1}{n}\{\sum_{j=k+1}^{n} X_{n:j} - (n-k)x\}. \tag{7.64}$$

Therefore, we have

$$\breve{\mu}_n(x) = \frac{1}{n-k}\sum_{j=k+1}^{n} X_{n:j} - x, \text{for } X_{n:k} \leqslant x < X_{n:k+1}, \tag{7.65}$$

for $x \in [X_{n:k}, X_{n:k+1})$ where $k = 0, 1, \ldots, n-1$; it follows from (7.62) that $\breve{\mu}_n(x)$ is undefined for $x \geqslant X_{n:n}$. Recall that by the fact that $0 < X_{n:1} < X_{n:2} < \cdots <$

$X_{n:n} < \infty$ (with probability one), we have

$$\bar{X}_n = \frac{1}{n}\sum_{i=1}^{n} X_i \leqslant \frac{1}{n-1}\sum_{j=2}^{n} X_{n:j} \leqslant \cdots \leqslant X_{n:n} < \infty, \tag{7.66}$$

so that on writing $d_{nj} = (n-j+1)(X_{n:j} - X_{n:j-1}), j = 1, \ldots, n$ (the *normalized spacings*) and letting conventionally $d_{n0} = 0 = d_{n(n+1)}$, we denote the *backward averages* as

$$_{nk}\bar{d} = \frac{1}{n-k+1}\sum_{j=k}^{n} d_{nj}, \quad k = 0, 1, \ldots, n.$$

Then, we have

$$_{nk}\bar{d} = \frac{1}{n-k+1}\sum_{j=k}^{n} X_{n:j} - X_{n:k-1} \text{ for } k = 1, \ldots, n, \tag{7.67}$$

and, as a result,

$$\breve{\mu}_n(x) =_{nk+1} \bar{d} - (x - X_{n:k}) \text{ for } X_{n:k} \leqslant x < X_{n:k+1} \tag{7.68}$$

for $k = 0, 1, \ldots, n-1$, so that

$$\breve{\mu}_n(X_{n:k+1}-) = \frac{1}{n-k}\sum_{j=k+1}^{n} X_{n:j} - X_{n:k+1} \tag{7.69}$$

$$\leqslant \frac{1}{n-k-1}\sum_{j=k+2}^{n} X_{n:j} - X_{n:k+1} = \breve{\mu}_n(X_{n:k+1}+), \tag{7.70}$$

for $k = 0, 1, \ldots, n-2$, and $\breve{\mu}_n(X_{n:n}-) = X_{n:n} - X_{n:n} = 0$ while $\breve{\mu}_n(x)$ is not properly defined for $x \geqslant X_{n:n}$.

Thus, $\breve{\mu}_n(x)$ is piecewise linear (negative slope) within $(X_{n:k}, X_{n:k+1})$, for $k = 0, \ldots, n-1$, having jump discontinuity at each $X_{n:k}, k \geqslant 1$, and is not properly defined for $x \geqslant X_{n:n}$ (see Fig. 7.1; Problem 7.11].

It is intuitively appealing to replace $S_n(\cdot)$ in (7.64) by its smoothed version $S_n^*(\cdot)$ obtained by using the weights in (7.23). Since the smoother is linear, it follows (Problem 7.12) that the smoothed version of $\breve{\mu}_n(\cdot)$ is

$$\breve{\mu}_n^*(x) = \frac{1}{S_n^*(x)}\sum_{k=1}^{n} S_n\left(\frac{n-k}{\lambda_n}\right)\int_x^{\infty} \omega_{n(n-k)}(t\lambda_n)dt. \tag{7.71}$$

In this respect, the choice of $\lambda_n = n/X_{n:n}$ may make the integral in (7.71) divergent, and hence, following Chaubey and Sen (2002), we use the Poisson weights, that is

$$\omega_{nk}(t\lambda_n) = e^{-t\lambda_n}(t\lambda_n)^k/k!, k = 0, 1, \ldots, \infty, \tag{7.72}$$

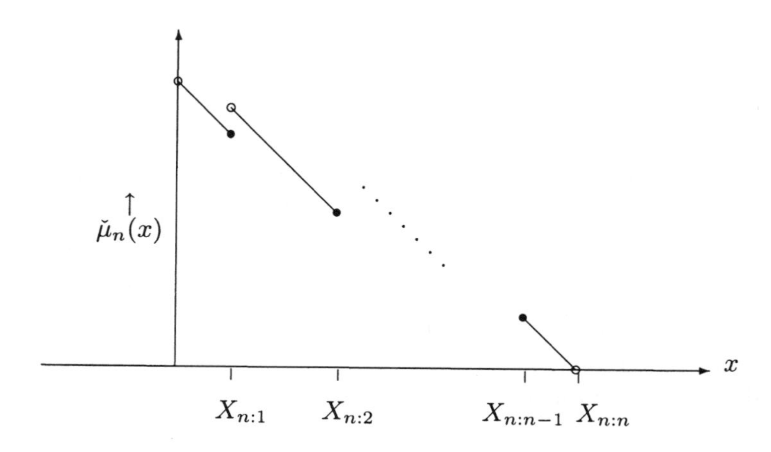

Fig. 7.1 Empirical MRL function.

so that via the resulting smoothed version of $S_n(\cdot)$, we arrive at the following smoothed estimator:

$$\breve{\mu}_n^0(x) = \frac{\sum_{k=0}^{n} \sum_{r=0}^{k} (x\lambda_n)^{k-r}}{\lambda_n \sum_{k=0}^{n} (x\lambda_n)^k} \frac{S_n(k/\lambda_n)/(k-r)!}{S_n(k/\lambda_n)/k!}, \tag{7.73}$$

which is defined for all $x \geqslant 0$. Problem 7.13 is set to verify this.

Unlike the case of the density or the hazard rate, for the MRL function, the NPMLE of $\mu(\cdot)$ may not universally exist. For the IMRL family, $\mu(x)$ is $\nearrow$ in x, and the NPMLE of $\mu(\cdot)$ does not exist. For the DMRL family it exists, but is computationally intractable when n is not very small (Kochar et al. (2000)). For the reason, we proceed to obtain suitable smooth estimators of $\mu(\cdot)$ along the same line as in the previous sections, i.e., isotonize the naive estimator via smoothing. Recall that the estimator $\breve{\mu}_n(\cdot)$ in (7.65) has jump discontinuities at each ordered $X_{n:k}, k = 1, \ldots, n$. We remove these discontinuities by considering the linear segmented estimator with the segments joining at $(X_{n:k}, \breve{\mu}_n(X_{n:k})), k = 1, \ldots, n$, where

$$\breve{\breve{\mu}}_n(X_{n:k} = \frac{1}{2}(\breve{\mu}_n(X_{n:k}-) + \breve{\mu}_n(X_{n:k}+)), k = 1, \ldots, n.$$

If we denote the solution of the minimization problem

$$\text{minimize} \sum_{k=1}^{n} (g_k - \breve{\breve{\mu}}_n(X_{n:k}))^2, \tag{7.74}$$

subject to the inequality constraint $g_0 \geqslant g_1 \geqslant \cdots \geqslant g_n = 0$, by $(\tilde{g}_{n0}, \ldots, \tilde{g}_{nn})$, then as in the previous sections, we have

$$\tilde{g}_{nk} = \min_{s \leqslant k-1} \max_{t \geqslant k} (\tilde{G}_{nt} - \tilde{G}_{ns})/(t-s), k = 1, \ldots, n, \tag{7.75}$$

where

$$\tilde{G}_{nk} = \sum_{i=0}^{k} \check{\tilde{\mu}}_n(X_{n:i}), k = 1, \ldots, n. \tag{7.76}$$

Thus, the $\tilde{g}_{nk}$ represent the left slope of the LCM of the $\tilde{G}_{nk}$. Using these isotonic values, we can define the isotonic segmented line estimator of the DMRL function as

$$\hat{\mu}_n^*(x) = \begin{cases} \tilde{g}_{nk} + \frac{(\tilde{g}_{n(k+1)} - \tilde{g}_{nk})(x - X_{n:k})}{X_{n:k+1} - X_{n:k}}, & X_{n:k} \leqslant x < X_{n:k+1}, 0 \leqslant k \leqslant n-1; \\ 0, & \text{for } x \geqslant X_{n:n}. \end{cases} \tag{7.77}$$

We may extend this method for the *random censoring* case as in the previous sections. We use the Poisson weights in conjunction with the Kaplan-Meier PLE in (7.30) and come up with the following naive estimator

$$\tilde{\mu}_n^0(x) = \begin{cases} \sum_{u=0}^{m_n - j} d_{j,u} Z_{n:j+u}^* - x, & Z_{n:j}^* \leqslant x < Z_{n:j+1}^*, 0 \leqslant j \leqslant m_n, \\ 0, & x > Z_{n:n}, \end{cases} \tag{7.78}$$

where the $d_{j,u}$ are derived from the values of the PLE at the $Z_{n:j+u}^*$. As a result, a smooth estimator of $\mu(x)$ in this case can be written as in (7.73) with $S_n(\cdot)$ replaced by the PLE $\tilde{S}_n(\cdot)$. With this modification, we may proceed as in (7.73) through (7.77) with this modified smooth estimator, and define the smooth isotonic estimator as in (7.77) with the $\tilde{g}_{nk}$ modified accordingly. For a numerical illustration see Chaubey and Sen (2002).

7.6 ISOTONIC NONPARAMETRIC REGRESSION: ESTIMATION

The term *nonparametric regression* has been used in more than one sense. We consider here the following scenarios.

Let $(X_i, Y_i), i = 1, \ldots, n$ be n *iid* random vectors, having a (bivariate) joint distribution $F(x, y)$, defined on $\mathbb{R}^2$. Let use denote by

$$m(x) = E(Y|X = x), \text{ for } x \in \mathbb{R}. \tag{7.79}$$

In a conventional parametric sense, it is assumed that $m(x) = \alpha + \beta x, x \in \mathbb{R}$ (or a subset of $\mathbb{R}$), where α (the intercept) and β (the slope) are the parameters of interest. In this simple linear regression model, it suffices to draw inference on the parameters (α, β). If we adopt this linear regression model then the regression is monotone increasing or decreasing according as β is positive or negative; for $\beta = 0$, there is no regression of Y on X.

In the simplest case, it is assumed that the conditional variance of Y, given $X = x$, denoted by σ_x^2, exists, and further that the model is *homoscedastic* in the sense that

$$\sigma_x^2 = \sigma^2 (< \infty), \quad \forall x \in \mathbb{R}. \tag{7.80}$$

Conditional *normality* of Y, given $X = x$, is also assumed for normal theory modeling and analysis. However, in many practical problems, the homoscedasticity or the normality condition may not be tenable. Even worse than that, generally, $m(x)$ may not be linear in x; in many situations it may not even be monotonic everywhere (in x). In view of this, we consider the following models, allowing increasing generality of the regression function. In all these developments, we denote the conditional distribution function of Y, given $X = x$, by $F(y|x), y \in \mathbb{R}, x \in \mathbb{R}$, and assume it to be continuous a.e. [see Sen (1996b)].

Model I (Homcoscedastic semiparametric model) We let

$$F(y|x) = F_0((y - m(x))/\sigma) \tag{7.81}$$

where F_0 is continuous and belongs to the location-scale family; $\sigma(0 < \sigma < \infty)$ is a scale parameter (not necessarily the conditional standard deviation of Y, given $X = x$). In fact, F_0 may not even have a finite second moment.

Model II (Heteroscedastic semiparametric model]). We let

$$F(y|x) = F_0^*((y - m(x))/\sigma_x), \tag{7.82}$$

where the scale factor σ_x may vary over x.

Model III (General nonparametric model]). We let

$$F(y|x) = F_x((y - m(x)), x \in \mathbb{R}, y \in \mathbb{R}, \tag{7.83}$$

where the form of the (unknown) d.f. F_x may vary over x, not necessarily only through a scale parameter σ_x (as in (7.82)), i.e., F_x may not belong to a location-scale family.

For all these models, $m(x)$ is nonparametric in the sense that apart from some local (in x) smoothness properties, no severe parametric form is presumed for $m(\cdot)$. Nonparametric estimation of $m(x)$, even in the isotonic case, can be formulated for the general model in (7.83), albeit for studying properties of such estimates, more specific regularity assumptions may be needed. In passing, we may remark that as $m(x)$ is nonparametric, a transform of $x \to x^* = h(x), h(\cdot)$ strictly increasing, would not affect the monotonicity of $m(x)$, if it is so. Hence, by making use of the (marginal) probability integral transformation on X, it is often assumed that X has a uniform distribution on $[0, 1]$. In practice, if the marginal d.f. of X is not known (as is usually the case), one may use the empirical d.f. of the X_i to reduce the range to $[0, 1]$, although it would not be strictly a uniform $(0, 1)$ d.f. for the transformed variables. Often, the random X_i are replaced by fixed equidistant points (on $[0, 1])t_{n,i} = i/(n + 1), 1 \leqslant i \leqslant n$. In the later case, it is taken that

$$Y_{ni} = m\{i/(n + 1)\} + \epsilon_{ni}, i = 1, \ldots, n, \tag{7.84}$$

where for each $n(\geqslant 1)$, the ϵ_{ni} are *iid* random variables with 0 mean, and possibly a finite variance $\sigma^2_{epsilon}$. In nonparametric analysis, even the existence of σ^2_ϵ may be necessary. The last model is termed the homoscedastic nonparametric regression (Eubank (1999) and Efromovich (1999)).

We provide here an elementary, intuitive treatise of the estimation problem; related asymptotics will be presented briefly along with. As in (7.26), we consider a kernel function $k(u), u \in \mathbb{R}$, such that $\int uk(u)du = 0$ and $\int u^2k(u)du = 1$. Also, let $\{h_n\}$ be a sequence of positive numbers such that $h_n \to 0$ but $nh_n \to \infty$ as $n \to \infty$; typically, $h_n \sim n^{-1/5}$. Let $F_n(x, y)$ be the (bivariate) empirical d.f. of the $(X_i, Y_i), 1 \leqslant i \leqslant n$, and let $F_{n1}(x)$ be the empirical d.f. of the $X_i, 1 \leqslant i \leqslant n$. Also, let $F_1(x)$ be the marginal d.f. of X, and denote the corresponding density and its derivatives by $f_1(x)$, $f_1'(x)$ and $f_1''(x)$, respectively. Likewise, the partial derivatives of $f(x, y)$ (with respect to x) are denoted by $f_{10}(x, y)$, $f_{20}(x, y)$, etc.

For a fixed $x \in (0, 1)$, consider the kernel estimator

$$\begin{aligned}\tilde{m}_n(x) &= \frac{(\int\int y\frac{1}{h_n}k(\frac{v-x}{h_n})dF_n(v,y))}{(\int\int \frac{1}{h_n}k(\frac{v-x}{h_n})dF_n(v,y))} \\ &= \frac{\frac{1}{nh_n}\sum_{i=1}^n Y_i k(\frac{X_i-x}{h_n})}{\frac{1}{nh_n}\sum_{i=1}^n k(\frac{X_i-x}{h_n})} = \frac{u_n}{v_n}, \text{ say}.\end{aligned} \tag{7.85}$$

Note that the $\{Y_i h_n^{-1} k\left(\frac{X_i - x}{h_n}\right), 1 \leqslant i \leqslant n\}$ form a triangular array of *iid* random variables with mean ξ_{n1} and variance γ_{n1}^2, where

$$\xi_{n1} = \int\int y h_n^{-1} k\left(\frac{v-x}{h_n}\right) f(v,y)dvdy \tag{7.86}$$

$$= \int\int yk(\omega)f(x + \omega h_n, y)d\omega dy \tag{7.87}$$

$$= \int\int yk(\omega)\{f(x,y) + \omega h_n f_{10}(x,y) + \frac{1}{2}\omega^2 h_n^2 f_{20}(x,y) + \ldots\}d\omega dy \tag{7.88}$$

$$= f_1(x)m(x) + h_n \int\int y f_{10}(x,y)\omega k(\omega)d\omega dy + \tag{7.89}$$

$$\frac{1}{2}h_n^2 \int\int y f_{20}(x,y)\omega^2 k^2(\omega)d\omega dy + O(h_n^3) \tag{7.90}$$

$$= f_1(x)m(x) + \frac{1}{2}h_n^2(\int y f_{20}(x,y)dy) + O(h_n^3). \tag{7.91}$$

Similarly, the $\{h_n^{-1}k\left(\frac{X_i - x}{h_n}\right), i = 1, \ldots, n\}$ forms a triangular array of *iid* random variables with mean ξ_{n2} and variance γ_{n2}^2, where

$$\xi_{n2} = Eh_n^{-1}k\left(\frac{X_i - x}{h_n}\right) = \int \frac{1}{h_n}k\left(\frac{v-x}{h_n}\right)f_1(v)dv \tag{7.92}$$

$$= \int k(\omega)f_1(x + h_n\omega)d\omega \tag{7.93}$$

$$= f_1(x) + 0 + \frac{1}{2}h_n^2 f_1''(x) + O(h_n^3). \tag{7.94}$$

Further, by similar manipulations, it can be shown (Chaubey and Sen (1999)) that the asymptotic variances γ_{nj}^2, $j = 1, 2$ as well as the asymptotic covariance of the two terms in the numerator and denominator of (7.6.7) corresponding to the ith observation (X_i, Y_i) are all functionals of the density f and its (partial) derivatives; we leave the details of verification as Problem 7.14. Problem 7.15 is set to verify that at any point x, the estimator in (7.85) is asymptotically normal; note that there is a bias term in this asymptotic normality result that is not ignorable if we let $h_n = O(n^{-1/5})$ (as was the case with the density estimators). Problem 7.16 is set to extend this result to a weak invariance principle for the process $\{n^{2/5}[\tilde{m}_n(x) - m(x)],\ x \in C\}$ where C is a compact interval on $\boldsymbol{R}$. Here also, the bias process needs to be noticed.

7.7 SHAPE CONSTRAINTS: HYPOTHESIS TESTING

In the preceding sections, we have discussed some important estimation problems relating to shape constraints of various types, mostly in nonparametric set-ups; the role of smoothing of constrained estimators has also been explored in this context. The complementary hypothesis testing problems will be discussed in this section. Like in estimation problems, in hypothesis testing problems as well, nonparametrics dominate the scenario. Further, in this combined CSI set-up, to a certain extent, asymptotics may be indispensable. However, for better motivation, we will explore the finite sample avenues, and then, whenever needed, appraise their statistical perspectives through asymptotic methods.

The basic theme in constrained hypothesis testing problems is related to characterizing appropriate *convex ordering*, which in turn characterizes inequality constraints on some functional parameters (or functionals of the underlying survival function) in a well-defined manner, so that *functional data analysis* tools are quite appropriate in this respect. In this context, a basic concept in reliability theory, namely, the *total time on test* (TTT) transformation plays the key role, and this will be adequately explored. Some of the (mostly asymptotic) tests in the context of monotone nonparametric regression problems are also discussed along with.

Consider two life distributions F and G, both defined on $\mathbb{R}^+$, and denote the corresponding survival functions by $S_F(.)$ and $S_G(.)$, respectively. A convex ordering of F and G, denoted by $\prec_c$, is defined by

$$F \prec_c G \text{ if } \int_x^\infty S_F(y)dy \leq \int_x^\infty S_G(y)dy,\ \forall x \geq 0. \tag{7.95}$$

To appraise fully the implications of this convex ordering, we consider first some simpler cases, some of which were discussed, albeit briefly, in earlier sections. The exponential survival function, having many characteristic aging properties (such as constant hazard rate, MRL, and others) forms the nucleus while aging property related constraints provide a nested network of convex nonparametric ordering of distributions. Nevertheless, we shall illustrate the convex ordering with some important parametric life distributions as well.

A life distribution F, defined on $\mathbb{R}^+$, is said to belong to the *new better (worse) than used* (NB(W)U) family if for every $x, y \geq 0$,

$$P\{X > x\} \geq (\leq) P\{X > x + y | X > y\}, \tag{7.96}$$

or equivalently,

$$S_F(x + y) \leq (\geq) S_F(x) S_F(y), \forall\, x, y \geq 0. \tag{7.97}$$

It has been observed in (7.41)–(7.42) that (7.97) is equivalent to the following inequality constraint in terms of the cumulative hazard function $H_F(x) (= \int_o^x h_F(y)dy)$:

$$H_F(x + y) \geq (\leq) H_F(x) + H_F(y),\ \forall\, x, y \geq 0, \tag{7.98}$$

so that a test for such an ordering may be based on either (7.97) or (7.98).

The d.f. F belongs to the class of *new better (worse) than used in expectation* (NB(W)UE), if

$$\int_0^\infty S_F(x + y)dy \leq (\geq) S_F(x) \int_0^\infty S_F(y)dy,\ \forall\, x \geq 0. \tag{7.99}$$

It is easy to verify that

$$NBU \Rightarrow NBUE\ (NWU \Rightarrow NWUE). \tag{7.100}$$

As in (7.61), the MRL at age x is defined as

$$\begin{aligned} \mu(x) &= \{S_F(x)\}^{-1} \int_0^\infty S_F(x + y)dy \\ &= \int_0^\infty e^{-\{H_F(x+y) - H_F(x)\}} dy,\ x \in \mathbb{R}^+. \end{aligned} \tag{7.101}$$

Using this notation, we may restate (7.99) as $\mu(x) \leq (\geq) \mu(0) = \mu$, $\forall x \geq 0$. On the other hand, the D(I)MRL property, introduced after (6.5.1) states that

$$\mu(x) \text{ is nonincreasing (nondecreasing) in } x \in \mathbb{R}^+. \tag{7.102}$$

Having defined the hazard and cumulative hazard functions as in Section 6.4, we introduce the I(D)FR class by the property that

$$h_F(x) \text{ is } \nearrow (\searrow) \text{ in } x \geq 0, \tag{7.103}$$

or, equivalently as $H_F(x)$ convex (concave) on $\mathbb{R}^+$. A life distribution F has *increasing (decreasing) failure rate average* (I(D)FRA) property if $x^{-1}H_F(x)$ is nondecreasing (nonincreasing) in $x \in \mathbb{R}^+$. There are some other aging properties of life distributions; we consider only another class that has come to the limelight of research in the recent past [Sen and Bhattacharjee (1997)].

For a d.f. F belonging to the class of life distributions with finite mean μ_F, we define the *first derived distribution* $TF(x)$ as

$$TF(x) = \mu_F^{-1} \int_0^x S_F(y)dy, \; x \in \mathbb{R}^+. \tag{7.104}$$

Note that by definition, $S_{TF}(x) = \mu_F^{-1} \int_x^\infty S_F(y)dy$, and the corresponding density function is

$$(dTF(x)/dx) = \mu_F^{-1} S_F(x) \searrow \text{ in } x \geq 0. \tag{7.105}$$

The MRL function corresponding to $TF(x)$ is

$$\mu_{TF}(x) = \{S_{TF}(x)\}^{-1} \int_x^\infty S_{TF}(y)dy, \; x \geq 0. \tag{7.106}$$

It is easy to verify (Problem 7.17) that $\mu_{TF}(x) < \infty$ whenever $\mu_F < \infty$. We say that F has *new better (worse) than renewal used in expectation* (NB(W)RUE) property if

$$\mu_F(0) \geq (\leq) \; \mu_{TF}(x), \; \forall \, x \geq 0. \tag{7.107}$$

For the NBRUE property, we have from above

$$\begin{aligned}
\mu_F(0) = \mu_F &= \int_0^\infty S_F(y)dy \geq \{S_{TF}(x)\}^{-1} \int_x^\infty S_{TF}(y)dy \\
&= \frac{\mu_F}{\int_x^\infty S_F(y)dy} \int_x^\infty S_{TF}(y)dy, \; x \geq 0,
\end{aligned} \tag{7.108}$$

so that the NBRUE property rests on the inequality constraint

$$\int_x^\infty S_{TF}(y)dy \; \leq \; \int_x^\infty S_F(y)dy, \; \forall x \geq 0. \tag{7.109}$$

Therefore, we conclude that

$$F \text{ NBRUE} \Rightarrow \; TF \prec_c F. \tag{7.110}$$

A similar characterization holds for the NWRUE class. Combining all these implications, we have the following

$$IFR \Rightarrow IFRA \Rightarrow NBU \Rightarrow NBUE \Rightarrow NBRUE$$
$$IFR \Rightarrow DMRL \Rightarrow NBUE \Rightarrow NBRUE.$$

A similar picture holds for the DFR class, etc. As such, in aging characterizations, the NBRUE class is of special interest. Tests for exponentiality against NBUE class as well as IFR and some other classes have been considered by Hollander and Proschan (1984) and others. Here, we present mainly the NBRUE class, and briefly mention some earlier works on other classes.

Let us introduce the TTT transformation, which plays a basic role in the contemplated constrained hypothesis testing problems. Consider n *iid* nonnegative random variables $X_1, \ldots, X_n$ drawn from a distribution F, defined on $\mathbb{R}^+$, and let $X_{n:1} < \ldots < X_{n:n}$ be the corresponding order statistics; conventionally, we let $X_{n:0} = 0$ and $X_{n:n+1} = \infty$. Note that the empirical survival function $S_n(x) = n^{-1} \sum_{i=1}^n I(X_i > x)$) assumes the value $(n-j)/n$ for $x \in [X_{n:j}, X_{n:j+1}), j = 0, 1, \ldots, n$. Also, define the *spacings* as the inter order statistics distances $X_{n:j} - X_{n:j_1}$, for $j = 1, \ldots, n$. Further, we define the *normalized spacings* as

$$d_{n,k} = (n-k+1)(X_{n:k} - X_{n:k-1}),\ k = 1, \ldots, n. \tag{7.111}$$

The cumulative normalized spacings are then denoted by

$$D_{n,k} = \sum_{j \le k} d_{n,j} = \sum_{j \le k-1} X_{n:j} + (n-k+1)X_{n:k},\ k = 1, \ldots, n; \tag{7.112}$$

conventionally, we let $d_{n,0} = D_{n,0} = 0$. We may then express the sample mean $\bar{X}_n$ as

$$\begin{aligned}\bar{X}_n &= \int_0^\infty S_n(x)dx = \sum_{k=0}^n \int_{X_{n:k}}^{X_{n:k+1}} S_n(x)dx \\ &= n^{-1} \sum_{k=0}^n (n-k)(X_{n:k+1} - X_{n:k}) = n^{-1} D_{n,n}. \end{aligned} \tag{7.113}$$

Thus, $D_{n,n} = n\bar{X}_n$ is the total life time of the n items with individual values $X_1, \ldots, X_n$.

Consider next an interval $[0, t), t \in \mathbb{R}^+$, and let

$$k_n(t) = \max\{k : X_{n:k} \le t\},\ t \in \mathbb{R}^+. \tag{7.114}$$

Then, noting that $X_{n:j}, j \le k_n(t)$ have already failed prior to time t, while the remaining $(n - k_n(t))$ items are still surviving, we conclude that up to the time t, the total time on test is given by

$$D_n(t) = D_{n,k_n(t)} + (n - k_n(t))(t - X_{n:k_n(t)}), \tag{7.115}$$

for all $t \le X_{n:n}$, and $D_n(t) = D_{n,n}$ for $t \ge X_{n:n}$. This observation leads us to define the TTT statistic (process) as

$$D_n(.) = \{D_n(t),\ t \in \mathbb{R}^+\}. \tag{7.116}$$

If the underlying d.f. F is exponential (with a constant hazard rate $\lambda > 0$), then the normalized spacings are *iid* random variables having the same exponential law (Problem 7.18). As a result, $D_{n,n}$ has the gamma (n, λ) distribution. This is a characterization of the exponentiality of F. As such, if the null hypothesis H_0 relates to F being exponential with unknown mean μ, exploiting the sufficiency and completeness of $D_{n,n}$ (under H_0), it seems natural to base tests on the $D_{n,k}$. Also exploiting the *iid* characterization of the $d_{n,k}$, distribution theory of such test statistics may be approximated by their permutation counterparts, although their unconditional form might be quite complicated. For some of the alternatives, such tests have tractable distribution theory for small n, although we will be relying more and more on asymptotic methods, to be discussed here. Such asymptotics provide good approximations for moderately large samples as well.

We consider first tests for exponentiality against NBUE alternatives. Note that the d.f. F belongs to the NBUE class if and only if

$$\xi_F(t) = \mu^{-1}\{\mu S_F(t) - \int_t^\infty S_F(u)du\} \geq 0, \ \forall t \geq 0, \tag{7.117}$$

with strict inequality on a set of measure nonzero; the opposite inequality holds for the NWUE class. Further, $\xi_F(t) = 0, \forall t \in \mathbb{R}^+$ if and only if F is exponential. Therefore, the functional $\xi_F = \{\xi_F(t), t \in \mathbb{R}^+\}$ leads us to formulate various test statistics, under complete or censored life-testing set-ups.

Using the notations for the normalized spacings and their cumulative versions, we see that (Problem 7.19) on replacing the d.f. F by its sample counterpart F_n,

$$\begin{aligned}\hat{\xi}_n(t) &= \xi_{F_n}(t) \\ &= \bar{X}_n^{-1}\{\bar{X}_n S_n(t) - \int_t^\infty S_n(u)du\} \\ &= \{D_n(t)/D_{n,n} - F_n(t)\}, \ t \in [0, X_{n:n}),\end{aligned} \tag{7.118}$$

and it vanishes for $t \geq X_{n:n}$. Note that

$$\hat{\xi}_n(X_{n:k}) = (D_{n,k}/D_{n,n} - k/n), \ k = 0, \ldots, n. \tag{7.119}$$

Noting that $F_n(t)$ is a step function with jumps of magnitude n^{-1} at each $X_{n:k}$ while $D_n(t)$ is continuous and piecewise linear (between successive order statistics $X_{n:k}$) though its first derivative is discontinuous at these order statistics, it seems quite motivating [Koul (1978), Kumazawa (1989); Bhattacharjee and Sen (1995)] to formulate Kolmogorov- Smirnov-type tests. Toward this end, taking into account the one-sided alternative, we let

$$\begin{aligned}K_n^+ &= \sup\{\hat{\xi}_n(t) : t \in \mathbb{R}^+\} \\ &= \max\{\hat{\xi}_n(X_{n:k}) : 0 \leq k \leq n\} \\ &= \max\{D_{n,k}/D_{n,n} - k/n : 1 \leq k \leq n\}.\end{aligned} \tag{7.120}$$

For NWUE alternatives, we use

$$K_n^- = \max\{k/n - D_{n,k}/D_{n,n} : 1 \leq k \leq n\}. \tag{7.121}$$

It was observed by Bhattacharjee and Sen (1995) that as the $d_{n,k}$ are *iid* exponential random variables (under H_0), the joint density of $\boldsymbol{U}_n = D_{n,n}^{-1}(d_{n,1}, \ldots, d_{n,n})'$ is given by

$$(n-1)! \, du_1 \cdots du_n, \ u_j \geq 0, \forall j, \ u_1 + \ldots + u_n = 1, \tag{7.122}$$

which does not depend on the unknown mean of the $d_{n,k}$ (Problem 7.20). Therefore, the tests based on K_n^+ or K_n^- are distribution-free; their distribution can be obtained by direct enumeration, albeit the task can be prohibitively laborious as n increases. For this reason, we present the asymptotic distribution theory briefly.

We introduce a stochastic process $\boldsymbol{W}_n = \{W_n(t),\ t \in [0,1]\}$ by letting

$$W_n(t) = \sqrt{n-1}\{D_{n[nt]} - tD_{nn}\}/D_{nn},\ t \in [0,1], \tag{7.123}$$

where $[s]$ denotes the largest integer $\leq s$. Then, we have

$$\sqrt{n-1}K_n^+ = sup\{W_n(t) : 0 \leq t \leq 1\}, \tag{7.124}$$

and a similar expression holds for K_n^-. Since the u_i are exchangeable random variables and are bounded, the classical permutational (functional) central limit theorem can be readily incorporated, yielding

$$\boldsymbol{W}_n \xrightarrow{\mathcal{D}} \boldsymbol{W}^o, \text{ as } n \to \infty, \tag{7.125}$$

where $\boldsymbol{W}^o$ is a standard Brownian Bridge on $[0,1]$; we leave the proof as Problem 7.21. This enables us to incorporate the general asymptotic theory of the classical Kolmogorov-Smirnov goodness-of-fit test statistic, and to conclude that

$$\lim_{n\to\infty} P\{\sqrt{n-1}K_n^+ \geq \lambda\} = exp\{-2\lambda^2\},\ \forall\, \lambda \geq 0. \tag{7.126}$$

These results have also been extended to Type I, II, and random censoring schemes. We refer to Bhattacharjee and Sen (1995).

Problems

Section 7.2

7.1

(a) Show that when F is continuous, the order statistics in (7.1) are all distinct, with probability one.

(b) Define the empirical d.f. F_n as in (7.2), and verify that F_n is a step function having jumps of the magnitude $1/n$ at each order statistic (and flat in between them). Hence, or otherwise verify (7.3).

(c) Show that for any fixed x, $E\{F_n(x)\} = F(x)$ and $\text{var}\{F_n(x)\} = n^{-1}F(x)[1-F(x)]$. Show further, pointwise, the empirical d.f. is the minimum variance unbiased estimator (MVUE) of F.

(d) Show that F_n has no derivative at the n order statistics, and elsewhere the derivative is identically equal to 0. (e) Viewed as a process on $\mathbb{R}$, is F_n the NPMLE of F?

7.2

(a) (Glivenko-Cantelli theorem). Show that if $F_n(x)$ is the empirical d.f. corresponding to the underlying d.f. $F(x)$, both defined on the real line, then $\sup_x |F_n(x) - F(x)| \to 0$, a.s., as $n \to \infty$..

(b) Use this result and the argument given in Section 7.2, and verify that whenever F is concave,

$$\sup_x \left|\hat{F}_n(x) - F(x)\right| \leqslant \sup_x |F_n(x) - F(x)| \to 0$$

a.s., as $n \to \infty$, where $\hat{F}_n$ is the least concave majorant of the empirical distribution function F_n.

(c) Let $a_n = (n/\log\log n)^{\frac{1}{2}}$. Show that

$$\overline{\lim}\, a_n \sup_x \left|\hat{F}_n(x) - F(x)\right| = 1/\sqrt{2} \quad \text{wp } 1. \tag{7.127}$$

[Hint: A result relating to the exact rate of convergence of $\sup_x |F_n(x) - F(x)|$ is (Csaki, 1968)

$$\overline{\lim}\, a_n \sup_x |F_n(x) - F(x)| = 1/\sqrt{2} \quad \text{wp } 1. \tag{7.128}$$

Use this and (7.9) to deduce

$$\overline{\lim}\, a_n \sup_x \left|\hat{F}_n(x) - F(x)\right| \leqslant 1/\sqrt{2} \quad \text{wp } 1. \tag{7.129}$$

Now, let x_0 be such that $F(x_0) = 1/2$, $Y_i = 1$ or 0 according as $X_i \leqslant x_0$ or $X_i > x_0$, and $Z_i = Y_i - F(x_0)$. By LIL, $\overline{\lim}\{2s_n^2(\log\log s_n)\}^{-1/2}\sum_{i=1}^n Z_i = 1$ wp 1, where $s_n^2 = \text{var}(\sum_{i=1}^n Z_i)$. Deduce that $\overline{\lim}\, a_n\{F_n(x_0) - F(x_0)\} = 1/2 \geqslant 1/\sqrt{2}$ with probability 1, where $a_n = (n/\log\log n)^{1/2}$, by substituting $s_n^2 = n/4$. Now,
$\overline{\lim}\, a_n \sup \left|\hat{F}_n(x) - F(x)\right| \geqslant \overline{\lim}\, a_n[\hat{F}_n(x_0) - F(x_0)] \geqslant$
$\overline{\lim}\, a_n[F_n(x_0) - F(x_0)] \geqslant 1/\sqrt{2}$ with probability 1.
Now combine this with (7.129).]

[Robertson et al. (1988), page 329].

7.3 Consider the histogram-type estimator of f in (7.12) and the isotonic estimator $\tilde{f}_n$ in (7.8). Show that $\tilde{f}_n$ is the isotonic regression of $\hat{f}_n$.

7.4 Integrate by parts the cumulative distribution function of a gamma distribution with shape parameter k, a positive integer, and unit scale parameter, and thereby verify the identity in (7.17). What happens if $k(> 0)$ is not an integer?

7.5 Use the Borel strong law of large numbers and show that if F is strictly monotone at T_p, then (7.29) holds.

Section 7.3

7.6 For a unimodal density with mode M, on the segment $(-\infty, M]$, the density is nondecreasing. Hence, consider a lateral inversion of this segment, i.e., take $g(x) = f(-x)$, for $x \in (-M, \infty)$, and apply the LCM algorithm on $g(.)$ to obtain the isotonic version, and show that it reduces to the GCM version on $f(x),\ x \in (-\infty, M]$.

7.7 Extend the results of the previous problem when the observations are subject to random censoring, and the Kaplan-Meier PLE is used instead of the empirical d.f. F_n.

Section 7.4

7.8 Show that the estimator $\hat{h}_n$ in (7.48) is the isotonic regression of the naive estimator in (7.47).

7.9 Provide a formal proof of the Hille's fundamental lemma. Show that if a function $u(.)$ is bounded and monotone (say, nondecreasing) on $\mathbb{R}$, then it is of bounded variation.

7.10 For the Gamma density $f(x) = \{\Gamma(p)\}^{-1}\lambda^p e^{\lambda x} x^{p-1}$, show that for $p > 1$, the density is strongly unimodal, and hence, $h(x)$ is nondecreasing. However, as $x \to \infty$, $h(x)$ attains its upper asymptote λ. [Hints: Show that the gamma density belongs to the exponential family, so that for large x, $h(x)$ behaves as $-f'(x)/f(x)$. Obtain the expression for $(d/dx)\log f(x)$ for large x, and verify the asymptote.]

Section 7.5

7.11 Consider the raw estimate in (7.70), and show that at each $X_{n:k}$, the left-hand slope is $\leq$ the right-hand slope, though they are both nonnegative. Also, show that for $x \geq X_{n:n}$, as the empirical survival function is equal to zero, the raw MRL is not defined.

7.12 Verify that the use of the smoothed version $S_n^*(.)$ leads to the estimator of the MRL in (7.71). Explain why it could be divergent (as $x \to \infty$) if λ_n is chosen as $n/X_{n:n}$.

7.13 Verify that the smoothed estimator in (7.73) is properly defined for all x

Section 7.6

7.14 Define (u_n, v_n) as in (7.85). Show that as $n \to \infty$, $(nh_n)^{1/2}(u_n - \xi_{n1}, v_n - \xi_{n2})$ has closely a bivariate normal law. [Hint. Apply the bivariate central limit theorem for a triangular scheme of *iid* random vectors.]

7.15 Make use of the asymptotics in the previous problem, and incorporate the Slutzky theorem to show that $(nh_n)^{1/2}(u_n/v_n - m(x))$ is asymptotically normal. By the delta method, obtain the expression for the bias and asymptotic mean squared error for the above asymptotic normal law.

7.16 Extend the asymptotic normality result in the previous problem to cover the weak convergence of the stochastic process $(nh_n)^{1/2}\{\tilde{m}_n(x) - m(x)\}$, $x \in \mathbb{R}^+$, to a Gaussian function. Show that this Gaussian function has the property that at any two distinct point x, y, the covariance term is zero. [Hint. You need to verify that all finite dimensional distributions converge, an extension of the result in the previous problem. Also, you need to verify the tightness or compactness of the processes, a result that requires more delicate stochastic analysis (Sen 1996)].

Section 7.7

7.17 Show that for a life distribution F, if μ_F is finite then defining $TF(.)$ as in (7.104) $\mu_{TF}(x) < \infty$, for every $x \geq 0$.

7.18 Show that if F is an exponential d.f., then defining the normalized spacings $d_{n,k}$ as in (7.111) they are *iid* having the same exponential d.f.

7.19 Define the sample function $\hat{\xi}_n(t)$ as in (7.118), and verify that it is the same functional as $\xi_F(.)$ but with F replaced by F_n.

7.20 Consider the multivariate beta density in (7.122) Show that it is free from the unknown mean of the $d_{n,k}$. Further, show that the elements of $\boldsymbol{U}_n$ are exchangeable, and, hence, the permutational central limit theorem applies to linear functions of functions of these elements.

7.21 Define the stochastic process $W_n(t), t \in [0, 1]$ as in (7.123) and show that it converges weakly to a standard Brownian bridge as n increases. [Hint. Use the permutational functional central limit theorem, Sen (1981).]

8

Bayesian Perspectives

8.1 INTRODUCTION

Bayes methods have permeated in almost all subfields of statistics, and constrained statistical inference (CSI) is no exception. The intricate support of *statistical decision theory* (SDT), nurturing *optimal Bayes* procedures, may, however, need a careful appraisal in the context of CSI. The primary roadblock in this respect stems from the fact that for restricted parametric models (or constrained parameter spaces), conventional *Bayes optimality* properties (including *admissibility*, *minimaxity*, *minimum risk estimability*) may not be universally true, and, hence, there is a genuine need to appraise the SDT aspects in a broader perspective.

In Chapter 5, while appraising the hypothesis testing problem for the multinormal mean vector against positive orthant restricted alternatives, when the dispersion matrix is *nuisance*, for the restricted likelihood ratio test (as well as its competitors), such optimality properties, resting intricately on SDT and Bayes methodology, could not be claimed; in fact, they may not be generally true for CSI (Tsai and Sen (2004)). A similar criticism may be labelled against restricted MLE and other estimators. We intend to present briefly such SDT based procedures in CSI, and discuss why in CSI problems, optimal Bayes procedures are often computationally intractable, and even sometimes, not existent.

There are, however, some dominance results for statistical tests and estimates that percolate through CSI. Variants of Bayes methods, particularly the *empirical Bayes* (EB) and *hierarchical Bayes* (HB) methodology have made significant impact in CSI. Improved estimation in nonregular (and restricted) cases, based on the *Stein phenomenon* (Stein 1956), has led to the evolution of *shrinkage* (or Stein-rule) estimation theory, which bears the support from the frequentists as well as the EB/HB advocates. Even the *Pitman measure of closeness* (PMC), which is often denied a berth in SDT, has created significant impact on CSI. Our treatise of CSI would remain somewhat incomplete without some dissemination of Bayes tests in a broad set-up, as well as SDT-based Bayes methodology and its variants; these are therefore outlined in the concluding sections of the present chapter.

8.2 STATISTICAL DECISION THEORY MOTIVATIONS

Both estimation and hypotheses testing problems are statistical decision problems. Given a (set of) observable random element(s), denoted by X, it is assumed that X has a probability law P that is at least partially unknown. In a parametric set-up, P is indexed by a parameter θ (possibly vector valued) that belongs to a set Θ, termed the *parameter space*, and, hence, a family $\mathcal{P} = \{P_\theta, \theta \in \Theta\}$ of probability laws is conceived. In a broader (possibly beyond parametric) set-up, P is assumed to belong to a class $\mathcal{P}$, the space of probability laws. Our objective is to gather information on Θ (or $\mathcal{P}$) when X is observed . The set of all possible outcomes (of X) is called a sample space $\mathcal{X}$ and realized values X are denoted by x $(\in \mathcal{X})$.

Our task is contemplated through a *decision rule* $d = d(X)$ whose domain is $\mathcal{X}$ and range is the set $\mathcal{A}$ of all possible decisions or *actions*, so that $\mathcal{A}$ is termed the *action* space. We denote by $\mathcal{N}$ the set of all true states of *nature*. [In estimation problems, it could be that $\mathcal{A} \equiv \mathcal{N} \equiv \Theta$.] If the action $d(X)$ chosen on the basis of X matches the true state of nature δ, there is no loss (or error) in our decision; otherwise, we have an incorrect decision made on the basis of X. Therefore, we conceive of a (nonnegative, real valued) *loss function*

$$L(a, b) \text{ on } \mathcal{N} \times \mathcal{A}, \; a \in \mathcal{N}, \; b \in \mathcal{A}, \tag{8.1}$$

such that $L(\delta, d(x))$ represents the cost of possibly incorrect decision made. Since X is a random element with probability law $P(= P_\theta)$, by allowing X to vary over $\mathcal{X}$, governed by P, we arrive at the *risk* or *expected loss* as

$$R(\delta, d(\cdot)) = E_P L(\delta, d(x)) = \int_{\mathcal{X}} L(\delta, d(x)) dP(x). \tag{8.2}$$

In an estimation problem, $d(x)$ is an estimator of θ, so that $\mathcal{A} = \mathcal{N} = \Theta$, and $L(a, b)$ is chosen as a suitable *norm* or *metric*, depicting the distance between a and b, both lying in Θ. For example, if θ is real valued, we may choose

$$L(a, b) = |a - b| \text{ (absolute distance)}, \tag{8.3}$$

$$L(a, b) = (a - b)^2 \text{ (squared distance)}, \tag{8.4}$$

$$L(a, b) = I(|a - b| > c), \text{ for some } c > 0, \tag{8.5}$$

the latter being a zero-one valued loss function. If θ is vector valued, one may take $L(\boldsymbol{a}, \boldsymbol{b}) = \|\boldsymbol{a} - \boldsymbol{b}\|$ (Euclidean norm), or a quadratic norm $(\boldsymbol{a} - \boldsymbol{b})^T \boldsymbol{Q} (\boldsymbol{a} - \boldsymbol{b})$ for a suitable Q, or analogous to (8.5), as $I((\boldsymbol{a}-\boldsymbol{b})^T\boldsymbol{Q}(\boldsymbol{a}-\boldsymbol{b}) \geq c^2)$, for some $c > 0$. [If θ is functional and so is $d(x)$ (viz., the true and sample d.f.), we may need to choose $L(\cdot,\cdot)$ based on suitable metric, such as the uniform metric.] For hypothesis testing and multiple decisions problems, usually $\mathcal{A}$ and $\mathcal{N}$ are taken to be finite sets, and zero-one valued loss functions are more commonly adopted.

We now introduce a third class, $\mathcal{D}$, the class of possible decision rules $\{d(X)\}$. Thus, the risk in (8.2), is now regarded as a function of $\delta \in \mathcal{N}$ and $d \in \mathcal{D}$, and it is generally strictly positive on this space $x \times \mathcal{D}$. A decision rule $d_0 (\in \mathcal{D})$ is said to be the *uniformly* (in $\delta \in \mathcal{N}$) *minimum risk* (UMR) rule, if

$$R(\delta, d_0) = \min_{d \in \mathcal{D}} R(\delta, d), \forall \delta \in \mathcal{N}. \tag{8.6}$$

Another way of seeking an optimal decision rule might be the following. Define

$$R^*(d) = \sup_{d \in \mathcal{N}} R(\delta, d), d \in \mathcal{D}. \tag{8.7}$$

Then, a decision rule d_0 is termed a *minimax rule* if

$$R^* = \min_{d \in \mathcal{D}} R^*(d). \tag{8.8}$$

Generally, UMR decision rules may not exist or may be hard to obtain, and minimax rules are relatively simpler to formulate. In the same vein, we say that, with respect to a chosen risk function $R(\cdot,\cdot)$, a decision rule d_1 is *R-better* than another rule d_2, if $R(\delta, d_1) \leq R(\delta, d_2), \forall \delta \in \mathcal{N}$, with strict inequality for some δ; d_1 is *R-equivalent* to d_2 if $R(\delta, d_1) = R(\delta, d_2), \forall \delta \in \mathcal{N}$. A decision rule d_0 is said to be *admissible* if there is no other decision rule that is R-better than d_0; if there exists an R-better decision rule then d_0 is inadmissible. We refer to Berger (1985) for a treatise of these aspects of SDT.

Bayesian decision rules occupy a focal point in SDT. For simplicity of presentation, we confine ourselves to the estimation problem. In a Bayesian approach, it is assumed that θ is itself a random element with a *prior* distribution $\pi(\theta)$ on Θ; note that θ is *unobservable* but often $\pi(\cdot)$ is assumed to be given (but not dependent on X). The joint probability law for (X, θ) is therefore $p_\theta(x)\pi(\theta)$ (on $\mathcal{X} \times \Theta$) where $p_\theta(\cdot)$ is the density corresponding to p_θ, defined in a meaningful way. The marginal density of X is obtained as $\int_\Theta p_\theta(X)d\pi(\theta) = p_\pi^*(X)$, say. Therefore, the conditional distribution of θ given X has the density

$$p_{\theta|X}^\pi = p_\theta(X)\pi(\theta)/p_\pi^*(X) \tag{8.9}$$

$$= \frac{p_\theta(X)\pi(\theta)}{\int_\Theta p_\theta(X)d\pi(\theta)}, \quad \theta \in \Theta, \ X \in \mathcal{X}. \tag{8.10}$$

This is termed the *posterior density* of θ given X, and, by analogy with the classical Bayes theorem, statistical inference based on such posterior laws, is termed *Bayes inference*.

The *Bayes risk* of an estimator $d(X)$ with respect to a prior $\pi(\cdot)$ on Θ is defined as

$$r(\pi, d) = E^{\pi}[R(\theta, d)] \tag{8.11}$$

$$= \int_{\Theta}\int_{\mathcal{X}} L(\theta, d(x))dP_{\theta}(x)d\pi(\theta). \tag{8.12}$$

An estimator, d_B, minimizing (8.12), is termed a *Bayes estimator* (with respect to the prior π). We make use of (8.10) and rewrite (8.12) as

$$r(\pi, d) = \int_{\Theta}\int_{\mathcal{X}} L(\theta, d(x))dP^{\pi}_{\theta|X}dP^{*}_{\pi}(x) \tag{8.13}$$

$$= \int_{\mathcal{X}}\left[\int_{\Theta} L(\theta, d(x))dP^{\pi}_{\theta|X}\right]dP^{*}_{\pi}(x) \tag{8.14}$$

$$= E\{E[L(\theta, d(x)|X]\} \tag{8.15}$$

where we may need to use the Fubini theorem to be able to interchange the integration order in the penultimate step from (8.15).

Several important results follow from (8.15) and the Rao-Blackwell theorem. First, if θ is real valued and we have the squared error loss: $L(a, b) = \gamma(a)(a - b)^2$, then letting

$$\delta_{\pi}(X) = E\{\gamma(\theta)\theta|X\}/E\{\gamma(\theta)|X\} \tag{8.16}$$

and noting that

$$\begin{aligned}
E^{\pi}\{\gamma(\theta)[\theta - d(X)]^2\} &= E[E\{\gamma(\theta)[\theta - d(X)]^2|\boldsymbol{X}\}] \\
&= E\{E[\gamma(\theta)[\theta - \delta_{\pi}(X)]^2|\boldsymbol{X}] + E[\gamma(\theta)[\delta_{\pi}(X)d(X)]^2|\boldsymbol{X}] \\
&\quad + 2E[\gamma(\theta)(\theta - \delta_{\pi}(X))(\delta_{\pi}(X) - d(X))|\boldsymbol{X}]\} \\
&= E\{E[\gamma(\theta)[\theta - \delta_{\pi}(X)]^2|\boldsymbol{X}]\} + E[E\{\gamma(\theta)[\delta_{\pi}(X) - d(X)]^2|\boldsymbol{X}\}] \\
&\geq E\{\gamma(\theta)(\theta - \delta_{\pi}(X))^2\},\ \forall\, d(\boldsymbol{X}),
\end{aligned}$$

we conclude that

$$\delta_{\pi}(\boldsymbol{X}) \text{ is the Bayes estimator of } \theta. \tag{8.17}$$

For absolute error loss, i.e., $L(a, b) = |a - b|$, it follows that similarly a median of the posterior distribution $p^{\pi}_{\theta|X}$ is the Bayes estimator of θ. It is clear from the above discussion that $\delta_{\pi}(X)$, the Bayes estimator of θ, depends on the prior π as well as the loss function $L(\cdot, \cdot)$. The characterization of the median of the posterior distribution as a Bayes estimator under absolute error loss has an intricate relationship with another notion of "closeness" of estimators, due to Pitman (1937), as extended to the notion of *posterior Pitman closeness* (PPC) by Ghosh and Sen (1991), and we introduce the latter briefly here.

Let $d_1(X)$ and $d_2(X)$ be two estimators of a parameter θ. Then $d_1(\cdot)$ is said to be *Pitman-closer* (PC) than $d_2(\cdot)$ if

$$P_{\theta}\{|d_1(X) - \theta| \leq |d_2(X) - \theta|\ \} \geq \frac{1}{2}, \forall\, \theta \in \Theta, \tag{8.18}$$

with strict inequality for some $\theta \in \Theta$. For a class $\mathcal{D}$ of estimators of θ, if (8.18) holds for all $d_2(X) \in \mathcal{D}$, then $d_1(X)$ is said to be the Pitman-closest estimator in the class $\mathcal{D}$. We define the Posterior distribution of θ, given X, $p^{\pi}_{\theta|X}$, as in (8.10). Then $d_1(\cdot)$, is said to be Posterior-Pitman-Closer than $d_2(\cdot)$, under π, if

$$P_{\pi}\{|d_1(X) - \theta| \leq |d_2(X) - \theta| \,|X\} \geq \frac{1}{2}, \; \forall\, X \in \mathcal{X}, \tag{8.19}$$

with strict inequality for some $X \in \mathcal{X}$. The usual PC criterion has sometimes been criticized from SDT perspectives; possible nontransitivity and nonuniqueness being the major target of this criticism. As a contrast, the PPC is free from both this shortcoming. Ghosh and Sen (1991) have shown that a posterior median is a PPC estimator of the parameter of interest, and further, if the posterior d.f. is continuous, this is the unique PP-closest estimator. Thus, the posterior median has not only the Bayes estimator characterization under absolute error loss but also is the posterior Pitman closest estimator whenever the posterior distribution of θ, given X, under π, is continuous.

For the characterization of uniformly minimum variance unbiased estimators (UMVUE), the classical Rao-Blackwell theorem incorporates sufficiency (and completeness) along with unbiased estimators (which need not be sufficient statistics). For the PPC characterization, in a similar set-up, *median-unbiased* estimators play a basic role. Recall that an estimator $\hat{\theta}$ is median-unbiased for θ, if

$$P_{\theta}\{\hat{\theta} < \theta\} \leq \frac{1}{2} \leq P_{\theta}\{\hat{\theta} \leq \theta\}, \forall\, \theta. \tag{8.20}$$

The following theorem is due to Ghosh and Sen (1989).

Theorem 8.2.1 *Let $\tilde{\theta}$ be a median-unbiased estimator of θ and consider the class C of all statistics of the form $V = \tilde{\theta} + Z$, where $\tilde{\theta}$ and Z are stochastically independent. Then for all $\theta \in \Theta$ and $V \in C$,*

$$P_{\theta}\{|V - \theta| \geq |\tilde{\theta} - \theta|\} \geq \frac{1}{2}\,, \tag{8.21}$$

i.e., $\tilde{\theta}$ is the PC estimator of θ within the class C.

Proof. Note that the left-hand side of (8.20) is equivalent to

$$P_{\theta}\{(\tilde{\theta} - \theta)^2 \leq (V - \theta)^2\} = P_{\theta}\{(\tilde{\theta} - \theta)^2 \leq (\tilde{\theta} + Z - \theta)^2\} \tag{8.22}$$

$$= P_{\theta}\{0 \leq Z^2 + 2Z(\tilde{\theta} - \theta)\} \tag{8.23}$$

$$\geq P_{\theta}\{Z(\tilde{\theta} - \theta) \geq 0\}. \tag{8.24}$$

Since Z and $\tilde{\theta}$ are assumed to be independent,

$$P_{\theta}\{\tilde{\theta} \geq \theta | Z\} = P_{\theta}\{\tilde{\theta} \geq \theta\} \geq \frac{1}{2}, \text{ by (8.20)}, \tag{8.25}$$

and hence the result follows from (8.24) and (8.25).

Since Theorem (8.1) is not confined to estimation problems that admit an equivariant structure, it follows that a Pitman-closest location-invariant estimator dominates every equivariant estimator, which is thus PC-inadmissible. Ghosh and Sen (1989) have shown that the independence of Z and $\tilde{\theta}$ is not crucial and can be replaced by a less restrictive condition that the conditional distribution of $\tilde{\theta}$, given Z, is symmetric about θ. As an illustration, consider the case of two $r.v.'s$ X_1, X_2, both having Cauchy density with location parameter θ and scale parameter 1. Let $\tilde{\theta} = \frac{1}{2}(X_1+X_2)$ and $Z = \frac{1}{2}(X_1 - X_2)$. Note that $\tilde{\theta} \sim$ Cauchy $(\theta, 1), X_1 \sim$ Cauchy $(\theta, 1)$ and $Z \sim$ Cauchy $(0, 1)$, and further $X_1 = \tilde{\theta} + Z$. It is easy to verify that the conditional distribution of $\tilde{\theta}$, given Z, is symmetric about θ, and $\tilde{\theta}$ is median-unbiased for θ. Problem 8.2.1 is set to verify this result. Hence, by Theorem (8.1), $\tilde{\theta}$ is Pitman closer than X_1, though marginally both have the same (Cauchy $(\theta, 1)$) density. Sen and Saleh (1992) have further relaxed the symmetry of the conditional distribution of $\tilde{\theta}$, given Z, to that of conditionally median-unbiased for θ, given Z. Thus, median-unbiasedness and median of the posterior distribution of θ, given Z, are quite characterizing for PC dominance results.

These findings are somewhat different in multiparameter settings, as well as, in CSI where equivalence, transitivity, etc. may not hold or play a basic role in Bayes estimation based on PCC. Nevertheless, PC dominance and quadratic risk, dominance results can be studied under appropriate regularity conditions, and comparison of estimators can be made in a meaningful manner. A basic observation of Stein (1956) (on inadmissibility of MLE) has an evolutionary impact in this field of research, and in the rest of this chapter, we provide an outline of these developments pertinent to CSI.

8.3 STEIN'S PARADOX AND SHRINKAGE ESTIMATION

For estimating the mean of a p-variate normal distribution under quadratic loss, Stein (1956) established that for $p \geq 3$, the sample mean (the best invariant and ML estimator) is inadmissible. This is known as the *Stein paradox*. Let us appraise the impact of the Stein paradox on improved estimation in a multiparameter set-up.

Let $\boldsymbol{X} \sim N_p(\boldsymbol{\theta}, \Sigma)$ and let $L(\boldsymbol{a}, \boldsymbol{b}) = (\boldsymbol{a} - \boldsymbol{b})^T\Sigma^{-1}(\boldsymbol{a} - \boldsymbol{b})$, the standardized quadratic loss. We let $\boldsymbol{\delta}_0 = \boldsymbol{X}$, and note that the risk of $\boldsymbol{\delta}_0$ (for estimating $\boldsymbol{\theta}$) is

$$R(\boldsymbol{\theta}, \boldsymbol{\delta}_0) = E_{\boldsymbol{\theta}}L(\boldsymbol{\theta}, \boldsymbol{\delta}_0) = p, \ \forall \boldsymbol{\theta} \in \mathbb{R}^p. \tag{8.26}$$

Stein (1956) showed that by considering estimators of the form

$$\boldsymbol{\delta}_{ab}(\boldsymbol{X}) = \{1 - b(a + \boldsymbol{X}^T\boldsymbol{\Sigma}^{-1}\boldsymbol{X})^{-1}\}\boldsymbol{X}, \ a > 0, b > 0, \tag{8.27}$$

for a suitable range of (a, b), the risk of $\boldsymbol{\delta}_{ab}$ is smaller than that of $\boldsymbol{\delta}_0$, so that the MLE $\boldsymbol{\delta}_0$ becomes *inadmissible* for $p \geq 3$. In particular, James and Stein (1961) considered a specific estimator $\boldsymbol{\delta}_{JS} = \{1 - (p-2)(\boldsymbol{X}^T\boldsymbol{\Sigma}^{-1}\boldsymbol{X})^{-1}\}\boldsymbol{X}$, and showed that under the standardized quadratic loss $L(\cdot, \cdot)$, $R(\boldsymbol{\theta}, \boldsymbol{\delta}_{JS}) \leq R(\boldsymbol{\theta}, \boldsymbol{\delta}_0)$, $\forall \, \boldsymbol{\theta} \in \mathbb{R}^p$, with strict inequality for small $||\boldsymbol{\theta}||$. We provide an outline of this dominance result for motivating subsequent findings in this chapter.

Note that by definition:

$$\boldsymbol{\theta} - \boldsymbol{\delta}_{JS} = \boldsymbol{\theta} - \boldsymbol{X} + (p-2)(\boldsymbol{X}^T\boldsymbol{\Sigma}^{-1}\boldsymbol{X})^{-1}\boldsymbol{X}. \tag{8.28}$$

Also, we let $\Delta = \boldsymbol{\theta}^T\boldsymbol{\Sigma}^{-1}\boldsymbol{\theta}$, and for simplicity of derivation (and without loss of generality) we let $\boldsymbol{\Sigma} = I$ and $\boldsymbol{\theta} = (\theta_1, \mathbf{0}^T)^T$ where $\theta_1 = \sqrt{\Delta}$. Then, we have

$$\begin{aligned}(\boldsymbol{\theta} - \boldsymbol{\delta}_{JS})^T(\boldsymbol{\theta} - \boldsymbol{\delta}_{JS}) =& (\boldsymbol{\theta} - \boldsymbol{X})^T(\boldsymbol{\theta} - \boldsymbol{X}) + (p-2)^2(\boldsymbol{X}^T\boldsymbol{X})^{-1} - \\ & 2(p-2)(\boldsymbol{X} - \boldsymbol{\theta})^T\boldsymbol{X}(\boldsymbol{X}^T\boldsymbol{X})^{-1}.\end{aligned} \tag{8.29}$$

Next, we note that $\boldsymbol{X}^T\boldsymbol{X}$ has noncentral chi square distribution with p degrees of freedom (DF) and noncentrality parameter Δ. Hence, invoking this distributional form (Problem 8.26), we obtain that for $p \geq 3$,

$$E_{\boldsymbol{\theta}}(\boldsymbol{X}^T\boldsymbol{X})^{-1} = e^{-\Delta/2}\Sigma_{r\geq 0}(\Delta/2)^r(r!)^{-1}(p-2+2r)^{-1}. \tag{8.30}$$

It is easy to verify that (8.30) is decreasing in $\Delta(\geq 0)$; at $\Delta = 0$, it is equal to $(p-2)^{-1}$ and it goes to 0 as $\Delta \to \infty$. As a matter of fact, if we denote by K a random variable having the Poisson distribution with parameter $\lambda = \Delta/2$, then we may rewrite

$$E_{\boldsymbol{\theta}}(\boldsymbol{X}^T\boldsymbol{X})^{-1} = E(p-2+2K)^{-1}. \tag{8.31}$$

Problem (8.27) is set to verify that for every integer $r \geq 0$ and $p \geq 1 + 2r$,

$$E_\theta(\boldsymbol{X}^T\boldsymbol{X})^{-r} = E\{(p-2+2K)\cdots(p-2r+2K)\}^{-1}. \tag{8.32}$$

To handle the last term on the right side of (8.29), we consider a simplified version of the famous *Stein identity*.

Lemma 8.3.1 *Let $\boldsymbol{X} \sim N_p(\boldsymbol{\theta}, I)$ and $K \sim$ Poisson$(\Delta/2)$, $\Delta = \boldsymbol{\theta}^T\boldsymbol{\theta}$. Without loss of generality, take $\boldsymbol{\theta} = (\theta_1, \mathbf{0}^T)^T$ where $\theta_1^2 = \Delta$. Then, for $p \geq 3$,*

$$E_{\boldsymbol{\theta}}\{(\boldsymbol{X}^T\boldsymbol{X})^{-1}\boldsymbol{X}^T(\boldsymbol{X} - \boldsymbol{\theta})\} = (p-2)E_{\boldsymbol{\theta}}(\boldsymbol{X}^T\boldsymbol{X})^{-1}. \tag{8.33}$$

Proof. Note that

$$(\boldsymbol{X}^T\boldsymbol{X})^{-1}\boldsymbol{X}^T(\boldsymbol{X} - \boldsymbol{\theta}) = 1 - \theta_1^2(\boldsymbol{X}^T\boldsymbol{X})^{-1} - \theta_1(X_1 - \theta_1)(\boldsymbol{X}^T\boldsymbol{X})^{-1}.$$

Further note that

$$\begin{aligned}&E((X_1 - \theta_1)(\boldsymbol{X}^T\boldsymbol{X})^{-1}\} = (\partial/\partial\theta_1)E\{(\boldsymbol{X}^T\boldsymbol{X})^{-1}\} \\ &= 2\theta_1(\partial/\partial\Delta)E\{(\boldsymbol{X}^T\boldsymbol{X})^{-1}\} \\ &= 2\theta_1\left\{-\frac{1}{2}E\left(\frac{1}{p-2+2K}\right) + \frac{1}{2}E\left(\frac{1}{p+2K}\right)\right\} \\ &= \theta_1 E\left\{\frac{1}{p+2K} - \frac{1}{p-2+2K}\right\}.\end{aligned}$$

Thus,

$$\begin{aligned}
&E_{\boldsymbol{\theta}}\left\{(\boldsymbol{X}^T\boldsymbol{X})^{-1}\boldsymbol{X}^T(\boldsymbol{X}-\boldsymbol{\theta})\right\}\\
&= 1-\Delta E\left(\frac{1}{p-2+2k}\right)-\Delta E\left\{\frac{1}{p+2k}-\frac{1}{p-2+2k}\right\}\\
&= 1-\Delta E\left\{\frac{1}{p+2K}\right\} = e^{-\frac{1}{2}\Delta}\Sigma_{r\geq 0}(\frac{1}{2}\Delta)^r\frac{1}{r!}\left\{1-\frac{\Delta}{p+2r}\right\}\\
&= e^{-\frac{1}{2}\Delta}\Sigma_{r\geq 0}\left(\frac{1}{2}\Delta\right)^r\frac{1}{r!}\left\{1-\frac{2r}{p+2r-2}\right\}\\
&= (p-2)e^{-\frac{1}{2}\Delta}\Sigma_{r\geq 0}\left(\frac{1}{2}\Delta\right)^r\frac{1}{r!}\frac{1}{p-2+2r}\\
&= (p-2)E(\boldsymbol{X}^TX)^{-1}.
\end{aligned}$$

■

From (8.29), (8.31), and (8.33), we obtain that

$$R(\boldsymbol{\theta},\boldsymbol{\delta}_{JS}) = p+(p-2)^2E\left(\frac{1}{p-2+2K}\right)-2(p-2)^2E\left(\frac{1}{p-2+2K}\right) \tag{8.34}$$

$$= p-(p-2)^2E\left\{\frac{1}{p-2+2K}\right\} \tag{8.35}$$

$$= 2+(p-2)\left[1-E\left\{\frac{p-2}{p-2+2K}\right\}\right]. \tag{8.36}$$

This shows that at $\Delta = 0$ (i.e., $K = 0$ with probability 1), (8.36) is equal to 2 and it monotonically increases with Δ increasing; its upper bound is p, the same as in (8.26). For $p \geq 3$, we may plot the relative risk picture for $\varphi_{p,\Delta} = R(\boldsymbol{\theta},\boldsymbol{\delta}_{JS})/R(\boldsymbol{\theta},\boldsymbol{\delta}_0)$ when $\boldsymbol{\theta}^T\boldsymbol{\theta} = \Delta$ (see Fig.8.1).

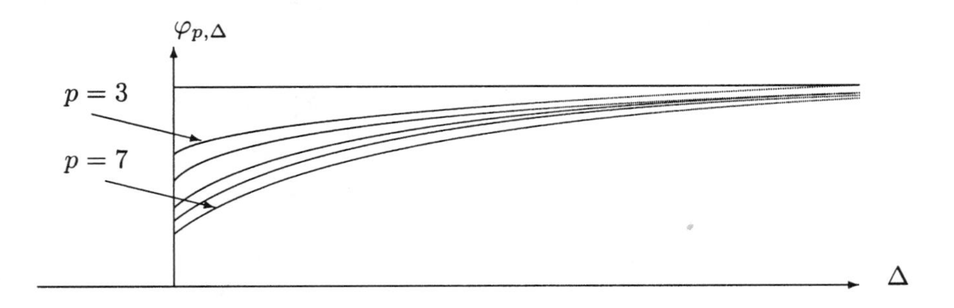

Fig. 8.1 The relative risk as a function of Δ

Clearly, for large p, the risk reduction is more, and the curves are clearly ordered. Actually, one may consider a $\boldsymbol{\delta}_{JS}$ type estimator wherein $(p-2)$ is replaced by

$c(\geq 0)$. As long as $c \in (0, 2(p-2))$, the estimator dominates $\boldsymbol{\delta}_0$ in risk, and the maximum dominance is perceived at $c = p - 2$. In any case, $\boldsymbol{\delta}_0$ being dominated by $\boldsymbol{\delta}_{JS}$, for $p \geq 3$, is inadmissible.

In a sense, the estimators of the type $\boldsymbol{\delta}_{JS}$ possess the property of *shrinking* the original estimator ($\boldsymbol{X}$) toward the origin $\mathbf{0}$, termed the *pivot*. For this reason, they are called *shrinkage estimators*. The dominance holds a.e., though it is more perceptible near the pivot. In fact, the pivot can be chosen to be any point, say $\boldsymbol{\theta}_0$, or even a lower dimensional subspace of $\mathbb{R}^p$. For example, one may take

$$\boldsymbol{\delta}_{JS}^{(1)} = \boldsymbol{\theta}_0 + \{1 - (p-2) \parallel \boldsymbol{X} - \boldsymbol{\theta}_0 \parallel_{\boldsymbol{\Sigma}}^{-1}\}(\boldsymbol{X} - \boldsymbol{\theta}_0) \tag{8.37}$$

or the Lindley's form

$$\boldsymbol{\delta}_{JS}^{(2)} = \bar{\boldsymbol{X}} + \{1 - (p-3) \parallel \boldsymbol{X} - \bar{\boldsymbol{X}} \parallel_{\boldsymbol{\Sigma}}^{-1}\}(\boldsymbol{X} - \bar{\boldsymbol{\theta}}), \qquad p \geq 4, \tag{8.38}$$

where $\bar{\boldsymbol{X}} = \{(X_1 + \cdots + X_p)/p\}\mathbf{1}$.

When $\boldsymbol{\Sigma}$ is unknown, one may use the following shrinkage estimator

$$\boldsymbol{\delta}_{JS}^{*} = \{1 - \frac{p-2}{n-p+1}(\boldsymbol{X}^T\boldsymbol{S}^{-1}\boldsymbol{X})^{-1}\}\boldsymbol{X} \tag{8.39}$$

where $\boldsymbol{S} \sim$ Wishart $(n-1, p, \boldsymbol{\Sigma})$ is independent of $\boldsymbol{X}$. The risk function of the above estimator is

$$R(\boldsymbol{\theta}, \boldsymbol{\delta}_{JS}^{*}) = p - \frac{n-p+1}{n-p+3}(p-2)^2 E_{\boldsymbol{\theta}}(\boldsymbol{X}^T\boldsymbol{\Sigma}^{-1}\boldsymbol{X})^{-1}. \tag{8.40}$$

The estimator in (8.39) may also be modified as in (8.37)–(8.38). In fact, Baranchik (1970) and Strawderman (1971) obtained a broader class of minimax estimators for the multivariate normal mean; for intended brevity, we omit these details.

Let us try to provide some other interpretations to the shrinkage estimators. Again, we confine ourselves to the normal mean vector $\boldsymbol{\theta}$ estimation problem, and consider first the case of $\boldsymbol{\Sigma} = \boldsymbol{I}$. Let $\pi(\boldsymbol{\theta})$ be a prior distribution for $\boldsymbol{\theta}$ and define the Bayes estimator $\boldsymbol{\delta}_\pi(\boldsymbol{X})$ of $\boldsymbol{\theta}$ as in (8.16)–(8.17) (but extended to the vector case). Under quadratic error loss function, we obtain that

$$\boldsymbol{\delta}_\pi(\boldsymbol{X}) = E^\pi(\boldsymbol{\theta}|\boldsymbol{X}). \tag{8.41}$$

Now consider the case where the prior $\pi(\boldsymbol{\theta})$ is itself a multinormal law with mean vector $\boldsymbol{\mu}$ and dispersion matrix $c\boldsymbol{I}$, for some $c > 0$. Then the marginal distribution $p_\pi^*(\boldsymbol{X})$ is the convolution of $p_{\boldsymbol{\theta}}(\boldsymbol{X})$ and $\pi(\boldsymbol{\theta})$ and is thereby multinormal with mean vector $\boldsymbol{\mu}$ and dispersion matrix $(1+c)\boldsymbol{I}$. As such, the posterior distribution of $\boldsymbol{\theta}$, given $\boldsymbol{X}$, is normal with mean vector $\boldsymbol{\mu} + \frac{c}{1+c}(\boldsymbol{X} - \boldsymbol{\mu})$ and dispersion matrix $\frac{c}{1+c}\boldsymbol{I}$; Problem 8.3.3 is set to verify this. As a result, by (8.41), we obtain that

$$\boldsymbol{\delta}_\pi(\boldsymbol{\mu}) = \boldsymbol{\mu} + (1 - \frac{1}{1+c})(\boldsymbol{X} - \boldsymbol{\mu}). \tag{8.42}$$

Since $\boldsymbol{\mu}$ and c for the prior $\pi(\cdot)$ may not be known, one may like to substitute some sensible estimators for them. These estimators may themselves be chosen from the data, possibly from the marginal distribution of $\boldsymbol{X}$. It follows that $\boldsymbol{X}$ has marginally multinormal law with mean vector $\boldsymbol{\mu}$ and dispersion matrix $(1+c)\boldsymbol{I}$, and hence, $\| \boldsymbol{X} - \boldsymbol{\mu} \|^2 /(1+c) \sim \chi_p^2$; Problem 8.3.4 is set to verify this. Therefore

$$E\{\| \boldsymbol{X} - \boldsymbol{\mu} \|^{-2}\} = \frac{1}{(p-2)(1+c)} \tag{8.43}$$

so that $\dfrac{(p-2)}{\| \boldsymbol{X} - \boldsymbol{\mu} \|^2}$ unbiasedly estimates $\dfrac{1}{1+c}$. On substitution in (8.42), we obtain an *empirical Bayes estimator* of $\boldsymbol{\theta}$;

$$\boldsymbol{\mu} + \{1 - \frac{p-2}{\| \boldsymbol{X} - \boldsymbol{\mu} \|^2}\}(\boldsymbol{X} - \boldsymbol{\mu}), \tag{8.44}$$

which is the James-Stein estimator in (8.37) when we take $\boldsymbol{\theta}_0 = \boldsymbol{\mu}$; $\boldsymbol{\delta}_{JS}$ actually takes $\boldsymbol{\mu} = \mathbf{0}$. If instead of $\boldsymbol{\Sigma} = I$, had we worked with $\boldsymbol{\Sigma}$ (known), the above argument would have given us the estimator in (8.37) with $\boldsymbol{\theta}_0 = \boldsymbol{\mu}$. The estimator in (8.39) would have similarly arisen from using the marginal distribution of $\boldsymbol{X}$ so that one could estimate $\boldsymbol{\Sigma}$ and c from the $(\boldsymbol{X} - \boldsymbol{\mu})(\boldsymbol{X} - \boldsymbol{\mu})^T$ and the distribution of $\boldsymbol{S}$. We refer to Efron and Morris (1973) for an excellent treatise of empirical Bayes methods in the light of shrinkage estimation theory.

In (8.18)–(8.19), we have outlined the basic motivation and rationality of the Pitman measure of closeness of estimators, in the frequentist as well as Bayesian perspectives. These measures have been developed for multiparameter models as well. So far in this section, we have motivated shrinkage estimation (also known as *Stein-rule* estimation) based on risk dominance with suitable quadratic loss functions. There is a natural query: does the dominance of the shrinkage estimators (over the MLE or conventional ones) hold under the PCC? Or, in other words, defining the distance between an estimator $\hat{\boldsymbol{\delta}}$ and the parameter $\boldsymbol{\theta}$ by a norm, say $L(\hat{\boldsymbol{\delta}}, \boldsymbol{\theta})$ can it be shown that

$$P\{L(\hat{\boldsymbol{\delta}}_{JS}, \theta) < L(\hat{\boldsymbol{\delta}}_0, \boldsymbol{\theta})\} + \frac{1}{2}P\{L(\hat{\boldsymbol{\delta}}_{JS}, \theta) = L(\hat{\boldsymbol{\delta}}_0, \boldsymbol{\theta})\} \text{ is (or not) } \geq \frac{1}{2}, \forall \boldsymbol{\theta}. \tag{8.45}$$

(and a similar PPC result for shrinkage estimators)? In some specific cases, an affirmative answer has been provided (viz., Sen, Kubokawa and Saleh (1989)). We intend to present some of these results for constrained statistical inference problems in a later section. The PCC may even allow $p \geq 2$ (instead of $p \geq 3$) for such dominance results to hold.

8.4 CONSTRAINED SHRINKAGE ESTIMATION

In the preceding chapters, excepting the one dealing with shape-constrained inference, a greater emphasis has been laid down on the hypothesis testing aspects. The general decision theoretic formulations, presented in the previous two sections, for unconstrained parameter spaces, enable us to present the estimation theory for constrained parameter spaces on an equal footing. This is explored systematically in this as well as later sections in this chapter.

We are primarily motivated by the risk dominance results in the preceding sections, albeit being aware that in a restricted parameter space, a parameter θ, to be estimated, may not belong to the interior of the space or satisfy certain regularity assumptions under which the classical MLE has some optimal properties, at least in appropriate asymptotic set-ups. The situation, as we have seen in shape-constrained problems, can be even more complex when dealing with nonparametric or semiparametric models. In multiparameter estimation theory, the advent of *Stein-rule estimators* (SRE) has clearly demonstrated the decision theoretic limitations of the classical MLE (which we shall term, in the sequel, as unrestricted MLE, or UMLE for short) and placed the corresponding SRE versions, to be termed SRMLE, in a risk-wise better standing. We would like to appraise this picture for constrained parameter spaces.

In the unrestricted case, a SRMLE of $\boldsymbol{\theta}$ dominates (in risk) the MLE; the choice of the pivot $\boldsymbol{\theta}_0$ (possibly a subspace of $\boldsymbol{\Theta}$) has an important role in SRE, and is taken to be an inner point (or subspace) of $\boldsymbol{\Theta}$ (quite comparable to the hypothesis testing problem where under the null as well as alternative hypotheses, $\boldsymbol{\theta}$ is taken to be an inner point of $\boldsymbol{\Theta}$). In a restricted parameter space, for hypothesis testing problems, we have seen that $\boldsymbol{\theta}$ may typically lie on (or near) the boundary of the restricted parameter space. As such, most of the optimality properties of *LRT*, etc., may not permeate in the restricted case. Similarly, in SRE in restricted parameter spaces, a pivot $\boldsymbol{\theta}_0$ may not be an inner point of the parameter space. On the same count, (restricted) RMLE may not automatically retain optimality properties even when their UMLE counterparts are optimal for the unrestricted model. Generally, RMLE performs better than the UMLE on the restricted parameter space, albeit it might not possess general optimality properties. In the same way as UMLE can be risk-dominated by SRMLE, one may like to explore the situation with RMLE versus SRRMLE for restricted parameter spaces. A comprehensive study of this dominance picture needs to focus on the risk of all the UMLE, RMLE, SRE as well as their restricted versions.

In Chapters 3 to 5, while considering *LRT* (and UIT) for a general class of restricted parameter space hypotheses, we have shown how the positive orthant space provides a general class covering a number of important practical applications. Along the same line, here also we will consider in detail the positive orthant parameter space in a multivariate normal mean estimation problem, and then briefly refer to some other cases. As such, we shall borrow the notation as far as possible from earlier chapters and avoid repetition. Specifically, we follow the notation introduced in Section 5.2.1 where the dual hypothesis testing problem has been considered.

Let $\boldsymbol{X}_1, \ldots, \boldsymbol{X}_n$ be n *iid* observations from a p-variate normal distribution with mean vector $\boldsymbol{\theta}$ and dispersion matrix $\boldsymbol{\Sigma}$, assumed to be positive definite. Although $\boldsymbol{\theta} \in \Theta = \mathbb{R}^p$, we are interested in a restricted parameter space Θ^+ and we choose (as explained earlier).

$$\Theta^+ = \mathbb{R}^{+p} = \{\boldsymbol{\theta} \in \mathbb{R}^p : \theta_j \geq 0, 1 \leq j \leq p\}. \tag{8.46}$$

Let $\bar{\boldsymbol{X}}_n = n^{-1}\sum_{i=1}^n \boldsymbol{X}_i$ and $\boldsymbol{S}_n = \sum_{i=1}^n (\boldsymbol{X}_i - \bar{\boldsymbol{X}}_n)(\boldsymbol{X}_i - \bar{\boldsymbol{X}}_n)$ be the sample mean vector and sum of product matrix. Thus, $\bar{\boldsymbol{X}}_n$ is the UMLE of $\boldsymbol{\theta}$; it is unbiased, but for $p \geq 3$, it is not admissible. For the RMLE, we follow the same partition system of P as in (5.9) through (5.12) in Section 5.2.1; see (5.9)–(5.12) for the notation $\bar{\boldsymbol{X}}_{na:a'}$, $\boldsymbol{S}_{naa:a'}$ and $\boldsymbol{I}_{na}$, for every $a : \phi \subset a \subset P$. Also, for every $a : \phi \subset a \subset P$ such that $|a| = r (0 \leq r \leq p)$, if $\boldsymbol{u}$ and $\boldsymbol{v}$ are r- and $(p-r)$-vectors, respectively, we denote by

$$[\mathcal{P}_a(\boldsymbol{u}, \boldsymbol{v})]_i = u_i \text{ or } v_i \text{ according as } i \in a \text{ or } i \in a'. \tag{8.47}$$

Then, the RMLE of $\boldsymbol{\theta}$, restricted to Θ^+, is given by

$$\hat{\boldsymbol{\theta}}^*_{RM} = \sum_{\phi \subset a \subset P} \mathcal{P}_a(\bar{\boldsymbol{X}}_{na:a'}, \mathbf{0}) I_{na}. \tag{8.48}$$

[If $\boldsymbol{\Sigma}$ were known, in the partitioning etc., $\boldsymbol{S}_n$ would be replaced by $\boldsymbol{\Sigma}$.]

Note that in (8.48) only *one* of the 2^p indicator functions I_{na} is 1 while the rest zero. Hence the RMLE $\hat{\boldsymbol{\theta}}^*_{RM}$ is actually expressible in terms of a single (random) term in this closed form. Further, the UMLE $\bar{\boldsymbol{X}}_n$ is affine equivariant (under any nonsingular transformation on the $\boldsymbol{X}_i$); the RMLE does not enjoy this affine equivariance property mainly due to the fact that $\boldsymbol{\Theta}^+$ is not invariant with respect to such arbitrary nonsingular transformations. The computational algorithms for the RMLE are the same as in the case of *LRT*/UIT discussed in Section 5.2.1. The main utility of (8.48) is that it provides a natural motivation for the SRE along the same line as in the previous section; we need to introduce the risk dominance for Θ^+ as well. For this, we consider the quadratic loss function $L(\boldsymbol{a}, \boldsymbol{b}) = (\boldsymbol{a} - \boldsymbol{b})^T \boldsymbol{\Sigma}^{-1} (\boldsymbol{a} - \boldsymbol{b})$ but allow $\boldsymbol{a}$ and $\boldsymbol{b}$ to be only in $\mathbb{R}^{+p}$.

If we look at the RMLE in (8.48), we may observe that the partition of $\mathbb{R}^p$ into 2^p subsets $a : \phi \subset a \subset P$ enables us to identify (in view) of possible dependence among the coordinates of the MLE $\bar{\boldsymbol{X}}$) the direction in which positivity of the components is highlighted. As such, it might be better to incorporate the same partitioning to trim the extent of shrinkage depending on the RMLE pertaining to that specific subset. Motivated by the James-Stein estimate and the above feature of the RMLE, Sengupta and Sen (1991) considered the following SRRMLE of $\boldsymbol{\theta}$ (on Θ^+):

$$\hat{\boldsymbol{\theta}}_{RSM} = \sum_{\{\phi \subset a \subset P\}} I_{na} \mathcal{P}_a(\bar{\boldsymbol{X}}_{na:a'}, \mathbf{0}) \{1 - \frac{c_a}{\| \hat{\boldsymbol{\theta}}^*_{RM} \|^2_{\boldsymbol{S}_n}}\} \tag{8.49}$$

where

$$0 \leq c_a \leq \frac{2(|a| - 2)^+}{n - p + 2}, q^+ = q \vee 0, \phi \subset a \subset P, \tag{8.50}$$

and by (8.48),

$$\| \hat{\theta}^*_{RM} \|^2_{\boldsymbol{S}_n} = (\bar{\boldsymbol{X}}_{na:a'})^T \boldsymbol{S}^{-1}_{naa:a'} (\bar{\boldsymbol{X}}_{na:a'}), \tag{8.51}$$

for $\phi \subset a \subset P$.

The SRRMLE in (8.49) may sometimes have an *overshrinking* feature (when $\| \hat{\boldsymbol{\theta}}^*_{RM} \|^2_{\boldsymbol{S}_n}$ is small, so that the shrinkage factor $1 - c_a \| \hat{\boldsymbol{\theta}}^*_{RM} \|^{-2}_{\boldsymbol{S}_n}$ can be negative, yielding possibly negative coordinates of $\hat{\theta}_{RSM}$—a contradiction to its placement in Θ^+. Problem 8.4.1 is set to verify this result. Hence, Sengupta and Sen (1991) advocated the use of the *positive rule shrinkage* version:

$$\hat{\boldsymbol{\theta}}^*_{RSM} = \sum_{\{\phi \subset a \subset P\}} I_{na} \mathcal{P}_a(\bar{\boldsymbol{X}}_{na:a'}, \mathbf{0}) \{1 - c_a \| \hat{\boldsymbol{\theta}}^*_{RM} \|^{-2}_{\boldsymbol{S}_n}\}^+. \tag{8.52}$$

In passing, we may remark that for the UMLE, such positive rule shrinkage estimators are known to have smaller risk [Sen, Kulokawa and Saleh (1989)], and hence, the motivation for $\hat{\boldsymbol{\theta}}^+_{RSM}$.

It may be remarked that in (8.50), the denominator $n - p + 2$ is based on the fact that $\boldsymbol{S}_n$ has the Wishart $(n-1, p, \boldsymbol{\Sigma})$ distribution with $n - 1$ degrees of freedom. In general, if $\boldsymbol{S}_n$ were $W(m, p, \boldsymbol{\Sigma})$ independently of $\bar{\boldsymbol{X}}_n$, then in (8.50) (and (8.52)) for the upper bound of c_a, the denominator should be replaced by $m - p + 3$. Also, in some special cases, as we have seen in Chapters 2–4, we would have $\boldsymbol{\Sigma} = \sigma^2 \boldsymbol{V}$ where $\boldsymbol{V}$ is a known p.d. matrix and $\sigma^2 (< \infty)$ is a scalar parameter, such that there exists an estimator S^2_n, independent of $\bar{\boldsymbol{X}}_n$ for which $mS^2_n/\sigma^2 \sim \chi^2_m$ for some $m \geq 1$. Then, we let $s^2 = m(m+2)^{-1} S^2_n$ and $\widehat{\boldsymbol{\Sigma}} = s^2 \boldsymbol{V}$. In that case, in the partitioning $\phi \leq a \leq P$, we replace $\boldsymbol{S}_n$ by $\widehat{\boldsymbol{\Sigma}}$ and get the positive-rule shrinkage estimator as in (8.52) wherein in the definition of the $\bar{\boldsymbol{X}}_{na:a'}$, $\boldsymbol{S}_n$ has been replaced by $\widehat{\boldsymbol{\Sigma}}$ and in (8.50), for $c_a, a \subset P$, for the upper bound, we would have simply $2(|a| - 2)^+, \phi \subset a \subset P$. Finally, if $\boldsymbol{\Sigma}$ were known, in the partitioning $\phi \subset a \subset P$, we replace $\boldsymbol{S}_n$ by $\boldsymbol{\Sigma}$, and in (8.50), for the upper bound for c_a, we would still have $2(|a| - 2)^+$.

The main risk-dominance result (Sengupta and Sen (1991)) can be summarized as follows; see Sengupta and Sen (1991) for a proof.

Theorem 8.4.1 *Under the quadratic loss* $L(\boldsymbol{a}, \boldsymbol{b}) = (\boldsymbol{a} - \boldsymbol{b})^T \boldsymbol{\Sigma}^{-1} (\boldsymbol{a} - \boldsymbol{b})$, *for every* $p \geq 3, \boldsymbol{\theta} \in \Theta^+$,

$$R(\boldsymbol{\theta}, \hat{\boldsymbol{\theta}}^+_{RSM}) \leq R(\boldsymbol{\theta}, \hat{\boldsymbol{\theta}}_{RSM}) \leq R(\boldsymbol{\theta}, \hat{\boldsymbol{\theta}}^*_{RM}), \tag{8.53}$$

where strict inequality holds in a neighborhood of the pivot $\mathbf{0}$ *(a vertex of* Θ^+*). Thus, the positive rule SRRMLE dominates the SRRMLE, which in turn dominates the RMLE when* $\boldsymbol{\theta} \in \Theta^+$.

Remarks. As in the case of UMLEs, confined to the set $a : \phi \subset a \subset P$, a maximal gain is achieved by letting $c_a = (|a| - 2)^+$. Also, the risk reduction due to shrinkage is maximum at the pivot $\mathbf{0}$. Further, in the unrestricted case, the shrinkage factor c_p does not depend on the partition $\phi \subset a \subset P$, while it does in the restricted case. Noting that $(|a| - 2)^+$ is a monotone nondecreasing function of $|a|$, and the risk reduction in the restricted case is some sort of average over the subsets $\phi \subset a \subset P$, we may have generally a smaller reduction of the risk of the SRRMLE over the RMLE than in the unrestricted case (i.e., SMLE over UMLE). This is not surprising as the RMLE are themselves adjusted estimators with due considerations on the set restraints for $\phi \leq a \leq P$. For this reason, it might of interest to compare the risk of SMLE and SRRMLE. We refer to Sengupta and Sen (1991), where the directional variation of the risk of the SRRMLE and RMLE as well as SMLE have been studied in detail; we will only discuss some of these results later in this section. Finally, the relative dominance picture in Theorem 8.4.1 is adapted to the particular loss function $L(\boldsymbol{a}, \boldsymbol{b}) = \| \boldsymbol{a} - \boldsymbol{b} \|^2_{\boldsymbol{\Sigma}}$. If instead of $\boldsymbol{\Sigma}^{-1}$, we use an arbitrary W (p.d.) then (8.53) may not hold. Problem 8.7 is set to verify this result. The situation is quite comparable to the unrestricted case where also the dominance of the SMLE rests on the adaptation of a particular loss function, and for different choice of W, shrinkage estimators having the desired dominance property may differ.

Motivated by the positive-rule shrinkage estimators and a general class of minimax estimators (in the unrestricted case) introduced by Baranchik (1971), we consider here some more general minimax estimators in the restricted case when $\boldsymbol{\Sigma}$ is known. In the unrestricted case, the MLE and its shrinkage versions are all equivariant under nonsingular transformations $\boldsymbol{X} \to \boldsymbol{Y} = \boldsymbol{B}\boldsymbol{X}, \boldsymbol{B}$ nonsingular. However, this equivariance is not generally true for $\mathcal{O}^+$ (when $\boldsymbol{\Sigma}$ is arbitrary p.d.) or for a positively homogeneous cone in $\mathbb{R}^p$. This makes it difficult to use a canonical reduction (on $\boldsymbol{\theta}, \boldsymbol{\Sigma}$) to establish the desired results for an arbitrary $\boldsymbol{\Sigma}$. However, under an additional condition on $(\boldsymbol{\theta}, \boldsymbol{\Sigma})$, such a general dominance result may be obtained. Let us define

$$\langle \boldsymbol{x}, \boldsymbol{y} \rangle_{\boldsymbol{\Sigma}} = \boldsymbol{x}^T \boldsymbol{\Sigma}^{-1} \boldsymbol{y}, \langle \boldsymbol{x}, \boldsymbol{x} \rangle_{\boldsymbol{\Sigma}} = \| \boldsymbol{x} \|^2_{\boldsymbol{\Sigma}}; \tag{8.54}$$

$$\psi_m(\boldsymbol{\theta}, \boldsymbol{\Sigma}) = (2\pi)^{p/2} |\boldsymbol{\Sigma}|^{-1/2} \int \cdots \int_{z>0} \| z \|^{-m}_{\boldsymbol{\Sigma}} \langle \boldsymbol{\theta}, z \rangle^m_{\boldsymbol{\Sigma}} \exp\left\{ -\frac{1}{2} \| z \|^2_{\boldsymbol{\Sigma}} \right\} dz, m \geq 0; \tag{8.55}$$

$$\Theta^*_+ = \{ \boldsymbol{\theta} \in \mathcal{O}^+ : \psi_m(\boldsymbol{\theta}_{a:a'}, \boldsymbol{\Sigma}_{a:a'}) \geq 0, \forall m \geq 0 \text{ and } a \subseteq N_m \}. \tag{8.56}$$

Note that a sufficient condition for Θ^*_+ to be nonempty is that $\boldsymbol{\Sigma}^{-1}_{a:a'} \boldsymbol{\theta}_{a:a'} \geq 0$, for every $a \in N^*_p (= \{ a \in N_p : |a| > 2 \}), p \geq 3$, and this is true in particular when $\boldsymbol{\Sigma}$ is diagonal. Thus, the results to follow hold for $\mathcal{O}^+$ when $\boldsymbol{\Sigma}$ is diagonal and $p \geq 3$.

For every a, $\phi \subseteq a \subseteq P$, let $t_a(y) : \mathbb{R}^+ \to (0, 2(|a| - 2)^+)$ be a nondecreasing, nonnegative and bounded function of y, and for $p \geq 3$, let $\boldsymbol{X} \sim N_p(\boldsymbol{\theta}, \boldsymbol{\Sigma})$,

$$\hat{\boldsymbol{\theta}}^* = \sum_{\{\phi \subseteq a \subseteq P\}} I_{na} \{ 1 - I(|a| > 2) [y^{-1} t_a(y)]_{y = \| \hat{\boldsymbol{\theta}}_{RM} \|^2_{\boldsymbol{\Sigma}}} \} \hat{\boldsymbol{\theta}}^*_{RM}, \tag{8.57}$$

where $\hat{\boldsymbol{\theta}}^*_{RM}$ refers to the RMLE in the case of known $\boldsymbol{\Sigma}$ (and $n = 1$). Note that the positive-rule estimator in (8.52) (with $\boldsymbol{S}_n$ replaced by $\boldsymbol{\Sigma}$) is a member of this class.

Theorem 8.4.2 *Let $\boldsymbol{X} \sim N_p(\boldsymbol{\theta}, \boldsymbol{\Sigma})$ where $\boldsymbol{\Sigma}$ is such that $\boldsymbol{\theta} \in \Theta^+$. Also assume that $t_a(y)$ is nonnegative, monotone nondecreasing, and bounded from above by $2(|a|-2)$, for every $a \subseteq P$. Then, for $p > 3$, any estimator of the form* (8.55) *dominates $\hat{\boldsymbol{\theta}}^*_{RM}$ (over Θ_+), and hence, is minimax (over Θ^+).*

See Sengupta and Sen (1991) for a proof of this theorem. In passing, we may remark that when $\boldsymbol{\Sigma}$ is diagonal, $\mathcal{X}_a(\boldsymbol{\Sigma}) = \mathcal{X}_a(\boldsymbol{I})$ and $\hat{\boldsymbol{\theta}}_{RM}$ does not depend on $\boldsymbol{\Sigma}$, but the factor $[y^{-1}t_a(y)]_{y-\|\hat{\boldsymbol{\theta}}_{RM}\|^2_{\boldsymbol{\Sigma}}}$ may depend on $\boldsymbol{\Sigma}$, and hence, (8.55) depends on $\boldsymbol{\Sigma}$, even if it is diagonal, although the conclusions in the theorem holds for the entire Θ^+.

Next, we consider an extension of the positive orthant model and some specific applications in linear models.

(I) *Suborthant model.* Consider the following:

$$\boldsymbol{X}_{(m+p)\times 1} = (\boldsymbol{X}_1^T, \boldsymbol{X}_2^T)^T \sim N_{m+p}(\boldsymbol{\theta}, \boldsymbol{\Sigma}); \boldsymbol{\theta} = (\boldsymbol{\mu}_1^T, \boldsymbol{\mu}_2^T)^T \tag{8.58}$$

where $\boldsymbol{\mu}_1$ and $\boldsymbol{\mu}_2$ are m and p-vectors, respectively, and $\boldsymbol{\Sigma}$ is a $(m+p) \times (m+p)$ matrix (p.d.). In this set-up, $\boldsymbol{\mu}_1$ is unrestricted while $\boldsymbol{\mu}_2$ belongs to the positive orthant $\mathbb{R}^{+p}$, i.e.,

$$\Theta_+^0 = \{\boldsymbol{\theta} = (\boldsymbol{\mu}_1^T, \boldsymbol{\mu}_2^T) : \boldsymbol{\mu}_1 \in \mathbb{R}^{+m}, \boldsymbol{\mu}_2 \in \mathbb{R}^{+p}\}, \tag{8.59}$$

and the *pivot* for $\boldsymbol{\mu}_2$ is 0. We denote by $N_p^0 = \{m+1, \ldots, m+p\}$ and for every $a : \phi \subseteq a \subseteq N_p^0$, the complementary subset (a') as well as the $\boldsymbol{X}_{1:a}(\boldsymbol{\Sigma}), \boldsymbol{X}_{a:a'}(\boldsymbol{\Sigma})$ and $\boldsymbol{\Sigma}_{a:a'}$ etc. are defined as in earlier sections. It can be shown that for this suborthant model, the RMLE of $\boldsymbol{\theta}$ is given by

$$\hat{\boldsymbol{\theta}}_{RM} = \sum_{\phi \subseteq a \subseteq N_p^0} \mathcal{P}_{N_m,a}(\boldsymbol{X}_{1:a'}, \boldsymbol{X}_{a:a'}, \boldsymbol{0})\boldsymbol{1}(\boldsymbol{\Sigma}_{a'a'}^{-1}\boldsymbol{X}_{a'} \leq \boldsymbol{0}, \boldsymbol{X}_{a:a'} > \boldsymbol{0}), \tag{8.60}$$

where $N_m = \{1, \ldots, m\}$. Actually, for the case of unknown $\boldsymbol{\Sigma}$, whenever we have a Wishart matrix $\boldsymbol{S}$ (with ν degrees of freedom), independent of $\boldsymbol{X}$, the RMLE of $\boldsymbol{\theta}$ is also given by (8.60) provided in the definition of the $\boldsymbol{X}_{1:a}, \boldsymbol{X}_{a:a'}$ etc., we replace $\boldsymbol{\Sigma}$ by $\boldsymbol{S}$.

Motivated by the RMLE in (8.60) and the Stein-rule estimators discussed earlier, we consider the following theorem on improved estimation for this suborthant model.

Theorem 8.4.3 *The shrinkage estimator*

$$\hat{\boldsymbol{\theta}}_{RSM} = \sum_{\phi \subseteq a \subseteq N_p^0} 1(\boldsymbol{X}_2 \in (\mathcal{X}_a(\boldsymbol{\Sigma}_2))\{1 - c_a \| \hat{\boldsymbol{\theta}}_{RM} \|_{\boldsymbol{\Sigma}}^{-2}\}\hat{\boldsymbol{\theta}}_{RM} \tag{8.61}$$

dominates the RMLE in (8.60) *[over Θ_+^0] whenever $p + m \geq 3$ and*

$$0 \leq c_a \leq 2(|a| + m - 2)^+, \textit{ for every } a : \phi \subseteq a \subseteq N_p^0. \tag{8.62}$$

See Sengupta and Sen (1991) for a proof. For the case of unknown $\boldsymbol{\Sigma}$, when $S \sim W(v, m+p; \boldsymbol{\Sigma})$, we replace $\boldsymbol{\Sigma}$ by S and in (8.62), the upper bound for c_a is taken as $2(|a| + m - 2)^+/(v - m - p + 3), a \in N_p^0$.

(II) *Ordered alternative model.* Suppose that $X_{ij}, j = 1, \ldots, n_i$ are *iid* random variables with the normal distribution $N(\mu_i, \sigma^2), i = 1, \ldots, r$; all these r samples being independent. The ordered alternative model relates to the following positively homogeneous subspace of $\mathbb{R}^r : \Theta^> = \{(\mu_1, \ldots, \mu_r) \epsilon \mathbb{R}^r : \mu_1 \leq \cdots \leq \mu_r\}$, so that the pivot is the line of equality $\mu_1 = \cdots = \mu_r \in \mathbb{R}$. If we write

$$\mu_i = \mu_1 \text{ for } i = 1 \text{ and } \mu_i = \mu_1 + \beta_2 + \cdots + \beta_i, i = 2, \ldots, r; \beta_2 = (\beta_2, \ldots, \beta_r)^T \tag{8.63}$$

then $\boldsymbol{\theta} = (\mu_1 : \boldsymbol{\beta}_2)$ relates to the suborthant model with $m = 1$ and $p = r - 1$. We may also consider a two-way layout: $X_{ij} = \mu_i + r_j + e_{ij}, 1 \leq j \leq n, i = 1, \ldots, n$ and characterize the order alternative model for the treatment effects $\tau_1, \ldots, \tau_r$ as a sub-orthant model. Since these are both particular cases of some linear models, we consider the latter in details.

(III) *SRRMLE for univariate linear models.* Consider the usual linear model

$$\boldsymbol{X}_n = (\boldsymbol{X}_1, \ldots, \boldsymbol{X}_n)^T = \boldsymbol{C}\boldsymbol{\beta} + e_n \sim N_n(\boldsymbol{0}, \boldsymbol{\sigma}^2 \boldsymbol{I}_n), 0 < \sigma^2 < \infty, \tag{8.64}$$

where $\boldsymbol{C}$ is a known matrix (of order $n \times p^*$) of regression constants, $\boldsymbol{\beta}^T = (\boldsymbol{\beta}_1^T, \boldsymbol{\beta}_2^T)^T$ is a p^*-vector of unknown regression parameters, $\boldsymbol{\beta}_j$ is a p_j-vector, $j = 1, 2$, and σ^2 is unknown. We consider the following suborthant model:

$$\boldsymbol{\beta} \in \Theta^> = \{\boldsymbol{\beta} : \boldsymbol{\beta}_1 \in \mathbb{R}^{p_1}, \boldsymbol{\beta}_2 \in \mathbb{R}^{+p}\}, p_2 = p \text{ and } p^* = p_1 + p_2. \tag{8.65}$$

Without any loss of generality (and allowing reparameterization if necessary), we may assume that $\boldsymbol{C}$ is of full rank $p^* < n$, and we desire to construct improved estimators of $\boldsymbol{\beta}$ under the set of restraints on $\boldsymbol{\beta}_2$ in (8.65). This model includes the K-sample location model as a special case of $p = K - 1$.

Note that the classical MLE for $(\boldsymbol{\beta}, \boldsymbol{\sigma}^2)$ are

$$\hat{\boldsymbol{\beta}} = (\hat{\boldsymbol{\beta}}_1 : \hat{\boldsymbol{\beta}}_2) = (\boldsymbol{C}^T\boldsymbol{C})^{-1}\boldsymbol{C}^T\boldsymbol{X}_n \text{ and } \hat{\sigma}^2 = (n - p^*)^{-1} \parallel \boldsymbol{X}_n - \boldsymbol{C}\hat{\boldsymbol{\beta}} \parallel^2, \tag{8.66}$$

and these are jointly sufficient for $(\boldsymbol{\beta}, \sigma^2)$. Moreover $\hat{\boldsymbol{\beta}}$ and $\hat{\sigma}^2$ are independent with

$$\hat{\boldsymbol{\beta}} \sim N(\boldsymbol{\beta}, \sigma^2(\boldsymbol{C}^T\boldsymbol{C}^{-1}) \text{ and } (n - p^*)\hat{\sigma}^2/\sigma^2 \sim \chi^2_{n-p^*}. \tag{8.67}$$

Problem 8.4.3 is set to verify this result. In this case, the RMLE of $\boldsymbol{\beta}$ is denoted by $\hat{\boldsymbol{\beta}}^*_{RM}$ and is defined by (8.58) with $\boldsymbol{X} = \hat{\boldsymbol{\beta}}$ and $\boldsymbol{\Sigma} = \hat{\sigma}^2(\boldsymbol{C}^T\boldsymbol{C})^{-1}$, and as in Theorem 8.2.1, we obtain the RSMLE as

$$\hat{\boldsymbol{\theta}}_{RSM} = \sum_{\phi \subseteq a \subseteq N_p^0} 1(\hat{\boldsymbol{\beta}}_2 \in \mathcal{X}_a(\hat{\sigma}^2(\boldsymbol{C}^T\boldsymbol{C})^{-1}))\{1 - c_a\hat{\sigma}^2[\hat{\boldsymbol{\beta}}^*_{RM}(\boldsymbol{C}^T\boldsymbol{C})\hat{\boldsymbol{\beta}}^*_{RM}]^{-1}\}\hat{\boldsymbol{\beta}}^*_{RM} \tag{8.68}$$

where

$$0 \leq c_a \leq 2(|a| + p_1 - 2)^+(n - p)/(n - p + 2), \forall a : \phi \subseteq a \subseteq N_p^0. \tag{8.69}$$

Then, under the quadratic error loss $(\boldsymbol{Y}-\boldsymbol{\beta})^T(\boldsymbol{C}^T\boldsymbol{C})(\boldsymbol{Y}-\boldsymbol{\beta})/\sigma^2$, we have

$$R(\hat{\boldsymbol{\beta}}_{RSM}, \boldsymbol{\beta}) \leq R(\hat{\boldsymbol{\beta}}_{RM}, \boldsymbol{\beta}), \forall \boldsymbol{\beta} \in \mathbb{R}^{p_1} \times \mathbb{R}^{+p}. \tag{8.70}$$

We may also consider a positive-rule version of the RSMLE (as in (3.4) with the obvious modifications) and verify that (4.1) holds as well in this model. Parallel dominance results hold for a general class of multivariate linear models too.

Let us make some comments on the relative risk picture of the SMLE and SR-RMLE. In spite of having some similarity, there is a basic difference. For the normal mean problem, positive orthant model, the SMLE is invariant under orthogonal transformations: $\boldsymbol{X} \to \boldsymbol{Y} = \boldsymbol{U}\boldsymbol{X}, \boldsymbol{U}^T\boldsymbol{U} = 1$, but the SRRMLE is not so. Thus, the risk function of the SMLE is constant on the contours $\delta = \boldsymbol{\theta}^T\boldsymbol{\Sigma}^{-1}\boldsymbol{\theta}$, and this characterization can be incorporated in the simplification of the risk picture of the SMLE. The risk function of the RMLE and SRRMLE may depend on the unknown $\boldsymbol{\theta}$ in a much more involved manner. This picture may actually depend on whether $\boldsymbol{\theta}$ belongs to the interior or Θ^+, or to its lower dimensional faces (edges). We refer to Sengupta and Sen (1991) for this study, and present the salient features as follows. The growth of the risk of these estimates, (RMLE and SRRMLE) is very much dependent on the direction of deviation of $\boldsymbol{\theta}$ from the pivot $\mathbf{0}$. Unlike the unrestricted parameter space case, the risk of the RMLE may not simply depend on Δ, while the MLE or the SMLE have risk function depending only through Δ. The risk is smaller on the boundaries of Θ^+ while it may be larger as $\boldsymbol{\theta}$ moves away to the interior of Θ^+. For $p = 3$, the RMLE dominates the SMLE while the SRRMLE dominates the SMLE. For $p \geq 4$, the SRRMLE dominates the SMLE on the edges of Θ^+.

Let us consider some alternative shrinkage estimator, for the positive orthant mean problem, due to Chang (1981, 1982). For $X \sim N(\mu, 1), \mu > 0$, Katz (1961) considered a Bayes estimator based on a uniform prior on $\mathbb{R}^+$. Thus, denoting the normal density by $f(x)$ (and noting that $f'(x) = -xf(x)$), we obtain that the posterior density of μ, given X, is $f(x-\mu)I(\mu > 0)/\int_0^\infty f(x-\mu)d\mu, \mu > 0$. The mean of this posterior law is equal to

$$\delta = \int_0^\infty f(x-\mu)\mu d\mu \Big/ \int_0^\infty f(x-\mu)d\mu \tag{8.71}$$

$$= x + \int_0^\infty (\mu - x)f(x-\mu)d\mu \Big/ \int_0^\infty f(x-\mu)d\mu \tag{8.72}$$

$$= x + \phi_0(x) \quad \text{where } \phi_0(x) = e^{-\frac{1}{2}x^2} \Big/ \int_{-\infty}^x e^{-\frac{1}{2}u^2}du, x \in \mathbb{R}. \tag{8.73}$$

Being a Bayes estimator, under squared error loss, δ is admissible. Motivated by this feature, Chang (1981, 1982) considered three shrinkage estimators, $\boldsymbol{\delta}^{(1)}, \boldsymbol{\delta}^{(2)}$, and $\boldsymbol{\delta}^{(3)}$, where $\boldsymbol{\delta}^{(r)} = (\delta_1^{(r)}, \ldots, \delta_p^{(r)})^T, r = 1, 2, 3$, and

$$\delta_j^{(1)} = X_j + \phi_0(X_j) - \frac{c\{X_j + \phi_0(X_j)\}}{\sum_{s=1}^p \{X_s + \phi_0(X_s)\}}, 1 \leq j \leq p, \tag{8.74}$$

and $0 < c < 2(p-2)$. Under quadratic loss, $\boldsymbol{\delta}^{(1)}$ dominates $\boldsymbol{X}$. Further, define $t(\cdot)$ as in (8.57), and let

$$\delta_j^{(2)} = \begin{cases} X_j + t(X_j) - cX_j / \sum_1^p X_j^2, & \text{if all the } X_i \geq 0 \\ X_j + t(X_j), & \text{if at least one } X_j < 0, \end{cases} \tag{8.75}$$

for $j = 1, \ldots, p$. Let us also introduce

$$\delta_j^{(3)} = X_j + t(X_j), 1 \leq j \leq p. \tag{8.76}$$

Chang (1982) has shown that for the estimation of $\boldsymbol{\theta}(\in \Theta^+)$ based on quadratic loss, $\boldsymbol{\delta}^{(2)}$ dominates $\boldsymbol{\delta}^{(3)}$. It is also possible to construct some other estimators that dominate either of these. Actually, Sengupta and Sen (1991) considered the following

$$\delta_j^* = X_j + t(X_j) - c_a X_j^+ \parallel \boldsymbol{X}_a \parallel^{-2} \text{ on } \mathcal{X}_a \phi \subseteq a \subseteq P, \tag{8.77}$$

for $j = 1, \ldots, p$ where $\mathcal{X}_a = \{\boldsymbol{x} \in \mathbb{R}^p : \boldsymbol{x}_a > 0, \boldsymbol{x}_{a'} \leq 0\}$. They showed that δ^* dominates $\boldsymbol{\delta}^{(2)}$. Similar modifications of $\boldsymbol{\delta}^{(1)}$ and $\boldsymbol{\delta}^{(3)}$ were considered by Sengupta and Sen (1991) and dominance results were studied. Very recently, Dunson and Neelon (2003) have considered Bayesian inference on order-constrained parameters in generalized linear models. One of their problems is the orthant problem treated above (albeit, they were apparently unaware of the earlier development as mentioned above). Their approach has been mostly to computing the Bayes factor and adjusting for the unconstrained posterior density using isotonic regression or other constrained algorithms. For intended brevity, we omit these details.

8.5 PCC AND SHRINKAGE ESTIMATION IN CSI

In the preceding section, it was observed that for the multivariate normal mean estimation problems when $\boldsymbol{\theta}$ belongs to a restricted domain (viz., positive orthant Θ^+), the RMLE fares better than the MLE, and the SRRMLE dominates the RMLE in the light of some quadratic risks. A parallel picture in the light of the Pitman closeness criterion (PCC) is presented here. In a general multiparameter setting, the PC dominance is to be interpreted as follows. Let $\boldsymbol{X} \sim N_p(\boldsymbol{\theta}, \boldsymbol{\Sigma})$ and let $\boldsymbol{\delta}_1$ and $\boldsymbol{\delta}_2$ be two estimators of $\boldsymbol{\theta}$. Also, for a given p.d. matrix $\boldsymbol{Q}$, define the quadratic loss as $\parallel \boldsymbol{X} - \boldsymbol{\theta} \parallel_{\boldsymbol{Q}}^2 = (\boldsymbol{x} - \boldsymbol{\theta})^T \boldsymbol{Q}^{-1} (\boldsymbol{x} - \boldsymbol{\theta}), \boldsymbol{x}, \boldsymbol{\theta} \in \mathbb{R}^p$. Then we say that $\boldsymbol{\delta}_1$ is *closer* to $\boldsymbol{\theta}$ than $\boldsymbol{\delta}_2$, under the norm $\parallel \cdot \parallel_{\boldsymbol{Q}}$, in the Pitman sense if

$$P_{\boldsymbol{\theta}, \boldsymbol{\Sigma}}\{\parallel \boldsymbol{\delta}_1 - \boldsymbol{\theta} \parallel_{\boldsymbol{Q}} \leq \parallel \boldsymbol{\delta}_2 - \boldsymbol{\theta} \parallel_{\boldsymbol{Q}}\} \geq \frac{1}{2} \tag{8.78}$$

for all $\boldsymbol{\theta}, \boldsymbol{\Sigma}$, with the strict inequality holding for some $\boldsymbol{\theta}$. Note that (8.78) is equivalent to saying

$$P_{\boldsymbol{\theta}, \boldsymbol{\Sigma}}\{\parallel \boldsymbol{\delta}_1 - \boldsymbol{\theta} \parallel_{\boldsymbol{Q}} < \parallel \boldsymbol{\delta}_2 - \boldsymbol{\theta} \parallel_{\boldsymbol{Q}}\} \geq P_{\boldsymbol{\theta}, \boldsymbol{\Sigma}}\{\parallel \boldsymbol{\delta}_1 - \boldsymbol{\theta} \parallel_{\boldsymbol{Q}} > \parallel \boldsymbol{\delta}_2 - \boldsymbol{\theta} \parallel_{\boldsymbol{Q}}\} \tag{8.79}$$

for all $\boldsymbol{\theta}, \boldsymbol{\Sigma}$, with strict inequality holding for some $\boldsymbol{\theta}$.

The evolution of shrinkage (or Stein-rule) estimators (SRE), sketched in Section 8.3, has primarily been under quadratic loss functions (along with other SDT augmentations). Sen, Kubokawa, and Saleh (1989) addressed the Stein-paradox in the sense of the Pitman measure of closeness, and this appraisal has further been extended to cover some restricted parameter spaces by Sen and Sengupta (1991). It has been demonstrated that a general class of shrinkage estimators, with a shrinkage factor especially suited for the PCC, possess the PC dominance property, and in that context, even p may be as small as 2. We intend to confine ourselves to restricted parameter cases (in line with our conformity to CSI) and, within this framework, explore the PC dominance of SRE over RMLE and MLE.

For simplicity of presentation, we consider explicitly the positive orthant model where $\boldsymbol{\Theta}^+ = \mathbb{R}^{+p} = \{\boldsymbol{x} \in \mathbb{R}^p : \boldsymbol{x} \geq \boldsymbol{\theta}\}$. As has been noted earlier, there is a basic difference between the unrestricted (i.e., $\boldsymbol{\Theta} = \mathbb{R}^p$) and restricted ($\boldsymbol{\Theta} = \boldsymbol{\Theta}^+$) cases. In the unrestricted case, the MLE as well as its shrinkage version are *equivariant* under the group of (affine) transformations:

$$\boldsymbol{X} \to \boldsymbol{Y} = \boldsymbol{a} + \boldsymbol{B}\boldsymbol{X}, \boldsymbol{B} \text{ nonsingular}, \tag{8.80}$$

so that one may choose $\boldsymbol{B}$ in such a way that $E\boldsymbol{Y} = \boldsymbol{\eta} = (\eta_1 : \mathbf{0})$, and $\eta_1 = \| \boldsymbol{\theta} \|^2$. Such a canonical reduction may not be possible for the restricted case: the primary difficulty stems from the fact that Θ^+ is not *invariant* under such a nonsingular (or even orthogonal) $\boldsymbol{B}$, although the invariance holds for all diagonal $\boldsymbol{B}$. In the negation of this equivariance property of the RMLE and its shrinkage versions, it is not surprising to see that the performance characteristics (be it in the quadratic risk or the P measure) of the RMLE and SRRMLE may depend on $\| \boldsymbol{\theta} \|$ but also on the direction cosines of the individual elements of $\boldsymbol{\theta}$. As such, this picture when $\boldsymbol{\theta}$ lies in the interior of $\boldsymbol{\Theta}^+$ (i.e., $\boldsymbol{\theta} > \mathbf{0}$) may not totally agree with the case when $\boldsymbol{\theta}$ lies on the boundary of $\boldsymbol{\Theta}^+$ (i.e., $\theta_j = 0$ for some $j = 1, \ldots, p$). However, the relative dominance picture remains the same, although the extent may differ from the edges to the interior of $\boldsymbol{\Theta}^+$.

For the general case $\boldsymbol{X} \sim N_p(\boldsymbol{\theta}, \boldsymbol{\Sigma}), \boldsymbol{\theta} \in \boldsymbol{\Theta}^+, \boldsymbol{\Sigma}$ unknown, and $\boldsymbol{S}_n \sim W(n-1, p, \boldsymbol{\Sigma})$, based on $\boldsymbol{X}_1, \ldots, \boldsymbol{X}_n$, the RMLE of $\boldsymbol{\theta}$ has been formulated in (8.48). If $\boldsymbol{\Sigma}$ were known, as has been indicated there, we need to replace $\boldsymbol{S}_n$ by $\boldsymbol{\Sigma}$ in the partitioning $a : \phi \subseteq a \subseteq P$. Consider an intermediate case where $\boldsymbol{\Sigma} = \sigma^2 \boldsymbol{V}, \boldsymbol{V}$ is a known p.d. matrix and σ^2 is unknown. In this case, in the partitioning $a : \phi \subseteq a \subseteq P$, we may replace $\boldsymbol{S}_n$ by V. In the simplest case $\boldsymbol{\Sigma} = \sigma^2 \boldsymbol{I}$, the solution in (8.48) simplifies considerably . To set the inquiries in the proper perspectives, first consider the case of $\boldsymbol{\Sigma} = \sigma^2 \boldsymbol{I}, \sigma^2$ known (and taken as 1, without loss of generality). In this special case, a natural choice of $\boldsymbol{Q}$ is $\boldsymbol{I}$. For the RMLE, we may observe that in this special case,

$$\hat{\boldsymbol{\theta}}_{RM} = \bar{\boldsymbol{X}}_n^+ = (\bar{X}_{1n} \vee 0, \ldots, \bar{X}_{pn} \vee 0)^T, \tag{8.81}$$

while $\hat{\boldsymbol{\theta}}_M = \bar{\boldsymbol{X}}_n$. Also, let $a(\boldsymbol{X}) = \sum_{j=1}^p I(X_j > 0)$ be the number of positive coordinates of $\boldsymbol{X}$, so that $0 \leq a(\boldsymbol{X}) \leq p, \forall \boldsymbol{X} \in \mathbb{R}^p$ and we let $[q]^+ = \max(q, 0)$. Then, for the model $\boldsymbol{X} \sim N_p(\boldsymbol{\theta}, \sigma^2 \boldsymbol{I}), \boldsymbol{\theta} \in \Theta$, the SRRMLE of $\boldsymbol{\theta}$ obtained in Section

8.4 reduces to

$$\hat{\boldsymbol{\theta}}^{+}_{RSM} = \{1 - [a(\boldsymbol{X}) - 2]^{+}\sigma^2 \parallel \boldsymbol{X}^{+} \parallel^{-2}\}^{+}\boldsymbol{X}^{+}, \tag{8.82}$$

where $\hat{\boldsymbol{\theta}}^{+}_{RSM}$ indicates the positive rule shrinkage estimator. We may express (8.82) also as

$$\hat{\boldsymbol{\theta}}_{RSM} = \{1 - [c_{a(\boldsymbol{X})}\sigma^2 \parallel \boldsymbol{X}^{+} \parallel^{-2}\}^{+}\boldsymbol{X}^{+}, \tag{8.83}$$

where $c_a : 0 \leq a \leq p$, is defined by (8.50). However, as has been remarked earlier, we may, under PCC, extend the domain of $c_a : 0 \leq a \leq p$ a bit more, enabling the comparison to be valid even for $p = 2$. We denote by $m_p (= \text{med}(\chi^2_p))$ the median of the chi-square distribution with p DF. Note that by the mean-median-model inequality for chi square distributions, for every $p \geq 2, p - 2 = \text{mode} < m_p < p = \text{mean}$; for $p = 2, m_2 = 2\log 2 \simeq 1.386$, while as p increases, m_p shrinks to $p - 1 + \cdot 3$. For $p = 1$, the median $m_1 = Q_3^2$, where Q_3 is the standard normal third quartile ($\simeq 0.6$). Problem 8.5.1 is set to verify these results. As such, motivated by the general results in Sen, Kubokawa, and Saleh (1989), we choose here

$$c^*_a = \begin{cases} 0, & \text{for } a = 0, 1; \\ m_a, & \text{for } a \geq 2. \end{cases} \tag{8.84}$$

Then, in (8.83), we replace $c_a(\boldsymbol{X})$, by $c^*_{a(\boldsymbol{X})}, \boldsymbol{X} \in \mathbb{R}^p$, and denote the resulting SRE by $\hat{\theta}^{*+}_{RSM}$. Our goal is to demonstrate the PC dominance of $\hat{\theta}^{*+}_{RSM}$ over $\hat{\theta}^{+}_{RM}$, for this simple model $\boldsymbol{X} \sim N_p(\boldsymbol{\theta}, \sigma^2\boldsymbol{I}), \boldsymbol{\theta} \in \boldsymbol{\Theta}^{+}$, and then to discuss more general cases along the same line.

Theorem 8.5.1 *For every $p \geq 2, \hat{\boldsymbol{\theta}}^{*+}_{RSM}$ dominates $\hat{\boldsymbol{\theta}}_{RM}$ in the PCC, and for $p \geq 3, \hat{\boldsymbol{\theta}}^{+}_{RSM}$ dominates $\hat{\boldsymbol{\theta}}_{RM}$ in the PCC.*

See Sen, Kubokawa, and Saleh (1989) for a proof. We consider now the case where $\boldsymbol{X} \sim N_p(\boldsymbol{\theta}, \sigma^2\boldsymbol{I}), \boldsymbol{\theta} \in \Theta^{+}$ and σ^2 is unknown: there exists a $r.v.S$ such that qS/σ^2 is distributed as $\chi^2_q, q \geq 1$, independently of $\boldsymbol{X}$. Then, we modify the definition of the estimates by substituting S for σ^2 in the respective definitions. The PC dominance results in Theorem 8.5.1 remain intact for this studentized case as well. Let us consider next a more general case where $\boldsymbol{X} \sim N_p(\boldsymbol{\theta}, \sigma^2\boldsymbol{V})$ with a known $\boldsymbol{V}$ (p.d.), σ^2 unknown, $qS/\sigma^2 \sim \chi^2_q$. In the unrestricted case, Sen, Kubokawa, and Saleh (1989) considered a general class of shrinkage estimators of the form

$$\boldsymbol{\delta}_\varphi = \boldsymbol{X} - \varphi(\boldsymbol{X}, S) \parallel \boldsymbol{X} \parallel^{-2}_{\boldsymbol{Q},V} \boldsymbol{Q}^{-1}\boldsymbol{V}^{-1}\boldsymbol{X}, \tag{8.85}$$

where

$$0 \leq \varphi(\boldsymbol{x}, s) \leq (p-1)(3p+1)/2p, \text{ for every } (\boldsymbol{x}, \beta), \tag{8.86}$$

$p \geq 2$ and $\parallel \boldsymbol{X} \parallel^2_{\boldsymbol{Q},\boldsymbol{V}} = \boldsymbol{X}^T\boldsymbol{V}^{-1}\boldsymbol{Q}^{-1}\boldsymbol{V}^{-1}\boldsymbol{X}$. They have shown that

$$P_{\boldsymbol{Q},\sigma}\{\parallel \boldsymbol{\delta}_\varphi - \boldsymbol{\theta} \parallel_{\boldsymbol{Q}} \leq \parallel \boldsymbol{X} - \boldsymbol{\theta} \parallel_{\boldsymbol{Q}}\} \geq \frac{1}{2}, \forall(\boldsymbol{\theta}, \sigma^2), \tag{8.87}$$

so that a SMLE $\boldsymbol{\delta}_{\varphi}$ of the form (8.85) dominates the MLE $\boldsymbol{X}$ in the Pitman closeness measure for all $p \geq 2$ and all shrinkage factor $\varphi(\cdot, \cdot)$ satisfying (8.86). The James-Stein (1961) estimator belongs to this class (for which $\varphi(\boldsymbol{X}, s) = b : 0 < b < (p-1)(3p+1)/2p$. The positive-rule version of the James-Stein (1961) estimator corresponds to

$$\varphi(\boldsymbol{x}, \sigma^2) = \begin{cases} \| \boldsymbol{x} \|^2 / \sigma^2, & \| \boldsymbol{x} \|^2 \leq b\sigma^2 \\ b, & \| \boldsymbol{x} \|^2 > b\sigma^2. \end{cases} \tag{8.88}$$

Although the results of Sen, Kubokawa, and Saleh (1989) can be used to establish PC dominance results to restricted parameter space models–it could be trifle harder to characterize their Pitman-Closest property. In the same vein, we may consider the estimators in (8.85) and obtain the shrinkage version, but characterization of "Pitman closest" property may be, in general, hard to establish.

Looking at Theorem 8.2.1 it seems very much desirable to have that result extended to the multiparameter case using $\| \cdot \|_{\boldsymbol{Q}}$ norm, and thereby to claim PC dominance in a more general way. In the first place, it needs a workable definition and interpretation of *multivariate median-unbiasedness*. One way of defining this multivariate median unbiasedness property for an estimator $\hat{\boldsymbol{\theta}}$ of a parameter $\boldsymbol{\theta}$ could be requiring that for every $\boldsymbol{\lambda} \in \mathbb{R}^p$, $\boldsymbol{\lambda}^T \tilde{\boldsymbol{\theta}}$ is median unbiased for $\boldsymbol{\lambda}^T \boldsymbol{\theta}$ (where, without loss of generality, one may take $\boldsymbol{\lambda}^T \boldsymbol{\lambda} = 1$). If $\hat{\boldsymbol{\theta}}$ has an ellipsoidally symmetric distribution around $\boldsymbol{\theta}$, then, of course, the above definition would fit well. However, for median-unbiasedness symmetry is not needed. For example, suppose that S^2 is an estimator of σ^2 such that qS^2/σ^2 is a χ_q^2 variable with $q \geq 2$. Define m_q as the median of χ_q^2 and let $S_0^2 = qS_q^2/m_q$. Then S_0^2 is median-unbiased for σ^2, though it might not have a symmetric distribution around σ^2. Consider a second example $\boldsymbol{X} \sim N_p(\boldsymbol{\theta}, \sigma^2 \boldsymbol{I}), \boldsymbol{\theta} \in \boldsymbol{\Theta}^+$. We have the RMLE for $\boldsymbol{\theta}$, given in (8.81). Although, $\bar{\boldsymbol{X}}_n$ has a multinormal law (and, hence, is symmetric), $\bar{\boldsymbol{X}}_n^+$ having a distribution defined over $\mathbb{R}^{+p}$, may not have a symmetric distribution unless $\boldsymbol{\theta}$ is away from $\boldsymbol{0}$ (inside $\boldsymbol{\Theta}^+$). For $\boldsymbol{\theta}$ close to the boundary of $\boldsymbol{\Theta}^+$, because of left truncation of $\bar{\boldsymbol{X}}_n$, this multivariate median-unbiasedness may not be true. Therefore, for CSI, to investigate PC dominance properties, direct verification may be necessary.

One discouraging feature of such shrinkage or Stein-rule estimators is that they do not amend directly for the confidence set estimation problems, even in unrestrained cases. The dominance of SRMLE over MLE is either in quadratic risk or PCC, as we have studied earlier. Even for the multinormal mean estimation problem, shrinkage estimators do not have multinormal law whereas the MLE has. The use of pivot and the Stein identity facilitate the verification of the risk dominance, but for the confidence set estimation problems, a knowledge of the sampling distribution of the shrinkage estimators is essential. Hwang and Peddada (1994) have studied the confidence interval estimation subject to order restrictions. The confidence set estimation problem becomes more unmanageable in the CSI cases. For this reason, we shall not provide a detailed discussion of constrained confidence set estimation problems based on shrinkage estimators.

8.6 BAYES TESTS IN CSI

In the previous sections, the emphasis has been mainly on Bayes estimators with justifications from SDT. We like to present the counterpart of CSI dealing with the hypothesis testing problems, albeit rather succinctly. In a frequentist set-up, as has been mostly considered in earlier chapters, the LRP or its variants have been incorporated to extend the Neyman-Pearson hypothesis testing theory in CSI as well. Type I (size) and Type II error (power) considerations dominate the scenario, and the so-called p-values or observed significance levels (OSL) play a basic role in this context. In a Bayesian set-up, in view of the emphasis on the prior(s) chosen (on the parameters or the probability laws), it is quite natural to formulate such tests based on appropriate posterior laws. We like to formulate such Bayes tests in an introductory fashion, and highlight their basic features.

We follow the same notation as in Section 8.2. For a random element $\boldsymbol{X}$, we denote the likelihood function for a given parameter $\boldsymbol{\theta}$ as $L_n(\boldsymbol{X}, \boldsymbol{\theta})$. Suppose now that $\boldsymbol{\theta}$ lies in a parameter space $\boldsymbol{\Theta}$. Keeping in mind the variety of models we have encountered earlier, we formulate the null and the alternative hypotheses as

$$H_0 : \boldsymbol{\theta} \in \boldsymbol{\Theta}_0 \quad \text{and} \quad H_1 : \boldsymbol{\theta} \in \boldsymbol{\Theta}_1,$$

where $\boldsymbol{\Theta}_0$ and $\boldsymbol{\Theta}_1$ are disjoint, and $\boldsymbol{\Theta}_0 \cup \boldsymbol{\Theta}_1 \subset \boldsymbol{\Theta}$. In a Bayes testing framework, the priors on $\boldsymbol{\theta}$ are therefore chosen on the appropriate domains under the two hypotheses. Keeping that in mind, we denote by $\Lambda_0(\boldsymbol{\theta})$ and $\Lambda_1(\boldsymbol{\theta})$ the priors under H_0 and H_1, respectively; these are defined exclusively on the two disjoint parameter spaces $\boldsymbol{\Theta}_0$ and $\boldsymbol{\Theta}_1$. They can also be defined on the entire parameter space $\boldsymbol{\Theta}$ with an understanding that these two priors are orthogonal. Let then

$$h_j(\boldsymbol{X}) = \int L_n(\boldsymbol{X}, \boldsymbol{\theta}) d\Lambda_j(\boldsymbol{\theta}), \ j = 0, 1.$$

Note that if we write

$$\Lambda(\boldsymbol{\theta}) = \Lambda_0(\boldsymbol{\theta}) I(\boldsymbol{\theta} \in \boldsymbol{\Theta}_0) + \Lambda_1(\boldsymbol{\theta}) I(\boldsymbol{\theta} \in \boldsymbol{\Theta}_1),$$

then the marginal likelihood of $\boldsymbol{X}$ is given by

$$\begin{aligned} p(\boldsymbol{X}) &= \int_{\boldsymbol{\Theta}} L_n(\boldsymbol{X}, \boldsymbol{\theta}) d\Lambda(\boldsymbol{\theta}) \\ &= \int_{\boldsymbol{\Theta}_0} L_n(\boldsymbol{X}, \boldsymbol{\theta}) d\Lambda(\boldsymbol{\theta}) + \int_{\boldsymbol{\Theta}_1} L_n(\boldsymbol{X}, \boldsymbol{\theta}) d\Lambda(\boldsymbol{\theta}) \\ &= \int_{\boldsymbol{\Theta}_0} L_n(\boldsymbol{X}, \boldsymbol{\theta}) d\Lambda_0(\boldsymbol{\theta}) + \int_{\boldsymbol{\Theta}_1} L_n(\boldsymbol{X}, \boldsymbol{\theta}) d\Lambda_1(\boldsymbol{\theta}) \\ &= h_0(\boldsymbol{X}) + h_1(\boldsymbol{X}). \end{aligned} \tag{8.89}$$

As a result, by the classical Bayes Theorem, the posterior distribution of $\boldsymbol{\theta}$, i.e., the conditional density of $\boldsymbol{\theta}$, given $\boldsymbol{X}$, is

$$g(\boldsymbol{\theta}|\boldsymbol{X}) = L_n(\boldsymbol{X}, \boldsymbol{\theta}) \Lambda(\boldsymbol{\theta}) / p(\boldsymbol{X}), \ \boldsymbol{\theta} \in \boldsymbol{\Theta}.$$

This gives us a way to evaluate the posterior probability of Θ_0 and Θ_1, the parameter space under the null and alternative hypothesis. These are

$$\begin{aligned} \alpha_0 &= \int_{\Theta_0} g(\boldsymbol{\theta}|\boldsymbol{X})d\boldsymbol{\theta} \\ &= h_0(\boldsymbol{X})/\{h_0(\boldsymbol{X}) + h_1(\boldsymbol{X})\}; \qquad (8.90) \\ \alpha_1 &= \int_{\Theta_1} g(\boldsymbol{\theta}|\boldsymbol{X})d\boldsymbol{\theta} \\ &= h_1(\boldsymbol{X})/\{h_0(\boldsymbol{X}) + h_1(\boldsymbol{X})\}. \qquad (8.91) \end{aligned}$$

Visualizing α_0 and α_1 as the *posterior confidence* or evidence in favor of H_0 and H_1, respectively, it is quite intuitive to consider the following (Bayes) test:

$$\begin{array}{c} \text{Reject } H_0 \text{ in favor of } H_1 \text{ when } \alpha_0 < \alpha_1 ; \\ \text{reject } H_1 \text{ in favor of } H_0 \text{ when } \alpha_1 < \alpha_0. \end{array}$$

Certain modifications may be needed if $\alpha_0 = \alpha_1$ can occur with a positive probability. Viewed from this Bayesian point, we may note that the Type I and Type II errors in a conventional Neyman-Pearson set-up are not that much incorporated in the formulation of the above test procedure. This can, however, be done with a slight adjustment. For example, if we consider a test function $\phi(\boldsymbol{X})$ (assuming values in the unit interval [0, 1]) defined by

$$\phi(\boldsymbol{X}) = 1 \text{ or } 0, \text{ according as } \frac{h_1(\boldsymbol{X})}{h_0(\boldsymbol{X})} > \text{ or } < k,$$

where $[\boldsymbol{X} : h_1(\boldsymbol{X}) = kh_0(\boldsymbol{X})]$ is of measure zero, then the number k can be so chosen to satisfy the Type I error-bound requirement. Further, in the presence of minimal sufficient statistics, admissibility and other properties of such Bayes tests can also be established (we refer to Chapter 6 in Lehmann (1994)) for some of these technicalities.

Let us now appraise this scenario in the context of CSI. For simplicity, let us consider the multinormal mean testing problem against the positive orthant alternative when the dispersion matrix is p.d. but unspecified. In this case, the parameter space Θ_0, under the null hypothesis is $\{(\mathbf{0}, \boldsymbol{\Sigma}) : \boldsymbol{\Sigma} \text{ is p. d.}\}$. Similarly, $\Theta_1 = \{\mathbb{R}^{p+}, \boldsymbol{\Sigma} \text{ p.d.}\}$. In the case of global alternatives, the role of the nuisance dispersion matrix is diffused by invoking affine invariance, and then appealing to the minimal sufficiency principle. In the present context, Θ_1 is not affine invariant, and hence, an invariant Bayes test may not be of much appeal.

The complexity of the prior on $\boldsymbol{\theta}$ in this testing problem is mainly due to the fact, for a given $\boldsymbol{\Sigma}$ (p.d.), the mean vector over $\mathbb{R}^{+p}$ and then the prior on $\boldsymbol{\Sigma}$ on the class of all p.d. matrices. For the second one, inverted Wishart distributional prior could be quite appealing. However, lacking affine-invariance (of the positive orthant), the product prior would not be simple to manipulate. Primarily, encountered with such roadblocks, Sugiura (1999) considered diffusion priors and formulated Bayes tests for such constrained inference problems. Nevertheless, it exhibits that some of the tests based on single contrasts (e.g., Tukey-Abelson test) could be interpreted as Bayes tests. We do not want to pursue this problem in greater depth here.

8.7 SOME DECISION THEORETIC ASPECTS: HYPOTHESIS TESTING

Admissibility and optimality (in a Bayes sense or not) and other desirable properties of statistical tests in a multidimensional parameter space have captured the attention of mathematical statisticians for more than six decades. Because of the fact that the genesis of such tests lies in the classical Neyman-Pearson (NP) lemma and its evolution to the gallery of likelihood based tests, particularly, the omnibus likelihood ratio test, attention has been mostly focused on such properties of the *LRT* and its siblings.

Even confined to the so called exponential family of distributions, in a multiparameter set-up, *uniformly most powerful* (UMP) tests may not generally exist. A classical example is the multinormal mean testing problem when the dispersion matrix is unspecified, albeit assumed to be of full rank. First, in the presence of such nuisance parameters (here the dispersion matrix), the classical NP lemma is not totally usable, and, hence, one may need to go through its generalizations. Secondly, the concept of UMP tests has to be amended to UMP similar regions, if they exist; the classical Neyman-structure of probability densities is a landmark in this respect. Finally, even if we look into the parameters under testing, there being a multidimensional alternative parameter space, UMP similar regions may not generally exist. If we go back to the multinormal mean testing problem, and, for simplicity, take the null hypothesis as the mean vector being null, against the global alternative that it is not null, for different rays in the alternative space, we may have different optimal similar regions, but none of them remain optimal over the entire parameter space under alternatives. For the one-dimensional case, by restricting to the class of unbiased tests, UMPU similar regions were advocated by Neyman and Pearson; we refer to Lehmann (1994) for a nice treatise of this subject matter.

Restricting by unbiasedness may not necessarily provide a resolution in a multiparameter hypothesis testing problem. Further restriction on the class of tests may therefore be necessary for a resolution. Simaika (1941), for the multinormal mean testing problem against global alternatives, made use of the affine invariance restraint, and showed that the Hotelling T^2-test is UMPI (invariant) test. If we invoke the (Bayesian) criterion of admissibility, Simaika's remarkable observation leads us to the basic property that the Hotelling T^2-test is *admissible* for global alternatives. Stein (1956) also provided a proof of the admissibility of Hotelling T^2-test by invoking the exponential structure of the parameter space. Alas, such a claim may not be valid for the CSI problems under consideration!

A related concept of *best average power* of a test has also been on statisticians' desk for more than 60 years; it involves the integration of the power function of a test over all possible direction in the alternative space wherein a contour is characterized by the very nature of the hypothesis testing problem. For example, in the multinormal mean $\boldsymbol{\theta}$ testing problem where $H_0 : \boldsymbol{\theta} = \mathbf{0}$ and $H_1 : \boldsymbol{\theta} \neq \mathbf{0}$, such a contour is taken as $\boldsymbol{\theta} \in \Theta_\Delta, \Delta = \boldsymbol{\theta}^T \Sigma^{-1} \boldsymbol{\theta} \geq 0$. While in this case, the affine invariance property may enable us to justify unequivocally this contour, in a general case, there remains the arbitrariness of the choice of a contour on which the best average power criterion rests.

Bayesians, some of whom, being the disciples of Charles Stein, have a handle over this tricky situation by formulating suitable *complete class theorems* for multiparameter hypothesis testing problems, and then using that for establishing admissibility and other desirable properties. Such a prescription may work out better for CSI where invariance constraints are of very limited use. We refer to Eaton (1970), where a complete class theorem for multidimensional one-sided alternatives has been elegantly established; it may still be a problem to verify his regularity conditions in a specific problem. Toward this, we consider the multinormal mean vector testing problem against the positive orthant alternative problem. The parameter $\boldsymbol{\theta}$ under the alternative is not affine invariant, and as a result, the UMPI criterion may not be tenable. In his seminal paper, Perlman (1969) considered this one-sided hypothesis testing problem in an elegant way, and yet his results show that the restricted *LRT* may not be unbiased, not to speak of being optimal (or at least, admissible) in a general sense. On the same track, the UIT may have the same drawbacks. There are other tests, discussed in Chapters 3–5, which are not invariant but unbiased for such restricted alternatives. A possible remedy to this undesirable property of *LRT* as well as UIT is to take recourse to a novel idea of Stein (1945) and to consider some *two-stage* tests; we refer to Sen and Tsai (1999) for an exploration of such two-stage *LRT* and UIT in multiparameter one-sided alternative hypothesis testing problems.

This CSI hypothesis testing problem has received considerable attention in the recent past; Perlman and Wu (1999) and Wang and McDermott (1998a) have attempted to bring more rationality to the use of *LRT* for CSI by invoking the principality of *conditionality*, although there could be some theoretical impasses in this detour. We refer to Perlman and Wu (2002a) and Perlman and Wu (2002b) for some further clarification of their philosophical stand in this respect, eliminating some of the minor inconsistencies in Perlman (1969). The issue of admissibility and other properties of *LRT* and UIT in CSI remains unresolved to a greater extent even now a days.

We conclude this section with a note on the Hotelling T^2-test in the context of the multinormal mean testing problem against the positive orthant alternatives. First, we note that T^2-test is unbiased and UMPI for the global alternatives; for this problem, it is equivalent to the *LRT*. But, for one-sided alternatives, the UIT and *LRT* exploit Hotelling T^2-statistics in different and possibly non-equivalent way (refer to Section 5.2). It is intuitively appealing to discard the Hotelling T^2 test for one-sided alternatives: for the one-dimensional case, T^2-statistic is isomorphic to the two-sided Student's t-test, and hence, is dominated by the one-sided t-test which is UMP similar. Given this clear inadmissibility of the T^2-test for the univariate case, it may be tempting to categorically label the same as inadmissible for the general multinormal mean testing problem against one-sided alternatives. The lack of invariance of the positive orthant, under affine transformations, adds more fuel to such a line of thinking. This is a challenging theoretical problem, and is left to time to come-up with a resolution in either way.

Tsai and Sen (2004) have considered a partial resolution. For the multinormal mean testing problem against the positive orthant alternatives, by using the elegant results of Eaton (1970), they have shown that the UIT and *LRT* can both be modified to a minor extent when the dispersion matrix belongs to a specific class (to be illustrated later), and such modified tests become members of Eaton's *essentially complete class* of tests. Further, it is shown there that the admissibility of Hotelling T^2-test is essentially tied down to the structure of the dispersion matrix $\mathbf{\Sigma}$. For a completely arbitrary (albeit p.d.) $\mathbf{\Sigma}$, the inadmissibility issue of the Hotelling T^2-test remains unresolved in a general CSI context. However, the inadmissibility of the Hotelling T^2-test has been established in a special class of the dispersion matrix.

A p.d. square matrix $\boldsymbol{A} = ((a_{ij}))$ with the $a_{ij} \leq 0,\ \forall i \neq j$ is termed an M-matrix, if and only if $\boldsymbol{A}^{-1} = ((a^{ij}))$ exists and further that $a^{ij} \geq 0,\ \forall i, j$. (see Tong (1990) page 78). If $\mathbf{\Sigma}^{-1}$ is an M-matrix, then all the elements of $\mathbf{\Sigma}$ are nonnegative (the diagonal ones being all positive), and this case arises in a class of important testing problems, as has been observed in Chapters 3–5. Tsai and Sen (2004) have exploited this M-matrix property and shown that if $\mathbf{\Sigma}^{-1}$ is an M-matrix, modified UIT and *LRT* are members of Eaton's essentially complete class of tests while the Hotelling T^2-test is inadmissible. An intuitive explanation for this is the following. If $\mathbf{\Sigma}^{-1} = ((\sigma^{ij}))$ is an M-matrix, then the off-diagonal σ^{ij} are all nonpositive, and this makes the noncentrality parameter $\boldsymbol{\theta}^T \mathbf{\Sigma}^{-1} \boldsymbol{\theta}$ of the Hotelling T^2 smaller in the positive orthant than in their image-points in quadrants 2 and 4. Thus, the average power gains its mass from alternatives in quadrants 2 and 4, and so when confined only to the positive orthant (quadrant 1), its noncentrality would be considerably smaller, adding to its inadmissibility. We refer to Tsai and Sen (2004) for derivation.

We have just touched on some decision theoretic aspects of CSI tests. In a more general CSI testing context, there remain some challenging tasks to innovate such Bayesian tools to fathom out the intricacies of optimal CSI. At the present, this is somewhat outside the scope of this monograph. However, we would like to draw the attention of serious, mature readers to this open area of statistical research.

Problems

8.1 Let X_1, X_2 be two independent r.v.'s, each having the Cauchy $(\theta, 1)$ density. Let $\tilde{\theta} = (X_1 + X_2)/2$ and $Z = (X_1 - X_2)/2$. Show that both $\tilde{\theta}, Z$ have marginally a Cauchy density; the density for Z being free from θ. Can you claim that $\tilde{\theta}$ and Z are stochastically independent? Show that the conditional density of $\tilde{\theta}$, given Z, is symmetric about θ, continuous and further that $\tilde{\theta}$ is median-unbiased for θ. Show that neither $\tilde{\theta}$ nor Z has finite first moment, and hence, the classical Bayes approach may not work out in this specific case.

8.2 Write down the expression for the density function corresponding to the noncentral chi-square distribution with p degrees of freedom and noncentrality parameter Δ. Using this result for $\boldsymbol{X}^T \boldsymbol{X}$, verify that for $p \geq 3$,

$$E_{\boldsymbol{\theta}}(\boldsymbol{X}^T \boldsymbol{X})^{-1} = e^{-\Delta/2} \sum_{r \geq 0} (\Delta/2)^r (r!)^{-1} (p + 2r - 2)^{-1}.$$

Show by differentiating with respect to Δ that the above expression is a monotone decreasing function of $\Delta(\geq 0)$, and at $\Delta = 0$, it is equal to $(p-2)^{-1}$. Further, let K be a r.v. having the Poisson distribution with mean $\Delta/2$. Write down the expression for the probability function of K, and, hence, or otherwise, show that $E_{\boldsymbol{\theta}}(\boldsymbol{X}^T\boldsymbol{X})^{-1}) = E(p-2+2K)^{-1}$.

8.3 For the previous problem, show that for any positive integer r, if $p \geq 2r+1$, then

$$E_{\boldsymbol{\theta}}(\boldsymbol{X}^T\boldsymbol{X})^{-r} = E\{(p-2+2K)...(p-2r+2K)\}^{-1}.$$

8.4 For the estimator in (8.42) if $\boldsymbol{X}$, given $\boldsymbol{\theta}$ is multinormal and $\boldsymbol{\theta}$ has also a multinormal law, obtain (i) the marginal law of $\boldsymbol{X}$ (by integrating the joint density of $\boldsymbol{X}$ and $\boldsymbol{\theta}$ over $\boldsymbol{\theta}$), and (ii) the conditional density of $\boldsymbol{\theta}$, given $\boldsymbol{X}$. Show that both are multinormal. Hence, or otherwise, verify the result in (8.43).

8.5 Consider the marginal density of $\boldsymbol{X}$ (in the previous problem), and verify that $||\boldsymbol{X} - \boldsymbol{\mu}||^2/(1+c)$, under the marginal law for $\boldsymbol{X}$, has the central chi-square distribution with p degrees of freedom. Hence, use the result in Problem 8.2 and verify that (8.44) holds.

8.6 Consider the SRRMLE in (8.49), and consider the case of $p \geq 3$. For any $a \subset P : c_a$ is positive (possible only for $|a| \geq 3$), show that

$$P_{\boldsymbol{\theta}}\{||\hat{\boldsymbol{\theta}}^*_{RM}||^2 < c_a\} > 0,$$

for all $\boldsymbol{\theta} \in \boldsymbol{\Theta}^+$. What can you say about this probability when $\boldsymbol{\theta}$ approaches the vertex $\mathbf{0}$?

8.7 Consider for simplicity the James-Stein estimator (worked out for $\boldsymbol{\Sigma} = \boldsymbol{I}$), and show that if actually the underlying dispersion matrix $\boldsymbol{\Sigma}$ has the property that $\boldsymbol{\Sigma}^{-1}$ is an M-matrix, then the dominance of the SR estimator may not hold for the positive orthant restricted problem. For simplicity, consider the case of $p = 3$ and a (negative) intraclass correlation pattern.

8.8 For the univariate linear model with errors normally distributed, show that the MLE $\hat{\boldsymbol{\beta}}$ being a linear estimator is also normally distributed. Also, use orthogonal transformation on the original observation matrix (of order $n \times p*$) and show that $(n-p*)\hat{\sigma}^2/\sigma^2$ has the central chi-square distribution with $n-p*$ degrees of freedom.

8.9 For $p \leq 2$, show that the density function of a χ^2_p variable is monotone nonincreasing, while for $p \geq 3$, it is bell-shaped. Does it belong to the class of log-concave densities or strongly unimodal densities? Verify that the mode of χ^2_p is equal to $p-2$ and its mean is p. Invoke the mean-median mode inequality for bell-shaped distribution (i.e., mean - mode is approximately equal to 3(mean − median) and mode $\leq$ median $\leq$ mean) and check how close are the values with these approximations.

9

Miscellaneous Topics

The general theory of likelihood-based inference for inequality-constrained problems was developed in Chapters 2 and 3. Other methods/approaches such as union intersection principle, intersection union test, and nonparametric methods were studied in Chapters 4 and 5. The results in the earlier chapters were used in Chapter 6 to formulate and solve specific problems that arise in the analysis of ordinal categorical data. There are several other areas in which the results of Chapters 2 to 5 have been found to be useful. In this chapter, we shall consider some developments that are of interest in the area of constrained statistical inference.

Section 9.1 considers the comparison of two treatments when the response is multivariate and the objective is to establish that one treatment is better than the other. This is the natural extension of the traditional two-sample problem with unrestricted alternative. This problem arises in many areas of applications, including clinical trials. In such trials, the multivariate response vector is sometimes referred to as *multiple end points*. Section 9.2 considers the closely related statistical inference problem for establishing that a particular treatment is the best among a group of treatments. The methods of Section 9.1 are mainly for settings in which the sample mean is approximately multivariate normal and the hypothesis testing problem would typically be formulated in terms of the mean of the response variable. By contrast, much of the attention in Section 9.2 is focused on univariate responses, although the ideas extend to multivariate settings as well. Section 9.3 provides a detailed account of the developments on *crossover interaction*; if one treatment is better than another under one set of conditions but worse under another, then we say that there is a crossover interaction. The methodological issues discussed in this section deals with some nonstandard problems as well. A general class of tests called *directed tests* is studied in Section 9.4. These are different from the likelihood ratio tests. Even for the simple problem of testing $\boldsymbol{\mu} = \mathbf{0}$ against $\boldsymbol{\mu} \geqslant \mathbf{0}$ based on a single observation of $\boldsymbol{X}$ where $\boldsymbol{X} \sim N(\boldsymbol{\mu}, \boldsymbol{V})$ and $\boldsymbol{V}$ is known, it is not possible to develop a test that has all the desirable properties. Consequently, there is no single best test. Recently there have been debates about the suitability of different tests for this and many other similar testing problems. Cohen and Sackrowitz (2004), Chaudhuri and Perlman (2004), and the references therein provide excellent coverage of the debate.

9.1 TWO-SAMPLE PROBLEM WITH MULTIVARIATE RESPONSES

Experiments are often carried out to establish that one treatment is better than another with respect to a multivariate response variable. Although different statistical procedures are available for this inference problem, none of them can be claimed to be the best or suitable for every problem. To formulate the problem, let $\boldsymbol{X}_1$ and $\boldsymbol{X}_2$ denote the p-variate responses associated with Treatments 1 and 2, respectively, $\boldsymbol{\mu}_i = E(\boldsymbol{X}_i)$ for $i = 1, 2$, and let $\boldsymbol{\theta} = \boldsymbol{\mu}_1 - \boldsymbol{\mu}_2$; thus, $\boldsymbol{\mu}_i = (\mu_{i1}, \ldots, \mu_{ip})^T$ and $\boldsymbol{\theta} = (\theta_1, \ldots, \theta_p)^T$ where $\theta_j = \mu_{1j} - \mu_{2j}$, $j = 1, \ldots, p$. Without loss of generality, assume that large values of μ_{ij} are preferred. Thus, if $\theta_1 \geqslant 0, \ldots, \theta_p \geqslant 0$ with at least one of them being a strict inequality, then Treatment 1 is better than Treatment 2. However, this is not the only possible formulation of "Treatment 1 is better than Treatment 2"; other formulations will be considered in this section.

Let $\boldsymbol{X}_{11}, \ldots, \boldsymbol{X}_{1n_1}$ be *iid* as $\boldsymbol{X}_1$ and let $\boldsymbol{X}_{21}, \ldots, \boldsymbol{X}_{2n_2}$ be *iid* as $\boldsymbol{X}_2$. Throughout this section it will be assumed that these two samples are independent. Let $\bar{\boldsymbol{X}}_i$ denote the sample mean of the n_i observations for sample i $(i = 1, 2)$. We are interested to test the claim that Treatment 1 is better than Treatment 2. Let us formulate this as test of

$$H_0 : \boldsymbol{\theta} = \mathbf{0} \qquad \text{vs} \qquad H_1 : \boldsymbol{\theta} \geqslant \mathbf{0}. \tag{9.1}$$

For simplicity, let us first consider the case when $p = 2$ and $\bar{\boldsymbol{X}}_1 - \bar{\boldsymbol{X}}_2 \sim N(\boldsymbol{\theta}, \boldsymbol{I})$ where $\boldsymbol{\theta}$ can be any point in $\mathbb{R}^2$. It may be verified that if $\bar{\boldsymbol{X}}_1 - \bar{\boldsymbol{X}}_2 = (-100, 100)^T$ then the *LRT* of H_0 against H_1 has a p-value of approximately zero, and hence would reject $H_0 : \boldsymbol{\theta} = \mathbf{0}$. Now, a question that arises is, can we conclude that $\boldsymbol{\theta} \gneqq \mathbf{0}$ despite the fact that the first component of $(\bar{\boldsymbol{X}}_1 - \bar{\boldsymbol{X}}_2)$ is large and negative? It would be appropriate to conclude that there is statistical evidence in favor of $\boldsymbol{\theta} \neq \mathbf{0}$, but not necessarily in favor of $\boldsymbol{\theta} \geqslant \mathbf{0}$, even though $\boldsymbol{\theta} \geqslant \mathbf{0}$ was the alternative hypothesis. Rejection of $\boldsymbol{\theta} = \mathbf{0}$ alone is inadequate to conclude that $\boldsymbol{\theta} \gneqq \mathbf{0}$. Now, let us also suppose that it is known *a priori* that $\boldsymbol{\theta} \geqslant \mathbf{0}$. Then, we can conclude that $\boldsymbol{\theta} \gneqq \mathbf{0}$. In this case, $\boldsymbol{\theta} \neq \mathbf{0}$ is based on statistical evidence and $\boldsymbol{\theta} \geqslant \mathbf{0}$ is based on *a priori* knowledge; together, they lead to $\boldsymbol{\theta} \gneqq \mathbf{0}$.

The foregoing example illustrates the following: if a test of hypotheses rejects the null then it does not necessarily follow that there is evidence to support the alternative hypothesis unless the latter is the complement of the null hypothesis. Unfortunately, it is not always possible to develop statistical methods when the null parameter space is the complement of the alternative parameter space. To illustrate this, suppose that it is not known *a priori* that $\boldsymbol{\theta} \geqslant \mathbf{0}$, and one of the objectives is to quantify the strength of statistical evidence in favor of $\boldsymbol{\theta} \gneq \mathbf{0}$. Since the alternative hypothesis is $H_1 : \boldsymbol{\theta} \geqslant \mathbf{0}$, let us consider the null hypothesis $H_0^* : \boldsymbol{\theta} \notin \mathcal{A}$ where $\mathcal{A} = \{(\theta_1, \theta_2) : \theta_1 = \theta_2 = 0,\ or\ \theta_1 < 0\ or\ \theta_2 < 0\}$. In this case, the null parameter space does not include its boundary, and therefore H_0^* is not a suitable null hypothesis for *LRT* and many other tests. One possible formulation to overcome this is

$$H_0 : \theta_1 \leq 0 \text{ or } \theta_2 \leq 0, \qquad \text{and} \qquad H_1 : \theta_1 > 0 \text{ and } \theta_2 > 0. \tag{9.2}$$

This says that Treatment 1 is better than Treatment 2 if the former is better than the latter with respect to every response. The class of intersection-union tests discussed in Chapter 5 offers a simple way of testing (9.2).

The formulation of H_1 in (9.2) may be too strong in some studies. A more liberal formulation that would be suitable for *LRT* and several other tests is the following (see Fig. 9.1): Let ϵ_1 and ϵ_2 be given nonnegative numbers and let

$$\Theta_1 = \{\boldsymbol{\theta} \in \mathbb{R}^2 : \theta_1 > -\epsilon_1, \theta_2 > -\epsilon_2 \text{ and } (\theta_1 > 0 \text{ or } \theta_2 > 0)\}. \tag{9.3}$$

Now, the null and alternative hypotheses are

$$H_0 : \boldsymbol{\theta} \notin \Theta_1 \quad \text{and} \quad H_1 : \boldsymbol{\theta} \in \Theta_1,$$

respectively. In this formulation, Treatment 1 is better than Treatment 2 if the former is better than the latter in terms of at least one response variable, and it is not inferior in terms of the other variable; here, *noninferiority* and *superiority*, with respect to the jth variable are formulated as $\theta_j > -\epsilon_j$ and $\theta_j > 0$, respectively. It is possible to modify the foregoing definition of Θ_1 in numerous other ways to capture different notions of what is meant by Treatment 1 is better than Treatment 2. The methods developed in this section can be applied to deal with several such different definitions of Θ_1.

In some studies, one is interested to detect differences between treatments that are of practical significance. In such cases, the problem may be formulated as follows. Let δ_1 and δ_2 be a given set of nonnegative numbers, and let the definition of Θ_1 in (9.3) be modified to

$$\Theta_1 = \{\boldsymbol{\theta} \in \mathbb{R}^2 : (\theta_1 > -\epsilon_1, \theta_2 > -\epsilon_2) \text{ and } (\theta_1 > \delta_1 \text{ or } \theta_2 > \delta_2)\}.$$

Now the different parts of the alternative hypothesis $H_1 : \boldsymbol{\theta} \in \Theta_1$ are interpreted as follows: If $\theta_j > \delta_j$ then Treatment 1 is better than Treatment 2 with respect to the jth variable and the difference is of practical significance; if $0 < \theta_j \leq \delta_j$ then Treatment 1 is still better than Treatment 2 but the difference may not be of practical significance. Without loss of generality, this inference problem can be seen to be the same as that in the previous paragraph by redefining θ_j as $\theta_j - \delta_j$ for every j.

It is quite possible for the response variables to be of unequal importance. For example, consider the comparison of two treatments for ulcer; Treatment 1, a new one, is being compared with an old one, Treatment 2. In such comparative studies, usually there are *primary variables* and *secondary variables*. The primary variables

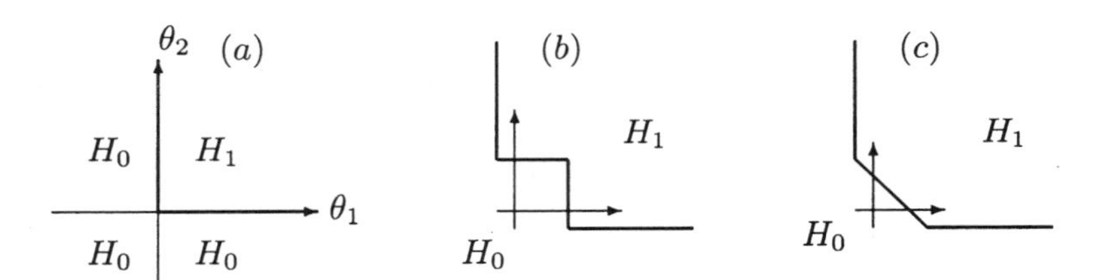

Fig. 9.1 Three possible formulations of the hypothesis that Treatment 1 is better than Treatment 2 when the response is bivariate. (a) Treatment 1 is superior with respect to each response variable: $H_1 : \theta_1 > 0$ and $\theta_2 > 0$; (b) Treatment 1 is noninferior with respect to any variable, and is superior with respect to at least one variable by a practically significant amount: $H_1 : (\theta_1 > -\epsilon_1, \theta_2 > -\epsilon_2)$ and $(\theta_1 > \delta_1$ or $\theta_2 > \delta_2)$; (c) Treatment 1 is noninferior with respect to any variable, and is superior overall by a practically significant amount: $H_1 : (\theta_1 > -\epsilon_1, \theta_2 > -\epsilon_2)$ and $(\theta_1 + \theta_2) > \delta)$.

may include direct effects on the ulcer, and secondary variables may include side effects of the treatment. The new treatment would be considered better than the old if the former is better in terms of the primary variables but not significantly worse in terms of the secondary variables. A small cost in terms of side effects may be tolerable if the treatment performs better in terms of its primary objective of treating the ulcer. One possible formulation of this testing problem is the following. Let $\boldsymbol{X}_1$ and $\boldsymbol{X}_2$ refer to the primary response variables corresponding to Treatments 1 and 2, respectively, and similarly let $\boldsymbol{Z}_1$ and $\boldsymbol{Z}_2$ denote the secondary response variables. Let $\boldsymbol{\beta}_i = E(\boldsymbol{Z}_i), i = 1, 2$. Now the null and alternative hypotheses may take the form

$$H_0 : \boldsymbol{\theta} \notin \Theta_1 \text{ or } \boldsymbol{\beta}_1 - \boldsymbol{\beta}_2 \leq -\boldsymbol{\eta}, \text{ and } H_1 : \boldsymbol{\theta} \in \Theta_1 \text{ and } \boldsymbol{\beta}_1 - \boldsymbol{\beta}_2 > -\boldsymbol{\eta}, \quad (9.4)$$

where Θ_1 could be as in (9.3), and $\boldsymbol{\eta}$ is to be chosen so that $\boldsymbol{\beta}_1 - \boldsymbol{\beta}_2 > -\boldsymbol{\eta}$ corresponds to Treatment 1 being not significantly worse than Treatment 2 in terms of the secondary variables. One way of testing this at level α is to apply an IUT as follows: test $\boldsymbol{\theta} \notin \Theta_1$ against $\boldsymbol{\theta} \in \Theta_1$ at level α, and test $\boldsymbol{\beta}_1 - \boldsymbol{\beta}_2 \leq -\boldsymbol{\eta}$ against $\boldsymbol{\beta}_1 - \boldsymbol{\beta}_2 > -\boldsymbol{\eta}$ at level α; if both of them reject their respective null hypotheses then accept H_1 in (9.4) at level α. Alternatively, if the p-values for the two tests are p_1 and p_2 then an upper bound for the p-value of the corresponding IUT of (9.4) is $\max\{p_1, p_2\}$.

In the foregoing discussions we assumed that there were no interim analyzes. Tang et al. (1989b, 1993) and Jennison and Turnbull (1993) have considered the two sample problem with interim analyzes.

Although we formulated the problem as one of testing a null against an alternative, it is also possible to formulate it differently as a multidecision problem. In this formulation, the parameter space is divided into more than two regions as follows.

Region 1: Treatment 1 is better than Treatment 2;

Region 2 : Treatment 1 is worse than Treatment 2;

Region 3 : Treatments 1 and 2 do not differ by a practically significant amount;

Region 4 : None of the treatments is the best and they are not equivalent either. Thall, Simon, and Shen (2000) studied a Bayesian approach to this problem.

In the subsequent subsections we shall consider test of $H_0 : \boldsymbol{\theta} = \mathbf{0}$ against $H_1 : \boldsymbol{\theta} \geqslant \mathbf{0}$ and of $H_0 : \boldsymbol{\theta} \notin \Theta_1$ against $H_1 : \boldsymbol{\theta} \in \Theta_1$ where Θ_1 is as in (9.3). We believe that the latter testing problem is likely to play an important role in applications. However, in what follows, we shall consider the former in greater details. This is a reflection of the status of the current literature on this topic.

9.1.1 One Degree of Freedom Tests of $H_0 : \boldsymbol{\mu}_1 - \boldsymbol{\mu}_2 = 0$ Against $H_1 : \boldsymbol{\mu}_1 - \boldsymbol{\mu}_2 \geqslant 0$

As in (9.1), consider the two-sample testing problem

$$H_0 : \boldsymbol{\theta} = \mathbf{0} \qquad \text{vs} \qquad H_1 : \boldsymbol{\theta} \geqslant \mathbf{0}, \tag{9.5}$$

where $\boldsymbol{\theta} = \boldsymbol{\mu}_1 - \boldsymbol{\mu}_2$. In this subsection, we simplify the problem by testing against an alternative that specifies a departure from $\boldsymbol{\theta} = \mathbf{0}$ in a prechosen direction of $\{\boldsymbol{\theta} : \boldsymbol{\theta} \geqslant \mathbf{0}\}$. Consequently, these tests tend to have good power for values of $\boldsymbol{\theta}$ close to the particular prechosen direction. These are essentially univariate methods. Multivariate methods of testing $H_0 : \boldsymbol{\theta} = \mathbf{0}$ against $H_1 : \boldsymbol{\theta} \geqslant \mathbf{0}$ will be discussed in a later subsection. As a general guide, the multivariate methods could be expected to have better average power over $\{\boldsymbol{\theta} \geqslant \mathbf{0}\}$ than the simpler methods of this subsection, which are likely to have good power in the chosen direction.

Let $\boldsymbol{a} = (a_1, \ldots, a_p)^T$ be a given direction in the positive orthant; thus, each a_i is nonnegative. Now consider the hypotheses

$$H_0^* : \boldsymbol{a}^T\boldsymbol{\theta} = 0 \qquad \text{and} \qquad H_1^* : \boldsymbol{a}^T\boldsymbol{\theta} > 0. \tag{9.6}$$

If $H_0^* : \boldsymbol{a}^T\boldsymbol{\theta} = 0$ is rejected then it follows that $H_0 : \boldsymbol{\theta} = \mathbf{0}$ can also be rejected. Clearly, a level-α test of H_0^* against H_1^* is a also a level-α test of H_0 against H_1. Therefore, a simple approach to testing H_0 against H_1 is to test H_0^* against H_1^*. Such a test may be developed as follows.

Let $\boldsymbol{U}$ be a statistic such that $(\boldsymbol{U} - \boldsymbol{\theta}) \approx N(\mathbf{0}, \boldsymbol{A})$; for example $\boldsymbol{U}$ can be $\bar{\boldsymbol{X}}_1 - \bar{\boldsymbol{X}}_2$. Then $(\boldsymbol{a}^T\boldsymbol{U} - \boldsymbol{a}^T\boldsymbol{\theta}) \approx N(\mathbf{0}, \boldsymbol{a}^T\boldsymbol{A}\boldsymbol{a})$. Therefore, a simple statistic for testing H_0^* against H_1^* is $T = \boldsymbol{a}^T\boldsymbol{U}/(\boldsymbol{a}^T\hat{\boldsymbol{A}}\boldsymbol{a})^{1/2}$ where $\hat{\boldsymbol{A}}$ is a consistent estimator of $\boldsymbol{A}$. Under H_0^*, T is approximately $N(0,1)$, and the test would reject H_0^* if T is large and positive. For testing H_0 against H_1, T would have good power for values of $\boldsymbol{\theta}$ close to the direction $\boldsymbol{a}$, but it is likely to decline quickly (compared with other multiparameter tests) as the direction of $\boldsymbol{\theta}$ moves away from $\boldsymbol{a}$. Thus, while $\boldsymbol{a}^T\boldsymbol{U}/(\boldsymbol{a}^T\hat{\boldsymbol{A}}\boldsymbol{a})^{1/2}$ is a simple statistic for testing H_0 against H_1, simplicity is likely to be achieved at the cost of loss of power overall for testing H_0 against H_1.

The foregoing procedure can be extended to tests concerning more general parameters. Let $\boldsymbol{\gamma} = (\gamma_1, \ldots, \gamma_p)^T$ be a general parameter of interest; for example, $\boldsymbol{\gamma}$ can be $\boldsymbol{\mu}_1 - \boldsymbol{\mu}_2$ or the difference between the two population medians. Let T_j be a statistic for testing $\gamma_j = 0$ against $\gamma_j > 0$ where large values of T_j support $\gamma_j > 0$.

Let $\boldsymbol{T} = (T_1, \ldots, T_p)^T$ and let the null and alternative hypotheses be $H_0^a : \boldsymbol{a}^T\boldsymbol{\gamma} = 0$ and $H_1^a : \boldsymbol{a}^T\boldsymbol{\gamma} > 0$. Suppose that, if $\boldsymbol{\gamma} = \boldsymbol{0}$ then $\boldsymbol{T} \xrightarrow{d} N(\boldsymbol{0}, \boldsymbol{B})$ for some $\boldsymbol{B}$. Let $\hat{\boldsymbol{B}}$ be a consistent estimator of $\boldsymbol{B}$. Then a statistic for testing H_0^a against H_1^a is

$$\boldsymbol{a}^T\boldsymbol{T}/(\boldsymbol{a}^T\hat{\boldsymbol{B}}\boldsymbol{a})^{1/2}. \tag{9.7}$$

In large samples, (9.7) is approximately $N(0, 1)$ under H_0^a.

If $T_1, \ldots, T_p$ have different units/scales then $T_1 + \ldots + T_p$ may not be a good test statistic. For example, if $T_1 \sim N(0, 1)$ and $T_2 \sim N(0, 100)$ under H_0, then a test based on $T_1 + T_2$ would be valid but it is likely to have poor power properties.

These types of tests are particularly useful when the response variables are similar or closely related. For example, consider the comparison of a new and expensive treatment program for depression with a standard one, in which (1) ethical and other cost considerations allow only a small sample size, (2) a large number of variables are measured on each patient to obtain as much information as possible to assess the efficacy of the treatments, and (3) sample size is too small to apply asymptotic multivariate results for statistical inference on a large number of free parameters. As an example, here we are thinking of samples of size around 20 with around 30 variables. In such cases, tests based on (9.7), with $\boldsymbol{a} = (1, \ldots, 1)^T$, for example, may offer reasonable solutions.

9.1.2 O'Brien's Test of $H_0 : \boldsymbol{\mu}_1 - \boldsymbol{\mu}_2 = 0$ vs $H_1 : \boldsymbol{\mu}_1 - \boldsymbol{\mu}_2 \geqslant 0$

O'Brien (1984) suggested two simple tests of $H_0 : \boldsymbol{\mu}_1 - \boldsymbol{\mu}_2 = \boldsymbol{0}$ against $H_1 : \boldsymbol{\mu}_1 - \boldsymbol{\mu}_2 \geqslant \boldsymbol{0}$; one is parametric and the other is nonparametric. Simplicity is achieved by reducing the multiparameter inference problem to one that involves only a scalar parameter, in a way that is very closely related to that in the previous subsection. These tests have attracted considerable attention in the area of multiple end points; these are discussed below.

Parametric Test:

Suppose that $\boldsymbol{X}_1 \sim N(\boldsymbol{\mu}_1, \boldsymbol{V}_1)$ and $\boldsymbol{X}_2 \sim N(\boldsymbol{\mu}_2, \boldsymbol{V}_2)$. Recall that $\boldsymbol{X}_{i1}, \ldots, \boldsymbol{X}_{in_i}$ are *iid* as $\boldsymbol{X}_i$, $\bar{\boldsymbol{X}}_i$ is the mean of these n_i observations $(i = 1, 2)$, and $\{\boldsymbol{X}_{11}, \ldots, \boldsymbol{X}_{1n_1}\}$ and $\{\boldsymbol{X}_{21}, \ldots, \boldsymbol{X}_{2n_2}\}$ are independent. In this subsection, we will assume that $\boldsymbol{V}_1 = \boldsymbol{V}_2$; let $\boldsymbol{V}$ denote the common value of $\boldsymbol{V}_1$ and $\boldsymbol{V}_2$, σ_j^2 denote the common variance V_{jj} of the jth response variable and let $\boldsymbol{\sigma} = (\sigma_1, \ldots, \sigma_p)^T$. Let $\boldsymbol{C}$ denote the correlation matrix corresponding to $\boldsymbol{V}$. Let $\boldsymbol{X}_{i\ell} = (X_{i\ell 1}, \ldots, X_{i\ell p})^T$; thus, $X_{i\ell j}$ corresponds to the ith treatment, ℓth observation and jth variable $(i = 1, 2, \ell = 1, \ldots, n_i, j = 1, \ldots, p)$. Let

$$\bar{X}_{ij} = n_i^{-1}\sum_{\ell=1}^{n_i} X_{i\ell j}, \quad s_j^2 = (n_1 + n_2 - 2)^{-1}\sum_{i=1}^{2}\sum_{\ell=1}^{n_i}(X_{i\ell j} - \bar{X}_{ij})^2,$$

$$T_j = (\bar{X}_{1j} - \bar{X}_{2j})/\{s_j(n_1^{-1} + n_2^{-1})^{1/2}\}, \qquad \boldsymbol{T} = (T_1, \ldots, T_p)^T,$$

$$\hat{C}_{qr} = (n_1 + n_2 - 2)^{-1} \sum_{i=1}^{2} \sum_{\ell=1}^{n_i} (X_{i\ell q} - \bar{X}_{iq})(X_{i\ell r} - \bar{X}_{ir})/\{s_q s_r\};$$

s_j^2 is the within sample pooled estimate of σ_j^2, $\hat{\boldsymbol{C}} = (\hat{C}_{qr})$ is the within sample pooled estimate of the correlation matrix and T_j is the two-sample t-statistic for testing $\mu_{1j} - \mu_{2j} = 0$ $(j = 1, \ldots, p)$.

To introduce O'Brien's statistic, let us first make the working assumption $\boldsymbol{\theta} = \lambda\boldsymbol{\sigma}$, where $\lambda \in \mathbb{R}$. This working assumption simplifies the task of developing a test of $H_0 : \boldsymbol{\theta} = \mathbf{0}$ against $H_1 : \boldsymbol{\theta} \geqslant \mathbf{0}$; however, the test is valid even if $\boldsymbol{\theta} \neq \lambda\boldsymbol{\sigma}$. Instead of $\boldsymbol{\sigma}$, we could have chosen any direction in the alternative parameter space of $\boldsymbol{\theta}$; the reason for choosing $\boldsymbol{\sigma}$ will be discussed later. Now, consider the hypotheses

$$H_0^* : \lambda = 0 \quad \text{and} \quad H_1^* : \lambda > 0.$$

Clearly, a test of H_0^* against H_1^* is also a valid test of H_0 against H_1. O'Brien's parametric test is essentially a generalized least squares-based test of H_0^* against H_1^*. This statistic may be expressed as

$$W = \boldsymbol{J}^T\hat{\boldsymbol{C}}^{-1}\boldsymbol{T}/\{\boldsymbol{J}^T\hat{\boldsymbol{C}}^{-1}\boldsymbol{J}\}^{1/2}, \quad \text{where} \quad \boldsymbol{J} = (1, \ldots, 1)^T. \tag{9.8}$$

The large sample null distribution of W is $N(0, 1)$ by central limit theorem, and the test rejects H_0^* if W is large.

In a simulation study, Pocock et al. (1987) observed that if the number of variables, p, is large, say 5 or more, then hundreds of observations may be needed for the null distribution of (9.8) to be close to $N(0, 1)$. They also observed that when $p = 2$, the distribution of (9.8) appears close to a t-distribution with $(n_1 + n_2 - 4)$ degrees of freedom for moderate sample sizes. The main point to note is that $N(0, 1)$ as a large sample approximation for the null distribution of W in (9.8) may not be good in moderate-sized samples if a high degree of accuracy is required. Simulation and bootstrap methods may be useful to improve on these approximations.

Nonparametric Test:

Since the components of $\boldsymbol{T}$ used in (9.8) are t-statistics and W is linear in $\boldsymbol{T}$, one would expect that the corresponding test would be sensitive to outliers. One way of ensuring that the test is robust against outliers is to replace $\bar{X}_{1j}$ and $\bar{X}_{2j}$ in the definition of T_j by robust estimators and make necessary changes to the denominator of W. Alternatively, one could replace T_j by a more robust two-sample test statistic.

O'Brien (1984) provided a nonparametric test of H_0 against H_1 based on ranks. Let $R_{i\ell j}$ denote the rank of $X_{i\ell j}$ among all values of variable j in the pooled set of data. Further, let $S_{i\ell} = \sum_j R_{i\ell j}$, the sum of the ranks assigned to the ℓ th unit in sample i $(i = 1, 2, \ell = 1, \ldots, n_i, j = 1, \ldots, p)$. Thus, $S_{i\ell}$ summarizes the overall effect of Treatment i on ℓ th observation. Now, a simple robust test of $H_0 : \boldsymbol{\theta} = \mathbf{0}$ against $H_1 : \boldsymbol{\theta} \geqslant \mathbf{0}$ is a one-sided univariate two-sample test of $\alpha = 0$ against $\alpha > 0$ based on the two independent samples, $\{S_{11}, \ldots, S_{1n_1}\}$ and $\{S_{21}, \ldots, S_{2n_2}\}$, where $\alpha = E(S_{1\ell}) - E(S_{2\ell})$.

Derivation of O'Brien's statistic W in (9.8):

Let us first suppose that $\boldsymbol{V}$, the common value of $\boldsymbol{V}_1$ and $\boldsymbol{V}_2$, is known, and let $\boldsymbol{Y}_i = (X_{i1}/\sigma_1, \ldots, X_{ip}/\sigma_p)^T, (i = 1, 2)$. Then $\boldsymbol{Y}_i \sim N(\boldsymbol{\beta}_i, \boldsymbol{C})$ where $\boldsymbol{\beta}_i = (\mu_{i1}/\sigma_1, \ldots, \mu_{ip}/\sigma_p)^T$ for i=1,2. Now, the null and alternative hypotheses in (9.1) are

$$H_0 : \boldsymbol{\beta}_1 - \boldsymbol{\beta}_2 = \boldsymbol{0} \qquad \text{and} \qquad H_1 : \boldsymbol{\beta}_1 - \boldsymbol{\beta}_2 \geqslant \boldsymbol{0},$$

respectively. Consider the null and alternative hypotheses

$$H_0^* : \lambda = 0 \qquad \text{and} \qquad H_1^* : \lambda > 0 \tag{9.9}$$

where $\boldsymbol{\beta}_1 - \boldsymbol{\beta}_2 = \lambda \boldsymbol{J}$ and $\boldsymbol{J}$ is the center direction, $(1, \ldots, 1)^T$. The main idea underlying O'Brien's approach for testing H_0 against H_1 is to test H_0^* against H_1^*; thus, this test attempts to detect departures from H_0 mainly in the center direction $\boldsymbol{J}$ in the parameter space $\{\boldsymbol{\beta}_1 - \boldsymbol{\beta}_2 \geqslant \boldsymbol{0}\}$. The choice of the center direction $\boldsymbol{J}$ to look for departure from the null appears reasonable based on symmetry resulting from equal variances for $Y_{i1}, \ldots, Y_{ip}$ and the assumed equal importance assigned to these variables.

Let $\hat{\lambda}$ denote the *mle* of λ. Then it may be verified that (see Exercise 9.1)

$$\hat{\lambda} = \boldsymbol{J}^T \boldsymbol{C}^{-1}(\bar{\boldsymbol{Y}}_1 - \bar{\boldsymbol{Y}}_2)/(\boldsymbol{J}^T \boldsymbol{C}^{-1} \boldsymbol{J}).$$

Therefore, the statistic, $\{\hat{\lambda}/se(\hat{\lambda})\}$, for testing H_0^* against H_1^* is

$$\boldsymbol{J}^T \boldsymbol{C}^{-1}(\bar{\boldsymbol{Y}}_1 - \bar{\boldsymbol{Y}}_2)/\{\boldsymbol{J}^T \boldsymbol{C}^{-1} \boldsymbol{J}(n_1^{-1} + n_2^{-1})\}^{1/2}; \tag{9.10}$$

this is distributed as $N(0, 1)$ under H_0^*, and is expected to be large under H_1^*.

Now, let us relax the assumption that $\boldsymbol{V}$ is known. Then (9.10) depends on unknown parameters. Let us substitute $\hat{\boldsymbol{C}}$ for $\boldsymbol{C}$ and s_j for σ_j $(j = 1, \ldots, p)$ in (9.10); consequently, $\bar{\boldsymbol{Y}}_i$ would be replaced by $(\bar{X}_{i1}/s_1, \ldots, \bar{X}_{ip}/s_p)^T$. Now (9.10) reduces to the O'Brien's statistic W in (9.8). If the observations are not assumed to be normally distributed, then the foregoing arguments based on maximum likelihood can be seen as those based on generalized least square (GLS), and the null distribution of (9.10) would be approximately $N(0, 1)$ in large samples.

It follows from the definition of $\boldsymbol{\beta}_1$ and $\boldsymbol{\beta}_2$ that $\boldsymbol{\beta}_1 - \boldsymbol{\beta}_2 = \lambda \boldsymbol{J}$ is equivalent to $\boldsymbol{\mu}_1 - \boldsymbol{\mu}_2 = \lambda \boldsymbol{\sigma}$ where $\boldsymbol{\sigma} = (\sigma_1, \ldots, \sigma_p)^T$. Thus, O'Brien's statistic W in (9.8) is essentially a GLS-type statistic for testing $\lambda = 0$ against $\lambda > 0$ where $\boldsymbol{\mu}_1 - \boldsymbol{\mu}_2 = \lambda \boldsymbol{\sigma}$.
■

9.1.3 A One Degree of Freedom Test for the Equality of General Parameters

The foregoing derivations leading to the statistic in (9.10) and the O'Brien's statistic W in (9.8) can be presented in a more general, concise and unified framework as follows. Let γ_1 and γ_2 be parameters associated with the distributions of $\boldsymbol{X}_1$ and $\boldsymbol{X}_2$, respectively. We wish to test

$$H_0 : \gamma_1 - \gamma_2 = \boldsymbol{0} \qquad \text{against} \qquad H_1 : \gamma_1 - \gamma_2 \geqslant \boldsymbol{0}. \tag{9.11}$$

Let T_j be a statistic for testing $\gamma_{1j} - \gamma_{2j} = 0$ against $\gamma_{1j} - \gamma_{2j} > 0$; assume that large values of T_j favor $\gamma_{1j} - \gamma_{2j} > 0$. Let $\boldsymbol{T} = (T_1, \ldots, T_p)^T$. Consider a sequence of local alternatives of the form $H_{1N} : \boldsymbol{\gamma}_1 - \boldsymbol{\gamma}_2 = \lambda \boldsymbol{a}/\sqrt{N}$ where N is a function of the sample sizes, for example $N = n_1 + n_2$, $\lambda(> 0)$ is an arbitrary fixed number, and $\boldsymbol{a}$ is a prechosen fixed direction in the positive orthant, the alternative parameter space of $\boldsymbol{\gamma}_1 - \boldsymbol{\gamma}_2$. Suppose that $\boldsymbol{T} \xrightarrow{d} N(\lambda \boldsymbol{b}, \boldsymbol{B})$ under H_{1N} for some $(\boldsymbol{b}, \boldsymbol{B})$. Let us write this in the form of a simple linear regression model with no intercept term: $\boldsymbol{T} = \lambda \boldsymbol{b} + \boldsymbol{E}$ where $\boldsymbol{E}$ is the error term that is approximately $N(\boldsymbol{0}, \boldsymbol{B})$. Now, if $\boldsymbol{B}$ is known and λ is estimated by minimizing the generalized sum of squares, $(\boldsymbol{T} - \lambda \boldsymbol{b})^T \boldsymbol{B}^{-1}(\boldsymbol{T} - \lambda \boldsymbol{b})$, then the resulting estimator of λ is $\boldsymbol{b}^T \boldsymbol{B}^{-1} \boldsymbol{T}/(\boldsymbol{b}^T \boldsymbol{B}^{-1} \boldsymbol{b})$ and its distribution is approximately $N\{\lambda, (\boldsymbol{b}^T \boldsymbol{B} \boldsymbol{b})^{-1}\}$ in large samples. Therefore, a suitable *global statistic* for testing $\lambda = 0$ against $\lambda > 0$ is

$$G = \boldsymbol{b}^T \hat{\boldsymbol{B}}^{-1} \boldsymbol{T}/(\boldsymbol{b}^T \hat{\boldsymbol{B}}^{-1} \boldsymbol{b})^{1/2}, \tag{9.12}$$

where $\hat{\boldsymbol{B}}$ is a consistent estimator of $\boldsymbol{B}$. This is approximately $N(0, 1)$ under $\lambda = 0$ and the test rejects $\lambda = 0$ and accepts $\lambda > 0$ for large values of G. Clearly, O'Brien's statistic W is of this form.

Instead of the generalized sum of squares, we may work with the ordinary least squares (OLS). The OLS estimate of λ is $(\boldsymbol{b}^T \boldsymbol{b})^{-1} \boldsymbol{b}^T \boldsymbol{T}$ and hence the corresponding statistic for testing $\lambda = 0$ against $\lambda > 0$ is $\boldsymbol{b}^T \boldsymbol{T}/(\boldsymbol{b}^T \hat{\boldsymbol{B}} \boldsymbol{b})^{1/2}$. This is of the form (9.7). Pocock, Geller, and Tsiatis (1987) suggested a special case of G; they assumed that $\boldsymbol{T}$ is standardized so that $\boldsymbol{B}$ is a correlation matrix and chose $\boldsymbol{b}$ to be the center direction $\boldsymbol{J}$. Pocock et al. (1987) studied the application of this in more general contexts than those in O'Brien (1984); two of them are discussed below briefly.

(i) Multivariate normal with unequal variance-covariance matrices:

Suppose that $\boldsymbol{X}_1 \sim N(\boldsymbol{\mu}_1, \boldsymbol{V}_1)$ and $\boldsymbol{X}_2 \sim N(\boldsymbol{\mu}_2, \boldsymbol{V}_2)$. Let us first suppose that $\boldsymbol{V}_1$ and $\boldsymbol{V}_2$ are known. Let $T_j = (\bar{X}_{1j} - \bar{X}_{2j})/(n_1^{-1}\sigma_{1j}^2 + n_2^{-1}\sigma_{2j}^2)^{1/2}$ and $\boldsymbol{C}$ be the correlation matrix of $(\bar{\boldsymbol{X}}_1 - \bar{\boldsymbol{X}}_2)$. Then, the statistic G in (9.12), with $\boldsymbol{b} = \boldsymbol{J}$, for testing $H_0 : \boldsymbol{\mu}_1 - \boldsymbol{\mu}_2 = \boldsymbol{0}$ against $H_1 : \boldsymbol{\mu}_1 - \boldsymbol{\mu}_2 \geqslant \boldsymbol{0}$, is

$$\boldsymbol{J}^T \boldsymbol{C}^{-1} \boldsymbol{T}/(\boldsymbol{J}^T \boldsymbol{C}^{-1} \boldsymbol{J})^{1/2}. \tag{9.13}$$

If $\boldsymbol{V}_1$ and $\boldsymbol{V}_2$ are unknown and consistent estimates are substituted for $\boldsymbol{\sigma}_1, \boldsymbol{\sigma}_2$ and $\boldsymbol{C}$, then (9.13) would be a suitable statistic for testing $H_0 : \boldsymbol{\mu}_1 - \boldsymbol{\mu}_2 = \boldsymbol{0}$ against $H_1 : \boldsymbol{\mu}_1 - \boldsymbol{\mu}_2 \geqslant \boldsymbol{0}$ (see Pocock et. al. (1987)).

(ii) Binary response data:

Suppose that X_{ij} is a binary random variable taking the values 0 and 1 ($i = 1, 2, j = 1, \ldots, p$). Let $\pi_{ij} = \text{pr}(X_{ij} = 1)$ and $\boldsymbol{\pi}_i = (\pi_{i1}, \ldots, \pi_{ip})^T$. Let the null and the alternative hypotheses be $H_0 : \boldsymbol{\pi}_1 = \boldsymbol{\pi}_2$ and $H_1 : \boldsymbol{\pi}_1 \geqslant \boldsymbol{\pi}_2$, respectively. Let $N = n_1 + n_2$, p_{ij} denote the sample proportion of successes for the j th component of the response variable in the i th treatment, $\bar{p}_j = (n_1 p_{1j} + n_2 p_{2j})/(n_1 + n_2)$, and

$$Z_j = (p_{1j} - p_{2j})\{\bar{p}_j(1 - \bar{p}_j)(n_1^{-1} + n_2^{-1})\}^{-1/2}; \tag{9.14}$$

Z_j is the usual two-sample statistic for testing the equality of two proportions, and it converges to $N(0,1)$ under H_0. An estimator of the correlation coefficient between Z_q and Z_r is $\hat{c}_{qr} = (s_{qr} - \bar{p}_q\bar{p}_r)/\{\bar{p}_q\bar{p}_r(1-\bar{p}_q)(1-\bar{p}_r)\}^{1/2}$, for $q \neq r$ where s_{qr} is the proportion of successes in both variables q and r (Pocock et al. (1987)). Now, the statistic G for testing $H_0 : \pi_1 = \pi_2$ against $H_1 : \pi_1 \geqslant \pi_2$ is $\boldsymbol{J}^T\hat{\boldsymbol{C}}^{-1}\boldsymbol{Z}/(\boldsymbol{J}^T\hat{\boldsymbol{C}}^{-1}\boldsymbol{J})^{1/2}$, which is approximately $N(0,1)$ under H_0 for large samples.

Pocock et al. (1987) also obtained explicit forms for the test statistic, G, in (9.12) when one of the response variable is a survival time subject to censoring and another is binary. This illustrates that the G in (9.12) is applicable even when some of the components of the multivariate responses are discrete and some are continuous.

9.1.4 Multiparameter Methods

The results in the previous subsections were all based on testing $\boldsymbol{\mu}_1 - \boldsymbol{\mu}_2 = \boldsymbol{0}$ against the alternative that $\boldsymbol{\mu}_1 - \boldsymbol{\mu}_2$ is parallel to a prechosen fixed direction in the alternative parameter space. This reduces the multiparameter testing problem to a test on a scalar parameter, which we denoted by λ. While this simplifies the testing problem considerably, it could also be expected to have low power at points away from the prechosen fixed direction. In this subsection we consider methods that do not reduce/simplify to a testing problem on a scalar parameter.

Methods that are asymptotically equivalent to *LRT*.

Suppose that $\boldsymbol{X}_1 \sim N(\boldsymbol{\mu}_1, \boldsymbol{V}_1)$ and $\boldsymbol{X}_2 \sim N(\boldsymbol{\mu}_2, \boldsymbol{V}_2)$ where $\boldsymbol{V}_1$ and $\boldsymbol{V}_2$ are known; later we shall relax the assumption that $\boldsymbol{V}_1$ and $\boldsymbol{V}_2$ are known. Then $\bar{\boldsymbol{X}}_1 - \bar{\boldsymbol{X}}_2 \sim N(\boldsymbol{\mu}_1 - \boldsymbol{\mu}_2, \boldsymbol{W})$ where $\boldsymbol{W} = (n_1^{-1}\boldsymbol{V}_1 + n_2^{-1}\boldsymbol{V}_2)$. Since $\boldsymbol{W}$ is known, we can apply *LRT* for testing

$$H_0 : \boldsymbol{\mu}_1 - \boldsymbol{\mu}_2 = \boldsymbol{0} \text{ against } H_1 : \boldsymbol{\mu}_1 - \boldsymbol{\mu}_2 \geqslant \boldsymbol{0};$$

the exact null distribution of the *LRT* is a $\bar{\chi}^2$. If $\boldsymbol{X}_1$ and $\boldsymbol{X}_2$ are not normal but $E(\boldsymbol{X}_i) = \boldsymbol{\mu}_i$ and $\text{cov}(\boldsymbol{X}_i) = \boldsymbol{V}_i$ for $i = 1, 2$ where $\boldsymbol{V}_1$ and $\boldsymbol{V}_2$ are known, then $\bar{\boldsymbol{X}}_1 - \bar{\boldsymbol{X}}_2$ is approximately $N(\boldsymbol{\mu}_1 - \boldsymbol{\mu}_2, \boldsymbol{W})$ for large n_1 and n_2. Therefore, the foregoing *LRT* developed under the assumption that $\boldsymbol{X}_1$ and $\boldsymbol{X}_2$ are exactly normal, is a valid large sample test. In practice, it is very unlikely that $\boldsymbol{V}_1$ and $\boldsymbol{V}_2$ would be known and hence a test of this type is unlikely to be of much use. However, these results are useful because they can be used as starting points for developing tests when $\boldsymbol{V}_1$ and $\boldsymbol{V}_2$ are unknown.

9.1.4.1 $\boldsymbol{X}_1 \sim N(\boldsymbol{\mu}_1, \sigma^2\boldsymbol{U}_1)$ and $\boldsymbol{X}_2 \sim N(\boldsymbol{\mu}_2, \sigma^2\boldsymbol{U}_2)$

Suppose that $\boldsymbol{X}_1 \sim N(\boldsymbol{\mu}_1, \sigma^2\boldsymbol{U}_1)$ and $\boldsymbol{X}_2 \sim N(\boldsymbol{\mu}_2, \sigma^2\boldsymbol{U}_2)$ where $\boldsymbol{U}_1$ and $\boldsymbol{U}_2$ are known but σ is unknown. In this case, an $\bar{F}$-test is applicable for testing H_0 against H_1. Note that $\bar{\boldsymbol{X}}_1 - \bar{\boldsymbol{X}}_2 \sim N(\boldsymbol{\mu}_1 - \boldsymbol{\mu}_2, \sigma^2\boldsymbol{W})$ where $\boldsymbol{W} = (n_1^{-1}\boldsymbol{U}_1 + n_2^{-1}\boldsymbol{U}_2)$. Let $\hat{\sigma}^2 = \nu^{-1}\sum_{i=1}^{2}\sum_{\ell=1}^{n_i}(\boldsymbol{X}_{i\ell} - \bar{\boldsymbol{X}}_i)^T\boldsymbol{U}_i^{-1}(\boldsymbol{X}_{i\ell} - \bar{\boldsymbol{X}}_i)$, where $\nu = (n_1 + n_2 - 2)p$.

Then, $\nu\hat{\sigma}^2/\sigma^2 \sim \chi^2_\nu$ and $\hat{\sigma}^2$ is independent of $(\bar{\boldsymbol{X}}_1, \bar{\boldsymbol{X}}_2)$. Let

$$\boldsymbol{Y} = (\bar{\boldsymbol{X}}_1 - \bar{\boldsymbol{X}}_2) \quad \text{and} \quad T(\boldsymbol{Y}, \boldsymbol{W}) = \hat{\sigma}^{-2}\|\Pi_{\boldsymbol{W}}(\boldsymbol{Y}; \mathbb{R}^{+p})\|^2_{\boldsymbol{W}} \quad \text{where}$$
$$\|\Pi_{\boldsymbol{W}}(\boldsymbol{Y}; \mathbb{R}^{+p})\|^2_{\boldsymbol{W}} = \boldsymbol{Y}^T\boldsymbol{W}^{-1}\boldsymbol{Y} - \min\{(\boldsymbol{Y} - \boldsymbol{a})^T\boldsymbol{W}^{-1}(\boldsymbol{Y} - \boldsymbol{a}) : \boldsymbol{a} \geqslant \boldsymbol{0}\}. \tag{9.15}$$

Now, the test rejects H_0 for large values of $T(\boldsymbol{Y}, \boldsymbol{W})$. By arguments similar to those for Theorem 3.9.2, the null distribution of $T(\boldsymbol{Y}, \boldsymbol{W})$ is the $\bar{F}$-distribution,

$$\text{pr}\{T(\boldsymbol{Y}, \boldsymbol{W}) \geqslant c\} = \sum_{i=0}^{p} w_i(p, \boldsymbol{W}, \mathbb{R}^{+p})\text{pr}(iF_{i,\nu} \geqslant c). \tag{9.16}$$

The expression on the RHS is the p-value with c denoting the sample value of $T(\boldsymbol{Y}, \boldsymbol{W})$. Even if $\boldsymbol{X}_1$ or $\boldsymbol{X}_2$ is not normal, $(\bar{\boldsymbol{X}}_1 - \bar{\boldsymbol{X}}_2)$ would be approximately normal and $\hat{\sigma}^2$ would be a consistent estimator of σ^2, and hence (9.16) holds approximately for large (n_1, n_2).

The asymptotic null distribution of $T(\boldsymbol{Y}, \boldsymbol{W})$, even if $\boldsymbol{X}_1$ and $\boldsymbol{X}_2$ are not normal, is given by

$$\text{pr}\{T(\boldsymbol{Y}, \boldsymbol{W}) \leqslant c\} \to \sum_{i=0}^{p} w_i(p, \boldsymbol{W}, \mathbb{R}^{+p})\text{pr}(\chi^2_i \leqslant c).$$

However, we believe that the p-values computed using the $\bar{F}$-distribution in (9.16) are likely to be closer to the true value than that corresponding to the limiting chi-bar-square distribution. This is based on our experience in some simulations involving the $\bar{F}$-test in linear regression; see Silvapulle (1992b).

9.1.4.2 $\boldsymbol{X}_1 \sim N(\boldsymbol{\mu}_1, \sigma_1^2\boldsymbol{U}_1)$ and $\boldsymbol{X}_2 \sim N(\boldsymbol{\mu}_2, \sigma_2^2\boldsymbol{U}_2)$

Suppose that $\boldsymbol{X}_1 \sim N(\boldsymbol{\mu}_1, \sigma_1^2\boldsymbol{U}_1)$ and $\boldsymbol{X}_2 \sim N(\boldsymbol{\mu}_2, \sigma_2^2\boldsymbol{U}_2)$ where $\boldsymbol{U}_1$ and $\boldsymbol{U}_2$ are known, but σ_1 and σ_2 may be unknown and unequal. Then $\bar{\boldsymbol{X}}_1 - \bar{\boldsymbol{X}}_2 \sim N(\boldsymbol{\mu}_1 - \boldsymbol{\mu}_2, \boldsymbol{B})$ where $\boldsymbol{B} = (n_1^{-1}\sigma_1^2\boldsymbol{U}_1 + n_2^{-1}\sigma_2^2\boldsymbol{U}_2)$. In this case, the exact result (9.16) does not hold. Let $\hat{\sigma}_i$ be a consistent estimator of σ_i for $i = 1, 2$ and let $\hat{\boldsymbol{B}} = (n_1^{-1}\hat{\sigma}_1^2\boldsymbol{U}_1 + n_2^{-1}\hat{\sigma}_2^2\boldsymbol{U}_2)$; for example, $\hat{\sigma}_i$ can be $\{(n_i - 1)p\}^{-1}\sum_{\ell=1}^{n_i}(\boldsymbol{X}_{i\ell} - \bar{\boldsymbol{X}}_i)^T\boldsymbol{U}_i^{-1}(\boldsymbol{X}_{i\ell} - \bar{\boldsymbol{X}}_i)$. Then $\hat{\boldsymbol{B}}$ is a consistent estimator of $\boldsymbol{B}$. Now, a statistic T for testing H_0 against H_1 and its asymptotic null distribution are given by

$$T = \|\Pi_{\hat{\boldsymbol{B}}}(\boldsymbol{Y}; \mathbb{R}^{+p})\|^2_{\hat{\boldsymbol{B}}}, \quad \text{and} \quad T \xrightarrow{d} \bar{\chi}^2(\boldsymbol{B}, \mathbb{R}^{+p}),$$

respectively, where Π is the projection as in (9.15) but now with respect to the matrix $\hat{\boldsymbol{B}}$. Note that $w_i(p, a\boldsymbol{U}_1 + b\boldsymbol{U}_2, \mathbb{R}^{+p}) = w_i(p, \boldsymbol{A}_\rho, \mathbb{R}^{+p})$ where $\rho = a/b$ and $\boldsymbol{A}_\rho = \boldsymbol{U}_1 + \rho\boldsymbol{U}_2$. Therefore, for large samples,

$$p\text{-value} \simeq \sup_\rho \text{pr}\{\bar{\chi}^2(\boldsymbol{A}_\rho, \mathbb{R}^{+p}) \geqslant t; \rho\}$$

where $\rho > 0$ and t is the sample value of T. This supremum is not difficult to compute approximately because $\text{pr}\{\bar{\chi}^2(\boldsymbol{A}_\rho, \mathbb{R}^{+p}) \geqslant t; \rho\}$ can be computed easily by simulation for values of ρ in a grid of $[0, \infty]$; $\rho = 0$ corresponds to $\boldsymbol{A}_\rho = \boldsymbol{U}_1$ and $\rho = \infty$ corresponds to $\boldsymbol{A}_\rho = \boldsymbol{U}_2$. Once this is done, the maximum over the grid would be a good approximation to the large sample p-value.

9.1.4.3 $\boldsymbol{X}_1 \sim N(\boldsymbol{\mu}_1, \boldsymbol{V}_1)$ and $\boldsymbol{X}_2 \sim N(\boldsymbol{\mu}_2, \boldsymbol{V}_2)$

Suppose that $\boldsymbol{X}_1 \sim N(\boldsymbol{\mu}_1, \boldsymbol{V}_1)$ and $\boldsymbol{X}_2 \sim N(\boldsymbol{\mu}_2, \boldsymbol{V}_2)$ where $\boldsymbol{V}_1$ and $\boldsymbol{V}_2$ are unknown and may be unequal. Then $\bar{\boldsymbol{X}}_1 - \bar{\boldsymbol{X}}_2 \sim N(\boldsymbol{\mu}_1 - \boldsymbol{\mu}_2, \boldsymbol{B})$ where $\boldsymbol{B} = (n_1^{-1}\boldsymbol{V}_1 + n_2^{-1}\boldsymbol{V}_2)$. Let $\hat{\boldsymbol{B}}$ be a consistent estimator of $\boldsymbol{B}$. Then a statistic for testing H_0 against H_1 is $\|\Pi_{\hat{\boldsymbol{B}}}(\bar{\boldsymbol{X}}_1 - \bar{\boldsymbol{X}}_2; \mathbb{R}^{+p})\|^2_{\hat{\boldsymbol{B}}}$, and its large sample null distribution is $\bar{\chi}^2(\boldsymbol{B}, \mathbb{R}^{+p})$. Corresponding to a sample value t for this statistic, a large sample p-value is $\sup \text{pr}\{\bar{\chi}^2(\boldsymbol{B}, \mathbb{R}^{+p}) \geqslant t : \boldsymbol{V}_1, \boldsymbol{V}_2\}$, where the supremum is taken over all possible $(\boldsymbol{V}_1, \boldsymbol{V}_2)$. This supremum is equal to

$$0.5\{\text{pr}(\chi^2_{p-1} \geqslant t) + \text{pr}(\chi^2_p \geqslant t)\}, \tag{9.17}$$

which is achieved when $(\boldsymbol{V}_1, \boldsymbol{V}_2)$ take extreme, often unrealistic, values. Consequently, the use of (9.17) as a p-value may be too conservative. In large samples, we can estimate the p-value by $\text{pr}\{\bar{\chi}^2(\boldsymbol{B}, \mathbb{R}^{+p}) \geqslant t\}$ evaluated at an estimate of $\boldsymbol{B}$. However, it is difficult to say how large the samples need to be to ensure that the error in the estimation is small.

9.1.4.4 An extension of Follmann's test

An extension of a test suggested by Follmann (1996b) can be applied to develop a test of H_0 against H_1. The large sample procedure is the following: Reject H_0 at level α if $T \geqslant \chi^2_{p,2\alpha}$ where

$$T = (\bar{\boldsymbol{X}}_1 - \bar{\boldsymbol{X}}_2)^T(n_1^{-1}\boldsymbol{S}_1 + n_2^{-1}\boldsymbol{S}_2)^{-1}(\bar{\boldsymbol{X}}_1 - \bar{\boldsymbol{X}}_2) I\{\sum_{j=1}^{p}(\bar{X}_{1j} - \bar{X}_{2j}) > 0\}, \tag{9.18}$$

$I\{\cdot\}$ is the indicator function and $\boldsymbol{S}_i$ is the sample variance-covariance matrix $\{(n_i - 1)^{-1}\sum_{\ell=1}^{n_i}(\boldsymbol{X}_{i\ell} - \bar{\boldsymbol{X}}_i)(\boldsymbol{X}_{i\ell} - \bar{\boldsymbol{X}}_i)^T$, $i = 1, 2$. The validity of this is stated in the next result.

Proposition 9.1.1 *Let $\boldsymbol{X}_1 \sim N(\boldsymbol{\mu}_1, \boldsymbol{V}_1)$ and $\boldsymbol{X}_2 \sim N(\boldsymbol{\mu}_2, \boldsymbol{V}_2)$. Assume that the two random samples on $\boldsymbol{X}_1$ and $\boldsymbol{X}_2$ are independent. Then, the asymptotic distribution of T in (9.18), under $H_0 : \boldsymbol{\mu}_1 - \boldsymbol{\mu}_2 = \boldsymbol{0}$, is given by*

$$pr(T \geqslant t \mid H_0) \rightarrow 0.5pr(\chi^2_p \geqslant t).$$

Proof: Let $\boldsymbol{Y}_1 = \boldsymbol{X}_1 - \mu_1, \boldsymbol{Y}_2 = \boldsymbol{X}_2 - \mu_2$, $\boldsymbol{Y}_{i\ell} = \boldsymbol{X}_{i\ell} - \boldsymbol{\mu}_i$ $(i = 1, 2, \ell = 1, \ldots, n_i)$ and $\boldsymbol{Z} = (\boldsymbol{Y}_{11} : \ldots : \boldsymbol{Y}_{1n_1} : \boldsymbol{Y}_{21} : \ldots : \boldsymbol{Y}_{2n_2})^T$. Then $\boldsymbol{Y}_i \sim N(\boldsymbol{0}, \boldsymbol{V}_i)$ and $\boldsymbol{Y}_i \sim -\boldsymbol{Y}_i$. Let

$$A = \{\boldsymbol{Z} : (\bar{\boldsymbol{Y}}_1 - \bar{\boldsymbol{Y}}_2)^T(n_1^{-1}\boldsymbol{S}_1 + n_2^{-1}\boldsymbol{S}_2)^{-1}(\bar{\boldsymbol{Y}}_1 - \bar{\boldsymbol{Y}}_2) \geqslant \chi^2_{p,2\alpha}\},$$

$B^+ = \{\boldsymbol{Z} : \sum_{j=1}^{p}(\bar{Y}_{1j} - \bar{Y}_{2j}) > 0\}$ and $B^- = \{\boldsymbol{Z} : \sum_{j=1}^{p}(\bar{Y}_{1j} - \bar{Y}_{2j}) < 0\}$. Let $f(\boldsymbol{z})$ denote the *pdf* of $\boldsymbol{Z}$ when $\boldsymbol{\mu}_1 = \boldsymbol{\mu}_2$. Then, we have that $\text{pr}(A \cap B^+) = \int_{\boldsymbol{z} \in A \cap B^+} f(\boldsymbol{z}) d\boldsymbol{z} = \int_{\boldsymbol{z} \in A \cap B^+} f(-\boldsymbol{z}) d\boldsymbol{z} = \int_{\boldsymbol{z} \in A \cap B^-} f(\boldsymbol{z}) d\boldsymbol{z} = \text{pr}(A \cap B^-)$. Now, the proof follows since $\text{pr}(A \cap B^+) + \text{pr}(A \cap B^-) = \text{pr}(A)$ and $\text{pr}(A) \to 2\alpha$ as $(n_1, n_2) \to \infty$ and $\lim(n_1/n_2) > 0$. ∎

It follows from the proposition that corresponding to a sample value t for T,

$$\text{a large sample } p\text{-value} = 0.5\text{pr}(\chi_p^2 \geqslant t).$$

This is very similar to the following well-known approach: in a well-behaved model that is known up to a scalar parameter β, a large sample p-value for testing $\beta = 0$ against $\beta > 0$ is half of that for the *LRT* for testing $\beta = 0$ against $\beta \neq 0$. For the statistic T in (9.18), the positive half-space $\{\boldsymbol{x} : x_1 + \ldots + x_p > 0\}$ plays the role of the positive axis $\{\beta : \beta > 0\}$ in the case of test on the scalar parameter β.

(V) *Approximate LRT of Tang, Gnecco, and Geller (1989):*

Tang et al. (1989a) introduced an Approximate *LRT* (ALR) for testing $H_0 : \boldsymbol{\mu} = \mathbf{0}$ against $H_1 : \boldsymbol{\mu} \geqslant \mathbf{0}$ based on an independently and identically distributed sample from $N(\boldsymbol{\mu}, \boldsymbol{V})$ where $\boldsymbol{V}$ may be known or unknown. The reason for introducing the ALR was to develop a procedure that is close to *LRT* but less demanding in terms of computations.

Let us introduce the essentials of ALR. Let $\boldsymbol{Y} = \bar{\boldsymbol{X}}_1 - \bar{\boldsymbol{X}}_2$, and assume that $\boldsymbol{Y} \sim N(\boldsymbol{\theta}, \boldsymbol{V})$ where $\boldsymbol{V}$ is known. Let the null and alternative hypotheses be

$$H_0 : \boldsymbol{\theta} = \mathbf{0} \quad \text{and} \quad H_1 : \boldsymbol{\theta} \geqslant \mathbf{0},$$

respectively. Let $\boldsymbol{V}^{-1} = \boldsymbol{A}^T\boldsymbol{A}$, $\boldsymbol{Z} = \boldsymbol{A}\boldsymbol{Y}$ and $\boldsymbol{\gamma} = \boldsymbol{A}\boldsymbol{\theta}$; the matrix $\boldsymbol{A}$ is not uniquely defined, and its choice will be considered soon. Now, the ALR of $H_0 : \boldsymbol{\theta} = \mathbf{0}$ against $H_1 : \boldsymbol{\theta} \geqslant \mathbf{0}$ is the *LRT* of $\boldsymbol{\gamma} = \mathbf{0}$ against $\boldsymbol{\gamma} \geqslant \mathbf{0}$ based on a single observation of $\boldsymbol{Z}$ and the fact that $\boldsymbol{Z} \sim N(\boldsymbol{\gamma}, \boldsymbol{I})$. Thus, the approximation of *LRT* is obtained by assuming that $\{\boldsymbol{\theta} : \boldsymbol{\theta} \geqslant \mathbf{0}\}$ is close to $\{\boldsymbol{\theta} : \boldsymbol{A}\boldsymbol{\theta} \geqslant \mathbf{0}\}$, and hence an approximate *LRT* of $H_0 : \boldsymbol{\theta} = \mathbf{0}$ against $H_1 : \boldsymbol{\theta} \geqslant \mathbf{0}$ is the *LRT* of $\boldsymbol{\theta} = \mathbf{0}$ against $\boldsymbol{A}\boldsymbol{\theta} \geqslant \mathbf{0}$.

Since the objective is to ensure that the ALR is close to *LRT*, the matrix $\boldsymbol{A}$ needs to be chosen so that $\{\boldsymbol{\theta} : \boldsymbol{\theta} \geqslant \mathbf{0}\}$ and $\{\boldsymbol{\theta} : \boldsymbol{A}\boldsymbol{\theta} \geqslant \mathbf{0}\}$ are close where $\boldsymbol{V}^{-1} = \boldsymbol{A}^T\boldsymbol{A}$; exactly, what is meant by "close" needs to be defined. Tang et al. (1989, 1993) provided a procedure for choosing $\boldsymbol{A}$ so that the center direction of $\{\boldsymbol{\theta} : \boldsymbol{\theta} \geqslant \mathbf{0}\}$ and that of $\{\boldsymbol{\theta} : \boldsymbol{A}\boldsymbol{\theta} \geqslant \mathbf{0}\}$ are close; they also provided the details for computing it as well. This is a good way of choosing the matrix $\boldsymbol{A}$. Once $\boldsymbol{A}$ is computed, the rest of the computations for ALR are quite easy.

It is easily seen that the null distribution of ALR is the chi-bar-square distribution given by

$$\text{pr}(ALR \geqslant c \mid \boldsymbol{\mu} = \mathbf{0}) = \sum_{i=0}^{p} 2^{-p}[p!/\{i!(p-i)!\}]\text{pr}(\chi_i^2 \geqslant c).$$

If $\boldsymbol{V}$ is unknown and is replaced by a consistent estimator of it, for example, the sample covariance matrix, then the foregoing results hold asymptotically (Tang et al. (1989)). Recall that, since $\boldsymbol{V}$ is unknown, the p-value for *LRT* is $\sup_{\boldsymbol{V}} \text{pr}(LRT \geqslant t_1 | H_0, \boldsymbol{V})$ where t_1 is the sample value of *LRT*. By contrast, the p-value for ALR does not require taking such a supremum because pr$(ALR \geqslant t|\boldsymbol{\mu} = \mathbf{0}, \boldsymbol{V})$ does not depend on $\boldsymbol{V}$. This is an attractive feature of ALR. However, it is worth noting that the alternative parameter space $\{\boldsymbol{\theta} : \boldsymbol{A\theta} \geqslant \mathbf{0}\}$ for ALR depends on the unknown parameter $\boldsymbol{V}$; its ramifications are not all that clear. In a simulation study, Tang et al. (1989) observed that differences between ALR and *LRT* were not major.

It is important to note that the ALR was not intended to overcome any deficiencies in the statistical properties of *LRT* as a method of inference in this particular problem. The main advantage of ALR is that it can be implemented without using a quadratic program and its asymptotic distribution is easy to use in computations.

Bloch et al. (2001) observed that, for testing $\boldsymbol{\theta} = \mathbf{0}$ against $\boldsymbol{\theta} \geqslant \mathbf{0}$ the probability that ALR rejects the null when the true value is outside $\boldsymbol{\theta} \geqslant \mathbf{0}$, was more than 90%. It is easily seen that *LRT* also has the same property; at a point which is very far from the origin and is on the boundary of the positive orthant $\{\boldsymbol{\theta} \geqslant \mathbf{0}\}$, the power of *LRT* is nearly 100%. Since the power function is continuous, the power would be approximately 100% at some points close to, but not necessarily in, the positive orthant as well. Although the same comment applies to O'Brien's test, its power at points outside $\{\boldsymbol{\theta} \geqslant \mathbf{0}\}$ is likely to be smaller than that for *LRT*. Therefore, ALR, *LRT*, O'Brien's test, and most of the other tests of $H_0 : \boldsymbol{\theta} = \mathbf{0}$ against $H_1 : \boldsymbol{\theta} \geqslant \mathbf{0}$ are not suitable for testing $H_0^a : \theta_i \leq 0$ *for some* i, against $H_1 : \theta_i > 0$ *for every* i. This testing problem arises in clinical trials with multiple end points. This is the topic of the next section.

9.1.5 H_1 : New Treatment is Noninferior With Respect to Every Variable and Superior With Respect to Some Variables

The foregoing methods are unlikely to be appropriate for establishing that Treatment 1 is better than Treatment 2 with respect to several variables if it is also required to provide statistical evidence simultaneously to establish that Treatment 1 is noninferior to Treatment 2 with respect to every variable. This would typically be the case in clinical trials with multiple end points where the number of end points is small compared with the number of observations; here an *end point* refers to a scalar response variable and *multiple end points* refers to a multivariate response. It was indicated in the introduction to this section that in such cases it is necessary to define the null and alternative hypotheses as $H_0 : \boldsymbol{\theta} \notin \Theta_1$ and $H_1 : \boldsymbol{\theta} \in \Theta_1$, where $\Theta_1 \subset \mathbb{R}^p$ is the set of values of $\boldsymbol{\theta}$ for which Treatment 1 would be considered to be better than Treatment 2. For example, the null and alternative hypothesis may take the form $H_0 : Not\ H_1$, and

> H_1 : Treatment 1 is better than Treatment 2 with respect to at least one response variable and is noninferior with respect to every response variable,

respectively. This inference problem is considered in this subsection.

Since the problem can be formulated in many different ways and Θ_1 may take unusual shapes, it is likely that computationally simple expressions for the null distribution of a test statistic would be difficult to obtain. Let us consider a setting for which closed-form expressions can be obtained.

Let us assume that Treatment 1 is *noninferior* to Treatment 2, with respect to jth response variable if $\theta_j > -\epsilon_j$, $(j = 1, \ldots, p)$; similarly, assume that Treatment 1 is *superior* to Treatment 2 if $\theta_j > \delta_j$. It is important to note that the choice of δ_js and ϵ_js may depend on sample sizes but not on the sample values. Now, let us define

$$\Theta_1 = \{\boldsymbol{\theta} \in \mathbb{R}^p : \theta_j > -\epsilon_j \text{ for every } j, \text{ and } \theta_j > \delta_j \text{ for some } j, j = 1, \ldots, p\} \tag{9.19}$$

$$\mathcal{C}_0 = \{\boldsymbol{\theta} : \theta_j \leq \delta_j, \text{ for } j = 1, \ldots, p\}, \text{ and } \mathcal{C}_1 = \{\boldsymbol{\theta} : \theta_j > -\epsilon_j, \text{ for } j = 1, \ldots, p\}.$$

Then $H_0 : \boldsymbol{\theta} \notin \Theta_1$ and $H_1 : \boldsymbol{\theta} \in \Theta_1$ may be stated as

$$H_0 : \boldsymbol{\theta} \notin \mathcal{C}_1 \text{ or } \boldsymbol{\theta} \in \mathcal{C}_0, \qquad \text{and} \qquad H_1 : \boldsymbol{\theta} \in \mathcal{C}_1 \text{ and } \boldsymbol{\theta} \notin \mathcal{C}_0, \tag{9.20}$$

respectively. This is in a form to which an IUT is applicable provided that we have a test of $\boldsymbol{\theta} \notin \mathcal{C}_1$ against $\boldsymbol{\theta} \in \mathcal{C}_1$, and another test of $\boldsymbol{\theta} \in \mathcal{C}_0$ against $\boldsymbol{\theta} \notin \mathcal{C}_0$. The first is a *sign testing* problem, which is discussed in detail in another section (see Section 9.2); the second is a *Type B* problem which was discussed in Chapter 3. Now, an approximate large sample IUT could be constructed as follows. Let $\boldsymbol{Y} = \bar{\boldsymbol{X}}_1 - \bar{\boldsymbol{X}}_2$. Assume that $\boldsymbol{Y} \approx N(\boldsymbol{\theta}, \boldsymbol{V})$ and $\hat{\boldsymbol{V}}$ is a consistent estimator of $\boldsymbol{V}$. For $j = 1, \ldots, p$, let p_j denote the large sample p-value for testing $\theta_j \leq -\epsilon_j$ against $\theta_j > -\epsilon_j$ based on $(Y_j + \epsilon_j)/se(Y_j) \approx N(0, 1)$. Let

$$U(\boldsymbol{Y}, \hat{\boldsymbol{V}}) = \min\{(\boldsymbol{Y} - \boldsymbol{a})^T \hat{\boldsymbol{V}}^{-1} (\boldsymbol{Y} - \boldsymbol{a}) : \boldsymbol{a} \in \mathcal{C}_0\};$$

this is an approximation to the *LRT* for testing $\boldsymbol{\theta} \in \mathcal{C}_0$ against $\boldsymbol{\theta} \notin \mathcal{C}_0$. A large sample upper bound for the p-value of this test is $p_0 = 0.5\{pr(\chi^2_{p-1} > c) + \text{pr}(\chi^2_p > c)\}$, where c is the sample value of $U(\boldsymbol{Y}, \hat{\boldsymbol{V}})$. Now, a large sample level-α IUT of (9.20) rejects H_0 if $\max\{p_0, p_1, \ldots, p_p\} \leq \alpha$.

If the shape of Θ_1 is changed, then it may not always be possible to obtain simple analytic expressions for computing critical values and/or p-value. In such cases, bootstrap may offer a practical solution; for example, see Bloch et al. (2001) and Reitmeir and Wassmer (1999). Let us note in passing that the bootstrap test of Bloch et al. (2001) assumes that the origin is the least favorable null value; since this is not always the case, some modifications may be required for this approach

To indicate other possible formulations of the testing problem, consider the case when the response vector has $p = 3$ components. Suppose that Treatment 1 is defined to be better than Treatment 2 if the former is better than the latter with respect to at least two of the three response variables and noninferior with respect to every variable. Let $\theta_{(1)} \geqslant \theta_{(2)} \geqslant \theta_{(3)}$ denote the ordered values of θ_1, θ_2 and θ_3. Now, the null and alternative hypotheses are

$$H_0 : \textit{not } H_1, \qquad \text{and} \qquad H_1 : \theta_j > -\epsilon_j \text{ for every } j = 1, 2, 3 \text{ and } \theta_{(2)} > 0;$$

it is easily seen that H_0 can also be expressed as $\theta_j \leq -\epsilon_j$ for some $j = 1, 2, 3$ or $\theta_{(2)} \leq 0$. To apply an IUT, we need a one-sided test on each of $\theta_1, \theta_2, \theta_3$ and $\theta_{(2)}$. For the first three, t-ratios can be applied; for $\theta_{(2)}$ a possible statistic is $Y_{(2)}$. Since analytic expressions for the relevant null distributions are likely to be complicated, bootstrap hypothesis testing appears to be a practically feasible way of implementing different tests. In this regard, the bootstrap test of Bloch et al. (2001) is worthy of note. This approach has considerable appeal because of its flexibility.

9.2 TESTING THAT AN IDENTIFIED TREATMENT IS THE BEST: THE MIN TEST

In this section, we consider a testing problem arising in the following context. There are k treatments, that we denote by $A_1, \ldots, A_k$. Further, there is a particular *identified* treatment that we denote by A_0. The objective is to establish that A_0 is better than $A_1, \ldots, A_k$. Therefore, the null and alternative hypotheses take the following forms:

$$H_0 : A_0 \text{ is not better than } A_i \text{ for some } i = 1, \ldots, k,$$
$$H_1 : A_0 \text{ is better than } A_i \text{ for every } i = 1, \ldots, k.$$

For simplicity, let us consider the case when the response is univariate, and let $\mu_0, \mu_1, \ldots, \mu_k$ denote the effects of the treatments, $A_0, A_1, \ldots, A_k$, respectively. Now, the null and alternative hypotheses may take the form,

$$H_0 : \theta_1 \leqslant 0, \text{ or } \ldots, \theta_k \leqslant 0, \quad \text{and} \quad H_1 : \theta_1 > 0, \ldots, \theta_k > 0, \tag{9.21}$$

respectively, where $\theta_i = \mu_0 - \mu_i, i = 1, \ldots, k$; some authors refer to this as the *sign testing* problem.

We have already noted that this type of inference problem arises in clinical trials involving combination therapies, where the objective is to establish that a combination of Treatments 1 and 2 is better than Treatment 1 and Treatment 2. In what follows, we consider the two cases, univariate response (single end point) and multivariate response (multiple end points), in turn. Most of the results of this section are based on intersection-union tests. Therefore, this section can be seen as a continuation of the section IUT in Chapter 4.

Let the null and alternative hypotheses be as in (9.21). For $i = 1, \ldots, k$ let X_i denote a statistic for testing $\theta_i \leqslant 0$ against $\theta_i > 0$. For any i, assume that the marginal distribution of X_i depends only on θ_i, and that larger values of X_i favor larger values of θ_i. Let a level-α critical region for testing $\theta_i \leqslant 0$ against $\theta_i > 0$ be $\{X_i \geqslant c_i\}$, for some c_i, $i = 1, \ldots, k$. Now, it follows from Proposition 5.3.1, that a level-α IUT for (9.21) is

$$\text{reject } H_0 \text{ if } \min\{X_1 - c_1, \ldots, X_k - c_k\} > 0. \tag{9.22}$$

This is also known as the *min test*. The term *min test* appears to be due to Laska and Meisner (1989); the basic idea has been known for some time (for example, see Lehmann (1952), Berger (1982), and Cohen et al. (1983)). If $\{X_1 \geqslant c_1\}$ is a size-α region and the conditions in Proposition 5.3.2 are satisfied, then the *min test* has size-α.

Clearly, the *min test* is an IUT. It has some properties that are important for inference problems arising in combination treatments. In this section, we shall address some of these issues. An example where the response variable is ordinal is given below; since this is an important example, we shall apply different methods and comment on the similarities and differences between such methods.

Example 9.2.1 *Combination drug for low back pain and spasm (Snappin (1987)).*

Dolobid (D) and Flexeril (F) are approved drugs for spasm and pain relief, respectively. The manufacturer of these drugs conjectured that a new drug that is a combination of Dolobid and Flexeril is likely to be better. A requirement of the regulatory authorities for the approval of a new combination drug such as DF is that it must be demonstrated that DF is better than its components D and F. A study was conducted to evaluate the efficacy of a Dolobid/Flexeril combination, DF, relative to Dolobid only, Flexeril only, and a placebo (P). Patients entering the study were randomly assigned to one of the four treatments, DF, D, F, and P. At the end of the study period, each patient provided an evaluation of the treatment on a five-point ordinal scale ranging from Marked Improvement to Worsening of the condition (see Table 9.1). The objective is to test the claim that DF is better than D, F, and P. Different ways of testing this will be discussed in this section; it will be noted that a multiple testing procedure with Bonferroni adjustment to control overall probability of Type I error is valid but inefficient. To ensure that the asymptotic results provide good approximations, the response categories *No Change* and *Worse* were amalgamated because the cell counts for *Worse* are small. ■

Table 9.1 Comparison of Dolobid/Flexeril with Dolobid, Flexeril, and Placebo

	Degree of improvement					
Treatment	Marked	Mod	Mild	No change	Worse	Total
Dolobid/Flexeril [DF]	50	29	19	11	5	114
Dolobid alone [D]	41	28	23	22	2	116
Flexeril alone [F]	38	27	18	27	3	113
Placebo [P]	35	20	23	37	4	119

Comparison of two or more treatments when the responses are ordinal arises frequently in practice. A *min test* can be used for testing the claim that a treatment is better than all the others when the responses are ordinal. It would be convenient to consider the two cases, two treatments and more than two treatments, separately. In the next example, we shall consider the comparison of the first two treatments; then in the following example, all four treatments are compared.

Example 9.2.2 *Comparison of two treatments with ordinal responses.*

Consider the data for Dolobid/Flexeril (DF) and Dolobid (D) in Example 9.2.1; in this example, we will need only the data for DF and D. Suppose that we wish to test that DF is better than D. Thus, the null and alternative hypotheses are

$$H_0 : \textit{DF is not better than D}, \quad \text{and} \quad H_1 : \textit{DF is better than D}. \tag{9.23}$$

We shall consider four different tests; they require different assumptions and hence complement each other.

Method 1: This is based on the well-known cumulative link models for ordinal data introduced by McCullagh (1980); for an excellent account of this and related methods, see Agresti (2002, Section 7.3). Let C denote the number of response categories; in this example $C = 4$ since the last two categories have been merged. This method *assumes* that

$$\text{pr(response category } \leq j) = \begin{cases} G(\alpha_j) & \text{for } DF \\ G(\alpha_j - \mu) & \text{for } D \end{cases} \tag{9.24}$$

for $j = 1, \ldots, C$ where G is a distribution function such as standard normal or logistic, and α_j and μ are unknown parameters. Now, the null and alternative hypotheses are

$$H_0^a : \mu \geqslant 0 \quad \text{and} \quad H_1^a : \mu < 0,$$

respectively. If G is assumed to be the standard normal distribution function then the p-value for testing H_0^a against H_1^a is 0.09; we used the ordinal regression procedure in MINITAB (version 13). Thus, it appears that there is inadequate evidence against $H_0^{(a)}$ and hence against H_0 in (9.24).

The model (9.24) essentially assumes that there is an unobserved response variable with its distribution function being G for every treatment; thus, the two distributions corresponding to the two treatments differ only by the location parameter μ. The range of the unobserved random variable is divided into C consecutive intervals so that the observed response takes category j whenever the unobserved response falls in the jth interval, $j = 1, \ldots, C$. ■

The assumption that the unobserved variable has the same shape for every treatment may not always be reasonable. Tests that do not require such an assumption are discussed below.

Let us first consider all possible 2×2 tables that can be formed by combining adjacent columns. The resulting tables are $(50, 64|41, 75)$, $(79, 35|69, 47)$, and $(98, 16|92, 2)$; the symbol '|' is used as a separator of rows. Let γ_1, γ_2, and γ_3 denote the population log (odds ratios) of these three collapsed tables. Then, $\gamma_1 = \gamma_2 = \gamma_3 = 0$ is equivalent to no difference between DF and D. One way of formulating the statement that DF is better than D is $\gamma_1 > 0, \gamma_2 > 0$ and $\gamma_3 > 0$; a less stringent one is $\gamma_1 \geqslant 0, \gamma_2 \geqslant 0$ and $\gamma_3 \geqslant 0$ with at least one strict inequality.

It is also possible to formulate the alternative hypothesis as *every local log (odds ratio) is positive*. This does not require any assumptions about the shape of the distribution as does Method 1, but it is more stringent than $\gamma_1 \geqslant 0, \gamma_2 \geqslant 0$ and $\gamma_3 \geqslant 0$.

Method 2: Let the null and the alternative hypothesis be

$$H_0^b : \gamma_i \leq 0, \text{ for some } i = 1, 2, 3, \quad \text{and} \quad H_1^b : \gamma_i > 0 \text{ for every } i = 1, 2, 3. \tag{9.25}$$

This is in a form to which a *min test* can be applied. The *mle* of γ_1 is $log\{(50 \times 75)/(64 \times 41)\} = 0.357$; an estimate of its standard error is $\{50^{-1} + 64^{-1} + 41^{-1} + 75^{-1}\}^{0.5} = 0.27$ (for example, see Agresti (1990, p 54)). Similarly, the *mle*s of γ_2 and γ_3 and their standard errors are 0.430 (se = 0.28) and 0.469 (se = 0.35), respectively. The corresponding one-sided p-values are 0.094, 0.061, and 0.093, respectively. Now, by Proposition 5.3.2, the p-value for the IUT of H_0^b against H_1^b is $max\{0.094, 0.061, 0.093\} = 0.094$. Therefore, there is insufficient evidence to reject H_0^b. An advantage of this method is that it does not require assumptions such as (9.24); all that it requires is that the observations be multinomial for each treatment.

Method 3: The formulation of the alternative hypothesis H_1^b in (9.25) may be too strong in some circumstances. A closely related but less stringent formulation of the alternative hypothesis is $\boldsymbol{\gamma} \geqslant \mathbf{0}, \boldsymbol{\gamma} \neq \mathbf{0}$. In this case, ideally the null hypothesis should be $\{\boldsymbol{\gamma} = \mathbf{0}\} \cup \{\boldsymbol{\gamma} \ngeqslant \mathbf{0}\}$, which is the complement of the alternative hypothesis. Unfortunately, this is not in a form that can be tested using the likelihood ratio because the boundary of the null hypothesis is in the alternative hypothesis. Suppose that we may assume *a priori* that $\boldsymbol{\gamma} \geqslant \mathbf{0}$. In this case, we may formulate the testing problem as

$$H_0^c : \boldsymbol{\gamma} = \mathbf{0} \quad \text{and} \quad H_1^c : \boldsymbol{\gamma} \geqslant \mathbf{0}.$$

This approach was discussed in an earlier chapter. Since $\hat{\gamma}_1, \hat{\gamma}_2$, and $\hat{\gamma}_3$ are positive, the maximum likelihood estimator of $\boldsymbol{\gamma}$ under H_1^c is the same as the unconstrained maximum likelihood estimator of $\boldsymbol{\gamma}$, and hence the sample value of the *LRT* for testing $\boldsymbol{\gamma} = \mathbf{0}$ against $\boldsymbol{\gamma} \geqslant \mathbf{0}$ is equal to that for testing $\boldsymbol{\gamma} = \mathbf{0}$ against $\boldsymbol{\gamma} \neq \mathbf{0}$. Therefore, it may be verified using a hand calculator that $LRT = 2.88$. The asymptotic distribution of LRT is a chi-bar square under H_0. A lower bound for the p-value is $0.5\{\text{pr}(\chi_1^2 \geqslant 2.88) + \text{pr}(\chi_2^2 \geqslant 2.88)\} = 0.16$. Again, there is inadequate evidence to reject the null hypothesis.

Method 4: Let us consider the three 2×2 tables that can be formed by using only adjacent columns; the tables are $(50, 29|41, 28)$, $(29, 19|28, 23)$, and $(19, 16|23, 24)$. Let α_1, α_2, and α_3 denote their population log (odds ratios); they are called local log (odds ratios). For a discussion of local odds ratios and their interpretation, see Chapter 6 and Agresti (1990, p 18). Now let us formulate the hypotheses as

$$H_0^d : \alpha_1 \leq 0, \text{ or } \alpha_2 \leq 0, \text{ or } \alpha_3 \leq 0, \quad \text{and} \quad H_1^d : \alpha_1 > 0, \alpha_2 > 0, \text{ and } \alpha_3 > 0.$$

Applying the procedure in Method 2, we have that $\hat{\alpha}_1 = \log\{(50 \times 28)/(41 \times 29)\} = 0.163$, $se(\hat{\alpha}_1) = \{50^{-1} + 29^{-1} + 41^{-1} + 28^{-1})^{0.5} = 0.34$ and a large sample one-sided p-value for this is 0.32. Similarly, large sample p-values for testing against $\alpha_2 > 0$ and $\alpha_3 > 0$ are 0.029 and 0.32, respectively. Now, an upper bound for the large sample p-value for the corresponding *min test* is $\max\{0.34, 0.29, 0.32\} = 0.34$. Therefore, there is insufficient evidence to reject H_0^d. ■

Thus, for comparing DF with D, the foregoing four methods lead to essentially the same conclusions because the p-values are large; however, we also note that p-values corresponding to different methods differ substantially. Now, let us compare the four treatments, DF, D, F, and P.

Example 9.2.3 *Combination drug for low-back pain and spasm (Snappin 1987).*

Let us consider Example 9.2.1. We wish to test the claim that the combination drug is better than each of its components. Therefore, the alternative hypothesis is DF *is better than* D, F, *and* P. For brevity, let "$DF > D$" and "$DF \not> D$" denote *DF is better than D* and *DF is not better than D*, respectively. Now, the null and the alternative hypotheses are

$$\begin{aligned} H_0 &: \quad DF \not> D \text{ or } DF \not> F \text{ or } DF \not> P, \\ H_1 &: \quad DF > D, DF > F, \text{ and } DF > P. \end{aligned}$$

To apply a *min test* we need to test $DF \not> D$ against $DF > D$, $DF \not> F$ against $DF > F$, and $DF \not> P$ against $DF > P$. For each of these, we will apply the first three methods in the previous example.

To apply Method 1, assume that the underlying response variable has the same shape for the four treatments; however, they may have different location parameters. For testing against each of $DF > D, DF > F$, and $DF > P$, the p-values are $0.094, 0.024$, and 0.000, respectively. It follows from Example 9.2.2 that the p-value for the *min test* of H_0 against H_1 is $\max\{0.094, 0.024, 0.000\} = 0.094$. Therefore, it appears that there is inadequate evidence to support the claim that the combination drug DF is better than each of D, F, and P.

Now, let us apply Methods 2 and 3 of the previous example.

First let us apply Method 2 for testing $DF \not> D$ against $DF > D$, $DF \not> F$ against $DF > F$, and $DF \not> P$ against $DF > P$. The corresponding p-valus are $0.094, 0.057$, and 0.011, respectively. By Propositions 5.3.1 and 5.3.2, the p-value for testing H_0 against H_1 is $\max\{0.094, 0.57, 0.011\} = 0.094$.

To apply Method 3 of the previous example, we need to assume that $DF \geqslant D, DF \geqslant F$ and $DF \geqslant P$. Under this assumption, the null and the alternative hypotheses are

$$\begin{aligned} &H_0^* : DF = D, or DF = For DF = P, \\ &\text{and } H_1^* : DF > D, DF > F \text{ and } DF > P, \end{aligned}$$

respectively. Now, test of $DF = D$ against $DF > D$ can be stated as test of $\gamma = \mathbf{0}$ against $\gamma \geqslant \mathbf{0}$; this was illustrated in Method 3 of the previous example. For testing $DF = D$ against $DF > D$, $DF = F$ against $DF > F$ and $DF = P$ against $DF > P$, the lower bounds for the p-values of the *LRT* are 0.16, 0.03, and 0.0002, respectively. Therefore, by Propositions 5.3.1 and 5.3.2, the p-value for testing H_0^* against H_1^* does not exceed $\max\{0.16, 0.03, 0.0002\} = 0.16$.

Irrespective of the method, there is insufficient evidence to support the claim that the combination drug is better than its components and the placebo. ■

Example 9.2.4 *Multiple dose combination drug in a Phase II clinical trial (Phillips, Cairns, and Koch (1992)).*

In clinical research of combination therapies, the appropriate dosages of components for a combination are usually unknown *a priori*. Therefore, such dosages need to be determined in a Phase II trial by comparing several doses.

This example is based on a clinical trial involving multiple doses of a combination drug for the treatment of hypertension. The estimates presented in Tables 9.2 and 9.3 are taken from Phillips, Cairns, and Koch (1992); readers are referred to their original paper for other related analyses/discussions of this study. The objective of this study was to compare the effects of Ramipril (R) and Hydrochlorothiazide (H) both as single therapies and in various combinations; the four levels of R (in mg) are 0, 2.5, 5, and 10 and the three levels of H (in mg) are 0, 12.5, and 25. The study design is a multicenter, randomized, double-blind, parallel group, and placebo-controlled based on a 3×4 factorial structure. The duration of the study was approximately 10 weeks with an initial placebo run-in phase followed by a treatment phase. The response variable considered here is the change in supine diastolic blood pressure (BP) from baseline to end point. There were 529 patients who were randomly assigned to the treatments; the number of patients per treatment group was between 42 and 48.

Table 9.2 Estimated mean reduction in blood pressure (mg) from baseline[a]

H (mg)	Ramipril dose (mg) 0	2.5	5	10
0	3.9 (1.3)	6.7 (1.2)	6.0 (1.1)	8.5 (1.2)
12.5	6.3 (1.1)	8.7 (2.4)	10.8 (1.1)	13.0 (1.2)
25	8.4 (1.6)	8.8 (1.4)	13.5 (1.0)	11.3 (1.6)

Table 9.3 p-Values for pairwise comparison of combination drug with components[b]

H (mg)	Ramipril (mg) 2.5	5	10
12.5	0.262	0.006	0.012
	0.185	0.012	0.001
	0.007	0.001	0.001
25	0.260	0.001	0.115
	0.828	0.005	0.104
	0.007	0.001	0.001

[a] The standard errors are in parentheses.

[b] The p-values in the first, second and third rows are for comparing with Ramipril only, hydrochlorothiazide only, and placebo, respectively.

The statistical model considered has the following form: for the kth patient receiving i th level of R and j th level of H, let

$$y_{ijk} = \mu + T_{ij} + \beta x_{ijk} + \epsilon_{ijk},$$

where y_{ijk} is the response (= change in blood pressure), T_{ij} is the effect of the treatment, x_{ijk} is a covariate to adjust for investigator effects, and the error terms $\{\epsilon_{ijk}\}$ are assumed to be independently and identically distributed . This model was estimated using the data for 529 patients. The estimates of the mean of the response variable are given in Table 9.2. Pairwise treatment group comparisons are given in Table 9.3. As an example, note that for comparing the combination $(R, H) = (10, 25)$ with Ramipril only, hydrochlorothiazide only, and placebo, the p-values are 0.115, 0.104, and 0.001, respectively. Application of *min test* to each cell in Table 9.3 shows that the combinations $(12.5, 5)$, $(25, 5)$, and $(12.5, 10)$ for (H, R) are significantly better than any one of the components in reducing the blood pressure. The other combinations of (H, R) do not appear to be significantly better than Ramipril only or hydrochlorothiazide only. ■

In the next subsection, we shall discuss the relevance of monotone tests, and an optimality property of the *min test*.

9.2.1 Monotone Critical Regions and Tests

In the foregoing examples, the alternative hypothesis was of the form $H_1 : \theta_i > 0$ for every $i = 1, \ldots, k$. When the alternative hypothesis takes such a one sided form, it is also desirable for the critical region to be one sided in some sense. In this subsection we shall discuss some of these issues.

Let us first introduce the following definitions: A region $C \subseteq \mathbb{R}^k$ is said to be *monotone nondecreasing* if

$$\boldsymbol{x} \in C, \boldsymbol{y} \in \mathbb{R}^k, x_i \leqslant y_i \text{ for every } i = 1, \ldots, k \Rightarrow \boldsymbol{y} \in C; \tag{9.26}$$

in this definition, if $\leqslant$ is replaced by $\geqslant$ then the term *nondecreasing* would be replaced by *nonincreasing*. A *monotone test* based on $(X_1, \ldots, X_k)$ is one for which the rejection region is monotone in the sample space of all possible values of $(X_1, \ldots, X_k)$.

Now, let us consider an example involving combination therapies to illustrate the importance of monotone tests; see also Laska and Meisner (1989) and Perlman and Wu (1999). Consider the combination drug problem in Example 1.1.9: A_1 and A_2 denote the two existing treatments, A_0 is the treatment obtained by combining A_1 and A_2, μ_i denotes the mean response for treatment $A_i (i = 1, 2), \theta_1 = \mu_0 - \mu_1$ and $\theta_2 = \mu_0 - \mu_2$. The objective is to test

$$H_0 : \theta_1 \leqslant 0 \text{ or } \theta_2 \leqslant 0 \quad \text{against } H_1 : \theta_1 > 0 \text{ and } \theta_2 > 0. \tag{9.27}$$

Let X_i denote a statistic for testing $\theta_i \leqslant 0$ against $\theta_i > 0$ $(i = 1, 2)$. For illustrative purposes, assume that $X_i \sim N(\theta_i, 1)$ for $i = 1, 2$. A critical region corresponding to a 5% level IUT for (9.27) is $\{(x_1, x_2) : x_1 \geqslant 1.645 \text{ and } x_2 \geqslant 1.645\}$; this is shown as region A in Fig. 9.2. Clearly, the critical region is monotone nondecreasing.

To illustrate that monotonicity of the critical region is essential for any reasonable test of the foregoing combination therapy example, let $P \equiv (2, 2)$, $Q \equiv (2, 4)$ and D be the acute cone in Fig. 9.3. Let us consider a test of (9.27) with critical region D; this is a nonmonotone critical region because the point Q is vertically above P and therefore each coordinate of Q is at least as large as that corresponding to P, yet P is in the critical region but not Q.

Our intuition says that evidence in favor of H_1 at Q is at least as strong as that at P, but the test D would reject H_0 when the outcome is P but not when it is Q. Thus the test D would do just the opposite of what we would normally expect it to do.

Laska and Meisner (1989) point out an interesting scenario in this context. Consider two drug companies evaluating two combination drugs similar to A_0 using the same design and sample sizes. Suppose that the outcomes reached by the Companies A and B are the points represented by P and Q, respectively. In this case, the test D would pass the drug for Company A but not for Company B, although Company

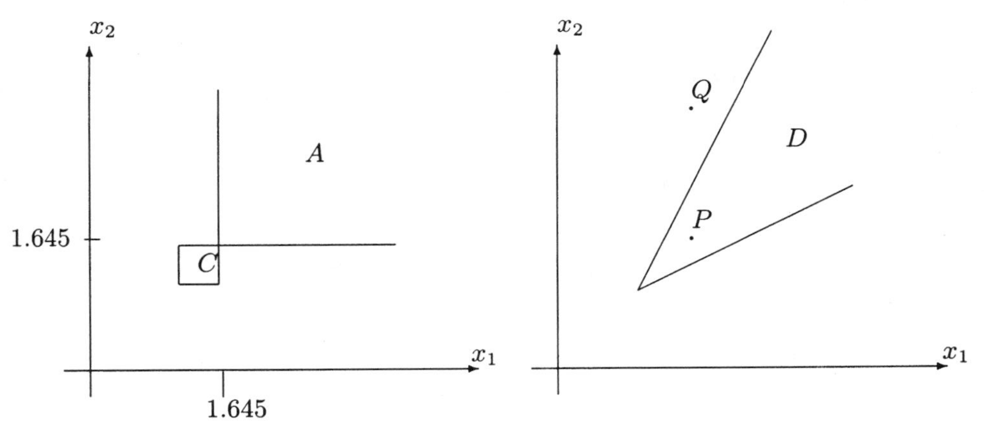

Fig. 9.2 Monotone (A) and nonmonotone ($A \cup C$) critical regions.

Fig. 9.3 A nonmonotone critical region, D.

B has stronger evidence in favor of its product than Company A. In view of this counterintuitive property of test D, it would be impossible to persuade a regulatory authority to adopt it.

Similarly, the critical region, $A \cup C$, in Figure 9.2 would not be reasonable because the point P may be inside the region C, and Q may be vertically above P but outside C. We conclude that any reasonable test of (9.27) for the combination drug problem must have a monotone nondecreasing critical region.

9.2.2 Optimality of Min Test

Let the testing problem be as in (9.27), and let X_i denote a statistic for testing $\theta_i \leq 0$ against $\theta_i > 0$; let us also assume that large values of X_i favor $\theta_i > 0$ $(i = 1, 2)$. Lehmann (1952), Cohen, Gatsonis, and Marden (1983a), and Sierra-Cavazos (1992) showed, under various conditions, that the *min test* is the uniformly most powerful monotone test of size-α based on (X_1, X_2). Laska and Meisner (1989) showed, under reasonable conditions, that the *min test* is the uniformly most powerful monotone test of size-α among the class of size-α tests of the form

$$\text{reject } H_0 \text{ if } g(X_1, X_2) \geqslant k \text{ for some } k,$$

where $g(x_1, x_2)$ is a nondecreasing function in each of its arguments. These optimality results do not assume that the joint distribution of (X_1, X_2) is known. It is important to note that the *min test* is not necessarily the uniformly most powerful among all possible tests of (9.21). This is further clarified in the next paragraph.

For any fixed i, it is possible that there may be a statistic other than X_i for testing $\theta_i \leqslant 0$ against $\theta_i > 0$; let Y_i be one such statistic and $\{Y_i \geqslant d_i\}$ be a size-α critical region, $i = 1, 2$. For example X_i, may be a t-ratio and Y_i may be based on ranks.

Now the *min test* based on (Y_1, Y_2) is

$$\text{reject } H_0 \text{ if } \min\{Y_1 - d_1, Y_2 - d_2\} > 0. \tag{9.28}$$

This *min test* is the uniformly most powerful monotone test of size-α among the class of tests of the form

$$\text{reject } H_0 \text{ if } h(Y_1, Y_2) \geqslant k \text{ for some } k,$$

where $h(y_1, y_2)$ is a nondecreasing function in each of its arguments. Thus, the *min tests* in (9.22) and (9.28) are the uniformly most powerful in different classes of tests. At this stage, we have not yet formulated the problem sufficiently precisely to say as to which of these two is better. For example, if robustness against outliers is desired, then (9.28) where Y_1 and Y_2 are nonparametric statistics such as the Wilcoxon rank sum statistic, is likely to be better than one based on sample means.

9.2.3 Multivariate Responses

Now, let us consider the case when the responses are multivariate (multiple end points). In this case, there is no unique way of formulating the testing problem. For simplicity, let us again consider the case when there are two treatments denoted by A_1 and A_2, and the identified treatment A_0. The results to be discussed below extend in a natural way to the case with more than two treatments.

Let $\boldsymbol{X}_i = (X_{i1}, \ldots, X_{ip})^T$ denote a $p \times 1$ random vector of responses corresponding to treatment A_i $(i = 0, 1, 2)$. Suppose that the only unknown parameter in the distribution of $\boldsymbol{X}_i$ is $\boldsymbol{\mu}_i = (\mu_{i1}, \ldots, \mu_{ip})^T$. For simplicity, assume that larger values of μ_{ij} correspond to better performance. The objective is to develop procedures for testing against

$$H_1 : A_0 \textit{ is better than } A_1 \textit{and } A_2.$$

A brief discussion of how this may be done is given below; some of these are based on Laska, Tang, and Meisner (1992).

Let us say that A_0 is *uniformly the best* if it is better than A_1 and A_2 with respect to every component of the parameter, $\boldsymbol{\mu}$. The corresponding null and alternative hypotheses are

$$H_0 : \mu_{0j} \leqslant \mu_{ij} \text{ for some } (i, j), \text{ and } H_1 : \mu_{0j} > \mu_{ij} \text{ for every } (i, j), \tag{9.29}$$

respectively. Here and in what follows it is assumed that $i \geqslant 1$. Now, it is a simple matter to apply a *min test*; all that we need is a test of $\mu_{0j} \leqslant \mu_{ij}$ against $\mu_{0j} > \mu_{ij}$, for every (i, j). The optimality of the *min test* that we referred to earlier for the univariate case also holds here (see Theorem 1 of Laska, Tang, and Meisner 1992).

The foregoing requirements for a treatment to be "uniformly the best" may be too strong in some situations, but not sufficiently strong in others. A more flexible formulation is the following. For a set of prechosen numbers, $\epsilon_1, \ldots, \epsilon_p$, let the testing problem be formulated as

$$H_0 : \delta_{ij} \leqslant 0 \text{ for some } (i, j), \quad \text{vs} \quad H_1 : \delta_{ij} > 0 \text{ for every } (i, j) \tag{9.30}$$

where $\delta_{ij} = (\mu_{0j} + \epsilon_j) - \mu_{ij}$ for every (i, j). This formulation would be relevant if we wish to accept the new treatment A_0 only if it is better than each A_i by a margin $(= \epsilon_i)$ of practical significance, $i \geqslant 1$. Again a *min test* is applicable for (9.30).

Another notion that may be useful here is admissibility of the treatment. Let us say that A_0 is *admissible* if, for each $i \geqslant 1$, A_0 is better than A_i with respect to at least one component of $\boldsymbol{\mu}$. In other words, A_0 is *inadmissible* if A_i is at least as good as A_0 in terms of every component of $\boldsymbol{\mu}$, for some i. This notion of inadmissibility is reasonable in the context of combination drug. The rationale is that if A_0 is the combination drug then it is likely to be more "expensive" (for example, in terms of side effects) than A_i for every $i \geqslant 1$. Therefore, A_0 would not be considered as desirable as A_i if the latter is at least as effective as the former.

The null and alternative hypotheses for testing against admissibility are,

$$H_0 : \delta_i \leqslant 0 \text{ for some } i, \quad \text{and} \quad H_1 : \delta_i > 0 \text{ for every } i, \tag{9.31}$$

respectively, where $\delta_i = \max_j(\mu_{0j} - \mu_{ij}), i \geqslant 1$. Again, a *min test* is applicable provided that a test of

$$H_0^{(i)} : \delta_i \leqslant 0 \quad \text{against} \quad H_1^{(i)} : \delta_i > 0 \tag{9.32}$$

is available for every $i \geqslant 1$. There are different ways of testing (9.32). With $\theta_{ij} = \mu_{0j} - \mu_{ij}$, the testing problem (9.32) can be restated as $H_0^{(i)} : \boldsymbol{\theta}_i \in \mathcal{C}$ and $H_1^{(i)} : \boldsymbol{\theta}_i \notin \mathcal{C}$, where $\boldsymbol{\theta}_i = (\mu_{01} - \mu_{i1}, \ldots, \mu_{0p} - \mu_{ip})$ and $\mathcal{C}$ is the closed convex cone, $\{\boldsymbol{x} : x_1 \leqslant 0, \ldots, x_p \leqslant 0\}$; this is a Type B testing problem discussed in Chapter 2 and hence the methods therein may be used. Alternatively, a many-to-one test as in Miller (1981) is also applicable for (9.32). For some optimality results concerning the *min test* for (9.31) see Laska, Tang, and Meisner (1992, Theorems 2 and 3).

9.2.4 Comparison of LRT with the Min Test

Let $\boldsymbol{\theta} \in \Theta$ denote the parameter of interest, $J = \{1, \ldots, k\}$ be an indexing set, Θ_j be s subset of Θ for $j \in J$, and let the testing problem be defined by

$$H_0 : \boldsymbol{\theta} \in \bigcup_{j \in J} \Theta_j \quad \text{and} \quad H_1 : \boldsymbol{\theta} \notin \bigcup_{j \in J} \Theta_j, \tag{9.33}$$

respectively. The IUT is based on marginal distributions for testing $\boldsymbol{\theta} \in \Theta_j$ against $\boldsymbol{\theta} \notin \Theta_j$ for one j at a time $(j \in J)$, while the *LRT* is based on the joint distribution of all the observations. For example, consider the following: $(X_1, X_2) \sim N(\boldsymbol{\theta}, V)$ where $V = (1, \rho|\rho, 1)$, and the null and alternative hypotheses are

$$H_0 : \theta_1 \leqslant 0 \text{ or } \theta_2 \leqslant 0, \quad \text{and} \quad H_1 : \theta_1 > 0 \text{ and } \theta_2 > 0, \tag{9.34}$$

respectively. For $i = 1, 2$, a test on θ_i could be based on the marginal distribution $N(\theta_i, 1)$ of X_i. These two tests can be used to construct an IUT for (9.34). Thus, as has been pointed out earlier, the IUT does not require the joint likelihood of X_1

and X_2, and it does not make use of the value of ρ even if it is known. Therefore, when ρ is unknown, this may be an attractive feature of IUT. On the other hand, if ρ is known, it is of interest to know whether or not the foregoing IUT is as good as the *LRT* even though the former does not make use of the known value of ρ but the latter does. In this subsection, some of these issues will be addressed.

Let $L(\boldsymbol{\theta}|\boldsymbol{x})$ denote the joint likelihood. Then, the *LRT* statistic $\Lambda(\boldsymbol{x})$ for (9.33) is given by

$$\Lambda(\boldsymbol{x}) = \min\{\Lambda_j(\boldsymbol{x}) : j \in J\},$$

where $\Lambda_j(\boldsymbol{x}) = \sup\{L(\boldsymbol{\theta}|\boldsymbol{x}) : \boldsymbol{\theta} \in \Theta\}/\sup\{L(\boldsymbol{\theta}|\boldsymbol{x}) : \boldsymbol{\theta} \in \Theta_j\}$, is the *LRT* for testing $H_0^{(j)} : \boldsymbol{\theta} \in \Theta_j$ against $H_1^{(j)} : \boldsymbol{\theta} \notin \Theta_j$, $j \in J$. Let $c_{j\alpha}$ denote the size-α critical value for $\Lambda_j(\boldsymbol{x}), j \in J$, and c_α denote the size-α critical value for $\Lambda(\boldsymbol{x})$. Then the IUT of H_0 against H_1 in (9.33) based on $\Lambda_j(\boldsymbol{x})$ for testing $\boldsymbol{\theta} \in \Theta_j$ against $\boldsymbol{\theta} \notin \Theta_j$, $j \in J$, is the following:

$$\text{IUT : Reject } H_0 \text{ if } \Lambda_j(\boldsymbol{x}) \geqslant c_{j\alpha} \text{ for every } j \in J.$$

The *LRT* of H_0 against H_1 is the following:

$$\textit{LRT} : \text{Reject } H_0 \text{ if } \Lambda_j(\boldsymbol{x}) \geqslant c_\alpha \text{ for every } j \in J.$$

Clearly, there is a close relationship between the foregoing IUT and the *LRT* because both use the statistics, $\{\Lambda_j(\boldsymbol{x}) : j \in J\}$. The critical value c_α of the *LRT* typically depends on the joint distribution of $\{\Lambda_j(\boldsymbol{x}) : j \in J\}$, but $c_{j\alpha}$ depends only on the marginal distribution of $\Lambda_j(\boldsymbol{x}), j \in J$. A feature of the IUT that makes it more flexible is that it allows different critical values for different j s. In Example 9.2.6 discussed below, this difference between IUT and *LRT* is exploited to construct an example in which the IUT is better than the *LRT*.

Now consider the special case when the $c_{j\alpha}$s are all equal for $j \in J$. Even in this case, it does not necessarily follow that c_α is the common value of $c_{j\alpha}$ because the IUT may not be size-α although we know that it is level-α. If the conditions in Proposition 5.3.2 or those in Corollary 5.3.3 are satisfied then the IUT would have size-α and hence the common value of $c_{j\alpha}$ would be equal to c_α. A more general result related to these is given below; its proof is similar to those of Propositions 5.3.1 and 5.3.2 (see Berger (1997)).

Proposition 9.2.1 *Suppose that $\{c_{j\alpha} : j \in J\}$ are not necessarily all equal. Without loss of generality, assume that $c_{1\alpha} = \max\{c_{j\alpha} : j \in J\}$. Further, assume that the conditions of Proposition 5.3.2 or those of Corollary 5.3.3 are satisfied. Then the following hold: (a) The size-α critical value, c_α, of LRT is equal to $c_{1\alpha}$. (b) The critical region of size-α LRT is contained in the critical region of the size-α IUT.*

It follows from part (b) of this result, that the IUT is uniformly at least as powerful as the *LRT*, and that

$$pr_{\boldsymbol{\theta}}(\text{Type I error with LRT}) \leqslant pr_{\boldsymbol{\theta}}\,(\text{Type I error with IUT}) \leqslant \alpha$$

for any $\boldsymbol{\theta} \in H_0$. Thus, the IUT achieves higher power than the *LRT* at the cost of higher probability of Type I error everywhere in H_0. Consequently, even though the

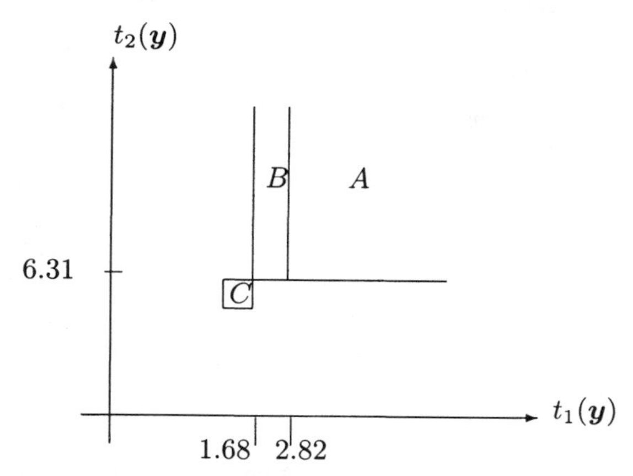

Fig. 9.4 Critical regions of an IUT ($= A \cup B$) and the *LRT*($= A$) for two independent samples.

IUT is uniformly more powerful than the LRT, it is not possible to claim that IUT is better than the *LRT* without any qualifications.

If X_1 and X_2 are independent and normal, and the number of observations on X_1 and X_2 are different, then the IUT turns out to be better than the *LRT*. This is illustrated in the next example due to Saikali and Berger (2001); for related discussions, see Perlman and Wu (1999), in particular, the discussion by R.L. Berger.

Example 9.2.6 to be discussed later in this section shows that a uniformly more powerful test is not necessarily a better test. Perlman and Wu (1999) provide excellent discussions of several issues related to this; we strongly recommend this article to those who are interested in these issues.

Example 9.2.5 *An example where IUT is better than LRT (Saikali and Berger 2001).*

Let $Y_{11,\dots,}Y_{1n_1}$ be a random sample from $N(\mu_1, \sigma_1^2)$ and $Y_{21,\dots,}Y_{2n_2}$ be an independent random sample from $N(\mu_2, \sigma_2^2)$. Let $n_1 = 40$ and $n_2 = 2$, $\boldsymbol{\theta} = (\mu_1, \mu_2, \sigma_1, \sigma_2)^T$ and $L(\boldsymbol{\theta}|\boldsymbol{Y})$ denote the likelihood. Suppose that we wish to test

$$H_0 : \mu_1 \leqslant 0 \text{ or } \mu_2 \leqslant 0, \quad \text{vs} \quad H_1 : \mu_1 > 0 \text{ and } \mu_2 > 0.$$

Let $T_i(\boldsymbol{Y}) = \bar{Y}_i/(S_i/\sqrt{n_i})$ be the t-ratio for testing $\mu_i \leq 0$ against $\mu_i > 0$.

It may be shown (see Problem 9.4) that the critical region for a 5% level *LRT* is $\{T_1(\boldsymbol{Y}) \geqslant 2.82 \text{ and } T_2(\boldsymbol{Y}) \geqslant 6.31\}$, and the critical region for a 5% level IUT is $\{T_1(\boldsymbol{Y}) \geqslant 1.68 \text{ and } T_2(\boldsymbol{Y}) \geqslant 6.31\}$. These are indicated in Fig. 9.4; the critical region for *LRT* is A, and that for the IUT is $A \cup B$. Since the critical region for *LRT* is contained in that for IUT, the latter is uniformly more powerful than the former. It also follows that the probability of Type I error for IUT is higher than that for *LRT* everywhere in H_0; however, the size of each of these tests is 0.05.

In this example, the IUT appears to be better than the *LRT*. This is because the former is uniformly more powerful than the latter and the increase in power is achieved without sacrificing important features of *LRT* such as monotonicity. Alternatively, we can say that the IUT would represent the statistical evidence better than the *LRT*, using a p-value. The fact that IUT is more powerful than the *LRT*, by itself, would not be sufficient to claim that the former is better than the latter. Finally, let us note that the IUT is better than the *LRT* despite the fact that the latter uses the known joint distribution of (X_1, X_2) but the former uses only the marginal distributions of X_1 and X_2. ■

In view of the fact that the *min test* is the uniformly most powerful monotone test based on (T_1, T_2), it is not possible to construct a test having uniformly more power without considering tests that are not monotone in (T_1, T_2). The next example shows that if monotonicity is not imposed then a test that is uniformly more powerful than the *min test* can be constructed. However, such a test is not necessarily better than the *min test* from a practical point of view; in fact, it is worse.

Example 9.2.6 *A test that is uniformly more powerful than the min test.*

Let us consider the same setting as in the previous example. First note that, for the IUT, the probability of Type I error at $(\mu_1, \mu_2) = (0, 0)$ is $0.05^2 = 0.0025$. Since, the power function is continuous, it follows that at points close to the origin and near the boundary of the first quadrant, the probability of Type I error would be smaller than 0.05. Consequently, it turns out that a rectangle similar to C in Fig. 9.4 can be included into the critical region of the IUT while ensuring that pr(Type I error $|H_0)$ does not exceed 0.05. Now, the test with critical region $A \cup B \cup C$ is uniformly more powerful than the IUT, which has critical region $A \cup B$, but the former is not monotone. If monotonicity is important, then the test with critical region $A \cup B \cup C$ is not as good as the IUT, which has the monotone critical region $A \cup B$. For results relating to these types of tests which are uniformly more powerful than the *LRT*, see Berger (1989), Liu and Berger (1995), and Wang and McDermott (1996, 2002); for arguments against this approach see Perlman and Wu (1999, 2002b). ■

9.3 CROSS-OVER INTERACTION

Consider an experiment to compare two treatments, T1 and T2, in two groups, G1 and G2. Let θ denote a scalar parameter of interest; for example, it could be the population mean of a response variable. Let θ_{ij} denote the θ parameter for treatment j in group i, $i = 1, 2, j = 1, 2$ (see Table 9.4). Without loss of generality, assume that a treatment with a larger mean is better. Three possible cell mean diagrams are shown in Fig. 9.5. In Fig. 9.5(a), the two lines are parallel and T2 is better than T1 in every group. In the standard terminology, the lines being parallel is the same as saying that there is no interaction between treatment and group. A particularly simplifying consequence of no interaction between treatment and group is that one can ignore the differences between groups when comparing treatments.

For the situation represented by Fig. 9.5(b) there is interaction, but T2 is still better than T1 in every group. For the situation represented by Fig. 9.5(c) there is interaction, but no treatment is better than the other in every group. If the orderings of

Table 9.4 Expected responses of two treatments in two groups

Group	*Treatment* T1	T2
G1	θ_{11}	θ_{12}
G2	θ_{21}	θ_{22}

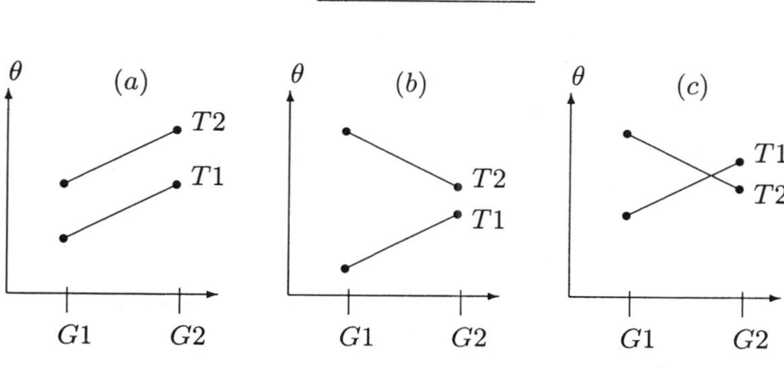

Fig. 9.5 Cell mean diagrams: (a) No interaction; (b) Quantitative interaction, but no cross-over interaction, (c) Cross-over interaction.

the treatments with respect to θ within the two groups are not the same, for example, as in Fig. 9.5(c), then we say that there is *cross-over interaction* between treatment and group. The term *qualitative interaction* is also used for cross-over interaction. For the situation represented by Fig. 9.5(b) we say that there is *quantitative interaction* but no cross-over interaction.

Cross-over interaction can be a more serious phenomenon than just quantitative interaction with no cross-over interaction. For example, let T1 be a placebo and T2 be a new treatment, and the two groups be males (= G1) and females (= G2). For the scenario depicted by Fig. 9.5(c), the new treatment is harmful for females but beneficial for males. For the scenario depicted by Fig. 9.5(b), while the new treatment is not as effective for females as for males, it is beneficial for both sexes.

In plant breeding, cross-over interaction between genotype and environment plays an important role (see Baker (1988)). If there is no cross-over interaction between genotype and environment then the need to breed new varieties capable of adapting to new environments does not arise. To give another example, consider comparing a treatment with a placebo in a multicenter clinical trial. In such a trial, if there is cross-over interaction between center and treatment then the treatment is better than the placebo in one center but worse in another. Consequently, the treatment cannot be claimed to better than the placebo irrespective of the center.

The roles of treatment and group are not interchangeable when we refer to "cross-over interaction between treatment and group." To illustrate this, consider the hypothetical example $(\theta_{11}, \theta_{12}, \theta_{21}, \theta_{22}) = (3, 4, 2, 1)$ where θ_{ij} is as in Table 9.4. These cell means are shown in Fig. 9.6; in the first diagram different lines correspond to different groups, and in the second different lines correspond to different treatments.

Note that the cell mean lines cross each other when group is on the x-axis but not when treatment is on the x-axis. In view of this asymmetry, cross-over interaction *between* treatment and group is not a particularly suitable phrase because the term "*between*" tends to imply that the roles of treatment and group are interchangeable. However, a more precise and informative phrase is likely to be cumbersome. Therefore, we shall not deviate from the foregoing terminology in the literature; as to which is the treatment and which is the group should be clear from the context.

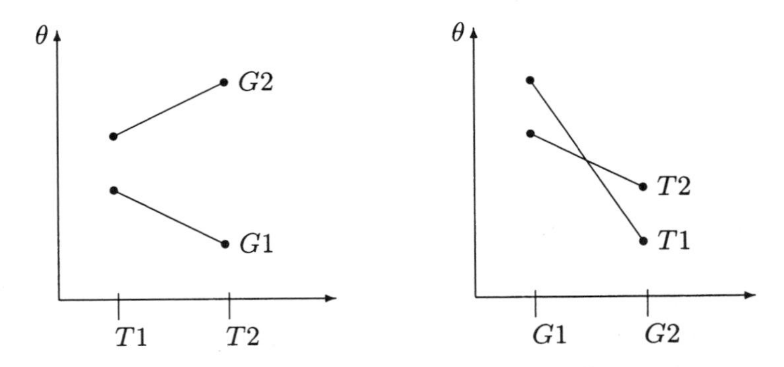

Fig. 9.6 Asymmetric nature of cross-over interaction; both diagrams correspond to $(\theta_{11}, \theta_{12}, \theta_{21}, \theta_{22}) = (3, 4, 2, 1)$.

The foregoing ideas on cross-over interaction extend in a natural way to several groups and several treatments. Consider J treatments being compared across I groups. If the ordering of the J treatments (with respect to the parameter of interest) within each group is the same for every group, then we say that there is no cross-over interaction. Fig. 9.7 shows three possibilities with three treatments and three groups. If the lines cross, then there is cross-over interaction; if they do not cross but are

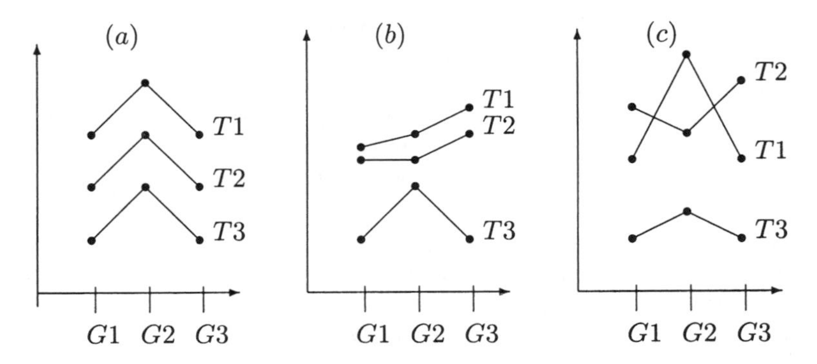

Fig. 9.7 Cell mean diagrams: (a) No interaction; (b) Quantitative interaction, but no cross-over interaction, (c) Cross-over interaction.

not parallel then there is quantitative interaction but not cross-over interaction. If the lines are parallel then there is no interaction. To state it more precisely let

$$\delta(j_1, j_2|i) = \theta_{ij_1} - \theta_{ij_2},$$

the difference between treatments j_1 and j_2 in group i. If, for any (j_1, j_2), $\delta(j_1, j_2|i)$ has the same sign for every group i, then we say that there is no cross-over interaction.

Various tests for and against cross-over interaction are available when there are only two treatments. However, for the case with more than two treatments, very little is available. We shall first consider the special case of two treatments in detail; the case with more than two treatments will be discussed later.

Let us present two examples to further illustrate the nature of the problem. At this stage we discuss only the nature of problem; solutions to the problems will be discussed later.

Example 9.3.1 *Cross-over interaction in a multicenter clinical trial. (Ciminera et al. (1993))*

A multicenter clinical trial was conducted to compare a drug with a placebo; the drug was a treatment to reduce blood pressure (see Table 1.10 on page 17)). This is an unpaired design (i.e., parallel group) within each center, and there were 12 centers. The issue that is of interest is whether or not there is any evidence that the drug is beneficial in some centers and harmful in some others. This requires a test against cross-over interaction between treatment and center.

Example 9.3.2 *Comparison of two treatments for breast cancer (Gail and Simon (1985)).*

This example is based on the National Surgical Adjuvant Breast and Bowel Project (Fisher et al. (1983)). The two treatments to be compared are PF and PFT:

PF: L-phenylalanine mustard, 5-fluorouracil;

PFT: L-phenylalamine, mustard, 5-fluorouracil, and tamoxifen.

The subjects of this trial are patients with primary operable breast cancer and positive nodes. The response variable is whether or not the patient is disease free at 3 years. Fisher et al. (1983) found evidence for heterogeneity in response to PFT therapy that is both age and progesterone receptor (PR) level dependent. They noted that young patients with progesterone receptor levels under 10 units do better in PF whereas other patients do better on PFT. In our terminology, this indicates presence of cross-over interaction. The summary data and relevant estimates are given in Table 9.5.

9.3.1 A Test Against Cross-over Interaction: Two Treatments and I Groups.

Consider a treatment and a control being compared across I groups. Let δ_i denote the treatment effect of interest, $i = 1, \ldots, I$. Let

$$\boldsymbol{\delta} = (\delta_1, \ldots, \delta_I), \mathcal{O}^+ = \{\boldsymbol{\delta} \in \mathbb{R}^I : \boldsymbol{\delta} \geqslant \mathbf{0}\} \text{ and } \mathcal{O}^- = \{\boldsymbol{\delta} \in \mathbb{R}^I : \boldsymbol{\delta} \leq \mathbf{0}\}. \quad (9.35)$$

Table 9.5 Comparison of PF and PFT for breast cancer

Group	Treatment[a]		$\hat{\delta}_i$	$\hat{\sigma}_i$	$\hat{\delta}_i/\hat{\sigma}_i$
	PF	PFT			
1. Age<50, PR<10	0.599 (0.0542)	0.436 (0.0572)	0.163	0.0788	2.07
2. Age ≥50, PR<10	0.526 (0.051)	0.639 (0.0463)	-0.144	0.0689	-1.65
3. Age<50, PR≥10	0.651 (0.0431)	0.698 (0.0438)	-0.047	0.0614	-0.765
4. Age≥50, PR ≥10	0.639 (0.0386)	0.790 (0.0387)	-0.151	0.0547	-2.76

[a] Columns 2 and 3 give estimates of proportion of patients free of breast cancer at 3 years. The standard errors are given in parentheses. The total number of observations is 1260.

Note that $\mathcal{O}^+$ and $\mathcal{O}^-$ are also the same as $\mathbb{R}^{+p}$ and $\mathbb{R}^{-p}$. In the cross-over interaction literature use of $\mathcal{O}^+$ and $\mathcal{O}^-$ is quite common, therefore, we shall adopt this slightly simpler notation. By definition, there is no cross-over interaction if $\boldsymbol{\delta} \geqslant \mathbf{0}$ or $\boldsymbol{\delta} \leq \mathbf{0}$. Therefore, the null and alternative hypotheses for testing against cross-over interaction are

$$H_0 : \boldsymbol{\delta} \leq \mathbf{0} \text{ or } \boldsymbol{\delta} \geqslant \mathbf{0} \quad \text{and} \quad H_1 : \text{ not } H_0. \tag{9.36}$$

Fig. 9.8(A) shows the null and alternative parameter spaces when $I = 2$. If $I \geqslant 3$ then H_0 consists of the two orthants, $\{\boldsymbol{\delta} : \boldsymbol{\delta} \geqslant \mathbf{0}\}$ and $\{\boldsymbol{\delta} : \boldsymbol{\delta} \leq \mathbf{0}\}$, and H_1 consists of the remaining $(2^I - 2)$ orthants.

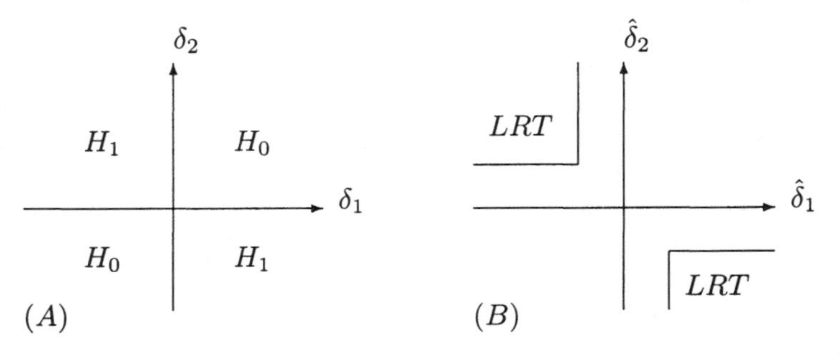

Fig. 9.8 Test of $H_0 : \boldsymbol{\delta} \leq \mathbf{0}$ or $\boldsymbol{\delta} \geqslant \mathbf{0}$ against H_1 : *not* H_0; (A) Parameter spaces, and (B) The critical region of the *LRT* when $\hat{\delta}_i \sim N(\delta_i, 1), i = 1, 2$.

Let $\hat{\boldsymbol{\delta}}$ be an estimator of $\boldsymbol{\delta}$ such that $\hat{\delta}_1, \ldots, \hat{\delta}_I$ are independent. Initially, we shall assume that

$$\hat{\delta}_i \sim N(\delta_i, \sigma_i^2), i = 1, \ldots, I \tag{9.37}$$

for some $\sigma_1, \ldots, \sigma_I$. It will be clear that if $\hat{\delta}_i$ is only approximately $N(\delta_i, \sigma_i^2)$ instead of being exactly $N(\delta_i, \sigma_i^2)$ for $i = 1, \ldots, I$, then the main results would hold approximately. The choice of a procedure for testing against cross-over interaction depends on what is known about $\sigma_1, \ldots, \sigma_I$. The following three cases are considered below.
(i) $\hat{\delta}_i \sim N(\delta_i, \sigma_i^2)$ where $\sigma_i^2 = \sigma^2 w_i$, σ^2 is unknown and w_i is known, $i = 1, \ldots, I$.
(ii) $\hat{\delta}_i \sim N(\delta_i, \sigma_i^2)$ where $\sigma_1, \ldots, \sigma_I$ are known.
(iii) $\hat{\delta}_i$ is approximately $N(\delta_i, \sigma_i^2)$ where $\sigma_1, \ldots, \sigma_I$ are unknown.

9.3.1.1 Case 1. $\hat{\delta}_i \sim N(\delta_i, \sigma_i^2)$ where $\sigma_i^2 = \sigma^2 w_i$, σ^2 is unknown and w_i is known, $i = 1, \ldots, I$.

First, let us point out two important settings in which this case arises.
(i) Let Z_{ijk} denote the kth observation for treatment j in group i. Suppose that the observations for the two treatments are paired. Then define $Y_{ik} = (Z_{i2k} - Z_{i1k})$, the difference between the responses for the kth pair in group i. Now, let $\hat{\delta}_i$ denote the mean $\bar{Y}_i$ where $\bar{Y}_i = n_i^{-1}(Y_{i1} + \ldots + \ldots Y_{in_i})$ and n_i is the number of matched pairs in group i. Assume that Y_{ik}'s are independent and are distributed as $N(\delta_i, \sigma^2)$. Then (9.37) holds with $\sigma_i^2 = \sigma^2 w_i$ and $w_i = n_i^{-1}, i = 1, \ldots, I$. In this case, σ^2 can be estimated by the error mean square, $S^2 = \sum_i \sum_k (Y_{ik} - \bar{Y}_i)^2/\nu$ where $\nu = \sum_i (n_i - 1)$.
(ii) Now, to indicate the second setting in which this case arises, suppose that the observations for the two treatments are independent, and hence unpaired. Let $\bar{Z}_{ij\cdot}$ be the mean of the n_{ij} observations, $\{Z_{ijk} : k = 1, \ldots, n_{ij}\}$, in cell (i, j). Assume that $Z_{ijk} \sim N(\mu_{ij}, \sigma^2)$. Let $\hat{\delta}_i = (\bar{Z}_{i2\cdot} - \bar{Z}_{i1\cdot}), i = 1, \ldots, I$. Then (9.37) holds with $\delta_i = (\mu_{i2} - \mu_{i1}), \sigma_i^2 = \sigma^2 w_i$, and $w_i = (n_{i1}^{-1} + n_{i2}^{-1})$. In this case, an estimator of σ^2 is the error mean square, $\nu^{-1} \sum_i \sum_j \sum_k (Z_{ijk} - \bar{Z}_{ij\cdot})^2$, where $\nu = \sum_i (n_{i1} + n_{i2} - 2)$.

To provide an insight into the derivation of the test statistics, let us temporarily assume that σ is known; this will be relaxed soon. Let

$$S(\boldsymbol{\delta}) = \sum_{i=1}^{I} (\hat{\delta}_i - \delta_i)^2 / w_i.$$

It is easily seen that -2 log (likelihood) is equal to $\{\sigma^{-2} S(\boldsymbol{\delta}) +$ *terms that depend on* σ *but not on* $\boldsymbol{\delta}\}$; thus $\sigma^{-2} S(\boldsymbol{\delta})$ is the kernel of the loglikelihood. Therefore, the *LRT* statistic is

$$T = \{\min_{H_0} S(\boldsymbol{\delta}) - \min_{H_0 \cup H_1} S(\boldsymbol{\delta})\} \sigma^{-2}.$$

Since the minimum of $S(\boldsymbol{\delta})$ over $H_0 \cup H_1$ is zero and H_0 is $\mathcal{O}^+ \cup \mathcal{O}^-$, it follows that $T = \sigma^{-2} \min\{S(\boldsymbol{\delta}) : \boldsymbol{\delta} \in \mathcal{O}^+ \cup \mathcal{O}^-\}$. Let $I(\cdot)$ denote the indicator function, and let

$$R^+ = \min\{S(\boldsymbol{\delta}) : \boldsymbol{\delta} \in \mathcal{O}^+\} \quad \text{and} \quad R^- = \min\{S(\boldsymbol{\delta}) : \boldsymbol{\delta} \in \mathcal{O}^-\}. \tag{9.38}$$

It may be verified that $R^+ = \sum_{i=1}^{I} w_i^{-1} \hat{\delta}_i^2 I(\hat{\delta}_i < 0)$, $R^- = \sum w_i^{-1} \hat{\delta}_i^2 I(\hat{\delta}_i > 0)$, and

$$T = \min(R^+, R^-)/\sigma^2. \tag{9.39}$$

Therefore, if σ is known, then the *LRT* against cross-over interaction is, reject H_0 when T is large. For the special case of two groups and $\sigma_1 = \sigma_2 = 1$, the shape of the critical region of *LRT* is shown in Fig. 9.8(B).

Now, let us relax the temporary assumption that σ is known. Clearly, T is not a usable test statistic because it depends on the unknown parameter σ^2. To construct an operational test statistic, we replace the σ in (9.39) by a consistent estimator of it. If the substituted estimator of σ^2 satisfies some conditions, then the exact finite sample distribution of the test statistic is known. Otherwise, only large sample results are available. In either case, the critical values and p-values can be computed easily.

Now, let us discuss the details for the case when exact results are available; large sample results under less-restrictive conditions will be discussed under Case 3. Let

$$F_Q = \min(R^+, R^-)/S^2, \quad \text{where} \quad S^2 = \text{ mean square for error.} \qquad (9.40)$$

This resembles the usual F-statistic in normal theory linear models. The test based on F_Q will be referred to as the F_Q*-test*. This test is not similar. Fortunately, the least favorable null distribution of F_Q has a simple form and therefore the application of F_Q to test against cross-over interaction is straight-forward. The required results are given in the next proposition; the proof is given in the Appendix.

Proposition 9.3.1 *(Silvapulle (2000, 2001)). Assume that* $\hat{\delta}_1, \ldots, \hat{\delta}_I$ *are independent, and* $\hat{\delta}_i \sim N(\delta_i, \sigma_i^2)$ *for* $i = 1, \ldots, I$. *Let* H_0 *and* H_1 *be as in (9.36),* $\boldsymbol{\delta}^* = (\infty, 0, \ldots, 0) \in H_0$, *and*

$$Bi(i; I-1) = (I-1)!\{i!(I-1-i)!\}^{-1}0.5^{I-1}, \qquad (9.41)$$

the binomial probability of i *successes out of* $(I-1)$ *trials when the probability of success in a single trial is 0.5.*

Assume that $\sigma_i^2 = \sigma^2 w_i$ *where* σ *is unknown and* w_i *is known for* $i = 1, \ldots, I$, S^2 *is an estimator of* σ^2 *such that* $\nu S^2/\sigma^2 \sim \chi^2_\nu$, *and* S^2 *and* $\{\hat{\delta}_1, \ldots, \hat{\delta}_I\}$ *are independent. Then, a test of* H_0 *against* H_1 *is, reject* H_0 *for large values of* F_Q *in (9.40). Further, the following results hold.*

(a) $pr(F_Q \geqslant c|\boldsymbol{\delta}) \leq pr(F_Q \geqslant c|\boldsymbol{\delta}^*)$ *for any* $\boldsymbol{\delta} \in H_0$. *Hence* $\boldsymbol{\delta}^*$ *is a least favorable null value; in fact any value of the form* $(0, \ldots, 0, \infty, 0, \ldots, 0)$ *for* $\boldsymbol{\delta}$ *is a least favorable null value.*

(b) The exact distribution of F_Q *at the least favorable null value* $\boldsymbol{\delta}^*$, *is given by*

$$pr(F_Q \geqslant c|\boldsymbol{\delta}^*) = \sum_{i=1}^{I-1} Bi(i; I-1)\, pr(iF_{i,\nu} \geqslant c). \qquad (9.42)$$

It follows from the foregoing proposition that an α-level test of H_0 against H_1 is, reject H_0 if $F_Q \geqslant c_\alpha$ where c_α is the value of c for which the RHS of (9.42) is equal to α. If the sample value of F_Q is c then the p-value is the expression on the right-hand side of (9.42). Computation of these critical and p-values using standard statistical software packages is straightforward. Values of c_α for $\alpha = 0.05$ are given in Silvapulle (2001); values of c_α for $\alpha = 0.01$ are given in Silvapulle (2000).

It is widely accepted that the F-critical values used in normal theory linear model are reasonably robust when the error distribution is not normal. The proof of Proposition 9.3.1 also suggests that the foregoing critical value, c_α, is also likely to be equally robust. Simulation results in Silvapulle (2001) support this conjecture.

Note that $iF_{i,\nu} \xrightarrow{d} \chi_i^2$ as $\nu \to \infty$. Therefore, if ν is large then an approximate version of Proposition 9.3.1 is obtained by replacing $\text{pr}(iF_{i,\nu} \geqslant c)$ by $\text{pr}(\chi_i^2 \geqslant c)$ in (9.42). As will be seen in Case 3, such an approximation holds if the distribution of $\hat{\delta}_i$ is only approximately $N(\delta_i, \sigma^2 w_i), i = 1, \ldots, I$.

Example 9.3.3 *Cross-over interaction in a multicenter clinical trial.*

Consider the clinical trial in Example 9.3.1. Let Z_{ijk} denote the change in baseline blood pressure after 4 weeks of treatment for the kth patient in the ith center receiving j^{th} treatment ($j = 1$ for drug and $j = 2$ for placebo). Assume that these observations are independent and normally distributed with common variance, σ^2. For center i, the effect of the drug compared with the placebo is estimated by $\hat{\delta}_i = (\bar{Z}_{i1\cdot} - \bar{Z}_{i2\cdot})$. Then $\hat{\delta}_i \sim N(\delta_i, \sigma^2 w_i)$ where $w_i = (n_{i1}^{-1} + n_{i2}^{-1})$.

For Center i, let ss_i denote the within center residual sum of squares,

$$\{\sum_k (z_{i1k} - \bar{z}_{i1\cdot})^2 + \sum_k (z_{i2k} - \bar{z}_{i2\cdot})^2\}.$$

For each center, the t-ratio for treatment is given in the last column of Table 1.10. The values, -2.29 for Center 2 and 3.04 for Center 5, are statistically significant at the 5% level. Since these effects have opposite signs, there is some indication that cross-over interaction may be present. However, since the Centers 2 and 5 were selected because they have the largest t-ratios with opposite signs, we cannot claim that this provides sufficient evidence of cross-over interaction at 5% level. We need to apply a test such as the foregoing F_Q.

To apply the F_Q-test, first note that an estimator of the unknown error variance σ^2 is the error mean square,

$$s^2 = \nu^{-1} \sum_i \sum_j \sum_k (z_{ijk} - \bar{z}_{ij\cdot})^2 = 125^{-1} \sum_i ss_i = 6534.16/125 = 52.27;$$

here we used the fact that the total error degrees of freedom is $\nu = \sum_i (n_{i1} + n_{i2} - 2) = 125$. Now, substituting the sample values of $\hat{\delta}_i$ and w_i, we have

$$R^+ = \sum I(\hat{\delta}_i < 0)\hat{\delta}_i^2/w_i = 144.9 \text{ and } R^- = \sum I(\hat{\delta}_i > 0)\hat{\delta}_i^2/w_i = 6543.16.$$

Therefore,

$$F_Q = \min(R^-, R^+)/s^2 = 144.9/52.27 = 2.77.$$

The table of critical values in Silvapulle (2001) does not provide the values for $\nu = 125$. However, since the critical values decrease with increasing error degrees of freedom, we will use the critical values on either side of $\nu = 125$. Because the sample value of 2.77 is much smaller than 12.85, the 5% level critical value corresponding to $\nu = 200$, H_0 cannot be rejected at the 5% level. By (9.42),

$$p\text{-value} = \sum_{i=1}^{11} 11!\{i!(11-i)!\}^{-1}\text{pr}(iF_{i,125} \geqslant 2.77) = 0.76.$$

Therefore, evidence in favor of treatment-by-center cross-over interaction is very weak. In other words, there is insufficient evidence to claim that the drug is beneficial in one center and harmful in another. ■

9.3.1.2 Case 2: $\hat{\delta}_i \sim N(\delta_i, \sigma_i^2), \hat{\delta}_1, \ldots, \hat{\delta}_I$ are independent, and $\sigma_1, \ldots, \sigma_I$ are known

Although it is unlikely that $\sigma_1, \ldots, \sigma_I$ would be known in practice, it is instructive to consider this case first. Let $\boldsymbol{\sigma} = (\sigma_1, \ldots, \sigma_I), \hat{\boldsymbol{\delta}} = (\hat{\delta}_1, \ldots, \hat{\delta}_I)$, $Q^+(\hat{\boldsymbol{\delta}}, \boldsymbol{\sigma}) = \min\{\sum\{(\hat{\delta}_i - \delta_i)/\sigma_i\}^2 : \boldsymbol{\delta} \in \mathcal{O}^+\}$, and $Q^-(\hat{\boldsymbol{\delta}}, \boldsymbol{\sigma}) = \min\{\sum\{(\hat{\delta}_i - \delta_i)/\sigma_i\}^2 : \boldsymbol{\delta} \in \mathcal{O}^-\}$. Then, it is easily shown that

$$Q^+(\hat{\boldsymbol{\delta}}, \boldsymbol{\sigma}) = \sum(\hat{\delta}_i/\sigma_i)^2 I(\hat{\delta}_i < 0) \text{ and } Q^-(\hat{\boldsymbol{\delta}}, \boldsymbol{\sigma}) = \sum(\hat{\delta}_i/\sigma_i)^2 \, I(\hat{\delta}_i > 0), \tag{9.43}$$

where $I(\cdot)$ is the indicator function. Further, it is also easily verified that the likelihood ratio statistic, $T(\hat{\boldsymbol{\delta}}, \boldsymbol{\sigma}) = -2[\max\{\text{loglik} : H_0\} - \max\{\text{loglik} : H_0 \cup H_1\}]$, is given by

$$T(\hat{\boldsymbol{\delta}}, \boldsymbol{\sigma}) = \min\{Q^+(\hat{\boldsymbol{\delta}}, \boldsymbol{\sigma}), Q^-(\hat{\boldsymbol{\delta}}, \boldsymbol{\sigma})\}. \tag{9.44}$$

Although the null distribution of $T(\hat{\boldsymbol{\delta}}, \boldsymbol{\sigma})$ depends on the particular value of $\boldsymbol{\delta}$ in the null parameter space, it turns out that a least favorable null value is $\boldsymbol{\delta}^* = (\infty, 0, \ldots, 0)$; at this value, the distribution of the test statistic $T(\hat{\boldsymbol{\delta}}, \boldsymbol{\sigma})$ does not depend on $\boldsymbol{\sigma}$. Therefore, if c is the sample value of $T(\hat{\boldsymbol{\delta}}, \boldsymbol{\sigma})$ then the p-value is $\text{pr}\{T(\hat{\boldsymbol{\delta}}, \boldsymbol{\sigma}) \geqslant c | \boldsymbol{\delta} = \boldsymbol{\delta}^*\}$. Therefore, the critical values and p-values can be computed at $\boldsymbol{\delta}^*$. The relevant result is given below.

Proposition 9.3.2 *(Gail and Simon (1985)). Assume that $\hat{\delta}_1, \ldots, \hat{\delta}_I$ are independent and that $\hat{\delta}_i \sim N(\delta_i, \sigma_i^2)$ where σ_i is known, $i = 1, \ldots, I$. Let $H_0, H_1, \boldsymbol{\delta}^*$, and $Bi(i; I-1)$ be as in Proposition 9.3.1. Then, the likelihood ratio test is, reject H_0 for large values of $T(\hat{\boldsymbol{\delta}}, \boldsymbol{\sigma})$ in (9.44). Further,*

$$(a) \quad pr\{T(\hat{\boldsymbol{\delta}}, \boldsymbol{\sigma}) \geqslant c | \boldsymbol{\delta} \in H_0\} \leq pr\{T(\hat{\boldsymbol{\delta}}, \boldsymbol{\sigma}) \geqslant c | \boldsymbol{\delta} = \boldsymbol{\delta}^*\}, \tag{9.45}$$

$$(b) \quad pr\{T(\hat{\boldsymbol{\delta}}, \boldsymbol{\sigma}) \geqslant c | \boldsymbol{\delta} = \boldsymbol{\delta}^*\} = \sum_{i=1}^{I-1} Bi(i; I-1) \, pr(\chi_i^2 \geqslant c). \tag{9.46}$$

This result was established by Gail and Simon (1985), and a similar result was also established by Berger (1984). They used results on majorization to prove results concerning the least favorable null values. Once this is done, critical values are obtained easily. The original ideas underlying Gail and Simon (1985) and Berger (1984) have been influential in the area of tests against cross-over interaction.

In practice, the distribution of $\hat{\delta}_i$ is unlikely to be exactly $N(\delta_i, \sigma_i^2)$ where σ_i is known, $i = 1, \ldots, I$. A more realistic scenario is considered in the next case.

9.3.1.3 Case 3: $\hat{\delta}_i$ is approximately $N(\delta_i, \sigma_i^2)$ and σ_i is unknown, $i = 1, \ldots, I$.

Since $\sigma_1, \ldots, \sigma_I$ are unknown, $T(\hat{\boldsymbol{\delta}}, \boldsymbol{\sigma})$ in (9.43) is not an operational test statistic. Let $\hat{\sigma}_i$ denote an estimator of σ_i, $i = 1, \ldots, I$. Now, $T(\hat{\boldsymbol{\delta}}, \hat{\boldsymbol{\sigma}})$ is a least squares based test statistic. A test of H_0 against H_1 is, reject H_0 when $T(\hat{\boldsymbol{\delta}}, \hat{\boldsymbol{\sigma}})$ is large.

In large samples, an approximate level-α critical value of $T(\hat{\boldsymbol{\delta}}, \hat{\boldsymbol{\sigma}})$ is the value of c for which the right-hand side of (9.46) is equal to α. Tables of these values are given in Silvapulle (2001) and Gail and Simon (1985). If the sample value of $T(\hat{\boldsymbol{\delta}}, \hat{\boldsymbol{\sigma}})$ is c, then an approximate expression for the p-value is the right-hand side of (9.46).

Let us point out that δ_i need not be the mean of a continuous random variable. For example, suppose that δ_i is a population proportion, and $\hat{\delta}_i$ is the sample proportion of n_i observations. Then the foregoing results for Case 3 are applicable with $\hat{\delta}_i \approx N(\delta_i, \sigma_i^2)$ where $\sigma_i^2 = \delta_i(1 - \delta_i)/n_i$ and $\hat{\sigma}_i^2 = \hat{\delta}_i(1 - \hat{\delta}_i)/n_i$.

Example 9.3.4 *Comparison of two treatments for breast cancer (Gail and Simon (1985)).*

Consider Example 9.3.2 in Chapter 1. The comment "young patients with progesterone receptor levels under 10 units do better in PF whereas other patients do better on PFT" of Fisher et al. (1983) points to possible cross-over interaction. Let us test against this phenomenon using the foregoing result for Case 3.

For a patient in group i under treatment j, let p_{ij} denote the probability of being disease free at three years. The parameter of interest is $\delta_i = (p_{i1} - p_{i2}), i = 1, \ldots, 4$. Since the sample sizes within each cell is reasonably large, the assumption that $\hat{\delta}_i$ is approximately $N(\delta_i, \sigma_i^2)$ is reasonable, for $i = 1, \ldots, 4$. Further, since $\hat{\delta}_i = \hat{p}_{i1} - \hat{p}_{i2}$, the condition, $\sigma_i^2 = \sigma^2 w_i$ where w_i is known, does not hold. Therefore, the F_Q-test is not applicable; however, $T(\hat{\boldsymbol{\delta}}, \hat{\boldsymbol{\sigma}})$ is applicable. Substituting directly into the formulas for $Q^+(\hat{\boldsymbol{\delta}}, \hat{\boldsymbol{\sigma}})$ and $Q^-(\hat{\boldsymbol{\delta}}, \hat{\boldsymbol{\sigma}})$, we have the following:

$$Q^+(\hat{\boldsymbol{\delta}}, \hat{\boldsymbol{\sigma}}) = \sum (\hat{\delta}_i/\hat{\sigma}_i)^2 I(\hat{\delta}_i < 0) = 1.65^2 + 0.765^2 + 2.76^2 = 10.9;$$

$$Q^-(\hat{\boldsymbol{\delta}}, \hat{\boldsymbol{\sigma}}) = \sum (\hat{\delta}_i/\hat{\sigma}_i)^2 I(\hat{\delta}_i > 0) = 2.07^2 = 4.3;$$

$$T(\hat{\boldsymbol{\delta}}, \hat{\boldsymbol{\sigma}}) = \min\{Q^+(\hat{\boldsymbol{\delta}}, \hat{\boldsymbol{\sigma}}), Q^-(\hat{\boldsymbol{\delta}}, \hat{\boldsymbol{\sigma}})\} = 4.3.$$

The 5% and 10% critical values are 5.43 and 4.01, respectively (using the table in Silvapulle (2001) , or that in Gail and Simon (1985)). In fact, the p-value is $\sum_{i=1}^{3} Bi(i, 3)\text{pr}(\chi_i^2 \geqslant 4.3) \approx 0.09$.

In these calculations, $\hat{\delta}_i$ was computed as $(\hat{p}_{i1} - \hat{p}_{i2})$. In principle, it is possible to define δ_i as a parameter that appears in a more elaborate model. For example, consider a separate Cox's proportional hazards model for each group; let the hazard be proportional to $\exp(\delta_i T)$ where $T = 0$ or 1 according as the treatment is PF or PFT. Then δ_i is log(relative hazard) for group i. Estimates of δ_i and standard error are given in Table 9.6. Substitution into the expressions for $Q^+(\hat{\boldsymbol{\delta}}, \hat{\boldsymbol{\sigma}}), Q^-(\hat{\boldsymbol{\delta}}, \hat{\boldsymbol{\sigma}})$ and $T(\hat{\boldsymbol{\delta}}, \hat{\boldsymbol{\sigma}})$ lead to the following: $Q^+(\hat{\boldsymbol{\delta}}, \hat{\boldsymbol{\sigma}}) = \sum(\hat{\delta}_i/\hat{\sigma}_i)^2 I(\hat{\delta}_i < 0) = 1.46^2 + 0.146^2 + 3.5^2 = 14.4$; $Q^-(\hat{\boldsymbol{\delta}}, \hat{\boldsymbol{\sigma}}) = \sum(\hat{\delta}_i/\hat{\sigma}_i)^2 I(\hat{\delta}_i > 0) = 2.55^2 + 0 + 0 + 0 = 6.5$; $T(\hat{\boldsymbol{\delta}}, \hat{\boldsymbol{\sigma}}) = \min\{Q^+(\hat{\boldsymbol{\delta}}, \hat{\boldsymbol{\sigma}}), Q^-(\hat{\boldsymbol{\delta}}, \hat{\boldsymbol{\sigma}})\} = \min\{14.4, 6.5\} = 6.5$. Since 6.5 exceeds the 5% level critical value 5.43 (corresponding to (9.46)), there is evidence to reject H_0 and accept that there is cross-over interaction when the parameter of interest is log(relative hazards). ■

Table 9.6 Estimates of log (relative hazards) for breast cancer data

Group	$\hat{\delta}_i$	$\hat{\sigma}_i^{(a)}$	$\hat{\delta}_i/\hat{\sigma}_i$
1. Age<50, PR< 10	0.531	0.208	2.55
2. Age⩾50, PR < 10	-0.266	0.182	-1.46
3. Age <50, PR⩾ 10	-0.030	0.206	-0.146
4. Age≥ 50, PR⩾ 10	-0.724	0.207	-3.50

$^{(a)}\hat{\sigma}_i = se(\hat{\delta}_i)$.

9.3.2 Robustness Against Outliers

Let us consider the experimental design in Example 9.3.3; this is an $I \times 2$ completely randomized block design with unequal numbers of observations in the cells. Let $\hat{\delta}_i$ be the difference between the sample means for the two treatments in block $i, i = 1, \ldots, I$. Let us consider the setting in Case 1 of Section 9.3.1.1, and the test statistic $T = \min\{R^+, R^-\}/S^2$. Note that $\min\{R^+, R^-\}$ is essentially a sum of squares of some of the $\hat{\delta}_i$ s. Thus, it is clear that T is likely to be just as sensitive to outliers in the response variable as the sample means, and that T is likely to be more robust against outliers if $\hat{\delta}_i$ is replaced by a more robust estimator of δ_i, for example, an M-estimator or a nonparametric estimator based on ranks; the denominator σ^2 in T also needs to be replaced by a suitable quantity. The relevant theoretical results are given in Silvapulle (2001). Here, we shall provide a brief statement of the results.

Let $\tilde{\delta}_i$ be a robust estimator of δ_i. Assume that $\tilde{\delta}_i$ is approximately $N(\delta_i, \xi^2 w_i)$, $i = 1, \ldots, I$; recall that $\hat{\delta}_i$ is approximately $N(\delta_i, \sigma^2 w_i)$, $i = 1, \ldots, I$. If the error distribution has long tails, then typically $\sigma > \xi$; in other words, the robust estimator $\tilde{\delta}_i$ is asymptotically more efficient than the least squares estimator $\hat{\delta}_i$, $i = 1, \ldots, I$. Now define the robust test statistic as

$$T_M = \min\{R_M^+, R_M^-\}/\tilde{\xi}^2,$$

where $\tilde{\xi}^2$ is an estimator ξ and $\{R_M^+, R_M^-\}$ is the same as $\{R^+, R^-\}$ except that $\hat{\delta}_i$ is replace by $\tilde{\delta}_i$ for every i. It turns out, as expected, that the robustness properties of $\tilde{\delta}_i$ carry over to T_M; consequently, the power of T_M is robust against outliers.

This qualitative result can be stated more precisely in terms of Pitman efficiency. Let $\boldsymbol{\delta}_0$ be a point on the boundary of the null parameter space and $\boldsymbol{a}$ be a fixed vector such that $\boldsymbol{\delta}_0 + n^{-1/2}\boldsymbol{a}$ is in the alternative parameter space, where n denotes the total sample size. Then, with respect to the sequence of local alternatives $H_n : \boldsymbol{\delta} = \boldsymbol{\delta}_0 + n^{-1/2}\boldsymbol{a}$, the asymptotic Pitman efficiency of the robust test relative to the least squares-based test is (σ^2/ξ^2), which is also the asymptotic efficiency of the robust estimator relative to the least squares estimator. Therefore, we conclude that the main asymptotic efficiency robustness properties of the robust estimator carry over to the test based on the same estimator.

In the foregoing results, we assumed that the error distributions within the I groups were identical. Now, we relax this assumption; in this case, the foregoing results generalize in a natural way. Suppose that δ_i is estimated using the data for Group i only, $i = 1, \ldots, I$; let $(\tilde{\delta}_i, \tilde{\xi}_i)$ be robust estimators of (δ_i, ξ_i). Now, the robust test statistic is the same as $T(\hat{\boldsymbol{\delta}}, \boldsymbol{\sigma})$ in (9.44) except that $\hat{\delta}_i$ and σ_i^2 are replaced by $\tilde{\delta}_i$ and $\tilde{\xi}_i^2 w_i$, respectively. Further, (9.45) and (9.46) hold asymptotically for $T(\tilde{\boldsymbol{\delta}}, \tilde{\boldsymbol{\xi}})$. Therefore, the asymptotic p-value and critical values are computed using (9.46).

The robustness properties of $(\tilde{\delta}_i, \tilde{\xi}_i)$ carry over to their corresponding test statistics in terms of Pitman efficiency. Therefore, the power of the test based on $(\tilde{\delta}_i, \tilde{\xi}_i)$ is robust against outliers. For more details and precise statements of these results, see Silvapulle (2001).

9.3.3 Dependent Treatment Effects

The tests against cross-over interaction studied thus far have been restricted to the case when the treatment effects for different groups are independent. However, in some cases such an independence condition may not be satisfied; the following example illustrates this.

Consider two treatments, A and B, for reducing blood pressure. Let y denote the response variable, the reduction in blood pressure. Let the population be divided into two groups: group 1 = $\{\text{age} \geqslant 30\}$ and group 2 = $\{\text{age} < 30\}$. Let δ_i denote the treatment effect for group $i, i = 1, 2$. Now, cross-over interaction is equivalent to δ_1 and δ_2 having opposite signs. If δ_i is estimated using data for group i only, $i = 1, 2$, then $\hat{\delta}_1$ and $\hat{\delta}_2$ are also independent, a condition that is required for the applicability of the results in the previous subsections.

Now, instead of treating age as a classification variable leading to two groups, let us consider a model of the form $E(y_j) = \alpha_j + \beta_j x$ where $j = A$ or B, and $x =$ age. A question of interest is whether or not one treatment is better than the other for every x in the range $x_0 \leq x \leq x_1$ where x_0 and x_1 are given; in other words, whether or not one regression line is above the other in the range $x_0 \leq x \leq x_1$. The question can be formulated as follows. Let

$$\delta_i = E(y_B|x_i) - E(y_A|x_i), \quad i = 1, 2.$$

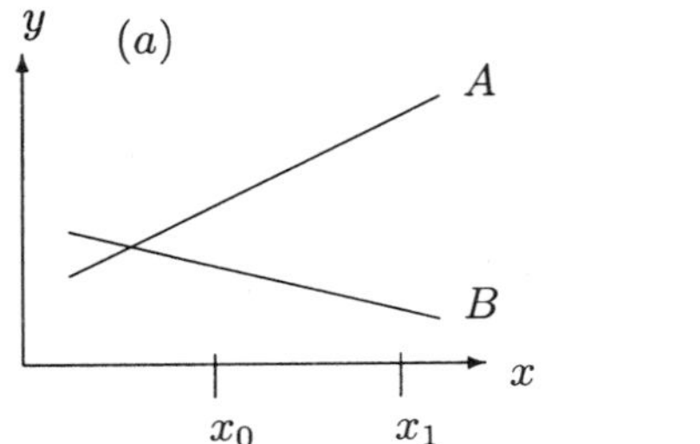

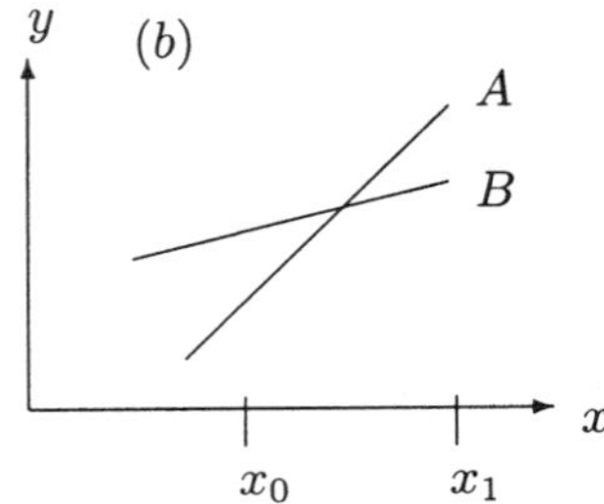

Fig. 9.9 Cross-over interaction in a regression context with treatments, A and B: (a) no cross-over interaction in the range $\hat{\delta}_0 \leq x \leq \hat{\delta}_1$; (b) cross-over interaction in the range $\hat{\delta}_0 \leq x \leq \hat{\delta}_1$.

Then, the null and alternative hypotheses are

$$H_0 : \boldsymbol{\delta} \geqslant \mathbf{0} \text{ or } \boldsymbol{\delta} \leq \mathbf{0} \qquad \text{and} \quad H_1 : \text{ not } H_0 \tag{9.47}$$

where $\boldsymbol{\delta} = (\delta_1, \delta_2)$. Note that H_0 is true if and only if the two regression lines do not intersect in the range $[x_0, x_1]$ (see Fig. 9.9). Let $(\hat{\alpha}_A, \hat{\alpha}_B, \hat{\beta}_A, \hat{\beta}_B)$ be an estimator of $(\alpha_A, \alpha_B, \beta_A, \beta_B)$. Then a natural estimator of δ_i is

$$\hat{\delta}_i = (\hat{\alpha}_B + \hat{\beta}_B x_i) - (\hat{\alpha}_A + \hat{\beta}_A x_i), i = 1, 2.$$

In general, $\hat{\delta}_1$ and $\hat{\delta}_2$ are stochastically dependent because each one of them is a function of $(\hat{\alpha}_A, \hat{\alpha}_B, \hat{\beta}_A, \hat{\beta}_B)$. Therefore, the methods developed in the earlier subsections are not applicable. In this case, the following results of Russek-Cohen and Simon (1993) are useful for testing H_0 against H_1; a limitation is that these are applicable for the special case of two groups only.

Proposition 9.3.3 *(Russek-Cohen and Simon (1993)). Let $\hat{\boldsymbol{\delta}}_{2\times 1} \sim N(\boldsymbol{\delta}, A)$ where A is in the correlation form $(1, \rho|\rho, 1)$ for some known correlation coefficient, ρ. Let the null and alternative hypothesis be as in (9.47). Then, we have the following for the likelihood ratio test of H_0 against H_1 based on a single observation of $\hat{\boldsymbol{\delta}}$: (a) The LRT is*

$$\textit{reject } H_0 \textit{ if } \{\hat{\delta}_1 \leq -c \textit{ and } \hat{\delta}_2 \geqslant c\} \textit{ or } \{\hat{\delta}_1 \geqslant c \textit{ and } \hat{\delta}_2 \leq -c\} \tag{9.48}$$

for some c. (b) The least favorable null value for the LRT is on the boundary $\{\boldsymbol{\delta} : \delta_1 = 0 \text{ or } \delta_2 = 0\}$. (c) If $\rho > 0$ then a least favorable null value of LRT is $(\infty, 0)$ and the critical value is the same as that for the case when $\hat{\delta}_1$ and $\hat{\delta}_2$ are independent. (d) If $\rho < 0$ then the least favorable null value moves from $(\infty, 0)$ to $(0, 0)$ as ρ reduces from 0 to -1. ■

In practice, it is very unlikely that an available $\hat{\boldsymbol{\delta}}$ would satisfy the condition, $\text{cov}(\hat{\delta}) = (1, \rho|\rho, 1)$, in the above proposition. However it is likely that a suitable test of (9.47) can be developed using the foregoing proposition. Let us suppose that $\hat{\boldsymbol{\delta}} \approx$

$N(\boldsymbol{\delta}, \sigma^2 U)$, where σ is unknown and U is known. Let $U = (u_{11}, u_{12}|u_{21}, u_{22})$. Then $\rho = \{u_{12}/\sqrt{u_{11}u_{22}}\}$ and the results in parts (a) to (d) of the foregoing proposition hold approximately if $\hat{\boldsymbol{\delta}}$ therein is replaced by $(\hat{\delta}_1/(s\sqrt{u_{11}}), \hat{\delta}_2/(s\sqrt{u_{22}}))$ where s is a consistent estimator of σ (see Russek-Cohen and Simon (1993)).

Now, consider the case when $\hat{\boldsymbol{\delta}} \approx N(\boldsymbol{\delta}, W)$, where W is unknown. Therefore, the correlation coefficient ρ corresponding to W is an unknown parameter. Let the $\hat{\boldsymbol{\delta}}$ in parts (a) to (d) of the foregoing proposition be replaced by $(\hat{\delta}_1/\sqrt{\hat{w}_{11}}, \hat{\delta}_2/\sqrt{\hat{w}_{22}})$, where $\hat{w}_{ii}$ is a consistent estimator of w_{ii} for $i = 1, 2$. In this case, the test in (9.48) is still reasonable although it is not the *LRT*. Since ρ is an unknown nuisance parameter, and the probability of Type I error for the test in (9.48) depends on ρ, the asymptotic critical value c_α for an α-level test is determined by

$$\sup_\rho \lim_n pr\{(\hat{\delta}_1 \leq -c_\alpha \text{ and } \hat{\delta}_2 \geqslant c_\alpha) \text{ or } (\hat{\delta}_1 \geqslant c_\alpha \text{ and } \hat{\delta}_2 \leq -c_\alpha)|H_0\} \leq \alpha. \quad (9.49)$$

Table 2 of Russek-Cohen and Simon (1993) shows that the critical values are monotonic in ρ, the supremum in (9.49) is achieved at $\rho = -1$, and the 5% level critical value is 1.96. The same table also suggests that if the range of possible values of ρ can be restricted to an interval of the form $[a, 1], a > -1$, then it is desirable to do so because the critical value would be smaller and the test would have higher power; it may be difficult to ensure that the choice of a is objective, and that it does not depend on the sample. The same table in Russek-Cohen and Simon (1993) also shows that the critical values tended to change substantially as a function of ρ only when ρ is close to -1.0, for example $\rho < -0.7$. Therefore, our conjecture is that the method of computing the p-value as the supremum over a confidence region of the nuisance parameter ρ is likely to be better than (9.49) (see Silvapulle (1996), Berger and Boss (1994)) because the confidence interval for ρ is likely to contain only a region away from -1.0. Finally, we note that nothing much is known about the case when there are more than two groups.

9.3.4 A Range Test Against Cross-over Interaction

A good competitor to the F_Q-test is the *Range Test* proposed by Piantadosi and Gail (1993) (PG). Of these two tests, none is uniformly more powerful. As to which one is better depends on the part of the parameter space in which higher power is desired.

The range test of PG is the following. Let $\hat{\delta}_i$ denote an estimator of $\delta_i, i = 1, \ldots, I$. Assume that $\hat{\delta}_i \sim N(\delta_i, \sigma_i^2), \;\; i = 1, \ldots, I$ and that $\hat{\delta}_1, \ldots, \hat{\delta}_I$ are independent. A test of $H_0 : \boldsymbol{\delta} \geqslant 0$ or $\boldsymbol{\delta} \leq \mathbf{0}$ against $H_1 :$ *not* H_0 is

$$\text{reject } H_0 \text{ if } \max_i(\hat{\delta}_i/\sigma_i) > c \text{ and } \min_i(\hat{\delta}_i/\sigma_i) < -c \quad (9.50)$$

for some c. This test can also be stated as follows: let

$$R_{PG} = \min\{\max_i(\hat{\delta}_i/\sigma_i), -\min_i(\hat{\delta}_i/\sigma_i)\}.$$

Now the test in (9.50), proposed by Piantadosi and Gail (1993), is reject H_0 for large values of R_{PG}. This test is not similar, and its null distribution is given in the next result.

Proposition 9.3.4 *(Piantadosi and Gail (1993)). The least favorable null distribution of R_{PG} occurs at $\boldsymbol{\delta} = \boldsymbol{\delta}^*$ where $\boldsymbol{\delta}^* = (\infty, 0, \ldots, 0)$. The α-level critical value c_α of R_{PG} is given by $1 - \{1 - \Phi(-c_\alpha)\}^{I-1} = \alpha$, where Φ is the standard normal distribution function.* ■

If σ_i is unknown and if it is replaced by $\hat{\sigma}_i$, a consistent estimator of σ_i, then the critical value obtained in this proposition is asymptotically valid.

The functional forms of F_Q and R_{PG} suggest that F_Q is likely to be more powerful than R_{PG} in directions such as $(1, 1, 1, 0, -1, -1, -1, 0)$ in which there are several positive and several negative components. On the other hand, R_{PG} is likely to be more powerful than F_Q in directions such as $(1, -1, 0, \ldots, 0)$ in which there is one large positive and one large negative component with the others being close to zero. The simulation results in Piantadosi and Gail (1993) corroborate this conjecture. If there are only two groups and (σ_1, σ_2) is known, then the range test and the likelihood ratio test are equivalent.

Example 9.3.5 *Example 9.3.2 continued: Comparison of two treatments for breast cancer.*

Let us test against cross-over interaction using the range test. It follows from Table 9.5 that max $\{\hat{\delta}_i/\hat{\sigma}_i\} = 2.07$ and min $\{\hat{\delta}_i/\hat{\sigma}_i\} = -2.76$. Since the sample sizes are large, let us assume that $\hat{\sigma}_i \approx \sigma_i$ for every i. Therefore, the range test statistic is

$$R_{PG} = \min\{2.07,\ 2.76\} = 2.07.$$

Using Table 1 in Piantadosi and Gail (1993), the 10% and 5% critical values for $I = 4$ groups are 1.82 and 2.12, respectively. Therefore, $0.10 > p$-value > 0.05. Essentially the same conclusion was reached in Example 9.3.2 by applying the test in Gail and Simon (1985). ■

9.3.5 A Test with Uniformly More Power and Less Bias

An unattractive feature of F_Q- and R_{PG}-tests is that they are both biased. At this stage, it is not clear how the available tests could be modified to have reduced bias without affecting their other desirable properties. To illustrate these let us consider a simple example with $I = 2$ groups.

Let $\hat{\delta}_i \sim N(\delta_i, 1)$ for $i = 1, 2$, and let $\hat{\delta}_1$ and $\hat{\delta}_2$ be independent. The size-α critical region, CR, of *LRT* is of the form,

$$\{(\hat{\delta}_1, \hat{\delta}_2) : \hat{\delta}_1 < -d, \hat{\delta}_2 > d\} \cup \{(\hat{\delta}_1, \hat{\delta}_2) : \hat{\delta}_1 > d, \hat{\delta}_2 < -d\} \text{ for some } d.$$

For illustrative purposes, let us choose $\alpha = 0.05$. Since, by Proposition 9.3.3, a least favorable null value is $(0, \infty)$, the critical value is $d = 1.645$; see Fig. 9.10. Note that pr(Type I error) is continuous on the null parameter space and is equal to $2 \times 0.05^2 = 0.005$ at the origin. Thus, pr(Type I error) is smaller than 0.05 near the origin. Therefore, a region of the sample space that is not in CR and close to the origin can be included into CR so that pr(Type I Error) would increase at the origin, but the size of the test remains at 0.05. The shape of one such critical region, CR*, is shown in Figure 9.10; for mathematical derivations leading to the shape of CR^*, see Zelterman (1990) or Berger (1989). In fact, the test that Zelterman (1990) constructed was a numerical approximation of an unbiased test and has critical region similar to CR*.

Note that (a) CR* and *LRT* have the same size, (b) CR* is less biased than *LRT*, (c) CR* is uniformly more powerful than *LRT*, and (d) CR* has higher probability of Type I error than *LRT* at any point in the null parameter space. Thus, in terms of some criteria, CR^* is better than the LRT. However, CR^* also has some features which may be of some concern.

First, note that CR^* is not monotone. To illustrate this, consider the possible outcomes P and Q indicated in Figure 9.10; P is in CR*, Q is vertically above P but it is not in CR*. Thus, CR* would reject the null hypothesis for the outcome P but not for Q despite the fact that the evidence in favor of H_1 at Q appears to be stronger than at P. Another feature of CR* that may be of some concern is that it contains points in the sample space that are very close to the origin. For example, the outcome $(\hat{\delta}_1, \hat{\delta}_2) = (-0.01, 0.01)$ would lead to a rejection of the null hypothesis. Since the standard deviation of $\hat{\delta}_1$ and of $\hat{\delta}_2$ is one, the outcome $(\hat{\delta}_1, \hat{\delta}_2) = (-0.01, 0.01)$ is too close to the origin to be seen as evidence against $\boldsymbol{\delta} = \mathbf{0}$ or in favor of cross-over interaction.

Therefore, although the test with critical region CR^* is uniformly more powerful and less biased than *LRT*, the latter has some features that are important in some practical applications. Nevertheless, if we consider size and power of a test as the main criteria of interest, Zelterman's construction is of interest.

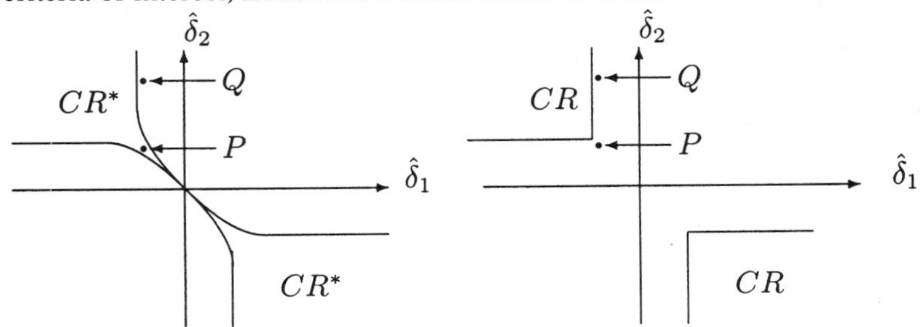

Fig. 9.10 A test with uniformly more power than the *LRT*; CR is the critical region for *LRT* and CR* is the union of CR and some region close to the origin.
[The figure on the left is reproduced from: *Statistics and Probability Letters*, Vol 10, D. Zelterman, On tests for qualitative interactions, Pages 59–63, Copyright(1990), with permission from Elsevier.]

9.3.6 Cross-over Interaction of Practical Significance

In the previous subsection, the hypotheses were formulated to establish the presence of a cross-over interaction, however small it may be. In some practical situations, the

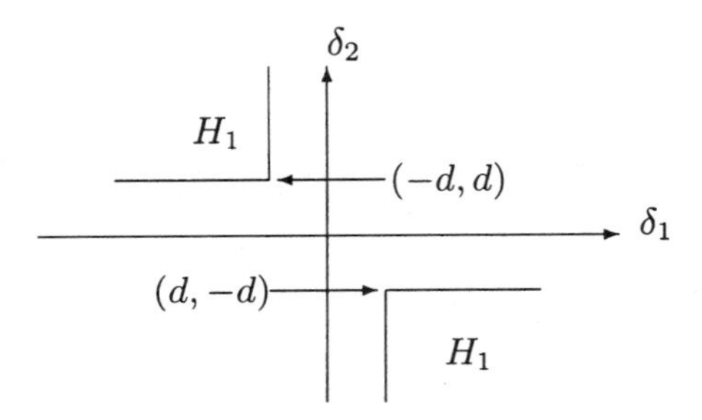

Fig. 9.11 Cross-over interaction of practical significance with two groups.

interest may be to detect interactions that are large enough to be practically significant. To illustrate this, consider a treatment to control pain. It is very unlikely that a single treatment for controlling pain would be uniformly superior (or inferior) to another across a range of different types of pain; let us treat the ith type of pain as the ith group. What may be of interest is to know whether or not Treatment A is substantially better than Treatment B in one group, but substantially worse in another. Therefore, the hypotheses of cross-over interaction of practical significance may be formulated as

$$\text{“There exist distinct groups } i \text{ and } j \text{ such that } \delta_i > d \text{ and } \delta_j < -d\text{,”} \qquad (9.51)$$

where d is a number chosen *a priori* to reflect what we mean by "practical significance" (see Fig. 9.11).

This formulation says that there is cross-over interaction of practical significance if Treatment A is better than Treatment B by d or more in one group but is worse by d or more in another. A test against cross-over interaction of practical significance was developed by Pan and Wolfe (1997). Their procedure is stated below.

Let the null and alternative hypothesis be stated as

$H_0(d) : \delta_i \geqslant -d$ for all $i = 1, \ldots, I$ or $\delta_i \leq d$ for all $i = 1, \ldots, I$.

$H_1(d) : \delta_i > d$ and $\delta_j < -d$ for some $i, j = 1, \ldots, I$ and $i \neq j$.

Suppose that $\hat{\delta}_i \sim N(\delta_i, \sigma_i^2)$ for $i = 1, \ldots, I$, $\sigma_1, \ldots, \sigma_I$ are known and $\hat{\delta}_1, \ldots, \hat{\delta}_I$ are independent. Now, the Pan-Wolfe test of $H_0(d)$ against $H_1(d)$ of size α is the following. Let $P_E = 2(1-\alpha)^{1/(I-1)} - 1$, and $[L_i, U_i]$ be the two-sided confidence interval for δ_i of confidence coefficient P_E, where $L_i = \hat{\delta}_i - z\sigma_i, U_i = \hat{\delta}_i + z\sigma_i$ and $pr\{N(0,1) \leq -z\} = \alpha/2$. Now, the test is, reject the null hypotheses if there exist distinct i and j such that the confidence interval for δ_i is entirely below $-d$ and that for δ_j is entirely above d. In other words, the test is

$$C(d) : \text{ reject } H_0 \text{ if there exist } i \text{ and } j \text{ such that } U_j < -d \text{ and } L_i > d. \qquad (9.52)$$

If $d = 0$ then this test is the same as the range test in (9.50). A practical advantage of this test is that we can tabulate the p-value for different values of d; such a table would be informative. It is easily shown that p-value $= [1 - \{(1+Q)/2\}^{I-1}$, where $Q = [1 - 2\text{pr}(Z \geqslant t_{PW})]$ and $t_{PW} = \min\{-\max_i(\hat{\delta}_i + d)/\hat{\sigma}_i, \min_i(\hat{\delta}_i - d)/\hat{\sigma}_i\}$.

Table 9.7 The p-values corresponding to $C(d)$

d	0.00	0.01	0.02	0.03	0.04
p-value	0.057	0.076	0.10	0.13	0.17

For the data in Table 9.5, p-values for different values of d are given in Table 9.7. If we consider that a p-value of 0.10 is small enough to reject the null hypothesis, then there is sufficient evidence to establish cross-over interaction of practical significance for values of d not smaller than 0.02.

Pan and Wolfe (1997) showed that a least favorable null value for $C(d)$ is $(\infty, -d, -d, \ldots, -d)$. Since this is far from the origin, it is clear that $C(d)$ is biased and has low power near the origin. Pan and Wolfe (1997) also provided a closed-form for the power of the test.

9.3.7 More than Two Treatments

The discussions so far have been limited to the case when there are only two treatments. Now, let us consider the case when there are J treatments and I groups; this was illustrated at the beginning of this section for the special case $(I, J) = (3, 3)$ (see Fig. 9.7). Let θ_{ij} denote the parameter of interest for group i and treatment j (i.e., i th row and j th column), $i = 1, \ldots, I$ and $j = 1, \ldots, J$. Let $\delta(j_1, j_2|i) = (\theta_{ij_2} - \theta_{ij_1})$. If there exist columns j_1 and j_2 and rows i_1 and i_2 such that $\delta(j_1, j_2|i_1)$ and $\delta(j_1, j_2|i_2)$ have opposite signs, then we say that there is cross-over (equivalently, qualitative) interaction. Therefore, to test against cross-over interaction, the null and alternative hypotheses may be stated as

H_0 : For any given pair, $(j_1, j_2), \delta(j_1, j_2|i) \geqslant 0$ for every $i = 1, \ldots, I$, or $\delta(j_1, j_2|i) \leq 0$ for every $i = 1, \ldots, I$.

H_1 : There exists (j_1, j_2) and (i_1, i_2) such that $\delta(j_1, j_2|i_1)\delta(j_1, j_2|i_2) < 0$.

It is easily seen that H_1 is the same as "not H_0". The methods of the previous subsections for the special case of two treatments do not extend easily to the present case of more than two treatments. Azzalini and Cox (1984) developed a test against H_1 but their null hypothesis is strictly smaller than (i.e., nested in) the foregoing H_0. Nevertheless, the main ideas therein are useful; these are indicated below.

Let Y_{ij} denote an estimator of θ_{ij} such that $Y_{ij} \sim N(\theta_{ij}, \sigma^2)$; σ^2 is assumed to be known for simplicity. In applications, a large sample approximation is obtained by substituting a consistent estimator for σ^2. The test proposed by Azzalini and Cox (1984) is, accept H_1 if, for a suitably chosen positive t,

$$Y_{i_1j_2} - Y_{i_1j_1} \geqslant t\sigma\sqrt{2} \text{ and } Y_{i_2j_2} - Y_{i_2j_1} \leq -t\sigma\sqrt{2} \qquad (9.53)$$

for some $\{i_1, i_2, j_1, j_2\}$. There is no completely satisfactory way of finding the critical value t because the least favorable configuration in H_0 is unknown. Azzalini and Cox

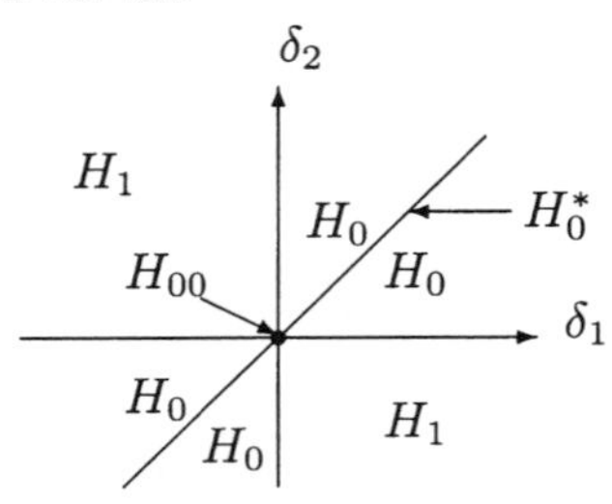

Fig. 9.12 Hypotheses relating to Azzalini-Cox test.

(1984) obtained the following results. Let H_{00} be defined as

$$H_{00} : \theta_{i1} = \cdots = \theta_{iJ}, \quad \text{for every } \ i = 1, \ldots I.$$

Thus, H_{00} is the same as saying "no difference between treatments"; it allows for differences between groups. It turns out that H_{00} is the least favorable null configuration when the null hypothesis is

$$H_0^* : \text{ No interaction between treatments and groups.}$$

For the case when the null hypothesis is H_{00}, Azzalini and Cox (1984) showed that an approximate α-level critical value, t_α, is given by

$$t_\alpha = \Phi^{-1}\left[\left\{-\frac{2\log(1-\alpha)}{I(I-1)J(J-1)}\right\}^{1/2}\right]. \tag{9.54}$$

Since H_{00} is least favorable in H_0^* but not in H_0, t_α is a level α critical value when the null hypothesis is H_0^*, but not necessarily when it is H_0. Ideally, to test against the cross-over interaction hypothesis H_1, what we need is a critical value corresponding to H_0 as the null hypothesis.

For the special case of two treatments and two groups, Fig. 9.12 shows the regions H_0^*, H_0 and H_1 where $\delta_1 = (\theta_{12} - \theta_{11})$ and $\delta_2 = (\theta_{22} - \theta_{21})$; H_{00} is the origin. Consequently, with H_0 as the null parameter space, the level of significance corresponding to the critical value t_α in (9.54) is higher than α. Therefore, (9.54) must be used with caution.

9.3.8 H_0 : Cross-over Interaction is Present

There may well be situations in which the interest is to establish that there is no cross-over interaction. In the previous subsections the interest was to show that there was cross-over interaction; thus the roles of null and alternative hypotheses were interchanged. As an example, it may be of interest to establish that the possibility where a treatment is better than placebo for males but not for females, does not arise. Similarly, in a multicenter clinical trial, the interest may be to show that the situation where a treatment is beneficial in one center but harmful in another, does not arise.

It is clear that what we essentially need is "no cross-over interaction" to be in the alternative hypothesis. In this section, we formulate the alternative hypothesis as, H_1 : *the lines corresponding to different treatments (for example, see Fig. 9.5) do not touch or cross.* Suppose that there are I groups and J treatments. Throughout this subsection, we shall assume that $1 \leq i_1 < i_2 \leq I$ and $1 \leq j_1 < j_2 \leq J$. Let the null and alternative hypotheses be stated as

$$H_0 : (\theta_{i_1 j_2} - \theta_{i_1 j_1})(\theta_{i_2 j_2} - \theta_{i_2 j_1}) \leq 0 \text{ for some}(i_1, i_2, j_1, j_2)$$
$$H_1 : (\theta_{i_1 j_2} - \theta_{i_1 j_1})(\theta_{i_2 j_2} - \theta_{i_2 j_1}) > 0 \text{ for every}(i_1, i_2, j_1, j_2).$$

Berger (1984) developed the *LRT* when the observations are independent and normal with equal variance; his main result is stated below without proof.

For group i and treatment j, let $\{Y_{ijk} : k = 1, \ldots, n_{ij}\}$ denote the observations that are assumed to be distributed as $N(\theta_{ij}, \sigma^2)$. Further, assume that $\{Y_{ijk} : i = 1, \ldots, I, j = 1, \ldots, J, k = 1, \ldots, n_{ij}\}$ are independent. Let S^2 denote the pooled within cell variance estimator $\nu^{-1} \sum\sum\sum(Y_{ijk} - \bar{Y}_{ij})^2$ where $\nu = (\sum\sum n_{ij} - IJ)$. Let $T(j_1, j_2|i) = (\bar{Y}_{ij_2} - \bar{Y}_{ij_1})/[S\{n_{ij_2}^{-1} + n_{ij_1}^{-1}\}^{1/2}]$; this is a two-sample t-statistic for comparing cells (i, j_1) with (i, j_2). Now, the main result concerning the *LRT* of H_0^A against H_1^A is the following:

Proposition 9.3.5 *(Berger (1984)). Let* $0 < \alpha < 0.5$. *Then, the level-*α *LRT of* H_0^A *against* H_1^A *is, reject* H_0^A *if, for every* $1 \leq j_1 < j_2 \leq J$, $\min\{T(j_1, j_2|i) : i = 1, \ldots, I\} > t_{\alpha,\nu}$ *or* $\max\{T(j_1, j_2|i) : i = 1, \ldots, I\} < -t_{\alpha,\nu}$, *where* $t_{\alpha,\nu}$ *is the upper* 100α *percentile of t-distribution with* ν *degrees of freedom.*

Clearly, the p-value corresponding to the *LRT* in the foregoing proposition is the smallest value of α at which the null hypothesis would be rejected. It may be computed as follows: Define Berger's statistic, B, as

$$B = \min_{1 \leq j_1 < j_2 \leq J} \{\min\{T(j_1, j_2|i) : i = 1, \ldots, I\}, -\max\{T(j_1, j_2|i) : i = 1, \ldots, I\}\}.$$

Now, the *LRT* is reject H_0 if $B > t$ for some t. Berger (1984) showed that a least favorable null configuration is achieved in the limit

$\theta_{11} = \theta_{12}$, and $\theta_{i(j+1)} - \theta_{ij} \to \infty$ for every (i, j) except $(i, j) = (1, 1)$.
Therefore, the critical value is computed at such a least favorable null value. Consequently, the level-α *LRT* is, reject H_0 if $B \geqslant t_{\alpha,\nu}$, and if the sample value of B is b then p-value = $\text{pr}(t_\nu \geqslant b)$.

9.3.9 Proof of Proposition 9.3.1

This proof is taken from Silvapulle (1999, unpublished); the approach is based on the proof of a similar result in Gail and Simon (1985). Let $Y_i = \hat{\delta}_i/\sqrt{w_i}$, where $\hat{\delta}_i$ is an estimator of treatment effect in group i and $\hat{\delta}_i \sim N(\delta_i, w_i\sigma^2)$. Then $Y_i \sim N(\theta_i, \sigma^2)$, where $\theta_i = \delta_i/\sqrt{w_i}$. Now, the null hypothesis can be stated as $H_0 : \boldsymbol{\theta} \leq \mathbf{0}$ or $\boldsymbol{\theta} \geqslant \mathbf{0}$.

Let $\boldsymbol{x} = (x_1, \ldots, x_I)$, $I(\cdot)$ denote the indicator function,

$$\phi_1(\boldsymbol{x}) = \sum_{i=1}^{I} x_i^2 I(x_i > 0), \phi_2(\boldsymbol{x}) = \sum_{i=1}^{I} x_i^2 I(x_i < 0),$$
$$\psi_1(\boldsymbol{x}) = I\{\phi_1(\boldsymbol{x}) > c\}, \psi_2(\boldsymbol{x}) = I\{\phi_2(\boldsymbol{x}) > c\}, \text{ and } \Psi(\boldsymbol{x}) = \psi_1(\boldsymbol{x})\psi_2(\boldsymbol{x}).$$

Then $R^+ = \phi_1(\boldsymbol{Y})$ and $R^- = \phi_2(\boldsymbol{Y})$. Let $Q_S^- = R^-/S^2$ and $Q_S^+ = R^+/S^2$. Then, we have that $Q_S^- = \phi_1(\boldsymbol{Y}/S)$ and $Q_S^+ = \phi_2(\boldsymbol{Y}/S)$. Let the events E_1 and E_2 be defined as $E_1 = [Q_S^- > c]$ and $E_2 = [Q_S^+ > c]$. Then we have that,

$$E_1 = I\{\phi_1(\boldsymbol{Y}/S) > c\} = \psi_1(\boldsymbol{Y}/S)$$
$$E_2 = I\{\phi_2(\boldsymbol{Y}/S) > c\} = \psi_2(\boldsymbol{Y}/S),$$
$$I\{E_1 \cap E_2\} = \psi_1(\boldsymbol{Y}/S)\psi_2(\boldsymbol{Y}/S) = \Psi(\boldsymbol{Y}/S).$$

Clearly, $\boldsymbol{Y} \sim N(\boldsymbol{\theta}, \sigma^2 I)$; let $\sigma^{-1} f\{(\boldsymbol{y}-\boldsymbol{\theta})/\sigma\}$ denote the density of $\boldsymbol{Y}$. Then, since $\boldsymbol{Y}$ and S are independent, their joint density is $\sigma^{-1} f\{(\boldsymbol{y}-\boldsymbol{\theta})/\sigma\}g(s)$ where $g(s)$ is the density of S. Let

$$\Psi^*(\boldsymbol{z}+\boldsymbol{\theta}) = \int \Psi\{(\boldsymbol{z}+\boldsymbol{\theta})/s\}g(s)ds.$$

Now,

$$\text{pr}(E_1 \cap E_2; \boldsymbol{\theta}) = \int I(E_1 \cap E_2)\sigma^{-1} f\{(\boldsymbol{y}-\boldsymbol{\theta})/\sigma\}g(s)d\boldsymbol{y}ds$$
$$= \int \Psi^*(\boldsymbol{y}/s)\sigma^{-1} f\{(\boldsymbol{y}-\boldsymbol{\theta})/\sigma\}g(s)d\boldsymbol{y}ds$$
$$= \sigma^{-1} \int \Psi^*\{(\boldsymbol{y}+\boldsymbol{\theta})/s\} f(\boldsymbol{y}/\sigma)d\boldsymbol{y} = E\{\Psi^*(\boldsymbol{Z}+\boldsymbol{\theta})\},$$

where $\sigma^{-1}\boldsymbol{Z}$ is distributed as multivariate normal with mean zero and identity covariance matrix.

Now, $\Psi^*(\boldsymbol{Z}+\boldsymbol{\theta})$ is Schur-convex by paragraph 1 in Section 5 of Marshal and Olkin (1974), and $E\{\Psi^*(\boldsymbol{Z}+\boldsymbol{\theta})\}$ is Schur-convex by E.5 on page 299 in Marshall and Olkin (1979). Let $\bar{\theta} = \sum \theta_i$, $\bar{\boldsymbol{\theta}} = (\bar{\theta}, 0, \ldots, 0)$, and $\bar{\boldsymbol{\theta}}^* = (\infty, 0, \ldots, 0)$. Since $\bar{\boldsymbol{\theta}}$ majorizes $\boldsymbol{\theta}$, we have that $\text{pr}(E_1 \cap E_2; \boldsymbol{\theta}) \leq \text{pr}(E_1 \cap E_2; \bar{\boldsymbol{\theta}})$. Let $V = \sum_2^I S^{-2} Y_i^2 I(Y_i > 0)$ and $W = \sum_2^I S^{-2} Y_i^2 I(Y_i < 0)$. Then by arguments similar to those in Gail and Simon (1985, Appendix), we have that

$$\text{pr}(E_1 \cap E_2; \bar{\boldsymbol{\theta}}) = \int_{w>c, v<c} \text{pr}(S^{-2}Y_1^2 > c - v)dG(w, v).$$

Since Y_1^2/S^2 has a noncentral F-distribution and $\text{pr}(Y_1^2/S^2 > c - v|\bar{\boldsymbol{\theta}})$ increases with $\bar{\theta}$, we have that

$$\text{pr}(E_1 \cap E_2; \boldsymbol{\theta}) \leq \text{pr}(E_1 \cap E_2; \bar{\boldsymbol{\theta}}) \leq \text{pr}(E_1 \cap E_2; \boldsymbol{\theta}^*).$$

Therefore, we conclude that $\boldsymbol{\theta}^* = (\infty, 0, \ldots, 0)$ is the least favorable null value.

Let pr* denote the probability evaluated at $\boldsymbol{\theta}^*$. Since $\text{pr}^*\{S^{-2}Y_1^2 I(Y_1 > 0) > c\} = 1$ we have that $\text{pr}^*(E_1) = 1$. Therefore, $\text{pr}^*(E_1 \cap E_2) = \text{pr}^*(E_2)$. Now,

$$\text{pr}^*(E_2) = \text{pr}^*\{\sum_1^I S^{-2}Y_i^2 I(Y_i < 0) > c\} = \text{pr}^*\{\sum_2^I S^{-2}Y_i^2 I(Y_i < 0) > c\},$$

the last equality holds because $\text{pr}^*(S^{-2}Y_1^2 I(Y_1 < 0) = 0) = 1$. Note that, since the Y_i's are *iid*, the probability that exactly j of the Y_i's are negative is given by the usual binomial probability. Further, $\text{pr}^*\{E_2 |$ exactly j of $Y_2, \ldots, Y_I$ are negative $\} = 1 - F_{j,\nu}(c/j)$. Now the proposition follows by applying the law of total probability. ■

9.4 DIRECTED TESTS

In Example 3.3.6 it was illustrated that the *LRT* of $H_0 : \boldsymbol{\mu} = \mathbf{0}$ against $H_1 : \boldsymbol{\mu} \geqslant \mathbf{0}$ based on a single observation of $\boldsymbol{X}$, where $\boldsymbol{X} \sim N(\boldsymbol{\mu}, \boldsymbol{V})$, and $\boldsymbol{V} = 0.2[1, 0.9 \mid 0.9, 1]$, could reject the null hypothesis even when every component of $\boldsymbol{X}$ is negative; for example, H_0 would be rejected at 5% level even when $\boldsymbol{X} = (-3, -2)$. At first, this may appear anomalous. However, simple calculations in that example provided clear explanation of the phenomenon (see Fig. 3.8). Essentially, because of the positive correlation between X_1 and X_2, the value $(-3, -2)$ is significantly more likely to have arisen from $\boldsymbol{\mu} = (0, 0.7)$ than $\boldsymbol{\mu} = (0, 0)$. Now, the question, should we expect a "good" test of H_0 against H_1 to reject H_0 for this value of $\boldsymbol{X}$? The answer would depend on how one would want to interpret the result, in particular how one wants to interpret a small p-value for $\boldsymbol{X} = (-3, -2)$. Perlman and Wu (1999, 2001) and Chaudhuri and Perlman (2004) argue in favor of *LRT*, but Cohen and Sackrowitz (1998, 2002, 2004) and Cohen et al. (2000) argue that *LRT* is deficiencient; interested readers are referred to these papers for detailed discussions.

Cohen and Sackrowitz (1998) introduced and developed the so-called class of directed tests, as an alternative to *LRT*. A fundamental property of a directed test of $\boldsymbol{\mu} = 0$ vs $\boldsymbol{\mu} \geq 0$ is that if it rejects H_0 for $\boldsymbol{X} = (a, b)$ then it would also reject H_0 for every value of $\boldsymbol{X} \geqslant (a, b)$; later, we will define a test with this property as cone order monotone. A suitable directed test of H_0 against H_1 would not reject H_0 when $\boldsymbol{X} = (-3, -2)$; if it did, then it has to reject H_0 for $\boldsymbol{X} = (0, 0)$ as well, which would not be acceptable for any test. In this section we will not discuss the philosophical issues relating to this and similar inference problems. Our objective in this section is to provide an introduction to the basic ideas of directed tests.

Cohen and Sackrowitz (1998) introduced an approach, which they call *directed test*, that is different from the *LRT* for constructing tests of $\boldsymbol{\theta} = \mathbf{0}$ against $\boldsymbol{\theta} \in \mathcal{C}$ where $\mathcal{C}$ is a closed convex cone and $\boldsymbol{\theta}$ is the natural parameter of an exponential family. This method of constructing tests is quite flexible; for example, in Example 3.3.6, a directed test of $H_0 : \boldsymbol{\theta} = \mathbf{0}$ against $H_1 : \boldsymbol{\theta} \geqslant \mathbf{0}$ can be constructed in such a way that the null hypothesis would not be rejected when the sample means $\bar{X}_1$ and $\bar{X}_2$ are both negative. For the simple case in Example 3.3.6, it is not difficult to construct a test by visually choosing the shape of the critical region to have the desired property. Such a visual approach could be difficult and subjective in more general cases such as contingency tables and the exponential families. The directed test of Cohen and Sackrowitz (1998) provides a rigorous theoretical way of developing tests with certain desired properties. For a given practical setting involving constrained inference, it would be possible to construct several different *LRT*s and Directed Tests by formulating the testing problem in different ways. However, the *LRT* and directed tests approach the testing problem from different directions and achieve different objectives.

First, let us define the terms *cone order* and *cone order monotone* test. Let $\mathcal{C}$ be a closed convex cone in $\mathbb{R}^p$. The *preorder*, $\preccurlyeq_{\mathcal{C}}$, with respect to $\mathcal{C}$ is defined by

$$\boldsymbol{x} \preccurlyeq_{\mathcal{C}} \boldsymbol{y} \quad \text{if and only if} \quad \boldsymbol{y} - \boldsymbol{x} \in \mathcal{C}.$$

Thus, $\boldsymbol{x} \preccurlyeq_{\mathcal{C}} \boldsymbol{y}$ is equivalent to any one of the following: (1) the direction from $\boldsymbol{x}$ to $\boldsymbol{y}$ is in $\mathcal{C}$, (2) $\boldsymbol{y} \in \boldsymbol{x} + \mathcal{C}$, and (3) if the cone $\mathcal{C}$ is translated by $\boldsymbol{x}$ then the point represented by $\boldsymbol{y}$ lies in the translated cone. Let the null and alternative hypotheses be denoted by H_0 and H_1, respectively, and let $\phi(\boldsymbol{x})$ denote a test function for testing H_0 against H_1; thus $\phi(\boldsymbol{x})$ is the probability of rejecting H_0 when $\boldsymbol{x}$ is the sample value. The test ϕ is said to be *cone order monotone* with respect to $\mathcal{C}$, denoted by COM[$\mathcal{C}$], if

$$\boldsymbol{x} \preccurlyeq_{\mathcal{C}} \boldsymbol{y} \Rightarrow \phi(\boldsymbol{x}) \leqslant \phi(\boldsymbol{y}).$$

Thus, a test is COM[$\mathcal{C}$] if $\boldsymbol{x} \preccurlyeq_{\mathcal{C}} \boldsymbol{y}$ and the test rejects H_0 when the sample outcome is $\boldsymbol{x}$ then it also rejects H_0 when the outcome is $\boldsymbol{y}$ as well.

9.4.1 Directed Test in a Simple Case

It would be instructive to consider the standard problem of testing $H_0 : \mu = 0$ against $H_1 : \mu > 0$ based on a single observation of X where X is a scalar and $X \sim N(\mu, 1)$. Let x denote the sample value of X. Then $[x, \infty)$ is the set of all possible outcomes that are at least as favorable to H_1 as x. In this case, we define the p-value as pr$(X \geqslant x \mid H_0)$. A different way of looking at this is the following. As an example, let $x = -5$. Now, a movement from -5 in the positive direction would take us in the direction of H_1 and also closer to H_0. Consequently, moving deeper into H_1 does not imply moving away from H_0. Therefore, we adopt the approach of rejecting H_0 only if every value in $[-5, \infty)$ is inconsistent with H_0. In this example, the null hypothesis would not be rejected because the value zero in $[-5, \infty)$ is consistent with it (i.e., the null hypothesis).

The procedure for developing the directed test are also similar. To introduce this, we first consider a bivariate example. Let $\boldsymbol{X} = (X_1, X_2)$ be a bivariate random variable distributed as $N(\boldsymbol{\mu}, \boldsymbol{V})$ where $\boldsymbol{V} = (1, \rho \mid \rho, 1)$ and ρ is known. Let us define

$$H_0 : \boldsymbol{\mu} = \boldsymbol{0}, \qquad H_1 : \boldsymbol{\mu} \geqslant \boldsymbol{0}, \quad E(\boldsymbol{x}) = \{\boldsymbol{y} : \boldsymbol{y} \geqslant \boldsymbol{x}\}$$

and consider test of H_0 against H_1. For a given observation, $\boldsymbol{x} = (x_1, x_2)$, suppose that we have decided to treat every point in $E(\boldsymbol{x})$ to be at least as favorable to H_1

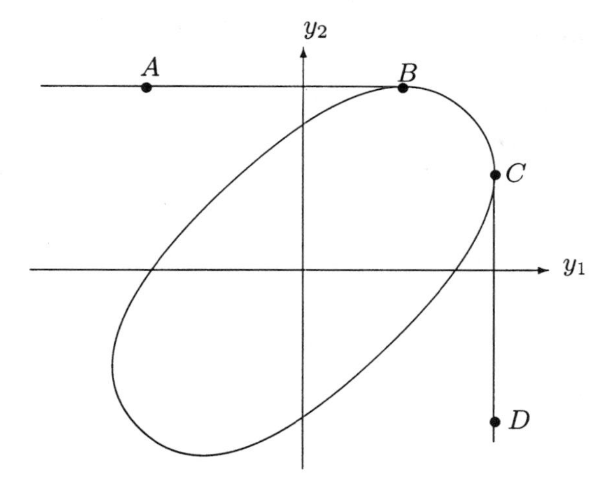

Fig. 9.13 Directed test of $\boldsymbol{\mu} = \mathbf{0}$ vs $\boldsymbol{\mu} \geqslant \mathbf{0}$ based on $\boldsymbol{Y} \sim N(\boldsymbol{\mu}, \boldsymbol{V})$; the acceptance region is the lower-left region bounded by $ABCD$.

as the observed one, $\boldsymbol{x}$. Therefore, if a test rejects H_0 when the observed value is $\boldsymbol{x}$ then we require it to reject H_0 when the observed value is $\boldsymbol{y}$ where $\boldsymbol{y}$ is any point in $E(\boldsymbol{x})$. If the test is COM[$\mathbb{R}^{+2}$] then this condition would be satisfied. A simple such test is one for which the critical region is $\{\boldsymbol{y} : \boldsymbol{y} \geqslant (c, c)\}$ for some positive c. The directed test provides a more formal approach to developing such tests.

Let $T(\boldsymbol{y})$ denote the *LRT*, $\boldsymbol{y}^T \boldsymbol{V}^{-1} \boldsymbol{y}$, for testing $H_0 : \boldsymbol{\mu} = \mathbf{0}$ against the unrestricted alternative $H_2 : \boldsymbol{\mu} \neq \mathbf{0}$ when the observed value is $\boldsymbol{y}$. Now, define the statistic $T_D(\boldsymbol{x})$ for testing $H_0 : \boldsymbol{\mu} = \mathbf{0}$ against $H_1 : \boldsymbol{\mu} \geqslant \mathbf{0}$ by

$$T_D(\boldsymbol{x}) = \inf\{T(\boldsymbol{y}) : \boldsymbol{y} \in E(\boldsymbol{x})\}; \tag{9.55}$$

the test rejects $H_0 : \boldsymbol{\mu} = \mathbf{0}$ in favor of $H_1 : \boldsymbol{\mu} \geqslant \mathbf{0}$ if T_D is large. This test says that once we have chosen a suitable set, $E(\boldsymbol{x})$, as the set of all outcomes that are at least as favorable to H_1 as the observed value $\boldsymbol{x}$, then reject H_0 if every outcome in $E(\boldsymbol{x})$ is inconsistent with H_0; the degree of inconsistency between $\boldsymbol{y}$ and H_0 is quantified by the test statistic $T(\boldsymbol{y})$.

To indicate the shape of the critical region of this test, let the inside of the oval shape in Fig. 9.13 be a typical acceptance region for a test of $\boldsymbol{\mu} = \mathbf{0}$ against $\boldsymbol{\mu} \neq \mathbf{0}$; for the *LRT* it is elliptical. Let AB be the horizontal tangent to the upper-right corner of the ellipse with B being the point of contact, and similarly let CD be the vertical tangent to the upper-right corner of ellipse with C being the point of contact. Then it may be verified that the lower-left region bounded by $ABCD$ is a typical acceptance region of T_D, and that the test is Com[$\mathbb{R}^{+2}$]. If the sample value $\boldsymbol{x}$ is the point represented by A then the p-value is the probability content, under H_0, of the upper-right region bounded by $ABCD$ (see Problem 9.8).

The foregoing test can also be defined by simply identifying the critical region directly rather than through a test statistic. This also provides a different insight because one can modify the critical region directly. Let $\mathcal{A}$ denote the elliptical region in Figure 9.13. This is the acceptance region for a test of $\boldsymbol{\mu} = \mathbf{0}$ against the unrestricted alternative, $\boldsymbol{\mu} \neq \mathbf{0}$; let α denote the level of significance of this unrestricted test. Suppose that we wish to construct a test that is COM[$\mathbb{R}^{+2}$]. To construct such a test, consider the acceptance region $\mathcal{B}$ defined by $\mathcal{B} = \cup_{\boldsymbol{t} \in -\mathbb{R}^{+2}} (\mathcal{A} + \boldsymbol{t})$. It is easily seen that $\mathcal{B}$ is the region swept by $\mathcal{A}$ when the center of $\mathcal{A}$ sweeps over negative orthant, $-\mathbb{R}^{+2}$. Since $\mathcal{B}$ is also the acceptance region (i.e., lower left region bounded by ABCD) in Fig. 9.13, these two tests are the same.

9.4.2 Directed Test Against Stochastic Order

Cohen et al. (2003) (CMS) proposed a class of tests for testing against stochastic orders in the analysis of ordinal response data. Their procedure for testing against simple stochastic order in $2 \times c$ table is discussed below.

Let us consider the comparison of two treatments when the response variable is ordinal. Let $\{n_{ij} : i = 1, 2, j = 1, \ldots, c\}$ denote the cell frequencies, and let the hypothesis testing problem be defined by

$$H_0 : \boldsymbol{\pi}_1 = \boldsymbol{\pi}_2 \quad \text{and} \quad H_1 : \boldsymbol{\pi}_1 \preccurlyeq_s \boldsymbol{\pi}_2,$$

where $\boldsymbol{\pi}_i = (\pi_{i1}, \ldots, \pi_{ic})^T$, $\pi_{ij} = \text{pr(Response} = j \mid \text{Row} = i)$ for every (i, j) and $\boldsymbol{\pi}_1 \preccurlyeq_s \boldsymbol{\pi}_2$ means that $\pi_{11} + \ldots + \pi_{1j} \geqslant \pi_{21} + \ldots + \pi_{2j}$ for $j = 1, \ldots, c$.

The directed test of Cohen, Madigan, and Sackrowitz (2003) suggests an alternative approach. To introduce this, let $e_{ij} = (n_{i+}n_{+j}/n_{++})$, the expected frequency for cell (i, j) under H_0 conditional on the marginal totals, $\{u_{ij}\}$ denote the frequencies for an arbitrary table, $y_j = u_{11} + \ldots + u_{1j}$ for $j = 1, \ldots, c$ and let

$$E(\boldsymbol{n}) = \{u_{11} + \ldots + u_{1j} \geqslant n_{11} + \ldots + n_{1j}, u_{i+} = n_{i+} \text{ and } u_{+j} = n_{+j}, \forall (i, j)\}.$$

Clearly $E(\boldsymbol{n})$ is also equal to $\{y_j \geqslant n_{11} + \ldots + n_{1j}, u_{i+} = n_{i+} \text{ and } u_{+j} = n_{+j}, \forall (i, j)\}$. Suppose that we need a test that is *concordant monotone*. It may be verified that this requires a test for which the p-value for any outcome in $E(\boldsymbol{n})$ to be at least as small as that for the observed one; equivalently, if $\boldsymbol{n}$ is in the critical region then $E(\boldsymbol{n})$ must be a subset of it. Cohen, Madigan, and Sackrowitz (2003) showed that this essentially requires the test to be COM[$\mathbb{R}^{+(c-1)}$].

The CMS test statistic is defined as

$$\chi_D^2 = \inf_{\boldsymbol{u} \in E(\boldsymbol{n})} \sum_i \sum_j (u_{ij} - e_{ij})^2 / e_{ij}. \tag{9.56}$$

The test rejects H_0 if χ_D^2 is large. Note that for every outcome in $E(\boldsymbol{n})$, the expected frequencies under H_0, given the marginal totals, are the same as those for the observed table. Therefore, the CMS test rejects H_0 if the minimum of the Pearson chi-square for the tables in $E(\boldsymbol{n})$ is large. Again, note that once we have chosen the set of all outcomes that are considered to be at least as extreme as the observed one, the procedure is to reject H_0 if every such outcome is inconsistent with H_0.

The exact finite sample distribution of χ_D^2 under H_0 and conditional on the marginal totals is unknown, but we know that it is does not depend on any nuisance parameters. Therefore, the p-value, pr($\chi_D^2 \geqslant obs \mid H_0$, marginal totals), can be estimated sufficiently precisely using simulation. To this end, the algorithm of Patefield (1981) for generating independent tables with a given margin can be used. The program RNTAB in IMSL offers a subroutine for generating pseudo random tables with fixed margins. The Website, http://stat.rutgers.edu/~madigan/dvp.html, offers a p-value very quickly upon entering the two rows of cell frequencies.

Now, we consider a hypothetical example to provide more insight into the shape of the critical region of the foregoing test. Suppose that there are two treatments, Treatment 1 and Treatment 2, to be compared. The response variable is ordinal with three categories, {poor, good, excellent}. The null and alternative hypotheses are $H_0 : (\pi_{11}, \pi_{12}, \pi_{13}) = (\pi_{21}, \pi_{22}, \pi_{23})$ and $H_1 : \pi_{11} \geqslant \pi_{21}, \pi_{11} + \pi_{12} \geqslant \pi_{21} + \pi_{22}$, respectively. Consider the two sets of data in Table 9.8; this table also provides the expected frequencies under the assumption that there is no difference between the treatments and conditional on the marginal totals being equal to the observed ones.

Table 9.8 Comparison of two treatments; two sets of data and expected values

	Set 1				Set 2			
	Poor	Good	Excellent	Total	Poor	Good	Excellent	Total
	Response				Response			
Treat 1	2	4	4	10	4	1	5	10
Treat 2	1	3	6	10	1	5	4	10
Total	3	7	10	20	5	6	9	20
	Expected Values				Expected Values			
Treat 1	1.5	3.5	5	10	2.5	3	4.5	10
Treat 2	1.5	3.5	5	10	2.5	3	4.5	10
Total	3	7	10	20	5	6	9	20

Source:

Fig. 9.14 shows the shape of the conditional critical region for data Set 2, in the (y_1, y_2) coordinates. The boundary of the oval shape corresponds to tables for which the Pearson chi-square statistic does not change. Let $\boldsymbol{y}_0$ denote the value of (y_1, y_2) corresponding to the expected frequencies $\{e_{ij}\}$; thus $\boldsymbol{y}_0 = (e_{11}, e_{11} + e_{12})$. Then the center of the oval shape corresponds to $\boldsymbol{y}_0$. These ovals are nested because as the value of the Pearson chi-square statistic increases the ovals become larger. Since $y_2 \geqslant y_1$, only the region above the 45 degree line in Figure 9.15 is relevant. By arguments similar to those for the normal mean presented earlier in this section, it may be verified that the critical region, conditional on the marginal totals, has the shape shown in Fig. 9.15 (i.e., the region inside the large triangle and bounded by ABCD). As in the example of the previous subsection, it can be verified that the every point on the curve $ABCD$ in Fig. 9.15 has the same value for χ_D^2; further, if the observed value of $\boldsymbol{y} = (n_{11}, n_{11} + n_{12})$ is on $ABCD$ then the p-value is the probability content of the upper-right corner bounded by $ABCD$.

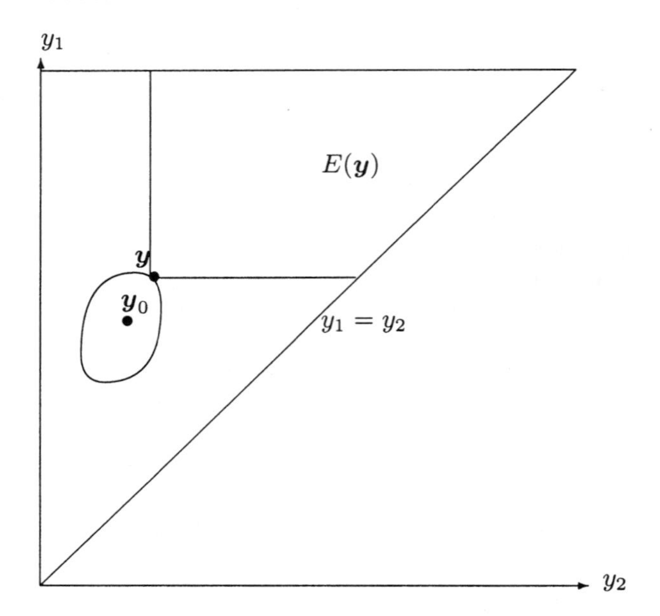

Fig. 9.14 Directed test for Set 2; χ_D^2 = Pearson χ^2 at $\boldsymbol{y}$.

Let us consider data Set 1 in Table 9.8. Let $\{\tilde{n}_{ij}\}$ denote the value of $\boldsymbol{u}$ at which the minimum in (9.56) is achieved. Thus, $\chi_D^2 = \sum_i \sum_j (\tilde{n}_{ij} - e_{ij})^2/e_{ij}$. For Set 1, it may be verified that $(\tilde{n}_{11}, \tilde{n}_{12}) = (2, 4)$, which is the same as the observed table; therefore, $\chi_D^2 = 0.876$, which is the Pearson chi-square statistic for Set 1. Thus, of all the tables in $E(\boldsymbol{n})$, the observed table is the most favorable to H_0 with respect to Pearson chi-square. This is consistent with the observation that the unrestricted estimator satisfies the constraints in H_1.

By contrast, for Set 2, the unrestricted estimator of $\boldsymbol{\pi}$ does not satisfy all the constraints in H_1. Consequently, as we move from the observed value, $\{\boldsymbol{y}\}$, vertically up by a small amount (see Fig. 9.16), we would move in the direction of H_1, but we would also move closer to $\{\boldsymbol{y}_0\}$ (i.e., closer to H_0) where closeness is measured in terms of Pearson chi-square. The outcome that is closest to H_0 is $(\tilde{n}_{11}, \tilde{n}_{12}) = (4, 2.4)$ and χ_D^2 is equal to the Pearson chi-statistic for the table $\tilde{\boldsymbol{n}}$; this leads to $\chi_D^2 = 2.4$.

9.4.3 $r \times c$ Tables

The discussions in CMS cover more general situations, including analysis of $r \times c$ tables with or without censored data and tests against other stochastic orders. Consider the comparison of r treatments when the response variable is ordinal with c categories. Let the null and alternative hypotheses be

$$H_0 : \boldsymbol{\pi}_1 = \ldots = \boldsymbol{\pi}_r \quad \text{and} \quad H_1 : \{\boldsymbol{\pi}_1, \ldots, \boldsymbol{\pi}_r\} \text{ satisfy a certain stochastic order.}$$

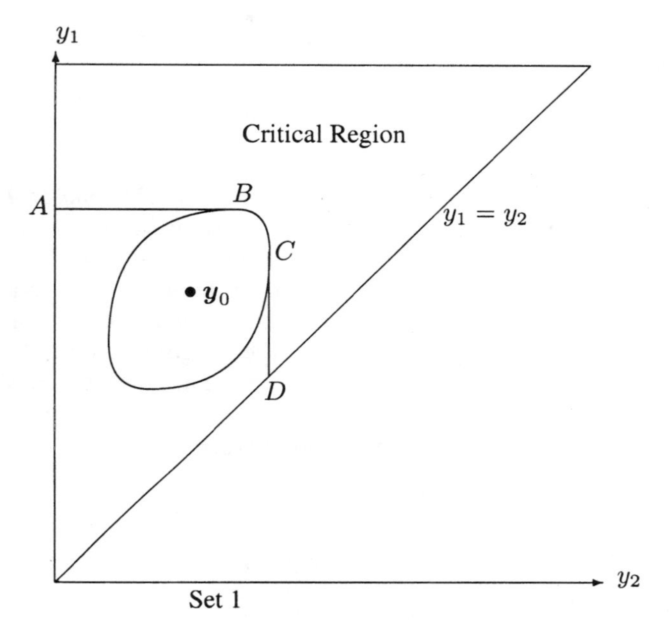

Fig. 9.15 The critical region for the directed test with margins as for Set 1

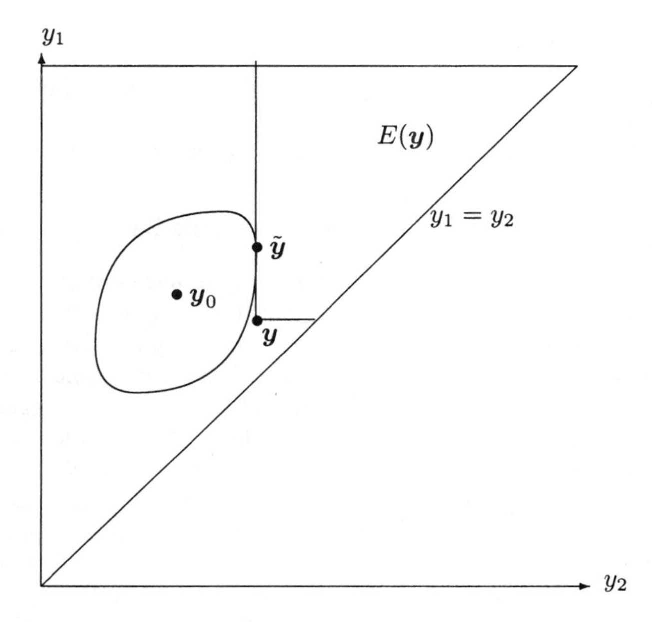

Fig. 9.16 Directed test for Set 1; χ^2_D = Pearson χ^2 at $\tilde{\boldsymbol{y}}$.

Generalizing the foregoing ideas for the comparison of two treatments, the directed test statistic is defined as

$$\chi_D^2 = \min_{u \in E(n)} \sum_i \sum_j (u_{ij} - e_{ij})^2 / e_{ij}$$

where $E(\boldsymbol{n}) = \{\boldsymbol{u} : \boldsymbol{B}(\boldsymbol{u} - \boldsymbol{n}) = \boldsymbol{0}, \boldsymbol{G}(\boldsymbol{u} - \boldsymbol{n}) \geqslant \boldsymbol{0}\}$, for suitably chosen $\boldsymbol{B}$ and $\boldsymbol{G}$. The matrix $\boldsymbol{B}$ is chosen to ensure that the marginal totals for $\{u_{ij}\}$ and $\{n_{ij}\}$ are equal; the matrix $\boldsymbol{G}$ is chosen such that the table of frequencies $\{u_{ij}\}$ with the same marginal totals as the observed one, would be considered to be at least as favorable to H_1 as the observed one if $\boldsymbol{G}(\boldsymbol{u} - \boldsymbol{n}) \geqslant \boldsymbol{0}$. It is of course possible that some of these are also more favorable to H_0 than the observed one. The foregoing test based on χ_D^2 as the test statistic amounts to rejecting H_0 if every outcome in $E(\boldsymbol{n})$ is inconsistent with H_0.

Let us consider testing against the simple stochastic order:

$$H_1 : \boldsymbol{\pi}_1 \preccurlyeq_s \ldots \preccurlyeq_s \boldsymbol{\pi}_r.$$

Let

$$\boldsymbol{G} = \begin{bmatrix} \boldsymbol{Q} & 0 & \ldots & \ldots & 0 & 0 \\ \boldsymbol{Q} & \boldsymbol{Q} & 0 & \ldots & 0 & 0 \\ \vdots & \vdots & \vdots & \cdots & \vdots & \vdots \\ \boldsymbol{Q} & \boldsymbol{Q} & \ldots & & \boldsymbol{Q} & 0 \end{bmatrix},$$

where

$$\boldsymbol{Q} = \begin{bmatrix} 1 & 0 & \ldots & \ldots & 0 & 0 \\ 1 & 1 & 0 & \ldots & 0 & 0 \\ \vdots & \vdots & \vdots & \cdots & \vdots & \vdots \\ 1 & 1 & \ldots & & 1 & 0 \end{bmatrix}.$$

Then, for testing $H_0 : \boldsymbol{\pi}_1 = \ldots = \boldsymbol{\pi}_r$ against $H_1 : \boldsymbol{\pi}_1 \preccurlyeq_s \ldots \preccurlyeq_s \boldsymbol{\pi}_r$, the χ_D^2 test is concordant monotone (see CMS for a proof). The acceptance/rejection regions are similar to those in Figure 9.15, except that they are in higher dimensions.

9.4.4 Improving Power by Enlarging the Achievable Set of p-Values

In many practical situations, the sample sizes involving ordinal response data are small and consequently the achievable set of p-values for a given sample size may not be sufficiently dense, which in turn leads to the test being conservative. This is a well-known problem in classical unconstrained analysis of categorical data; for example see Agresti (2002, p 20). One of the methods that has been suggested is to use the so called *mid p-value,* which works as follows: If T is a test statistic, then instead of defining the p-value as $\text{pr}(T \geqslant obs)$, define the *mid p-value,* as $\{\text{pr}(T > obs) + 0.5\text{pr}(T = obs)\}$, and the test rejects the null hypothesis if this mid p-value is small. Cohen and Sackrowitz (1992) proposed a related procedure in which the term $0.5\text{pr}(T = obs)$ was replaced by pr($T = obs$, *and the table has smaller probability of being observed than the observed one*). CMS suggested a procedure that is also appealing. Their procedure is to use a back-up statistic, L, which is itself a suitable statistic for testing H_0 against H_1, and define the p-value as

$$\text{pr}(\chi_D^2 > obs) + \text{pr}(\chi_D^2 = obs, L \geqslant obs);$$

CMS chose $L = \sum_{j=1}^{c} \sum_{\ell=1}^{j} n_{1\ell}(n_{+j} + n_{+(j+1)})$, the Wilcoxon-Mann-Whitney statistic for L.

A consequence of using a back-up statistic is that the achievable set of p-values become more dense and hence the test becomes less conservative (consequently, more powerful). Cohen and Sackrowitz (1998) suggested a different method of obtaining a finer grid of p-values. They called it *Directed Vertex Peeling* (DVP); for a simpler account of this method, see Cohen, Sackrowitz, and Sackrowitz (2000).

9.4.5 Exponential Families

In the example on testing against stochastic order in $2 \times c$ tables, concordant monotonicity was achieved by ensuring that the test is COM[$\mathbb{R}^{+(c-1)}$]. Cohen, Sackrowitz, and Samuel-Cahn (1995) showed that tests that are COM[$\mathcal{C}^*$], where $\mathcal{C}^* = \{\boldsymbol{x} : \boldsymbol{x}^T\boldsymbol{y} \geqslant 0 \text{ for any } \boldsymbol{y} \in \mathcal{C}\}$ is the positive dual of $\mathcal{C}$, have some desirable properties; in particular, that it is a *complete class*. These suggest that there may be instances in which it may be of interest to ensure that the test is COM[K] for some K; the choice of K would depend on the particular problem. The directed test of Cohen and Sackrowitz (1998) offers a general approach to constructing such tests in exponential families.

Let $\boldsymbol{X} = (X_1, \ldots, X_k)$ be a random vector having an exponential family density given by $h(\boldsymbol{x})g(\boldsymbol{\nu})\exp(\boldsymbol{x}^T\boldsymbol{\nu})$. Let the hypothesis testing problem be $H_0 : \boldsymbol{\nu} = \boldsymbol{0}$ and $H_1 : \boldsymbol{\nu} \in \mathcal{C}$, respectively, where $\mathcal{C}$ is a closed convex cone. In general, the hypothesis testing problem would involve nuisance parameters. However, often the nuisance parameter can be eliminated by conditioning on sufficient statistics and a testing problem would reduce to the foregoing form without nuisance parameters. The approach to constructing a test of $\boldsymbol{\nu} = \boldsymbol{0}$ against $\boldsymbol{\nu} \in \mathcal{C}$ is to start with a test of $\boldsymbol{\nu} = \boldsymbol{0}$ against the unrestricted alternative $\boldsymbol{\nu} \neq \boldsymbol{0}$ and then directionalize it by an extension of the procedure presented earlier for the location and the contingency table models.

Let us consider a test of $\boldsymbol{\nu} = \boldsymbol{0}$ against the unrestricted alternative $\boldsymbol{\nu} \neq \boldsymbol{0}$; for example, it could be the *LRT* or the Pearson chi-square test. Let A denote an acceptance region of this test; it is assumed that A is convex. Now consider a test with acceptance region C defined by $C = A - K$. Thus, $C = \cup_{k \in K}(A - k)$. This is the directed test of $\boldsymbol{\nu} = \boldsymbol{0}$ against $\boldsymbol{\nu} \in \mathcal{C}$. It has a convex acceptance region, is COM[K], and lies in the complete class. The fact that the test has convex acceptance region is useful because it is a requirement for admissibility. Finally, let us note that the *projected tests* introduced by Cohen, Kemperman and Sackrowitz (1994) is a special case of this general class of directed tests.

Problems

Multiple end points

9.1 Let $\boldsymbol{X}_i$ be $p \times 1$ random vector, $\boldsymbol{X}_i \sim N(\boldsymbol{\mu}_i, \boldsymbol{V}_i)$, and $\boldsymbol{X}_{i1}, \ldots, \boldsymbol{X}_{in_i}$ be independently and identically distributed as $\boldsymbol{X}_i, (i = 1, 2)$. Let $\boldsymbol{\delta} = (\delta_1, \ldots, \delta_p)^T$ be a given fixed vector, $\boldsymbol{\delta} \geqslant \boldsymbol{0}$, $\boldsymbol{\mu}_2 = \boldsymbol{\mu}_1 + \lambda\boldsymbol{\delta}$ for some $\lambda \in \mathbb{R}$. Let $L(\boldsymbol{\mu}_1, \lambda)$ denote the loglikelihood of the samples on $\boldsymbol{X}_1$ and $\boldsymbol{X}_2$. Prove the following assuming that $\boldsymbol{V}_1 = \boldsymbol{V}_2 = \boldsymbol{V}$.

9.1.1. The kernel of $L(\boldsymbol{\mu}_1, \lambda)$ is

$$\sum_{\ell=1}^{n_1}(\boldsymbol{X}_{1\ell} - \boldsymbol{\mu}_1)^T\boldsymbol{V}^{-1}(\boldsymbol{X}_{1\ell} - \boldsymbol{\mu}_1) + \sum_{\ell=1}^{n_2}(\boldsymbol{X}_{2\ell} - \boldsymbol{\mu}_1 - \lambda\boldsymbol{\delta})^T\boldsymbol{V}^{-1}(\boldsymbol{X}_{2\ell} - \boldsymbol{\mu}_1 - \lambda\boldsymbol{\delta}).$$

9.1.2. The maximum likelihood estimating equations are

$$(n_1\bar{\boldsymbol{X}}_1 + n_2\bar{\boldsymbol{X}}_2) - (n_1 + n_2)\hat{\boldsymbol{\mu}}_1 - \hat{\lambda}(n_2\boldsymbol{\delta}) = \mathbf{0}$$

$$\boldsymbol{\delta}^T\boldsymbol{V}^{-1}\bar{\boldsymbol{X}}_2 - \boldsymbol{\delta}^T\boldsymbol{V}^{-1}\hat{\boldsymbol{\mu}}_1 - \hat{\lambda}(\boldsymbol{\delta}^T\boldsymbol{V}^{-1}\boldsymbol{\delta}) = 0.$$

9.1.3. Deduce that $\hat{\lambda} = \boldsymbol{\delta}^T\boldsymbol{V}^{-1}(\bar{\boldsymbol{X}}_1 - \bar{\boldsymbol{X}}_2)/(\boldsymbol{\delta}^T\boldsymbol{V}^{-1}\boldsymbol{\delta})$, and $se(\hat{\lambda}) = (\boldsymbol{\delta}^T\boldsymbol{V}^{-1}\boldsymbol{\delta})^{-1}$ $(n_1^{-1} + n_2^{-1})^{1/2}$.

9.1.4. Let σ_j^2 denote V_{jj} the variance of the ith component of $\boldsymbol{X}_1$. Define $\boldsymbol{D}$ and $\boldsymbol{C}$ by $\boldsymbol{D} = diag\{\sigma_1, \ldots, \sigma_p\}$ and $\boldsymbol{V} = \boldsymbol{DCD}$. Verify that $\boldsymbol{C}$ is the correlation matrix of $\boldsymbol{X}_1$. Show that

$$\hat{\lambda}/se(\hat{\lambda}) = \boldsymbol{a}^T\boldsymbol{C}^{-1}\boldsymbol{Z}/(\boldsymbol{a}^T\boldsymbol{C}^{-1}\boldsymbol{a})^{1/2}$$

where $\boldsymbol{a} = (a_1, \ldots, a_p)^T$, $a_j = \delta_j/\sigma_j$, $\boldsymbol{Z} = (Z_1, \ldots, Z_p)^T$, and $Z_j = (\bar{X}_{1j} - \bar{X}_{2j})/\{\sigma_j(n_1^{-1} + n_2^{-1})^{1/2}\}$.

9.2 Let the setting be as in Problem 9.1. Now prove the following without assuming that $\boldsymbol{V}_1$ and $\boldsymbol{V}_2$ are equal.

9.2.1. The kernel of the loglikelihood is $\sum_{k=1}^{n_1}(\boldsymbol{X}_{1k} - \boldsymbol{\mu}_1)^T\boldsymbol{V}_1^{-1}(\boldsymbol{X}_{1k} - \boldsymbol{\mu}_1) + \sum_{k=1}^{n_2}(\boldsymbol{X}_{2k} - \boldsymbol{\mu}_1 - \lambda\boldsymbol{\delta})^T\boldsymbol{V}_2^{-1}(\boldsymbol{X}_{2k} - \boldsymbol{\mu}_1 - \lambda\boldsymbol{\delta})$.

9.2.2. The maximum likelihood equations are

$$\boldsymbol{\delta}^T\boldsymbol{V}_2^{-1}(\bar{\boldsymbol{X}}_2 - \hat{\boldsymbol{\mu}}_1 - \hat{\lambda}\boldsymbol{\delta}) = \mathbf{0}$$

$$(n_1\boldsymbol{V}_1^{-1} + n_2\boldsymbol{V}_2^{-1})\hat{\boldsymbol{\mu}}_1 + n_2\hat{\lambda}\boldsymbol{\delta} = (n_1\boldsymbol{V}_1^{-1}\bar{\boldsymbol{X}}_1 + n_2\boldsymbol{V}_2^{-1}\bar{\boldsymbol{X}}_2).$$

9.2.3. The *mle* of λ is $c\boldsymbol{\delta}^T\boldsymbol{V}_2^{-1}(n_1\boldsymbol{V}_1^{-1} + n_2\boldsymbol{V}_2^{-1})^{-1}\boldsymbol{V}_1^{-1}(\bar{\boldsymbol{X}}_1 - \bar{\boldsymbol{X}}_2)$ for some positive c that does not depend on the data.

9.2.4. $\hat{\lambda}/se(\hat{\lambda}) = \boldsymbol{a}^T\boldsymbol{C}^{-1}\boldsymbol{Z}/(\boldsymbol{a}^T\boldsymbol{C}^{-1}\boldsymbol{a})^{1/2}$ where $\boldsymbol{a} = \boldsymbol{D}^{-1}\boldsymbol{\delta}, \boldsymbol{D} = diag\{d_1, \ldots, d_p\}$, $d_j^2 = (n_1^{-1}V_{1jj} + n_2^{-1}V_{2jj})$, $\boldsymbol{C} = corr(\bar{\boldsymbol{X}}_1 - \bar{\boldsymbol{X}}_2)$, and $\boldsymbol{Z} = \boldsymbol{D}^{-1}(\bar{\boldsymbol{X}}_1 - \bar{\boldsymbol{X}}_2)$.

Intersection Union Test

9.3 Suppose that $Y_{ij} = \alpha_i + \beta_i x_{ij} + \epsilon_{ij}$ where $i = 1, 2$ and $j = 1, \ldots, n_i$, $(\alpha_1, \beta_1, \alpha_2, \beta_2)$ is an unknown parameter, $\{x_{ij}\}$ are fixed, $\{\epsilon_{ij}\}$ are independent and identically distributed as $N(0, \sigma^2)$ where σ is unknown. It is desired to compare the two regression lines $(\alpha_1 + \beta_1 x)$ and $(\alpha_2 + \beta_2 x)$ on the finite interval $[x_a, x_b]$. More specifically the objective is to test

$$H_0 : \alpha_1 + \beta_1 x \leqslant \alpha_2 + \beta_2 x \text{ for some } x \in [x_a, x_b]$$

against

$$H_1 : \alpha_1 + \beta_1 x > \alpha_2 + \beta_2 x \text{ for every } x \in [x_a, x_b].$$

Let $\theta_1 = \{(\alpha_1 + \beta_1 x_a) - (\alpha_2 + \beta_2 x_a)\}$ and $\theta_2 = \{(\alpha_1 + \beta_1 x_b) - (\alpha_2 + \beta_2 x_b)\}$. Let $(\hat{\alpha}_1, \hat{\beta}_1, \hat{\alpha}_2, \hat{\beta}_2)$ denote the least squares estimator of $(\alpha_1, \beta_1, \alpha_2, \beta_2)$ and let s^2 denote the estimator $\nu^{-1} \sum_i \sum_j (y_{ij} - \hat{\alpha}_i - \hat{\beta}_i x_{ij})^2$ of σ^2 where $\nu = (n_1 + n_2 - 4)$. Show that

9.3.1. $\nu s^2/\sigma^2 \sim \chi^2_\nu$

9.3.2. $T_j = (\hat{\theta}_j - \theta_j)/\{s\sqrt{\sigma_{jj}}\} \sim t_\nu$ for $j = 1, 2$ where

$$\sigma_{11}^2 = \sum_{i=1}^{2} \left\{ \sum_{j=1}^{n_i} (x_{ij} - \bar{x}_i)^2 \right\}^{-1} \left\{ \sum_{j=1}^{n_i} n_i^{-1} x_{ij}^2 - 2x_a \bar{x}_i + x_a^2 \right\},$$

and σ_{22}^2 has the same expression as for σ_{11}^2 except that x_a is replaced by x_b.

9.3.3. Show that the IUT which rejects H_0 if $T_1 > t_{1-\alpha,\nu}$ and $T_2 > t_{1-\alpha,\nu}$, has exact size α. [Hint: Use Corollary 5.3.3.]

[(Tsutakawa and Hewett (1978); Example 4.1 in Berger (1982)]

9.4 Consider the Example 9.2.5. Let us define, $H_0^{(i)} : \mu_i \leqslant 0$ for $i = 1, 2$. Let the null hypothesis be

$$H_0 : H_0^{(1)} \text{ or } H_0^{(2)}.$$

Let $\Lambda_i(\boldsymbol{Y})$ denote the likelihood ratio statistic for testing $H_0^{(i)}$ against $H_1^{(i)}, i = 1, 2$.

9.4.1. Show that

$$\Lambda_i(\boldsymbol{Y}) = \sup_{\boldsymbol{\theta}} L(\boldsymbol{\theta}|\boldsymbol{Y}) / \sup\{L(\boldsymbol{\theta}|\boldsymbol{Y}) : \mu_i \leqslant 0\}$$
$$= \begin{cases} \{1 + (n_i - 1)^{-1} T_i^2\}^{n_i/2} & \text{if } \bar{Y}_i > 0 \\ 1 & \text{if } \bar{Y}_i < 0, \end{cases}$$

where $T_i = \bar{Y}_i/(S_i/\sqrt{n_i})$ is the t-ratio for testing $\mu_i \leqslant 0$ against $\mu_i > 0$. Thus, the likelihood ratio and the t-ratio for testing $H_0^{(i)}$ against $H_1^{(i)}$ are monotonically related.

9.4.2. Let a size-α critical region for testing $\mu_i \leqslant 0$ against $\mu_i > 0$ be $\{\Lambda_i(\boldsymbol{Y}) \geqslant c_{i\alpha}\}$ for some $c_{i\alpha}, i = 1, 2$. Deduce from Proposition 5.3.2 that the level-α *LRT* of H_0 against H_1 is

$$\text{reject } H_0 \text{ if } \Lambda(\boldsymbol{Y}) \geqslant \max\{c_{1\alpha}, c_{2\alpha}\}$$

where

$$\Lambda(\boldsymbol{Y}) = \min\{\Lambda_1(\boldsymbol{Y}), \Lambda_2(\boldsymbol{Y})\}.$$

9.4.3. Using the fact that $\Lambda_i(\boldsymbol{Y})$ is an increasing function of $T_i(\boldsymbol{Y})$, deduce that $c_{i\alpha} = \{1 + (n_{i-1})^{-1} t^2_{\alpha,n_{i-1}}\}^{n_i/2}$ where $t_{\alpha,\nu}$ is the upper 100α-percentile of a t-distribution with ν degrees of freedom.

9.4.4. Deduce that the IUT

$$\text{reject } H_0 \text{ if } T_i \geqslant t_{\alpha,n_i-1} \text{ for every } i$$

of H_0 against H_1 has size-α.

9.4.5. To compare the IUT with the *LRT*, consider the case when $\alpha = 0.05, n_1 = 40$ and $n_2 = 2$. From standard statistical tables, we have that $t_{\alpha,n_1-1} = 1.68, t_{\alpha,n_2-1} = 6.31$. Verify that $c_{1\alpha} = \{1 + (40-1)^{-1}1.68^2\}^{40/2} = 4.044, c_{2\alpha} = \{1 + (2-1)^{-1}6.31^2\}^{2/2} = 40.816, \max\{c_{1\alpha}, c_{2\alpha}\} = 40.816$ and

$$\Lambda_i(\boldsymbol{Y}) \geqslant 40.816 \Leftrightarrow \{1 + (n_i - 1)^{-1}T_i^2(\boldsymbol{Y})\}^{n_i/2} \geqslant 40.816$$
$$\Leftrightarrow T_i(\boldsymbol{Y}) \geqslant \{40.816^{2/n_i} - 1\}(n_i - 1)^{1/2}.$$

9.4.6. Deduce from the previous part that the size-0.05 critical regions of IUT and the *LRT* of H_0 against H_1 are $\{T_1(\boldsymbol{Y}) \geqslant 1.68 \text{ and } T_2(\boldsymbol{Y}) \geqslant 6.31\}$ and $\{T_1(\boldsymbol{Y}) \geqslant 2.82 \text{ and } T_2(\boldsymbol{Y}) \geqslant 6.31\}$, respectively.

9.4.7. Consider the special case when $n_1 = n_2 = n$. Then $c_{1\alpha} = c_{2\alpha}$, and the IUT and the *LRT* are identical. The t-statistic T_i has a noncentral t-distribution with noncentrality parameter $\nu_i = \sqrt{n}\mu_i/\sigma_i$. Therefore, the testing problem $H_0 : \mu_1 \leq 0$ or $\mu_2 \leq 0$ against $H_1 : \mu_1 > 0$ and $\mu_2 > 0$ is equivalent to testing

$$H_0^* : \nu_i = 0 \text{ for some } i, \text{ against } H_1^* : \nu_i > 0 \text{ for every } i.$$

Deduce from Theorem 3.3 of Cohen, et al. (1983 b) that the foregoing IUT is uniformly most powerful size-α monotone test [Hint: Use the fact that t-distribution has a monotone likelihood ratio in the noncentrality parameter, and that T_1 and T_2 are independent.] [Saikali and Berger (2001)]

Cross-over Interaction

9.5 Consider m_2 treatments being compared across m_1 groups. Let $Y_{ij} \sim N(\theta_{ij}, \sigma^2)$ for $i = 1, \ldots, m_1$ and $j = 1, \ldots, m_2$. Assume that Y_{ij}'s are independent and σ^2 is known. Let

H_0^* : No interaction between treatments and groups.
H_0: No cross-over interaction.
H_1: There is cross-over interaction.

Let

$$I^t_{j_1 j_2 k_1 k_2} = \begin{cases} 1 & \text{if } Y_{j_1 k_2} - Y_{j_1 k_1} \geqslant t\sigma\sqrt{2} \text{ and } Y_{j_2 k_2} - Y_{j_2 k_1} \leq -t\sigma\sqrt{2} \\ 1 & \text{if } Y_{j_1 k_2} - Y_{j_1 k_1} \leq -t\sigma\sqrt{2} \text{ and } Y_{j_2 k_2} - Y_{j_2 k_1} \geqslant t\sigma\sqrt{2} \\ 0 & \text{otherwise,} \end{cases}$$

and

$$S = \sum_{j_1 < j_2} \sum_{k_1 < k_2} I^t_{j_1 j_2 k_1 k_2}.$$

Thus, S is the total number of "large" reversals of treatment effects. The test is, accept H_1 if $S \geqslant 1$.

Assume that t is large enough so that pr$(S \geqslant 2)$ is negligible. Therefore, assume that pr$(S = 1) = E(S)$ in the following, except for the first part.

9.5.1. Show that $E(I^t_{j_1 j_2 k_1 k_2})$ is equal to

$$\Phi(-t+A)\Phi(-t-B)+\Phi(-t-A)\Phi(-t+B) \tag{9.57}$$

where $A = A(j_1, k_1, k_2) = (\theta_{j_1 k_2} - \theta_{j_1 k_1})/(\sigma\sqrt{2})$ and $B = Bi(j_2, k_1, k_2) = (\theta_{j_2 k_2} - \theta_{j_2 k_1})/(\sigma\sqrt{2})$.

9.5.2. Show that pr(accept $H_1|H_0$) is a sum of terms of the form of (9.57). Deduce that pr(accept $H_1|H_0^*$) is a sum of terms of the form (9.57) with $A(j_1, k_2, k_1) = A(j_2, k_2, k_1)$.

9.5.3. Show that (a) $(d^2/dt^2)\log\Phi(-t) < 0$ and hence $\log\Phi(-t)$ is concave, (b) $\log\Phi(-t-a)+\log\Phi(-t+a) \leq 2\log\Phi(-t)$, (c) $E(S)$ reaches its maximum over H_0^* when there is no main effect for treatment, and (d) the least favorable value in H_0^* is obtained when there is no main effect for treatment.

9.5.4. Assume that the null hypothesis is H_0: *no cross-over interaction.* Does the least favorable null distribution still correspond to the case when there is no main effect for treatment and no interaction, as in (b) of the previous part?
[Hint: Consider the special case when $m_2 = 2$. For this special case, the Azzalini-Cox test is the same as the range test of Piantadosi and Gail (1993). In this case, the least favorable null value is obtained when the treatment effect in one group is ∞ and is zero in the others (see Piantadosi and Gail (1993)].

[Azzalini and Cox (1984)]

9.6 Let $\boldsymbol{X}_{2\times 1} \sim N(\boldsymbol{\theta}, V)$ where $V = (1, \rho|\rho, 1), \Theta_0 = \{\boldsymbol{\theta} \in \mathbb{R}^2 : \boldsymbol{\theta} \geqslant \mathbf{0}$ or $\boldsymbol{\theta} \leq \mathbf{0}\}$. Consider the *LRT* of $H_0 : \boldsymbol{\theta} \in \Theta_0$ against $H_1 : \boldsymbol{\theta} \notin \Theta_0$ based on a single observation of $\boldsymbol{X}$.

9.6.1. Let $\boldsymbol{x} = (x_1, x_2)^T \in \mathbb{R}^2$. Show that the point, $\boldsymbol{a}$, on the x_1-axis for which $(\boldsymbol{x}-\boldsymbol{a})^T V^{-1}(\boldsymbol{x}-\boldsymbol{a})^T$ is a minimum has coordinates $(x_1 - \rho x_2, 0)^T$. Similarly, the point, $\boldsymbol{a}$, on the x_2-axis for which $(\boldsymbol{x}-\boldsymbol{a})^T V^{-1}(\boldsymbol{x}-\boldsymbol{a})$ is a minimum has coordinates $(0, x_2 - \rho x_1)^T$.

9.6.2. Show that the distance, $\|\boldsymbol{x} - x_1 mbox{-}axis\|_V$, between $\boldsymbol{x}$ and the x_1-axis is $|x_2|$; similarly, the distance, $\|\boldsymbol{x} - x_2\text{-axis}\|_V$, between $\boldsymbol{x}$ and the x_2-axis is $|x_1|$. $(\|\boldsymbol{x}\|^2 = \boldsymbol{x}^T V^{-1}\boldsymbol{x})$.

9.6.3. Show that the *LRT* statistic for testing $H_0 : \boldsymbol{\theta} \in \Theta_0$ against $H_1 : \boldsymbol{\theta} \notin \Theta_0$ is

$$\min\{\sum X_i^2 I(X_i > 0),\ \sum X_i^2 I(X_i < 0)\}$$

where $I(\cdot)$ is the indicator function.

9.7 Three different types of adhesives were compared for three different types of assembly; the adhesives are "Treatments" and the type of assembly is "Group." The response variable is a bonding strength (in unspecified units). There are five observations in each of the nine cells. The cell means are given in Table 9.9. The within-cell error mean square is 17.485 with 36 degrees of freedom. Assume that cell means are normally distributed with variance (17.485/5); although this is only an estimate of error variance, we can obtain a large sample approximation by assuming that they are equal. Test against cross-over interaction using the Azzalini-Cox test. [see Azzalini and Cox (1984); the data are from Johnson and Leone (1964, p 94].

Table 9.9 Mean bond strength

Assembly type	Adhesive 1	2	3
1	17.2	20.6	22.4
2	180	18.8	21.8
3	16.8	24.6	16.4

Reprinted from: *Statistics and Experimental Design*, Vol. 2, N. L. Johnson and F. Leone, Copyright (1964), with permission John Wiley and Sons.

Directed test

9.8 Consider the statistic T_D in (9.55) for the same bivariate example as in Fig. 9.13. Let $\boldsymbol{x}$ be the sample value of $\boldsymbol{X}$ and let $\boldsymbol{y}$ be an arbitrary point in $\mathbb{R}^2$.

1. Suppose that $\boldsymbol{x}$ is represented by A in Figure 9.13. Show that as $\boldsymbol{y}$ moves to the right horizontally along AB, $T(\boldsymbol{y})$ decreases first, reaches a minimum at B and then increases.

2. Suppose that $\boldsymbol{x}$ is represented by D in the same figure. Show that as $\boldsymbol{y}$ moves vertically up, $T(\boldsymbol{y})$ decreases first, reaches a minimum at C and then increases.

3. Show that $T(\boldsymbol{y})$ has the same value at every point $\boldsymbol{y}$ on the arc BC.

4. Deduce that $T_D(\boldsymbol{y})$ has the same value at every point on $ABCD$.

5. Deduce that if the sample value $\boldsymbol{x}$ is any point on $ABCD$ then the p-value is $\{1 - \text{pr}(\mathcal{A} \mid H_0)\}$ where $\mathcal{A}$ is the lower-left region bounded by $ABCD$.

6. Verify that T_D is COM[$\mathbb{R}^{+2}$]. [Hint: Let $\boldsymbol{x} \leq \boldsymbol{y}$; draw two curves similar to $ABCD$, one through $\boldsymbol{x}$ and one through $\boldsymbol{y}$. verify that the curve through $\boldsymbol{y}$ is to upper-right of that through $\boldsymbol{x}$.]

Bibliography

Aaberge, R. (1999). UMP unbiased tests for multiparameter testing problems with restricted alternatives. *Metrika, 50*, 179–193.

Abel, U. (1986). A nonparametric test against ordered alternatives for data defined by intervals. *Statistica Neerlandica, 40*, 87–91.

Abelson, R. B. and Tukey, J. W. (1963). Efficient utilization of non-numerical information in quantitative analysis: General theory and the case of simple order. *Annals of Mathematical Statistics, 34*, 1347–1369.

Adegboye, O. S. and Gupta, A. K. (1989). On testing against restricted alternatives for the variances of Gaussian models. *The Australian Journal of Statistics, 31*, 409–415.

Agresti, A. (1990). *Categorical Data Analysis.* John Wiley and Sons, New York.

Agresti, A. (1999). Modelling ordered categorical data: Recent advances and future challenges. *Statistics in Medicine, 18*, 2191–2207.

Agresti, A. (2002). *Categorical Data Analysis.* John Wiley and Sons, New York.

Agresti, A. and Chuang, C. (1986). Bayesian and maximum likelihood approaches to order restricted inference for models from categorical data. In Dykstra et al., editor, *Advances in order restricted statistical inference*, pages 6–27. Springer-Verlag, New York.

Agresti, A. and Coull, B. A. (1996). Order-restricted tests for stratified comparisons of binomial proportions. *Biometrics, 52*, 1103–1111.

Acknowledgment: This bibliography was prepared in part using the Current Index to Statistics (2002), by permission of the Current Index to Statistics Management Committee.

Note: This bibliography contains mainly the publications during the past sixteen years; for references to earlier ones, see Robertson et al. (1988) and Barlow et al. (1972).

Agresti, A. and Coull, B. A. (1998a). An empirical comparison of inference using order-restricted and linear logit models for a binary response. *Communications in Statistics, Part B – Simulation and Computation*, *27*, 147–166.

Agresti, A. and Coull, B. A. (1998b). Order-restricted inference for monotone trend alternatives in contingency tables. *Computational Statistics and Data Analysis*, *28*, 139–155.

Agresti, A. and Coull, B. A. (2002). The analysis of contingency tables under inequality constraints. *Journal of Statistical Planning and Inference*, *107*(1-2), 45–73.

Agresti, A., Wackerly, D., and Boyett, J. M. (1979). Exact conditional tests for cross-classifications: Approximation of attained significance levels. *Psychometrika*, *44*, 75–84.

Agresti, A., Chuang, C., and Kezouh, A. (1987). Order-restricted score parameters in association models for contingency tables. *Journal of the American Statistical Association*, *82*, 619–623.

Ahmad, I. A. (1994). A class of statistics useful in testing increasing failure rate average and new better than used life distributions. *Journal of Statistical Planning and Inference*, *41*, 141–149.

Ahmad, I. A. (2000). Testing exponentiality against positive ageing using Kernel methods. *Sankhya, Series A, Indian Journal of Statistics*, *62*(2), 244–257.

Ahmad, I. A. (2001). Testing stochastic ordering in tails of distribution. *Journal of Nonparametric Statistics*, *13*(6), 775–790.

Ahmad, I. A. and Kochar, S. C. (1990). Testing whether F is more IFR than G. *Metrika*, *37*, 45–58.

Aitchison, J. and Silvey, S. D. (1958). Maximum likelihood estimation of parameters subject to restraints. *Annals of Mathematical Statistics*, *29*, 813–828.

Akkerboom, J. C. (1990). *Testing Problems with Linear or Angular Inequality Constraints*. Springer-Verlag, New York.

Al-Rawwash, H. M. (1990). On dominating the χ^2-test in the case of multivariate normal distribution with one-sided alternatives. *Sankhya, Series B, Indian Journal of Statistics*, *52*, 174–182.

Albert, J. H. (1994). A Bayesian approach to estimation of GPAs of University of Iowa freshmen under order restrictions. *Journal of Educational Statistics*, *19*, 1–21.

Alvo, M. and Cabilio, P. (1995). Testing ordered alternatives in the presence of incomplete data. *Journal of the American Statistical Association*, *90*, 1015–1024.

Aly, E.-E. A. A., Kochar, S. C., and McKeague, I. W. (1994). Some tests for comparing cumulative incidence functions and cause-specific hazard rates. *Journal of the American Statistical Association*, *89*, 994–999.

Anderson, T. W. (1955). The integral of a symmetric unimodal function over a symmetric convex set and some probaility inequalities. *Proceedings of the American Mathematical Society*, *6*, 170–176.

Anderson, T. W. (1984). *An introduction to multivariate statistical analysis*. John Wiley and Sons, New York.

Andrews, D. W. K. (1998). Hypothesis testing with a restricted parameter space. *Journal of Econometrics*, *84*, 155–199.

Andrews, D. W. K. (1999). Estimation when a parameter is on a boundary. *Econometrica, 67*, 1341–1383.

Andrews, D. W. K. (2000). Inconsistency of the bootstrap when a parameter is on the boundary of the parameter space. *Econometrica, 68*(2), 399–405.

Andrews, D. W. K. (2001). Testing when a parameter is on the boundary of the maintained hypothesis. *Econometrica, 69*(3), 683–734.

Anevski, D. and O., H. (2002). Monotone regression and density function estimation at a point of discontinuity. *Journal of Nonparametric Statistics, 14*, 279–294.

Anraku, K. (1989). Approximately somewhere most powerful tests for marginal homogeneity of a square table under restricted alternatives. *Behaviormetrika, Journal of the Behaviormetric Society of Japan, 25*, 1–13.

Anraku, K. (1994). Estimation of odds ratios under order restrictions. *Communications in Statistics, Part A – Theory and Methods, 23*, 3257–3272.

Anraku, K. (1999). An information criterion for parameters under a simple order restriction. *Biometrika, 86*, 141–152.

Anraku, K., Nishi, A., and Yanagawa, T. (1988). Tests for the marginal probabilities in the two-way contingency table under restricted alternatives. *Annals of the Institute of Statistical Mathematics, 40*, 149–163.

Aras, G., Jammalamadaka, S. R., and Kafai, M. (1989). A rank test for stochastically ordered alternatives. *Communications in Statistics, Part A – Theory and Methods, 18*, 2263–2277.

Arcones, M. A. and Samaniego, F. J. (2000). On the asymptotic distribution theory of a class of consistent estimators of a distribution satisfying a uniform stochastic ordering constraint. *Annals of Statistics, 28*(1), 116–150.

Arcones, M. A., Kvam, P. H., and Samaniego, F. J. (2002). Nonparametric estimation of a distribution subject to a stochastic precedence constraint. *Journal of the American Statistical Association, 97*(457), 170–182.

Arjas, E. and Gasbarra, D. (1996). Bayesian inference of survival probabilities, under stochastic ordering constraints. *Journal of the American Statistical Association, 91*, 1101–1109.

Arnold, S. F. (1981). *The theory of linear models and multivariate analysis*. John Wiley and Sons, New York.

Arnold, S. F. (1988). Union-intersection principle. In S. Kotz and N. L. Johnson, editors, *Encyclopedia of Statistical Sciences (Vol. 1)*, pages 417–420. John Wiley and Sons, New York.

Avriel, M. (1976). *Nonlinear Programming*. Prentice-Hall, Englewood Cliffs, N.J.

Bacchetti, P. (1989). Additive isotonic models. *Journal of the American Statistical Association, 84*, 289–294.

Bagai, I. and Rao, P. B. L. S. (1995). Kernel-type density and failure rate estimation for associated sequences. *Annals of the Institute of Statistical Mathematics, 47*, 253–266.

Bagai, I., Deshpandé, J. V., and Kochar, S. C. (1989). Distribution free tests for stochastic ordering in the competing risks model. *Biometrika, 76*, 775–781.

Bagui, S. C., Bhaumik, D. K., and Parnes, M. (1996). One-sided tolerance limits for unbalanced m-way random-effects ANOVA models. *Journal of Applied Statistical Science, 3*, 135–147.

Bandyopadhyay, D. and Basu, A. P. (1991). A class of tests for bivariate exponentiality against bivariate increasing failure rate alternatives. *Journal of Statistical Planning and Inference*, *29*, 337–349.

Bandyopadhyay, U. and Chattopadhyay, G. (1992). Inverse sampling for bivariate non-parametric two-sample problems against restricted alternatives. *Calcutta Statistical Association Bulletin*, *42*, 221–236.

Banerjee, M. and Wellner, J. A. (2001). Likelihood ratio tests for monotone functions. *Annals of Statistics*, *29*(6), 1699–1731.

Bar-Lev, S. K. and Fygenson, M. (2000). Tests for exponentiality versus increasing failure rate alternatives. *Far East Journal of Theoretical Statistics*, *4*(1), 73–86.

Baranchik, A. J. (1970). A family of minimax estimators of the mean of a multivariate normal distribution. *Annals of Mathematical Statistics*, *41*, 642–645.

Barker, L., Rolka, H., Rolka, D., and Brown, C. (2001). Equivalence testing for binomial random variables: Which test to use? *The American Statistician*, *55*(4), 279–287.

Barlow, R. E., Bartholomew, D. J., Bremner, J. M., and Brunk, H. D. (1972). *Statistical Inference Under Order Restrictions*. John Wiley and Sons, New York.

Barnhart, H. X. and Sampson, A. R. (1995). Multiple population models for multivariate random length data – with applications in clinical trials. *Biometrics*, *51*, 195–204.

Barrett, G. F. and Donald, S. G. (2003). Consistent tests for stochastic dominance. *Econometrica*, *71(1)*, 71–104.

Bartolucci, F. and Forcina, A. (2000). A likelihood ratio test for MTP_2 within binary variables. *Annals of Statistics*, *28*(4), 1206–1218.

Bartolucci, F., Forcina, A., and Dardanoni, V. (2001). Positive quadrant dependence and marginal modeling in two-way tables with ordered margins. *Journal of the American Statistical Association*, *96*(456), 1497–1505.

Basawa, I. V. (1985). Neyman-Lecam testsbased on estimating functions. In L. M. LeCam and R. A. Olshen, editors, *Proceedings of the Berkeley Conference in Honour of Jerzey Neyman and Jack Keifer*, volume II, pages 811–825. Institute of Mathematical Statistics, Hayward, CA.

Basu, A. P. and Habibullah, M. (1987). A test for bivariate exponentiality against BIFRA alternative. *Calcutta Statistical Association Bulletin*, *36*, 79–84.

Bauer, P. (1997). A note on multiple testing procedures in dose finding. *Biometrics*, *53*, 1125–1128.

Bauer, P. and Budde, M. (1994). Multiple testing for detecting efficient dose steps. *Biometrical Journal*, *36*, 3–15.

Bauer, P., Hackl, P., Hommel, G., and Sonnemann, E. (1986). Multiple testing of pairs of one-sided hypotheses. *Metrika*, *33*, 121–127.

Bazaraa, M. S., Sherali, H. D., and Shetty, C. M. (1993). *Nonlinear Programming : Theory and Algorithms*. Springer-Verlag, Berlin.

Beg, A. B. M. R. A., Silvapulle, M. J., and Silvapulle, P. (2001). Tests against inequality constraints when some nuisance parameters are present only under the alternative: Test of ARCH in ARCH-M models. *Journal of Business and Economic Statistics*, *19*(2), 245–253.

Beier, F. and Büning, H. (1997). An adaptive test against ordered alternatives. *Computational Statistics and Data Analysis*, *25*, 441–452.

Bellout, D. (1989). Order restricted estimation of distributions with censored data. *Journal of Statistical Planning and Inference*, *21*, 27–39.

Belzunce, F., Candel, J., and Ruiz, J. M. (1998). Testing the stochastic order and the IFR, DFR, NBU, NWU ageing classes. *IEEE Transactions on Reliability*, *47*, 285–296.

Benjamini, Y. and Hochberg, Y. (1995). Controlling the false discovery rate: A practical and powerful approach to multiple testing. *Journal of the Royal Statistical Society, Series B, Methodological*, *57*, 289–300.

Benjamini, Y. and Hochberg, Y. (1997). Multiple hypothesis testing with weights. *Scandinavian Journal of Statistics*, *24*, 407–418.

Benjamini, Y. and Hochberg, Y. (1999). More on simes' test. *Department of Statistics and OR, Tel Aviv University*.

Benjamini, Y. and Hochberg, Y. (2000). Adaptive control of false discovery rate in multiple hypothesis testing. *Jour. Behav. Educ. Statist.*, *25*, 60–83.

Benjamini, Y. and Liu, W. (1999). A step-down multiple hypotheses testing procedure that controls the false discovery rate under independence. *Journal of Statistical Planning and Inference*, *82*, 163–170.

Benjamini, Y. and Yekutieli, D. (2001). The control of the false discovery rate in multiple testing under dependency. *Annals of Statistics*, *29*(4), 1165–1188.

Beran, R. and Dümbgen, L. (1998). Modulation of estimators and confidence sets. *Annals of Statistics*, *26*(5), 1826–1856.

Berger, J. O. (1985). *Statistical Decision Theory and Bayesian Analysis*. Springer-Verlag, New York.

Berger, R. L. (1982). Multiparameter hypothesis testing and acceptance sampling. *Technometrics*, *24*, 295–300.

Berger, R. L. (1984a). Testing for the same ordering in several groups of means. In *Design of experiments: ranking and selection: essays in honor of Robert E. Bechhofer*, pages 241–249.

Berger, R. L. (1984b). Testing whether one regression function is larger than another. *Communications in Statistics, Part A – Theory and Methods*, *13*, 1793–1810.

Berger, R. L. (1988). A nonparametric, intersection-union test for stochastic order. In *Statistical decision theory and related topics IV, Volume 2*, pages 253–264. Springer-Verlag, New York.

Berger, R. L. (1989). Uniformly more powerful tests for hypotheses concerning linear inequalities and normal means. *Journal of the American Statistical Association*, *84*, 192–199.

Berger, R. L. (1997). Likelihood ratio tests and intersection union tests. In S. Panchapakesan and N. Balakrishnan, editors, *Advances in Statistical Decision Theory and Applications*, pages 225–237. Birkhäuser, Boston.

Berger, R. L. and Boos, D. D. (1994). P values maximized over a confidence set for the nuisance parameter. *Journal of the American Statistical Association*, *89*, 1012–1016.

Berger, R. L. and Hsu, J. C. (1996). Bioequivalence trials, intersection-union tests and equivalence confidence sets (Disc: p303-319). *Statistical Science*, *11*, 283–302.

Berger, R. L. and Sinclair, D. F. (1984). Testing hypotheses concerning unions of linear subspaces. *Journal of the American Statistical Association*, *79*, 158–163.

Berger, V. and Sackrowitz, H. (1997). Improving tests for superior treatment in contingency tables. *Journal of the American Statistical Association, 92*, 700–705.

Berger, V. W., Permutt, T., and Ivanova, A. (1998). Convex hull test for ordered categorical data. *Biometrics, 54*, 1541–1550.

Berk, R. and Marcus, R. (1996). Dual cones, dual norms, and simultaneous inference for partially ordered means. *Journal of the American Statistical Association, 91*, 318–328.

Bhattacharjee, M. C. and Sen, P. K. (1995). *Analysis of censored data*, volume 27 of *IMS Lecture Monograph Series*, chapter On Kolmogorov-Smirnov type tests for NB(W)UE alternates under censoring schemes, pages 25–38.

Bhattacharjee, M. C. and Sen, P. K. (1997). TTT-transform characterization of the NBRUE property and tests for exponentiality. In *Frontiers in Reliability, (Ed: A. P. Basu et al.)*, pages 71–82. World Science Press, London.

Bhattacharjee, M. C. and Sen, P. K. (1998). TTT-transformation characterization of the NBRUE property and tests for exponentiality. In A. P. Basu et al., editor, *Frontiers in Reliability*, pages 71–81. World Science Publications, London.

Bhattacharya, B. (1995). Restricted tests for and against the increasing failure rate ordering on multinomial parameters. *Statistics and Probability Letters, 25*, 309–316.

Bhattacharya, B. (1996). Tests of bivariate symmetry with a one-sided alternative in the analysis of variance. *Biometrical Journal, 38*, 791–808.

Bhattacharya, B. (1997a). On testing diagonal homogeneity with a one-sided alternative in the analysis of variance. *Sankhya, Series A, Indian Journal of Statistics, 59*, 198–214.

Bhattacharya, B. (1997b). On tests of symmetry against one-sided alternatives. *Annals of the Institute of Statistical Mathematics, 49*, 237–254.

Bhattacharya, B. (1997c). Testing multinomial parameters under order restrictions. *Communications in Statistics, Part A – Theory and Methods, 26*, 1839–1865.

Bhattacharya, B. (1998). Testing conditional symmetry against one-sided alternatives in square contingency tables. *Metrika, 47*, 71–84.

Bhattacharya, B. and Nandram, B. (1996). Bayesian inference for multinomial populations under stochastic ordering. *Journal of Statistical Computation and Simulation, 54*, 145–163.

Bhattacharya, G. K. and Johnson, R. A. (1970). Aayer rank test for ordred bivariate altrnative. *Annals of Mathematical Statistics, 41*, 1296–1310.

Bhaumik, D. K. and Kulkarni, P. M. (1991). One-sided tolerance limits for unbalanced one-way ANOVA random effects model. *Communications in Statistics, Part A – Theory and Methods, 20*, 1665–1675.

Bickel, P. J. and Doksum, K. (1977). *Mathematical Statistics*. Holden-Day, San Francisco, CA.

Birgé, L. (1987a). Estimating a density under order restrictions: Nonasymptotic minimax risk. *Annals of Statistics, 15*, 995–1012.

Birgé, L. (1987b). On the risk of histograms for estimating decreasing densities. *Annals of Statistics, 15*, 1013–1022.

Birgé, L. (1997). Estimation of unimodal densities without smoothness assumptions. *Annals of Statistics, 25*, 970–981.

Black, S. and Mansouri, H. (1995). On exact distributions of rank tests for ordered alternatives in block designs. *Computational Statistics and Data Analysis*, *20*, 265–274.

Blair, R. C., Troendle, J. F., and Beck, R. W. (1996). Control of familywise errors in multiple endpoint assessments via stepwise permutation tests. *Statistics in Medicine*, *15*, 1107–1121.

Bloch, D. A., Lai, T. L., and Tubert-Bitter, P. (2001). One-sided tests in clinical trials with multiple endpoints. *Biometrics*, *57*(4), 1039–1047.

Block, H. W., Qian, S., and Sampson, A. R. (1996). Isotonic regression on permutations. In *Distributions with fixed marginals and related topics*, pages 45–64.

Bofinger, E. (1992). Expanded confidence intervals, one-sided tests, and equivalence testing. *Journal of Biopharmaceutical Statistics*, *2*, 181–188.

Bohrer, R. and Chow, W. (1978). Weights for one-sided multivariate inference. *Applied Statistics*, *27*, 100–104.

Boyd, M. N. and Sen, P. K. (1983). Union-intersection rank tests for ordered alternatives in some simple linear models. *Communications in Statistics, Part A – Theory and Methods*, *12*, 1737–1753.

Boyd, M. N. and Sen, P. K. (1984). Union-intersection rank tests for ordered alternatives in a complete block design. *Communications in Statistics, Part A – Theory and Methods*, *13*, 285–303.

Boyd, M. N. and Sen, P. K. (1986). Union-intersection rank tests for ordered alternatives in ANOCOVA. *Journal of the American Statistical Association*, *81*, 526–532.

Boyer, J. E. J. (1990). Ordered alternatives: A means of improving power. In *Proceedings of the 1990 Kansas State University Conference on Applied Statistics in Agriculture*, pages 196–205.

Boyles, R. A., Marshall, A. W., and Proschan, F. (1985). Inconsistency of the maximum likelihood estimator of a distribution having increasing failure rate average. *Annals of Statistics*, *13*, 413–417.

Bregenzer, T. and Lehmacher, W. (1998). Directional tests for the analysis of clinical trials with multiple endpoints allowing for incomplete data. *Biometrical Journal*, *40*, 911–928.

Bremner, J. M. (1993). A new approach to subset selection for normal means. *Journal of Statistical Computation and Simulation*, *44*, 187–208.

Bretz, F. and Hothorn, L. A. (2000). A powerful alternative to Williams' test with application to toxicological dose-response relationships of normally distributed data. *Environmental and Ecological Statistics*, *7*(2), 135–154.

Bricker, D. L., Kortanek, K. O., and Xu, L. (1997). Maximum likelihood estimates with order restrictions on probabilities and odds ratios: A geometric programming approach. *Journal of Applied Mathematics and Decision Sciences*, *1*, 53–65.

Bristol, D. R. (1992). One-sided multiple comparisons of response rates with a control. In F. M. Hoppe, editor, *Multiple Comparisons, Selection, and Applications in Biometry. A Festschrift in Honor of Charles W. Dunnett*, pages 77–96. Marcel Dekker, New York.

Bristol, D. R. (1993). Probabilities and sample sizes for the two one-sided tests procedure. *Communications in Statistics, Part A – Theory and Methods*, *22*, 1953–1961.

Bryant, J. and Day, R. (1995). Incorporating toxicity considerations into the design of two-stage phase II clinical trials. *Biometrics*, *51*, 1372–1383.

Budde, M. and Bauer, P. (1989). Multiple test procedures in clinical dose finding studies. *Journal of the American Statistical Association, 84*, 792–796.

Büning, H. (1999). Adaptive Jonckheere-type tests for ordered alternatives. *Journal of Applied Statistics, 26*, 541–551.

Büning, H. and Kössler, W. (1996). Robustness and efficiency of some tests for ordered alternatives in the c-sample location problem. *Journal of Statistical Computation and Simulation, 55*, 337–352.

Büning, H. and Kössler, W. (1999). The asymptotic power of Jonckheere-type tests for ordered alternatives. *The Australian and New Zealand Journal of Statistics, 41*, 67–77.

Calvin, J. A. (1994). One-sided test of a covariance matrix with a known null value. *Communications in Statistics, Part A – Theory and Methods, 23*, 3121–3140.

Calvin, J. A. and Dykstra, R. L. (1991). Maximum likelihood estimation of a set of covariance matrices under Löwner order restrictions with applications to balanced multivariate variance components models. *Annals of Statistics, 19*, 850–869.

Calvin, J. A. and Dykstra, R. L. (1995). REML estimation of covariance matrices with restricted parameter spaces. *Journal of the American Statistical Association, 90*, 321–329.

Capizzi, T., Survill, T. T., Heyse, J. F., and Malani, H. (1992). An empirical and simulated comparison of some tests for detecting progressiveness of response with increasing doses of a compound. *Biometrical Journal, 34*, 275–289.

Cardoso-Neto, J. and Paula, G. A. (2001). Wald one-sided test using generalized estimating equation. *Computational Data Analysis, 36*, 475–495.

Carolan, C. and Dykstra, R. (1999). Asymptotic behavior of the Grenander estimator at density flat regions. *The Canadian Journal of Statistics, 27*, 557–566.

Carolan, C. and Dykstra, R. (2001). Marginal densities of the least concave majorant of Brownian motion. *Annals of Statistics, 29*(6), 1732–1750.

Carriere, K. C. and Kochar, S. C. (2000). Comparing sub-survival functions in a competing risks model. *Lifetime Data Analysis, 6*(1), 85–97.

Casella, G. and Berger, R. L. (2002). *Statistical Inference*. Thomson Learning, Pacific Grove, CA.

Castillo, E., Hadi, A. S., Lacruz, B., and Sarabia, J. M. (2002). Constrained mixture distributions. *Metrika, 55*(3), 247–269.

Ceesay, P., Sarkar, S. K., and Snapinn, S. (1997). Assessing the superiority of a combination drug from a Bayesian perspective. In *ASA Proceedings of the Biopharmaceutical Section*, pages 263–271. American Statistical Association, Alexandria, VA.

Chacko, V. J. (1963). Testing homogeneity against ordered alternatives. *Annals of Mathematical Statistics, 34*, 945–956.

Chacko, V. J. (1966). Modified chi-square test for ordered alternatives. *Sankhya, Series B, Indian Journal of Statistics, 28*, 185–190.

Chakraborti, S. (1990). A one-sided test of homogeneity against simple tree alternative for right-censored data. *Communications in Statistics, Part B – Simulation and Computation, 19*, 879–889.

Chakraborti, S. and Desu, M. M. (1988). A class of distribution-free tests for testing homogeneity against ordered alternatives. *Statistics and Probability Letters, 6*, 251–256.

Chakraborti, S. and Gibbons, J. D. (1992). One-sided nonparametric comparison of treatments with a standard for unequal sample sizes. *Journal of Experimental Education*, *60*, 235–242.

Chakraborti, S. and Hettmansperger, T. P. (1996). Multi-sample inference for the simple-tree alternative based on one-sample confidence intervals. *Communications in Statistics, Part A – Theory and Methods*, *25*, 2819–2837.

Chakraborti, S. and Sen, P. K. (1992). Order restricted quantile tests under unequal right-censorship. *Sankhya, Series B, Indian Journal of Statistics*, *54*, 150–164.

Chang, M. N. (1996). On the asymptotic distribution of an isotonic window estimator for the generalized failure rate function. *Communications in Statistics, Part A – Theory and Methods*, *25*, 2239–2249.

Chang, M. N. and Chung, D. (1998). Isotonic window estimators of the baseline hazard function in cox's regression model under order restriction. *Scandinavian Journal of Statististics*, *25*, 151–161.

Chang, Y. T. (1981). Stein-type estimators for parameters restricted by linear inequalities. *Keito Science Technical Report*, *34*, 83–95.

Chang, Y. T. (1982). Stein-type estimators for parameters in truncated spaces. *Keito Science Technical Report*, *35*, 185–193.

Chant, D. (1974). On asymptotic tests of complete hypotheses in nonstandard conditions. *Biometrika*, *61*, 291–298.

Charnes, A., Cooper, W. W., and Tyssedal, J. (1983). Khincin-Kullback-Leibler estimation with inequality constraints. *Mathematische Operationsforschung und Statistik, Series Operations Research*, *14*, 377–380.

Chatterjee, S. K. (1962). Sequential inference procedures of Stein's type for a class of multivariate regression problems. *Annals of Mathematical Statistics*, *33*, 1039–1064.

Chatterjee, S. K. (1984). Restricted alternatives. In P.R. Krishnaiah and P.K. Sen, editor, *Handbook of Statistics (Vol. 4): Nonparametric Methods*, pages 327–345. North-Holland, New York.

Chatterjee, S. K. and Sen, P. K. (1964). Nonparametric tests for the bivariate two-sample location problem. *Calcatta Statistical Association Bulletin*, *13*, 18–58.

Chaubey, Y. and Sen, P. K. (2002a). Smooth estimation of multivariate survival and density function. *Journal of Statistical Planning and Inference*, *102*, 349–358.

Chaubey, Y. and Sen, P. K. (2003). Tail-behaviour of survival functions. *Journal of Statistical Planning and Inference*, *90*, 223–232.

Chaubey, Y. P. and Sen, P. K. (1996). On smooth estimation of survival and density functions. *Statistics and Decisions*, *14*, 1–22.

Chaubey, Y. P. and Sen, P. K. (1997). On smooth estimation of hazard and cumulative hazard functions. In S. P. Mukherjee et al., editor, *Frontiers in Probability and Statistics*, pages 92–100. Narosa, New Delhi.

Chaubey, Y. P. and Sen, P. K. (1999). On smooth estimation of mean residual life. *Journal of Statistical Planning and Inference*, *75*, 223–236.

Chaubey, Y. P. and Sen, P. K. (2002b). Smooth isotonic estimation of density, hazard and MRL functions. *Calcutta Statistical Association Bulletin*, *52*, 99–116.

Chaudhuri, S. and Perlman, M. D. (2004). The role of reversals in order restricted inference. *Canadian Journal of Statistics (in press).*

Chen, S.-Y. (2000). One-sided range test for testing against an ordered alternative under heteroscedasticity. *Communications in Statistics, Part B – Simulation and Computation, 29*(4), 1255–1272.

Chen, S.-Y. and Chen, H. J. (1999). A single-stage procedure for testing homogeneity of means against ordered alternatives under unequal variances. In *ASA Proceedings of the Statistical Computing Section*, pages 244–249. American Statistical Association, Alexandria, VA.

Chen, T. T. and Simon, R. M. (1994). Extension of one-sided test to multiple treatment trials. *Controlled Clinical Trials, 15*, 124–134.

Chen, Y. I. (1991). Notes on the Mack-Wolfe and Chen-Wolfe tests for umbrella alternatives. *Biometrical Journal, 33*, 281–290.

Chen, Y.-I. (1993a). Nonparametric comparisons of umbrella pattern treatment effects with a control in a one-way layout. *Communications in Statistics, Part B – Simulation and Computation, 22*, 749–764.

Chen, Y. I. (1993b). On the comparison of umbrella pattern treatment means with a control mean. *Biometrical Journal, 35*, 689–700.

Chen, Y. I. and Wolfe, D. A. (1990a). Modifications of the Mack-Wolfe umbrella tests for a generalized Behrens-Fisher problem. *The Canadian Journal of Statistics, 18*, 245–253.

Chen, Y. I. and Wolfe, D. A. (1990b). A study of distribution-free tests for umbrella alternatives. *Biometrical Journal, 32*, 47–57.

Chen, Y.-I. and Wolfe, D. A. (1993). Nonparametric procedures for comparing umbrella pattern treatment effects with a control in a one-way layout. *Biometrics, 49*, 455–465.

Chen, Y.-I. and Wolfe, D. A. (2000). Umbrella tests for right-censored survival data. *Statistica Sinica, 10*(2), 595–612.

Chen, Y.-J., Hewett, J., Wang, Y., and Johnson, J. (1998). A rank test for equality of two multivariate populations vs a particular ordered alternative. *Computational Statistics and Data Analysis, 29*, 129–144.

Cheng, M.-Y. and Hall, P. (1998). Calibrating the excess mass and dip tests of modality. *Journal of the Royal Statistical Society, Series B, 60*, 579–89.

Cheng, M.-Y., Gasser, T., and Hall, P. (1999). Nonparametric density estimation under unimodality and monotonicity constraints. *Journal of Computational and Graphical Statistics, 8*, 1–21.

Cheng, R. C. H. and Iles, T. C. (1988). One-sided confidence bands for cumulative distribution functions. *Technometrics, 30*, 155–159.

Chepoi, V., Cogneau, D., and Fichet, B. (1997). Polynomial algorithms for isotonic regression. In Y. Dodge, editor, *L_1-Statistical Procedures and Related Topics*, pages 147–160. Institute of Mathematical Statistics, Hayward, CA.

Chernoff, H. (1954). On the distribution of the likelihood ratio. *Annals of Mathematical Statistics, 25*, 573–578.

Chernoff, H. and Lander, E. (1995). Asymptotic distribution of the likelihood ratio test that a mixture of two binomials is a single binomial. *Journal of Statistical Planning and Inference, 43*, 19–40.

Cheung, S. H. and Zhu, H. T. (1998). Simultaneous one-sided pairwise comparisons in a two-way design. *Biometrical Journal*, *40*, 613–625.

Chinchilli, V. M. and Sen, P. K. (1981a). Multivariate linear rank statistics and the union-intersection principle for hypothesis testing under restricted alternatives. *Sankhya, Series B, Indian Journal of Statistics*, *43*, 135–151.

Chinchilli, V. M. and Sen, P. K. (1981b). Multivariate linear rank statistics and the union-intersection principle for the orthant restriction problems. *Sankhya, Series B, Indian Journal of Statistics*, *43*, 152–171.

Chinchilli, V. M. and Sen, P. K. (1982). Multivariate linear rank statistics for profile analysis. *Journal of Multivariate Analysis*, *12*, 219–229.

Chuang-Stein, C. and Agresti, A. (1997). A review of tests for detecting a monotone dose-response relationship with ordinal response data. *Statistics in Medicine*, *16*, 2599–2618.

Chung, D. and Chang, M. N. (1994). An isotonic estimator of the baseline hazard function in Cox's regression model under order restriction. *Statistics and Probability Letters*, *21*, 223–228.

Ciminera, J. L., Heyse, J. F., Nguyen, H. H., and Tukey, J. W. (1993a). Evaluation of multicentre clinical trial data using adaptations of the Mosteller-Tukey procedure. *Statistics in Medicine*, *12*, 1047–1061.

Ciminera, J. L., Heyse, J. F., Nguyen, H. H., and Tukey, J. W. (1993b). Tests for qualitative treatment-by-centre interaction using a "pushback" procedure. *Statistics in Medicine*, *12*, 1033–1045.

Cohen, A. and Fygenson, M. (1995). Testing homogeneity of uniform scale distributions against two-sided and one-sided alternatives. *Journal of the American Statistical Association*, *90*, 1062–1067.

Cohen, A. and Sackrowitz, H. B. (1990). Unbiased tests for some one-sided testing problems. *The Canadian Journal of Statistics*, *18*, 337–346.

Cohen, A. and Sackrowitz, H. B. (1992a). An evaluation of some tests of trend in contingency tables. *Journal of the American Statistical Association*, *87*, 470–475.

Cohen, A. and Sackrowitz, H. B. (1992b). Improved tests for comparing treatments against a control and other one-sided problems. *Journal of the American Statistical Association*, *87*, 1137–1144.

Cohen, A. and Sackrowitz, H. B. (1992c). Some remarks on a notion of positive dependence, association, and unbiased testing. In *Stochastic Inequalities*, pages 33–37. Institute of Mathematical Statistics, Hayward, CA.

Cohen, A. and Sackrowitz, H. B. (1993a). Evaluating tests for increasing intensity of a Poisson process. *Technometrics*, *35*, 446–448.

Cohen, A. and Sackrowitz, H. B. (1993b). Inadmissibility of Studentized tests for normal order restricted models. *Annals of Statistics*, *21*, 746–752.

Cohen, A. and Sackrowitz, H. B. (1995). Inadmissibility of some tests for order-restricted alternatives. *Statistics and Probability Letters*, *24*, 153–156.

Cohen, A. and Sackrowitz, H. B. (1996a). Cone order association and stochastic cone ordering with applications to order-restricted testing. *Annals of Statistics*, *24*, 2036–2048.

Cohen, A. and Sackrowitz, H. B. (1996b). Lower confidence bounds using pilot samples with an application to auditing. *Journal of the American Statistical Association*, *91*, 338–342.

Cohen, A. and Sackrowitz, H. B. (1996c). Lower confidence regions for restricted parameter spaces. *Mathematical Methods of Statistics*, *5*, 113–123.

Cohen, A. and Sackrowitz, H. B. (1996d). Tests for the umbrella alternative under normality. *Communications in Statistics, Part A – Theory and Methods*, *25*, 2807–2817.

Cohen, A. and Sackrowitz, H. B. (1998). Directional tests for one-sided alternatives in multivariate models. *Annals of Statistics*, *26*, 2321–2338.

Cohen, A. and Sackrowitz, H. B. (2000a). Properties of Bayes testing procedures in order restricted inference. *Statistics and Probability Letters*, *49*(2), 205–209.

Cohen, A. and Sackrowitz, H. B. (2000b). Testing whether treatment is "better" than control with ordered categorical data: Definitions and complete class theorems. *Statistics and Decisions*, *18*(1), 1–25.

Cohen, A. and Sackrowitz, H. B. (2001). Wherefore similar tests? *Statistics and Probability Letters*, *54*(3), 283–290.

Cohen, A. and Sackrowitz, H. B. (2002). Directed likelihood ratio tests for ordered alternatives and inference results for the star-shaped restriction. *Journal of Statistical Planning and Inference*, *103*(1-2), 181–189.

Cohen, A. and Sackrowitz, H. B. (2004). A discussion of some inference issues in order restricted models. *Canadian Journal of Statistics (in press)*.

Cohen, A., Gatsonis, C., and Marden, J. I. (1983a). Hypothesis testing for marginal probabilities in a $2 \times 2 \times 2$ contingency table with conditional independence. *Journal of the American Statistical Association*, *78*, 920–929.

Cohen, A., Gatsonis, C., and Marden, J. I. (1983b). Hypothesis tests and optimality properties in discrete multivariate analysis. In S. Karzin, T. Amemiya, and L. A. Goodman, editors, *Studies in Econometrics, Time Series, and Multivariate Statistics*, pages 379–405. Academic Press, New York.

Cohen, A., Kemperman, J. H. B., and Sackrowitz, H. B. (1993). Unbiased tests for normal order restricted hypotheses. *Journal of Multivariate Analysis*, *46*, 139–153.

Cohen, A., Kemperman, J. H. B., and Sackrowitz, H. B. (1994). Projected tests for order restricted alternatives. *Annals of Statistics*, *22*, 1539–1546.

Cohen, A., Sackrowitz, H. B., and Samuel-Cahn, E. (1995a). Cone order association. *Journal of Multivariate Analysis*, *55*, 320–330.

Cohen, A., Sackrowitz, H. B., and Samuel-Cahn, E. (1995b). Constructing tests for normal order-restricted inference. *Bernoulli*, *1*, 321–333.

Cohen, A., Kemperman, J. H. B., and Sackrowitz, H. B. (2000). Properties of likelihood inference for order restricted models. *Journal of Multivariate Analysis*, *72*(1), 50–77.

Cohen, A., Kemperman, J. H. B., and Sackrowitz, H. (2002). On the bias in estimating genetic length and other quantities in simplex constrained models. *Annals of Statistics*, *30*(1), 202–219.

Cohen, A., Madigan, D., and Sackrowitz, H. B. (2003). Effective directed tests for models with ordered categorical data. *Australian and New Zealand Journal of Statistics*, pages 285–300.

Colombi, R. and Forcina, A. (2000). Modelling discrete data by equality and inequality constraints. *Statistica*, *60*(2), 195–214.

Conaway, M., Pillers, C., Robertson, T., and Sconing, J. (1990). The power of the circular cone test: A noncentral chi-bar-squared distribution. *The Canadian Journal of Statistics, 18*, 63–70.

Conaway, M., Pillers, C., Robertson, T., and Sconing, J. (1991). A circular-cone test for testing homogeneity against a simple tree order. *The Canadian Journal of Statistics, 19*, 283–296.

Conover, W. J. (1967). A k-sample extension of the one-sided two sample smirnov test statistic. *Annals of Mathematical Statistics, 38*, 1726–1730.

Considine, T. J. (1990). Symmetry constraints and variable returns to scale in logit models. *Journal of Business and Economic Statistics, 8*, 347–353.

Cox, D. R. and Hinkley, D. V. (1974). *Theoretical Statistics*. Chapman and Hall, London.

Cuesta, J. A., Dominguez, J. S., and Matràn, C. (1993). On the consistency of the L_p-isotonic regression. *Theory of Probability and its Applications, 37*, 129–132.

Cutler, J. A., Follmann, D., Elliott, P., and Suh, I. (1991). An overview of randomized trials of sodium reduction and blood pressure. *Hypertension, 17*(1), 27–33.

Dardanoni, V. and Forcina, A. (1998). A unified approach to likelihood inference on stochastic orderings in a nonparametric context. *Journal of the American Statistical Association, 93*, 1112–1123.

Das, S. and Sen, P. K. (1994). Restricted canonical correlations. *Linear Algebra and its Applications, 210*, 29–47.

Das, S. and Sen, P. K. (1995). Simultaneous spike-trains and stochastic dependence. *Sankhya, Series B, Indian Journal of Statistics, 57*, 32–47.

Das, S. and Sen, P. K. (1996). Asymptotic distribution of restricted canonical correlations and relevant resampling methods. *Journal of Multivariate Analysis, 56*, 1–19.

Datta, S. (1995). A minimax optimal estimator for continuous monotone densities. *Journal of Statistical Planning and Inference, 46*, 183–193.

Davidson, R. and Duclos, J.-Y. (2000). Statistical inference for stochastic dominance and for the measurement of poverty and inequality. *Econometrica, 68*(6), 1435–1464.

Davis, L. J. (1989). Intersection union tests for strict collapsibility in three-dimensional contingency tables. *Annals of Statistics, 17*, 1693–1708.

Davis, W. W. (1978). Bayesian analysis of the linear model subject to linear inequality constraints. *Journal of the American Statistical Association, 73*, 573–579.

Deaton, L. W. (1980). An empirical Bayes approach to polynomial regression under order restrictions. *Biometrika, 67*, 111–117.

Delecroix, M., Simioni, M., and Thomas-Agnan, C. (1996). Functional estimation under shape constraints. *Nonparametric Statistics, 6*, 69–89.

Demos, A. and Sentana, E. (1998). Testing for GARCH effects: A one-sided approach. *Journal of Econometrics, 86*, 97–127.

Denby, L. and Vardi, Y. (1986). The survival curve with decreasing density. *Technometrics, 28*, 359–367.

DeSarbo, W. S. and Mahajan, V. (1984). Constrained classification: The use of a priori information in cluster analysis. *Psychometrika, 49*, 187–215.

Desarbo, W. S., Hausman, R. E., Lin, S., and Thompson, W. (1982). Constrained canonical correlations. *Psychometrika, 47*, 489–516.

Deshpande, J. V. (1983). A class of tests for exponentiality against increasing failure rate average alternatives. *Biometrika*, *70*, 514–518.

Dhawan, A. K. and Gill, A. N. (1997). Simultaneous one-sided confidence intervals for the ordered pairwise differences of exponential location parameters. *Communications in Statistics, Part A – Theory and Methods*, *26*, 247–262.

Di Castelnuovo, A., Mazzaro, D., Pesarin, F., and Salmaso, L. (2000). Multivariate permutation tests for isotonic inference problems in genetics. *Statistica*, *60*(4), 691–700.

Diaz, M. M. and Salvador, G. B. (1988). The validity of the "Pool-Adjacent-Violator" algorithm. *Statistics and Probability Letters*, *6*, 143–145.

DiCiccio, T. J., Martin, M. A., and Stern, S. E. (2001). Simple and accurate one-sided inference from signed roots of likelihood ratios. *The Canadian Journal of Statistics*, *29*(1), 67–76.

Dietz, E. J. (1989). Multivariate generalizations of Jonckheere's test for ordered alternatives. *Communications in Statistics, Part A – Theory and Methods*, *18*, 3763–3783.

Diewert, W. E. and Wales, T. J. (1987). Flexible functional forms and global curvature conditions. *Econometrica*, *55*, 43–68.

Diggle, P., Morris, S., and Morton-Jones, T. (1999). Case-control isotonic regression for investigation of elevation in risk around a point source. *Statistics in Medicine*, *18*, 1605–1613.

Dilleen, M., Heimann, G., and Hirsch, I. (2003). Nonparametric estimators of a monotonic dose-response curve and bootstrap confidence interval. *Statistics in Medicine*, *22*, 869–882.

Dinh, K. T. and Nguyen, T. T. (1994). Maximum likelihood estimators of binomial parameters under an order restriction (Ack: 94V48 p354). *The American Statistician*, *48*, 29–30.

Dobler, C. P. (2002). The one-way layout with ordered parameters: a survey of advances since 1988. *Journal of Statistitical Planning and Inference*, *107*, 75–88.

Doll, R. and Pygott, F. (1952). Factors influencing the rate of healing of gastric ulcers. *Lancet*, *259*, 171–175.

Douglas, R., Fienberg, S. E., Lee, M.-L. T., Sampson, A. R., and Whitaker, L. R. (1991). Positive dependence concepts for ordinal contingency tables. In H. W. Block, A. R. Sampson, and T. H. Savits, editor, *Topics in Statistical Dependence*, IMS Lecture Notes Series, Hayward, CA, pages 189–202.

Drees, H. and Milbrodt, H. (1994). The one-sided Kolmogorov-Smirnov test in signal detection problems with Gaussian white noise. *Statistica Neerlandica*, *48*, 103–116.

Dubey, S. D. (1991). Some thoughts on the one-sided and two-sided tests. *Journal of Biopharmaceutical Statistics*, *1*, 139–150.

Dubnicka, S. R. (2002). Testing against ordered alternatives for multivariate censored survival data. *Statistics and Probability Letters*, *56*(2), 207–216.

Dufour, J.-M. (1989). Nonlinear hypotheses, inequality restrictions, and non-nested hypotheses: Exact simultaneous tests in linear regressions. *Econometrica*, *57*, 335–355.

Dümbgen, L. (1995). Minimax tests for convex cones. *Annals of the Institute of Statistical Mathematics*, *47*, 155–165.

Dunnett, C. W. (1997). Comparisons with a control. In N. L. J. Samuel Kotz, editor, *Encyclopedia of Statistical Sciences*, pages 126–134. John Wiley and Sons, New York.

Dunson, D. and Neelon, B. (2003a). Bayesian inference on order-constrained parameters in generalized linear models. *Biometrics*, *59*, 286–295.

Dunson, D. B. (2001). Bayesian modeling of the level and duration of fertility in the menstrual cycle. *Biometrics*, *57*(4), 1067–1073.

Dunson, D. B. (2003). Bayesian inferences in the cox model for order-restricted hypotheses. *Biometrics*, *59*, 916–923.

Dunson, D. B. and Neelon, B. (2003b). Bayesian inferences in the cox model for order-restricted hypothesess. *Biometrics*, *59*, 916–923.

Dupačová, J. and Wets, R. (1988). Asymptotic behavior of statistical estimators and of optimal solutions of stochastic optimization problems. *Annals of Statistics*, *16*, 1517–1549.

Durot, C. (2002). Sharp asymptotics for isotonic regression. *Probability Theory and Related Fields*, *122*(2), 222–240.

Durot, C. and Tocquet, A.-S. (2001). Goodness of fit test for isotonic regression. *ESAIM P and S: Probability and Statistics*, *5*, 119–140.

Durot, C. and Tocquet, A. S. (2003). On the distance between the empirical process and its concave majorant in a monotone regression framework. *Annales De L Institut Henri Poincare-Probabilites Et Statistiques*, *39(2)*, 217–240.

Dykstra, R. (1991). Asymptotic normality for chi-bar-square distributions. *The Canadian Journal of Statistics*, *19*, 297–306.

Dykstra, R. and Carolan, C. (1999). The distribution of the argmax of two-sided Brownian motion with quadratic drift. *Journal of Statistical Computation and Simulation*, *63*, 47–58.

Dykstra, R., Kochar, S., and Robertson, T. (1991). Statistical inference for uniform stochastic ordering in several populations. *Annals of Statistics*, *19*, 870–888.

Dykstra, R., Kochar, S., and Robertson, T. (1995a). Inference for likelihood ratio ordering in the two-sample problem. *Journal of the American Statistical Association*, *90*, 1034–1040.

Dykstra, R., Kochar, S., and Robertson, T. (1995b). Likelihood based inference for cause specific hazard rates under order restrictions. *Journal of Multivariate Analysis*, *54*, 163–174.

Dykstra, R., Kochar, S., and Robertson, T. (1995c). Likelihood ratio tests for symmetry against one-sided alternatives. *Annals of the Institute of Statistical Mathematics*, *47*, 719–730.

Dykstra, R., Kochar, S., and Rojo, J. (1997). Stochastic comparisons of parallel systems of heterogeneous exponential components. *Journal of Statistical Planning and Inference*, *65*, 203–211.

Dykstra, R., Kochar, S., and Robertson, T. (1998). Restricted tests for testing independence of time to failure and cause of failure in a competing-risks model. *The Canadian Journal of Statistics*, *26*, 57–68.

Dykstra, R., Hewett, J., and Robertson, T. (1999). Nonparametric, isotonic discriminant procedures. *Biometrika*, *86*, 429–438.

Dykstra, R. L. and El Barmi, H. (1997). Chi-bar-square distributions. In N. L. J. Samuel Kotz, editor, *Encyclopedia of Statistical Sciences*, pages 89–93. John Wiley and Sons, New York.

Dykstra, R. L. and Feltz, C. J. (1989). Nonparametric maximum likelihood estimation of survival functions with a general stochastic ordering and its dual. *Biometrika*, *76*, 331–341.

Dykstra, R. L. and Lee, C.-I. C. (1991). Multinomial estimation procedures for isotonic cones. *Statistics and Probability Letters, 11*, 155–160.

Dykstra, R. L. and Lemke, J. H. (1988). Duality of I projections and maximum likelihood estimation for log-linear models under cone constraints. *Journal of the American Statistical Association, 83*, 546–554.

Dykstra, R. L., Lee, C.-I. C., and Yan, X. (1996). Multinomial estimation procedures for two stochastically ordered distributions. *Statistics and Probability Letters, 30*, 353–361.

Eaton, M. L. (1970). A complete class theorem for multidimensional one-sided alternatives. *Annals of Mathematical Statistics, 41*, 1884–1888.

Ebrahimi, N. and Kirmani, S. N. U. A. (1996). Some results on ordering of survival functions through uncertainty. *Statistics and Probability Letters, 29*, 167–176.

Efromovich, S. (1999). *Nonparametric Curve Estimation : Methods, Theory and Applications*. Springer-Verlag, New York.

Efromovich, S. (2001). Density estimation under random censorship and order restrictions: From asymptotic to small samples. *Journal of the American Statistical Association, 96*(454), 667–684.

Efron, B. and Morris, C. (1973a). Combining possibly related estimation problems (with discussion). *Journal of the Royal Statistical Society, Series B, Methodological, 35*, 379–421.

Efron, B. and Morris, C. (1973b). Stein's estimation rule and its competitors – An empirical Bayes approach. *Journal of the American Statistical Association, 68*, 117–130.

El Barmi, H. (1996). Empirical likelihood ratio test for or against a set of inequality constraints. *Journal of Statistical Planning and Inference, 55*, 191–204.

El Barmi, H. (1997). Testing for or against a trend in odds ratios. *Communications in Statistics, Part A – Theory and Methods, 26*, 1877–1891.

El Barmi, H. and Dykstra, R. (1995). Testing for and against a set of linear inequality constraints in a multinomial setting. *The Canadian Journal of Statistics, 23*, 131–143.

El Barmi, H. and Dykstra, R. (1998). Maximum likelihood estimates via duality for log-convex models when cell probabilities are subject to convex constraints. *Annals of Statistics, 26*(5), 1878–1893.

El Barmi, H. and Dykstra, R. L. (1999). Likelihood ratio test against a set of inequality constraints. *Journal of Nonparametric Statistics, 11*, 233–250.

El Barmi, H. and Kochar, S. C. (1995). Likelihood ratio tests for bivariate symmetry against ordered alternatives in a square contingency table. *Statistics and Probability Letters, 22*, 167–173.

El Barmi, H. and Nelson, P. I. (2000). Three monotone density estimators from selection biased samples. *Journal of Statistical Computation and Simulation, 67*(3), 203–217.

El Barmi, H. and Pontius, J. S. (2000). Testing ordered hypotheses in animal resource selection studies. *Journal of Agricultural, Biological, and Environmental Statistics, 5*(1), 88–101.

El Barmi, H. and Rojo, J. (1997). Likelihood ratio tests for peakedness in multinomial populations. *Journal of Nonparametric Statistics, 7*, 221–237.

El Barmi, H. and Rothmann, M. D. (1999). Estimation of weighted multinomial probabilities under log-convex constraints. *Journal of Statistical Planning and Inference, 81*, 1–11.

El Barmi, H. and Zimmerman, D. (1999). Likelihood ratio tests for and against decreasing in transposition in $K \times K \times K$ contingency tables. *Statistics and Probability Letters*, *45*, 1–10.

El Barmi, H., Harris, I. R., and Basu, A. (1996). Statistical inference concerning weighted Poisson rates under some natural order restrictions. *Journal of Applied Statistics*, *23*, 507–514.

Elfessi, A. and Pal, N. (1992). A note on the common mean of two normal populations with order restricted variances. *Communications in Statistics, Part A – Theory and Methods*, *21*, 3177–3184.

Engelhardt, M., Guffey, J. M., and Wright, F. T. (1990). Tests for positive jumps in the intensity of a Poisson process: A power study. *IEEE Transactions on Reliability*, *39*, 356–360.

Eubank, R. (1988a). Comments on "Monotone regression splines in action.". *Statistical Science*, *3*, 446–450.

Eubank, R. L. (1988b). *Spline Smoothing and Nonparametric Regression*. Marcel Dekker, New York.

Eubank, R. L. (1999). *Nonparametric Regression and Spline Smoothing*. Marcel Dekker, New York.

Evans, M., Gilula, Z., Guttman, I., and Swartz, T. (1997). Bayesian analysis of stochastically ordered distributions of categorical variables. *Journal of the American Statistical Association*, *92*(437), 208–214.

Fahrmeir, L. and Kaufmann, H. (1985). Consistency and asymptotic normality of the maximum likelihood estimator in generalized linear models (Corr: V14 p1643). *Annals of Statistics*, *13*, 342–368.

Fahrmeir, L. and Klinger, J. (1994). Estimating and testing generalized linear models under inequality restrictions. *Statistical Papers*, *35*, 211–229.

Fahrmeir, L., Gieger, C., and Heumann, C. (1999). An application of isotonic longitudinal marginal regression to monitoring the healing process. *Biometrics*, *55*, 951–956.

Fairley, D., Pearl, D. K., and Verducci, J. S. (1987). The penalty for assuming that a monotone regression is linear. *Annals of Statistics*, *15*, 443–448.

Farebrother, R. W. (1995). Simpler tests of linear inequality constraints in the standard linear model. In *New Trends in Probability and Statistics Volume 3: Multivariate Statistics and Matrices in Statistics (Proceedings of the 5th Tartu Conference)*, pages 67–73.

Fearn, D. H. and Nebenzahl, E. (1992). Tests for exponentiality against increasing failure rate average alternatives with censored data. *Communications in Statistics, Part A – Theory and Methods*, *21*, 2557–2567.

Feder, P. I. (1968). On the distribution of the loglikelihood ratio statistic when the true parameter is near the boundaries of the hypothesis region. *Annals of Mathematical Statististics*, *39*, 2044–2055.

Fernàndez, M. A., Rueda, C., and Salvador, B. (1997). On the maximum likelihood estimator under order restrictions in uniform probability models. *Communications in Statistics, Part A – Theory and Methods*, *26*, 1971–1980.

Fernàndez, M. A., Rueda, C., and Salvador, B. (1998). Simultaneous estimation by isotonic regression. *Journal of Statistical Planning and Inference*, *70*, 111–119.

Fernàndez, M. A., Rueda, C., and Salvador, B. (1999). The loss of efficiency estimating linear functions under restrictions. *Scandinavian Journal of Statistics*, *26*, 579–592.

Fernàndez, M. A., Rueda, C., and Salvador, B. (2000). Parameter estimation under orthant restrictions. *The Canadian Journal of Statistics*, *28*(1), 171–181.

Finkelstein, D. M. (1991). Modeling the effect of dose on the lifetime tumor rate from an animal carcinogenicity experiment. *Biometrics*, *47*, 669–680.

Finkelstein, D. M. and Wolfe, R. A. (1986). Isotonic regression for interval censored survival data using an E-M algorithm. *Communications in Statistics, Part A – Theory and Methods*, *15*, 2493–2505.

Fiorentini, G., Sentana, E., and Calzolari, G. (2003). Maximum likelihood estimation and inference in multivariate conditionally heteroscedastic dynamic regression models with student t innovations. *Journal of Business and Economic Statistics*, *21*, 532–546.

Fisher, L. D. (1991). The use of one-sided tests in drug trials: An FDA Advisory Committee member's perspective. *Journal of Biopharmaceutical Statistics*, *1*, 151–156.

Fisher, R. A. (1932). *Statistical Methods for Research Workers*. Oliver and Boyd, London.

Folks, J. L. (1984). Combination of independent tests. In *Handbook of Statistics (Volume 4): Nonparametric Methods*, pages 113–121.

Follmann, D. (1995). Multivariate tests for multiple endpoints in clinical trials. *Statistics in Medicine*, *14*, 1163–1175.

Follmann, D. (1996a). Discussion of "Monitoring clinical trials with multiple endpoints". *Statistics in Medicine*, *15*, 2367–2370.

Follmann, D. (1996b). A simple multivariate test for one-sided alternatives. *Journal of the American Statistical Association*, *91*, 854–861.

Forster, J. J., McDonald, J. W., and Smith, P. W. F. (1996). Montecarlo exact conditional tests for loglinear and logistic models. *Journal of the Royal Statistical Society, Series B*, *58*, 445–453.

Franck, W. E. (1984). A likelihood ratio test for stochastic ordering. *Journal of the American Statistical Association*, *79*, 686–691.

Francom, S. F., Chuang-Stein, C., and Landis, J. R. (1989). A log-linear model for ordinal data to characterize differential change among treatments. *Statistics in Medicine*, *8*, 571–582.

Fraser, D. A. S. and Massam, H. (1989). A mixed primal-dual bases algorithm for regression under inequality constraints. Application to concave regression. *Scandinavian Journal of Statistics*, *16*, 65–74.

Frick, H. (1994). A maxmin linear test of normal means and its application to Lachin's data. *Communications in Statistics, Part A – Theory and Methods*, *23*, 1021–1029.

Frick, H. (1995). Comparing trials with multiple outcomes: The multivariate one-sided hypothesis with unknown covariances. *Biometrical Journal*, *37*, 909–917.

Frick, H. (1996). On the power behaviour of Läuter's exact multivariate one-sided tests. *Biometrical Journal*, *38*, 405–414.

Frick, H. (1997). Computing projections into cones generated by a matrix. *Biometrical Journal. Journal of Mathematical Methods in Biosciences*, *39*, 975–987.

Friedman, J. and Tibshirani, R. (1984). The monotone smoothing of scatterplots. *Technometrics*, *26*, 243–250.

Futschik, A. and Pflug, G. C. (1998a). Distance testing for selecting the best population. *The Australian and New Zealand Journal of Statistics*, *40*, 443–464.

Futschik, A. and Pflug, G. C. (1998b). The likelihood ratio test for simple tree order: A useful asymptotic expansion. *Journal of Statistical Planning and Inference*, *70*, 57–68.

Gail, M. and Simon, R. (1985). Testing for qualitative interactions between treatment effects and patient subsets. *Biometrics*, *41*, 361–372.

Gallant, A. R. and Golub, G. H. (1982). Imposing curvature restrictions on flexible functional forms. *Journal of Econometrics*, *26*, 295–322.

Garren, S. T. (2000). On the improved estimation of location parameters subject to order restrictions in location-scale families. *Sankhya, Series B, Indian Journal of Statistics*, *62*(2), 189–201.

Gatto, R. and Jammalamadaka, S. R. (2002). A saddlepoint approximation for testing exponentiality against some increasing failure rate alternatives. *Statistics and Probability Letters*, *58*(1), 71–81.

Gaylord, C. K. and Ramirez, D. E. (1991). Monotone regression splines for smoothed bootstrapping. *Computational Statistics*, *6*, 85–97.

Gelfand, A. E. and Carlin, B. P. (1990). Bayesian inference for hard problems using the Gibbs sampler. In *Computing Science and Statistics: statistics of many parameters: curves, images, spatial models*, pages 29–37.

Gelfand, A. E. and Kottas, A. (2001). Nonparametric Bayesian modeling for stochastic order. *Annals of the Institute of Statistical Mathematics*, *53*(4), 865–876.

Gelfand, A. E. and Kuo, L. (1991). Nonparameteric Bayesian bioassay including ordered polytomous response. *Biometrika*, *78*, 657–66.

Geller, N. L., Gnecco, C., and Tang, D.-I. (1992). Group equential designs for trials with multiple endpoints. In *Biopharmaceutical Sequential Statistical Applications*, pages 29–33.

Gerlach, B. (1987). Testing exponentiality against increasing failure rate with randomly censored data. *Statistics*, *18*, 275–286.

Gerlach, B. (1989). Tests for increasing failure rate average with randomly right censored data. *Statistics*, *20*, 287–295.

Geyer, C. J. (1991). Constrained maximum likelihood exemplified by isotonic convex logistic regression. *Journal of the American Statistical Association*, *86*, 717–724.

Geyer, C. J. (1994). On the asymptotics of constrained M-estimation. *Annals of Statistics*, *22*, 1993–2010.

Ghiassi, S. H. M. and Govindarajulu, Z. (1986). An asymptotically distribution-free test for ordered alternatives in two-way layouts. *Journal of Statistical Planning and Inference*, *13*, 239–249.

Ghosal, S., Sen, A., and Van Der Vaart, A. W. (2000). Testing monotonicity of regression. *Annals of Statistics*, *28*(4), 1054–1082.

Ghosh, J. K. and Sen, P. K. (1985). On the asymptotic performance of the log likelihood ratio statistic for the mixture model and related results. In L. LeCam and R. A. Olshen, editors, *Proceedings of the Berkeley Coference in Honor of J. Neyman and J. Kiefer*, volume 2, pages 789–806. Wadsworth, Monterey, CA.

Ghosh, M. and Sen, P. K. (1989). Median unbiasedness and Pitman closeness. *Journal of the American Statistical Association*, *84*, 1089–1091.

Ghosh, M. and Sen, P. K. (1991). Bayesian Pitman closeness (Disc: p3679-3695). *Communications in Statistics, Part A – Theory and Methods*, *20*, 3659–3678.

Gill, A. N. and Dhawan, A. K. (1996). Testing homogeneity of scale parameters against ordered alternatives using Hodges-Lehmann estimators. *Statistics*, *28*, 159–170.

Gill, A. N. and Dhawan, A. K. (1999). A one-sided test for testing homogeneity of scale parameters against simple ordered alternative. *Communications in Statistics, Part A – Theory and Methods*, *28*, 2417–2439.

Giltinan, D. M., Capizzi, T. P., and Malani, H. (1988). Diagnostic tests for similar action of two compounds. *Applied Statistics*, *37*, 39–50.

Gilula, Z., Krieger, A. M., and Ritov, Y. (1988). Ordinal association in contingency tables: Some interpretive aspects. *Journal of the American Statistical Association*, *83*, 540–545.

Gleser, L. J. (1973). On a theory of intersection-union tests (preliminary report). *IMS Bulletin*, *2*, 233.

Godfrey, L. G. (1988). *Misspecification Tests in Econometrics: The Lagrange multiplier principle and other approaches*. Cambridge University Press, Cambridge, U.K.

Gogoi, B. and Kakoty, S. (1987). A distribution-free test for ordered alternatives in randomised block design (STMA V30 1574). *Journal of the Indian Statistical Association*, *25*, 37–43.

Goodman, L. A. (1991). Measures, models, and graphical displays in the analysis of cross-classified data. *Journal of the American Statistical Association*, *86*, 1085–1111.

Gordon, L. and Pollak, M. (1995). A robust surveillance scheme for stochastically ordered alternatives. *Annals of Statistics*, *23*, 1350–1375.

Gore, A. P., Rao, K. S. M., and Sahasrabudhe, M. N. (1986). Median tests for ordered alternatives. *Gujarat Statistical Review*, *13*(1), 55–63.

Gourieroux, C. (1997). *ARCH Models and Financial Applications*. Springer-Verlag, New York.

Gourieroux, C. and Monfort, A. (1995). *Statistics and Econometric Models. Volumes 1 and 2*. Cambridge University Press, Cambridge, England.

Gourieroux, C., Holly, A., and Monfort, A. (1982). Likelihood ratio test, Wald test, and Kuhn-Tucker test in linear models with inequality constraints on the regression parameters. *Econometrica*, *50*, 63–80.

Graubard, B. I. and Korn, E. L. (1987). Choice of column scores for testing independence in ordered $2 \times k$ contingency tables. *Biometrics*, *43*, 471–476.

Grenander, U. (1956). On the theory of mortality measurement. part ii. *Skand. Aktuar.*, *39*.

Grenander, U. and Rosenblatt, M. (1956). Some problems in estimating the spectrum of a time series. In *Proceedings of the Third Berkeley Symposium on Mathematical Statistics and Probability, Volume 1*, pages 77–93. University of California Press, Berkeley, CA.

Groeneboom, P. (1985). Estimating a monotone density. In L. M. Le Cam and R. A. Olshen, editors, *Proceedings of the Berkeley conference in honor of Jerzy Neyman and Jack Kiefer (Vol. 2)*, pages 539–555. Institute of Mathematical Statistics, Hayward, CA.

Groeneboom, P. and Jongbloed, G. (1995). Isotonic estimation and rates of convergence in Wicksell's problem. *Annals of Statistics*, *23*, 1518–1542.

Groeneboom, P. and Lopuhaä, H. P. (1993). Isotonic estimators of monotone densities and distribution functions: Basic facts. *Statistica Neerlandica*, *47*, 175–183.

Groeneboom, P., Hooghiemstra, G., and Lopuhaä, H. P. (1999). Asymptotic normality of the L_1 error of the Grenander estimator. *Annals of Statistics*, *27*(4), 1316–1347.

Groeneboom, P., Jongbloed, G., and Wellner, J. A. (2001a). A canonical process for estimation of convex functions: The "invelope" of integrated Brownian motion $+t^4$. *Annals of Statistics*, *29*(6), 1620–1652.

Groeneboom, P., Jongbloed, G., and Wellner, J. A. (2001b). Estimation of a convex function: Characterizations and asymptotic theory. *Annals of Statistics*, *29*(6), 1653–1698.

Grove, D. M. (1980). A test of independence against a class of ordered alternatives in a $2 \times C$ contingency table. *Journal of the American Statistical Association*, *75*, 454–459.

Guess, F., Holander, M., and Proschan, F. (1986). Testing exponentiality versus a trend change in mean residual life. *Annals of Statistics*, *14*, 1388–98.

Guffey, J. M. and Wright, F. T. (1996). Testing for monotonicity in the intensity of a nonhomogeneous Poisson process. *Statistics and Probability Letters*, *28*, 195–202.

Gutmann, S. (1987). Tests uniformly more powerful than uniformly most powerful monotone tests. *Journal of Statistical Planning and Inference*, *17*, 279–292.

Gutmann, S., Kemperman, J. H. B., Reeds, J. A., and Shepp, L. A. (1991). Existence of probability measures with given marginals. *Annals of Probability*, *19*, 1781–1797.

Hájek, J. (1961). Some extensions of the Wald-Wolfowitz-Noether theorem. *Annals of Mathematical Statistics*, *32*, 506–523.

Hájek, J., Šidák, J., and Sen, P. K. (1999). *Theory of Rank Tests*. Academic Press, New York.

Hall, D. B. and Præstgaard, J. T. (2001). Order-restricted score tests for homogeneity in generalized linear and nonlinear mixed models. *Biometrika*, *88*, 739–751.

Hall, P. and Huang, L.-S. (2001). Nonparametric kernel regression subject to monotonicity constraints. *Annals of Statistics*, *29*(3), 624–647.

Hall, P. and Yao, Q. W. (2003). Data tilting for time series. *Journal of the Royal Statistical Society Series B*, *65*, 425–442.

Hall, P., Huber, C., and Speckman, P. L. (1997). Covariate-matched one-sided tests for the difference between functional means. *Journal of the American Statistical Association*, *92*, 1074–1083.

Hall, W. J. and Mathiason, D. (1990). On large-sample estimation and testing in parametric models. *Internatioal Statistical Review*, *58*, 77–97.

Hallahan, C. B. (1993). Model estimation with inequality constraints: A Bayesian approach using SAS/IML software. In *Proceedings of the SAS Users Group International Conference*, volume 18, pages 1043–1049. SAS Institute, Cary, NC.

Hansen, M. B. and Lauritzen, S. L. (2002). Nonparametric Bayes inference for concave distribution functions. *Statistica Neerlandica*, *56*(1), 110–127.

Hanson, D. L. and Mukerjee, H. (1990). Isotonic estimation in stochastic approximation. *Statistics and Probability Letters*, *9*, 279–287.

Hartigan, J. A. (1985a). A failure of likelihood asymptotics for normal mixtures. In L. LeCam and R. A. Olshen, editors, *Proceedings of the Berkeley conference in honor of Jerzy Neyman and Jack Kiefer (Vol. 2)*, pages 807–810. Wadsworth, Monterey, CA.

Hartigan, J. A. (1988). The span test for unimodality. In *Classification and related methods of data analysis*, pages 229–236. North-Holland, New York.

Hartigan, J. A. and Hartigan, P. M. (1985). The dip test of unimodality. *Annals of Statistics*, *13*, 70–84.

Hartigan, P. M. (1985b). [Algorithm AS 217] Computation of the dip statistic to test for unimodality. *Applied Statistics*, *34*, 320–325.

Hartlaub, B. A. and Wolfe, D. A. (1999). Distribution-free ranked-set sample procedures for umbrella alternatives in the m-sample setting. *Environmental and Ecological Statistics*, *6*, 105–118.

Hartmann, W. (1986). Canonical decomposition with linear equality and inequality constraints on parameters. In *Classification as a tool of research*, pages 191–199. North-Holland, NewYork.

Hasegawa, H. (1989). On some comparisons between Bayesian and sampling theoretic estimators of a normal mean subject to an inequality constraint. *Journal of the Japan Statistical Society*, *19*, 167–177.

Hayter, A. J. (1990). A one-sided studentized range test for testing against a simple ordered alternative. *Journal of the American Statistical Association*, *85*, 778–785.

Hayter, A. J. and Liu, W. (1996). Exact calculations for the one-sided studentized range test for testing against a simple ordered alternative. *Computational Statistics and Data Analysis*, *22*, 17–25.

Hayter, A. J. and Liu, W. (1999). A new test against an umbrella alternative and the associated simultaneous confidence intervals. *Computational Statistics and Data Analysis*, *30*, 393–401.

Hayter, A. J., Miwa, T., and Liu, W. (2000). Combining the advantages of one-sided and two-sided procedures for comparing several treatments with a control. *Journal of Statistical Planning and Inference*, *86*(1), 81–99.

He, X. and Shi, P. (1998). Monotone B-spline smoothing. *Journal of the American Statistical Association*, *93*, 643–650.

Headrick, T. C. and Rotou, O. (2001). An investigation of the rank transformation in multiple regression. *Computational Statistics and Data Analysis*, *38*(2), 203–215.

Hengartner, N. W. (1999). A note on maximum likelihood estimation. *The American Statistician*, *53*, 123–125.

Hengartner, N. W. and Stark, P. B. (1995). Finite sample confidence envelopes for shape-restricted densities. *Annals of Statistics*, *23*, 525–550.

Hettmansperger, T. P. and Norton, R. M. (1987). Tests for patterned alternatives in k-sample problems. *Journal of the American Statistical Association*, *82*, 292–299.

Heyde, C. C. (1997). *Quasi-likelihood and its Application : A general approach to optimal parameter estimation*. Springer Verlag, New York.

Higgins, J. J. and Bain, P. T. (1999). Nonparametric tests for lattice-ordered alternatives in unreplicated two-factor experiments. *Journal of Nonparametric Statistics*, *11*, 307–318.

Hill, N. J., Padmanabhan, A. R., and Puri, M. L. (1988). Adaptive nonparametric procedures and applications. *Applied Statistics*, *37*, 205–218.

Hille, E. (1948). *Functional Analysis and Semigroups*. Amer. Math. Colloq., Pub. 31, New York.

Hiriart-Urruty, J.-B. and Lemaréchal, C. (1993). *Convex Analysis and Minimization Algorithms*. Springer-Verlag, Berlin.

Hirotsu, C. (1982). Use of cumulative efficient scores for testing ordered alternatives in discrete models. *Biometrika*, *69*, 567–577.

Hochberg, Y. (1988). A sharper Bonferroni procedure for multiple tests of significance. *Biometrika*, *75*, 800–802.

Hochberg, Y. (1995). On assessing multiple equivalence with reference to bioequivalence. In H. N. Nagaraja, P. K. Sen, and D. F. Morrison, editors, *Statistical Theory and Applications: Papers in Honor of Herbert A. David*, pages 267–278. Springer-Verlag, New York.

Hochberg, Y. and Liberman, U. (1994). An extended Simes' test. *Statistics and Probability Letters*, *21*, 101–105.

Hochberg, Y. and Rom, D. (1995). Extensions of multiple testing procedures based on Simes' test. *Journal of Statistical Planning and Inference*, *48*, 141–152.

Hoferkamp, C. and Peddada, S. D. (2002). Parameter estimation in linear models with heteroscedastic variances subject to order restrictions. *Journal of Multivariate Analysis*, *82*, 65–87.

Hoff, P. D. (2000). Constrained nonparametric maximum likelihood via mixtures. *Journal of Computational and Graphical Statistics*, *9*(4), 633–641.

Hoffman, W. P. and Leurgans, S. E. (1990). Large sample properties of two tests for independent joint action of two drugs. *Annals of Statistics*, *18*, 1634–1650.

Hogg, R. V. (1965). On models and hypotheses with restricted alternatives. *Journal of the American Statistical Association*, *60*, 1153–1162.

Holford, T. R. and Bracken, M. B. (1992). A model for estimating level and net severity of spinal cord injuries. *Statistics in Medicine*, *11*, 1171–1186.

Hollander, M. (1988). Wilcoxon-type tests for ordered alternatives in randomized blocks. In S. Kotz and C. B. R. Norman L. Johnson (associate editor, editors, *Encyclopedia of Statistical Sciences (Vol. 1)*, pages 619–627. John Wiley and Sons, New York.

Hollander, M. and Proschan, F. (1975). Tests for the mean residual life (Corr: V63 p412; V67 p259). *Biometrika*, *62*, 585–594.

Hollander, M. and Proschan, F. (1984). Nonparametric concepts and methods in reliability. In *Handbook of Statistics (Volume 4): Nonparametric Methods*, pages 613–655.

Hommel, G. (1988). A stagewise rejective multiple test procedure based on a modified bonferroni test. *Biometrika*, *75*, 383–386.

Hommel, G. (1989). A comparison of two modified Bonferroni procedures. *Biometrika*, *76*, 624–625.

Hong, Y. (1997). One-sided testing for conditional heteroskedasticity in time series models. *Journal of Time Series Analysis*, *18*, 253–277.

Hoshino, N., Miyazaki, H., and Seki, Y. (1995). On the level probabilities for useful partially ordered alternatives in the analysis of variance. *Communications in Statistics, Part A – Theory and Methods*, *24*, 2059–2071.

Hotelling, H. (1936). Relations between two sets of variates. *Biometrika*, *28*, 321–377.

Hothorn, L. (1989a). On the behaviour of Fligner-Wolfe-trend test "control versus k treatments' with application in toxicology. *Biometrical Journal*, *31*, 767–780.

Hothorn, L. (1989b). Robustness study on Williams- and Shirley-procedure, with application in toxicology. *Biometrical Journal*, *31*, 891–903.

Hothorn, L. and Lehmacher, W. (1991). A simple testing procedure "Control versus k treatments" for one-sided ordered alternatives, with application in toxicology. *Biometrical Journal*, *33*, 179–189.

Hothorn, L. A. and Bretz, F. (2000). One-sided simultaneous confidence intervals for effective dose steps in unbalanced designs. *Biometrical Journal*, *42*(8), 995–1006.

Hothorn, L. A. and Hauschke, D. (2000). Identifying the maximum safe dose: A multiple testing approach. *Journal of Biopharmaceutical Statistics*, *10*(1), 15–30.

Hothorn, L. A., Hilton, J. F., and Neuhäuser, M. (1998). Stratified trend tests for dichotomous endpoints, with bio-pharmaceutical applications. In *ASA Proceedings of the Biopharmaceutical Section*, pages 91–94. American Statistical Association, Alexandria, VA.

Hsueh, H.-M., Liu, J.-P., and Chen, J. J. (2001). Unconditional exact tests for equivalence or noninferiority for paired binary endpoints. *Biometrics*, *57*(2), 478–483.

Hu, X. (1995). A note on the symmetry of normal mean hypotheses and its implications. *Statistics and Probability Letters*, *25*, 15–20.

Hu, X. (1997). Maximum-likelihood estimation under bound restriction and order and uniform bound restrictions. *Statistics and Probability Letters*, *35*, 165–171.

Hu, X. (1998). An exact algorithm for projection onto a polyhedral cone. *The Australian and New Zealand Journal of Statistics*, *40*, 165–170.

Hu, X. (1999). Application of the limit of truncated isotonic regression in optimization subject to isotonic and bounding constraints. *Journal of Multivariate Analysis*, *71*, 56–66.

Hu, X. (2000). Tests for monotonic intensity in a Poisson process. *The Australian and New Zealand Journal of Statistics*, *42*(3), 359–365.

Hu, X. and Wright, F. T. (1994a). Likelihood ratio tests for a class of non-oblique hypotheses. *Annals of the Institute of Statistical Mathematics*, *46*, 137–145.

Hu, X. and Wright, F. T. (1994b). Monotonicity properties of the power functions of likelihood ratio tests for normal mean hypotheses constrained by a linear space and a cone. *Annals of Statistics*, *22*, 1547–1554.

Huang, J. and Rossini, A. J. (1997). Sieve estimation for the proportional-odds failure-time regression model with interval censoring. *Journal of the American Statistical Association*, *92*, 960–967.

Huang, J. and Wellner, J. A. (1995). Estimation of a monotone density or monotone hazard under random censoring. *Scandinavian Journal of Statistics*, *22*, 3–33.

Huang, Y. and Zhang, C.-H. (1994). Estimating a monotone density from censored observations. *Annals of Statistics*, *22*, 1256–1274.

Hung, H. M. J. (1996a). Global tests for combination drug studies in factorial trials. *Statistics in Medicine*, *15*, 233–247.

Hung, H. M. J. (1996b). On evaluation of multiple-dose combination drugs in factorial clinical trials. In *ASA Proceedings of the Biopharmaceutical Section*, pages 25–32. American Statistical Association, Alexandria, VA.

Hung, H. M. J. (2000). Evaluation of a combination drug with multiple doses in unbalanced factorial design clinical trials. *Statistics in Medicine*, *19*(16), 2079–2087.

Hung, H. M. J., Chi, G. Y. H., and Lipicky, R. (1994). On some statistical methods for analysis of combination drug studies. *Communications in Statistics, Part A – Theory and Methods*, *23*, 361–376.

Huque, M. F. and Sankoh, A. J. (1997). A reviewer's perspective on multiple endpoint issues in clinical trials. *Journal of Biopharmaceutical Statistics*, *7*, 545–564.

Huque, M. F., Sankoh, A. J., and Rashid, M. M. (1997). Large sample inference for non-inferiority and clinical equivalence trials and the impact of multiple endpoints. In *ASA Proceedings of the Biopharmaceutical Section*, pages 224–229. American Statistical Association, Alexandria, VA.

Huynh, H. (1981). Testing the identity of trends under the restriction of monotonicity in repeated measures designs. *Psychometrika*, *46*, 295–305.

Hwang, J. T. G. and Peddada, S. D. (1994). Confidence interval estimation subject to order restrictions. *Annals of Statistics*, *22*, 67–93.

Iliopoulos, G. and Kourouklis, S. (2000). Interval estimation for the ratio of scale parameters and for ordered scale parameters. *Statistics and Decisions*, *18*(2), 169–184.

Iman, R. and Conover, J. (1989). Monotone regression utilizing ranks. In *Proceedings of the SAS Users Group International Conference*, volume 14, pages 1310–1312. SAS Institute, Cary, NC.

Intriligator, M. D. (1971). *Mathematical Optimization and Economic Theory*. Prentice-Hall, Englewood Cliffs, N.J.

Iwasa, M. (1991). Admissibility of unbiased tests for a composite hypothesis with a restricted alternative. *Annals of the Institute of Statistical Mathematics*, *43*, 657–665.

Iwasa, M. (1996). An extension of a convolution inequality for G-monotone functions and an approach to Bartholomew's conjectures. *Journal of Multivariate Analysis*, *59*, 249–271.

Iwasa, M. (1998). Directional monotonicity properties of the power functions of likelihood ratio tests for cone-restricted hypotheses of normal means. *Journal of Statistical Planning and Inference*, *66*, 223–233.

James, W. and Stein, C. (1961). Estimation with quadratic loss. In *Proceedings of the Fourth Berkeley Symposium on Mathematical Statistics and Probability, Volume 1*, pages 361–379. University of California Press, Berkeley, CA.

Jammalamadaka, S. R., Tiwari, R. C., Sen, P. K., and Ebneshahrashoob, M. (1991). A rank test based on the number of "near-matches" for ordered alternatives in randomized blocks. *Sankhya, Series A, Indian Journal of Statistics*, *53*, 183–193.

Jennison, C. and Turnbull, B. W. (1993). Group sequential tests for bivariate response - interim analyses of clinical trials with both efficacy and safety endpoints. *Biometrics*, *49*, 741–752.

Jewitt, I. (1991). Applications of likelihood orderings in economics. In K. Mosler and M. Scarsini, editors, *Stochastic orders and decision under risk*, pages 174–189. Institute of Mathematical Statistics, Lecture Notes-Monograph Series, Vol 19, Hayward, CA.

Jin, C. and Pal, N. (1991). A note on the location parameters of two exponential distributions under order restrictions. *Communications in Statistics, Part A – Theory and Methods*, *20*, 3147–3158.

Jin, K. and Chi, G. Y. H. (1997). Application of bootstrap in handling multiple endpoints. In *ASA Proceedings of the Biopharmaceutical Section*, pages 150–155. American Statistical Association, Alexandria, VA.

Johnson, N. L. and Kotz, S. (1972). *Continuous Multivariate Distributions*. John Wiley and Sons, New York.

Johnson, N. L. and Leone, F. (1964). *Statistics and Experimental Design*. John Wiley and Sons, New York.

Johnson, R. A. and Mehrota, K. G. (1971). Some c-sample nonparametric tests for ordered alternative. *Joural of Indian Statistical Association*, *9*, 8–23.

Johnson, R. A., Sim, S., Klein, B. E., and Klein, R. (1998). A multivariate multisample quantile test for ordered alternatives. *Journal of the American Statistical Association*, *93*, 807–818.

Jones, M. P. (2001). Unmasking the trend sought by Jonckheere-type tests for right-censored data. *Scandinavian Journal of Statistics*, *28*(3), 527–535.

Jongbloed, G. (2001). Sieved maximum likelihood estimation in Wicksell's problem related deconvolution problems. *Scandinavian Journal of Statistics*, *28*(1), 161–183.

Jonkheere, A. R. (1954a). A distribution free k-sample test against ordered alternatives. *Biometrika*, *41*, 133–145.

Jonkheere, A. R. (1954b). A test for significance of the relation between m rankings and k ranked categories. *The British Journal of Statistical Psychology*, *7*, 93–100.

Joshi, S. N. and MacEachern, S. N. (1997). Isotonic maximum likelihood estimation for the change point of a hazard rate. *Sankhya, Series A, Indian Journal of Statistics*, *59*, 392–407.

Kao, L. H. and Chakraborti, S. (1997). One-sided sign-type non-parametric procedures for comparing treatments with a control in a randomized complete block design. *Journal of Applied Statistics*, *24*, 251–264.

Kaplan, E. L. and Meier, P. (1958). Nonparametric estimation from incomplete observations. *Journal of the American Statistical Association*, *53*, 457–481.

Karabatsos, G. and Sheu, C. F. (2004). Order-constrained bayes inference for dichotomous models of unidimensional nonparametric irt. *Applied Psychological Measurement*, *28*, 110–125.

Kariya, T. and Cohen, A. (1992). On the invariance structure of the one-sided testing problem for a multivariate normal mean (Corr: p619). *Statistica Sinica*, *2*, 221–236.

Karmous, A. R. (1994). M-estimation of k-sample location parameters: A two-sided preliminary test approach. *Communications in Statistics, Part A – Theory and Methods*, *23*, 3063–3073.

Karmous, A. R. and Sen, P. K. (1988a). Isotonic M-estimation of location: Union-intersection principle and preliminary test versions. *Journal of Multivariate Analysis*, *27*, 300–318.

Karmous, A. R. and Sen, P. K. (1988b). On preliminary test isotonic M-estimation of multivariate location parameters. *Calcutta Statistical Association Bulletin*, *37*, 1–16.

Karmous, A. R. and Sen, P. K. (1991). Preliminary test isotonic M-estimation in some multivariate linear models. *Calcutta Statistical Association Bulletin*, *41*, 1–17.

Karrison, T. G. and O'Brien, P. C. (1999). A rank-sum-type test for paired data with multiple endpoints. In *ASA Proceedings of the Biometrics Section*, pages 246–251. American Statistical Association, Alexandria, VA.

Katz, B. P. and Brown, M. B. (1989). A comparison of methods to detect adulteration in natural products under a restricted alternative. *Biometrical Journal*, *31*, 639–648.

Katz, M. W. (1961). Admissible and minimax estimates of parameters in truncated spaces. *Annals of Mathematical Statistics*, *32*, 136–142.

Kazarian, L. (1998). A consistent bootstrapped GMM estimator for the linear model with arbitrary inequality constraints on parameters. *Statistical Papers*, *39*, 325–333.

Kelly, C. and Rice, J. (1990). Monotone smoothing with application to dose-response curves and the assessment of synergism. *Biometrics*, *46*, 1071–1085.

Kelly, R. E. (1989). Stochastic reduction of loss in estimating normal means by isotonic regression. *Annals of Statistics*, *17*, 937–940.

Kepner, J. L. and Robinson, D. H. (1984). A distribution-free rank test for ordered alternatives in randomized complete block designs. *Journal of the American Statistical Association*, *79*, 212–217.

Kiefer, J. (1982). Optimum rates for non-parametric density and regression estimates, under order restrictions. In J. G. G. Kallianpur, P.R. Krishnaiah, editor, *Statistics and probability: Essays in honour of C. R. Rao*, pages 419–428. North-Holland, Amsterdam.

Kim, D. and Agresti, A. (1997). Nearly exact tests of conditional independence and marginal homogeneity for sparse contingency tables. *Computational Statistics and Data Analysis*, *24*, 89–104.

Kim, D. H. and Kim, Y. C. (1992). Distribution-free tests for umbrella alternatives in a randomized block design. *Journal of Nonparametric Statistics*, *1*, 277–285.

Kim, D. H. and Lim, D. H. (1995). Rank tests for parallelism of regression lines against umbrella alternatives. *Journal of Nonparametric Statistics*, *5*, 289–302.

Kim, K. and Demets, D. L. (1987). Confidence intervals following group sequential tests in clinical trials. *Biometrics*, *43*, 857–864.

Kim, K. and Foutz, R. V. (1997). Kolmogorov-type tests of goodness-of-fit against stochastic ordering. *Communications in Statistics, Part A – Theory and Methods*, *26*, 885–897.

Kim, S. W. (1999). A Bayesian test for simple tree ordered alternative using intrinsic priors. *Journal of the Korean Statistical Society*, *28*(1), 73–92.

Kimeldorf, G., Sampson, A. R., and Whitaker, L. R. (1992). Min and max scorings for two-sample ordinal data. *Journal of the American Statistical Association*, *87*, 241–247.

King, A. J. and Rockafellar, R. T. (1993). Asymptotic theory for solutions in statistical estimation and stochastic programming. *Mathematics of Operations Research*, *18*, 148–162.

King, M. L. and Smith, M. D. (1986). Joint one-sided tests of linear regression coefficients. *Journal of Econometrics*, *32*, 367–383.

King, M. L. and Wu, P. X. (1997). Locally optimal one-sided tests for multiparameter hypotheses. *Econometric Reviews*, *16*, 131–156.

Kirchner, H. L. and Lemke, J. H. (2002). Simultaneous estimation of intrarater and interrater agreement for multiple raters under order restrictions for a binary trait. *Statistics in Medicine*, *21*(12), 1761–1772.

Kitchens, L. J. (1998). *Exploring Statistics: A modern introduction to data analysis and inference*. Brook/Cole, Pacific Grove, CA.

Knautz, H. (2000). Inference in linear models with inequality constrained parameters. *Discussiones Mathematicae – Probability and Statistics*, *20*(1), 135–161.

Koch, G. G. (1991). One-sided and two-sided tests and p values. *Journal of Biopharmaceutical Statistics*, *1*, 161–170.

Koch, G. G. and Gillings, D. B. (1988). One-sided versus two-sided tests. In Samuel Kotz, Norman L. Johnson, Campbell B. Read, editor, *Encyclopedia of Statistical Sciences (Vol. 1)*, pages 218–222. John Wiley and Sons, New York.

Kochar, A. and Kochar, S. C. (1989). Some distribution-free tests for testing homogeneity of location parameters against ordered alternatives. *Journal of the Indian Statistical Association*, *27*, 1–8.

Kochar, S. C. and Gupta, R. P. (1988). A Monte Carlo study of some asymptotically optimal tests of exponentiality against positive aging. *Communications in Statistics, Part B – Simulation and Computation*, *17*, 803–811.

Kochar, S. C., Mukerjee, H., and Samaniego, F. J. (2000). Estimation of a monotone mean residual life. *Annals of Statistics*, *28*(3), 905–921.

Kodde, D. A. and Palm, F. C. (1986). Wald criteria for jointly testing equality and inequality restrictions (STMA V28 2359). *Econometrica*, *54*, 1243–1248.

Koebel, B., Falk, M., and Laisney, F. (2003). Imposing and testing curvature conditions on a box-cox cost function. *Journal of Business and Economic Statistics*, *21(2)*, 319–335.

Korkhin, A. S. (1988). Parameter estimation in multiple linear regression with fuzzy inequality constraints. *Cybernetics*, *24*, 203–211.

Kortanek, K. O. (1993). Semi-infinite programming duality for order restricted statistical inference models. *Zeitschrift für Operations Research*, *37*, 285–301.

Kössler, W. and Büning, H. (2000). The efficacy of some c-sample rank tests of homogeneity against ordered alternatives. *Journal of Nonparametric Statistics*, *13*(1), 95–106.

Kössler, W. and Büuning, H. (2000). The asymptotic power and relative efficiency of some c-sample rank tests of homogeneity against umbrella alternatives. *Statistics*, *34*(1), 1–26.

Koul, H. L. (1978a). A class of tests for testing 'new is better than used'. *The Canadian Journal of Statistics*, *6*, 249–271.

Koul, H. L. (1978b). Testing for new is better than used in expectation. *Communications in Statistics, Part A – Theory and Methods*, *7*, 685–702.

Koyak, R. A., Schmehl, R. L., Cox, D. C., Dewalt, F. G., Haugen, M. M., Schwemberger, J. G., J., and Scalera, J. V. (1998). Statistical models for the evaluation of portable lead measurement technologies — Part I: Chemical test kits. *Journal of Agricultural, Biological, and Environmental Statistics*, *3*, 451–465.

Kriener, S. (1987). Analyis of multidimensional contigency tables by exact conditional tests: tests and strategies. *Scandinavian Journal of Statistics*, *14*, 97–112.

Krishnaiah, P. R. e. (1982). *Handbook of Statistics, Volume 1: Analysis of Variance*. North-Holland, Amsterdam].

Krogen, B. L. and Magel, R. C. (2000). Proposal of k sample tests for bivariate censored data for nondecreasing ordered alternatives. *Biometrical Journal*, *42*(4), 435–455.

Krummenauer, F. and Hommel, G. (1999). The size of Simes' global test for discrete test statistics. *Journal of Statistical Planning and Inference*, *82*, 151–162.

Kübler, J. and Schumacher, M. (1990). Survival time models for analysing drug combination treatments. *Statistics in Medicine*, *9*, 1527–1539.

Kubokawa, T. and Saleh, A. K. M. E. (1994). Estimation of location and scale parameters under order restrictions. *Journal of Statistical Research*, *28*, 41–51.

Kudo, A. (1963). A multivariate analogue of the one sided test. *Biometrika*, *50*, 403–418.

Kuhlmann, D. W. and Wright, F. T. (1993). A property of projection residuals with applications to concave regression. *Statistics and Probability Letters*, *16*, 313–319.

Kulatunga, D. D. S., Asai, M., and Sasabuchi, S. (1996). A simulation study of several tests of the equality of binomial probabilities against ordered alternatives. *Journal of Statistical Computation and Simulation*, *56*, 53–75.

Kumar, N., Gill, A. N., and Dhawan, A. K. (1994a). A class of distribution-free statistics for homogeneity against ordered alternatives. *South African Statistical Journal*, *28*, 55–65.

Kumar, N., Gill, A. N., and Mehta, G. P. (1994b). Distribution-free test for homogeneity against ordered alternatives. *Communications in Statistics, Part A – Theory and Methods*, *23*, 1247–1256.

Kumazawa, Y. (1992). Tests for increasing failure rate with randomly censored data. *Statistics*, *23*, 17–25.

Kuriki, S. (1993). One-sided test for the equality of two covariance matrices. *Annals of Statistics*, *21*, 1379–1384.

Kuriki, S. and Takemura, A. (2000a). Shrinkage estimation towards a closed convex set with a smooth boundary. *Journal of Multivariate Analysis*, *75*(1), 79–111.

Kuriki, S. and Takemura, A. (2000b). Some geometry of the cone of nonnegative definite matrices and weights of associated $\bar{\chi}^2$ distribution. *Annals of the Institute of Statistical Mathematics*, *52*(1), 1–14.

Kuriki, S., Shimodaira, H., and Hayter, T. (2002). On the isotonic range statistic for testing against an ordered alternative. *Journal of Statistical Planning and Inference*, *105*(2), 347–362.

Kusum, K. and Bagai, I. (1988). A new class of distribution free procedures for testing homogeneity of scale parameters against ordered alternatives. *Communications in Statistics, Part A – Theory and Methods*, *17*, 1365–1376.

Kvam, P. H. and Singh, H. (1998). Estimating reliability of components with increasing failure rate using series system data. *Naval Research Logistics: An International Journal*, *45*, 115–123.

Kvam, P. H. and Singh, H. (2001). On non-parametric estimation of the survival function with competing risks. *Scandinavian Journal of Statistics*, *28*(4), 715–724.

Kvam, P. H., Singh, H., and Whitaker, L. R. (2002). Estimating distributions with increasing failure rate in an imperfect repair model. *Lifetime Data Analysis*, *8*(1), 53–67.

La Velle, J. M. (1986). Potasium chromate potentiates frameshift mutagenesis in e. coli and s. typhimurium. *Mutation Research*, *171*, 1–10.

Laitila, T. (1998). Adaptation of Honda's one-sided test of random effects to repeated measurements experiments. *Communications in Statistics, Part A – Theory and Methods*, *27*, 3015–3033.

Laska, E. M. and Meisner, M. J. (1989). Testing whether identified treatment is best. *Biometrics*, *45*, 1139–1151.

Laska, E. M., Tang, D.-I., and Meisner, M. J. (1992). Testing hypotheses about an identified treatment when there are multiple endpoints. *Journal of the American Statistical Association*, *87*, 825–831.

Laska, E. M., Meisner, M., and Tang, D.-I. (1996). Combination treatments with multiple endpoints. In *ASA Proceedings of the Biopharmaceutical Section*, pages 33–39. American Statistical Association, Alexandria, VA.

Laska, E. M., Meisner, M., and Tang, D.-I. (1997). Classification of the effectiveness of combination treatments. *Statistics in Medicine*, *16*, 2211–2228.

Laslett, G. M. (1982). The survival curve under monotone density constraints with applications to two-dimensional line segment processes. *Biometrika*, *69*, 153–160.

Lavine, M. and Mockus, A. (1995). A nonparametric Bayes method for isotonic regression. *Journal of Statistical Planning and Inference*, *46*, 235–248.

Le, C. T. (1988). A new rank test against ordered alternatives in K-sample problems. *Biometrical Journal*, *30*, 87–92.

Lee, C.-I. C. (1987). Chi-squared tests for and against an order restriction on multinomial parameters. *Journal of the American Statistical Association*, *82*, 611–618.

Lee, C.-I. C. (1988). Quadratic loss of order restricted estimators for treatment means with a control. *Annals of Statistics*, *16*, 751–758.

Lee, C.-I. C. (1996). On estimation for monotone dose-response curves. *Journal of the American Statistical Association*, *91*, 1110–1119.

Lee, C.-I. C. and Yan, X. (1996). On the consistency of rank transformed regression. *Communications in Statistics, Part A – Theory and Methods*, *25*, 2159–2168.

Lee, C.-I. C., Robertson, T., and Wright, F. T. (1989). On the testing of marginal homogeneity with a one-sided alternative in the analysis of variance. *Annals of the Institute of Statistical Mathematics*, *41*, 149–167.

Lee, C.-I. C., Robertson, T., and Wright, F. T. (1993). Bounds on distributions arising in order restricted inferences with restricted weights. *Biometrika*, *80*, 405–416.

Lee, J. H. H. and King, M. L. (1993). A locally most mean powerful based score test for ARCH and GARCH regression disturbances (Corr: 94V12 p139). *Journal of Business and Economic Statistics*, *11*, 17–27.

Lee, W.-C. and Chen, Y.-I. (2001). Weighted Kaplan-Meier tests for umbrella alternatives. *Annals of the Institute of Statistical Mathematics*, *53*(4), 835–852.

Lee, Y. J. (1977). Maximin tests of randomness against ordered alternatives: The multinomial distribution case. *Journal of the American Statistical Association*, *72*, 673–675.

Lehmacher, W., Wassmer, G., and Reitmeir, P. (1991). Procedures for two-sample comparisons with multiple endpoints controlling the experimentwise error rate. *Biometrics*, *47*, 511–521.

Lehmann, E. L. (1952). Testing multiple hypotheses. *Annals of Mathematical Statistics*, *23*, 541–552.

Lehmann, E. L. (1991). *Theory of Point Estimation*. Wadsworth, Monterey, CA.

Lehmann, E. L. (1994). *Testing Statistical Hypotheses*. Chapman and Hall, New York.

Lei, X., Peng, Y. B., and Wright, F. T. (1995). Testing homogeneity of normal means with a simply ordered alternative and dependent observations. *Computational Statistics and Data Analysis*, *20*, 173–183.

Leurgans, S. (1982). Asymptotic distributions of slope-of-greatest-convex-minorant estimators. *Annals of Statistics*, *10*, 287–296.

Leurgans, S. (1986). Isotonic M-estimation. In *Advances in order restricted statistical inference*, pages 48–68. Springer Verlag, New York.

Liang, T. (1991). Empirical Bayes selection for the highest success probability in Bernoulli processes with negative binomial sampling. *Statistics and Decisions*, *9*, 213–234.

Lim, D. H. and Wolfe, D. A. (1997). Nonparametric tests for comparing umbrella pattern treatment effects with a control in a randomized block design. *Biometrics*, *53*, 410–418.

Lin, D. Y. and Liu, P. Y. (1992). Nonparametric sequential tests against ordered alternatives in multiple-armed clinical trials. *Biometrika*, *79*, 420–425.

Lin, Y. and Lindsay, B. G. (1997). Projections on cones, chi-bar squared distributions, and Weyl"s formula. *Statistics and Probability Letters*, *32*, 367–376.

Lindqvist, B. H. (1988). Estimation and testing in Markov models for repairable systems with positively dependent components. *Scandinavian Journal of Statistics*, *15*, 243–257.

Liu, H. (2000). Uniformly more powerful, two-sided tests for hypotheses about linear inequalities. *Annals of the Institute of Statistical Mathematics*, *52*(1), 15–27.

Liu, H. and Berger, R. L. (1995). Uniformly more powerful, one-sided tests for hypotheses about linear inequalities. *Annals of Statistics*, *23*, 55–72.

Liu, P. Y. and Tsai, W. Y. (1999). A modified logrank test for censored survival data under order restrictions. *Statistics and Probability Letters*, *41*, 57–63.

Liu, P. Y., Green, S., Wolf, M., and Crowley, J. (1993). Testing against ordered alternatives for censored survival data. *Journal of the American Statistical Association*, *88*, 153–160.

Liu, P.-Y., Tsai, W. Y., and Wolf, M. (1998). Design and analysis for survival data under order restrictions with a modified logrank test. *Statistics in Medicine*, *17*, 1469–1479.

Liu, Q. (1998). An order-directed score test for trend in ordered $2 \times K$ tables. *Biometrics*, *54*, 1147–1154.

Liu, W. (1999). A one-sided confidence set hidden under a two-sided sequential test of a normal mean. *Statistics and Probability Letters*, *41*, 435–438.

Lloyd, C. J. (1988). Doubling the one-sided p-value in testing independence in 2×2 tables against a two-sided alternative. *Statistics in Medicine*, *7*, 1297–1306.

Lloyd, C. J. (1999). *Statistical Analysis of Categorical Data*. John Wiley and Sons, New York.

Lo, S.-H. (1986). Estimation of a unimodal distribution function. *Annals of Statistics*, *14*, 1132–1138.

Lo, S.-H. (1987). Estimation of distribution functions under order restrictions. *Statistics and Decisions*, *5*, 251–262.

Lochner, R. H. and Basu, A. P. (1977). A Bayesian approach for testing increasing failure rate. *The theory and applications of reliability with emphasis on Bayesian and nonparametric methods*, *1*, 67–84.

Long, T. and Gupta, R. D. (1998). Alternative linear classification rules under order restrictions. *Communications in Statistics, Part A – Theory and Methods*, *27*, 559–575.

Louv, W. C. and Littell, R. C. (1986). Combining one-sided binomial tests. *Journal of the American Statistical Association, 81*, 550–554.

Lu, Y. and Bean, J. A. (1995). On the sample size for one-sided equivalence of sensitivities based upon McNemar's test. *Statistics in Medicine, 14*, 1831–1839.

Lu, Z. Q. J. and Clarkson, D. B. (1999). Monotone spline and multidimensional scaling. In *ASA Proceedings of the Statistical Computing Section*, pages 185–190. American Statistical Association, Alexandria, VA.

Lucas, L. A. and Wright, F. T. (1991). Testing for and against a stochastic ordering between multivariate multinomial populations. *Journal of Multivariate Analysis, 38*, 167–186.

Luceno, A. and Box, G. E. P. (2000). Influence of the sampling interval, decision limit and autocorrelation on the average run length in Cusum charts. *Journal of Applied Statistics, 27*(2), 177–183.

Lumley, T. (1995). Efficient execution of Stone's likelihood ratio tests for disease clustering. *Computational Statistics and Data Analysis, 20*, 499–510.

Magel, R. C. (1989). A nonparametric test for ordered alternatives in two-way layouts with censored data. *Biometrical Journal, 31*, 609–618.

Magel, R. C. and Degges, R. (1998). Tests for ordered alternatives with right censored data. *Biometrical Journal, 40*, 495–518.

Mammen, E. (1991). Estimating a smooth monotone regression function. *Annals of Statistics, 19*, 724–740.

Mammen, E. and Thomas-Agnan, C. (1999). Smoothing splines and shape restrictions. *Scandinavian Journal of Statistics, 26*, 239–252.

Mammen, E., Marron, J. S., Turlach, B. A., and Wand, M. P. (2001). A general projection framework for constrained smoothing. *Statistical Science, 16*(3), 232–248.

Mansouri, H. (1989). Linear rank tests for homogeneity against ordered alternatives in anova and anocova. *Communications in Statistics, Part A – Theory and Methods, 18*, 4321–4334.

Mansouri, H. (1990). Rank tests for ordered alternatives in analysis of variance. *Journal of Statistical Planning and Inference, 24*, 107–117.

Mansouri, H. (1991). Distribution-free tests in randomized blocks. *Communications in Statistics, Part A – Theory and Methods, 20*, 2753–2773.

Mansouri, H. and Zheng, X. (1991). Small sample distributions of some nonparametric tests for testing treatments vs control. *Communications in Statistics, Part B – Simulation and Computation, 20*, 969–979.

Marcus, R. and Genizi, A. (1987). Nonparametric analysis of covariance with ordered alternatives. *Journal of the Royal Statistical Society, Series B, Methodological, 49*, 102–111.

Marcus, R. and Genizi, A. (1991). Simultaneous one-sided tolerance intervals for sets of contrasts from normal populations. *Communications in Statistics, Part A – Theory and Methods, 20*, 1861–1870.

Marcus, R. and Genizi, A. (1994). Simultaneous confidence intervals for umbrella contrasts of normal means. *Computational Statistics and Data Analysis, 17*, 393–407.

Marden, J. I. (1982). Minimal complete classes of tests of hypotheses with multivariate one-sided alternatives. *Annals of Statistics, 10*, 962–970.

Marden, J. I. (1985). Combining independent one-sided noncentral t or normal mean tests. *Annals of Statistics*, *13*, 1535–1553.

Marshall, A. W. (1991). Multivariate stochastic orderings and generating cones of functions. In K. Mosler and M. Scarsini, editors, *Stochastic orders and decision under risk*, pages 231–247. Institute of Mathematical Statistics, Hayward, CA.

Marshall, A. W. and Proschan, F. (1965). Maximum likelihood estimation for distributions with monotone failure rate. *Annals of Mathematical Statistics*, *36*, 69–77.

Mau, J. (1985). Isotonic regression trend tests for counting processes. *Statistics and Decisions Supplement Issue*, *2*, 187–191.

McCullagh, P. and Nelder, J. A. (1989). *Generalized Linear Models*. Chapman and Hall, London.

McDermott, M. and Wang, Y. (2002). Construction of uniformly more powerful tests for hypotheses about linear inequalities. *Journal of Statistical Planning and Inference*, *107*, 207–217.

McDermott, M. P. (1999). Generalized orthogonal contrast tests for homogeneity of ordered means. *The Canadian Journal of Statistics*, *27*, 457–470.

McDonald, J. W. and Diamond, I. D. (1990). On the fitting of generalized linear models with nonnegativity parameter constraints. *Biometrics*, *46*, 201–206.

McNichols, D. T. and Padgett, W. J. (1982a). Maximum likelihood estimation of unimodal and decreasing densities based on arbitrary right censored data. *Communications in Statistics: Theory and Methods*, *A11*, 2259–2270.

McNichols, D. T. and Padgett, W. J. (1982b). Maximum likelihood estimation of unimodal and decreasing densities based on arbitrarily right-censored data. *Communications in Statistics, Part A – Theory and Methods*, *11*, 2259–2270.

Mehta, C. R., Patel, N. R., and Senchauduri, P. (1988). Importance sampling for estimating exact probabilities in permutational inference. *Journal of the American Statistical Association*, *83*, 999–1005.

Menendez, J. A. and Salvador, B. (1987). An algorithm for isotonic median regression. *Computational Statistics and Data Analysis*, *5*, 399–406.

Menéndez, J. A. and Salvador, B. (1987). Deficiencies of the likelihood ratio test for certain hypothesis with order restrictions (Spanish). *Trabajos de Estadistica*, *2*(2), 71–80.

Menéndez, J. A., Rueda, C., and Salvador, B. (1991). Conditional test for testing a face of the tree order cone. *Communications in Statistics, Part B – Simulation and Computation*, *20*, 751–762.

Menéndez, J. A., Rueda, C., and Salvador, B. (1992). Testing non-oblique hypotheses. *Communications in Statistics, Part A – Theory and Methods*, *21*, 471–484.

Menendez, M. L., Pardo, L., and Zografos, K. (2002). Test of hypotheses for and against order restrictions on multinomial parameters based on phi-divergences. *Utilitas Mathematica*, *61*, 209–223.

Meyer, M. and Woodroofe, M. (1999). An extension of the mixed primal-dual based algorithm to the case of more constraints than dimension. *Journal of Statistical Planning and Inference*, *81*, 13–31.

Meyer, M. and Woodroofe, M. (2000). On the degrees of freedom in shape restricted regression. *Annals of Statistics*, *28*, 1083–1104.

Meyer, M. C. (2003a). An evolutionary algorithm with applications to statistics. *Journal of Computational and Graphical Statistics, 12(2)*, 265–281.

Meyer, M. C. (2003b). A test for linear versus convex regression function using shape-restricted regression. *Biometrika, 90*, 223–232.

Misra, N. and van der Meulen, E. C. (1997). On estimation of the common mean of k (≥ 2) normal populations with order restricted variances. *Statistics and Probability Letters, 36*, 261–267.

Miwa, T. and Hayter, T. (1999). Combining the advantages of one-sided and two-sided test procedures for comparing several treatment effects. *Journal of the American Statistical Association, 94*, 302–307.

Miyazaki, H. and Hoshino, N. (1991). An useful $\bar{E}_k^2$ test for partially ordered alternatives in the analysis of variance. *Communications in Statistics, Part B – Simulation and Computation, 20*, 281–307.

Monti, K. L. and Sen, P. K. (1976). The locally optimal combination of independent test statistics. *Journal of the American Statistical Association, 71*, 903–911.

Moonesinghe, R. and Wright, F. T. (1994). Likelihood ratio tests involving a bivariate trend in two-factor designs: The level probabilities. *Communications in Statistics, Part B – Simulation and Computation, 23*, 143–155.

Moonesinghe, R. and Wright, F. T. (1995). Testing homogeneity with an ordered alternative in a two-factor layout by combining p-values. *Annals of the Institute of Statistical Mathematics, 47*, 505–523.

Morais, M. and Natàrio, I. (1998). Improving an upper one-sided c-chart. *Communications in Statistics, Part A – Theory and Methods, 27*, 353–364.

Morais, M. C. and Pacheco, A. (1998). Two stochastic properties of one-sided exponentially weighted moving average control charts. *Communications in Statistics, Part B – Simulation and Computation, 27*, 937–952.

Moran, P. A. P. (1971). Maximum likelihood estimators in nonstandard conditions. In *Proceedings of the Cambridge Philosophical Society*, volume 70, pages 441–50. Cambridge Philosophical Society, Cambridge, UK.

Morris, M. D. (1988). Small-sample confidence limits for parameters under inequality constraints with application to quantal bioassay. *Biometrics, 44*, 1083–1092.

Morton-Jones, T., Diggle, P., Parker, L., Dickinson, H. O., and Binks, K. (2000). Additive isotonic regression models in epidemiology. *Statistics in Medicine, 19*(6), 849–859.

Mosler, K. and Scarsini, M. (1991a). Some theory of stochastic dominance. In *Stochastic orders and decision under risk*, pages 261–284. Institute of Mathematical Statistics, Hayward, CA.

Mosler, K. C. and Scarsini, M. (1991b). *Stochastic orders and decision under risk - [edited by Mosler and Scarsini]*. Institute of Mathematical Statistics, Hayward, CA.

Mosler, K. C. and Scarsini, M. (1993). *Stochastic Orders and Applications: A classified bibliography*. Springer-Verlag, New York.

Mudholkar, G. S. and McDermott, M. P. (1989). A class of tests for equality of ordered means. *Biometrika, 76*, 161–168.

Mudholkar, G. S. and Subbaiah, P. (1980). Testing significance of a mean vector – A possible alternative to Hotelling's T^2. *Annals of the Institute of Statistical Mathematics, 32*, 43–52.

Mudholkar, G. S., McDermott, M. P., and Mudholkar, A. (1995). Robust finite-intersection tests for homogeneity of ordered variances. *Journal of Statistical Planning and Inference*, *43*, 185–195.

Mukerjee, H. (1988a). Monotone nonparametric regression. *Annals of Statistics*, *16*, 741–750.

Mukerjee, H. (1988b). Order restricted inference in a repeated measures model. *Biometrika*, *75*, 616–617.

Mukerjee, H. (1995). A simple combinatorial proof of a result by Robertson and Pillers. *Statistics and Probability Letters*, *24*, 173–175.

Mukerjee, H. (1996). Estimation of survival functions under uniform stochastic ordering. *Journal of the American Statistical Association*, *91*, 1684–1689.

Mukerjee, H. and Stern, S. (1994). Feasible nonparametric estimation of multiargument monotone functions. *Journal of the American Statistical Association*, *89*, 77–80.

Mukerjee, H. and Tu, R. (1995). Order-restricted inferences in linear regression. *Journal of the American Statistical Association*, *90*, 717–728.

Mukerjee, H., Robertson, T., and Wright, F. T. (1986). A probability inequality for elliptically contoured densities with applications in order restricted inference. *Annals of Statistics*, *14*, 1544–1554.

Munk, A. (2000). An unbiased test for the average equivalence problem – the small sample case. *Journal of Statistical Planning and Inference*, *87*(1), 69–86.

Mykytyn, S. and Santner, T. (1981). Maximum likelihood estimation of the survival function based on censored data under hazard rate assumptions. *Communications in Statistics: Theory and Method*, *A10*, 1369–1387.

Nadaraya, E. A. (1964). On estimating regression. *Theory of Probbaility and Application*, *15*, 134–137.

Nadaraya, E. A., Nadaraya, E. A., and Seckler, B. e. (1964a). On estimating regression. *Theory of Probability and its Applications*, *9*, 141–142.

Nadaraya, E. A., Nadaraya, E. A., and Seckler, B. e. (1964b). Some new estimates for distribution functions. *Theory of Probability and its Applications*, *9*, 497–500.

Nair, V. N. (1987). Chi-squared-type tests for ordered alternatives in contingency tables. *Journal of the American Statistical Association*, *82*, 283–291.

Nakahira, M. (1989). A new approach to testing a linear hypothesis against a restricted alternative by spectral decomposition (Japanese). *Japanese Journal of Applied Statistics*, *18*, 163–180.

Nandram, B. and Sedransk, J. (1993). Bayesian inference for the mean of a stratified population when there are order restrictions. In *ASA Proceedings of the Section on Bayesian Statistical Science*, pages 79–84. American Statistical Association, Alexandria, VA.

Nandram, B., Sedransk, J., and Smith, S. J. (1997). Order-restricted Bayesian estimation of the age composition of a population of Atlantic cod. *Journal of the American Statistical Association*, *92*, 33–40.

Neel, J. V. and Schull, W. J. (1954). *Human Heredity*. University of Chicago Press, Chicago, IL.

Nelson, D. B. and Cao, C. Q. (1992). Inequality constraints in the univariate GARCH model. *Journal of Business and Economic Statistics*, *10*, 229–235.

Nelson, L. S. (1992). A randomization test for ordered alternatives (Corr: p251). *Journal of Quality Technology*, *24*, 51–53.

Nelson, L. S. (2000). A distribution-free test for ordered alternatives. *Journal of Quality Technology*, *32*(4), 464–467.

Nelson, P. L. and Toothaker, L. E. (1975). An empirical study of Jonckheere's non-parametric test of ordered alternatives. *The British Journal of Mathematical and Statistical Psychology*, *28*, 167–176.

Neto, J. C. and Paula, G. A. (2001). Wald one-sided test using generalized estimating equations approach. *Computational Statistics and Data Analysis*, *36*(4), 475–495.

Nettleton, D. (1999a). Convergence properties of the EM algorithm in constrained parameter spaces. *The Canadian Journal of Statistics*, *27*, 639–648.

Nettleton, D. (1999b). Order restyricted hypothesis testing in a variation of the normal mixture model. *Canadian Journal of Statistics*, *27*, 383–394.

Nettleton, D. and Præstgaard, J. (1998). Interval mapping of quantitative trait loci through order-restricted inference. *Biometrics*, *54*, 74–87.

Neuhäuser, M. and Hothorn, L. A. (1997). Trend tests for dichotomous endpoints with application to carcinogenicity studies. *Drug Information Journal*, *31*, 463–469.

Neuhäuser, M. and Hothorn, L. A. (1999). An exact Cochran-Armitage test for trend when dose-response shapes are a priori unknown. *Computational Statistics and Data Analysis*, *30*, 403–412.

Nomakuchi, K. and Sakata, T. (1987). A note on testing two-dimendsional normal mean. *Annals of the Institute of Statistical Mathematics*, *39*, 489–495.

Nomakuchi, K. and Shi, N.-Z. (1988). A test for a multiple isotonic regression problem. *Biometrika*, *75*, 181–184.

Nyquist, H. (1991). Restricted estimation of generalized linear models. *Applied Statistics*, *40*, 133–141.

O'Brien, P. C. (1984). Procedures for comparing samples with multiple endpoints. *Biometrics*, *40*, 1079–1087.

O'Brien, P. C. and Geller, N. L. (1997). Interpreting tests for efficacy in clinical trials with multiple endpoints. *Controlled Clinical Trials*, *18*, 222–227.

Odeh, R. E. (1989). Simultaneous one-sided prediction intervals to contain all of k future means from a normal distribution. *Communications in Statistics, Part B – Simulation and Computation*, *18*, 1557–1585.

Odeh, R. E. and Mee, R. W. (1990). One-sided simultaneous tolerance limits for regression. *Communications in Statistics, Part B – Simulation and Computation*, *19*, 663–680.

O'Donnell, C. J., Rambaldi, A. N., and Doran, H. E. (2001). Estimating economic relationships subject to firm- and time-varying equality and inequality constraints. *Journal of Applied Econometrics*, *16*(6), 709–726.

Oh, M. (1995). On maximum likelihood estimation of cell probabilities in $2 \times k$ contingency tables under negative dependence restrictions with various sampling schemes. *Communications in Statistics, Part A – Theory and Methods*, *24*, 2127–2143.

Oh, M. (1998a). Tests for and against a positive dependence restriction in two-way ordered contingency tables. *Journal of the Korean Statistical Society*, *27*, 205–220.

Oh, M.-S. (1998b). A Bayes test for simple versus one-sided hypothesis on the mean vector of a multivariate normal distribution. *Communications in Statistics, Part A – Theory and Methods*, *27*, 2371–2389.

O'Hagan, A. and Leonard, T. (1976). Bayes estimation subject to uncertainty about parameter constraints. *Biometrika*, *63*, 201–203.

Oluyede, B. O. (1993). A modified chi-square test for testing equality of two multinomial populations against an order restricted alternative. *Biometrical Journal*, *35*, 997–1012.

Oluyede, B. O. (1994a). Estimation and testing of multinomial parameters under order restrictions. In *ASA Proceedings of the Section on Quality and Productivity*, pages 276–280. American Statistical Association, Alexandria, VA.

Oluyede, B. O. (1994b). A modified chi-square test of independence against a class of ordered alternatives in an $r \times c$ contingency table. *The Canadian Journal of Statistics*, *22*, 75–87.

Oluyede, B. O. (1994c). Test of independence against a class of ordered alternatives in an $R \times C$ contingency table. *Biometrical Journal*, *36*, 935–951.

Oluyede, B. O. (1994d). Tests for equality of several binomial populations against an order restricted alternative and model selection for one-dimensional multinomials. *Biometrical Journal*, *36*, 17–32.

Oluyede, B. O. and Patterson, R. (1996). On analyzing dichotomous data under order restrictions. In *ASA Proceedings of the Biometrics Section*, pages 334–339. American Statistical Association, Alexandria, VA.

Ouassou, I. and Strawderman, W. E. (2002). Estimation of a parameter vector restricted to a cone. *Statistics and Probability Letters*, *56*(2), 121–129.

Overall, J. E. (1991). A comment concerning one-sided tests of significance in new drug applications. *Journal of Biopharmaceutical Statistics*, *1*, 157–160.

Öztürk, O. (2001). A generalization of Ahmad's class of mann-whitney-wilcoxon statistics. *The Australian and New Zealand Journal of Statistics*, *43*(1), 67–74.

Pace, L. and Salvan, A. (1997). *Principles of Statistical Inference: from a neo Fisherian perspective*. World Scientific Publishing, River Edge, NJ.

Padgett, W. J. and Wei, L. J. (1980). Maximum likelihood estimation of a distribution function with increasing failure rate based on censored observations. *Biometrika*, *67*, 470–474.

Padgett, W. J. and Wei, L. J. (1981). A Bayesian nonparametric estimator of survival probability assuming increasing failure rate. *Communications in Statistics, Part A – Theory and Methods*, *10*, 49–63.

Page, E. B. (1963). Order hypotheses for multiple treatments: A significance test for linear ranks. *Journal of the American Statistical Association*, *58*, 216–230.

Pal, N. and Kushary, D. (1992). On order restricted location parameters of two exponential distributions. *Statistics and Decisions*, *10*, 133–152.

Pan, G. (1996a). Distribution-free confidence procedure for umbrella orderings. *The Australian Journal of Statistics*, *38*, 161–172.

Pan, G. (1996b). Distribution-free tests for umbrella alternatives. *Communications in Statistics, Part A – Theory and Methods*, *25*, 3185–3194.

Pan, G. (1996c). Subset selection with additional order information. *Biometrics*, *52*, 1363–1374.

Pan, G. (1997). Confidence subset containing the unknown peaks of an umbrella ordering. *Journal of the American Statistical Association*, *92*, 307–314.

Pan, G. (2000). Nonparametric methods for checking the validity of prior order information. *Annals of the Institute of Statistical Mathematics*, *52*(4), 680–697.

Pan, G. and Wolfe, D. A. (1995). Distribution-free test based on ranks for comparing groups with umbrella orderings. *Journal of Nonparametric Statistics*, *5*, 409–416.

Pan, G. and Wolfe, D. A. (1996a). Comparing groups with umbrella orderings. *Journal of the American Statistical Association*, *91*, 311–317.

Pan, G. and Wolfe, D. A. (1996b). Likelihood ratio tests for qualitative interaction when variances are unknown. In *ASA Proceedings of the Biopharmaceutical Section*, pages 151–155. American Statistical Association, Alexandria, VA.

Pan, G. and Wolfe, D. A. (1997). Test for qualitative interaction of clinical significance (Corr: 1998V17 p2015-2016). *Statistics in Medicine*, *16*, 1645–1652.

Park, C. (1994). Testing for failure rate ordering between survival distributions. *Journal of the Korean Statistical Society*, *23*, 349–365.

Park, C. G. (1998a). Least squares estimation of two functions under order restriction in isotonicity. *Statistics and Probability Letters*, *37*, 97–100.

Park, C. G. (1998b). Testing for unimodal dependence in an ordered contingency table with restricted marginal probabilities. *Statistics and Probability Letters*, *37*, 121–129.

Park, C. G. (2000a). Likelihood ratio test for homogeneity of steady-state availabilities against order restrictions. *Journal of Statistical Planning and Inference*, *84*(1-2), 159–169.

Park, C. G. (2001). An adaptive test for ordered interquartile ranges among several distributions. *Journal of the Korean Statistical Society*, *30*(1), 63–76.

Park, C. G. (2002). Testing for ordered trends of binary responses between contingency tables. *Journal of Multivariate Analysis*, *81*(2), 229–241.

Park, C. G., Park, T., and Shin, D. W. (1997). Analysis of ordered covariate effects among groups with repeated measurements. *Communications in Statistics, Part A – Theory and Methods*, *26*, 2291–2301.

Park, C. G., Lee, C.-I. C., and Robertson, T. (1998a). Goodness-of-fit test for uniform stochastic ordering among several distributions. *The Canadian Journal of Statistics*, *26*, 69–81.

Park, E. and Lee, Y. J. (2000). Non-parametric test of ordered alternatives in incomplete blocks. *Statistics in Medicine*, *19*(10), 1329–1337.

Park, S.-G. (2000b). Weighted logrank testing procedures for the simple tree alternatives. *Biometrical Journal*, *42*(2), 145–160.

Park, T., Shin, D. W., and Park, C. G. (1998b). A generalized estimating equations approach for testing ordered group effects with repeated measurements. *Biometrics*, *54*, 1645–1653.

Parsian, A. and Farsipour, N. S. (1997). Estimation of parameters of exponential distribution in the truncated space using asymmetric loss function. *Statistical Papers*, *38*, 423–443.

Parzen, E. (1962). On estimating the probability density function and mode. *Annals of Mathematical Statistics*, *33*, 1065–1076.

Patefield, W. M. (1981). Algorithm as159. an efficient method of generating random $r \times c$ tables with given row and column totals. *Journal of the Royal Statistical Society, Series C*, *30*, 91–97.

Patefield, W. M. (1982). Exact tests for trends in ordered contingency tables. *Journal of the Royal Statistical Society, Series C*, *31*, 32–43.

Paula, G. A. (1996). On approximation of the level probabilities for testing ordered parallel regression lines. *Statistics and Probability Letters*, *30*, 333–338.

Paula, G. A. (1999). One-sided tests in generalized linear dose-response models. *Computational Statistics and Data Analysis*, *30*, 413–427.

Paula, G. A. and Artes, R. (2000). One-sided test to assess correlation in linear logistic models using estimating equations. *Biometrical Journal*, *42*(6), 701–714.

Paula, G. A. and Sen, P. K. (1994). Tests of ordered hypotheses in linkage in heredity. *Statistics and Probability Letters*, *20*, 395–400.

Paula, G. A. and Sen, P. K. (1995). One-sided tests in generalized linear models with parallel regression lines (Corr: 96V52 p1164). *Biometrics*, *51*, 1494–1501.

Peace, K. E. (1991). One-sided or two-sided p values: Which most appropriately address the question of drug efficacy? *Journal of Biopharmaceutical Statistics*, *1*, 133–138.

Peddada, S. D. (1997). Confidence interval estimation of population means subject to order restrictions using resampling procedures. *Statistics and Probability Letters*, *31*, 255–265.

Peddada, S. D., Prescott, K. E., and Conaway, M. (2001). Tests for order restrictions in binary data. *Biometrics*, *57*(4), 1219–1227.

Peers, H. W. (1995). Invariant hypothesis testing in order-restricted inference. *Revista Brasileira de Probabilidade e Estatistica (The Brazilian Journal of Probability and Statistics)*, *9*, 99–118.

Pepe, M. S. and Fleming, T. R. (1989). Weighted Kaplan-Meier statistics: A class of distance tests for censored survival data. *Biometrics*, *45*, 497–507.

Perlman, M. D. (1969). One-sided problems in multivariate analysis (Corr: V42 p1777). *Annals of Mathematical Statistics*, *40*, 549–567.

Perlman, M. D. (1971). Multivariate one-sided testing problems involving Fisher's discriminant function. *Sankhya, Series A, Indian Journal of Statistics*, *33*, 19–34.

Perlman, M. D. and Wu, L. (1999). The emperor's new tests (Disc: p370-381). *Statistical Science*, *14*(4), 355–369.

Perlman, M. D. and Wu, L. (2002a). A class of conditional tests for a multivariate one-sided alternatives. *Journal of Statistical Planning and Inference*, *107*, 155–171.

Perlman, M. D. and Wu, L. (2002b). A defense of the likelihood ratio test for one-sided and order-restricted alternatives. *Journal of Statistical Planning and Inference*, *107*, 173–186.

Permutt, T. and Berger, V. W. (2000). A new look at rank tests in ordered $2 \times k$ contingency tables. *Communications in Statistics, Part A – Theory and Methods*, *29*(5-6), 989–1003.

Petroni, G. R. and Wolfe, R. A. (1994). A two-sample test for stochastic ordering with interval-censored data. *Biometrics*, *50*, 77–87.

Pfanzagl, J. (1982). *Contributions to general asymptotic statistical theory*, volume 13 of *Lecture Notes in Statistics*. Springer-Verlag, New York.

Phillips, J. A., Cairns, V., and Koch, G. G. (1992). The analysis of a multiple-dose combination-drug clinical trial using response surface methodology. *Journal of Biopharmaceutical Statistics*, *2*(1), 49–67.

Piegorsch, W. W. (1990). One-sided significance tests for generalized linear models under dichotomous response. *Biometrics*, *46*, 309–316.

Pirie, W. (1983). Jonckheere tests for ordered alternatives. In C. B. R. Samuel Kotz, Norman L. Johnson ; associate editor, editor, *Encyclopedia of Statistical Sciences (Vol. 1)*, pages 315–318. John Wiley and Sons, New York.

Pirie, W. (1985). Page test for ordered alternatives. In C. B. R. Samuel Kotz, Norman L. Johnson; associate editor, editor, *Encyclopedia of Statistical Sciences (Vol. 1)*, pages 553–555. John Wiley and Sons, New York.

Pitman (1936). The closest estimates of statistical estimators. *Proceedings of the Cambridge Philosophical Society, Cambridge Philosophical Society, Cambridge*, *32*, 567–579.

Pitman, E. J, G. (1937). The closest estimates of statistical parameters. *Proceedings of the Cambridge Philosophical Society*, *33*, 212–222.

Pocock, S. J., Geller, N. L., and Tsiatis, A. A. (1987). The analysis of multiple endpoints in clinical trials. *Biometrics*, *43*, 487–498.

Poiraud-Casanova, S. and Thomas-Agnan, C. (2000). About monotone regression quantiles. *Statistics and Probability Letters*, *48*(1), 101–104.

Polonik, W. (1998). The silhouette, concentration functions and ML-density estimation under order restrictions. *Annals of Statistics*, *26*, 1857–1877.

Præstgaard, J. T. and Huang, J. (1996). Asymptotic theory for nonparametric estimation of survival curves under order restrictions. *Annals of Statistics*, *24*, 1679–1716.

Prakasa Rao, B. (1983). Estimation of unimodal density. *Sankya, Ser A, Indian Journal of Statistics*, *31*, 23–36.

Preston, S. R. and Ryan, Thomas A., J. (1995). Sign-scored testing for ordered alternatives in the one-way layout. In *Statistics and manufacturing with subthemes in environmental statistics, graphics and imaging: Computing science and statistics. Proceedings of the 27th Symposium on the Interface*, pages 506–510. Interface Foundation of North America, Fairfax Station, VA.

Puri, M. L. (1965). Some distribution free k-sample rank tests of homogeneity against ordered alternatives. *Communications in Pure and Applied Mathematics*, *18*, 51–63.

Puri, M. L. and Sen, P. K. (1971). *Nonparametric Methods in Multivariate Analysis*. John Wiley, New York.

Puri, M. L. and Sen, P. K. (1985). *Nonparameteric Methods in General Linear Models*. John Wiley and Sons, New York.

Puri, P. S. and Singh, H. (1990). On recursive formulas for isotonic regression useful for statistical inference under order restrictions. *Journal of Statistical Planning and Inference*, *24*, 1–11.

Pyke, R. (1994). Minimax one-sided Kolmogorov-type distribution-free tests. In *Studies in applied probability. Essays in honour of Lajos Takàcs (Journal of Applied Probability, Volume 31A)*, pages 291–308. Applied probability trust, Sheffield.

Qian, S. (1992). Minimum lower sets algorithms for isotonic regression. *Statistics and Probability Letters*, *15*, 31–35.

Qian, S. (1994a). Efficient programs for simulating chi-bar square distributions. In *Computing Science and Statistics; Computationally Intensive Statistical Methods; Proceedings of the*

26th Symposium on the Interface, pages 232–236. Interface Foundation of North America, Fairfax Station, VA.

Qian, S. (1994b). Generalization of least-square isotonic regression. *Journal of Statistical Planning and Inference*, *38*, 389–397.

Qian, S. (1994c). The structure of isotonic regression class for LAD problems with quasi-order constraints. *Computational Statistics and Data Analysis*, *18*, 389–401.

Qian, S. (1996). An algorithm for tree-ordered isotonic median regression. *Statistics and Probability Letters*, *27*, 195–199.

Qian, S. and Eddy, W. F. (1996). An algorithm for isotonic regression on ordered rectangular grids. *Journal of Computational and Graphical Statistics*, *5*, 225–235.

Quan, H., Bolognese, J., and Yuan, W. (2001). Assessment of equivalence on multiple end-points. *Statistics in Medicine*, *20*(21), 3159–3173.

Rachev, S. T., Yakovlev, A. Y., Kadyrova, N. O., and Myasnikova, E. M. (1988). On the statistical inference from survival experiments with two types of failure. *Biometrical Journal*, *30*, 835–842.

Ramirez, D. E. and Smith, P. W. (1998). Applications of smoothed monotone regression splines and smoothed bootstrapping in survival analysis. In *COMPSTAT – Proceedings in Computational Statistics, 13th Symposium*, pages 425–430. Physica-Verlag, Heidelberg.

Ramsay, J. O. (1988). Monotone regression splines in action (diss/res: p442-461). *Statistical Science*, *3*, 425–441.

Ramsay, J. O. and Abrahamowicz, M. (1989). Binomial regression with monotone splines: A psychometric application. *Journal of the American Statistical Association*, *84*, 906–915.

Ramsey, F. L. (1972). A Bayesian approach to bioassay (Com: V29 p225-226, V29 p830). *Biometrics*, *28*, 841–858.

Rao, B. L. S. P. (1969). Estimation of a unimodal density. *Sankhya, Series A, Indian Journal of Statistics*, *31*, 23–36.

Rao, B. L. S. P. (1970). Estimation for distributions with monotone failure rate. *Annals of Mathematical Statistics*, *41*, 507–519.

Rao, B. L. S. P. (1983). *Nonparametric Functional Estimation*. Academic Press, Orlando.

Rao, C. R. (1972). *Linear Statistical Infernce*. John Wiley and Sons, New York.

Raubertas, R. F., Lee, C.-I. C., and Nordheim, E. V. (1986). Hypothesis tests for normal means constrained by linear inequalities. *Communications in Statistics, Part A – Theory and Methods*, *15*, 2809–2833.

Rayner, J. C. W. and Best, D. J. (1999). Nonparametric tests for data in randomised blocks with ordered alternatives. *Journal of Applied Mathematics and Decision Sciences*, *3*, 143–153.

Reiss, R.-D. (1973). On the measurability and consistency of maximum likelihood estimates for unimodal densities. *Annals of Statistics*, *1*, 888–901.

Reiss, R.-D. (1976). On minimum distance estimators for unimodal densities. *Metrika*, *23*, 7–14.

Reitmeir, P. and Wassmer, G. (1996). One-sided multiple endpoint testing in two-sample comparisons. *Communications in Statistics, Part B – Simulation and Computation*, *25*, 99–117.

Reitmeir, P. and Wassmer, G. (1999). Resampling-based methods for the analysis of multiple endpoints in clinical trials. *Statistics in Medicine, 18*, 3453–3462.

Restrepo, A. and Bovik, A. C. (1993). Locally monotonic regression. *IEEE Transactions on Acoustics, Speech, and Signal Processing, 41*, 2796–2810.

Reynolds, Marion R., J. and Arnold, J. C. (1989). Optimal one-sided Shewhart control charts with variable sampling intervals. *Sequential Analysis, 8*, 51–77.

Rice, W. R. and Gaines, S. D. (1994a). Extending nondirectional heterogeneity tests to evaluate simply ordered alternative hypotheses. *Proceedings of the National Academy of Sciences, 91*, 225–226.

Rice, W. R. and Gaines, S. D. (1994b). The ordered-heterogeneity family of tests. *Biometrics, 50*, 746–752.

Richardson, M., Richardson, P., and Smith, T. (1992). The monotonicity of the term premium. *Journal of Financial Economics, 31*, 97–105.

Rindskopf, D. (1983). Parameterizing inequality constraints on unique variances in linear structural models. *Psychometrika, 48*, 73–83.

Rivas Moya, T. (2000). Goodness of fit measure based on sample isotone regression of Mokken double monotonicity model. In H.A.L. Kiers et al., editor, *Data Analysis, Classification, and Related Methods*, pages 267–272. Springer Verlag, Berlin.

Robert, C. P. and Hwang, J. T. G. (1996). Maximum likelihood estimation under order restrictions by the prior feedback method. *Journal of the American Statistical Association, 91*, 167–172.

Robertson, T., Wright, F. T., and Dykstra, R. (1988). *Order Restricted Statistical Inference*. John Wiley and Sons, New York.

Rockafellar, R. T. and Wets, R. (1998). *Variational analysis*. Springer, New York.

Rödel, E. (1996). Projections on convex cones with applications in statistics. In *COMPSTAT 1996. Proceedings in Computational Statistics. 12th Symposium*, pages 423–428. Physica-Verlag, Heidelberg.

Rogers, A. J. (1986). Modified Lagrange multiplier tests for problems with one-sided alternatives. *Journal of Econometrics, 31*, 341–361.

Röhmel, J. and Mansmann, U. (1999). Unconditional non-asymptotic one-sided tests for independent binomial proportions when the interest lies in showing non-inferiority and/or superiority. *Biometrical Journal, 41*, 149–170.

Rojo, J. (1998). Estimation of the quantile function of an IFRA distribution. *Scandinavian Journal of Statistics, 25*, 293–310.

Rojo, J. and Samaniego, F. J. (1993). On estimating a survival curve subject to a uniform stochastic ordering constraint. *Journal of the American Statistical Association, 88*, 566–572.

Rosenberger, W. F., Lachin, J. M., and Bain, R. P. (1995). Nonparametric test of stochastic ordering for multiple longitudinal measures. *Journal of Biopharmaceutical Statistics, 5*, 235–243.

Rosenblatt, M. (1956a). Remarks on some nonparameteric estimates of a density function. *Annals of Mathematical Statistics, 27*, 832–837.

Rosenblatt, M. (1956b). Some regression problems in time series analysis. In *Proceedings of the Third Berkeley Symposium on Mathematical Statistics and Probability, Volume 1*, pages 165–186. University of California Press, Berkeley, CA.

Rothe, G. (1989). Some rank tests for one-sided many-one comparisons. *Statistica Neerlandica, 43*, 91–108.

Roy, J. (1958). Step-down procedures in multivariate analysis. *Annals of Mathematical Statistics*, *29*, 1177–1188.

Roy, S. N. (1953). On a heuristic method of test construction and its use in multivariate analysis. *Annals of Mathematical Statistics*, *24*, 220–238.

Roy, S. N. (1957). *Some Aspects of Multivariate Analysis*. John Wiley and Sons, Asia Publishing, Bombay.

Roy, S. N., Gnanadesikan, R., and Srivastava, J. N. (1971). *Analysis and Design of Certain Quantitative Multiresponse Experiments*. Pergamon Press, New York.

Rueda, C., Salvador, B., and Fernàndez, M. A. (1997). Simultaneous estimation in a restricted linear model. *Journal of Multivariate Analysis*, *61*, 61–66.

Rueda, C., Menéndez, J. A., and Salvador, B. (2002). Bootstrap adjusted estimators in a restricted setting. *Journal of Statistical Planning and Inference*, *107*, 123–131.

Sagara, N. and Fukushima, M. (1994). A hybrid method for solving the nonlinear least squares problem with linear inequality constraints. *Journal of the Operations Research Society of Japan*, *38*, 55–69.

Sakata, T. (1987). Likelihood ratio test for one-sided hypothesis of covariance matrices of two normal populations. *Communications in Statistics, Part A – Theory and Methods*, *16*, 3157–3168.

Sakata, T. (1988). Asymptotic expansion of the null distribution of the likelihood ratio test for the one-sided hypothesis of covariance matrices (Two sample problem). In K. Matusita, editor, *Statistical theory and data analysis II*, pages 379–384. North-Holland, New York.

Saleh, A. K. M. E. and Sen, P. K. (1992). Estimation under uncertain prior information on the Cox proportional hazard model. In *Order Statistics and Nonparametrics: Theory and Applications*, pages 289–303.

Sall, J. (1991). A monotone regression smoother based on ordinal cumulative logistic regression. In *ASA Proceedings of the Statistical Computing Section*, pages 276–281. American Statistical Association, Alexandria, VA.

Sampson, A. R. and Singh, H. (2002). Min and max scorings for two sample partially ordered categorical data. *Journal of Statistical Planning and Inference*, *107*, 219–236.

Sampson, A. R. and Whitaker, L. R. (1989). Estimation of multivariate distributions under stochastic ordering. *Journal of the American Statistical Association*, *84*, 541–548.

Samuel-Cahn, E. (1996). Is the Simes' improved Bonferroni procedure conservative? *Biometrika*, *83*, 928–933.

Samuel-Cahn, E. (1999). A note about a curious generalization of Simes' theorem. *Journal of Statistical Planning and Inference*, *82*, 147–149.

Sarkar, S. K. (1998). Some probability inequalities for ordered MTP_2 random variables: a proof of the Simes conjecture. *Annals of Statistics*, *26*(2), 494–504.

Sarkar, S. K. and Chang, C.-K. (1997). The Simes method for multiple hypothesis testing with positively dependent test statistics. *Journal of the American Statistical Association*, *92*, 1601–1608.

Sarkar, S. K., Snapinn, S., and Wang, W. (1995). On improving the *min test* for the analysis of combination drug trials (Corr: 1998V60 p180-181). *Journal of Statistical Computation and Simulation*, *51*, 197–213.

Sasabuchi, S. (1980). A test of a multivariate normal mean with composite hypotheses determined by linear inequalities. *Biometrika*, *67*, 429–439.

Sasabuchi, S. (1988a). A multivariate one-sided test with composite hypotheses when the covariance matrix is completely unknown. *Memoirs of the Faculty of Science, Kyushu University, Series A, Math*, *42*, 37–46.

Sasabuchi, S. (1988b). A multivariate test with composite hypotheses determined by linear inequalities when the covariance matrix has an unknown scale factor. *Memoirs of the Faculty of Science, Kyushu University, Series A, Math*, *42*, 9–19.

Sasabuchi, S., Inutsuka, M., and Kulatunga, D. D. S. (1992). An algorithm for computing multivariate isotonic regression. *Hiroshima Mathematical Journal*, *22*, 551–560.

Sasabuchi, S., Kulatunga, D. D. S., and Saito, H. (1998). Comparison of powers of some tests for testing homogeneity under order restrictions in multivariate normal means. *American Journal of Mathematical and Management Sciences*, *18*, 131–158.

Sasabuchi, S., Tanaka, K., and Tsukamoto, T. (2003). Testing homogeity of multivariate normal mean vectors under an order restriction when the covaraince matrices are common but unknown. *Annals of Statistics*, *31*, 1517–1536.

Scheffé, H. (1943). On solutions of the behrens-fisher problem based on the t-distribution. *Annals of Mathematical Statistics*, *14*, 35–44.

Scheffé, H. (1953). A method for judging all contrasts in the analysis of variance (Corr: V56 p229). *Biometrika*, *40*, 87–104.

Scheiblechner, H. (1995). Isotonic ordinal probabilistic models. *Psychometrika*, *60*, 281–304.

Scheiblechner, H. (1999). Additive conjoint isotonic probabilistic models (ADISOP). *Psychometrika*, *64*, 295–316.

Schell, M. J. and Singh, B. (1997). The reduced monotonic regression method. *Journal of the American Statistical Association*, *92*, 128–135.

Schervish, M. J. (1983). Multivariate normal probabilities with error bound. *Applied Statistics*, *32*, 81–87.

Schervish, M. J. (1996). P values: What they are and what they are not. *The American Statistician*, *50*, 203–206.

Schoenfeld, D. A. (1986). Confidence bounds for normal means under order restrictions, with application to dose-response curves, toxicology experiments, and low-dose extrapolation. *Journal of the American Statistical Association*, *81*, 186–195.

Scholz, F. W. (1980). Towards a unified definition of maximum likelihood. *Canadian Journal of Statistics*, *8*, 193–203.

Sedransk, J., Monahan, J., and Chiu, H. Y. (1985). Bayesian estimation of finite population parameters in categorical data models incorporating order restrictions. *Journal of the Royal Statistical Society, Series B, Methodological*, *47*, 519–527.

Self, S. G. and Liang, K.-Y. (1987). Asymptotic properties of maximum likelihood estimators and likelihood ratio tests under nonstandard conditions. *Journal of the American Statistical Association*, *82*, 605–610.

Sen, K. and Jain, M. B. (1991). A test for bivariate exponentiality against BIFR alternative. *Communications in Statistics, Part A – Theory and Methods*, *20*, 3139–3145.

Sen, P. K. (1966). On a distribution-free method of estimating asymptotic efficiency of a class of nonparametric tests. *Annals of Mathematical Statistics*, *37*, 1759–1770.

Sen, P. K. (1968). On a class of aligned rank order tests in two-way layouts. *Annals of Mathematical Statistics*, *39*, 1115–1124.

Sen, P. K. (1981). *Sequential Nonparametrics: Invariance Principles and Statistical Inference*. John Wiley and Sons, New York.

Sen, P. K. (1982). The UI-principle and LMP rank tests. In B. V. Gnedenko et al., editor, *Nonparametric Statistical Inference (Colloquia Mathematica Societatis János Bolyai, V 32)*, pages 843–858. North-Holland, Amsterdam.

Sen, P. K. (1983a). A Fisherian detour of the step-down procedure. In *Contributions to Statistics: Essays in honour of Norman L. Johnson*, pages 367–377.

Sen, P. K. (1984). Subhypotheses testing against restricted alternatives for the Cox regression model. *Journal of Statistical Planning and Inference*, *10*, 31–42.

Sen, P. K. (1985). Nonparametric testing against restricted alternatives under progressive censoring. *Sequential Analysis*, *4*, 247–273.

Sen, P. K. (1988a). Combination of statistical tests for multivariate hypotheses against restricted alternatives. In *Proceedings of the International Conference on Advances in Multivariate Statistical Analysis*, pages 377–402.

Sen, P. K. (1988b). Combination of statistical tests for multivariate hypotheses against restricted alternatives. In S. Dasgupta and J. K. Ghosh, editors, *Advances in Multivariate Statistical Analysis*, pages 377–402. Indian Statistical Institute, Calcutta.

Sen, P. K. (1996a). Design and analysis of experiments: Nonparamentric methods with applications to clinical trials. In *Handbook of Statistics (Volume 13): Design and Analysis of Experiments*, pages 91–150. North-Holland, Amsterdam.

Sen, P. K. (1996b). Regression rank scores estimation in ANOCOVA. *Annals of Statistics*, *24*, 1586–1601.

Sen, P. K. (1999a). Multiple comparisons in interim analysis. *Journal of Statistical Planning and Inference*, *82*, 5–23.

Sen, P. K. (1999b). Robust nonparametrics in mixed-MANOCOVA models. *Journal of Statistical Planning and Inference*, *75*, 433–451.

Sen, P. K. (1999c). Some remarks on Simes-type multiple tests of significance. *Journal of Statistical Planning and Inference*, *82*, 139–145.

Sen, P. K. and Bhattacharjee, M. C. (1998). Tests for a property of aging under renewals: Rationality and general asymptotics. In *Frontiers in Probability and Statistics*, pages 328–340. Narosa Publishers, New Delhi.

Sen, P. K. and Puri, M. L. (1967). On the theory of rank order tests for location in the multivariate one sample problem. *Annals of Mathematical Statistics*, *38*, 1216–1228.

Sen, P. K. and Sengupta, D. (1991). On characterizations of Pitman closeness of some shrinkage estimators. *Communications in Statistics, Part A – Theory and Methods*, *20*, 3551–3580.

Sen, P. K. and Silvapulle, M. J. (2002). An appraisal of some aspects of statistical inference under inequality constraints. *Journal of Statistical Planning and Inference*, *107*, 3–43.

Sen, P. K. and Singer, J. M. (1993). *Large Sample Methods in Statistics: An introduction with applications*. Chapman and Hall, London.

Sen, P. K. and Tsai, M.-T. (1999). Two-stage likelihood ratio and union-intersection tests for one-sided alternatives multivariate mean with nuisance dispersion matrix. *Journal of Multivariate Analysis*, *68*, 264–282.

Sen, P. K., Kubokawa, T., and Saleh, A. K. M. E. (1989). The Stein paradox in the sense of the Pitman measure of closeness. *Annals of Statistics*, *17*, 1375–1386.

Sen, P. K. e. (1983b). *Contributions to Statistics: Essays in honour of Norman L. Johnson*. Elsevier, North-Holland.

Sengupta, D. and Sen, P. K. (1991). Shrinkage estimation in a restricted parameter space. *Sankhya, Series A, Indian Journal of Statistics*, *53*, 389–411.

Shaddick, G. and Elliott, P. (1996). Use of Stone's method in studies of disease risk around point sources of environmental pollution. *Statistics in Medicine*, *15*, 1927–1934.

Shaffer, J. P. (1986). Modified sequentially rejective multiple test procedures. *Journal of the American Statistical Association*, *81*, 826–831.

Shanubhogue, A. (1988). Distribution-free test for homogeneity against stochastic ordering. *Metrika*, *35*, 109–119.

Shapiro, A. (1985). Asymptotic distribution of test statistics in the analysis of moment structures under inequality constraints. *Biometrika*, *72*, 133–144.

Shapiro, A. (1987). A conjecture related to chi-bar-squared distributions. *The American Mathematical Monthly*, *94*, 46–48.

Shapiro, A. (1988). Towards a unified theory of inequality constrained testing in multivariate analysis. *International Statistical Review*, *56*, 49–62.

Shapiro, A. (1989). Asymptotic properties of statistical estimators in stochastic programming. *Annals of Statistics*, *17*, 841–858.

Shapiro, A. (2000a). On the asymptotics of constrained local M-estimators. *Annals of Statistics*, *28*(3), 948–960.

Shapiro, A. (2000b). Statistical inference of stochastic optimization problems. In S. P. Uryasev, editor, *Probabilistic constrained optimization: Theory and applications*, pages 91–116. Kluwer Academic Publishers, Boston.

Shea, G. (1979). Monotone regression and covariance structure. *Annals of Statistics*, *7*, 1121–1126.

Shen, Y. and Fisher, L. (1999). Statistical inference for self-designing clinical trials with a one-sided hypothesis. *Biometrics*, *55*, 190–197.

Shi, N.-Z. (1988a). Rank test statistics for umbrella alternatives. *Communications in Statistics, Part A – Theory and Methods*, *17*, 2059–2073.

Shi, N.-Z. (1988b). A test of homogeneity for umbrella alternatives and tables of the level probabilities. *Communications in Statistics, Part A – Theory and Methods*, *17*, 657–670.

Shi, N.-Z. (1988c). A test of homogeneity of regression coefficients against ordered alternatives. In *Statistical theory and data analysis II*, pages 397–408. North-Holland, Amsterdam.

Shi, N.-Z. (1989). Testing for umbrella order restrictions on multinomial parameters. *Sankhya, Series B, Indian Journal of Statistics*, *51*, 13–23.

Shi, N.-Z. (1991). A test of homogeneity of odds ratios against order restrictions. *Journal of the American Statistical Association*, *86*, 154–158.

Shi, N.-Z. (1994). Maximum likelihood estimation of means and variances from normal populations under simultaneous order restrictions. *Journal of Multivariate Analysis*, *50*, 282–293.

Shi, N.-Z. (1995). The minimal L_1 isotonic regression. *Communications in Statistics, Part A – Theory and Methods*, *24*, 175–189.

Shi, N.-Z., Gao, W., and Zhang, B.-X. (2001). One-sided estimating and testing problems for location models from grouped samples. *Communications in Statistics, Part B – Simulation and Computation*, *30*(4), 885–898.

Shin, D. W., Park, C. G., and Park, T. (1996). Testing for ordered group effects with repeated measurements. *Biometrika*, *83*, 688–694.

Shin, D. W., Park, C. G., and Park, T. (2001). Testing for one-sided group effects in repeated measures study. *Computational Statistics and Data Analysis*, *37*(2), 233–247.

Shirley, A. G. (1992). Is the minimum of several location parameters positive? *Journal of Statistical Planning and Inference*, *31*, 67–79.

Shorack, G. R. (1967). Testing against ordered alternatives in Model I analysis of variance; normal theory and nonparametric. *Annals of Mathematical Statistics*, *38*, 1740–1752.

Sikand, Y., Hewett, J. E., and Wright, F. T. (1993). On the performance of some tests for the equality of two discrete distributions with an ordered alternative. *Journal of Statistical Computation and Simulation*, *46*, 201–215.

Silvapulle, M. J. (1981). On the existence of maximum likelihood estimators for the binomial response models. *Journal of the Royal Statistical Society, Series B, Methodological*, *43*, 310–313.

Silvapulle, M. J. (1985). Asymptotic behavior of robust estimators of regression and scale parameters with fixed carriers. *Annals of Statistics*, *13*, 1490–1497.

Silvapulle, M. J. (1991). On limited dependent variable models: Maximum likelihood estimation and test of one-sided hypothesis. *Econometric Theory*, *7*, 385–395.

Silvapulle, M. J. (1992a). On M-methods in growth curve analysis with asymmetric errors. *Journal of Statistical Planning and Inference*, *32*, 303–309.

Silvapulle, M. J. (1992b). Robust tests of inequality constraints and one-sided hypotheses in the linear model. *Biometrika*, *79*, 621–630.

Silvapulle, M. J. (1992c). Robust Wald-type tests of one-sided hypotheses in the linear model. *Journal of the American Statistical Association*, *87*, 156–161.

Silvapulle, M. J. (1994). On tests against one-sided hypotheses in some generalized linear models. *Biometrics*, *50*, 853–858.

Silvapulle, M. J. (1995). A Hotelling's T^2-type statistic for testing against one-sided hypotheses. *Journal of Multivariate Analysis*, *55*, 312–319.

Silvapulle, M. J. (1996a). On an F-type statistic for testing one-sided hypotheses and computation of chi-bar-squared weights. *Statistics and Probability Letters*, *28*, 137–141.

Silvapulle, M. J. (1996b). A test in the presence of nuisance parameters. *Journal of the American Statistical Association*, *91*, 1690–1693.

Silvapulle, M. J. (1997a). A curious example involving the likelihood ratio test against one-sided hypotheses. *The American Statistician*, *51*, 178–180.

Silvapulle, M. J. (1997b). On order restricted inference in some mixed linear models. *Statistics and Probability Letters*, *36*, 23–27.

Silvapulle, M. J. (1997c). Robust bounded influence tests against one-sided hypotheses in general parametric models. *Statistics and Probability Letters*, *31*(1), 45–50.

Silvapulle, M. J. (2000). Tests against qualitative interaction: Exact critical values and robust tests. *Unpublished manuscript.*

Silvapulle, M. J. (2001). Tests against qualitative interaction: Exact critical values and robust tests. *Biometrics*, *57*(4), 1157–1165.

Silvapulle, M. J. and Burridge, J. (1986). Existence of maximum likelihood estimates in regression models for grouped and ungrouped data. *Journal of the Royal Statistical Society, Series B, Methodological*, *48*, 100–106.

Silvapulle, M. J. and Sen, P. K. (1993). Robust tests in group sequential analysis: One- and two-sided hypotheses in the linear model. *Annals of the Institute of Statistical Mathematics*, *45*, 159–171.

Silvapulle, M. J. and Silvapulle, P. (1995). A score test against one-sided alternatives. *Journal of the American Statistical Association*, *90*, 342–349.

Silvapulle, M. J., Silvapulle, P., and Basawa, I. V. (2002). An order restricted semiparametric score test. *Journal of Statistical Planning and Inference*, *107*, 307–320.

Silverman, B. W. (1993). *Density Estimation for Statistics and Data Analysis*. Chapman and Hall, London.

Silvey, S. D. (1959). The Lagrangian multiplier test. *Annals of Mathematical Statistics*, *30*, 389–407.

Simes, R. J. (1986). An improved Bonferroni procedure for multiple tests of significance. *Biometrika*, *73*, 751–754.

Simpson, D. G. and Margolin, B. H. (1986). Recursive nonparametric testing for dose-response relationships subject to downturns at high doses. *Biometrika*, *73*, 589–596.

Singh, B. and Wright, F. T. (1989). The power functions of the likelihood ratio tests for a simply ordered trend in normal means. *Communications in Statistics, Part A – Theory and Methods*, *18*, 2351–2392.

Singh, B. and Wright, F. T. (1990). Testing for and against an order restriction in mixed-effects models. *Statistics and Probability Letters*, *9*, 195–200.

Singh, B. and Wright, F. T. (1993). The level probabilities for a simple loop ordering. *Annals of the Institute of Statistical Mathematics*, *45*, 279–292.

Singh, B. and Wright, F. T. (1996). Testing order restricted hypotheses with proportional hazards. *Lifetime Data Analysis*, *2*, 363–389.

Singh, B. and Wright, F. T. (1998). Comparing survival times for treatments with those for a control under proportional hazards. *Lifetime Data Analysis*, *4*, 265–279.

Singh, B., Schell, M. J., and Wright, F. T. (1993). The power functions of the likelihood ratio tests for a simple tree ordering in normal means: Unequal weights. *Communications in Statistics, Part A – Theory and Methods*, *22*, 425–449.

Singh, H., Singh, R. S., and Qian, S. (2000). Testing uniform stochastic ordering for survival functions. *Communications in Statistics, Part A – Theory and Methods*, *29*(7), 1649–1661.

Small, K. A. (1987). A discrete choice model for ordered alternatives (STMA V29 954). *Econometrica*, *55*, 409–424.

Smith, T. J. (2000). L_1 optimization under linear inequality constraints. *Journal of Classification*, *17*(2), 225–242.

Snapinn, S. M. and Sarkar, S. K. (1996). A note on assessing the superiority of a combination drug with a specific alternative. *Journal of Biopharmaceutical Statistics*, *6*, 241–251.

Snappin, S. M. (1987). Evaluating the efficacy of a combination therapy. *Statistics in Medicine*, *6*, 656–665.

Snidjers, T. (1980). Asymptotic optimality theory for testing problems with restricted alternatives. *Journal of the American Statistical Association*, *75*.

Snijders, T. A. B. (1979). *Asymptotic Optimality Theory for Testing Problems with Restricted Alternatives.* CWI, Math. Centrum [Centrum voor Wiskunde en Informatica], Amsterdam.

Solorzano, E. and Spurrier, J. D. (1999). One-sided simultaneous comparisons with more than one control [corr: vol 64, pp 387–387]. *Journal of Statistical Computation and Simulation*, *63*, 37–46.

Solow, A. R. (2000). Comment on "Multiple comparisons of entropies with application to dinosaur biodiversity" (1999V55 p1300-1305). *Biometrics*, *56*(4), 1272–1273.

Solow, A. R. and Costello, C. J. (2001). A test for declining diversity. *Ecology*, *82*(8), 2370–2372.

Somerville, P., Miwa, T., Liu, W., and Hayter, A. (2001). Combining one-sided and two-sided confidence interval procedures for successive comparisons of ordered treatment effects. *Biometrical Journal*, *43*(5), 533–542.

Stablein, D. M., Novak, J. W., Peace, K. E., Laska, E. M., and Meisner, M. J. (1990). Optimization in clinical trials and combination drug development. In K. E. Peace, editor, *Statistical issues in drug research and development*, pages 263–303. Marcel Dekker, New York.

Stefanski, L. A. and Bay, J. M. (1996). Simulation extrapolation deconvolution of finite population cumulative distribution function estimators. *Biometrika*, *83*, 407–417.

Stein, C. (1956). Inadmissibility of the usual estimator for the mean of a multivariate normal distribution. In *Proceedings of the Third Berkeley Symposium on Mathematical Statistics and Probability, Volume 1*, pages 197–206. University of California Press, Berkeley, CA.

Stepanavage, M., Quan, H., Ng, J., and Zhang, J. (1993). A review of statistical methods for multiple endpoints in clinical trials. In *ASA Proceedings of the Biopharmaceutical Section*, pages 42–47. American Statistical Association, Alexandria, VA.

Stoer, J. and Witzgall, C. (1970). *Convexity and Optimization in Finite Dimensions.* John Wiley and Sons, New York.

Stone, R. A. (1988). Investigations of excess environmental risks around putative sources: Statistical problems and a proposed test. *Statistics in Medicine*, *7*, 649–660.

Stram, D. O. and Lee, J. W. (1994). Variance components testing in the longitudinal mixed effects model (Corr: 95V51 p1196). *Biometrics*, *50*, 1171–1177.

Strand, M. (2000). A generalized nonparametric test for lattice-ordered means. *Biometrics*, *56*(4), 1222–1226.

Strawderman, W. E. (1971). Proper Bayes minimax estimators of the multivariate normal mean. *Annals of Mathematical Statistics*, *42*, 385–388.

Strömberg, U. (1991). An algorithm for isotonic regression with arbitrary convex distance function. *Computational Statistics and Data Analysis*, *11*, 205–219.

Strömberg, U. (1993a). Controlling for a confounder monotonically related to exposure by means of isotonic quantile regression. *Statistics in Medicine*, *12*, 1989–1998.

Strömberg, U. (1993b). Distribution-free estimation of a monotonic p-th quantile function. *Biometrical Journal*, *35*, 601–613.

Stylianou, M. and Flournoy, N. (2002). Dose finding using the biased coin up-and-down design and isotonic regression. *Biometrics*, *58*(1), 171–177.

Sugiura, N. (1999). Generalized Bayes and multiple contrasts tests for simple loop-ordered normal means. *Communications in Statistics, Part A – Theory and Methods*, *28*, 549–580.

Sun, H.-J. (1988a). A FORTRAN subroutine for computing normal orthant probabilities of dimensions up to nine. *Communications in Statistics, Part B – Simulation and Computation*, *17*, 1097–1111.

Sun, H.-J. (1988b). A general reduction method for n-variate normal orthant probability. *Communications in Statistics, Part A – Theory and Methods*, *17*, 3913–3921.

Takemura, A. and Kuriki, S. (1997). Weights of $\bar{\chi}^2$ distribution for smooth or piecewise smooth cone alternatives. *Annals of Statistics*, *25*, 2368–2387.

Takemura, A. and Kuriki, S. (1999). Shrinkage to smooth non-convex cone: Principal component analysis as Stein estimation. *Communications in Statistics, Part A – Theory and Methods*, *28*, 651–669.

Takemura, A. and Kuriki, S. (2002). On the equivalence of the tube and Euler characteristic methods for the distribution of the maximum of Gaussian fields over piecewise smooth domains. *Annals of Applied Probability*, *12*(2), 768–796.

Takeuchi, H. (2001). On the likelihood ratio test for a single model against the mixture of two known densities. *Communications in Statistics, Part A – Theory and Methods*, *30*(5), 931–942.

Tang, D. (1994). Uniformly more powerful tests in a one-sided multivariate problem (Corr: 96V91 p1757). *Journal of the American Statistical Association*, *89*, 1006–1011.

Tang, D.-I. (1998). Testing the hypothesis of a normal mean lying outside a convex cone. *Communications in Statistics, Part A – Theory and Methods*, *27*, 1517–1534.

Tang, D.-I. and Geller, N. L. (1999). Closed testing procedures for group sequential clinical trials with multiple endpoints. *Biometrics*, *55*, 1188–1192.

Tang, D.-I. and Lin, S. P. (1991). Extension of the pool-adjacent-violators algorithm. *Communications in Statistics, Part A – Theory and Methods*, *20*, 2633–2643.

Tang, D.-I. and Lin, S. P. (1994). On improving some methods for multiple endpoints. In *ASA Proceedings of the Biopharmaceutical Section*, pages 466–474. American Statistical Association, Alexandria, VA.

Tang, D.-I. and Lin, S. P. (1997). An approximate likelihood ratio test for comparing several treatments to a control. *Journal of the American Statistical Association, 92*, 1155–1162.

Tang, D.-I., Gnecco, C., and Geller, N. L. (1989a). An approximate likelihood ratio test for a normal mean vector with nonnegative components with application to clinical trials. *Biometrika, 76*, 577–583.

Tang, D.-I., Gnecco, C., and Geller, N. L. (1989b). Design of group sequential clinical trials with multiple endpoints. *Journal of the American Statistical Association, 84*, 776–779.

Tang, D.-I., Geller, N. L., and Pocock, S. J. (1993). On the design and analysis of randomized clinical trials with multiple endpoints. *Biometrics, 49*, 23–30.

Tantiyaswasdikul, C. and Woodroofe, M. B. (1994). Isotonic smoothing splines under sequential designs. *Journal of Statistical Planning and Inference, 38*, 75–87.

Tebbs, J. and Swallow, W. (2003). Estimating ordered binomial proportions with the use of group testing. *Biometrika, 90(2)*, 471–477.

Terpstra, T. J. (1985). One-sided and multi-sided rank tests against trend in the case of concordant observers. *Metrika, 32*, 109–123.

Thakur, A. K. (1984). A FORTRAN program to perform the nonparametric Terpstra-Jonckheere test. *Computer Methods and Programs in Biomedicine, 18*, 235–240.

Ting, N., Burdick, R. K., Graybill, F. A., and Gui, R. (1989). One-sided confidence intervals on nonnegative sums of variance components. *Statistics and Probability Letters, 8*, 129–135.

Tiwari, R. C., Jammalamadaka, S. R., and Zalkikar, J. N. (1989a). A class of large sample tests for bivariate exponentiality versus BIFRA and BNBU-(t_0, t_0) alternatives. *Statistics, 20*, 297–304.

Tiwari, R. C., Jammalamadaka, S. R., and Zalkikar, J. N. (1989b). Testing an increasing failure rate average distribution with censored data. *Statistics, 20*, 279–286.

Tong, Y. L. (1990). *The Multivariate Normal Distribution*. Springer-Verlag, New York.

Tripathi, R. C. and Kale, B. K. (1991). Combined sample scores versus two-sample scores under censoring. *Communications in Statistics, Part A – Theory and Methods, 20*, 2095–2118.

Troendle, J. F. and Legler, J. M. (1998). A comparison of one-sided methods to identify significant individual outcomes in a multiple outcome setting: Stepwise tests or global tests with closed testing. *Statistics in Medicine, 17*, 1245–1260.

Tryon, P. V. and Hettmansperger, T. P. (1973). A class of non-parametric tests for homogeneity against ordered alternatives. *Annals of Statistics, 1*, 1061–1070.

Tsai, K.-T. and Koziol, J. A. (1994). Assessing multiple endpoints with ordered alternatives. *Communications in Statistics, Part A – Theory and Methods, 23*, 533–544.

Tsai, M. (1992). On the power superiority of likelihood ratio tests for restricted alternatives. *Journal of Multivariate Analysis, 42*, 102–109.

Tsai, M.-T. (1993). UI score tests for some restricted alternatives in exponential families. *Journal of Multivariate Analysis, 45*, 305–323.

Tsai, M.-T. (1995). Estimation of covariance matrices under Löwner order restrictions. *Sankhya, Series A, Indian Journal of Statistics, 57*, 433–439.

Tsai, M.-T. (2004). Maximum likelihood estimation of covariance matrices under simple tree ordering. *Journal of Multivariate Analysis, 89*, 292–303.

Tsai, M.-T. and Sen, P. K. (1989). Asymptotic optimality of nonparametric tests for restricted alternatives in some multivariate mixed models. *Journal of Multivariate Analysis*, *29*, 292–307.

Tsai, M.-T. and Sen, P. K. (1990). Asymptotically efficient rank MANOVA tests for restricted alternatives in randomized block designs. *Annals of the Institute of Statistical Mathematics*, *42*, 375–385.

Tsai, M.-T. and Sen, P. K. (1991a). Asymptotic optimality and distribution theory of UI-LMP rank tests for restricted alternatives. *Sankhya, Series B, Indian Journal of Statistics*, *53*, 151–175.

Tsai, M.-T. and Sen, P. K. (1991b). Tests for independence in two-way contingency tables based on canonical analysis. *Calcutta Statistical Association Bulletin*, *40*, 109–123.

Tsai, M.-T. and Sen, P. K. (1993). On the local optimality of optimal linear tests for restricted alternatives. *Statistica Sinica*, *3*, 103–115.

Tsai, M.-T. and Sen, P. K. (1995). A test of quasi-independence in ordinal triangular contingency tables. *Statistica Sinica*, *5*, 767–780.

Tsai, M.-T. and Sen, P. K. (2004). On the inadmissibility of hotelling T^2 test for restricted alternative. *Journal of Multivariate Analysis*, *89*, 87–96.

Tsai, M.-T., Sen, P. K., and Yang, Y.-H. (1994). Power-robustness of likelihood ratio (union-intersection score) tests for some restricted alternative problems. *Statistics and Decisions*, *12*, 231–244.

Tsai, W.-Y. (1988). Estimation of the survival function with increasing failure rate based on left truncated and right censored data. *Biometrika*, *75*, 319–324.

Tsubaki, M. (1995). New weighted sum of χ^2 test for testing a linear hypothesis against a restricted alternative hypothesis. *Communications in Statistics, Part B – Simulation and Computation*, *24*, 987–1004.

Tu, D. (1997). Two one-sided tests procedures in establishing therapeutic equivalence with binary clinical endpoints: Fixed sample performances and sample size determination. *Journal of Statistical Computation and Simulation*, *59*, 271–290.

Turner, T. R. and Wollan, P. C. (1997). Locating a maximum using isotonic regression. *Computational Statistics and Data Analysis*, *25*, 305–320.

Uesaka, H. (1998). Validity and applicability of tests for ordered alternatives with binary response for a clinical dose-response study. *Journal of the Japanese Society of Computational Statistics*, *11*, 121–136.

van der Vaart, A. W. (2000). *Asymptotic Statistics*. Cambridge University Press, Cambridge, England.

van Eaden, C. and Zidek, J. V. (2002). Combining sample information in estimating ordered normal means. *Sankhya Ser A, Indian Journal of Statistics*, *64*, 588–610.

van Es, B., Jongbloed, G., and van Zuijlen, M. (1998). Isotonic inverse estimators for non-parametric deconvolution. *Annals of Statistics*, *26*(6), 2395–2406.

van Soest, A. and Kooreman, P. (1990). Coherency of the indirect translog demand system with binding nonnegativity constraints. *Journal of Econometrics*, *44*, 391–400.

van Soest, A., Kapteyn, A., and Kooreman, P. (1993). Coherency and regularity of demand systems with equality and inequality constraints. *Journal of Econometrics*, *57*, 161–188.

VanEs, B., Jongbloed, G., and vanZuijlen, M. (1998). Isotonic inverse estimators for nonparametric deconvolution. *Annals of Statistics*, *26*(6), 2395–2406.

Vardi, Y. (1989). Multiplicative censoring, renewal process, deconvolution and decreasing density: nonparametric estimation. *Biometrika*, *76*, 751–761.

Vermunt, J. K. (2001). The use of restricted latent class models for defining and testing nonparametric and parametric item response theory models. *Applied Psychological Measurement*, *25*(3), 283–294.

Vu, H. T. V. and Maller, R. A. (1996). The likelihood ratio test for Poisson versus binomial distributions. *Journal of the American Statistical Association*, *91*, 818–824.

Vu, H. T. V. and Zhou, S. (1997). Generalization of likelihood ratio tests under nonstandard conditions. *Annals of Statistics*, *25*, 897–916.

Wada, C. Y. and Hotta, L. K. (2000). Restricted alternatives tests in a bivariate exponential model with covariates. *Communications in Statistics, Part A – Theory and Methods*, *29*(1), 193–210.

Wada, C. Y. and Sen, P. K. (1992). Restricted alternatives test in a competing risk logistic model. *Revista Brasileira de Probabilidade e Estatistica (The Brazilian Journal of Probability and Statistics)*, *6*, 69–84.

Wada, C. Y. and Sen, P. K. (1995). Restricted alternative test in a parametric model with competing risk data. *Journal of Statistical Planning and Inference*, *44*, 193–203.

Wada, C. Y., Sen, P. K., and Tarumoto, M. (1993). Restricted alternative tests in a proportional hazard model with competing risk data. *Revista Brasileira de Probabilidade e Estatistica (The Brazilian Journal of Probability and Statistics)*, *7*, 21–35.

Wan, A. T. K. (1996). On the bias and mean square error of the least square estimator in a regression model with two inequality constraints and multivariate t error terms. *Communications in Statistics, Part A – Theory and Methods*, *25*, 2079–2091.

Wan, A. T. K. and Ohtani, K. (2000). Minimum mean-squared error estimation in linear regression with an inequality constraint. *Journal of Statistical Planning and Inference*, *86*(1), 157–173.

Wang, J. (1996a). Asymptotics of least squares estimators for constrained nonlinear regression. *Annals of Statistics*, *24*, 1316–1326.

Wang, J.-L. (1986). Asymptotically minimax estimators for distributions with increasing failure rate. *Annals of Statistics*, *14*, 1113–1131.

Wang, J.-L. (1987). Estimators of a distribution function with increasing failure rate average. *Journal of Statistical Planning and Inference*, *16*, 415–427.

Wang, S.-J. (1998). A closed procedure based on Follmann's test for the analysis of multiple endpoints. *Communications in Statistics, Part A – Theory and Methods*, *27*, 2461–2480.

Wang, S.-J. and Hung, H. M. J. (1997). Large sample tests for binary outcomes in fixed-dose combination drug studies. *Biometrics*, *53*, 498–503.

Wang, W. (1997). Optimal unbiased tests for equivalence in intrasubject variability. *Journal of the American Statistical Association*, *92*, 1163–1170.

Wang, Y. (1994). A Bartlett-type adjustment for the likelihood ratio statistic with an ordered alternative. *Statistics and Probability Letters*, *20*, 347–352.

Wang, Y. (1995a). Asymptotic expansions of the likelihood ratio statistic with ordered hypotheses. *Sankhya, Series A, Indian Journal of Statistics*, *57*, 410–423.

Wang, Y. (1995b). The L_1 theory of estimation of monotone and unimodal densities. *Journal of Nonparametric Statistics*, *4*, 249–261.

Wang, Y. (1996b). The L_2 risk of an isotonic estimate. *Communications in Statistics, Part A – Theory and Methods*, *25*, 281–294.

Wang, Y. (1996c). A likelihood ratio test against stochastic ordering in several populations. *Journal of the American Statistical Association*, *91*, 1676–1683.

Wang, Y. and McDermott, M. P. (1998a). Conditional likelihood ratio test for a nonnegative normal mean vector. *Journal of the American Statistical Association*, *93*, 380–386.

Wang, Y. and McDermott, M. P. (1998b). A conditional test for a non-negative mean vector based on a Hotelling's T^2-type statistic. *Journal of Multivariate Analysis*, *66*, 64–70.

Wang, Y. and McDermott, M. P. (2001). Uniformly more powerful tests for hypotheses about linear inequalities when the variance is unknown. *Proceedings of the American Mathematical Society*, *129*(10), 3091–3100.

Wassmer, G., Reitmeir, P., Kieser, M., and Lehmacher, W. (1999). Procedures for testing multiple endpoints in clinical trials: An overview. *Journal of Statistical Planning and Inference*, *82*, 69–81.

Wegman, E. J. (1970a). Maximum likelihood estimation of a unimodal density function. *Annals of Mathematical Statistics*, *41*, 457–471.

Wegman, E. J. (1970b). Maximum likelihood estimation of a unimodal density, II. *Annals of Mathematical Statistics*, *41*, 2169–2174.

Wei, L. J. and Knuiman, M. W. (1987). A one-sided rank test for multivariate censored data. *The Australian Journal of Statistics*, *29*, 214–219.

Werner, H. J. and Yapar, C. (1995). More on partitioned possibly restricted linear regression. In T. K. E.-M. Tiit and H. Niemi, editors, *New Trends in Probability and Statistics Volume 3: Multivariate Statistics and Matrices in Statistics (Proceedings of the 5th Tartu Conference)*, pages 57–66. Utrecht, The Netherlands.

Westfall, P. H., Ho, S.-Y., and Prillaman, B. A. (2001). Properties of multiple intersection-union tests for multiple endpoints in combination therapy trials. *Journal of Biopharmaceutical Statistics*, *11*(3), 125–138.

Wijsmann, R. A. (1979). Constructing all smallest simultaneous confidence sets in a general class with applications to MANOVA. *Annals of Statistics*, *7*, 1003–1018.

Winsberg, S. and Ramsay, J. O. (1982). Monotone splines: A family of transformations useful for data analysis. In *COMPSTAT 1982, Proceedings in Computational Statistics*, pages 451–456. Physica-Verlag, Heidelberg.

Winsberg, S. and Ramsay, J. O. (1983). Monotone spline transformations for dimension reduction. *Psychometrika*, *48*, 575–595.

Wolak, F. A. (1987a). Discussion on "regression topics: Nonnested hypothesis tests, constrained estimation, and extrapolation". In *ASA Proceedings of the Business and Economic Statistics Section*, pages 399–401. American Statistical Association, Alexandria, VA.

Wolak, F. A. (1987b). An exact test for multiple inequality and equality constraints in the linear regression model. *Journal of the American Statistical Association*, *82*, 782–793.

Wolak, F. A. (1988). Duality in testing multivariate hypotheses. *Biometrika*, *75*, 611–615.

Wolak, F. A. (1989a). Local and global testing of linear and nonlinear inequality constraints in nonlinear econometric models. *Econometric Theory*, *5*, 1–35.

Wolak, F. A. (1989b). Testing inequality constraints in linear econometric models. *Journal of Econometrics*, *41*, 205–235.

Wolak, F. A. (1991). The local nature of hypothesis tests involving inequality constraints in nonlinear models. *Econometrica*, *59*, 981–995.

Wollan, P. C. and Dykstra, R. L. (1987). [Algorithm AS 225] Minimizing linear inequality constrained Mahalanobis distances. *Applied Statistics*, *36*, 234–240.

Woodroofe, M. and Sun, J. (1993). A penalized maximum likelihood estimate of $f(0+)$ when f is nonincreasing. *Statistica Sinica*, *3*, 501–515.

Woodroofe, M. and Sun, J. (1999). Testing uniformity versus a monotone density. *Annals of Statistics*, *27*, 338–360.

Woodroofe, M. and Zhang, R. (1999). Isotonic estimation for grouped data. *Statistics and Probability Letters*, *45*, 41–47.

Wright, F. T. (1988). The one-way analysis of variance with ordered alternatives: A modification of Bartholomew's E^2 test. *The Canadian Journal of Statistics*, *16*, 75–85.

Wright, F. T. (1990). Bounds on distributions arising in order restricted inference: The partially ordered case. *Sankhya, Series B, Indian Journal of Statistics*, *52*, 158–173.

Wu, P. X. and King, M. L. (1994). One-sided hypothesis testing in econometrics: A survey. *Pakistan Journal of Statistics, Series A*, *10*, 261–300.

Wynn, H. P. (1975). Integrals for one-sided confidence bounds: A general result. *Biometrika*, *62*, 393–396.

Xiong, C. and Milliken, G. A. (2000). Changepoints in stochastic ordering. *Communications in Statistics, Part A – Theory and Methods*, *29*(2), 381–400.

Yancey, T. A., Judge, G. G., and Bohrer, R. (1989). Sampling performance of some joint one-sided preliminary test estimators under squared error loss. *Econometrica*, *57*, 1221–1228.

Zarantonello, E. H. (1998). Projections on convex sets in Hilbert spaces and spectral theory. In E. H. Zarantonello, editor, *Contributions to Nonlinear Functional Analysis*, pages 237–424. Academic Press, New York.

Zelazo, P. R., Zelazo, N. A., and Kolb, S. (1972). Walking in the newborn. *Science*, *176*, 314–315.

Zelterman, D. (1990). On tests for qualitative interactions. *Statistics and Probability Letters*, *10*, 59–63.

Zelterman, D. and Kershner, R. P. (1987). Tests for qualitative interactions. *Proceedings of the Biopharmaceutical Section, Alexandria, Virginia: American Statistical Association*, pages 131–133.

Zeng, Q. and Davidian, M. (1997). Testing homogeneity of intra-run variance parameters in immunoassay. *Statistics in Medicine*, *16*, 1765–1776.

Zequeira, R. and Valdés, J. (2001). An approach for the Bayesian estimation in the case of ordered piecewise constant failure rates. *Reliability Engineering and System Safety*, *72*(3), 227–240.

Zhang, C. H. (2002). Risk bounds in isotonic regression. *Annals of Statistics*, *30*, 528–555.

Zhang, R., Kim, J., and Woodroofe, M. (2001). Asymptotic analysis of isotonic estimation for grouped data. *Journal of Statistical Planning and Inference*, *98*(1-2), 107–117.

Zhao, L. and Peng, L. (2002). Model selection under order restriction. *Statistics and Probability Letters*, *57*(4), 301–306.

Zhou, X.-H. and Gao, S. (2000). One-sided confidence intervals for means of positively skewed distributions. *The American Statistician*, *54*(2), 100–104.

Zielinski, J. M., Wolfson, D. B., Nilakantan, L., and Confavreux, C. (1993). Isotonic estimation of the intensity of a nonhomogeneous Poisson process: The multiple realization setup. *The Canadian Journal of Statistics*, *21*, 257–268.

Index

Active constraints, 191
Adaptive test, 180
  See also semiparametric test, 180
Admissibility, 381, 401, 403
  of Hotelling's T^2, 402
Admissible treatment, 431
Aging property, 357, 369–370
Agresti-Coull test for marginal homogeneity, 331
Aligned rank statistic, 251
Angle
  between vectors, 115
  obtuse, 76, 115
Approximate LRT, see Tang-Gnecco-Geller test, 419
Approximating cone, 188
  see also tangent cone, 188
ARCH effect
  Lee-King test for, 181
  local score test for, 182
  testing for the presence of, 182
ARCH effect, local score test, 18
Azzalini-Cox test, see cross-over interaction, 451
B4, 23
Backward average, 364
Bahadur
  deficiency, 274
  efficiency, 269
Ballot theorem, 270
Bandwidth, 351
Bayes
  estimator, 382
    of binomial parameters, 298
    under absolute error loss, 382
    under squared error loss, 395
    with uniform prior on the positive orthant, 395
  optimality, 379
  risk, 382
  test, 400–401
  theorem, 400
Best average power, 402
Binomial parameters
  Bayes estimator, 298
  bootstrap test, 297
  ordered hypothesis, 11, 160, 286
Binomial samples
  comparison of several populations, 293
  comparison using exact conditional test, 327
  connection to isotonic regression, 293
  LRT for Type A testing problem, 293
  LRT for Type B testing problem, 294
  matrix hypothesis, 296
  one-way layout, 293
  pairwise contrasts, 296
Bioassay, 229, 346
Bioinformatics, 266, 275
Biological assay; see bioassay, 346
Block
  see PAVA, 47
Bootstrap test
  on binomial parameters, 297

ordered normal means, 34
Bounded influence tests under constraints, 27
Bounds test, 151
Canonical analysis, 276
Canonical correlation, 223
Censoring
Type I, 352
Type II, 352
Central limit theorem
permutational, 246
Chernoff regular, 187
See also approximating cone, 187
Chi-bar-square
bounds on the tail probability, 80
distribution, 75, 77
computing tail probability, 78
statistic ($\bar{\chi}^2$), 78
weights, 77, 81
closed forms for $p < 4$, 80
computation for polyhedral, 79
computation when the cone is not a polyhedral, 80
Chi-square with zero degrees of freedom, 66
Clark regularity, 214
Classification, 276
Cohen-Sackrowitz directed test, 455
COM[$\mathcal{C}$], 456
see cone order monotone, 456
Comparison of treatments
formulation in terms of odds ratio, 12, 301
noninferiority, 409
primary variables, 409
secondary variables, 409
superiority, 409
with interim analyzes, 410
Comparison of two means
O'Brien's tests, 412
with unequal covariance matrices, 415
Comparison of two treatments, 412
bootstrap test of superiority, 421
multivariate binary response, 415
Pocock-Geller-Tsiatis global test, 415
testing for departure in a specified direction, 411
Complete class
essentially, 404
theorem, 403
Condition A, 173
Condition LS, 180
Condition Q, 146
Condition Q2, 198
Condition Q3, 199
Conditional p-value, 326
Conditional test
exact, 328
in $r \times c$ tables, 328
multinormal mean, 105
see Perlman-Wu test, 106
see Wang-McDermott test, 105
Conditionality principal, 403
Cone of tangents, 184
Cone order monotone, 456
Cone
dual, 67
face of, 124
polar, 67, 76, 113
polyhedral, 122
with vertex at $\boldsymbol{x}_0$, 60
Confidence interval for a scalar parameter, 298
Consistency
of a constrained estimator, 146
of a test, 107
Consistency, see optimality, 107
Constrained hypothesis, 26
Constrained minimum of a quadratic function, 36
Constrained mle, 64
Constraint qualification
Cottle, 218
Kuhn-Tucker, 218
Mangasarian-Fromowitz, 190
Constraint
active, 191
pairwise, 42
Contingency tables
pseudo random, 459
Continuation odds ratio
see odds ratio, 302
Convex
hull, 122
set, 60
Convex;See also near convexity, 214
Coupling method, 231
Covariance matrix
permutation, 280
Cross-over interaction, 435
Azzalini-Cox test, 451
Berger's test, 453
Gail-Simon test, 442
Pan-Wolfe test, 450
Piantadosi-Gail range test, 447
robust test, 444
testing for the presence of, 452
with more than two treatments, 451
Zelterman's test, 449
Cross-product ratio
see odds ratio, 299
CSD (cumulative sum diagram), 48
Cumulative sum diagram, 48
Data mining, 276
Decision rule, 380–381
R-better, 381
Bayesian, 381
minimax, 381

optimal, 381
uniformly minimum risk, 381
Decision theory, 1
multiple, 270
statistical, 1, 379
Decreasing mean residual life, 358
Density estimation, 351
bandwidth, 351
Density
log-concave, 349
strongly unimodal, 349
Diagonal symmetry, 254
Direct sum of linear spaces, 113
Directed test of Cohen and Sackrowitz, 455
Directed test
against stochastic order, 458
exponential family, 463
Directed vertex peeling[DVP], 463
Distance statistic, 154, 176
Distance
$\boldsymbol{V}$-distance, 70
between a point and a plane, 112
between a point and a set, 188
between two points, 112
with respect to $\|,\|_{\boldsymbol{V}}$, 69
with respect to $\langle \boldsymbol{x},\boldsymbol{y}\rangle_{\boldsymbol{V}}$, 69
Distribution
$\bar{F}$, 29, 99
chi-bar-square, 75
chi-square with zero degrees of freedom, 66
hypergeometric, 324
see also $\bar{F}$-test, 29
Distributional robustness, 345
DMRL (decreasing mean residual life), 358, 362
Dual cone, see cone, 67
DVP, see directed vertex peeling, 463
E-bar square
$\bar{E}^2$, 35
connection to LRT, 35
difference between $\bar{F}$ and $\bar{E}^2$, 96
E-bar Square
equivalence to likelihood ratio test, 35
E-bar square
equivalence to likelihood ratio test, 97
EB (Empirical Bayes), 379
Elliptically symmetric, 110
Empirical distribution function, 347
Empirical likelihood, 194, 199
Emprical Bayes, 379
End point, see also multiple end points, 420
Estimating equation, 173
Estimating function, 173
Euclidean space, 61
Exact conditional p-value
computation by simulation, 327
Exact conditional test
Fisher's, 324
in 2×2 tables, 324
of stochastic order in $2 \times c$ tables, 325
Expected loss, 380
Exponential family
directed test, 463
ordered means, 50
Face
of a cone, 124
relative interior of, 124
False discovery rate, 270
Family-wise error, 270
FDR (false discovery rate), 270
Finite union-intersection test, 224
Fisher's exact test, 324
Follman's test
extension to two samples, 418
FUIT (finite union-intersection test), 224
Functional parameter, 1
Functional parameter, see shape constraint, 1
FWE (family-wise error), 270
Gail-Simon test, see cross-over interaction, 442
GCM (Greatest Convex Minorant), 48
Generalized additive model, 276
Generalized Estimating Equation, 144
Generalized linear model, 11, 15, 143, 156, 159, 173, 285
ordered Bayesian inference, 396
Generalized Method of Moments, 144
Generator of a polyhedral, 122
Global statistic, 415
Greatest convex minorant, 48
Group-sequential procedure, 274
Half-space, 61
Hausdorff distance, 188
Hazard rate order, 302
HB (hierarchical Bayes), 379
Hierarchical Bayes, 379
Hilbert space
projections onto convex sets, 112
Hille's lemma, 359–360, 376
Hotelling's T^2, 101, 222
extension to constrained hypotheses, 101
inadmissibility, 404
relationship to LRT, 234
Hull
conical, 122
positive, 122
Hypothesis
matrix, 296
IFR (increasing failure rate), 357
Inadmissibility of mle, 384
Increasing failure rate, 357
Independence of direction and length of N(0 I), 125
Independent binomial samples

asymptotics of LRT, 294
nonsimilarity of LRT, 294
Inequality constraint, 25
Inner product, 112
Interaction
cross-over, 435
qualitative, 435
quantitative, 435
Interim analysis, 13, 274
Intersection union test, 236
UMP monotone property, 466
Intrablock ranks, 249, 278
Invariant
orthogonally, 125
Isomorphism, 113
Isotonic, 45
Isotonic regression, 42–43
closed form for simple order, 47
computation, 46
connection to quadratic program, 46
for simple order, 43
least squares, 46
ordered means of exponential family, 50
relationship to pairwise constraints, 43
relationship to quasi-order, 44
relationship to weighted least squares, 43
simple order, 47
with respect to L_p, 46
with respect to weights, 46
Jonkheere Test, 250
Kühn-Tücker multiplier test, 166
Kaplan-Meier estimator, 353
Kernel
multinomial loglikelihood, 316
of loglikelihood, 63
Knowledge discovery, 276
Lagrange multiplier
statistic, 166
test, 166
Largest root criterion, 223
Laska-Tang-Meisner tests, 430
Latin square, 267
Layer alternative, 258
LCM (least concave majorant), 348
Least concave majorant, 348
Least favorable
null distribution, 91
null value for Type B problem, 91
tail probability near the null value, 310
Level probabilities, 77
Life distribution, 354
Likelihood principle, 24
Likelihood ratio test
Type A problem on normal mean, 83
Type B problem on normal mean, 90
against simple stochastic order, 309
least favorable null value, 309
asymptotic distribution for Type A problems, 149
binomial parameters, 292
equivalence to a distance statistic, 149
equivalence to score and Wald tests, 155
equivalence to score and Wald tests for constrained hypotheses, 176
in $2 \times c$ table, 307
in $r \times c$ tables, 317
normal theory in two dimensions, 63
Likelihood
concentrated, 97
nonparametric, 346
profile, 97, 346
pseudo, 346
Linear model
normal theory, 95
Type A test, 97
Type B test, 97
Lipchitz continuity, 149
LMPR (locally most powerful rank test), 253
Local optimality, 244
Local score-type test, 173–175, 179
connection to $C(\alpha)$-test, 179
Local score test, 169, 175–177
equivalence to LRT, 178
Locally most mean powerful score test for ARCH, see Lee-King test, 181
Locally most powerful rank test, 253
Log odds ratio
see odds ratio, 314
Logically related hypothesis testing, 270
Loglikelihood
kernel, 63, 67
quadratic approximation, 147
Loss function, 380
quadratic error, 387
zero-one valued, 380
LRHT (logically related hypothesis testing), 270
LRT; See likelihood ratio test, 59
Maintained hypothesis, 4
Mangasarian-Fromowitz constraint qualification, 190
Marginal homogeneity
computation of exact p-value by simulation, 332
exact conditional test, 331
in square contingency tables, 322
Matched pairs
connection to square tables, 322
ordinal response, 331
Matrix hypothesis, 296
Mean remaining life
see mean residual life, 362
Mean residual life, 362
Measure of agreement, 249

Median unbiased estimator, 383
Min test, 422, 430
  extension to multivariate response, 430
  optimality of, 429
  see also Laska-Tang-Meisner tests, 430
  uniformly most powerful monotone property, 430
Minimaxity, 379
Minimum risk estimability, 379
Mode, 354
Monotone
  cone order, 456
  critical region, 428
  density, 346
  dependence, 298
  failure rate, 346
  regression, 44
  test, 428
MRL (mean residual life), 362
MTP (multiple testing procedure), 270
Multiparameter hypothesis testing, 222, 402–403
Multiple comparisons procedure, 270
Multiple decision theory, 270
Multiple end points, 15, 407, 420
Multiple regression
  restricted alternative, 261
Multiple testing, 89, 239, 270, 273
  conservative property of, 271
  procedure, 270
NBU (new better than used), 357
NBUE (new better than used in expectation), 357
Near convexity, 214
Neural network, 276
New better than used, 357
New better than used in expectation, 357
New worse than used, 357
Neyman-Scott problem, 221
Noninferiority of a treatment, 409
Nonlinear model, 156, 180, 345
Nonparametric mle, 348
Nonparametrics, 243, 345
Norm[$\|.\|_V$], 112
Normal mean
  test when covariance matrix is known, 63
  test when covariance matrix is unknown, 100
Nuisance parameter space
  sensitivity of the p-value to changes in, 310
Null hypothesis
  constrained, 38
NWU (new worse than used), 357
O'Brien's test
  comparison of two normal means, 412
  nonparametric test, 413
  parametric test, 412
Observed significance level, 267
Odds ratio, 143, 162, 298–299
  asymptotic distribution of, 300
  confidence interval, 300
  continuation, 302–303
  cumulative, 300, 303
  generalized, 303
  global, 303
  local, 12, 162, 300, 303
  log, 314
  relationship among, 302
Optimal Bayes procedure, 379
Optimality
  consistency of test, 107
  King-Wu locally optimal test, 181
  of LP, 346
  of mle, 346
  test on normal mean, 107
Order
  antisymmetric, 45
  convex, 369
  hazard rate, 302
  likelihood ratio, 303–304
  partial, 45
  quasi, 44–45
  reflexive, 45
  restriction, 25
  simple, 43
  simple stochastic, 303
  stochastic, 304
  transitive, 45
  tree, 43
  uniform stochastic, 303–304
Ordered alternative
  linear model, 97
  one-way layout, 27
  two-way layout, 248
Ordered categorical data, 275
Ordered hypothesis, 26
Ordinal response
  test against nonlinear constraints, 307
Orthogonal, 134
Orthogonal projection, see projection, 116
Orthogonal with respect to $\boldsymbol{V}$
  complement of a set, 113
  sum of sets, 113
Orthogonal
  $\boldsymbol{V}$-orthogonal, 70
  complement, 113
  decomposition, 113
  sum of two sets, 113
  vectors, 70, 112
  with respect to $\langle . \rangle_{\boldsymbol{V}}$, 70
Orthogonally invariant, 125
OSL (observed significance level), 267
Page's Test, 250
Paired sample, 229, 235, 277, 343
Pan-Wolfe test, see cross-over interaction, 450

Partial likelihood, 194
PAVA (pool adjacent violators algorithm), 47
PCC (Pitman closeness criterion), 396
PCLT (permutational central limit theorem), 246
PCLT
  a primitive approach to proof, 278
Perlman-Wu conditional test, 106
Perlman
  $\mathcal{U}$-statistic, 151
  test, 100–101, 151
Permutation covariance matrix, 280
Permutational central limit theorem, 246
Piantadosi-Gail range test, see cross-over interaction, 447
Pitman closeness criterion, 396
Pitman measure of closeness, 380
PMC (Pitman measure of closeness), 380
Polar cone, 76, 113
  see cone, 67
Polar coordinates, transformation to, 125
Polyhedral
  cone, 122
  defined by tight(non-redundant) constraints, 122
  definition, 61
  explicit formulas for the polar cone, 122
  generator, 122
  Minkowski's theorem, 123
  Weyl's theorem, 123
Pool adjacent violators algorithm, 47
Positive rule estimator, 391
Posterior
  confidence, 401
  density, 381, 395
  median, 383
  Pitman closeness, 382
PPC (posterior Pitman closeness), 382
Preorder, 456
Primary variables, see comparison of treatments, 409
Principal component models, 276
Prior distribution, 242
Pro-regular
  see also near convexity, 213
Pro regularity of a set, 213
Probability integral transformation, 267
Product limit estimator, 353
Profile likelihood
  see likelihood, 346
Projected test of Cohen-Kemperman-Sackrowitz, 463
Projected test
  See also directed test, 463
Projection
  $\boldsymbol{V}$-projection, 70
  least square, 46
  matrix, 113–114
  onto a cone, 64
  onto a cone inside a linear space, 118
  onto a set, 112
  onto convex cone, 113
  onto convex sets in Hilbert space, 112
  onto linear spaces, 112
  orthogonal with respect to $\boldsymbol{V}$, 116
  pursuit, 276
  with respect to $\boldsymbol{V}$, 70
  with respect to inner product($\langle\cdot\rangle_{\boldsymbol{V}}$), 46
Publication bias, 265
Pythagoras theorem in an inner product space, 112
Quadratic function
  constrained minimum, 36
Quadratic program, 37
Qualitative data models, 275
Qualitative interaction, 435
Quantitative interaction, 435
Quasi-likelihood, 194
Randomized block design, 253
Rank collection matrix, 257, 259
Regression trees, 276
Regular models, 144
Relative interior, 124
Reliability theory, 357–358
Residual sum of squares, 31
Restricted MLE, 389
Restricted alternative
  multisample, 233
Risk, 379–381, 384
  Bayes, 382
  dominance in, 387
  minimum, 379
  see also expected loss, 380
Robust test
  bounded influence, 27
  in general parametric models, 27
Robust tests in the linear model, 27
  Pitman efficiency, 27
Robust
  against outliers, 444
  test, 160
Rotational symmetry, 254
Sample mean, 63
Score test
  a general local, 179
  an overview, 168
  application to time series, 180
  equivalence to LRT and Wald tests, 155
  local, 177
Score
  function, 145
  global statistic, 155
  statistic, 153, 155
  test, 153
SDT (statistical decision theory), 1, 379

Secondary variables, see comparison of treatments, 409
Semiparametric test
  efficient, 180
Semiparametrics, 179, 345
Separating plane, 115
Set containment argument, 91, 109
Shape constraint, 346
Shape constraint on hazard, 357
Shape constraint
  hypothesis testing, 369
  increasing failure rate, 358
  on density, 44
  on regression function, 44
Shrinkage estimation, 379, 387–388
  connection to PCC, 396
  dominance over MLE, 388
Sign testing, 422
  for log-odds ratio in $2 \times c$ tables, 314
Significance
  practical, 409
Simes-type procedure, 264, 269
Simes theorem
  relationship to Ballot theorem, 270
Simple independent action model, 11
Simple stochastic order
  LRT for, 311
  test against, 308
Simultaneous confidence interval, 226
Simultaneous confidence set, 222–223, 233
Smoothing the isotonic estimator, 349
Smoothing
  a general result on, 351
  kernel, 347
  regression, 345
Spacing, 372
  cumulative normalized, 372
  for exponential distribution, 372
  normalized, 364, 372, 377
Spending function, 273
Spherical symmetry, 254
SRE (Stein-rule estimator), 389
SRMLE (Stein-rule MLE), 389
Statistical decision theory, see decision theory, 379
Stein's paradox, 384
Stein-rule
  MLE, 389
  estimation, see shrinkage estimation, 380
  estimator, 389
Stein paradox, 384
Step-down procedure, 227
Stochastic order, 245
  hazard rate, 302
  simple, 289
Stochastically larger
  see stochastic order, 303
Strongly unimodal, 357
Suborthant model, 393
Superiority of a treatment, 409
Superiority
  test for, see Laska-Tang-Meisner tests, 430
Symmetry
  elliptical, 254
  rotational, 254
  spherical, 254
  total, 254
Synergism
  test against, 11
Tang-Gnecco-Geller test, 419
Tangent cone
  Boulingard, 186
  derived, 186
  examples, 191
  ordinary, 186
  see also approximating cone, 188
  see also cone of attainable directions, 186
  see cone of tangents, 184
Tangent
  derivable, 186
  examples, 186
Test on normal mean
  unknown covariance, 100
Test
  $\bar{F}$-test, 29, 96
  F_Q-test, 440
  an identified treatment is the best, 422
  consistency, 107
  distance, 176
  global score, 166
  Hausman-Wald-type, 166
  Kühn-Tücker multiplier, 166
  local score, 169
  min test, 422
  Rao's score, 174
  sign testing, 422
  similar, 84
  Type A and Type B problems, 61
  uniformly more powerful, 448
  Wald, 170
Tight constraints, 122
Tilted distribution, 255
Total time on test, 369
Tree order, 43
TTT (total time on test), 369
Two-sample problem
  global statistic, 415
Two-way layout
  ordered alternative, 248
  aligned-rank statistic for, 251
Two sample, 232, 243
  layer alternative, 258
  nonparametric test, 258

problem, 231, 279, 408
rank statistic, 244
Type A testing problem, 61
simple examples, 64
Type B testing problem, 61
simple examples, 61
UIP (union-intersection principle), 222
Umbrella order, 43
UMLE (unrestricted MLE), 389
UMPI (uniformly most powerful invariant), 402
UMR (uniformly minimum risk), 381
Uniformly minimum risk, see also UMR, 381
Uniformly more powerful test, 448
Uniformly most powerful, 222, 243, 402
invariant test, 222
monotone test, 429
similar test, 324
Unimodal, 110, 354, 357
Unimodal density, 354
estimator of Birge, 356
Union-intersection principle, 24, 101, 222
equivalence to LRT, 224
Unrestricted MLE, 389
Wald statistic, 153
Wang-McDermott conditional test, 106
Zelterman's test, see cross-over interaction, 449

WILEY SERIES IN PROBABILITY AND STATISTICS

ESTABLISHED BY WALTER A. SHEWHART AND SAMUEL S. WILKS

The ***Wiley Series in Probability and Statistics*** is well established and authoritative. It covers many topics of current research interest in both pure and applied statistics and probability theory. Written by leading statisticians and institutions, the titles span both state-of-the-art developments in the field and classical methods.

Reflecting the wide range of current research in statistics, the series encompasses applied, methodological and theoretical statistics, ranging from applications and new techniques made possible by advances in computerized practice to rigorous treatment of theoretical approaches.

This series provides essential and invaluable reading for all statisticians, whether in academia, industry, government, or research.

ABRAHAM and LEDOLTER · Statistical Methods for Forecasting
AGRESTI · Analysis of Ordinal Categorical Data
AGRESTI · An Introduction to Categorical Data Analysis
AGRESTI · Categorical Data Analysis, *Second Edition*
ALTMAN, GILL, and McDONALD · Numerical Issues in Statistical Computing for the Social Scientist
AMARATUNGA and CABRERA · Exploration and Analysis of DNA Microarray and Protein Array Data
ANDĚL · Mathematics of Chance
ANDERSON · An Introduction to Multivariate Statistical Analysis, *Third Edition*
*ANDERSON · The Statistical Analysis of Time Series
ANDERSON, AUQUIER, HAUCK, OAKES, VANDAELE, and WEISBERG · Statistical Methods for Comparative Studies
ANDERSON and LOYNES · The Teaching of Practical Statistics
ARMITAGE and DAVID (editors) · Advances in Biometry
ARNOLD, BALAKRISHNAN, and NAGARAJA · Records
*ARTHANARI and DODGE · Mathematical Programming in Statistics
*BAILEY · The Elements of Stochastic Processes with Applications to the Natural Sciences
BALAKRISHNAN and KOUTRAS · Runs and Scans with Applications
BARNETT · Comparative Statistical Inference, *Third Edition*
BARNETT and LEWIS · Outliers in Statistical Data, *Third Edition*
BARTOSZYNSKI and NIEWIADOMSKA-BUGAJ · Probability and Statistical Inference
BASILEVSKY · Statistical Factor Analysis and Related Methods: Theory and Applications
BASU and RIGDON · Statistical Methods for the Reliability of Repairable Systems
BATES and WATTS · Nonlinear Regression Analysis and Its Applications
BECHHOFER, SANTNER, and GOLDSMAN · Design and Analysis of Experiments for Statistical Selection, Screening, and Multiple Comparisons
BELSLEY · Conditioning Diagnostics: Collinearity and Weak Data in Regression

*Now available in a lower priced paperback edition in the Wiley Classics Library.

* BELSLEY, KUH, and WELSCH · Regression Diagnostics: Identifying Influential Data and Sources of Collinearity
BENDAT and PIERSOL · Random Data: Analysis and Measurement Procedures, *Third Edition*
BERRY, CHALONER, and GEWEKE · Bayesian Analysis in Statistics and Econometrics: Essays in Honor of Arnold Zellner
BERNARDO and SMITH · Bayesian Theory
BHAT and MILLER · Elements of Applied Stochastic Processes, *Third Edition*
BHATTACHARYA and WAYMIRE · Stochastic Processes with Applications
BILLINGSLEY · Convergence of Probability Measures, *Second Edition*
BILLINGSLEY · Probability and Measure, *Third Edition*
BIRKES and DODGE · Alternative Methods of Regression
BLISCHKE AND MURTHY (editors) · Case Studies in Reliability and Maintenance
BLISCHKE AND MURTHY · Reliability: Modeling, Prediction, and Optimization
BLOOMFIELD · Fourier Analysis of Time Series: An Introduction, *Second Edition*
BOLLEN · Structural Equations with Latent Variables
BOROVKOV · Ergodicity and Stability of Stochastic Processes
BOULEAU · Numerical Methods for Stochastic Processes
BOX · Bayesian Inference in Statistical Analysis
BOX · R. A. Fisher, the Life of a Scientist
BOX and DRAPER · Empirical Model-Building and Response Surfaces
*BOX and DRAPER · Evolutionary Operation: A Statistical Method for Process Improvement
BOX, HUNTER, and HUNTER · Statistics for Experimenters: An Introduction to Design, Data Analysis, and Model Building
BOX and LUCEÑO · Statistical Control by Monitoring and Feedback Adjustment
BRANDIMARTE · Numerical Methods in Finance: A MATLAB-Based Introduction
BROWN and HOLLANDER · Statistics: A Biomedical Introduction
BRUNNER, DOMHOF, and LANGER · Nonparametric Analysis of Longitudinal Data in Factorial Experiments
BUCKLEW · Large Deviation Techniques in Decision, Simulation, and Estimation
CAIROLI and DALANG · Sequential Stochastic Optimization
CASTILLO, HADI, BALAKRISHNAN, and SARABIA · Extreme Value and Related Models with Applications in Engineering and Science
CHAN · Time Series: Applications to Finance
CHATTERJEE and HADI · Sensitivity Analysis in Linear Regression
CHATTERJEE and PRICE · Regression Analysis by Example, *Third Edition*
CHERNICK · Bootstrap Methods: A Practitioner's Guide
CHERNICK and FRIIS · Introductory Biostatistics for the Health Sciences
CHILÈS and DELFINER · Geostatistics: Modeling Spatial Uncertainty
CHOW and LIU · Design and Analysis of Clinical Trials: Concepts and Methodologies, *Second Edition*
CLARKE and DISNEY · Probability and Random Processes: A First Course with Applications, *Second Edition*
*COCHRAN and COX · Experimental Designs, *Second Edition*
CONGDON · Applied Bayesian Modelling
CONGDON · Bayesian Statistical Modelling
CONOVER · Practical Nonparametric Statistics, *Third Edition*
COOK · Regression Graphics
COOK and WEISBERG · Applied Regression Including Computing and Graphics
COOK and WEISBERG · An Introduction to Regression Graphics
CORNELL · Experiments with Mixtures, Designs, Models, and the Analysis of Mixture Data, *Third Edition*

*Now available in a lower priced paperback edition in the Wiley Classics Library.

COVER and THOMAS · Elements of Information Theory
COX · A Handbook of Introductory Statistical Methods
*COX · Planning of Experiments
CRESSIE · Statistics for Spatial Data, *Revised Edition*
CSÖRGŐ and HORVÁTH · Limit Theorems in Change Point Analysis
DANIEL · Applications of Statistics to Industrial Experimentation
DANIEL · Biostatistics: A Foundation for Analysis in the Health Sciences, *Eighth Edition*
*DANIEL · Fitting Equations to Data: Computer Analysis of Multifactor Data, *Second Edition*
DASU and JOHNSON · Exploratory Data Mining and Data Cleaning
DAVID and NAGARAJA · Order Statistics, *Third Edition*
*DEGROOT, FIENBERG, and KADANE · Statistics and the Law
DEL CASTILLO · Statistical Process Adjustment for Quality Control
DEMARIS · Regression with Social Data: Modeling Continuous and Limited Response Variables
DEMIDENKO · Mixed Models: Theory and Applications
DENISON, HOLMES, MALLICK and SMITH · Bayesian Methods for Nonlinear Classification and Regression
DETTE and STUDDEN · The Theory of Canonical Moments with Applications in Statistics, Probability, and Analysis
DEY and MUKERJEE · Fractional Factorial Plans
DILLON and GOLDSTEIN · Multivariate Analysis: Methods and Applications
DODGE · Alternative Methods of Regression
*DODGE and ROMIG · Sampling Inspection Tables, *Second Edition*
*DOOB · Stochastic Processes
DOWDY, WEARDEN, and CHILKO · Statistics for Research, *Third Edition*
DRAPER and SMITH · Applied Regression Analysis, *Third Edition*
DRYDEN and MARDIA · Statistical Shape Analysis
DUDEWICZ and MISHRA · Modern Mathematical Statistics
DUNN and CLARK · Basic Statistics: A Primer for the Biomedical Sciences, *Third Edition*
DUPUIS and ELLIS · A Weak Convergence Approach to the Theory of Large Deviations
*ELANDT-JOHNSON and JOHNSON · Survival Models and Data Analysis
ENDERS · Applied Econometric Time Series
ETHIER and KURTZ · Markov Processes: Characterization and Convergence
EVANS, HASTINGS, and PEACOCK · Statistical Distributions, *Third Edition*
FELLER · An Introduction to Probability Theory and Its Applications, Volume I, *Third Edition,* Revised; Volume II, *Second Edition*
FISHER and VAN BELLE · Biostatistics: A Methodology for the Health Sciences
FITZMAURICE, LAIRD, and WARE · Applied Longitudinal Analysis
*FLEISS · The Design and Analysis of Clinical Experiments
FLEISS · Statistical Methods for Rates and Proportions, *Third Edition*
FLEMING and HARRINGTON · Counting Processes and Survival Analysis
FULLER · Introduction to Statistical Time Series, *Second Edition*
FULLER · Measurement Error Models
GALLANT · Nonlinear Statistical Models
GHOSH, MUKHOPADHYAY, and SEN · Sequential Estimation
GIESBRECHT and GUMPERTZ · Planning, Construction, and Statistical Analysis of Comparative Experiments
GIFI · Nonlinear Multivariate Analysis
GLASSERMAN and YAO · Monotone Structure in Discrete-Event Systems
GNANADESIKAN · Methods for Statistical Data Analysis of Multivariate Observations, *Second Edition*

*Now available in a lower priced paperback edition in the Wiley Classics Library.

GOLDSTEIN and LEWIS · Assessment: Problems, Development, and Statistical Issues
GREENWOOD and NIKULIN · A Guide to Chi-Squared Testing
GROSS and HARRIS · Fundamentals of Queueing Theory, *Third Edition*
*HAHN and SHAPIRO · Statistical Models in Engineering
HAHN and MEEKER · Statistical Intervals: A Guide for Practitioners
HALD · A History of Probability and Statistics and their Applications Before 1750
HALD · A History of Mathematical Statistics from 1750 to 1930
HAMPEL · Robust Statistics: The Approach Based on Influence Functions
HANNAN and DEISTLER · The Statistical Theory of Linear Systems
HEIBERGER · Computation for the Analysis of Designed Experiments
HEDAYAT and SINHA · Design and Inference in Finite Population Sampling
HELLER · MACSYMA for Statisticians
HINKELMAN and KEMPTHORNE: · Design and Analysis of Experiments, Volume 1: Introduction to Experimental Design
HOAGLIN, MOSTELLER, and TUKEY · Exploratory Approach to Analysis of Variance
HOAGLIN, MOSTELLER, and TUKEY · Exploring Data Tables, Trends and Shapes
*HOAGLIN, MOSTELLER, and TUKEY · Understanding Robust and Exploratory Data Analysis
HOCHBERG and TAMHANE · Multiple Comparison Procedures
HOCKING · Methods and Applications of Linear Models: Regression and the Analysis of Variance, *Second Edition*
HOEL · Introduction to Mathematical Statistics, *Fifth Edition*
HOGG and KLUGMAN · Loss Distributions
HOLLANDER and WOLFE · Nonparametric Statistical Methods, *Second Edition*
HOSMER and LEMESHOW · Applied Logistic Regression, *Second Edition*
HOSMER and LEMESHOW · Applied Survival Analysis: Regression Modeling of Time to Event Data
HUBER · Robust Statistics
HUBERTY · Applied Discriminant Analysis
HUNT and KENNEDY · Financial Derivatives in Theory and Practice
HUSKOVA, BERAN, and DUPAC · Collected Works of Jaroslav Hajek—with Commentary
HUZURBAZAR · Flowgraph Models for Multistate Time-to-Event Data
IMAN and CONOVER · A Modern Approach to Statistics
JACKSON · A User's Guide to Principle Components
JOHN · Statistical Methods in Engineering and Quality Assurance
JOHNSON · Multivariate Statistical Simulation
JOHNSON and BALAKRISHNAN · Advances in the Theory and Practice of Statistics: A Volume in Honor of Samuel Kotz
JOHNSON and BHATTACHARYYA · Statistics: Principles and Methods, *Fifth Edition*
JOHNSON and KOTZ · Distributions in Statistics
JOHNSON and KOTZ (editors) · Leading Personalities in Statistical Sciences: From the Seventeenth Century to the Present
JOHNSON, KOTZ, and BALAKRISHNAN · Continuous Univariate Distributions, Volume 1, *Second Edition*
JOHNSON, KOTZ, and BALAKRISHNAN · Continuous Univariate Distributions, Volume 2, *Second Edition*
JOHNSON, KOTZ, and BALAKRISHNAN · Discrete Multivariate Distributions
JOHNSON, KOTZ, and KEMP · Univariate Discrete Distributions, *Second Edition*
JUDGE, GRIFFITHS, HILL, LÜTKEPOHL, and LEE · The Theory and Practice of Econometrics, *Second Edition*
JUREČKOVÁ and SEN · Robust Statistical Procedures: Aymptotics and Interrelations

*Now available in a lower priced paperback edition in the Wiley Classics Library.

JUREK and MASON · Operator-Limit Distributions in Probability Theory
KADANE · Bayesian Methods and Ethics in a Clinical Trial Design
KADANE AND SCHUM · A Probabilistic Analysis of the Sacco and Vanzetti Evidence
KALBFLEISCH and PRENTICE · The Statistical Analysis of Failure Time Data, *Second Edition*
KASS and VOS · Geometrical Foundations of Asymptotic Inference
KAUFMAN and ROUSSEEUW · Finding Groups in Data: An Introduction to Cluster Analysis
KEDEM and FOKIANOS · Regression Models for Time Series Analysis
KENDALL, BARDEN, CARNE, and LE · Shape and Shape Theory
KHURI · Advanced Calculus with Applications in Statistics, *Second Edition*
KHURI, MATHEW, and SINHA · Statistical Tests for Mixed Linear Models
*KISH · Statistical Design for Research
KLEIBER and KOTZ · Statistical Size Distributions in Economics and Actuarial Sciences
KLUGMAN, PANJER, and WILLMOT · Loss Models: From Data to Decisions, *Second Edition*
KLUGMAN, PANJER, and WILLMOT · Solutions Manual to Accompany Loss Models: From Data to Decisions, *Second Edition*
KOTZ, BALAKRISHNAN, and JOHNSON · Continuous Multivariate Distributions, Volume 1, *Second Edition*
KOTZ and JOHNSON (editors) · Encyclopedia of Statistical Sciences: Volumes 1 to 9 with Index
KOTZ and JOHNSON (editors) · Encyclopedia of Statistical Sciences: Supplement Volume
KOTZ, READ, and BANKS (editors) · Encyclopedia of Statistical Sciences: Update Volume 1
KOTZ, READ, and BANKS (editors) · Encyclopedia of Statistical Sciences: Update Volume 2
KOVALENKO, KUZNETZOV, and PEGG · Mathematical Theory of Reliability of Time-Dependent Systems with Practical Applications
LACHIN · Biostatistical Methods: The Assessment of Relative Risks
LAD · Operational Subjective Statistical Methods: A Mathematical, Philosophical, and Historical Introduction
LAMPERTI · Probability: A Survey of the Mathematical Theory, *Second Edition*
LANGE, RYAN, BILLARD, BRILLINGER, CONQUEST, and GREENHOUSE · Case Studies in Biometry
LARSON · Introduction to Probability Theory and Statistical Inference, *Third Edition*
LAWLESS · Statistical Models and Methods for Lifetime Data, *Second Edition*
LAWSON · Statistical Methods in Spatial Epidemiology
LE · Applied Categorical Data Analysis
LE · Applied Survival Analysis
LEE and WANG · Statistical Methods for Survival Data Analysis, *Third Edition*
LePAGE and BILLARD · Exploring the Limits of Bootstrap
LEYLAND and GOLDSTEIN (editors) · Multilevel Modelling of Health Statistics
LIAO · Statistical Group Comparison
LINDVALL · Lectures on the Coupling Method
LINHART and ZUCCHINI · Model Selection
LITTLE and RUBIN · Statistical Analysis with Missing Data, *Second Edition*
LLOYD · The Statistical Analysis of Categorical Data
MAGNUS and NEUDECKER · Matrix Differential Calculus with Applications in Statistics and Econometrics, *Revised Edition*
MALLER and ZHOU · Survival Analysis with Long Term Survivors
MALLOWS · Design, Data, and Analysis by Some Friends of Cuthbert Daniel
MANN, SCHAFER, and SINGPURWALLA · Methods for Statistical Analysis of Reliability and Life Data

*Now available in a lower priced paperback edition in the Wiley Classics Library.

MANTON, WOODBURY, and TOLLEY · Statistical Applications Using Fuzzy Sets
MARCHETTE · Random Graphs for Statistical Pattern Recognition
MARDIA and JUPP · Directional Statistics
MASON, GUNST, and HESS · Statistical Design and Analysis of Experiments with Applications to Engineering and Science, *Second Edition*
McCULLOCH and SEARLE · Generalized, Linear, and Mixed Models
McFADDEN · Management of Data in Clinical Trials
* McLACHLAN · Discriminant Analysis and Statistical Pattern Recognition
McLACHLAN, DO, and AMBROISE · Analyzing Microarray Gene Expression Data
McLACHLAN and KRISHNAN · The EM Algorithm and Extensions
McLACHLAN and PEEL · Finite Mixture Models
McNEIL · Epidemiological Research Methods
MEEKER and ESCOBAR · Statistical Methods for Reliability Data
MEERSCHAERT and SCHEFFLER · Limit Distributions for Sums of Independent Random Vectors: Heavy Tails in Theory and Practice
MICKEY, DUNN, and CLARK · Applied Statistics: Analysis of Variance and Regression, *Third Edition*
*MILLER · Survival Analysis, *Second Edition*
MONTGOMERY, PECK, and VINING · Introduction to Linear Regression Analysis, *Third Edition*
MORGENTHALER and TUKEY · Configural Polysampling: A Route to Practical Robustness
MUIRHEAD · Aspects of Multivariate Statistical Theory
MULLER and STOYAN · Comparison Methods for Stochastic Models and Risks
MURRAY · X-STAT 2.0 Statistical Experimentation, Design Data Analysis, and Nonlinear Optimization
MURTHY, XIE, and JIANG · Weibull Models
MYERS and MONTGOMERY · Response Surface Methodology: Process and Product Optimization Using Designed Experiments, *Second Edition*
MYERS, MONTGOMERY, and VINING · Generalized Linear Models. With Applications in Engineering and the Sciences
*NELSON · Accelerated Testing, Statistical Models, Test Plans, and Data Analyses
NELSON · Applied Life Data Analysis
NEWMAN · Biostatistical Methods in Epidemiology
OCHI · Applied Probability and Stochastic Processes in Engineering and Physical Sciences
OKABE, BOOTS, SUGIHARA, and CHIU · Spatial Tesselations: Concepts and Applications of Voronoi Diagrams, *Second Edition*
OLIVER and SMITH · Influence Diagrams, Belief Nets and Decision Analysis
PALTA · Quantitative Methods in Population Health: Extensions of Ordinary Regressions
PANKRATZ · Forecasting with Dynamic Regression Models
PANKRATZ · Forecasting with Univariate Box-Jenkins Models: Concepts and Cases
*PARZEN · Modern Probability Theory and Its Applications
PEÑA, TIAO, and TSAY · A Course in Time Series Analysis
PIANTADOSI · Clinical Trials: A Methodologic Perspective
PORT · Theoretical Probability for Applications
POURAHMADI · Foundations of Time Series Analysis and Prediction Theory
PRESS · Bayesian Statistics: Principles, Models, and Applications
PRESS · Subjective and Objective Bayesian Statistics, *Second Edition*
PRESS and TANUR · The Subjectivity of Scientists and the Bayesian Approach
PUKELSHEIM · Optimal Experimental Design
PURI, VILAPLANA, and WERTZ · New Perspectives in Theoretical and Applied Statistics
PUTERMAN · Markov Decision Processes: Discrete Stochastic Dynamic Programming
*RAO · Linear Statistical Inference and Its Applications, *Second Edition*

*Now available in a lower priced paperback edition in the Wiley Classics Library.